Chromatographic Processes

Addressing all aspects of the design, modeling, and simulation of chromatographic processes, this result-oriented primer provides a practical guide to all the necessary approaches, methodologies, and tools.

Beginning with key definitions and concepts, it builds up from the simplest to the most complex situations, including multicomponent systems, non-uniform velocity profiles, bed instability, particle size distributions, and the influence of complex environments on chromatographic process design. In addition to covering classical approaches, it introduces efficient tools for investigating chromatographic processes, such as the "Russian-Lego" approach for linear systems, phenomenological models, and specific shortcuts for deriving the key properties of industrial processes.

With an emphasis on real-world problems and applications, step-by-step modeling design guidelines, and detailed exercises for self-assessment, this is a must-have guide for practitioners and researchers working in chemical, biochemical, food, and pharmaceutical engineering.

Roger-Marc Nicoud is the founder and former CEO of Novasep, which has installed over 2000 chromatographic units worldwide and pioneered the use of large-scale continuous chromatographic processes in the life science industry. He is President of École Nationale Supérieure des Industries Chimiques (ENSIC), and in 2014, founded Ypso-Facto (www.ypsofacto.com).

Cambridge Series in Chemical Engineering

Books in Series

Baldea and Daoutidis, *Dynamics and Nonlinear Control of Integrated Process Systems*
Chau, *Process Control: A First Course with MATLAB*
Cussler, *Diffusion: Mass Transfer in Fluid Systems, Third Edition*
Cussler and Moggridge, *Chemical Product Design, Second Edition*
De Pablo and Schieber, *Molecular Engineering Thermodynamics*
Denn, *Chemical Engineering: An Introduction*
Denn, *Polymer Melt Processing: Foundations in Fluid Mechanics and Heat Transfer*
Duncan and Reimer, *Chemical Engineering Design and Analysis: An Introduction*
Fan and Zhu, *Principles of Gas-Solid Flows*
Fox, *Computational Models for Turbulent Reacting Flows*
Franses, *Thermodynamics with Chemical Engineering Applications*
Leal, *Advanced Transport Phenomena: Fluid Mechanics and Convective Transport Processes*
Lim and Shin, *Fed-Batch Cultures: Principles and Applications of Semi-Batch Bioreactors*
Marchisio and Fox, *Computational Models for Polydisperse Particulate and Multiphase Systems*
Mewis and Wagner, *Colloidal Suspension Rheology*
Morbidelli, Gavriilidis, and Varma, *Catalyst Design: Optimal Distribution of Catalyst in Pellets, Reactors, and Membranes*
Nicoud, *Chromatographic Processes*
Noble and Terry, *Principles of Chemical Separations with Environmental Applications*
Orbey and Sandler, *Modeling Vapor-Liquid Equilibria: Cubic Equations of State and their Mixing Rules*
Petyluk, *Distillation Theory and its Applications to Optimal Design of Separation Units*
Rao and Nott, *An Introduction to Granular Flow*
Russell, Robinson, and Wagner, *Mass and Heat Transfer: Analysis of Mass Contactors and Heat Exchangers*
Schobert, *Chemistry of Fossil Fuels and Biofuels*
Shell, *Thermodynamics and Statistical Mechanics*
Sirkar, *Separation of Molecules, Macromolecules and Particles: Principles, Phenomena and Processes*
Slattery, *Advanced Transport Phenomena*
Varma, Morbidelli, and Wu, *Parametric Sensitivity in Chemical Systems*

Chromatographic Processes

Modeling, Simulation, and Design

ROGER-MARC NICOUD

CAMBRIDGE
UNIVERSITY PRESS

CAMBRIDGE
UNIVERSITY PRESS

University Printing House, Cambridge CB2 8BS, United Kingdom

One Liberty Plaza, 20th Floor, New York, NY 10006, USA

477 Williamstown Road, Port Melbourne, VIC 3207, Australia

314-321, 3rd Floor, Plot 3, Splendor Forum, Jasola District Centre, New Delhi - 110025, India

79 Anson Road, #06-04/06, Singapore 079906

Cambridge University Press is part of the University of Cambridge.

It furthers the University's mission by disseminating knowledge in the pursuit of education, learning and research at the highest international levels of excellence.

www.cambridge.org
Information on this title: www.cambridge.org/9781107082366

First published 2015

A catalogue record for this publication is available from the British Library

Library of Congress Cataloging in Publication data
Nicoud, Roger-Marc, 1962–
Chromatographic processes : modeling, simulation, and design / Roger-Marc Nicoud, École Nationale Supérieure des Industries Chimiques (ENSIC).
pages cm. – (Cambridge series in chemical engineering)
ISBN 978-1-107-08236-6 (Hardback)
1. Chromatographic analysis. 2. Chemistry, Analytic. 3. Chemical engineering. I. Title.
QD79.C45N53 2015
543′.8–dc23 2014035285

ISBN 978-1-107-08236-6 Hardback

Contents

Preface and acknowledgments *page* xiii

1 Our approach to modeling chromatographic processes 1

1.1 System description 3
1.2 Adsorption equilibria 7
1.2.1 Equal concentrations in fluid mobile phase and intragranular fluid 9
1.2.2 Linear equilibria 9
1.3 Mass balances: retention times 10
1.4 Kinetics 14
1.5 Hydrodynamics 14
1.6 Modeling moving beds 15
1.7 Terminology choices 16
References 18

2 Linear chromatography: the Russian Lego 20

2.1 Basic definitions 21
2.2 Modeling hydrodynamic dispersion 27
2.3 Adding adsorption at equilibrium to the hydrodynamic model 34
2.4 Adding diffusional limitations to hydrodynamics and adsorption 37
2.5 Dead volumes, injected volumes and sensors: a Lego approach 42
2.6 Modeling: the Russian Lego 46
2.7 A warning on modeling non-linear systems 49
2.8 Conclusion 50
2.9 Solutions of the exercises 51
References 61

3 Non-linear chromatography: equilibrium theory 62

3.1 Introduction 62
3.2 Elementary definitions for adsorption isotherms 63
3.3 Single-solute systems 66
3.3.1 The equilibrium model 66
3.3.1.1 Propagation of waves and shocks 68

3.3.1.2 Construction of breakthrough curves 71
3.3.1.3 Construction of chromatograms 75
3.3.1.4 Adsorption isotherms with inflection points 81
3.3.2 More realistic chromatograms: the MC–$\overline{\text{Equil}}$ model 83
3.4 Binary-solute systems: competition between solutes 84
3.4.1 The equilibrium model 85
3.4.2 More realistic chromatograms: the MC–$\overline{\text{Equil}}$ model 102
3.5 Multi-solute systems: competition between solutes 103
3.5.1 The equilibrium model 103
3.5.2 An orthogonalization procedure: the ω transform 106
3.5.3 Application of the $\boldsymbol{\omega}$ transform to binary systems 108
3.5.4 More realistic chromatograms: the MC–$\overline{\text{Equil}}$ model 112
3.6 Similarities and differences between ion exchange and adsorption 113
3.7 Qualitative rules and complex multi-solute systems 115
3.8 Conclusion 120
3.9 Solutions of the exercises 121
References 137

4 Fluid–solid phase equilibria 139

4.1 Introduction 139
4.2 Porosity 140
4.2.1 Total porosity 141
4.2.2 Extragranular porosity 142
4.2.3 Intragranular porosity 143
4.3 Modeling adsorption isotherms 144
4.3.1 General considerations 144
4.3.2 Single-solute systems 146
4.3.3 Multi-solute systems 152
4.4 The case of supercritical fluids 159
4.5 The case of ion exchange 161
4.5.1 General considerations 161
4.5.2 The lumped model: ideal situations 161
4.5.3 The lumped model: non-ideal situations 166
4.5.4 The porous model 167
4.6 The case of size exclusion 177
4.7 Phenomenological models 178
4.7.1 Introductory concepts 178
4.7.2 Some general properties of a phenomenological mechanism 181
4.7.3 Illustrative example: amino acids 185
4.7.4 Illustrative example: induced pH gradients on weak cationic resins 190
4.8 Experimental methods 192
4.8.1 General considerations 192
4.8.2 Single-solute systems 193

4.8.2.1 Static methods 193
4.8.2.2 Dynamic methods 194
4.8.3 Multi-solute systems 198
4.8.3.1 Static methods 199
4.8.3.2 Dynamic methods 199
4.8.4 The case of ion exchange 203
4.8.5 Sensitivity 204
4.9 Conclusion 206
4.10 Solutions of the exercises 207
References 213

5 Mass transfer 216

5.1 External transfer 217
5.2 Internal transfer: the lumped $\overline{\text{Fick}}$ model 221
5.2.1 Using Fick's law 221
5.2.2 The $\overline{\text{LDF}}$ approximation 225
5.2.3 Typical experimental results 228
5.2.3.1 Influence of Henry's coefficient 229
5.2.3.2 Influence of solute concentration 231
5.3 Internal transfer: the $\overline{\overline{\text{Fick}}}$ model 234
5.3.1 General description 235
5.3.2 Linear adsorption 239
5.3.3 Non-linear adsorption 241
5.3.4 $\overline{\overline{\text{LDF}}}$ approximations for the non-linear porous model 247
5.3.5 Comparison of experimental results with the $\overline{\overline{\text{Fick}}}$ model 250
5.3.6 Special situation: the biporous model 253
5.3.7 Special situation: internal convection 256
5.3.8 Special situation: adsorption kinetics 257
5.3.9 Special situation: size exclusion 258
5.4 Mass transfer: multiple solutes 260
5.4.1 General considerations on Fick's law 260
5.4.2 External transfer 261
5.4.3 Internal transfer 262
5.4.4 The $\overline{\text{LDF}}$ and $\overline{\overline{\text{LDF}}}^{\text{var}}$ approximations 265
5.5 The case of ion exchange 266
5.5.1 Internal transfer 266
5.5.2 External transfer 270
5.6 The rate-controlling step 271
5.7 Expression of the column transfer function (linear adsorption isotherms) 276
5.8 The Van Deemter equation (linear adsorption isotherms) 280
5.9 Influence of particle size distributions 287
5.10 Experimental methods 292
5.10.1 Batch methods 292

5.10.2 Shallow-bed method (1) 294
5.10.3 Shallow-bed method (2) 294
5.10.4 The Van Deemter method 296
5.10.5 Curve fitting (the inverse method) 297
5.11 Influence of mass transfer on chromatograms 298
5.12 Perspectives: from Fick to Maxwell–Stefan? 301
5.13 Conclusion 304
5.14 Solutions of the exercises 306
References 314

6 Hydrodynamics of chromatographic columns 317

6.1 Ideal systems 317
6.1.1 Pressure-drop modeling 318
6.1.2 Hydrodynamic dispersion 321
6.1.3 Fluid compressibility 324
6.1.3.1 Gas chromatography 326
6.1.3.2 Supercritical fluid chromatography 327
6.2 Bed compressibility 330
6.2.1 General considerations 330
6.2.2 Simplistic model 332
6.2.3 Incompressible chromatographic media 342
6.2.4 Compressible chromatographic media 346
6.2.5 Swelling and shrinking media 350
6.2.6 Conclusion 351
6.3 Radial profiles 352
6.3.1 Temperature profiles 353
6.3.2 RTD in the presence of radial profiles 360
6.4 Particle size distribution 370
6.5 Viscosity and gravity effects 372
6.6 Conclusion 377
6.7 Solutions of the exercises 378
References 385

7 Simulating chromatographic columns 388

7.1 Introduction 388
7.2 Modeling hydrodynamics and mass transfer with a single parameter 396
7.3 Modeling mass transfer via the LDF approximation 402
7.4 Modeling mass transfer by solving Fick's law in spherical geometry 407
7.5 Influence of a non-uniform radial velocity profile 410
7.6 The case of ion exchange 416
7.7 Influence of reactions at equilibrium 419
7.8 Miscellaneous effects 422

7.8.1 Fluid compressibility 423
7.8.2 Bed compressibility 424
7.8.3 Heat effects 424
7.8.4 Viscous fingering and multi-phase effects 425
7.9 Numerical methods 425
7.9.1 The Craig model 426
7.9.2 Solving ordinary differential equations 427
7.9.3 Solving partial differential equations 429
7.9.3.1 Finite-difference methods 431
7.9.3.2 Weighted-residual methods 432
7.10 Conclusion 434
References 435

8 Counter-current systems 438

8.1 Introduction 438
8.2 Single-zone counter-current systems: the equilibrium model 439
8.3 Single-zone counter-current systems: the MC and PD models 455
8.4 Two-zone counter-current systems: the equilibrium model 461
8.5 Four-zone counter-current systems: the equilibrium model 468
8.6 Four-zone counter-current systems: the MC and PD models 474
8.7 Simulated moving beds 476
8.8 Analogy between TMB/SMB and elution chromatography 481
8.9 Imperfect implementations of simulated moving beds 483
8.9.1 Asymmetry in the loop 483
8.9.2 Differences between the columns 483
8.10 Conclusion 485
8.11 Solutions of the exercises 486
References 495

9 Chromatographic modes and their optimization 497

9.1 Introduction 497
9.2 Presentation of chromatographic modes 497
9.2.1 Single-column systems 497
9.2.1.1 Adapting piping and control 498
9.2.1.2 Recycling of partially purified fractions 499
9.2.1.3 Chemical modulation 500
9.2.1.4 Physical modulation 501
9.2.2 Multi-column systems 502
9.2.2.1 Adapting piping and control 503
9.2.2.2 Recycling of partially purified fractions and implementing gradients 504
9.2.2.3 Physical modulation 506

9.2.3 Combining different modes or steps 506
9.2.4 Conclusion 506
9.3 Presentation of the optimization problem 507
9.3.1 Variables, tuning parameters, constrained parameters and results 508
9.3.2 Objective function (simplified) 512
9.3.3 Objective function (detailed) 513
9.4 Algorithmic part of the optimization procedure 518
9.4.1 General considerations 518
9.4.2 Type I problems 520
9.4.2.1 Structure 520
9.4.2.2 Scaling the system 524
9.4.2.3 Reference to the equilibrium model 525
9.4.3 Type 0 shortcuts: estimating tuning parameters from scratch 528
9.4.3.1 Estimating all tuning parameters except column length and one velocity 529
9.4.3.2 Determining length and velocity 533
9.5 Application to elution chromatography 534
9.5.1 Type I problems 534
9.5.2 Type 0 shortcuts: influence of pressure drop and particle size 538
9.6 Application to TMB 541
9.6.1 Type I problems 541
9.6.2 Type 0 shortcuts: influence of pressure drop and particle size 545
9.7 Application to SMB and multi-column systems 547
9.7.1 Type I problems 547
9.7.2 Type 0 shortcuts: influence of pressure drop and particle size 548
9.8 Conclusion 549
9.9 Solutions of the exercises 550
References 556

10 Addressing a few industrial problems 561

10.1 Sugar separation 561
10.1.1 Problem description 562
10.1.2 Laboratory data 562
10.1.3 Process design: linear 564
10.1.4 Process design: non-linear 568
10.1.5 Conclusion 572
10.2 Optical isomer separation 573
10.2.1 Problem description 574
10.2.2 Laboratory data 574
10.2.3 Equilibrium model analysis 575
10.2.4 MC–$\overline{\text{Equil}}$ model analysis 576
10.2.5 Comparison of elution and SMB: practical considerations 578
10.2.6 Conclusion 582

10.3 Chromatography as a Lego piece 584
10.3.1 Problem description 584
10.3.2 Process design: HFCS and by-pass strategy 584
10.3.3 Process design: elimination of radioactive pollutants 588
10.3.4 Conclusion 593
10.4 Complex mixtures 595
10.4.1 Gradient operations 595
10.4.2 Flow reversal 597
10.4.3 Multi-step strategies 599
10.5 Conclusion 602
References 602

11 Conclusion 604

Appendix A Some important properties of the Laplace transform 606

Appendix B Inlet and outlet boundary conditions 614

Appendix C Equilibrium theory: single-solute chromatograms 618

Appendix D Equilibrium theory: binary chromatograms 622

Appendix E The influence of the porosity determination on chromatographic modeling 625

Appendix F Useful physico-chemical data and orders of magnitude 628

Appendix G Fick and Maxwell–Stefan approaches to diffusion 631

Appendix H Non-linear LDF for multi-solute systems 638

Appendix I Situations that make the use of the MC model problematic 640

Appendix J Typical industrial chromatographic processes 642

Notation 644
Index 656

Preface and acknowledgments

No three inventions have had greater impact on the working day of the physical scientist than the computer, the photocopier and the chromatograph.
Friedrich Helfferich, 1986

The dramatic increase in purification needs, from the development to the production stage, especially in the pharmaceutical and biopharmaceutical industries, has led preparative chromatography to become a widely used technology. There are now so many companies, teams and individuals using or developing preparative chromatographic processes that a real chromatographic world has appeared during the last decades.

Experience shows that very diverse people are living in the chromatographic world. Some of them come from the "Chemists" planet and they speak about polarity, adsorbent deactivation or resolution. The ones from the "Biochemists" planet speak about protein denaturation, isoelectric points or viral contamination. Others come from the "Chemical Engineers" planet and speak about mass transfer, dispersion coefficients and adsorption isotherms. Those from the "Pharmacists" planet speak about contamination or good manufacturing practices. One can even find people from the "Mathematicians" planet, speaking about Riemann invariants or Laplace transforms. Obviously, since all of these people use their own language, the chromatographic world may sound very dissonant, not propitious for discussion.

My experience, skills and passion are certainly on the chemical engineering side. The chemical engineering methodology, "La voie Royale" in the terminology of Jacques Villermaux, sounds like a rational way to approach complex problems. My confidence in the ability of this approach to efficiently design chromatographic processes is at the origin of my decision to create Novasep. Twenty years later, I know how fruitful the approach is, but I also know how hard it is to explain and to diffuse it. Rational chemical engineering approaches are too seldom used in the industry, and the emergence of Quality by Design (QbD) and associated Design of Experiments (DoE) approaches, encouraged by the authorities and embraced by the industry, sounds too often like a refusal to understand the physics of chromatography.

I decided to write this book in order to present a chemical engineering approach in a comprehensive manner, and to help people coming from different origins to appreciate what can be expected from this approach. The content is not exhaustive, as for instance chromatographic media are not presented, and perhaps not even objective: only information considered useful according to my experience has been included. I readily apologize if important contributions are not mentioned; this would simply be a measure of my ignorance.

This book has been written in order to show how concepts may help in solving practical problems ("the real world"), and how practical problems may give some very interesting research subjects for which new concepts are probably required. Instead of giving recipes, I propose a methodology that will, I believe, help the reader to solve even the most difficult problems. I know that some chromatographers may find some parts difficult because of the mathematics involved. I have done my best to illustrate any part that could be of specific difficulty, but I believe that this complexity is the price for access to the methodology. If somebody wants to learn guitar, he has to accept that the instrument has six chords, even if initially this number seems way too high.

The first chapters of the book are devoted to understanding and modeling chromatographic processes. Starting with simple models of linear chromatography with incompressible fluids, various simplifying assumptions are progressively relaxed, and models of increasing complexity are proposed for describing, with the required degree of precision, the thermodynamic, hydrodynamic and mass-transfer processes that determine a chromatogram's shape.

The theoretical concepts presented in the first chapters are then used for selecting and optimizing different chromatographic processes. The theoretical models are used as tools to predict the behavior of chromatographic systems in unexplored regions, assuming that an optimization criterion is known for selecting the most appropriate system.

The approach presented in this book is the result of 30 years of involvement in chromatography, starting with research work on ion exchange initiated in 1984 in the Laboratoire des Sciences du Génie Chimique, Nancy. Professor J. Villermaux was the Director of the Laboratory, and Dr. D. Schweich my research director. I will never be able to express how these first years, and the opportunity to benefit from the exceptional scientific skills of Professor Villermaux and Dr. Schweich, have been critical for my entire career.

The work continued at Separex, and my early cooperation with Professor M. Perrut has been instrumental in making things happen and building the confidence that concepts will eventually lead to design, design to reality.

Many ideas presented in this book have been developed, tested or improved during more than two decades with my coworkers, and especially P. Adam, J. Blehaut, Dr. F. Charton, Dr. J. Y. Clavier, Dr. H. Colin, Dr. F. Denet, Dr. W. Hauck, Dr. M. Holzer, Dr. O. Ludemann-Hombourger and Dr. E. Valery. Some partners from academia, with whom I co-authored scientific publications or co-directed theses, also played important roles, and the names of Dr. M. Bailly, Professor M. Morbidelli, Professor M. Mazzotti, Professor A.E. Rodrigues and Professor A. Seidel-Morgenstern immediately cross my mind. More recently, I had some discussions with Professor M. Kaspereit on a preliminary version of the book, and he then contributed to improving the presentation of the single-solute equilibrium model; Professor J.P. Mota and Professor A.E. Rodrigues agreed to read and comment on the manuscript; D. Pfister was brave enough to perform a final detailed check. Even prior to the creation of Novasep, when using counter-current chromatography for producing active pharmaceutical molecules was no more than a dream, I benefited from the unconditional support of Professor Kinkel, at that time head of the preparative chromatography laboratory at Merck (Darmstadt). I would like to express to all of them my sincere gratitude.

As a consequence of these historical cooperations, I will normally use the first person plural for presenting models and results, and the first person singular for expressing an opinion.

The roots of this book can probably be found in the early 1990s, when I received support from the European Union program COMETT for producing materials aimed at teaching a methodology for designing chromatographic processes. I then, from time to time, worked with Dr. Michel Bailly, in order to extend the initial material, and to propose a book. My professional duties did not allow me to allocate the time required for moving forward efficiently, so the writing was put on hold for many years. When I took a break from operational management, my first desire was to complete what was initiated so long ago. In writing this book, organizing things and reading recent contributions, I realized that I could have made better use of some interesting concepts and tools. These are mentioned, although briefly, in order to share some possible directions with those who are now in charge of developing the existing knowledge and methods.

In order to produce a document of the highest possible quality, I asked Dr. Daniel Schweich to check and critically review the content. He accepted immediately, and I have to admit that I feel some guilt because I did not realize how big this task was going to be. I would like to express to him my gratitude for having taken on the challenge. This book owes him a lot.

Roger-Marc NICOUD

1 Our approach to modeling chromatographic processes

Let's start the journey . . .

The title of this book contains the words *processes*, *modeling*, *simulation* and *design*. Let us start by briefly commenting on their respective meanings.

According to the dictionary, the word *process* has a double meaning as it can refer to either:

- a series of actions that produce a change or development (e.g. the process of digestion)
- a method of doing or producing something (e.g. the Bessemer process for the mass production of steel).

In order to clarify the distinction, let us consider that the first definition refers to *elementary processes* while the second refers to *manufacturing processes*. When speaking of *chromatographic processes*, one can refer either to elementary processes (the physico-chemical mechanisms at the origin of the separation) or to manufacturing processes (like the Parex© process for producing para-xylene).

The thesis of this book is that understanding elementary processes is of primary importance for properly designing manufacturing processes.

The difference between *modeling* and *simulation* may be less obvious for many chromatographers. By modeling, we mean the task of understanding and predicting processes (either elementary or manufacturing). By simulation, we mean the task of actually representing and quantitatively evaluating a process; this is typically associated with more or less complex numerical tools. A good simulator based on a poor model will give wrong predictions (e.g. simulating ion-exchange processes without taking into account electroneutrality constraints). A good model without a simulator can give useful information and trends (e.g. the equilibrium model neglecting all sources of hydrodynamic dispersion and mass transfer limitations, as presented in Chapter 3). With a good simulator in hand, one can then address the final step, consisting of designing the manufacturing process.

A good example of what modeling can do for the chromatographer is the determination of flow rates in simulated moving beds. I used to say that finding these parameters is like finding a small piece of blue in a dark sky: for the vast majority of flow rates, the SMB delivers "purified" fractions having the composition of the feed, and is thus totally inefficient. There are, however, some special combinations of flow rate for which the system is very efficient and delivers pure products. When the "piece of blue" has been found, one can then use more sophisticated simulators for precisely predicting performance and finally designing the machine.

Many articles and books have been devoted to the tasks of modeling and simulating chromatographic processes; some outstanding contributions were made in the second part of the 20th century. My personal selection would include the book by (Helfferich, 1962) for a rational understanding of ion exchange, the books of (Rhee *et al.*, 1989) and (Helfferich and Klein, 1970) for the development of multi-component non-linear chromatography, the books of (Ruthven, 1984), (Wankat, 1986a), (Wankat, 1986b) for bridging between theory and implementation and the book edited by (Rodrigues and Tondeur, 1981) containing a collection of articles from authors who contributed outstandingly to the development of chromatography understanding (Broughton, Klein, Rhee, Tondeur, Villermaux, Wankat). Many references to these contributions will be made in the subsequent chapters.

I don't believe, however, that a book containing a comprehensive presentation of a methodology based on chemical engineering tools and associated with a broad industrial experience is available. Delivering this comprehensive scheme and methodology is our task, and achieving this goal will require clear definitions, and choices that are sometimes not those generally accepted by chromatographers. This chapter, aimed at presenting our key definitions and choices, is thus fundamental for understanding the development presented in the following chapters.

Prior to starting our modeling presentation and system description, let us mention that the book neither describes nor presents chromatographic media and chromatographic equipment. For these matters the reader is referred to (Schmidt-Traub *et al.*, 2012) and (Carta and Jungbauer, 2011) for pharmaceutical and biopharmaceutical applications, and to (Wankat, 1986a), (Wankat, 1986b) and (Ruthven, 1984) for large-scale adsorption processes.

It is certainly widely accepted that modeling chromatographic systems requires taking into account thermodynamic, hydrodynamic and kinetic processes, which are involved in mass- and possibly heat-balance equations. The way to address the above-mentioned processes is probably less widely accepted, and the literature proposes many different approaches. Prior to presenting the one that I believe is the most pertinent, let us take a bit of distance from the detailed modeling activity.

As mentioned by Octave Levenspiel, citing a sentence attributed to German mathematician Friedrich Gauss, "Give me four parameters and I will draw an elephant for you, with five I will have him raise and lower his trunk and his tail."[1] The idea was that better representing reality by adding parameters is not proof of a better model. If the problem is to estimate the weight of the animal, probably a four-parameter model is adequate (we could even assume the elephant to be spherical for a rough approximation!). If the problem is to propose a drawing that could have been produced by a gifted artist, things are different. Making this choice, adapting the complexity of the modeling to the needs,

[1] (Wei, 1975) then showed that a fairly decent elephant's silhouette can be obtained with 15 terms in a Fourier series.

is the art of the scientist. When the elephant is perfectly represented with a high number of parameters, we could even see him from his back moving away. Miracle, the elephant is symmetrical! One can divide the number of parameters by two: here starts modeling![2]

1.1 System description

We consider a system containing a *chromatographic medium*, made of particles at the origin of the retention of the different species, and a fluid in which solutes are dissolved. Our goal is to describe the different zones of the system in which the solute can be located. Only simple basics are presented here; more detailed descriptions are given in the subsequent chapters.

The chromatographic medium can be silica (modified or not), polymer resins, zeolites or generally any type of solid having adsorbing, ion-exchange or exclusion properties. The associated particles are described using the following simple assumptions:

- The particles are "well-formed" particles having a well-identified skeleton and well-defined pores.
- The particles are totally rigid so that no swelling or shrinking can occur.

These two assumptions are often legitimate and widely used, but one needs to have in mind that concepts such as pore, intraparticular pore fluid and wall surface become vague in a gel or with zeolites. We also know that resins can swell; taking into account this possibility would, however, introduce additional complexity that is not required at this introductory level.

Solutes contained in a chromatographic system can be located in at least three different zones:

1. The fluid outside the particles contained in the *external* (or *extragranular*) *porosity*, i.e. the solvent or carrier fluid. Later on, this fluid will systematically be called the *fluid mobile phase*.
2. The fluid contained in the *internal* (or *intragranular*) *porosity*. This does not move like the carrier fluid and will be called the *intragranular fluid*.
3. The *solid-phase skeleton*, which is at the origin of the separation process. It can be the adsorbing medium or it can be coated with a suitable adsorbent. We will speak of the solid-phase skeleton of the chromatographic medium, even if the term is not truly appropriate when considering a coated adsorbent like reversed-phase silica. This, however, allows us to make an unambiguous distinction from the intragranular fluid.

The model which differentiates the solid-phase skeleton and the intragranular fluid phase inside the particles will be called the *Porous Model*.

The different zones that are present in a chromatographic column are schematically represented in Figure 1.1. The column volume is the sum of two fluid volumes and one

[2] From a discussion with Daniel Schweich.

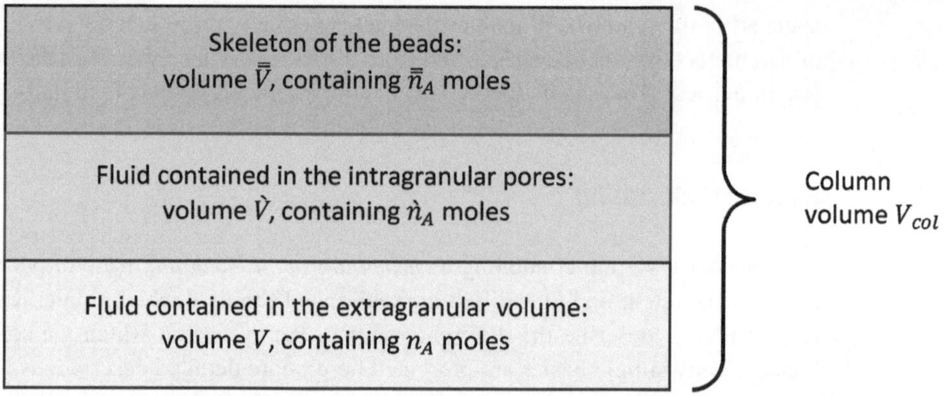

Figure 1.1 Schematic representation of the different solute locations in a chromatographic column according to the porous model.

solid volume:

$$V_{col} = V + \grave{V} + \overline{\overline{V}} \tag{1.1}$$

where V, $\grave{V}$ and $\overline{\overline{V}}$ represent respectively the volume of the fluid mobile phase, the volume of the intragranular fluid and the volume of the solid-phase skeleton.

Similarly, the total number of moles of solute in the system is also the sum of three contributions, two being in a fluid phase, one being in a solid phase:

$$n_A^{tot} = n_A + \grave{n}_A + \overline{\overline{n}}_A \tag{1.2}$$

where n_A, $\grave{n}_A$ and $\overline{\overline{n}}_A$ represent respectively the number of moles of solute A present in the fluid mobile phase, in the intragranular fluid and on the solid-phase skeleton.

Throughout this book, for a given variable X, we will call X its value in the fluid mobile phase, $\grave{X}$ its value in the intragranular fluid, $\overline{\overline{X}}$ its value on the solid-phase skeleton (thus excluding intragranular liquid). Additionally, we will denote by $\overline{X}$ the sum of $\grave{X}$ and $\overline{\overline{X}}$ for extensive variables (like volumes) or the volume average of $\grave{X}$ and $\overline{\overline{X}}$ for intensive variables (like concentrations).

It is important to understand this notation, as it will be used for many variables and parameters, including concentrations, adsorption isotherm expressions and coefficients, diffusion coefficients and temperature.

Using this notation convention, the two equations above can be further detailed:

- For volume (extensive quantity):

$$\begin{aligned} V_{col} &= V + \grave{V} + \overline{\overline{V}} \\ V_{col} &= V + \overline{V} \\ \overline{V} &= \grave{V} + \overline{\overline{V}} \end{aligned} \tag{1.3}$$

- For the number of moles (extensive quantity):

$$
\begin{aligned}
n_A^{tot} &= n_A + \grave{n}_A + \bar{\bar{n}}_A \\
n_A^{tot} &= n_A + \bar{n}_A \\
\bar{n}_A &= \grave{n}_A + \bar{\bar{n}}_A
\end{aligned}
\tag{1.4}
$$

We can define concentration (intensive quantity) in the same way, so that one has:

$$
\begin{aligned}
C_A &= \frac{n_A}{V} \\
\grave{C}_A &= \frac{\grave{n}_A}{\grave{V}} \\
\bar{\bar{C}}_A &= \frac{\bar{\bar{n}}_A}{\bar{\bar{V}}}
\end{aligned}
\tag{1.5}
$$

Three different concentrations are thus considered to fully describe the system: the concentration in the fluid mobile phase, C_A, the concentration in the intragranular fluid, $\grave{C}_A$, and the concentration on the solid-phase skeleton, $\bar{\bar{C}}_A$.

A fourth concentration, $\bar{C}_A$, in the fictitious pseudo-homogeneous solid of volume $\bar{V}$, is related to the others:

$$
\bar{C}_A = \frac{\bar{n}_A}{\bar{V}} = \frac{\grave{n}_A \grave{V}}{\grave{V}\bar{V}} + \frac{\bar{\bar{n}}_A \bar{\bar{V}}}{\bar{\bar{V}}\bar{V}} = \grave{C}_A \frac{\grave{V}}{\bar{V}} + \bar{\bar{C}}_A \frac{\bar{\bar{V}}}{\bar{V}}
\tag{1.6}
$$

The concentration $\bar{C}_A$ will be named the *lumped solid-phase concentration* or *lumped concentration* for short, because the solute molecules located both in the intragranular fluid and on the solid skeleton are lumped together to define the average concentration that would prevail should the particle be a homogeneous phase. We will later speak of lumped solid phase, lumped diffusion coefficients, lumped adsorption isotherms, and so on.

The combined intragranular fluid phase and solid-phase skeleton are generally called the *stationary phase*. This is a misleading term since two phases of different concentration (solid and fluid) are involved and also because the stationary phase can be moving in a true moving bed. It is certainly a convenient concept provided it is not misused, but we prefer the term *lumped solid phase* in order to avoid ambiguities.

The concentrations defined above are volume-averaged concentrations. While non-uniform concentration profiles generally appear inside the particles during the saturation/desaturation process, we will see that the main chromatogram's characteristics depend primarily on volume-averaged concentrations.

Choices for concentration references and definitions call for the following remarks:

- A first practical remark: Instead of referring the concentrations to the volume of particles, one could choose the mass of the solid skeleton. The two approaches are strictly equivalent for rigid particles. Using the mass of the solid skeleton as a reference has some advantages in the case of possible swelling/shrinking since this mass is independent of experimental conditions, as opposed to the particle volume, which varies. The drawback of using the mass of the solid skeleton as a reference is that modeling chromatographic columns involves mass balances written on volumes, so that the particle density must be determined. Instead of having to consider a volume of particles possibly varying with experimental conditions one has thus to consider a density of particles varying with experimental conditions. Consequently, the two approaches are equivalent, and I believe that using volume-averaged concentration is in general a bit easier to manipulate.

- A second practical remark: Chemical considerations favor the expression of concentrations in mol/m^3 (molar concentration), while chromatographers normally use kg/m^3 (massic concentration), perceived to be more practical. We plead guilty as we will often use massic concentration instead of molar concentration. The differences are in general purely academic (especially when one considers the separation of optical isomers having the same molecular mass) and we will warn the reader when this may not be the case. There is one situation for which massic concentration must be avoided: where ion exchange is concerned, molar or, even better, equivalent concentrations must systematically be used as this allows straightforward expression for the electroneutrality equation.

Nothing prohibits modeling of the chromatographic process using only lumped solid-phase concentrations, provided these concentrations are properly related to the fluid mobile phase concentrations and adequate mass balances are written. Under these conditions, the detailed structure of the particle is ignored and the chromatographic medium is approximated by a pseudo-homogeneous (lumped) phase of volume $\overline{V}$ containing $\bar{n}_A$ moles at concentration $\overline{C}_A$. A model considering only lumped solid-phase concentration will be referred to as a *Lumped Model*, in contrast to the porous model.

The choice between the lumped model and the porous model is largely a matter of convenience and objectives. When the objective is to model pressure drop, the lumped model is certainly adequate. When looking for a detailed description of the transport of solutes inside the particles in order to understand or improve these transport properties, the detailed porous model is necessary. The fact that the porous model is more detailed than the lumped model should not lead one to conclude that the former is more rigorous than the latter. Let us recall that the definitions of solid skeleton and pores are pretty fuzzy in some cases (gel ion-exchange resins, zeolites, etc.). Similarly or consequently, the definition of adsorbed phase versus fluid phase is not always as precise as it may appear: in ion-exchange resins, there is more of a continuum than two clearly different physical states, as shown in Section 4.5.4, and for adsorption, interpreting experimental results for diffusion coefficients can require assuming that adsorbed species can diffuse and are thus not as fixed as anticipated (Chapter 5). We are now approaching the profound meaning of the elephant joke ...

Finally, let us mention that, provided the *intragranular volume* is defined, the concentrations associated with the two models are related by Eq. (1.6), so that one can move from a more global description to a more detailed one or vice versa. This approach, consisting in lumping different sub-domains in a larger domain, allows for the construction of models of different complexity. This is a sort of Lego® approach, to be further presented in Chapters 2, 4 and 5.

By defining ratios between the different volumes, one can define different porosities (i.e. void fractions). The *external* or *extragranular porosity* is defined by:

$$\begin{aligned}\varepsilon_e &= \frac{V}{V_{col}} \\ 1-\varepsilon_e &= \frac{V_{col}-V}{V_{col}} = \frac{\overline{V}}{V_{col}}\end{aligned} \tag{1.7}$$

The *internal* or *intragranular porosity* is defined by:

$$\begin{aligned}\varepsilon_i &= \frac{\grave{V}}{\overline{V}} \\ 1-\varepsilon_i &= \frac{\overline{V}-\grave{V}}{\overline{V}} = \frac{\overline{\overline{V}}}{\overline{V}}\end{aligned} \tag{1.8}$$

The *total porosity* is a combination of the two:

$$\begin{aligned}\varepsilon_T &= \frac{V+\grave{V}}{V_{col}} = \frac{V_{col}-\overline{\overline{V}}}{V_{col}} = \varepsilon_e + (1-\varepsilon_e)\varepsilon_i \\ 1-\varepsilon_T &= \frac{\overline{\overline{V}}}{V_{col}} = \frac{\overline{\overline{V}}}{\overline{V}}\frac{\overline{V}}{V_{col}} = (1-\varepsilon_i)(1-\varepsilon_e)\end{aligned} \tag{1.9}$$

Introducing the intragranular porosity in the definition of the lumped solid-phase concentration $\overline{C}_A$ given in Eq. (1.6), one obtains:

$$\overline{C}_A = \varepsilon_i \grave{C}_A + (1-\varepsilon_i)\overline{\overline{C}}_A \tag{1.10}$$

Equation (1.10) allows a simple connection between the concentrations of the lumped model and those of the porous model. The only assumption is that intragranular pore volume, and thus porosities, can be defined.

1.2 Adsorption equilibria

Going further in the system description requires relating fluid to solid concentrations via so-called *adsorption isotherms*. The definition of the adsorption isotherm is affected by the choice of the structure model, be it the lumped model or the porous model. Again, both options are equally valid in theory and one can move from one definition to the other. The objective is to build the correct associated theoretical framework and to understand how these choices are connected with practical determination and modeling.

This section is aimed at defining equilibrium situations. This does not mean that the two phases are in general at equilibrium, but it will give us the state that the system is going to reach, given sufficient time.

- Lumped Model: The two phases considered are the fluid mobile phase and the lumped solid phase. At equilibrium, the concentrations of a single solute A in the two phases are related by:

$$\overline{C}_A = \overline{f}_A(C_A) \tag{1.11}$$

Strictly speaking, the equation $\overline{C}_A = \bar{f}_A(C_A)$ relating the lumped solid-phase concentration to the fluid mobile phase concentration of the single solute A is not the "phase equilibrium law" (or adsorption isotherm) as usually defined by thermodynamicists. We will name it the *lumped adsorption isotherm.*

Notice that when more than one solute is present, the adsorption of one solute is in general influenced by the presence of the others, as shown in Chapter 4.

- Porous Model: The three phases considered are the fluid mobile phase, the intragranular fluid and the solid skeleton. The fluid–solid equilibrium is assumed to be reached between the intragranular fluid phase and the solid skeleton phase, so the concentrations of a single solute A in these two phases must be related by:

$$\overline{\overline{C}}_A = \bar{\bar{f}}_A(\grave{C}_A) \tag{1.12}$$

Notice again that when more than one solute is present, the adsorption of one solute is in general influenced by the presence of the others, as shown in Chapter 4.

$\bar{\bar{f}}_A$ is the "phase equilibrium law" according to the thermodynamicists.[3] Equation (1.12) is awkward to use as it involves the concentration of solute in the intragranular fluid, which is unknown *a priori.*

At first glance, since the solvent is the "same" in the extragranular fluid and in the intragranular fluid, one may consider that $\grave{C}_A = C_A$. This is the case in many situations – with some meaningful exceptions such as Donnan ion exclusion and size exclusion, for which $\grave{C}_A \neq C_A$.

In general, one can assume that the intragranular fluid and the extragranular fluid concentrations are related by:

$$\grave{C}_A = \grave{f}_A(C_A) \tag{1.13}$$

Equation (1.13) is the partition law between intragranular fluid and the extragranular fluid. It will later be called the *partition law.*

The concentrations on the solid skeleton and in the fluid mobile phase are then related by:

$$\overline{\overline{C}}_A = \bar{\bar{f}}_A(\grave{C}_A) = \bar{\bar{f}}_A\left(\grave{f}_A(C_A)\right) \tag{1.14}$$

The lumped solid concentration being given by Eq. (1.10), the *lumped equilibrium law* $\bar{f}_A$ connecting the lumped solid-phase concentration to the fluid mobile phase concentration is related to the equilibrium law $\bar{\bar{f}}_A$ by:

$$\overline{C}_A = \bar{f}_A(C_A) = \varepsilon_i \grave{f}_A(C_A) + (1 - \varepsilon_i)\bar{\bar{f}}_A\left(\grave{f}_A(C_A)\right) \tag{1.15}$$

The two models are finally equivalent if one uses Eq. (1.15) to relate the different concentrations.

[3] This assumes that solute A is the only component present in the system, which is impossible in liquid chromatography. Defining a single-component adsorption isotherm is not that simple; more information will be given in Chapter 4.

The two approaches are thus theoretically connected. However, determining independently the partition law $\grave{C}_A = \grave{f}_A(C_A)$ and the adsorption law $\bar{\bar{C}}_A = \bar{\bar{f}}_A(\grave{C}_A)$ is not an easy task, even in the simplest situations.

We will further illustrate the models by successively assuming that:

- there are no exclusion-like processes, so that $\grave{C}_A = C_A$
- the equilibria are linear, or the solutes so dilute that the equilibrium law "reduces to its initial slope".

1.2.1 Equal concentrations in fluid mobile phase and intragranular fluid

Under the assumption $\grave{C}_A = C_A$, Eq. (1.15) connecting the adsorption law to the lumped adsorption law becomes:

$$\bar{f}_A(C_A) = \varepsilon_i C_A + (1 - \varepsilon_i)\bar{\bar{f}}_A(C_A) \tag{1.16}$$

This relationship shows that the lumped equilibrium law and the equilibrium law are simply related by a linear relation involving the concentration in the fluid mobile phase. The two models can thus be used in very similar ways, but, again, this assumes that the fluid mobile phase and intragranular fluid have the same concentration. In this case, besides purely academic considerations, the two models are indistinguishable at equilibrium.

1.2.2 Linear equilibria

We now relax the assumption of identity between concentrations in the fluid mobile phase and the intragranular fluid, but we assume that the concentrations in the different phases are linearly related. This may or may not be true, but non-linearity will not change our conclusions.

Under the linearity assumption, the concentration of solute A in the intragranular fluid is related to the concentration of solute A in the fluid mobile phase by:

$$\grave{C}_A = \grave{K}_A C_A \tag{1.17}$$

In the absence of exclusion, the coefficient $\grave{K}_A$ equals 1, so the concentrations in the intragranular fluid and fluid mobile phase are identical.

The concentration of the adsorbed phase is also assumed to be linearly related to the concentration in the intragranular fluid by a linear adsorption law:

$$\bar{\bar{C}}_A = \bar{\bar{K}}_A \grave{C}_A \tag{1.18}$$

The coefficient $\bar{\bar{K}}_A$ will be called *Henry's coefficient*. The reader may be more familiar with the expression "Henry's constant". However $\overline{K}_A$ is not constant: it varies with temperature, composition of the solvent and so on, so we prefer the term "coefficient".

The two above equilibrium relations together with Eq. (1.15) allow an estimate of the lumped solid-phase concentration:

$$\overline{C}_A = \varepsilon_i \grave{C}_A + (1-\varepsilon_i)\overline{\overline{C}}_A = \left[\varepsilon_i \grave{K}_A + (1-\varepsilon_i)\grave{K}_A \overline{\overline{K}}_A\right] C_A \tag{1.19}$$

which can also be written:

$$\overline{C}_A = \overline{K}_A C_A \tag{1.20}$$

with

$$\overline{K}_A = \varepsilon_i \grave{K}_A + (1-\varepsilon_i)\grave{K}_A \overline{\overline{K}}_A \tag{1.21}$$

The lumped solid-phase concentration is thus linearly related to the extragranular concentration via the *lumped Henry's coefficient* $\overline{K}_A$. The lumped Henry's coefficient is a linear combination of unrelated equilibrium constants. This means that the initial slope of an experimentally determined adsorption isotherm is a combination of intragranular porosity, partition law and adsorption. For readers familiar with the subject, this means that there is no way to investigate the adsorption law using retention time measurements unless intragranular porosity and partition law are known.

If the solutes are not excluded from the particles so that $\grave{K}_A = 1$, Eq. (1.21) expressing the lumped Henry's coefficient $\overline{K}_A$ becomes:

$$\overline{K}_A = \varepsilon_i + (1-\varepsilon_i)\overline{\overline{K}}_A \tag{1.22}$$

so that the lumped Henry's coefficient should at least equal the intragranular porosity even if the solute does not adsorb on the solid.

1.3 Mass balances: retention times

An important part of chromatography modeling consists in determining retention times. These retention times can be associated with a pulse injection of a given solute at trace level or with large frontal injections of a complex multi-component mixture. Their prediction systematically requires writing and solving mass balances. The way these mass balances are written is largely influenced by the choice of the lumped model or the porous model.

General methodologies for simulating chromatograms will be given in the following chapters. For the time being, we simply consider a column of infinite efficiency (equivalent to an infinite number of plates), initially solute-free, and fed by an *incompressible* carrier fluid at constant flow rate containing a solute A at concentration C_A^F. Additionally, we assume the establishment of a linear equilibrium. The concentration at the outlet of the column is thus expected to stay at zero until the so-called *retention time* t_R is attained, and then to abruptly increase to C_A^F. At the end of the saturation process, one assumes that, everywhere in the column, the extragranular fluid concentration equals C_A^F and that the concentrations $\overline{\overline{C}}_A$ and $\grave{C}_A$ are in equilibrium with C_A^F. The situation is schematically represented in Figure 1.2. At the very time t_R the outlet concentration reaches the inlet

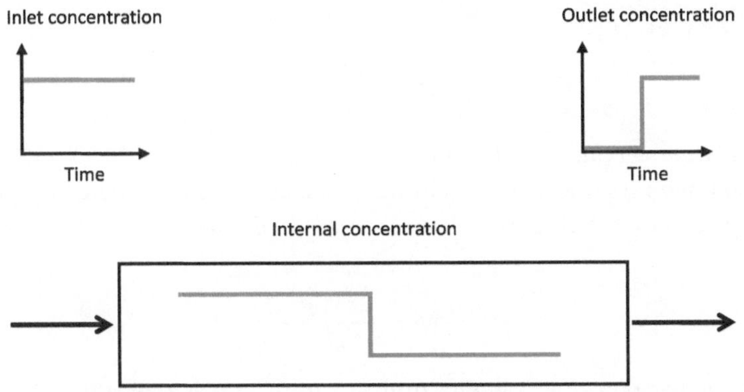

Figure 1.2 Schematic representation of a breakthrough curve and bed saturation.

concentration, the amount of solute fed into the column equals the amount of solute contained in the column.

Knowing that the solute concentration in the fluid mobile phase is C_A^F, the mass balance can be written for the porous model:

$$Qt_R C_A^F = C_A^F V + \grave{C}_A \grave{V} + \overline{\overline{C}}_A \overline{\overline{V}} \tag{1.23}$$

and for the lumped model:

$$Qt_R C_A^F = C_A^F V + \overline{C}_A \overline{V} \tag{1.24}$$

concentrations $\grave{C}_A$, $\overline{\overline{C}}_A$ and $\overline{C}_A$ being in equilibrium with C_A^F.

Assuming a linear equilibrium between the different phase concentrations and introducing the different porosities, the retention time t_R of the front can be calculated for the porous model:

$$t_R = \frac{V_{col}}{Q}\left[\varepsilon_e + (1-\varepsilon_e)\varepsilon_i \grave{K}_A + (1-\varepsilon_e)(1-\varepsilon_i)\grave{K}_A \overline{\overline{K}}_A\right] \tag{1.25}$$

and for the lumped model:

$$t_R = \frac{V_{col}}{Q}\left[\varepsilon_e + (1-\varepsilon_e)\overline{K}_A\right] \tag{1.26}$$

It must be noted that t_R also gives the retention of a narrow chromatographic peak: the linear assumption ensures that the saturation and desaturation retention times are identical.

In order to make a distinction between the time spent in the column by the solutes because of their access to void volumes and the time spent in the column due to adsorption, it is convenient to define a retention time reference that would be obtained for a non-retained solute. This retention time reference is normally called the *zero retention time*.

A first option (option E) is to consider that a non-retained solute has access to the *extragranular* fluid only (the mobile phase) and we thus define a *zero (excluded) retention time* (Z_eRT):

$$t_0 = \frac{\varepsilon_e V_{col}}{Q} \tag{1.27}$$

ε_e is typically in the range 0.35 to 0.4.

A second option (option T) is to consider that a non-retained solute has access to the *total* fluid phase contained in the column, so that we define a *zero (total) retention time* (Z_tRT):

$$t_{0R} = \frac{\varepsilon_T V_{col}}{Q} \tag{1.28}$$

With intragranular porosities ranging from 0.3 to 0.7, ε_T is typically in the range 0.5 to 0.8.

In case one elects to work with the lumped model, only option E that uses the Z_eRT as reference makes sense, porosities ε_i and thus ε_T not being defined.

When the porous model is selected, either option E or option T can be used. This choice, however, produces some fundamental differences in the system description, and these differences will show that the extragranular porosity choice is better suited to modeling. We are aware that this is not the choice generally made by chromatographers, and we know from many discussions that this choice is a source of surprise or even discomfort. The reasons for our preference for option E are presented below.

If one uses option E, thus the Z_eRT as a reference, Eq. (1.25) expressing the retention time t_R gives, after simple rearrangement:

$$\begin{aligned} t_R &= t_0[1 + \overline{K}'_A] \\ \overline{K}'_A &= \frac{1-\varepsilon_e}{\varepsilon_e}\overline{K}_A = \frac{1-\varepsilon_e}{\varepsilon_e}\grave{K}_A\left[\varepsilon_i + (1-\varepsilon_i)\overline{\overline{K}}_A\right] \end{aligned} \tag{1.29}$$

If one uses option T, thus the Z_tRT as a reference, Eq. (1.25) leads to:

$$\begin{aligned} t_R &= t_{0R}\left[\frac{1-\varepsilon_e}{\varepsilon_T}\varepsilon_i(\grave{K}_A - 1) + 1 + k'_A\right] \\ k'_A &= \frac{1-\varepsilon_T}{\varepsilon_T}\grave{K}_A\overline{\overline{K}}_A \end{aligned} \tag{1.30}$$

These equations have a bit of an unfamiliar shape for chromatographers because possible solute exclusion is taken into account.

If one considers situations in which no solute exclusion occurs, so that $\grave{K}_A = 1$, Eq. (1.29) and Eq. (1.30) become:

$$t_R = t_0(1 + \overline{K}'_A) \quad \text{with} \quad \overline{K}'_A = \frac{1-\varepsilon_e}{\varepsilon_e}\overline{K}_A \tag{1.31}$$

and

$$t_R = t_{0R}(1 + k'_A) \quad \text{with} \quad k'_A = \frac{1-\varepsilon_T}{\varepsilon_T}\overline{\overline{K}}_A \tag{1.32}$$

The coefficients $\overline{K}'_A$ and k'_A are called *retention factors*.

The shape indicated by $t_R = t_0(1 + k')$ is very familiar to chromatographers, but we stress that Eq. (1.31) is based on the Z_eRT (the less common choice for chromatographers) whereas Eq. (1.32) is based on the Z_tRT (the more common choice for chromatographers).

Choosing between option E and option T, and thus between Eq. (1.31) and Eq. (1.32), seems to be a question of taste. This is true, at equilibrium, and in the absence of solute exclusion.

If we consider possible solute exclusion, the choice becomes obvious:

- The retention time given by Eq. (1.29) shows additive contributions (time spent in the extragranular fluid, in the intragranular fluid and on the solid skeleton), as expected.
- The retention time given by Eq. (1.30) again shows additive contributions, but one is unexpectedly negative (when $\grave{K}_A < 1$) as if a "negative adsorption" were taking place. This is obviously counterintuitive and misleading. The retention time of the tracer (Z_tRT) being greater than the retention time of the totally excluded and non-adsorbed solute, Eq. (1.30) must include a negative contribution.

Let us finally note a fundamental difference between the two options. The porous model, when using the Z_tRT as a reference time, emphasizes the difference between a *solid phase* (the intragranular fluid is not taken into account in this zone) and a *fluid phase* (including the fluid contained in the pores). The lumped model and the porous model, when using the Z_eRT as a reference time, emphasize the difference between a lumped solid phase (the intragranular fluid is taken into account in this zone) and a fluid *mobile* phase (thus excluding the fluid contained in the pores). This is schematically represented in Figure 1.3.

Focusing on the difference between lumped solid and fluid *mobile* phases using the Z_eRT as a reference thus has the big advantage of describing situations from total

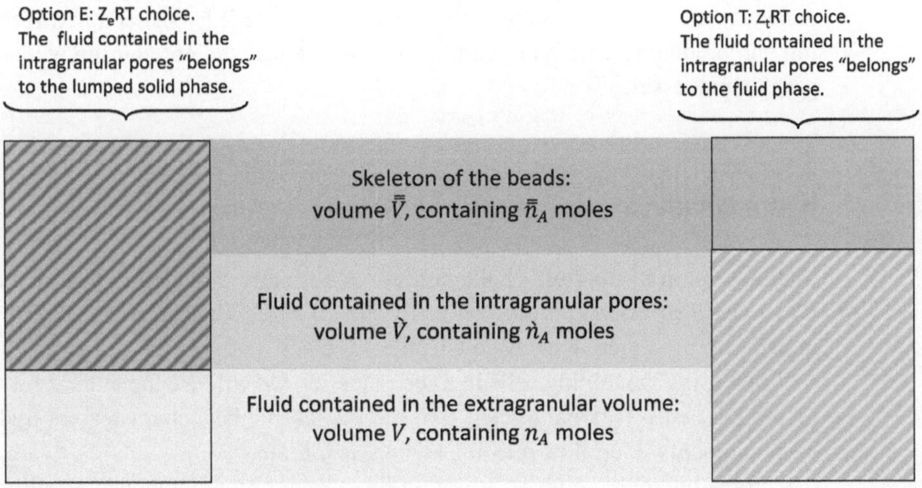

Figure 1.3 Two options for describing the fluid–solid system.

exclusion to adsorption within the same theoretical framework. That is a reason why the Z_eRT reference is the one selected by chemical engineers. The reader can easily check this assertion by reading the numerous references given in this book. Other reasons favoring the use of the Z_eRT as a reference will be given in subsequent sections.

1.4 Kinetics

The porous model and the lumped model can be extended to describe non-equilibrium situations. As chromatographic processes are frequently limited by diffusional transport occurring inside the particles, modeling frequently involves Fick's law, associated with a more or less complex geometrical description of the beads.

We show in Chapter 5 how to relate the microscopic descriptions provided by more or less refined porous models to the lumped model, and for that we will define lumped diffusion coefficients, just as we defined lumped concentrations.

Modeling diffusional transport inside particles requires describing the evolution of internal concentration profiles. Assuming that a porous model is used to describe the particles, this especially requires the description of the spatial evolution of the concentration $\grave{C}_A$ in the intragranular fluid, which is obviously incompatible with the assumption $\grave{C}_A = C_A$.

Consequently, when associating the porous model with mass balances written at the column level, using the Z_tRT as a reference is not really an option as it implicitly assumes $\grave{C}_A = C_A$ to be practical. If one uses the Z_tRT as a reference for modeling chromatographic columns in which kinetic limitations occur, one has to accept impractical expressions.

This is another important argument in favor of using the Z_eRT as a reference which offers a fully consistent framework for modeling situations ranging from equilibrium to strong diffusional limitations. Once again, we appreciate that this choice is not the one generally made by chromatographers, but we believe it is best adapted for understanding and predicting column dynamics, i.e. designing efficient and innovative chromatographic processes.

1.5 Hydrodynamics

The dispersion of the fluid in the column is normally modeled either by the *plug flow plus axial dispersion* (PD) model or by the *mixing cells in series* (MC) model. These models are presented in detail in Chapters 2, 6 and 7.

When using the mixing cells in a series model, we will use the letter J for the number of mixing cells. This parameter, like the Péclet number, characterizes hydrodynamic dispersion only. Plug flow prevails when J is infinite.

Chromatographic columns are often represented by a number of equilibrium stages in series, so-called "plates". We will use letter N for the number of plates.

J and N may appear to represent similar concepts, leading to some confusion. The concepts are in fact different: the number of plates (N) is a parameter in which the effects of hydrodynamic dispersion and mass-transfer kinetics are lumped together, while the number of mixing cells (J) characterizes hydrodynamics only. In general, one has $J > N$.

The PD and MC models share the common assumption that velocity and concentration profiles are radially uniform. This assumption is at the origin of a widely accepted assumption that chromatography can be scaled up linearly with column cross section, namely: "in order to scale up a chromatographic process, it is sufficient to keep bed length and fluid velocity constant". This assertion is true – until it is not true!

The influence of radial non-uniformity will be studied specifically in Chapters 6 and 7. This is an important topic as it gives the main (only?) reason why chromatographic processes are not always as easily scalable as expected.

1.6 Modeling moving beds

Let us now address the general mass balance associated with a single-zone true moving bed, in which the solid and fluid phases are contacted counter-currently. The fluid is assumed to move upwards, the chromatographic medium downwards. According to our terminology, we then speak about *lumped solid moving phase* and *fluid moving phase*. Assuming steady state and incompressible fluids, as described in Chapter 8, the general expression for the mass balance, assuming no dispersion, is given by:

$$Q\frac{\partial C}{\partial z} - \grave{Q}\frac{\partial \grave{C}}{\partial z} - \bar{\bar{Q}}\frac{\partial \bar{\bar{C}}}{\partial z} = 0 \tag{1.33}$$

where Q is the extragranular fluid volumic flow rate, $\grave{Q}$ the volumic flow rate of the intragranular fluid and $\bar{\bar{Q}}$ the volumic flow rate of the bead skeleton.

It is important to note that the fluid concentrations C and $\grave{C}$ are associated with different signs in the mass balance as the former moves upwards while the latter moves downwards with the particles.

The intragranular fluid and skeleton flow rates can be written:

$$\begin{aligned} \grave{Q} &= \varepsilon_i \bar{Q} \\ \bar{\bar{Q}} &= (1 - \varepsilon_i)\bar{Q} \end{aligned} \tag{1.34}$$

where $\bar{Q}$ is the volumic flow rate of the beads.

Working with an extragranular porosity reference (option E) and thus lumped concentration $\bar{C}$, Eq. (1.33) allows us to readily obtain the classical conservation equation used for TMB modeling in numerous references (see Chapter 8):

$$Q\frac{\partial C}{\partial z} - \bar{Q}\frac{\partial \bar{C}}{\partial z} = 0 \tag{1.35}$$

Now let us describe our system with option T, thus differentiating between the solid skeleton and fluid phases. For simplicity, we assume that no exclusion occurs so that $\grave{C}_A = C_A$.

The mass balance (1.33) becomes:

$$(Q - \varepsilon_i \overline{Q}) \frac{\partial C}{\partial z} - (1 - \varepsilon_i) \overline{Q} \frac{\partial \overline{\overline{C}}}{\partial z} = 0 \tag{1.36}$$

Equation (1.36) is a bit less symmetrical than Eq. (1.35) obtained with option E. The underlying reason is that making a distinction according to motion as in option E is more appropriate than making a distinction according to a physical state (solid or liquid) as in option T.

It seems that this issue is not always fully recognized in the literature, as option T (the classical one for chromatographers) is sometimes used to describe fluid–solid equilibria when Eq. (1.35) instead of Eq. (1.36) is used for the mass balance. This is a source of confusion that can lead to an incorrect allocation of the pore liquid motion.

The importance of proper solute allocation is simply understood with the important desalting processes involving ion exclusion. In these processes, some solutes can enter the particles, others cannot. If one uses option T, how do we properly write mass balances for excluded solutes? Improper definition of concentrations, due to an incorrect porosity reference, would lead the excluded products to move in an incorrect direction and the desalted solution to be ... pretty salty.

1.7 Terminology choices

To paraphrase F. Helfferich when he said "the world is full of eigenvalues", we can say that "the chromatographic world is full of misleading terms" and proper modeling certainly requires recognizing some terminology traps.

We have insisted in previous sections on the definition of terms as simple as *solid phase* and *fluid phase*, the risk of confusion with *phase mobility* being real. These definitions have direct consequences on concentration definitions.

We will see in subsequent chapters that terms as common as adsorption isotherm and diffusion coefficient can be confusing because they depend on the model selected. We will systematically use the adjective *lumped* when the variable or parameter is associated with the lumped model of particles.

The higher the number of components involved, the higher the risk of being confused or even misled. Consider the expressions "single-component" and "multi-component". They are very often used to refer to situations where the system contains in fact a single *solute* or multiple *solutes*: a chromatographic system cannot be single-component as it contains at least one solute (at the very least) and one solvent. The difference may look semantic, but it is not: realizing that diffusion is the motion of one species relative to others so that a diffusion problem cannot be single-component is essential for properly modeling this process. Similar considerations could be reported for adsorption, let alone ion exchange. The term "single-component" should preferably be banished from chromatography and replaced by "single-solute". As a consequence, we will also use the expression "multiple-solute" rather than "multi-component".[4]

Column modeling is in fact "no more" than incorporating hydrodynamic, kinetic and thermodynamic processes in mass-balance equations and then solving them. The variety of models available for describing hydrodynamics, thermodynamics and kinetics makes the number of possible combinations very significant, so that many column models are available.

The way various models are named may vary from one author to another, even though some terminology rationalization has been proposed, for instance by (Guiochon *et al.*, 2006) and (Seidel-Morgenstern *et al.*, 2012).

For a decade or two, chromatographers used the term "ideal model" for models neglecting hydrodynamic and kinetic dispersion and focusing on the impact of the adsorption isotherms on chromatograms. This approach was called the *Equilibrium Model* by their promoters (De Vault, Glüeckauf, Helfferich, Klein, Rhee and Amundson etc.). We believe that this earlier terminology is more appropriate as this model focuses on the influence of equilibrium properties on chromatogram characteristics. This being said, we acknowledge that equilibrium is also assumed to be reached in the plate model proposed by (Martin and Synge, 1941).

A model can be simple or complex, precise or very approximate, useful or useless, but in this domain, "ideality" is largely a matter of personal appreciation. For some mathematicians, the equilibrium model bears probably some "ideality" as some assumptions allow the use of specific mathematical tools. The point of view of other experts, investigating hydrodynamics via transfer functions, could very well be that an "ideal" model is a linear model because the Laplace toolbox can be used. As a chemical engineer – and this is a strictly personal point of view – the MC model is pretty close to being ideal. It is easy to understand and to program, shows flexibility and normally delivers adequate accuracy for design purposes.

Instead of using terms found in the literature such as "ideal model", "general rate model", "transport model", "dispersive model" that can potentially lead to confusion, we prefer naming the models precisely using the format *hydro–kinetics*.

Hydro refers to the hydrodynamic sub-model. It can thus be MC (mixing cells), PF (plug flow), PD (plug flow + dispersion), possibly completed with subscript R (radial) when a radial profile is taken into account.

Kinetics refers to the kinetic sub-model. It can be Fick, LDF or Equil when the mass-transfer contribution is taken into account according to, respectively, a Fickian mechanism, a Linear Driving Force approximation or is neglected so that the moving fluid and lumped solid phases are at equilibrium. In order to specify the type of particle description involved, we will specify the kinetics as $\overline{\text{Fick}}$, $\overline{\text{LDF}}$ or $\overline{\text{Equil}}$ if the lumped model is used and as $\overline{\overline{\text{Fick}}}$, $\overline{\overline{\text{LDF}}}$ or $\overline{\overline{\text{Equil}}}$ if the porous model is used.

[4] Notice that an n-solute system contains at least $n + 1$ components when one takes into account the solvent.

Thus the equilibrium model presented above, assuming plug flow in the column and equilibrium between the moving fluid phase and the lumped solid phase, will be referred to as PF–$\overline{\text{Equil}}$.

A model describing the column with mixing cells, using lumped solid concentrations, taking into account kinetic limitations (external and internal) modeled via LDF approximations will be referred to as MC–$\overline{\text{LDF}}_{\text{E+I}}$.

The model describing the column in the plug flow + dispersion representation, describing the diffusion of the species inside the pores of the particle with Fick's law and taking into account possible external mass transfer limitations will be referred to as PD–$\overline{\overline{\text{Fick}}}_{\text{E}}$.

Although this naming may seem a bit cumbersome, it has the merit of avoiding many ambiguities, and telling concisely the main assumptions of the model.

I am certainly not a big fan of using the qualifier "ideal" for models. My experience allows me to be a bit flexible about chromatographic systems. The definition of an "ideal" system from the perspective of the executive suite is probably: "An ideal system is one that is cheaper, more productive and more flexible than the one you just installed." Objectives are (normally) more realistic when interacting with scientists... When chromatography is one option amongst many, it is normally heavily challenged before being selected; the concept of avoiding chromatography as much as possible is still present in many chemists' heads. Chromatography can also be *the* option, other technologies being unable to deliver the required purity. When a decision is made to go with chromatography, failure is not an option due to big economic impact and possibly a lack of alternatives. Chromatography being normally presented as linearly scalable in cross section (a key advantage over other technologies), the scale-up must be no surprise. I saw often that, for users, "an ideal chromatographic process is a process that allows a direct scale-up from the laboratory apparatus", the unavoidable stress being kept for other matters. I will from time to time use the term "ideal situation" when the scale-up is linear in cross section, and "non-ideal situation" when it is more complex for temperature, compressibility or whatever reasons. Chromatography being generally linearly scalable in cross section, this definition has a big merit: it makes chromatography an ideal process ... in general!

We are now ready to dig into the details.

References

Carta, G. and Jungbauer, A. (2011), *Protein Chromatography*, Wiley-VCH.

Guiochon, G., Felinger, A., Golshan-Shirazi, S. and Katti, A. M. (2006), *Fundamentals of Preparative and Nonlinear Chromatography*, Elsevier Academic Press.

Helfferich, F. G. (1962), *Ion Exchange*, McGraw-Hill.

Helfferich, F. G. and Klein, G. (1970), *Multicomponent Chromatography*, Marcel Dekker.

Martin, A. and Synge, R. (1941), *Biochem. J.* **35**, 1358–1368.

Rhee, H.-K., Aris, R. and Amundson, N. R. (1989), *First-Order Partial Differential Equations. Vol. II: Theory and Applications of Hyperbolic Systems of Quasilinear Equations*, Prentice-Hall.

Rodrigues, A. E. and Tondeur, D. (1981), *Percolation Processes: Theory and Applications*, vol. 33 of *NATO ASI Series, Series E: Applied Sciences*, Sijthoff & Noordhoff Publishers.

Ruthven, D. M. (1984), *Principles of Adsorption and Adsorption Processes*, John Wiley & Sons.

Schmidt-Traub, H., Kaspereit, M., Engell, S., Susanto, A., Epping, A. and Jupke, A. (2012). *In* H. Schmidt-Traub, M. Schulte and A. Seidel-Morgenstern, eds., *Preparative Chromatography* 2nd edn., Wiley-VCH, p. 425.

Seidel-Morgenstern, A., Schmidt-Traub, H., Michel, M., Epping, A. and Jupke, A. (2012). In H. Schmidt-Traub, M. Schulte and A. Seidel-Morgenstern, eds., *Preparative Chromatography*, 2nd edn., Wiley-VCH, p. 321.

Wankat, P. C. (1986a), *Large-Scale Adsorption and Chromatography, Vol I*, CRC Press.

Wankat, P. C. (1986b), *Large-Scale Adsorption and Chromatography, Vol II*, CRC Press.

Wei, J. (1975), *Chemtech* **5**(2), 128–129.

2 Linear chromatography: the Russian Lego

This chapter is dedicated to Jacques Villermaux.

This chapter presents the basic concepts that allow the modeling of linear chromatographic systems, i.e. systems where the outlet signals (peaks, breakthrough curves) are proportional to the inlet concentration signal.

A chromatographic system can be considered linear at "sufficiently small" concentrations. As a rule of thumb, this is usually the case for concentrations lower than a few g/l for columns packed with silica or reversed phase and lower than 1 g/l on chiral media, whereas the chromatography of sugars is linear up to about 100 g/l on cationic ion-exchange resins. Conversely, when one considers separation of ions on resins (cationic or anionic), the system usually behaves linearly if the ionic fraction of the solute is lower than about 0.05.

In order to increase productivity, the inlet concentrations in preparative chromatography are normally made as high as possible, and the system usually does not behave linearly. Consequently, the theory of linear chromatography is usually not adequate for the comprehensive modeling of preparative chromatography. So why should we spend a fair number of pages describing this matter?

Probably most readers expect to find a rewriting of known analytical solutions for linear chromatography. Those who have looked at a few pages of this chapter have seen Laplace transforms and may fear the worse.

Let me explain why I believe this matter is possibly one of the least described and one of the most important. The theory of linear systems and consequently of linear chromatography mainly provides an efficient investigation tool for the study of hydrodynamics, kinetics and fluid–solid interactions. That is already a lot!

In order to benefit from this approach, one must accept working with the Laplace transform. This will most likely look strange to many chromatographers, as the Laplace transform is, to say the least, not their primary tool. I am aware of this risk, but I believe that spending some time understanding the concept is extremely rewarding as this will give access quickly to the significant properties of solutions of non-trivial problems.

When working in the real-time domain, modeling is quite cumbersome: some analytical solutions can be derived in specific situations, but changes as small as adding a dead volume will make the initial solution inoperative. By contrast, working with Laplace transforms (in fact with the so-called transfer functions) allows the building of models of increasing complexity, just as a child builds cars, planes or houses with *Lego*.

Similarly, models can be built in order to contain sub-models, a little bit like Russian dolls. That is why I use the term "Russian Lego" to tease the reader.

When this technique is mastered, modeling the influence of any possible dead volume, non-uniform velocity profile, particle size distribution or chromatographic media heterogeneity, or assessing different kinetic models, becomes a non-event. In short, deriving the key characteristics of a peak can be done without obtaining cumbersome expressions for this peak.

There is a price to pay: for the Lego pieces or Russian dolls to fit together, one needs to use precise definitions and formalism. That is the objective of the next section.

Readers willing to investigate these tools in greater detail are referred to articles written by (Villermaux, 1981) and, in a shorter version, by (Sardin *et al.*, 1991), describing the migration of solutes in porous media.

2.1 Basic definitions

A chromatographic column of volume V_{col} is eluted with a *constant* flow rate Q of a carrier fluid. A single solute is injected at the inlet of the column with the inlet signal $S^{inl}(t)$, while the time evolution of the outlet signal is $S^{out}(t)$ (Figure 2.1). The $S(t)$ signals can be the solute concentrations or not, and they may be expressed in arbitrary units, as long as they are proportional to the concentration.

The aim of modeling is to relate the outlet signal to:

- well-identified physico-chemical processes and their parameters;
- the operating conditions
- the shape of the inlet signal.

The first important step is to normalize the inlet and outlet signals and to define:

$$X(t) = \frac{S^{inl}(t)}{\int_0^\infty S^{inl}(t)dt}$$
$$Y(t) = \frac{S^{out}(t)}{\int_0^\infty S^{out}(t)dt} \tag{2.1}$$

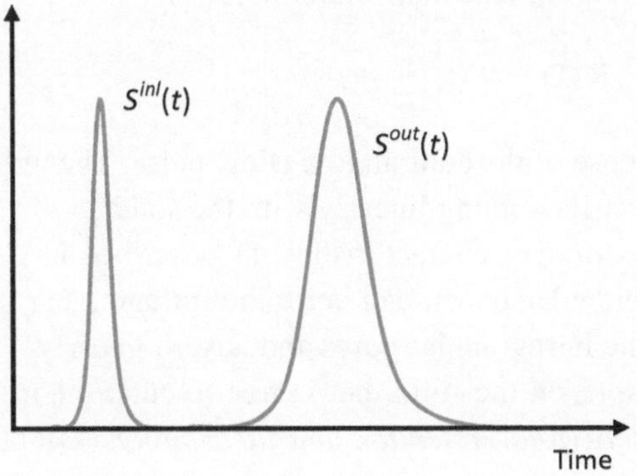

Figure 2.1 Time evolution of detected signals at the inlet and outlet of a chromatographic column.

The "mathematical monster" $\int_0^\infty S^{inl}(t)dt$ in Eq. (2.1) is no more than the area under the injected solute peak. Provided that the eluent flow rate is constant, $\int_0^\infty S^{inl}(t)dt$ is proportional to the total amount of solute injected. Assuming that all molecules of the solute exit the column sooner or later, $\int_0^\infty S^{inl}(t)dt$ should equal $\int_0^\infty S^{out}(t)dt$. If these two integrals are different, something is wrong! Reasons can include calculation mistakes, column leakage, solute being subject to some chemical reaction or non-constant eluent flow rate.

Note that the normalized signals $X(t)$ and $Y(t)$ are dimensional and that they have the dimension of reciprocal time, s^{-1}.

In practice, a precise estimate of the areas below the signals may require some expertise and care, especially with peaks exhibiting noise, when baseline drifts or when only discrete (sampled) information is available. This step is, however, very important as the theory presented later requires this area normalization. Our experience shows that an efficient way to measure these areas can be to fit the curve with an analytical expression (or simulated curve) and then to calculate the areas from the analytical expression (or simulated curve).

By definition, the areas under the $X(t)$ and $Y(t)$ signals equal unity:

$$\int_0^\infty X(t)dt = \int_0^\infty Y(t)dt = 1 \tag{2.2}$$

We can now introduce two relatively intuitive definitions:

Definition 1 *Two inlet signals are proportional if they have the same X(t).*

Definition 2 *The system is linear if proportional inlet signals lead to identical responses Y(t).*

As an illustration, in the case of a rectangular pulse injection of volume V_{inj} and concentration C_A^{inl}, the system is linear if the outlet signal $Y(t)$ is independent of C_A^{inl}.

It must, however, be noticed that $Y(t)$ depends on V_{inj}, even very slightly for very small V_{inj}, irrespective of the linearity of the system.

If the injected volume is very small (typically less than 1% of the column volume), the injection $X(t)$ can be approximated by a Dirac injection (pulse), and the normalized outlet signal $E(t)$ is called the *retention time distribution* (RTD):

$$\text{RTD} = E(t) = Y(t) \tag{2.3}$$

The RTD is the normalized response of the column to a Dirac pulse. The term RTD will be used in any situation of a tracer interacting linearly with the solid.

The injected solute can have different characteristics, as described in Chapter 1: it can be excluded from the intragranular pores, can enter the intragranular pores freely without adsorbing or can enter the intragranular pores and adsorb linearly.

When the solute *does not adsorb* on the solid, but is free to enter the *intragranular* pores or not, we will speak of a *zero (total) retention time distribution* (Z_tRTD), denoted by $E_t(t)$.

When the solute *does not adsorb* on the solid and *is excluded* from the *intragranular* pores (thus having access to the *extragranular* porosity only), we will speak of a *zero (excluded) retention time distribution* (Z_eRTD), denoted by $E_e(t)$.

RTD is the most general term, Z_eRTD the less. For instance, Z_tRTD and Z_eRTD are both RTDs. Additionally, a Z_tRTD can be a Z_eRTD but not the reverse. The distinction between these terms is based on considerations of porosity and adsorption presented in Chapter 1:

- The Z_eRTD contains all the information characterizing hydrodynamics only, i.e. the carrier fluid flow in the *extragranular* porosity.
- The Z_tRTD is impacted by hydrodynamics, and by mass-transfer processes outside and inside the beads, to the exclusion of adsorption on the solid.
- The RTD is impacted by hydrodynamics, mass-transfer processes and adsorption on the solid. Incidentally, let us mention that adsorption linearity must be checked prior to using the concept of RTD.

Notice that in the literature RTD is normally the acronym used for *residence* (not retention) time distribution for a non-retained tracer. Because of the ambiguity of the concept of non-retained solute in chromatography, we decided to be precise about things and define three types of RTD (both having the dimension of reciprocal time, s^{-1}). We will use RTD as a generic term and specify Z_eRTD or Z_tRTD when necessary. Finally, for the chemical engineers, notice that the *residence* time distribution presented by (Wen and Fan, 1975) and (Levenspiel, 1972), aimed at investigating reactor mixing patterns, is our Z_eRTD.

The RTD $E(t)$ is of paramount importance in chromatography (and more generally in chemical engineering) and has a simple physical meaning: $\int_0^{t_s} E(t)dt$ represents the fraction of molecules of solute that have been in the column for a time less than t_s. In a differential form, $E(t)dt$ represents the fraction of solute molecules that have been in the column for a time between t and $t + dt$.

A simple relation can be given between the Z_tRTD and the outlet concentration signal. We assume that n_A^{tot} moles of a non-adsorbing solute A are injected into the system during the Dirac injection. The number of moles of solute having spent less than time t_s in the system is given by $n_A^{tot} \int_0^{t_s} E_t(t)dt$. This is also the number of moles having left the system prior to time t_s which is given by $Q \int_0^{ts} C^{out}(t)dt$, where Q is the fluid flow rate, assumed to be constant.

As a consequence of the above:

$$E_t(t) = \frac{Q}{n_A^{tot}} C^{out}(t) \tag{2.4}$$

The shape of $E_t(t)$, containing all the information describing the time spent by the carrier fluid molecules moving across the column, is an invaluable tool for investigating models describing column hydrodynamics.

Most often the whole RTD does not need to be used and a sketchy description is sufficient. A good way to extract information from $E(t)$ consists in working with the moments of the distribution, defined by:

$$\mu_n = \int_0^\infty E(t) t^n dt \tag{2.5}$$

By definition of the RTD, μ_0 equals unity.

Similarly, one can define the centered moments of the distribution by:

$$\mu_{nc} = \int_0^\infty E(t)\,(t-\mu_1)^n\,dt \tag{2.6}$$

After simple math, one obtains:

$$\begin{aligned} \mu_{1c} &= 0 \\ \mu_{2c} &= \mu_2 - \mu_1^2 \\ \mu_{3c} &= \mu_3 + 2\mu_1^3 - 3\mu_1\mu_2 \end{aligned} \tag{2.7}$$

Practical difficulties in determining moments of a given experimental distribution increase with the order of the moment. This is mainly associated with uncertainties at large times due to limits in experimental precision. Even more advantageous than for the peak area normalization, fitting the curve with an analytical (or simulated) expression and using the moment of the analytical (or simulated) expression can help a lot.

When evaluating the centered moments of orders greater than 1 by direct numerical integration, it is recommended that Eq. (2.6) and not Eq. (2.7) be used because subtractions are involved in the latter equations, and they may induce uncertainties (including negative results).

The most commonly used moments are μ_1, which is the mean retention time t_R, and μ_2 and μ_3, which are associated with the variance σ^2 and the asymmetry of the output signal. According to definitions (2.5) and (2.6), one obtains:

$$\begin{aligned} t_R &= \mu_1 \\ \sigma^2 &= \mu_2 - \mu_1^2 \end{aligned} \tag{2.8}$$

Later on, like any chromatographer, we will use the term "retention time" instead of "mean retention time".

The mean retention time is also given by the ratio of the fluid volume accessible to the solute to the volumetric flow rate (assumed to be constant):

$$\begin{aligned} t_R &= \frac{\varepsilon_e V_{col}}{Q} = t_0 \quad \text{for the } \mathrm{Z_eRTD} \\ t_R &= \frac{\varepsilon_T V_{col}}{Q} = t_{0R} \quad \text{for the } \mathrm{Z_tRTD} \end{aligned} \tag{2.9}$$

In general, t_R is affected by adsorption, as discussed in Chapter 1; a more detailed account will be given in Section 2.3.

The asymmetry of the signal is typically characterized by its skewness S_k, defined by:

$$S_k = \frac{\mu_{3c}}{\mu_{2c}^{3/2}} \tag{2.10}$$

S_k is a non-dimensional number which is zero for a symmetrical distribution, positive if the dispersion exhibits a peak tailing and negative in the case of peak fronting.

In order to give orders of magnitude, and as shown later in this chapter, the skewness of a column equivalent to 10 000 plates is 0.02, whereas it is 0.2 for a 100-plate column and 2 for a 1-plate column.

Knowledge of the first two or three moments is normally considered to be sufficient to characterize classical chromatographic peaks. For less classical situations, analysis of the complete shape of the RTD is required.

Knowledge of $E(t)$ allows a calculation of the response of the system to any inlet signal, as shown below.

The inlet signal $X(\tau)$ can be subdivided into successive thin injections $X(\tau)d\tau$, each representing the fraction of solute injected between times τ and $\tau + d\tau$. By definition of the RTD, this fraction will give a contribution to the outlet elution profile given by $X(\tau)E(t-\tau)d\tau$.

By adding the contributions of all successive thin injections, one obtains the response of the column to the whole inlet signal $X(t)$:

$$Y(t) = \int_0^t X(\tau)E(t-\tau)d\tau \tag{2.11}$$

so that the outlet signal is the convolution product of the inlet signal by the RTD.

Though elegant, the above equation is a bit impractical, so that moving into the Laplace domain will prove to be extremely fruitful.

The Laplace transform is a suitable mathematical tool for solving linear chromatographic models, as shown by (Villermaux, 1974, 1981, 1987, 1990). By definition, the Laplace transform of a time-dependent function $x(t)$ is (Varma and Morbidelli, 1997):

$$\hat{x}(s) = \mathscr{L}(x(t)) = \int_0^\infty e^{-st}x(t)dt \tag{2.12}$$

The most interesting properties of Laplace transforms for chromatography are summarized in Appendix A.

We now define the transfer function of the column as the Laplace transform of the RTD:

- *The transfer function $\hat{E}(s)$ of the column is the Laplace transform of the* RTD *$E(t)$.*
- *More specifically, we will define $\hat{E}_e(s)$ as the Laplace transform of the* Z_eRTD *$E_e(t)$, and $\hat{E}_t(s)$ as the Laplace transform of the* Z_tRTD *$E_t(t)$*

By the definitions of the Laplace transform and of $E(t)$, one has $\hat{E}(0) = 1$.

As a consequence of the property of the Laplace transform regarding the convolution product (Appendix A), the Laplace transform of the outlet signal $\hat{Y}(s)$ is related to the Laplace transform of the inlet signal $\hat{X}(s)$ by:

$$\hat{Y}(s) = \hat{X}(s)\hat{E}(s) \tag{2.13}$$

If the transfer function $\hat{E}(s)$ is known, the response $\hat{Y}(s)$ of the system to any given inlet signal is thus easily calculated in the Laplace domain as the ordinary product of the transfer function $\hat{E}(s)$ by the Laplace transform of the inlet signal $\hat{X}(s)$.

In addition, as shown in Appendix A, the Laplace transform of a Dirac pulse is $\hat{X}(s) = 1$, while the Laplace transform of a step of height a is $\hat{X}(s) = a/s$, so that one can write:

- $\hat{Y}(s) = 1 \times \hat{E}(s) = \hat{E}(s)$ response to a Dirac injection in the Laplace domain
- $\hat{Y}(s) = \frac{a}{s} \times \hat{E}(s)$ response to a step of height a in the Laplace domain

All the information required to calculate the output signal (the chromatogram of a single solute) for any input signal is thus contained in the transfer function $\hat{E}(s)$ of the column.

Another very important property of the Laplace transform is given by Van der Laan's theorem. This allows calculation of the moments of the RTD by simple successive differentiations of the transfer function $\hat{E}(s)$:

$$\mu_n = (-1)^n \left. \frac{\partial^n \hat{E}(s)}{\partial s^n} \right|_{s=0} \tag{2.14}$$

We will show that the transfer functions can usually be derived easily. The theoretical expressions for the moments of the RTD are thus directly calculated using Van der Laan's theorem and thus without requiring knowledge of an analytical expression for the RTD and calculating the infinite integrals involved in Eq. (2.5).

Additionally, the simple observation of the shape of the RTD gives information on possible unwanted behaviors such as by-pass, non-uniform velocity profiles or inadequate distribution. Illustrations will be given in Chapter 6 when hydrodynamics is addressed.

Important note: due to the properties of the Laplace transform, and especially the one giving the Laplace transform of a time derivative (see Appendix A), the most efficient way to model a system is to use deviation variables. A deviation variable is the difference between a time-dependent variable (typically concentration in chromatography, possibly temperature) and the value of this variable at steady state for time $t = 0$. Very often, modeling chromatography aims at describing a transient behavior starting from a state at which the variables of interest equal zero. In this situation, the deviation variable equals the variable itself. This is especially the case when chromatograms obtained on an initially solute-free column are simulated. In order to keep our mathematical description as simple as possible, we will normally assume that the initial steady state is associated with a zero value of the concentration so that the deviation variable equals the variable. Specific warning (Exercise 2.2, Exercise 2.8 and Appendix A) will be given when this is not the case.

EXERCISE 2.1 *$F(t)$ is the response of an initially solute-free column (linear system) to a step input of magnitude 1. Show that $E(t)$ is the derivative of $F(t)$.*

EXERCISE 2.2 *Using the Laplace transform, derive the response of a solute-free, perfectly stirred cell to a step input of concentration C_A^F. Answer the same question for a perfectly stirred vessel at initial concentration $C_A^I \neq 0$ and flushed with a solution at concentration C_A^F.*

Figure 2.2 Representation of a column as a series of mixing cells (MC model).

2.2 Modeling hydrodynamic dispersion

In this section, we focus on investigating the flow inside a conventional fixed bed. For this purpose we consider a column packed with beads having *no* intragranular porosity and *no* interaction with the solutes (i.e. there is no retention of the solutes). We will show that a proper description of this "uncommon" situation is of primary importance for modeling more complex chromatographic processes.

Porosities were defined in Chapter 1, and typical values of the extragranular porosity usually lie in the range 0.35–0.40 for normally packed columns.

In addition to knowledge of porosity, modeling a chromatographic column requires selecting a physical representation for the bed. As shown in Figure 2.2, following the historical model of (Martin and Synge, 1941), a simple option consists in viewing the column as a series of J mixing cells, in each of which the concentrations are assumed to be uniform.[1] In what follows we will denote this as the mixing cells (MC) model. If the column is eluted at a flow rate Q, the following transient mass balance can be written for a solute A in each cell k:

$$QC_A^{k-1} = QC_A^k + \frac{\varepsilon_e V_{col}}{J}\frac{dC_A^k}{dt} \tag{2.15}$$

where C_A^k represents the concentration of solute A in cell k.

Introducing the Z_eRT t_0, Eq. (2.15) becomes:

$$C_A^{k-1} = C_A^k + \frac{t_0}{J}\frac{dC_A^k}{dt} \tag{2.16}$$

Moving into the Laplace domain, assuming a constant t_0 (thus a constant flow rate) and assuming $C_A^k\,(t=0)=0$, Eq. (2.16) becomes:

$$\hat{C}_A^{k-1} = \hat{C}_A^k + \frac{t_0}{J}s\hat{C}_A^k \tag{2.17}$$

We have thus transformed an ordinary differential equation (ODE) into a simple algebraic equation. This immediately leads to:

$$\hat{C}_A^{out} = \hat{C}_A^J = \hat{C}_A^{inl}\left[1+\frac{t_0}{J}s\right]^{-J} \tag{2.18}$$

where $\hat{C}_A^{out}$ and $\hat{C}_A^{inl}$ are the Laplace transforms of the outlet and inlet concentrations, respectively.

[1] As mentioned in Chapter 1, J is not yet the well-known plate number N as it is only related to the flow and does not contain elements related to mass-transfer or equilibrium relations between the fluid and the solid phases.

By comparison with Eq. (2.13), one obtains from Eq. (2.18) an expression for the transfer function $\hat{E}_e(s)$ for the MC model:

$$\hat{E}_e(s) = \left[1 + \frac{t_0}{J}s\right]^{-J} \tag{2.19}$$

Recall that $\hat{E}_e(s)$ is the transfer function of the MC model for a non-adsorbed solute injected into a column packed with non-porous particles (or for a non-adsorbed solute prevented from entering inside porous particles).

Another model consists in viewing the column as a continuous medium in which the fluid flow is governed by:

- an ideal plug flow, characterized by a superficial velocity $u = Q/\Omega_{col}$, assumed to be uniform over the column cross section
- a dispersive flow, described by a Fickian mechanism and characterized by an axial dispersion coefficient D_{ax}.

We will name this the plug flow + dispersion (PD) model; it is schematically represented in Figure 2.3.

A mass balance on solute A can be written over a slice of thickness dz:

$$\begin{aligned} QC_A(z,t) &= QC_A(z+dz,t) + \varepsilon_e \Omega_{col} dz \frac{\partial C_A(z,t)}{\partial t} \\ &\quad + \varepsilon_e \Omega_{col} D_{ax} \left\{ \left[\frac{\partial C_A(z,t)}{\partial z}\right]_z - \left[\frac{\partial C_A(z,t)}{\partial z}\right]_{z+dz} \right\} \end{aligned} \tag{2.20}$$

After some simple manipulations, one obtains:

$$\frac{u}{\varepsilon_e}\frac{\partial C_A(z,t)}{\partial z} + \frac{\partial C_A(z,t)}{\partial t} = D_{ax}\frac{\partial^2 C_A(z,t)}{\partial^2 z} \tag{2.21}$$

As $t_0 = \dfrac{\varepsilon_e V_{col}}{Q} = \dfrac{\varepsilon_e L_{col}}{u}$, Eq. (2.21) can be rewritten:

$$\begin{aligned} &\frac{\partial C_A(\xi,t)}{\partial \xi} + t_0 \frac{\partial C_A(\xi,t)}{\partial t} = \frac{1}{Pe}\frac{\partial^2 C_A(\xi,t)}{\partial^2 \xi} \\ &\xi = \frac{z}{L_{col}} \\ &Pe = \frac{u}{\varepsilon_e}\frac{L_{col}}{D_{ax}} \quad \text{Péclet number} \end{aligned} \tag{2.22}$$

which is a normalized version of the mass-balance equation for the PD model.

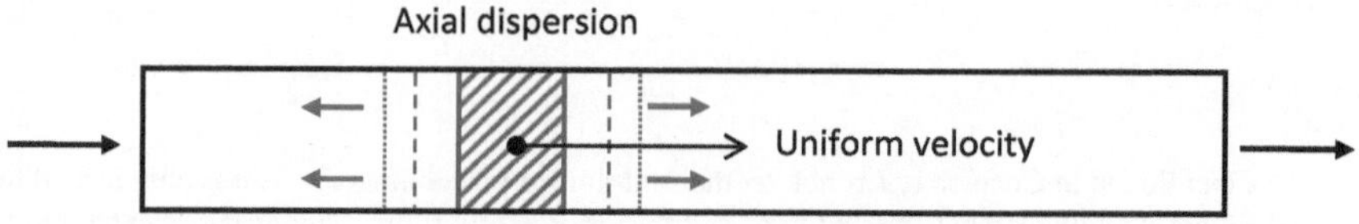

Figure 2.3 Representation of a column by the plug flow + dispersion (PD) model.

In the Laplace domain, and assuming $C_A\,(\xi,t=0)=0$, Eq. (2.22) gives:

$$\frac{d\hat{C}_A(\xi,s)}{d\xi}+t_0 s\hat{C}_A(\xi,s)=\frac{1}{Pe}\frac{d^2C_A(\xi,s)}{d^2\xi} \tag{2.23}$$

This is a homogeneous second-order linear ODE in $\hat{C}_A$. It can be integrated via classical methods provided that two boundary conditions are imposed. A simple way to look at these boundary conditions is the neglect of possible concentration discontinuities at the extremities of the column. Under this assumption, one has:

$$C_A^{inl}=C_A|_{\xi=0}\ \text{and}\ C_A^{out}=C_A|_{\xi=1}\ \text{are finite} \tag{2.24}$$

We will name these conditions the *simplified boundary conditions* (simplified BC).

It can be shown that these conditions are normally acceptable for *Pe* greater than about 20. At lower *Pe* values, they should be replaced by the so-called Danckwerts conditions. This is discussed in detail in Appendix B.

Using the simplified BC, we show in Appendix B that Eq. (2.24) has the solution:

$$\hat{C}_A^{out}=\hat{C}_A^{inl}\exp\left[\frac{Pe}{2}\left(1-\sqrt{1+\frac{4t_0s}{Pe}}\right)\right] \tag{2.25}$$

Consequently the transfer function for the PD model (simplified BC) is:

$$\hat{E}_e(s)=\exp\left[\frac{Pe}{2}\left(1-\sqrt{1+\frac{4t_0s}{Pe}}\right)\right] \tag{2.26}$$

$\hat{E}_e(s)$ is the transfer function of the PD model (simplified *BC*) for a non-adsorbed solute injected into a column packed with non-porous particles (or for a non-adsorbed solute prevented from entering inside porous particles).

EXERCISE 2.3 *Use the MC and PD transfer functions to calculate the response of a column of infinite efficiency (J or Pe being infinite) to a pulse.*

Van der Laan's theorem (2.14) allows determination of the moments of the distribution $E_e(t)$ by differentiation of the transfer function $\hat{E}_e(s)$. Key results are given in Table 2.1.[2] The Laplace transform allowed us to estimate the mean retention times, variances and skewness of both models without any complex calculations and without even obtaining an analytical expression for the RTD.

The mean retention time (based on moment definition) appears to be independent of the dispersion parameters and is given by t_0 (assuming simplified BC for the PD model). It is proportional to the column length.

[2] The moments of Table 2.1 are valid for the selected boundary conditions only. Other conditions would lead to different moments. For instance, the first moment associated with the PD model can be t_0, $t_0\left(1+\frac{1}{Pe}\right)$ or $t_0\left(1+\frac{2}{Pe}\right)$, corresponding to simplified, open–closed, closed–closed boundary conditions, respectively. Definitions of the different boundary conditions are given in Appendix B.

Table 2.1 Expressions for moment, variance and skewness associated with the MC and PD (simple BC) models.

	MC model	PD model (simplified BC)
$\mu_1 (= t_R)$	t_0	t_0
μ_2	$\frac{J+1}{J} t_0^2$	$\left(1+\frac{2}{Pe}\right) t_0^2$
μ_3	$\frac{(J+1)(J+2)}{J^2} t_0^3$	$\left[\left(1+\frac{2}{Pe}\right)\left(1+\frac{4}{Pe}\right)+\frac{4}{Pe^2}\right] t_0^3$
$\mu_{2c} (= \sigma^2)$	$\frac{1}{J} t_0^2$	$\frac{2}{Pe} t_0^2$
μ_{3c}	$\frac{2}{J^2} t_0^3$	$\frac{12}{Pe^2} t_0^3$
S_k	$\frac{2}{\sqrt{J}}$	$\frac{3}{\sqrt{Pe/2}}$

As shown in Table 2.1, the variances given by the two models are proportional to t_0^2 and to the reciprocal of J (resp. Pe), so that the two models give the same variance if:

$$Pe = 2J \tag{2.27}$$

This well-known relation between Pe and J defines an equivalence between the MC and PD models. The quality of this equivalence will be assessed later in this section.

Let us now spend a bit of time on the skewness factor, which is not often considered. For skewness different from zero (Table 2.1), the RTD is never symmetrical, even in the case of the very simple models dealt with up to now. Positive skewness means that peaks are tailing; however, the larger J (or Pe), the weaker the asymmetry and the more inconspicuous the tail.

Additionally, using the $Pe = 2J$ equivalence in the expressions for the skewness factors given in Table 2.1 does not allow us to obtain rigorously identical skewness for the MC and PD models. In practice the impact of this difference is insignificant for sufficiently high $Pe = 2J$. Yet it is worth keeping in mind the intrinsic difference between the two models and thus to be prepared to identify situations where special care must be taken.

We want to stress the fact that the retention times and variances presented above are rigorously defined from the moments of the RTD. For convenience, however, approximations are often introduced by chromatographers. As shown in Figure 2.4, the idea is to roughly characterize the peak by the position of its maximum, the width W at mid-height or the projection on the time axis of the inflection tangents b. Retention times and variance are then calculated with formulas given in Table 2.2.

Using the approximations in the table one obtains:

$$\begin{aligned} t_R &= t_R^{max} && \text{instead of } \mu_1 \\ J &= \frac{Pe}{2} \approx 5.54\left(\frac{t_R^{max}}{W}\right)^2 \approx 16\left(\frac{t_R^{max}}{b}\right)^2 && \text{instead of } \frac{\mu_1^2}{\mu_{2c}} \end{aligned} \tag{2.28}$$

Table 2.2 Rigorous and approximate definitions of the mean retention time and variance of the RTD.

	Rigorous definition	Generally used
t_R	μ_1	t_R^{max}
σ^2	$\mu_2 - \mu_1^2$	$\frac{W^2}{5.54} \simeq \frac{b^2}{16}$

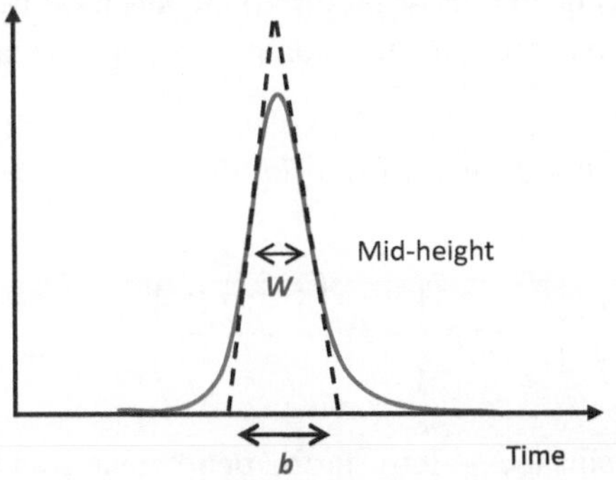

Figure 2.4 Approximate chromatogram characteristics.

It is emphasized that care must be taken when using these approximations, since they are strictly valid only for Gaussian peaks. For instance (and as illustrated later), the position of the peak maximum is affected by hydrodynamics and mass transfer limitations, while the first moment is not. This is not a big concern for relatively symmetrical chromatograms obtained on efficient columns but can be odd for asymmetrical peaks. Moreover, for peaks showing some tailing, the efficiency measured with the moments can be significantly lower than the result obtained with mid-height measurement. Using optimistic efficiencies based on mid-height measurement can certainly lead to disappointing performance. This is especially true when implementing a process aimed at delivering high-purity fractions with not very well-separated solutes.

Finally, by inverting the transfer functions $\hat{E}_e(s)$, thus returning from the Laplace domain to the time domain, one obtains the RTD. For the MC model one obtains:

$$E_e(t) = \frac{1}{t_0} \frac{J^J}{(J-1)!} \left(\frac{t}{t_0}\right)^{J-1} \exp\left(-J\frac{t}{t_0}\right) \tag{2.29}$$

When J is high, the factorial in Eq. (2.29) can be replaced by the Stirling relation, leading to an expression a bit more convenient for numerical estimates:

$$E_e(t) = \frac{1}{t_0} \frac{J^J}{(J-1)^{J-1}} \left(\frac{t}{t_0}\right)^{J-1} \frac{\exp\left(-J\frac{t}{t_0} + J - 1\right)}{\sqrt{2\pi(J-1)}} \tag{2.30}$$

For the PD model associated with the simplified BC, the inversion leads to analytical solutions presented by (Aris, 1959), (Gibilaro, 1978):

$$E_e(t) = \frac{1}{t_0}\frac{1}{2\frac{t}{t_0}}\left[\frac{Pe}{\pi\frac{t}{t_0}}\right]^{1/2} \exp\left(-Pe\frac{\left(1-\frac{t}{t_0}\right)^2}{4\frac{t}{t_0}}\right) \tag{2.31}$$

Let us stress again the fact that Eq. (2.31) assumes the simplified BC and is thus valid for *Pe* greater than about 20.

Solutions to the PD (open–closed) or PD (closed–closed) models have been derived by (Villermaux and Swaay, 1969) but they are too complex to be of real practical interest.

Note: an expression looking similar but different from Eq. (2.31) can be found in the literature:

$$E_e(t) = \frac{1}{2t_0}\left[\frac{Pe}{\pi\frac{t}{t_0}}\right]^{1/2} \exp\left(-Pe\frac{\left(1-\frac{t}{t_0}\right)^2}{4\frac{t}{t_0}}\right) \tag{2.32}$$

This other expression does not contain the $\frac{t}{t_0}$ term in the denominator and is in fact associated with open–open boundary conditions (see definition in Appendix B) and is possibly less relevant for chromatography.

For comparison, the expression for a Gaussian distribution which is symmetrical, with the same first-order moment and variance, would be:

$$E_e^{Gauss}(t) = \frac{1}{2t_0}\left[\frac{Pe}{\pi}\right]^{1/2} \exp\left(-Pe\frac{\left(1-\frac{t}{t_0}\right)^2}{4}\right) \tag{2.33}$$

EXERCISE 2.4 *Using the MC model, estimate the time at which the maximum of the RTD occurs. Compare this to the mean retention time. Perform the same comparison for the peak variance measured either by mid-height width or by the rigorous formula. Assume that the column is represented by 10 and 100 mixing cells. The reader is then strongly encouraged to go to Exercise 2.5 to realize how bad this approximation can be.*

EXERCISE 2.5 *Using the graphical method (peak maximum, and mid-height width), estimate the mean retention time and variance (thus, number of plates) of the peak given in Figure 2.5. Compare the experimental peak with the peak for the MC model associated with the correct mean retention time and number of plates determined above.*

Equations (2.29) through (2.33) allow one to easily compute chromatogram shapes for various retention and dispersion parameters.

The theoretical RTD curves obtained with both models are compared in Figure 2.6. The MC and PD models lead to very similar curves if $Pe = 2J$ is greater than about 20. It must be kept in mind that these results are rigorous only for linear chromatography. In practice we can use either the PD or the MC model for efficient columns, i.e. $Pe \approx 2J > 20$. It is largely a matter of personal preference.

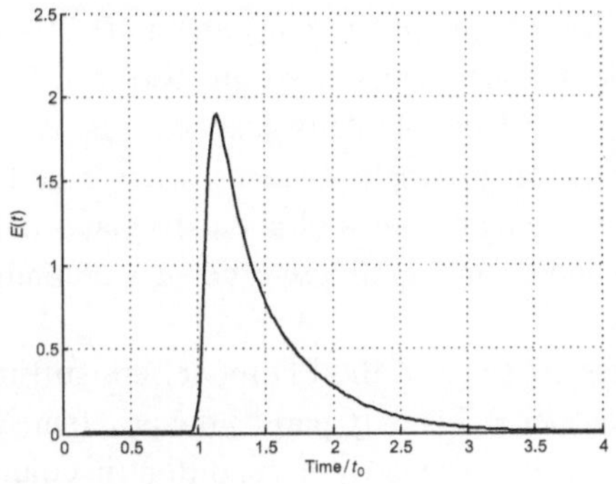

Figure 2.5 For Exercise 2.5

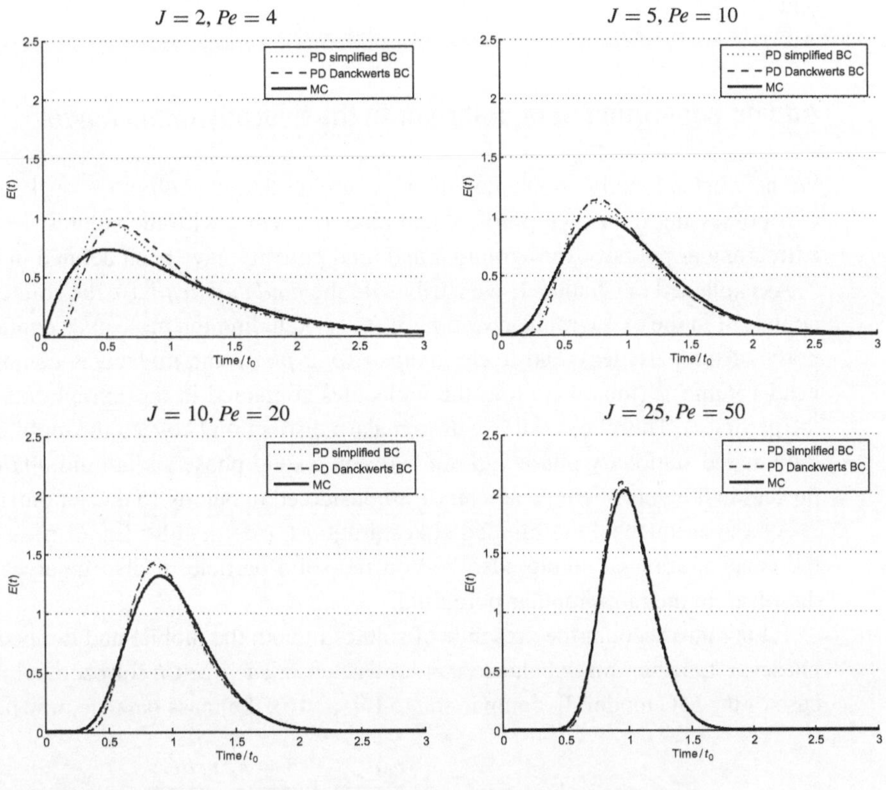

Figure 2.6 Comparison of RTD curves obtained with MC and PD models.

When $Pe = 2J$ is small, the Danckwerts conditions are normally considered to be relevant (at least more relevant than the simplified BC). In such a case the transfer function of the PD model must be inverted numerically, as shown for instance by (Seidel-Morgenstern, 1991). Taking into account the Danckwerts BC instead of the simplified BC gives different RTD shapes for low-efficiency columns, as shown in Figure 2.6.

All curves being similar if $Pe = 2J$ is high enough (say, about 10–20), this discussion of BC may look a bit academic as many separations are performed on columns equivalent to 100 plates or more. This is in fact not totally academic, as we will show in Chapter 10 that optimum continuous chromatography systems, when used for performing relatively easy separations, may be using columns of about 10 plates only! Adding plates may be very costly, so precise understanding of low-efficiency columns is indeed of interest.

It is, however, unlikely that the assumptions of the PD model are still met in such dispersive environments, so setting proper BC is only part of the problem. What is the real meaning of the dispersion concept and of Danckwerts conditions in columns having wide and complex distributors? Under low-efficiency conditions especially, using a model as close as possible to reality will be of utmost importance when addressing modeling of non-linear behaviors. More detailed information is provided in Chapters 6 and 7.

2.3 Adding adsorption at equilibrium to the hydrodynamic model

We now extend the previous models by assuming that the column is packed with particles possessing their own porosity and able to interact with the solutes. In addition to extragranular porosity, intragranular and total porosity have been defined in Chapter 1.

As explained in Chapter 1, we attribute to the *fluid mobile phase* the solute molecules contained in the extragranular volume V (thus excluding the molecules contained in the pores of the particles), and to the *lumped solid phase* the molecules contained in the bead volume $\overline{V}$ (thus including the molecules contained in the intragranular liquid of the particles). This choice allows us to make a distinction between the mobile phase and the lumped stationary phase and not between a solid phase and a fluid phase: the fluid present in the pores being stagnant, it is considered to belong to the stationary phase.

As a consequence, the lumped concentration $\overline{C}_A$ defined by Eq. (1.6) is not simply the concentration of solute adsorbed on the solid particle: it also involves the solute dissolved in the intragranular pore fluid.

Taking into account the presence of solutes in both the mobile and lumped stationary phases, a transient mass balance can be written in each cell k for each solute A in the case of the MC model. In comparison to Eq. (2.16), the mass balance now becomes

$$C_A^{k-1} = C_A^k + \frac{t_0}{J}\frac{dC_A^k}{dt} + \frac{1-\varepsilon_e}{\varepsilon_e}\frac{t_0}{J}\frac{d\overline{C}_A^k}{dt} \tag{2.34}$$

Similarly, the mass-balance equation (2.22) associated with the PD model becomes:

$$\frac{\partial C_A(\xi,t)}{\partial \xi} + t_0\frac{\partial C_A(\xi,t)}{\partial t} + \frac{1-\varepsilon_e}{\varepsilon_e}t_0\frac{\partial \overline{C}_A(\xi,t)}{\partial t} = \frac{1}{Pe}\frac{\partial^2 C_A(\xi,t)}{\partial^2 \xi} \tag{2.35}$$

If the flow rate is sufficiently small, one can expect equilibrium to be reached between the solid and the fluid phase. If in addition the concentrations are small enough, one can assume that they are linearly related:

$$\overline{C}_A = \overline{K}_A C_A \tag{2.36}$$

$\overline{K}_A$ being the lumped Henry's coefficient.

Note again that the coefficient $\overline{K}_A$ takes account of all solute molecules in the intragranular pores and on the solid skeleton of the stationary phase.

Because equilibrium is assumed, the MC and PD models are now named more precisely as MC$-\overline{\text{Equil}}$ and PD$-\overline{\text{Equil}}$.

Inserting Eq. (2.36) into Eq. (2.34) or Eq. (2.35), we obtain:

$$C_A^{k-1} = C_A^k + \frac{t_0}{J}\left(1 + \frac{1-\varepsilon_e}{\varepsilon_e}\overline{K}_A\right)\frac{dC_A^k}{dt} \quad \text{MC}-\overline{\text{Equil}} \tag{2.37}$$

$$\frac{\partial C_A(\xi,t)}{\partial \xi} + t_0\left(1 + \frac{1-\varepsilon_e}{\varepsilon_e}\overline{K}_A\right)\frac{\partial C_A(\xi,t)}{\partial t} = \frac{1}{Pe}\frac{\partial^2 C_A(\xi,t)}{\partial \xi^2} \quad \text{PD}-\overline{\text{Equil}} \tag{2.38}$$

By introducing:

$$\begin{aligned} t_R &= t_0\left[1 + \frac{1-\varepsilon_e}{\varepsilon_e}\overline{K}_A\right] = t_0\left[1 + \overline{K}_A'\right] \\ \overline{K}_A' &= \frac{1-\varepsilon_e}{\varepsilon_e}\overline{K}_A \end{aligned} \tag{2.39}$$

Eq. (2.37) and Eq. (2.38) can be rewritten as:

$$C_A^{k-1} = C_A^k + \frac{t_R}{J}\frac{dC_A^k}{dt} \tag{2.40}$$

$$\frac{\partial C_A(\xi,t)}{\partial \xi} + t_R\frac{\partial C_A(\xi,t)}{\partial t} = \frac{1}{Pe}\frac{\partial^2 C_A(\xi,t)}{\partial \xi^2} \tag{2.41}$$

Noting that Eq. (2.40) and Eq. (2.41) have exactly the same form as Eq. (2.16) and Eq. (2.22) obtained for an excluded solute, the transfer functions are readily obtained:

$$\hat{E}(s) = \left[1 + \frac{t_R}{J}s\right]^{-J} \quad \text{MC}-\overline{\text{Equil}} \tag{2.42}$$

$$\hat{E}(s) = \exp\left[\frac{Pe}{2}\left(1 - \sqrt{1 + \frac{4t_R s}{Pe}}\right)\right] \quad \text{PD}-\overline{\text{Equil}} \text{ (simplified BC)} \tag{2.43}$$

The immediate derivation of Eq. (2.42) and Eq. (2.43) on the basis of Eq. (2.16) and Eq. (2.21) is a first simple illustration of the Lego approach to modeling that we referred to in the chapter introduction. Starting from a well-defined first brick (modeling the RTD of an excluded product) one can then add bricks of different complexities (intragranular porosity for now) while keeping the same model structure. This will be further illustrated in the following sections and chapters. The approach is made possible by the precise definitions progressively introduced in the chapters. Based on these definitions the various processes can be hierarchically organized as compartments and subcompartments that make modeling easy.

Table 2.3 Expressions for moments, variance and skewness associated with the MC and PD (simplified BC) models. Linear adsorption at equilibrium is assumed. $\overline{K}'_A = \frac{1-\varepsilon_e}{\varepsilon_e}\overline{K}_A$.

	MC−$\overline{\text{Equil}}$	PD−$\overline{\text{Equil}}$ (simplified BC)
μ_1	$t_R = t_0\left(1+\overline{K}'_A\right)$	$t_R = t_0\left(1+\overline{K}'_A\right)$
μ_2	$\frac{J+1}{J}t_R^2$	$\left(1+\frac{2}{Pe}\right)t_R^2$
μ_3	$\frac{(J+1)(J+2)}{J^2}t_R^3$	$\left[\left(1+\frac{2}{Pe}\right)\left(1+\frac{4}{Pe}\right)+\frac{4}{Pe^2}\right]t_R^3$
$\mu_{2c}\ (\equiv\sigma^2)$	$\frac{1}{J}t_R^2$	$\frac{2}{Pe}t_R^2$
μ_{3c}	$\frac{2}{J^2}t_R^3$	$\frac{12}{Pe^2}t_R^3$
S_k	$\frac{2}{\sqrt{J}}$	$\frac{3}{\sqrt{Pe/2}}$

Using Van der Laan's theorem (2.14), the moments of the distribution $E(t)$ are obtained from the transfer function $\hat{E}(s)$. The useful expressions obtained for the moments are summarized in Table 2.3. In comparing these to Table 2.1, one recognizes that, in replacing t_0 by t_R, one gets exactly the expressions obtained on a column packed with non-porous beads having no interaction with the solutes.

Again, the transfer function approach allows us to demonstrate that the MC and PD models (assuming the simplified BC for the PD model):

- lead to the same retention times, independent of the parameters Pe and J
- give the same variance, provided $Pe = 2J$.

As with the excluded solute and the Z_eRTD, the identity of the two leading moments does not guarantee the identity of the two RTDs. In practice, the difference between the MC and PD models is, however, small enough so as to make them indistinguishable, provided J is greater than about 20.

From Eq. (2.42) and Eq. (2.43) one concludes that the analytical expressions for the RTD are still given by Eq. (2.29) to Eq. (2.31), replacing t_0 by t_R.

EXERCISE 2.6 *Show that the retention time associated with a step input (frontal chromatography) is still given by the expression for Table 2.3, provided linear conditions prevail.*

EXERCISE 2.7 $\overline{V} = 3$ *ml of porous silica at incipient wetness is mixed in a batch vessel with* $V = 10$ *ml of hexane initially containing 1 g/l of solute A. After equilibration, the concentration of A in hexane is 0.2 g/l. A small injection of A is performed on a column of 1 l packed with the same silica and eluted with a flow rate of 1 l/min of hexane. Which physico-chemical parameter or parameters are identical in the preliminary and chromatographic experiments? Estimate the retention time of A (hint: this exercise uses some orders of magnitude given earlier).*

2.4 Adding diffusional limitations to hydrodynamics and adsorption

In the previous section, adsorption equilibrium between the phases was presumed to be reached. This assumption might not always be valid, especially for small retention times.

It is well accepted that chromatographic processes are normally kinetically controlled by diffusional processes. A "rigorous" modeling would involve solving a law of diffusion (Fick, Nernst–Planck or Maxwell–Stefan according to the situation) in an adequate geometry. It will be shown in Chapter 5 that this model can be replaced by a simpler one, which only takes into account the volume-averaged concentration inside the beads. According to this model, denoted as the linear driving force (LDF) approximation, the uptake of molecules by the beads is approximated by a first-order dynamic system, such that the rate at which the system goes towards equilibrium is proportional to the deviation from equilibrium in the lumped solid phase:

$$t_A^m \frac{d\overline{C}_A}{dt} = \overline{K}_A C_A - \overline{C}_A \tag{2.44}$$

where t_A^m is the characteristic time for mass transfer. The time t_A^m gives an order of magnitude for the time needed to saturate a bead.[3] Its dependence on particle diameter, shape and diffusion coefficient will be discussed in Chapter 5.

As a limit assumption, if $t_A^m = 0$, Eq. (2.44) becomes

$$0\frac{d\overline{C}_A}{dt} = 0 = C_A - \overline{C}_A/\overline{K}_A$$

so that the system is at equilibrium.

The comparison of characteristic times provides a simple assessment of the importance of kinetic limitations in a chromatographic column: if $t_R \gg t_A^m$, the retention time, i.e. the time spent in the column, is much greater than the time needed for diffusion to reach equilibrium; as a consequence the system behaves as if t_A^m were close to or equal to 0, i.e. as if the phases were close to or at equilibrium. We will show later in this chapter and in Chapter 5 that important chromatogram characteristics are directly dependent upon different characteristic time ratios.

Moving into the Laplace domain will drastically simplify column modeling. In the Laplace domain, assuming $C_A = 0$ initially, the kinetic equation (2.44) becomes:

$$\hat{\overline{C}}_A = \frac{\overline{K}_A \hat{C}_A}{1 + t_A^m s} \tag{2.45}$$

Once again, if $t_A^m = 0$, diffusion is infinitely fast, and Eq. (2.45) becomes $\hat{\overline{C}}_A = \overline{K}_A \hat{C}_A$, so that equilibrium is reached.

[3] For instance, one can show from Eq. (2.44) that the mean concentration of solute in a bead initially free of A and put in contact with a solution of concentration C_A^I varies according to $\overline{C}_A = \overline{K}_A C_A^I \left[1 - \exp\left(-t/t_A^m\right)\right]$. From this equation it is concluded that equilibrium is "roughly reached" after $5t_A^m$. Note that this would be the time required to obtain the equilibration required in Exercise 2.7.

Equations (2.44) or (2.45) hold anywhere in the column, so the MC and PD models can easily be modified to account for mass transfer limitations modeled with the LDF approximation. These models will be denoted MC$-\overline{\text{LDF}}$ and PD$-\overline{\text{LDF}}$, respectively. The bar on top of LDF is a reminder that a lumped model is used.

Starting with the MC model, insertion of Eq. (2.44) into the Laplace transform of the mass-balance equation (2.34), which holds whether phase equilibrium is reached or not, leads to:

$$\hat{C}_A^{k-1} = \hat{C}_A^k + \frac{t_0}{J}s\hat{C}_A^k + \frac{1-\varepsilon_e}{\varepsilon_e}\frac{t_0}{J}s\frac{\overline{K}_A\hat{C}_A^k}{1+t_A^m s} = \left(1+\frac{t_0}{J}s\left(1+\frac{1-\varepsilon_e}{\varepsilon_e}\frac{\overline{K}_A}{1+t_A^m s}\right)\right)\hat{C}_A^k \tag{2.46}$$

so that the transfer function associated with the MC$-\overline{\text{LDF}}$ model is:

$$\hat{E}(s) = \frac{\hat{C}_A^J}{\hat{C}_A^0} = \frac{1}{\left(1+\frac{t_0 s}{J}\left(1+\frac{1-\varepsilon_e}{\varepsilon_e}\frac{\overline{K}_A}{1+t_A^m s}\right)\right)^J} \tag{2.47}$$

Using a similar procedure, one obtains the transfer function of the PD$-\overline{\text{LDF}}$ model (with simplified BC):

$$\hat{E}(s) = \exp\left[\frac{Pe}{2}\left(1-\sqrt{1+\frac{4t_0}{Pe}s\left(1+\frac{1-\varepsilon_e}{\varepsilon_e}\frac{\overline{K}_A}{1+t_A^m s}\right)}\right)\right] \tag{2.48}$$

Here again, the immediate derivation of equations (2.47) and (2.48) illustrates the Lego approach. The Van der Laan Eq. (2.14) then allows us to obtain the moments of the distributions which are given in Table 2.4.

The expressions given in Table 2.4 reveal two important results:

- The retention time (first moment) is not affected by mass transfer limitations (again assuming simplified BC for the PD$-\overline{\text{LDF}}$ model).
- The variance appears as the sum of a hydrodynamic contribution and a kinetic contribution.

The MC$-\overline{\text{LDF}}$ and PD$-\overline{\text{LDF}}$ models are clearly more sophisticated than the MC and PD models, since kinetic limitations are taken into account. It is, however, convenient, and often sufficient, to lump hydrodynamic and kinetic contributions into a single parameter: the number of theoretical stages or simply the plate number N.

The idea is to mimic both mass transfer (i.e. t_A^m) and hydrodynamic dispersion (i.e. J or Pe) by an equivalent pseudo-dispersion which is accounted for by an equivalent number of N mixing cells where the phases are at equilibrium. This leads to the MC$-\overline{\text{Equil}}$ model. Since the width, and thus the variance, of a peak is larger in the presence of mass transfer limitations, we must have $N < J$. The equivalence between the MC$-\overline{\text{Equil}}$ and MC$-\overline{\text{LDF}}$ models is then obtained by matching their variances, since the retention times are already identical. Using the expression for the variance in Table 2.2 with N instead

Table 2.4 Expressions for moments, variance and skewness associated with the MC−$\overline{\text{LDF}}$ and PD−$\overline{\text{LDF}}$ (simplified BC) models.

	MC−$\overline{\text{LDF}}$	PD−$\overline{\text{LDF}}$
μ_1	$t_R = t_0\left(1+\overline{K}'_A\right)$	$t_R = t_0\left(1+\overline{K}'_A\right)$
μ_2	$\frac{J+1}{J}t_R^2 + 2\overline{K}'_A t_A^m t_0$	$\left(1+\frac{2}{Pe}\right)t_R^2 + 2\overline{K}'_A t_A^m t_0$
μ_3	$\frac{(J+1)(J+2)}{J^2}t_R^3 + 6\overline{K}'_A t_A^m t_0\left[\frac{J+1}{J}t_R + t_A^m\right]$	not given
$\mu_{2c}\ (\equiv \sigma^2)$	$t_R^2\left[\frac{1}{J} + 2\frac{\overline{K}'_A}{1+\overline{K}'_A}\frac{t_A^m}{t_R}\right]$	$t_R^2\left[\frac{2}{Pe} + 2\frac{\overline{K}'_A}{1+\overline{K}'_A}\frac{t_A^m}{t_R}\right]$
μ_{3c}	$\frac{2}{J^2}t_R^3 + 6\overline{K}'_A t_A^m t_0\left[\frac{t_R}{J} + t_A^m\right]$	not given
S_k	$\dfrac{\frac{2}{J^2} + 6\overline{K}'_A\frac{t_A^m t_0}{t_R^2}\left[\frac{1}{J} + \frac{t_A^m}{t_R}\right]}{\left[\frac{1}{J} + 2\frac{\overline{K}'_A}{1+\overline{K}'_A}\frac{t_A^m}{t_R}\right]^{3/2}}$	not given

of J, and the one in Table 2.4, one obtains:

$$\frac{1}{N} = \frac{1}{J} + 2\frac{\overline{K}'_A}{1+\overline{K}'_A}\frac{t_A^m}{t_R} \tag{2.49}$$

According to Eq. (2.49), one can represent a column either by J mixing cells, where J is connected to hydrodynamic dispersion only and in which mass transfer limitations occur with a characteristic time t_A^m, or by a series of N equilibrium stages (mixing cells in which equilibrium is reached). Although J and N are "numerically equivalent", they should not be confused. Let us remark that J is independent of $\overline{K}'_A$ whereas N depends on this retention factor. This means that when a plate number is indicated, the associated retention factor (i.e. solute) should be given. Unfortunately, this is not often the case in practice.

Equation (2.49) allows one to express the height equivalent to a theoretical plate ($HETP = L_{col}/N$), which is a classical parameter for column efficiency characterization:

$$HETP = \frac{L_{col}}{N} = \frac{L_{col}}{J} + 2\frac{\overline{K}'_A}{\left(1+\overline{K}'_A\right)^2}t_A^m u_e \tag{2.50}$$

L_{col}/J is typically about 2 particle diameters for well-packed columns.

According to Eq. (2.50), one can expect the equivalent number of plates to vary linearly with the fluid velocity, which is approximately true in many cases. More sophisticated models and explanations will be given in Chapter 5 presenting mass transfer.

If one identifies the variance obtained by the MC$-\overline{\text{Equil}}$ model with the variance obtained by the PD$-\overline{\text{LDF}}$ model, one obtains, similarly to Eq. (2.50):

$$HETP = \frac{2L_{col}}{Pe} + 2\frac{\overline{K}'_A}{\left(1+\overline{K}'_A\right)^2} t_A^m u_e \tag{2.51}$$

Similarly to the approach followed for the MC model, one can lump mass transfer and hydrodynamic dispersion in the PD model. Such a model, characterized by an *apparent Péclet number* Pe^{app}, will be called PD$-\overline{\text{Equil}}$. Just as we had $Pe = 2J$ we have $Pe^{app} = 2N$. For the sake of completeness, the apparent Péclet number Pe^{app} can be related to an *apparent axial dispersion coefficient* D_{ax}^{app}, in which the effects of mass transfer and hydrodynamic dispersion are lumped.

Recall that equations (2.50) and (2.51) have been derived on the basis of identical variances for the $\overline{\text{Equil}}$ and $\overline{\text{LDF}}$ models. Having identified the variances is certainly a necessary step, but this does not ensure the identity of the RTDs.

In order to quantify possible differences between the $\overline{\text{Equil}}$ and $\overline{\text{LDF}}$ models, let us consider skewness. We know that the skewness of the MC$-\overline{\text{Equil}}$ model equals $2/\sqrt{N}$ by analogy with the result given in Table 2.1 for the MC model. The skewness of the MC$-\overline{\text{LDF}}$ model is given in Table 2.4.

Introducing N as expressed by Eq. (2.49) in the skewness S_k(MC$-\overline{\text{LDF}}$) allows us to relate the skewness S_k(MC$-\overline{\text{LDF}}$) to S_k(MC$-\overline{\text{Equil}}$) and to a correction factor depending on the ratio N/J and the retention factor $\overline{K}'_A$:

$$\begin{aligned} &S_k\left(\text{MC}-\overline{\text{LDF}}\right) = S_k\left(\text{MC}-\overline{\text{Equil}}\right) \times (\text{correction factor}) \\ &S_k\left(\text{MC}-\overline{\text{Equil}}\right) = 2/\sqrt{N} \\ &\text{correction factor} = \left(\frac{N}{J}\right)^2 + \frac{3}{2}\left(1-\frac{N}{J}\right)\left(\frac{N}{J} + \frac{1+\overline{K}'_A}{2\overline{K}'_A}\left(1-\frac{N}{J}\right)\right) \end{aligned} \tag{2.52}$$

In the absence of kinetic limitations, according to Eq. (2.49) N and J are identical, so the correction factor equals 1. When kinetic limitations prevail, the correction factor can then be greater or smaller than 1 as shown in Figure 2.7. The lower the ratio N/J, the higher the influence of kinetic limitations, the higher the possible difference between the correction factor and 1, and thus the higher the possible difference between the two models.

When N is sufficiently high, let us say above 100, the skewness factor of the MC$-\overline{\text{Equil}}$ model is very small (let's say 0.1 or less) so that the peaks are practically symmetrical. While the correction factor is different from 1, its impact is negligible in practice.

The situation is different for lower values of N. The skewness factor of the MC$-\overline{\text{Equil}}$ model becomes significant, such that the RTD shows a fair amount of asymmetry. As the correction factor can differ significantly from 1, the RTD can be substantially different for the two models.

For a better view of possible situations, chromatograms associated with different parameters have been simulated via numerical inversion of the transfer function and are presented in Figure 2.8.

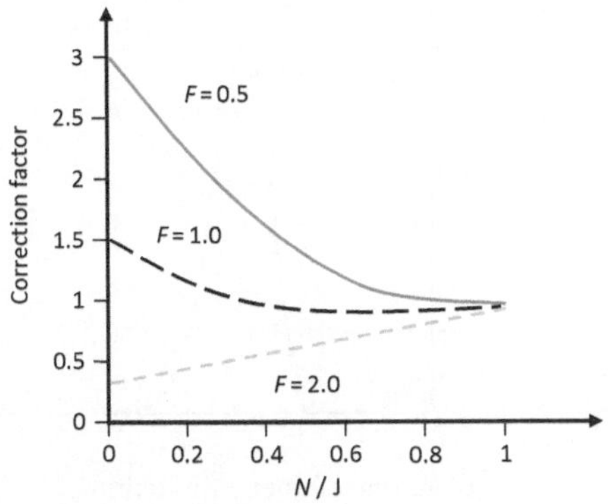

Figure 2.7 Correction factor for Eq. (2.52) for three values of $F = 2\frac{\overline{K}'_A}{\left(1+\overline{K}'_A\right)}$.

When kinetic limitations are "moderate", the curves predicted by the two models are extremely similar, as shown in Figures 2.8a and 2.8b. When kinetic limitations increase significantly, the peaks predicted by the MC$-\overline{\text{LDF}}$ model become strongly asymmetrical and thus very different from the MC$-\overline{\text{Equil}}$ peaks, as shown in Figures 2.8c and 2.8d. According to the MC$-\overline{\text{LDF}}$ model, a single chromatogram can even exhibit two maxima, as shown in Figure 2.8d and already reported by (Villermaux, 1974) or (Chen and Hsu, 1987) for instance.

We can give a simple and qualitative explanation for this last observation. When internal diffusion is slow enough (large t_A^m) some solute molecules may never enter the particles. As a result, these molecules are eluted as if they were flowing in a column packed with non-porous beads. This explains the narrow and high sub-peak observed on the leading edge of the chromatogram of Figure 2.8d. For the solute molecules which were able to enter the beads, slow diffusion means that they will have difficulty diffusing back into the flowing fluid. This explains their high retention and the tailing edge of the chromatogram.

Lumping kinetics and hydrodynamics into one single parameter is thus an excellent approximation in the case of "efficient" columns for which the kinetic contribution is not dramatically reducing the efficiency. Strong kinetic limitations can lead to very asymmetrical peaks that cannot be accounted for by the classical plates model. We will show in Chapter 5 that the impact is even more dramatic under strong non-linear conditions.

EXERCISE 2.8 *Using the MC$-\overline{\text{LDF}}$ model, estimate in the Laplace domain the response of a column initially equilibrated at concentration C_A^I to a step input (positive or negative) of concentration C_A^F.*

EXERCISE 2.9 *A 30 cm-long column is packed with 100 μm particles. At a superficial fluid velocity of 1 cm/sec, the column is equivalent to 300 plates. For a given separation a minimum number of 1000 plates is required. Which fluid velocity should be used? What is the skewness in both conditions for a product having $\overline{K}'_A = 0.5$?*

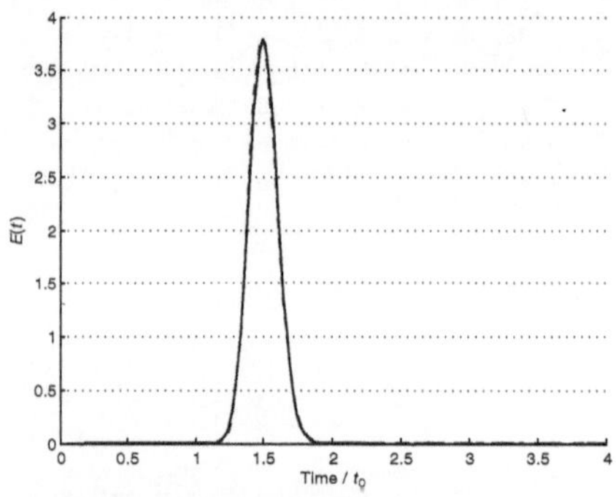

(a) Moderate kinetic limitations:
$t_A^m = 0.01$ min so $N = 202.2$
$S_k(\text{MC}-\overline{\text{LDF}}) = 0.28$
$S_k(\text{MC}-\overline{\text{Equil}}) = 0.14$

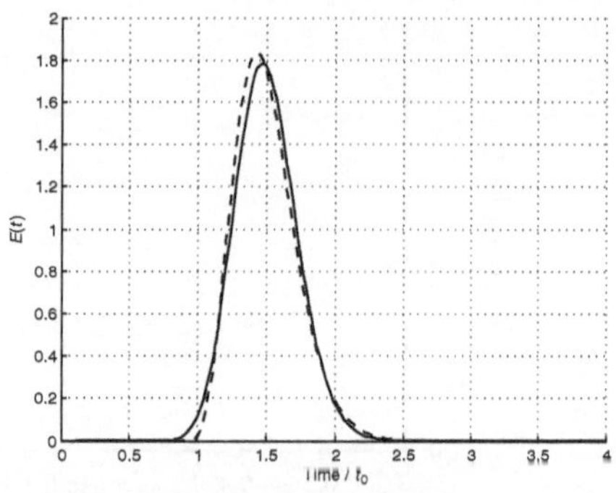

(b) Average kinetic limitations:
$t_A^m = 0.05$ min so $N = 44.0$
$S_k(\text{MC}-\overline{\text{LDF}}) = 0.66$
$S_k(\text{MC}-\overline{\text{Equil}}) = 0.30$

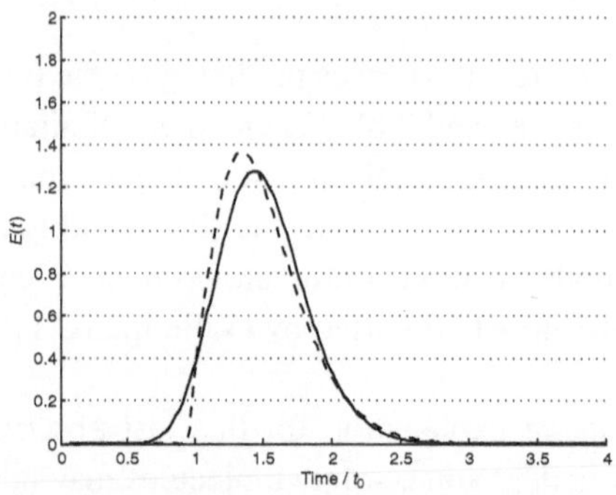

(c) Strong kinetic limitations:
$t_A^m = 0.1$ min so $N = 22.25$
$S_k(\text{MC}-\overline{\text{LDF}}) = 0.95$
$S_k(\text{MC}-\overline{\text{Equil}}) = 0.42$

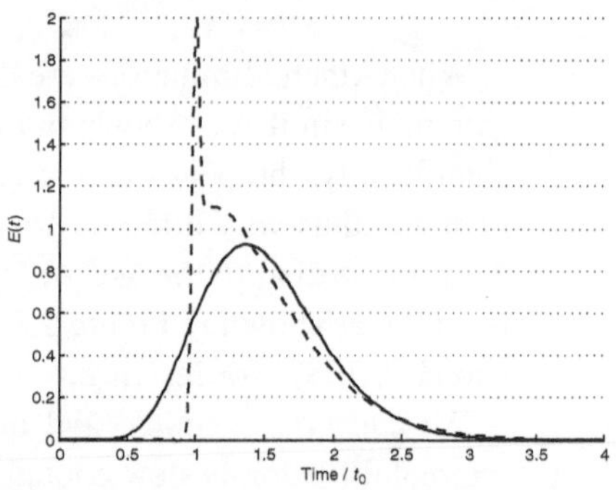

(d) Very strong kinetic limitations:
$t_A^m = 0.2$ min so $N = 11.8$
$S_k(\text{MC}-\overline{\text{LDF}}) = 1.3$
$S_k(\text{MC}-\overline{\text{Equil}}) = 0.58$

Figure 2.8 Chromatograms for the MC$-\overline{\text{Equil}}$ (continuous) and MC$-\overline{\text{LDF}}$ (discontinuous) models, simulated by inversion of the transfer function. N is always calculated via Eq. (2.49), so the two models have identical variances. $J = 2000$, $t_0 = 1$ min, $\overline{K}'_A = 0.5$, for all simulations. Note that non-integer values of N can be used in the transfer function (though they have a questionable physical meaning).

2.5 Dead volumes, injected volumes and sensors: a Lego approach

The results presented in the previous sections rely on several technical assumptions that are not necessarily true:

- The inlet signal is ideal (Dirac or step change); what is the effect of deviations from these ideal signals?
- The column is the only source of peak broadening; what happens when fittings, filters and other connecting tubes, i.e. extra column dead volumes, are taken into account?

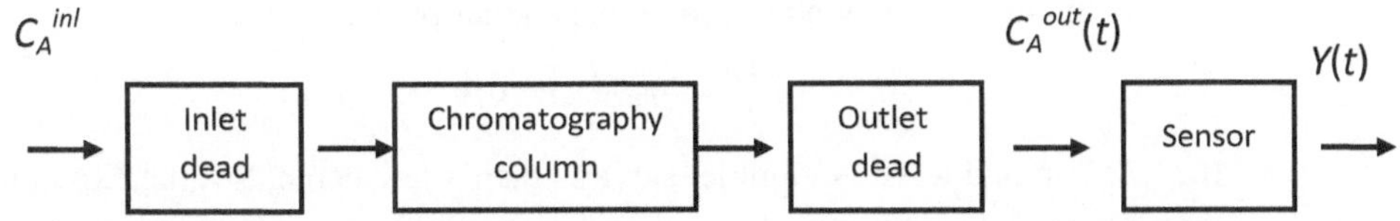

Figure 2.9 Schematic representation of a column with inlet and outlet dead volumes.

- The detection device is ideal and does not contribute to signal broadening; what happens if the detector dynamics is taken into account, or if a remote detector is used?

System dynamics and Laplace transforms will help in answering these questions and in proposing simple strategies to avoid or at least evaluate the problems. This is an illustration of the Lego concept, others being for instance presented in Chapter 6 when non-uniform velocity profiles are considered.

In general, a chromatographic setup can be represented by an inlet dead volume, a column and an outlet dead volume, as shown in Figure 2.9.

Since the various components are in series, the Laplace transform of the measured signal $Y(t)$ is given by:

$$\hat{Y}(s) = \hat{E}_{dead}^{inl}(s)\hat{E}^{col}(s)\hat{E}_{dead}^{out}(s)\hat{E}^{sens}(s)\hat{C}_A^{inl}(s) \tag{2.53}$$

The first observation is that, under the linearity assumption, the ordering of the dead volumes, column and sensor is of no importance. This means that it is useless to distinguish the "inlet" and "outlet" dead volumes, and we will now consider:

$$\left.\begin{aligned} \hat{Y}(s) &= \hat{E}^{glob}\hat{C}_A^{inl} \\ \hat{E}^{glob} &= \hat{E}^{col}\hat{E}^{sens}\hat{E}^{dead} \\ \hat{E}^{dead} &= \hat{E}_{dead}^{inl}\hat{E}_{dead}^{out} \end{aligned}\right\} \tag{2.54}$$

where $\hat{E}^{dead}$ is the transfer function of extra-column effects excluding the sensor, and $\hat{E}^{glob}$ is the Laplace transform of the apparent RTD that accounts for these effects.

The problem is to know whether the properties of $E^{glob}(t)$ differ significantly from those of $E(t)$. The main properties are again given by Van der Laan's theorem, which leads to:

$$\begin{aligned} t_R^{glob} &= t_R^{dead} + t_R^{col} + t_R^{sens} + \mu_1^{inl} \\ {\sigma_{glob}}^2 &= {\sigma_{dead}}^2 + {\sigma_{col}}^2 + {\sigma_{sens}}^2 + {\sigma_{inl}}^2 \end{aligned} \tag{2.55}$$

The additivity rule in Eq. (2.55) is not valid for centered moments of order higher than 2.

We can now state a general rule of thumb: *extra-column effect and injection signal effect are negligible provided that t_R^{col} and σ_{col} are the main contributors in Eq. (2.55).*

It remains to check whether or not this rule is satisfied.

The injection signal. In general this is well represented by a rectangular pulse of width t_{inj} and height C_A^F. This can be delivered by an injection valve equipped with a sample loop. It can then be shown that, for a rectangular pulse:

$$\mu_1^{inl} = t_{inj}/2 \qquad \sigma_{inl}^{\ 2} = t_{inj}^2/12 \tag{2.56}$$

If $t_{inj}/2$ is found to be negligible, say less than a few percent of the experimental t_R^{glob}, then it is expected that the inlet contribution to the retention time will be negligible. Let us stress, however, that the choice of the percentage deviation is subjective. This will also hold for similar comparisons of other contributions to global retention time and variance.

EXERCISE 2.10 *Derive the formulas in Eq. (2.56) using the definition of moments given by Eqs. (2.5) and (2.6). If you are courageous, derive the Laplace transform of the rectangular pulse (refer to Appendix A), and obtain Eq. (2.56) using Van der Laan's theorem.*

The sensor. Analytical chromatographs are equipped with fast-response and small volume sensors. In general, their contribution to Eq. (2.55) is negligible. However, in preparative systems this is not that obvious, especially when the detector is remotely located on a sampling line. Thus, it is advisable to determine its dynamic properties. The method consists in using system dynamics again: the detector is excited with a known inlet signal generated by an appropriate setup. The best choice is to connect the sampling line to two reservoirs of fluid at different compositions through a three-way valve. When the valve is opened, the sensor is excited by a step-like signal. Interpreting the response leads to t_R^{sens} and σ^{sens}. Finally, the weight of these contributions to the global chromatogram is evaluated as in the case of the injection signal (see "The injection signal" above).

EXERCISE 2.11 *In order to characterize a sensor, a sampling line is connected to two reservoirs of fluid at different compositions through a three-way valve. When the valve is opened, the sensor is excited by a step-like signal. The signal given by the detector is given in Figure 2.10. Propose a transfer function for the sensor.*

Dead volumes. These are those fluid volumes that are not used for separation. They are located in fittings, distributors and collectors of large diameter columns, etc. When the contributions of the injection signal and of the sensor are known, the dead volumes

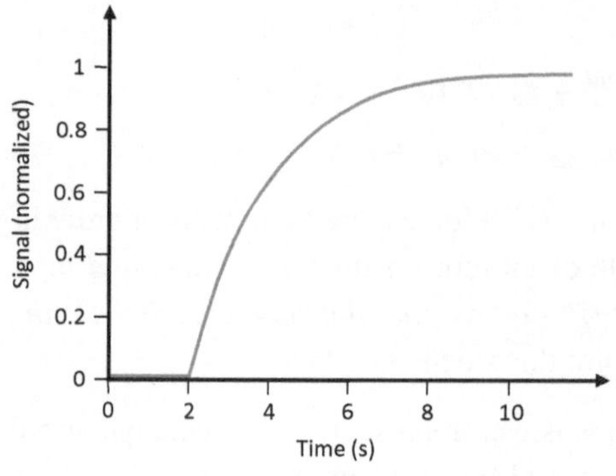

Figure 2.10 For Exercise 2.11.

are easy to characterize: suffice to perform an experiment without the column, and Eq. (2.55) gives the result after subtracting the injection and sensor contributions. Let us, however, mention that the possible distributor and collector must be kept in the system whereas ideally only the column bed is withdrawn: a suitable "zero-length" connection between the distributor and collector must be available.

It would be dangerous to assume $E(t) = E^{glob}(t)$ without any validation. This would imply that:

$$N_{app} = \left(\frac{t_R^{glob}}{\sigma_{glob}}\right)^2$$

is assumed to be the plate number of the column. This is in general wrong, and most often N_{app} is much smaller than N. This is illustrated in the following exercise.

EXERCISE 2.12 *A column ($t_{col} = 60$ s, $N = 500$ plates) is included in a more complex system having two dead volumes. The first one can be approximated by a perfectly stirred reservoir ($t_{st} = 1$ s) while the second one is equivalent to a pipe characterized by $t_p = 3$ s and $J = 10$ mixing cells. An injection of 5 s duration is made. Estimate the apparent plate number associated with the complete system.*

System dynamics provides another smart method for eliminating the sensor and dead volume effects. Let us assume that two identical experiments are performed successively with and without the chromatographic column. The respective Laplace transforms of the responses are (see Eq. (2.54)):

$$\begin{aligned} \hat{Y}(s) &= \hat{C}_A^F \hat{E}^{col} \hat{E}^{dead} \hat{E}^{sens} && \text{with the column} \\ \hat{Y}^*(s) &= \hat{C}_A^F \hat{E}^{dead} \hat{E}^{sens} && \text{without the column} \end{aligned} \tag{2.57}$$

It is then deduced that:

$$\frac{\hat{Y}(s)}{\hat{Y}^*(s)} = \hat{E}^{col} \tag{2.58}$$

The transfer function of the column, $\hat{E}^{col}$, is simply given by the ratio of the Laplace transforms of the two experiments.

As a consequence of Eq. (2.58), one has:

$$t_R^{col} = \mu_{1Y} - \mu_{1Y^*} \quad {\sigma_{col}}^2 = {\sigma_Y}^2 - {\sigma_{Y^*}}^2 \tag{2.59}$$

When the experimental curves $Y(t)$ and $Y^*(t)$ are available, the characteristics of the column can be determined, irrespective of the exact pulse shape, the quality of the sensor (provided it is linear!) or the presence of dead volumes. Let us stress that, since two successive experiments are required, *they must be as reproducible as possible*, and the *flow pattern must be independent of the presence of the column*. The latter condition precludes using this technique in gas-phase chromatography, owing to the compressibility of the fluid, unless the pressure drop in the column is negligible with respect to the mean working pressure.

2.6 Modeling: the Russian Lego

The first sections of this chapter were devoted to the basic models of linear chromatography. These models involved several assumptions about thermodynamics, hydrodynamics and kinetics. The methodology presented in this chapter and based on the concept of transfer function can be used for more complex models in which some of the above-mentioned assumptions are relaxed.

The chromatographic system can be modeled as an association of "boxes" that are in series (see for instance Section 2.5), in parallel (see for instance Section 6.3.2) or embedded in one another. The latter concept is for instance illustrated by a particle description embedded in a column model. Each box has its own characteristics and a succession of boxes allows the building of models of increasing complexity. When a particular box is considered, system dynamics allows us to forget its internal structure, and one is interested only in input and output variables. An illustration is briefly described in the following.

We focus on a small volume of column where the composition is assumed to be uniform. It can be a mixing cell of the MC model or a slice of volume δV of the PD model. The amounts of solute in the mobile and stationary phases are respectively δn_A and $\delta \bar{n}_A$.

In the Laplace domain, these amounts are related by a given transfer function:

$$\delta \hat{\bar{n}}_A = \hat{\mathscr{L}}(s) \delta \hat{n}_A \tag{2.60}$$

$\hat{\mathscr{L}}(s)$ is presumed to be independent of the position in the column.

If $\hat{E}_e(s)$ is the Laplace transform of the $\mathrm{Z_eRTD}$, then it can be shown ((Villermaux, 1981), (Villermaux *et al.*, 1993)) that the Laplace transform of the RTD describing the behavior of a solute A is:

$$\hat{E}(s) = \hat{E}_e \left[s \left(1 + \hat{\mathscr{L}}(s) \right) \right] \tag{2.61}$$

Equation (2.61) is a very powerful relation, allowing us to associate any type of hydrodynamic model with any type of fluid–solid interaction description in order to build column models. The transfer function of the column in the presence of a linearly interacting solute is obtained by replacing s by $s\left(1 + \hat{\mathscr{L}}(s)\right)$ in the expression for the transfer function $\hat{E}_e$ of an excluded solute.

As an illustration, let us use Eq. (2.61) for the MC$-\overline{\text{LDF}}$ model. We have shown with Eq. (2.45) that the concentrations in the solid and liquid phases are related by:

$$\hat{\bar{C}}_A = \frac{\bar{K}_A}{1 + t_m^A s} \hat{C}_A \tag{2.62}$$

In any given cell the compositions are uniform and one has:

$$\delta n_A = \varepsilon_e C_A \frac{V_{col}}{J} \text{ and } \delta \bar{n}_A = (1 - \varepsilon_e) \bar{C}_A \frac{V_{col}}{J} \tag{2.63}$$

Consequently one has $\hat{\bar{n}}_A = \frac{1-\varepsilon_e}{\varepsilon_e}\frac{\overline{K}_A}{1+t_m^A s}\hat{n}_A$ so that:

$$\hat{\mathscr{L}}(s) = \frac{1-\varepsilon_e}{\varepsilon_e}\frac{\overline{K}_A}{1+t_m^A s} \tag{2.64}$$

The Laplace transform of the Z$_e$RTD for the MC model is given by:

$$\hat{E}_e(s) = \left[1 + \frac{t_0}{J}s\right]^{-J} \tag{2.65}$$

By replacing the argument s by the product $s\left(1+\hat{\mathscr{L}}(s)\right)$ in Eq. (2.19) one obtains:

$$\hat{E}(s) = \frac{\hat{C}_A^J}{\hat{C}_A^{inl}} = \frac{1}{\left(1 + \frac{t_0 s}{J}\left(1 + \frac{1-\varepsilon_e}{\varepsilon_e}\frac{\overline{K}_A}{1+t_A^m s}\right)\right)^J} \tag{2.66}$$

which is the transfer function of the MC$-\overline{\text{LDF}}$ model given by Eq. (2.47).

Equation (2.61) can be applied to the simple systems cases already dealt with in the previous sections, and the results are presented in Table 2.5. The classical results are recognized.

Equation (2.61) can be easily applied to more sophisticated situations such as heterogeneity in retention site, heterogeneity in internal porosity and non-instantaneous adsorption ((Villermaux, 1981), (Villermaux *et al.*, 1993)). Its usefulness will be shown in Chapters 5 and 6 where complex mass transfer and hydrodynamic situations are addressed. Note that the approach has even been extended to model chromatographic media exhibiting a continuous distribution of site properties with the concept of transfer time distribution (TTD) (Villermaux, 1981).

By simple application of Eq. (2.61) and successive derivations, one obtains, thanks to Van der Laan's theorem:

$$\begin{aligned} t_R &= t_0\left[1 + \mathscr{L}(0)\right] \\ \left(\frac{\sigma}{t_R}\right)^2 &= \left(\frac{\sigma_0}{t_0}\right)^2 + \frac{2}{t_R}\frac{-\mathscr{L}'(0)}{1+\mathscr{L}(0)} \end{aligned} \tag{2.67}$$

where σ_0 is the variance obtained for a non-retained (excluded) solute.

Equation (2.61) is a complete generalization of the Van Deemter equation to any type of hydrodynamic or mass-transfer model.

Equations (2.61) and (2.67) reveal extraordinary properties of a peak in linear chromatography. First, (2.67) tells us that deviation from plug flow (σ_0/t_0 term) and mass transfer (second term) are always *additive* contributions to the peak variance whatever the deviation from plug flow (dispersion, radially non-uniform axial velocity) and the mass-transfer process (external or internal diffusion, adsorption kinetics etc.).

Second, in Eq. (2.67), $\mathscr{L}(0)$ and $-\mathscr{L}'(0)$ are the zeroth- and first-order moments of the mass-transfer sub-model that relates the stationary and mobile phase compositions. Equation (2.67) is general and independent of the exact nature of the phase equilibrium

Table 2.5 Expressions for different transfer functions.

Model	$\hat{\mathscr{L}}(s)$	$\hat{E}_e(s)$	$\hat{E}(s)$
MC–$\overline{\text{Equil}}$	$\frac{1-\varepsilon_e}{\varepsilon_e}\overline{K}_A$	$\frac{1}{\left(1+\frac{t_0}{J}s\right)^J}$	$\frac{1}{\left(1+\frac{t_0}{J}s\left(1+\frac{1-\varepsilon_e}{\varepsilon_e}\overline{K}_A\right)\right)^J}$
PD–$\overline{\text{Equil}}$	$\frac{1-\varepsilon_e}{\varepsilon_e}\overline{K}_A$	$\exp\left(\frac{Pe}{2}\left(1-\sqrt{1+4\frac{t_0}{Pe}s}\right)\right)$	$\exp\left(\frac{Pe}{2}\left(1-\sqrt{1+4\frac{t_0}{Pe}s\left(1+\frac{1-\varepsilon_e}{\varepsilon_e}\overline{K}_A\right)}\right)\right)$
MC–$\overline{\text{LDF}}$	$\frac{1-\varepsilon_e}{\varepsilon_e}\frac{\overline{K}_A}{1+t_A^m.s}$	$\frac{1}{\left(1+\frac{t_0}{J}s\right)^J}$	$\frac{1}{\left(1+\frac{t_0}{J}s\left(1+\frac{1-\varepsilon_e}{\varepsilon_e}\frac{\overline{K}_A}{1+s.t_A^m}\right)\right)^J}$
PD–$\overline{\text{LDF}}$	$\frac{1-\varepsilon_e}{\varepsilon_e}\frac{\overline{K}_A}{1+t_A^m.s}$	$\exp\left(\frac{Pe}{2}\left(1-\sqrt{1+4\frac{t_0}{Pe}s}\right)\right)$	$\exp\left(\frac{Pe}{2}\left(1-\sqrt{1+4\frac{t_0}{Pe}s\left(1+\frac{1-\varepsilon_e}{\varepsilon_e}\frac{\overline{K}_A}{1+s.t_A^m}\right)}\right)\right)$

and of the mass-transfer mechanism, provided that they are both linear. The extraordinary result is that the peak retention time and variance depend only on these leading moments of the mass-transfer sub-model. We may thus *define*:

$$\begin{aligned} \overline{K}'_A &\equiv \mathscr{L}(0) \\ \overline{K}'_A t_A^m &\equiv -\mathscr{L}'(0) \end{aligned} \tag{2.68}$$

While the first equation in (2.68) is no more than a self-evident redefinition of the retention factor, the second truly *defines* the mass transfer time constant used in Eq. (2.44), whatever the exact nature of the mass transfer process and particle geometry. This is a result of prime importance that shows that detailed and cumbersome mass-transfer submodels are useless for estimating peak variance, number of plates and the last term of the van Deemter equation (see (2.50) or (2.51)).

The exact nature of the mass-transfer process, i.e. the expression for $\mathscr{L}(s)$, is involved in higher-order moments (skewness) and detailed peak shape.

EXERCISE 2.13 *A chromatographic column is packed with a mixture of two different chromatographic media characterized by different Henry's coefficients and diffusion times t_A^m. Express the transfer function of the column using the MC model. Calculate the mean retention time and variance. Show that a column in which plug flow takes place behaves as two columns in series, one packed with the first adsorbent, the other with the second.*

2.7 A warning on modeling non-linear systems

If the system is linear (according to the definition given at the beginning of this chapter), then the relationship between $\overline{C}_A$ and C_A must be linear. The reverse may not be true.

Deviation from linearity is mainly due to the phase equilibrium law(s) and to column "overloading" owing to productivity constraints. Other non-linear processes may, however, be involved:

- Non-isothermal behavior owing to the heat of adsorption of the solute, temperature programmed separation or even temperature controller failure
- Non-isobaric behavior (pressure swing adsorption)
- Gravity-induced natural convection in concentrated solution flowing in vertical columns; transient flow pattern modification owing to concentration-dependent viscosity
- Viscosity and gravity effects
- Exotic phenomena such as chemical reaction, degassing or phase separation.

As mentioned already, it should be emphasized that knowledge of the RTD is not sufficient to predict the behavior of non-linear systems. Many results valid in linear chromatography may be wrong when non-linearity is building up. A simple illustration is given by the positioning of dead volumes:

- If linear conditions prevail, then the exact location of the dead volumes is of no importance.
- If non-linear conditions prevail, then the chromatographic peaks depend on whether the dead volumes are upstream or downstream of the column (see Exercise 3.3.)

The main reason for this is that $E(t)$ contains all the information describing the time spent by the fluid molecules in moving across the column, but it does not describe how the molecules have moved across the column or, for instance, when or where a broadening process (dispersion, kinetics, dead volumes etc.) took place (for example, very close to the inlet, to the outlet or over the whole column).

We will show in Chapter 5 that two almost identical RTDs can lead to very different non-linear chromatograms. For instance, lumping hydrodynamic dispersion and kinetics into a single parameter may give an excellent approximation in the case of a linear system, while it can lead to unacceptable errors in the case of a strongly non-linear situation.

We will even show in Chapter 7 that two columns having exactly the same RTD may lead to very different non-linear chromatograms if velocity heterogeneity is involved.

The RTD is a fantastic tool for investigating models and obtaining valuable predictions. However, the models must make sense and be as close as possible to physical reality. In particular, understanding the specific effects of hydrodynamics, kinetics, thermodynamics, boundary conditions, etc. is in general mandatory.

2.8 Conclusion

The concepts of RTD (Z_eRTD and Z_tRTD included), transfer functions and moments have enabled us to derive the main properties of chromatographic peaks *under the linearity assumption*. Let us immediately stress that some of the simple properties symbolized by the Russian Lego expression are intimately connected with the definitions in Chapter 1 concerning the notions of mobile phase, lumped stationary phase, extragranular and intragranular porosities.

Using the first-order moment of the RTD, we showed that the mean retention time of a chromatogram is normally affected by thermodynamics only, and not by kinetics or hydrodynamic dispersion. Exceptions can be found in low-efficiency columns due to boundary-condition issues.

The same methodology allowed us to show in general that the contributions of hydrodynamics and kinetics to dispersion are additive, irrespective of the models selected for hydrodynamics and kinetics. This shows that lumping together dispersion and mass transfer (plate concept) is trustworthy for linear chromatography in most situations. This is a validation and a generalization of the Van Deemter curve.

We also showed, and this is by far less known, that mass transfer can only impact the variance of a chromatographic peak by the zeroth- and first-order moments of the mass-transfer sub-model that relates the stationary and mobile phase compositions. This is a result of prime importance, showing that detailed and cumbersome mass-transfer submodels are useless for estimating peak variance.

Extracting the maximum information from a chromatogram requires using the moments of the distribution instead of graphical approximations that can easily destroy valuable information.

Identical peak variances in alternate models do not ensure the equivalence of the corresponding peaks. This is especially illustrated by the concept of theoretical plates, which encapsulates in a single dispersion-like parameter all the broadening processes that define the shape of the peak. All the models are not equivalent! Strong differences between simulated peaks are typically seen when kinetic limitations are very strong.

We have presented the effect of the flow pattern (Z_eRTD) and of the mass-transfer process with a very basic and academic approach. It remains to know how column design, chromatographic medium properties and operating conditions may determine more detailed and practical features:

- What are the contributions to t_A^m? Is mass transfer dependent on fluid flow or not, and why? How does t_A^m depend on particle shape and size, and which of these characteristics is of greater importance? The answers to these questions will be dealt with in Chapter 5.
- How does axial dispersion depend on particle shape and size, and on fluid velocity? Can a radially non-uniform axial velocity profile (another deviation to plug flow) be represented by an axial dispersion coefficient? The answers to these questions will be dealt with in Chapter 6.

A main assumption of this chapter is *linearity*. In non-linear cases the RTD concept loses its predictive ability and its knowledge is, *a priori*, not sufficient to predict non-linear situations. This is because it describes the time spent by the solute molecules inside the column, but does not describe where various processes take place. This is shown by the location of the dead volumes: the location, either upstream or downstream of the column, is of no importance in the linear case, while it matters in the non-linear case. Similarly, lumping together dispersion and mass transfer (plate concept) is trustworthy for linear chromatography in most situations, but is hazardous in the *non-linear* case.

We will see how to deal with non-linear chromatography in the next chapters.

2.9 Solutions of the exercises

Exercise 2.1

Let us denote by $\hat{F}(s)$ the Laplace transform of $F(t)$. As the Laplace transform of a step input 1 is $1/s$, Eq. (2.13) gives: $\hat{F}(s) = \hat{E}(s)/s$, thus $\hat{E}(s) = s\hat{F}(s)$. The column being initially solute-free, one has $F(0) = 0$. Consequently $E(t)$ is the first derivative of $F(t)$ (see Eq. (A.4) in Appendix A with $F(t = 0) = 0$).

Exercise 2.2

Consider a cell of volume V and a constant inlet flow rate Q. If C_A is the concentration of the tracer in the cell, one has:

$$QC_A^F = QC_A + V\frac{dC_A}{dt} \quad \text{thus} \quad C_A^F = C_A + \tau\frac{dC_A}{dt} \quad \text{with} \quad \tau = V/Q$$

Start with the situation:

$$C_A^F = cst \text{ and } C_A\left(t=0\right) = 0$$

From the property of the Laplace transform, one has:

$$\mathscr{L}\left(\frac{dC_A}{dt}\right) = s\,\hat{C}_A - C_A\left(t=0\right) = s\,\hat{C}_A$$

Since C_A^F is constant, the Laplace transform of the mass-balance equation is:

$$\frac{C_A^F}{s} = \hat{C}_A + \tau s\hat{C}_A \quad \text{thus} \quad \hat{C}_A = \frac{C_A^F}{s} - \tau\frac{C_A^F}{1+\tau s}$$

After inversion (see the table of inverse Laplace transforms given in Appendix A) one has:

$$C(t) = C_A^F\left(1 - e^{-t/\tau}\right)$$

Let us now consider the situation:

$$C_A^F = cst \text{ and } C_A\left(t=0\right) = C_A^I \neq 0$$

The methodology presented above can be generalized if one defines deviation variables:

$$\Delta C_A = C_A - C_A\left(t=0\right) = C_A - C_A^I$$

$$\Delta C_A^F = C_A^F - C_A^I$$

Using deviation variables, the mass-balance equation becomes:

$$\Delta C_A^F = \Delta C_A + \tau\frac{d\Delta C_A}{dt}$$

As ΔC_A equals zero at $t = 0$, the above equation gives, in the Laplace domain:

$$\frac{\Delta C_A^F}{s} = \Delta\hat{C}_A + \tau s\Delta\hat{C}_A \quad \text{thus} \quad \Delta\hat{C}_A = \frac{\Delta C_A^F}{s} - \tau\frac{\Delta C_A^F}{1+\tau s}$$

After inversion one has:

$$\Delta C(t) = \Delta C_A^F\left(1 - e^{-t/\tau}\right)$$

and coming back to the variables:

$$C(t) = C_A^I + \left(C_A^F - C_A^I\right)\left(1 - e^{-t/\tau}\right)$$

Deviation variables is the general method that must be used unless one is sure that the value of the variables at $t = 0$ equals zero.

Exercise 2.3

The transfer functions of the MC and PD models are respectively given by

$$\hat{E}_e(s) = \left[1 + \frac{t_0}{J}s\right]^{-J} \quad \text{and} \quad \hat{E}_e(s) = \exp\left[\frac{Pe}{2}\left(1 - \sqrt{1 + \frac{4t_0 s}{Pe}}\right)\right]$$

When Pe reaches infinity, the PD transfer function becomes

$$\hat{E}_e(s) = \exp\left[-t_0 s\right]$$

Similarly, noting that

$$\hat{E}_e(s) = \left[1 + \frac{t_0}{J}s\right]^{-J} = \exp\left[-J \ln\left(1 + \frac{t_0}{J}s\right)\right]$$

as J goes to infinity, the MC transfer function gives

$$\hat{E}_e(s) = \exp\left[-t_0 s\right]$$

The two models lead to $\hat{E}_e(s) = \exp\left[-t_0 s\right]$, which, as mentioned in Appendix A, is the Laplace transform of a pure time delay t_0. This is also the Z_eRTD of an ideal plug flow.

Exercise 2.4

The time of the peak maximum is obtained for: $\frac{dE_e(t)}{dt} = 0$. Equation (2.29) gives:

$$\frac{dE_e(t)}{dt} = \frac{J^J}{(J-1)!}\left\{\frac{J-1}{t_0}\left(\frac{t}{t_R}\right)^{J-2} + \left(\frac{t}{t_0}\right)^{J-1}\left(\frac{-J}{t_0}\right)\right\}\exp\left(-J\frac{t}{t_0}\right)$$

$\frac{dE_e(t)}{dt} = 0$ leads to $t_{max} = t_0\left(1 - \frac{1}{J}\right)$.

The time of the peak maximum is thus affected by the column efficiency even if the effect is limited for efficient columns.

In order to assess the precision of the mid-height measurement on the number of plates determination, let us consider the two situations $N = 10$ and $N = 100$ and use Eq. (2.29).

Mid-height widths are respectively 0.72 and 0.23 (normalized time) for $N = 10$ and $N = 100$ so that:

	$N = 10$	$N = 100$
Peak maximum	0.905	0.99
Mid-height width	0.72	0.23
$\sigma^2 = W^2/5.54$	0.126	0.0298
$N = \left(\frac{t_R}{\sigma}\right)^2$	9	103

Number of plates determined with the approximative method are very close to the rigorous plates numbers because the peaks are practically symmetrical.

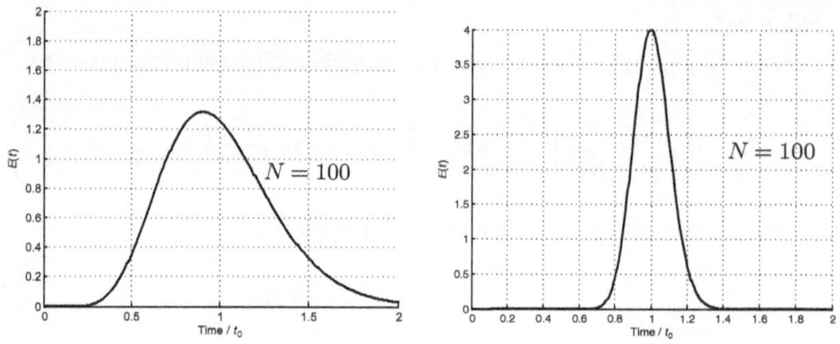

Figure 2.11 For solution of Exercise 2.4.

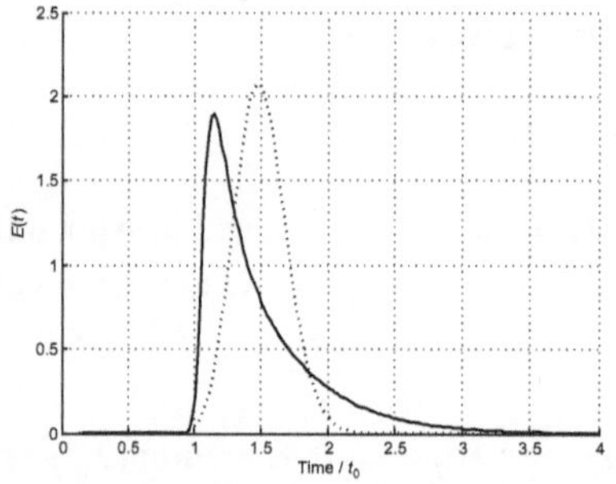

Figure 2.12 For solution of Exercise 2.5.

Exercise 2.5

The peak is clearly asymmetrical, but that type of asymmetry may not look totally unfamiliar to many preparative chromatographers.

From the chromatogram, one obtains $t_R^{max} = 1.15$ and $W = 0.35$, so that $J = 60$.

By calculating the first-order moment (or in this case with a simple geometrical approximation) one can see that the mean positioning of the peak equals $t_0 = 1.5$. The simulated RTD curve with $t_0 = 1.5$ and $J = 60$ is given in Figure 2.12. The agreement is very poor!

Needless to say, using one or the other option may lead to very different conclusions in terms of process performance!

Exercise 2.6

A first option has been given in Chapter 1: a mass balance leads to deriving Eq. (1.31), which is identical to the expression given in Table 2.3.

Another solution is based on system dynamics; it is aimed at exercising the skills of the reader in that domain.

Since the chromatography problem is assumed to be linear, the response to a positive step (from $C_A = 0$ to C_A^F) has the same properties and thus the same retention time as the response to a negative step (from $C_A = C_A^F$ to 0). Let $F(t)$ be the response to a unit step ($C_A^F = 1$ unit of concentration) and $F^-(t)$ be the response to a negative unit step. The response to the negative unit step is a rounded rectangle starting at $t = 0$; the more efficient the column (high J), the sharper the corners of the rectangle. In the limit of an infinitely efficient column, the response is thus a rectangle starting at $t = 0$ and ending at time θ (the retention time of the step input), the surface area below the rectangle being θ.

The Laplace transform of $F(t)$ is $\hat{E}(s)/s$. The response $F^-(t)$ to the negative unit step is $F^-(t) = 1 - F(t)$. The Laplace transform of $F^-(t)$ is thus $\hat{F}^-(s) = 1/s - \hat{E}(s)/s = \left[1 - \hat{E}(s)\right]/[0 - s]$.

From the definition of the Laplace transform with $s = 0$, $\hat{F}^-(0)$ equals the surface area under the response to the unit negative step. Consequently, $\hat{F}^-(0) = \theta$.

Moreover, when s goes to 0, $\hat{F}^-(s) = \left[1 - \hat{E}(s)\right]/[0 - s]$ becomes exactly the derivative of $\hat{E}(s)$ at $s = 0$, i.e. the mean retention time t_R of the RTD according to Van der Laan's theorem. Consequently $\hat{F}^-(0) = t_R$.

We have thus shown that $\hat{F}^-(0) = \theta = t_R$: the mean retention time of the step input θ is thus identical to the mean retention time t_R of the RTD.

Exercise 2.7

In the batch and chromatographic experiments, only the equilibrium parameters are identical. Conversely, the ratios of mobile- and stationary-phase volumes are not necessarily identical. One must thus first look for the equilibrium parameter from the batch experiment.

A mass balance associated with the batch experiment leads to:

$$\overline{V}\overline{C}_A + VC_A = VC_A^I$$

At incipient wetness, the intragranular porosity of the silica is saturated with hexane; as a consequence, the external volume of the solution does not vary when silica is immersed in the 10 ml of solution of A. At equilibrium, since $\overline{C}_A = \overline{K}_A C_A$ (assuming we are in the linear region), one obtains:

$$\overline{K}_A = \frac{VC_A^I - VC_A}{\overline{V}C_A} = \frac{V}{\overline{V}}\frac{C_A^I - C_A}{C_A}$$

$$\overline{K}_A = \frac{10}{3}\frac{1 - 0.2}{0.2} = 13.3$$

Assuming an extragranular porosity of the column of 0.4 (cf. Chapter 1), the zero retention time is given by:

$$t_0 = \frac{0.4 \times 1}{1} = 0.4 \text{ min}$$

The retention time is thus:

$$t_R = t_0\left(1 + \frac{1 - \varepsilon_e}{\varepsilon_e}\overline{K}_A\right) = 8.38 \text{ min}$$

Note: If dry silica of intragranular porosity $\varepsilon_i = 0.5$ were used in the batch experiment, 1.5 ml of liquid would be sucked inside the particles owing to capillary force. The volume of external solution would thus be 8.5 ml instead of 10 ml. This would result in $\overline{K}_A = 11.3$, i.e. 15% deviation and $t_R = 7.2$ min instead of 8.4. This clearly shows that interpretation of preliminary experiments must rely on a rigorous and careful analysis of the problem.

Exercise 2.8

The two mass-balance equations to be taken into account are:

$$C_A^{k-1} = C_A^k + \frac{t_0}{J}\frac{dC_A^k}{dt} + \frac{1-\varepsilon_e}{\varepsilon_e}\frac{t_0}{J}\frac{d\overline{C}_A^k}{dt}$$

$$t_A^m \frac{d\overline{C}_A}{dt} = \overline{K}_A C_A - \overline{C}_A$$

Using the same approach as in Exercise 2.2, we define deviation variables:

$$\Delta C_A^k = C_A^k - C_A^k\,(t=0) = C_A^k - C_A^I$$

$$\Delta \overline{C}_A^k = \overline{C}_A^k - \overline{C}_A^k\,(t=0) = \overline{C}_A^k - \overline{K}_A C_A^I$$

$$\Delta C_A^F = C_A^F - C_A^I$$

With deviation variables, the mass balances are written:

$$\Delta C_A^{k-1} = \Delta C_A^k + \frac{t_0}{J}\frac{d\Delta C_A^k}{dt} + \frac{1-\varepsilon_e}{\varepsilon_e}\frac{t_0}{J}\frac{d\Delta\overline{C}_A^k}{dt}$$

$$t_m \frac{d\Delta\overline{C}_A}{dt} = \overline{K}_A \Delta C_A - \Delta\overline{C}_A \qquad (2.69)$$

These have exactly the same form as Eq. (2.34) and Eq. (2.44), the variables being replaced by deviation variables.

The deviation variables ΔC_A^k and $\Delta\overline{C}_A^k$ equaling zero at time zero, one can move into the Laplace domain just like in Section 2.4 so that one obtains:

$$\frac{\Delta\hat{C}_A^J}{\Delta\hat{C}_A^F} = \frac{1}{\left(1 + \frac{t_0 s}{J}\left(1 + \frac{1-\varepsilon_e}{\varepsilon_e}\frac{\overline{K}_A}{1+t_A^m s}\right)\right)^J} = \hat{E}(s)$$

The deviation variables are thus related by the same transfer function as the variables when initial concentrations equal zero.

The above equation leads to:

$$\hat{C}_A^J = \frac{C_A^I}{s} + \hat{E}(s)\frac{\left(C_A^F - C_A^I\right)}{s}$$

Coming back to the concentrations, one obtains:

$$C_A^J = C_A^I + F(t)\left(C_A^F - C_A^I\right)$$

$F(t)$ is the response of the column initially free of solute to a step change of concentration.

This generalizes the expression obtained when the column is initially free of product. Once again, unless one is absolutely certain that variables equal zero at initial time, the deviation variables must be used to express mass balances before moving into the Laplace domain.

Exercise 2.9

According to Eq. (2.50) the HETP can be written:

$$HETP = \frac{L_{col}}{N} = \frac{L_{col}}{J} + Cu \text{ (cm)}$$

As $L_{col}/J \approx 2 \times d_p \approx 2 \times 10^{-2}$ (cm) one has $J = 1500$, so:

$$HETP = \frac{L_{col}}{N} = 2 \times 10^{-2} + Cu \text{ (cm)}$$

For $u = 1$ cm/s, one has:

$$\frac{L_{col}}{N} = \frac{30}{300} = 2 \times 10^{-2} + C \text{ (cm) thus } C = 8 \times 10^{-2} \text{ s/cm}$$

1000 plates are thus obtained at $u = \frac{\frac{30}{1000} - 2\times10^{-2}}{8\times10^{-2}} = 0.125$ cm/s.

We have thus two sets of conditions:

- for $u = 0.125$ cm/s: $J = 1500$, $N = 1000$
- for $u = 1.0$ cm/s: $J = 1500$, $N = 300$

For the MC$-\overline{\text{Equil}}$ model, skewness is given by $2/\sqrt{N}$. For the MC$-\overline{\text{LDF}}$ model, one has to use Eq. (2.52) with $\overline{K}'_A = 0.5$. The skewness increases when N decreases, as expected, and in this situation the asymmetry is higher with the MC$-\overline{\text{LDF}}$ model than with the MC$-\overline{\text{Equil}}$ model.

Table 2.6 Skewness factors for the two models (Exercise 2.9).

	MC$-\overline{\text{Equil}}$	MC$-\overline{\text{LDF}}$
$u = 0.125$ cm/s, $J = 1500$, $N = 1000$	0.0632	0.115
$u = 1.0$ cm/s, $J = 1500$, $N = 300$	0.0649	0.1978

Exercise 2.10

The total area under the rectangular pulse of intensity C_A^F and duration t_{inj} is $t_{inj}C_A^F$. This is the zeroth-order moment of the inlet signal. The surface normalized inlet signal $X(t)$ is thus given by:

$$X(t) = \frac{1}{t_{inj}} \text{ if } 0 \leq t \leq t_{inj}$$
$$X(t) = 0 \text{ if } t \geq t_{inj}$$

Consequently one obtains:

$$\mu_1 = \int_0^{\infty} X(t)t^1 dt = \int_0^{t_{inj}} \frac{t}{t_{inj}} dt = \frac{1}{2} t_{inj}$$

$$\mu_2 = \int_0^{\infty} X(t)t^2 dt = \int_0^{t_{inj}} \frac{t^2}{t_{inj}} dt = \frac{1}{3} t_{inj}^2$$

Finally one obtains: $\sigma^2 = \mu_2 - \mu_1^2 = \frac{1}{12} t_{inj}^2$.

Now let us solve the same problem in the Laplace domain. A rectangular pulse of intensity C_A^F and duration t_{inj} can be considered as a combination of two step functions:

$$g(t) = C_A^F \left[u\,(t-0) - u(t - t_{inj}) \right]$$

function u being defined in Appendix A.

We will use a function normalized to the zeroth-order moment of this function ($C_A^F t_{inj}$):

$$h(t) = \frac{1}{t_{inj}} \left[u\,(t-0) - u(t - t_{inj}) \right]$$

Recalling that the Laplace transform is a linear operator, in the Laplace domain one has:

$$\hat{h}(s) = \frac{1}{st_{inj}} \left[e^{-0} - e^{-s.t_{inj}} \right] = \frac{1}{st_{inj}} \left[1 - e^{-s.t_{inj}} \right]$$

For small values of s, $\hat{h}(s)$ can be expanded as:

$$\hat{h}(s) = \left[1 - \frac{1}{2!} st_{inj} + \frac{1}{3!} s^2 t_{inj}^2 + \cdots \right]$$

Equating this to the Maclaurin expansion

$$f(x) = f(0) + xf'(0) + \frac{1}{2!} x^2 f''(0) + \cdots$$

and using Van der Laan's theorem, one obtains:

$$\mu_0 = 1$$

$$\mu_1 = \frac{1}{2} t_{inj}$$

$$\mu_2 = \frac{1}{3} t_{inj}^2$$

So $\sigma_{inj}^2 = \mu_2 - {\mu_1}^2 = \dfrac{t_{inj}^3}{12}$.

Exercise 2.11

Looking at Figure 2.10, it is apparent that the signal given by the detector contains a delay of about 2 s. It is then followed by a curve that can be represented by $(1 - \exp(-t/3))$.

From the property of the Laplace transform presented in Appendix A, we obtain the Laplace transform of the detector's response to the step input:

$$\hat{Y}(s) = \exp(-2s)\left(\frac{1}{s} - \frac{3}{1+3s}\right)$$

The Laplace transform of the input signal being $1/s$, one obtains the transfer function of the sensor:

$$\hat{E}^{sens} = \frac{\hat{Y}(s)}{1/s} = \exp(-2s)\left(1 - \frac{3s}{1+3s}\right) = \exp(-2s)\left(\frac{1}{1+3s}\right)$$

One can use this transfer function as part of the global transfer function described in Eq. (2.54), or can simply determine the contribution of the sensor to the retention time and variance using Van der Laan's theorem.

Exercise 2.12

All times are given in s, variances in s^2.

For the column one has $t_{col} = 60$, $\sigma^2 = \frac{60^2}{500} = 7.2$.

For the first dead volume one has $t_{st} = 1$, $\sigma^2 = \frac{1^2}{1} = 1$.

For the second dead volume one has $t_p = 3$, $\sigma^2 = \frac{3^2}{10} = 0.9$.

For the rectangular pulse injection one has $\mu_1^{inl} = 5/2$, $\sigma_{inl}^2 = \frac{5^2}{12} = 2.08$.

Consequently, for the entire system one has:

$$t_R^{glob} = 60 + 1 + 3 + 2.5 = 66.5$$

$$\sigma_{glob}{}^2 = 7.2 + 1 + 0.9 + 2.08 = 11.18$$

So the apparent plate number N_{app} is given by:

$$N = \left(\frac{t_R^{glob}}{\sigma_{glob}}\right)^2 = \frac{66.5^2}{11.18} = 395$$

Exercise 2.13

For each chromatographic medium, the Laplace transforms of the concentrations contained in the lumped solid and fluid phases are related by:

$$\hat{\overline{C}}_{A1} = \frac{\overline{K}_{A1}}{1 + t_A^{m1} s}\hat{C}_A$$

$$\hat{\overline{C}}_{A2} = \frac{\overline{K}_{A2}}{1 + t_A^{m2} s}\hat{C}_A$$

If the volumetric fraction of medium type 1 is θ, we can define a global mean solid-phase concentration by:

$$\hat{\overline{C}}_A = \theta\hat{\overline{C}}_{A1} + (1-\theta)\,\hat{\overline{C}}_{A2}$$

Consequently, the number of moles of solute in each phase are related by:

$$\hat{\bar{n}}_A = \mathscr{L}(s)\hat{n}_A$$

with:

$$\mathscr{L}(s) = \frac{1-\varepsilon_e}{\varepsilon_e}\left(\theta\frac{\overline{K}_{A1}}{1+t_A^{m1}s} + (1-\theta)\frac{\overline{K}_{A2}}{1+t_A^{m2}s}\right)$$

The global transfer function is given by:

$$\hat{E}(s) = \hat{E}_e\left[s\left(1+\hat{\mathscr{L}}(s)\right)\right]$$

Consequently, if the column is modeled with the MC model, one has:

$$\hat{E}(s) = \left[1+\frac{t_0}{J}s\left(1+\frac{1-\varepsilon_e}{\varepsilon_e}\left(\theta\frac{\overline{K}_{A1}}{1+t_A^{m1}s}+(1-\theta)\frac{\overline{K}_{A2}}{1+t_A^{m2}s}\right)\right)\right]^J$$

Noting that $\hat{\mathscr{L}}(0) = \frac{1-\varepsilon_e}{\varepsilon_e}\left(\theta\overline{K}_{A1}+(1-\theta)\overline{K}_{A2}\right)$, Eq. (2.67) gives:

$$t_R = t_0\left(1+\mathscr{L}(0)\right) = t_0\left(1+\frac{1-\varepsilon_e}{\varepsilon_e}\left(\theta\overline{K}_{A1}+(1-\theta)\overline{K}_{A2}\right)\right)$$

Noting that $(\sigma_0)^2 = t_0^2/J$ and that $\mathscr{L}'(0) = -\frac{1-\varepsilon_e}{\varepsilon_e}\left(\theta\overline{K}_{A1}t_A^{m1}+(1-\theta)\overline{K}_{A2}t_A^{m2}\right)$, Eq. (2.67) gives:

$$\left(\frac{\sigma}{t_R}\right)^2 = \left(\frac{\sigma_0}{t_0}\right)^2 + \frac{2}{t_R}\frac{\left(-L'(0)\right)}{1+L(0)} = \frac{1}{J}+\frac{2}{t_R}\frac{\frac{1-\varepsilon_e}{\varepsilon_e}\left(\theta\overline{K}_{A1}t_A^{m1}+(1-\theta)\overline{K}_{A2}t_A^{m2}\right)}{1+\frac{1-\varepsilon_e}{\varepsilon_e}\left(\theta\overline{K}_{A1}+(1-\theta)\overline{K}_{A2}\right)}$$

Notice that if the two sites are identical, the above equation gives:

$$\left(\frac{\sigma}{t_R}\right)^2 = \frac{1}{J}+\frac{2}{t_R}\frac{\frac{1-\varepsilon_e}{\varepsilon_e}\overline{K}_A t_A^m}{1+\frac{1-\varepsilon_e}{\varepsilon_e}\overline{K}_A}$$

which is identical to (2.49).

Finally, when plug flow prevails, the Laplace transform of the Z_eRTD is

$$\hat{E}_e(s) = \exp\left[-t_0 s\right]$$

Using Eq. (2.67) one obtains immediately:

$$\hat{E}(s) = \exp\left[-t_0 s\left(1+\frac{1-\varepsilon_e}{\varepsilon_e}\left(\theta\frac{\overline{K}_{A1}}{1+t_A^{m1}s}+(1-\theta)\frac{\overline{K}_{A2}}{1+t_A^{m2}s}\right)\right)\right]$$

so that:

$$\hat{E}(s) = \exp\left[-t_0 s\theta\left(1+\frac{1-\varepsilon_e}{\varepsilon_e}\frac{\overline{K}_{A1}}{1+t_A^{m1}.s}\right)\right] \times \exp\left[-t_0 s(1-\theta)\left(1+\frac{1-\varepsilon_e}{\varepsilon_e}\frac{\overline{K}_{A2}}{1+t_A^{m2}.s}\right)\right]$$

The latter expression means that the column behaves as two columns in series, one packed with the first adsorbent, the other with the second. If each column is sliced into several subcolumns, the order of the subcolumns is of no importance in the chromatogram, owing to the linearity of the phenomena.

References

Aris, R. (1959), *Chem. Eng. Sci.* **9**, 266.

Chen, L. and Hsu, J. (1987), *AIChE J.* **33**(8), 1387.

Gibilaro, L. (1978), *Chem. Eng. Sci.* **33**, 487.

Levenspiel, O. (1972), *Chemical Reaction Engineering*, 2nd edn., John Wiley & Sons.

Martin, A. and Synge, R. (1941), *Biochem. J.* **35**, 1358–1368.

Sardin, M., Schweich, D., Leij, F. and Van Genuchten, M. (1991), *Water Resour. Res.* **27**(9), 2287–2307.

Seidel-Morgenstern, A. (1991), *Chem. Eng. Sci.* **46**(10), 2567–2571.

Varma, A. and Morbidelli, M. (1997), *Mathematical Methods in Chemical Engineering*, Oxford University Press.

Villermaux, J. (1974), *J. Chromatogr. Sci.* **12**, 822–831.

Villermaux, J. (1981). In A. E. Rodrigues and D. Tondeur, eds., *Percolation Processes Theory and Applications*, vol. 33 of *NATO ASI Series, Series E: Applied Sciences*, Sijthoff & Noordhof, pp. 83–140.

Villermaux, J. (1987), *J. Chromatogr. A* **406**, 11–26.

Villermaux, J. (1990), *J. Pet. Sci. Eng.* **4**(1), 21–30.

Villermaux, J., Schweich, D. and Sardin, M. (1993). In G. Ganetsos and P. E. Barker, eds., *Preparative and Production Scale Chromatography*, vol. 61 of *Chromatographic Science Series*, Marcel Dekker, pp. 701–739.

Villermaux, J. and Swaay, V. (1969), *Chem. Eng. Sci.* **24**, 1097.

Wen, C. and Fan, L. (1975), *Chemical Processes and Engineering, Vol. 3: Models of Flow Systems and Chemical Reactors*, Marcel Dekker.

3 Non-linear chromatography: equilibrium theory

Sort of a naked version of chromatography . . .

3.1 Introduction

The previous chapters introduced fundamental principles assuming that chromatographic processes can be described as *linear* systems. However, most if not all industrial chromatographic systems are *non-linear* due to the high concentrations applied there. Addressing the key consequences of non-linearity is the objective of this chapter.

Non-linear phenomena can arise from several sources including the nature of the distribution equilibria of the solutes between phases, chemical reactions or equilibria in the column, solubility limitations, sorption effects in gas chromatography and viscosity effects in liquids. We will speak of thermodynamic non-linearity later on; this chapter will stress the fundamental difference between the effects of thermodynamics and kinetics on the shape of chromatograms.

In this chapter we will restrict our study to the influence of essential features of fluid–solid equilibria (specifically adsorption isotherms) because this is usually the main reason for non-linearity in chromatography. Just as "linear chromatography" refers to systems with linear lumped adsorption isotherms in Chapter 2, "non-linear chromatography" refers to non-linear lumped adsorption isotherms in this chapter.

Non-linearity gives rise to various intriguing phenomena and sometimes non-intuitive behavior of chromatographic processes. Well-known effects are *peak tailing* and *fronting*, which are related to the formation of self-sharpening and self-broadening fronts (another frequent term is *compressive* and *dispersive waves*). While peak tailing and fronting are already observed in single-solute systems, more complex situations can arise when separating mixtures, for instance, so-called *displacement* (a strongly retained solute pushing and concentrating less-retained solutes) and *tag-along* (appearance of long plateaus of the more retained species) effects, and *system peaks* (extraneous peaks not directly associated with a species contained in the feed).

In order to extract the essence of thermodynamic non-linearity we will assume a column of infinite efficiency in which neither hydrodynamic dispersion nor kinetic limitation occurs. Using the terminology introduced in Chapters 1 and 2, this means that the number of mixing cells (J), the number of plates (N) and the Péclet number (Pe) all equal infinity. This is certainly an ideal representation of a complex reality, but this will allow the derivation of important features of chromatograms or fronts that are exclusively due to thermodynamic non-linearity. This will give us a kind of asymptotic behavior that chromatographic systems tend to reach if the column is efficient and long enough.

The fact that many of the key features of chromatograms can be obtained from simple knowledge of the adsorption isotherms was recognized decades ago in contributions by (De Vault, 1943) and (Glueckauf, 1949), which still provide a good introduction to the subject. The development was then generalized by (Rhee *et al.*, 1970) and (Helfferich and Klein, 1970). The associated framework is referred to as *equilibrium theory*. The reader interested in a broader view is referred to, for example, a series of papers by Helfferich (Helfferich and Carr, 1993; Helfferich and Whitley, 1996; Helfferich, 1997). While the mathematics are relatively accessible for simple systems, they can quickly become very unfamiliar and exceed the comfort level of the chromatographer. The objective of this chapter is to give enough background to share the key results and methodology of equilibrium theory, knowing that the "real world" will have to be approached by the unavoidable numerical resolution of more complex models. The understanding of some non-linear behavior as predicted by equilibrium theory will then help in understanding the simulated results – or in detecting bugs in the numerical model!

Finally, as seen in Chapters 1 and 2, the relevant information for positioning chromatograms is the *distribution equilibrium* of the solutes between the mobile and lumped stationary phases. This is why the lumped model will be used in this chapter. A consequence is that all adsorption isotherms mentioned in this chapter are in fact lumped adsorption isotherms, even if the qualificative *lumped* is sometimes omitted for convenience.

3.2 Elementary definitions for adsorption isotherms

Prior to presenting equilibrium theory, we need to introduce some basic terminology and definitions related to adsorption isotherms. A detailed presentation of this topic and more generally of fluid–solid phase equilibria is given in Chapter 4.

As presented in Chapter 1, the lumped adsorption isotherm of a single solute A describes the thermodynamic equilibrium between its concentration in the extragranular fluid phase, C_A, and the corresponding concentration in the lumped solid phase, $\overline{C}_A$, at constant temperature T (hence the name "adsorption isotherm", which is otherwise ambiguous). Most generally, this is a functional relationship of the form:

$$\overline{C}_A = \bar{f}_A(C_A) \quad \text{lumped adsorption isotherm at a given temperature} \tag{3.1}$$

Recalling the definitions given in Chapter 1, the lumped concentration $\overline{C}_A$ subsumes all molecules located on the solid skeleton at concentration $\overline{\overline{C}}_A$ together with those located in the intragranular pore fluid at concentration $\grave{C}_A$.

It is important to know that adsorption isotherms must satisfy certain constraints in order to be thermodynamically acceptable. This aspect is further discussed in Sections 3.7, 4.3.1, 4.3.3 and 4.7.

In the simplest case, an adsorption isotherm (resp. lumped) reduces to the linear expression $\overline{\overline{C}}_A = \overline{\overline{K}}_A C_A$ (resp. $\overline{C}_A = \overline{K}_A C_A$) used in Chapters 1 and 2. However, $\overline{\overline{f}}_A(C_A)$ (resp. $\bar{f}_A(C_A)$) is usually a non-linear function, and many corresponding adsorption isotherm models will be reviewed in Chapter 4.

It is relatively frequent (but not systematic) for adsorption isotherms to be linear at small concentrations, then to exhibit a continuously declining slope with increasing fluid concentration until the adsorbed concentration reaches a plateau. An important expression for representing this type of single-solute adsorption isotherms is the well-known Langmuir model, having the form $\overline{\overline{C}}_A = \dfrac{a_1 C_A}{1 + a_2 C_A}$ – thus with an initial slope a_1 and reaching plateau a_1/a_2 at large fluid concentrations. As the lumped solid concentrations subsume solutes contained in the intragranular pores, it is often expedient to represent the lumped adsorption isotherm by a linearly modified Langmuir expression like:

$$\overline{C}_A = \overline{\lambda}_A C_A + \frac{\overline{N}_A \tilde{K}_A C_A}{1 + \tilde{K}_A C_A} \tag{3.2}$$

linearly modified Langmuir lumped adsorption isotherm

For the time being, we will simply consider $\overline{\lambda}_A$, $\overline{N}_A$ and $\tilde{K}_A$ as adjustable parameters. More physical sense will be given to these parameters in Chapter 4.

The derivative of the lumped adsorption isotherm (3.2), which will be shown to play an important role, is given by:

$$\frac{d\overline{C}_A}{dC_A} = \overline{\lambda}_A + \frac{\overline{N}_A \tilde{K}_A}{\left(1 + \tilde{K}_A C_A\right)^2} \tag{3.3}$$

which immediately leads to the initial slope (lumped Henry's coefficient):

$$\overline{K}_A = \overline{\lambda}_A + \overline{N}_A \tilde{K}_A \tag{3.4}$$

Lumped adsorption isotherms having a derivative $\bar{f}'_A(C_A)$ which decreases with increasing concentration C_A, thus $\bar{f}''_A(C_A) \leq 0$, are named *favorable lumped adsorption isotherms*. Lumped adsorption isotherms having a derivative $\bar{f}'_A(C_A)$ increasing with increasing concentration C_A thus $\bar{f}''_A(C_A) \geq 0$ are named *unfavorable lumped adsorption isotherms*. Linearly modified Langmuir lumped adsorption isotherms are favorable assuming $\tilde{K}_A > 0$.

In order to represent favorable and unfavorable Langmuir lumped adsorption isotherms with a unique expression, following (Mazzotti, 2006) we propose:

$$\overline{C}_A = \overline{\lambda}_A C_A + \frac{\overline{N}_A \tilde{K}_A C_A}{1 + p_A \tilde{K}_A C_A} \tag{3.5}$$

where p_A is a parameter whose value can be 1 for favorable lumped adsorption isotherms and -1 for unfavorable lumped adsorption isotherms. Using the terminology proposed by (Mazzotti, 2006), we will refer to the adsorption isotherms described by Eq. (3.5) as linearly modified *generalized* lumped adsorption isotherms. This certainly allows representation of lumped adsorption isotherms having $\bar{f}''_A(C_A) \geq 0$, but this representation is dangerous as it can, for instance, lead to negative $\overline{C}_A$ when fluid concentration exceeds a certain value.

Let us now consider a mixture of n different solutes of concentrations C_i in the extragranular fluid and $\overline{C}_i$ in the lumped solid. In most cases the solutes are competing for the adsorption/exchanging/complexing sites so that $\overline{C}_i$ depends on C_i but also on C_j for $j \neq i$. In other words, the lumped concentrations in the lumped solid phase $\overline{C}_i$ can be related to the fluid-phase concentrations via n adsorption laws:

$$\overline{C}_i = f_i(C_1, \ldots, C_n) \quad 1 \leq i \leq n \tag{3.6}$$

As pointed out already, the initial slopes of the adsorption isotherms (lumped Henry's coefficients $\overline{K}_i$) play a very important role:

$$\overline{K}_i = \left.\frac{\partial \bar{f}_i}{\partial C_i}\right|_{C_j=0} \quad 1 \leq i \leq n, j \in [1,n] \tag{3.7}$$

The values $\overline{K}_i$ characterize single-solute behavior under very dilute conditions.

As with single-solute systems, adsorption isotherms for multi-solute mixtures are often represented with a linearly modified Langmuir expression:

$$\overline{C}_i = \bar{\lambda}_i C_i + \frac{\overline{N}_i \tilde{K}_i C_i}{1 + \sum\limits_{k=1}^{n} \tilde{K}_k C_k} \tag{3.8}$$

linearly modified Langmuir lumped adsorption isotherm

which immediately leads to the lumped Henry's coefficients $\overline{K}_i$:

$$\overline{K}_i = \bar{\lambda}_i + \overline{N}_i \tilde{K}_i \tag{3.9}$$

For binary-solute systems containing solutes A and B, one writes:

$$\begin{aligned} \overline{C}_A &= \bar{\lambda}_A C_A + \frac{\overline{N}_A \tilde{K}_A C_A}{1 + \tilde{K}_A C_A + \tilde{K}_B C_B} \\ \overline{C}_B &= \bar{\lambda}_B C_B + \frac{\overline{N}_B \tilde{K}_B C_B}{1 + \tilde{K}_A C_A + \tilde{K}_B C_B} \end{aligned} \tag{3.10}$$

and the partial derivatives which will be shown to play an important role are given by:

$$\begin{aligned} \frac{\partial \overline{C}_A}{\partial C_A} &= \bar{\lambda}_A + \frac{\overline{N}_A \tilde{K}_A \left(1 + \tilde{K}_B C_B\right)}{\left(1 + \tilde{K}_A C_A + \tilde{K}_B C_B\right)^2} \\ \frac{\partial \overline{C}_A}{\partial C_B} &= \frac{-\overline{N}_A \tilde{K}_A \tilde{K}_B C_A}{\left(1 + \tilde{K}_A C_A + \tilde{K}_B C_B\right)^2} \\ \frac{\partial \overline{C}_B}{\partial C_A} &= \frac{-\overline{N}_B \tilde{K}_A \tilde{K}_B C_B}{\left(1 + \tilde{K}_A C_A + \tilde{K}_B C_B\right)^2} \\ \frac{\partial \overline{C}_B}{\partial C_B} &= \bar{\lambda}_B + \frac{\overline{N}_B \tilde{K}_B \left(1 + \tilde{K}_A C_A\right)}{\left(1 + \tilde{K}_A C_A + \tilde{K}_B C_B\right)^2} \end{aligned} \tag{3.11}$$

Finally, the extension of Eq. (3.10) to *generalized* lumped adsorption isotherms is straightforward:

$$\begin{aligned}\overline{C}_A &= \overline{\lambda}_A C_A + \frac{\overline{N}_A \tilde{K}_A C_A}{1 + p_A \tilde{K}_A C_A + p_B \tilde{K}_B C_B} \\ \overline{C}_B &= \overline{\lambda}_B C_B + \frac{\overline{N}_B \tilde{K}_B C_B}{1 + p_A \tilde{K}_A C_A + p_B \tilde{K}_B C_B}\end{aligned} \tag{3.12}$$

Again, following (Mazzotti, 2006), we define four cases for binary systems:

- Case L (Langmuir): $p_A = 1$ and $p_B = 1$
- Case A (anti-Langmuir): $p_A = -1$ and $p_B = -1$
- Case M_A (mixed A): $p_A = 1$ and $p_B = -1$
- Case M_B (mixed B): $p_A = -1$ and $p_B = 1$

Conditions for the thermodynamic consistency of Eq. (3.12) are discussed by (Mazzotti, 2006). While this expression is not always thermodynamically consistent, we present it as it has undoubtedly some practical merits, provided it is used in full awareness of possible limitations.

3.3 Single-solute systems

Professor Malte Kaspereit helped to make the presentation of single-solute systems as pedagogical as possible. His contribution is greatly acknowledged.

Although dealing with a chromatographic process involving one single solute may seem useless since no separation is needed, this simple situation will help illustrate the method and the main behaviors observed in multi-solute systems.

3.3.1 The equilibrium model

This section provides the background that allows estimation of the main features of chromatograms. In order to set the focus exclusively on the influence of adsorption isotherms, let us recall that we consider an *ideal column, without hydrodynamic dispersion or mass transfer limitations*. For such systems, the material balance for the single solute A can be written:

$$u_e \frac{\partial C_A}{\partial z} + \frac{\partial C_A}{\partial t} + F_e \frac{\partial \overline{C}_A}{\partial t} = 0 \tag{3.13}$$

where $u_e = u/\varepsilon_e$ is the interstitial velocity and $F_e = (1-\varepsilon_e)/\epsilon_e$ is the ratio of the lumped volumes of the particles to that of the extragranular fluid. This equation is similar to (2.35) given in Chapter 2 with $Pe = \infty$. Notice that this mass balance, assuming fluid velocity to be constant, in fact assumes the solute to be sufficiently diluted in an inert solvent.

Since we neglect mass transfer resistances, the system is permanently at adsorption equilibrium and Eq. (3.1) holds throughout the column. This allows replacement of the derivative for $\overline{C}_A$ in Eq. (3.13) using the chain rule:

$$\frac{\partial \overline{C}_A}{\partial t} = \frac{d\overline{C}_A}{dC_A}\frac{\partial C_A}{\partial t} = \bar{f}'_A(C_A)\frac{\partial C_A}{\partial t}$$

After introduction of the chain rule into Eq. (3.13) one obtains:

$$u_e\frac{\partial C_A}{\partial z} + \left[1 + F_e\bar{f}'_A(C_A)\right]\frac{\partial C_A}{\partial t} = 0 \tag{3.14}$$

Equation (3.14), a quasi-linear partial differential equation (PDE) of first order, is the basis for the equilibrium model of chromatography. It is sometimes also referred to as the "ideal model", a terminology that we will not use for reasons explained in Chapter 1. According to the terminology introduced in Chapter 1, the equilibrium model corresponds to the PF–$\overline{\text{Equil}}$ model (or to the PF–$\overline{\overline{\text{Equil}}}$ model if reference is made to total porosity instead of extragranular porosity). We will in general use the term "equilibrium model" in deference to those who were at the origin of this breakthrough in the understanding of chromatography.

Despite its simplicity, this model will turn out to be very powerful for describing the essential features of chromatographic processes. For solving this type of model together with appropriate initial and boundary conditions, there exists a powerful theoretical framework, commonly referred to as equilibrium theory, which is presented below.

In essence, let us remark that a chromatographic separation is no more than a race: each species is propagating in the column at its own velocity; the velocity is thus the key concept. We can show that the solutions are dependent on a single parameter σ having the dimensions of velocity:

$$\sigma = \frac{z}{t} \tag{3.15}$$

We can demonstrate this result by replacing z and t in Eq. (3.14) by σ. For the partial derivatives:

$$\frac{\partial C_A(\sigma)}{\partial z} = \frac{\partial\sigma}{\partial z}\frac{dC_A}{d\sigma} = \frac{1}{t}\frac{dC_A}{d\sigma} \quad \text{and} \quad \frac{\partial C_A(\sigma)}{\partial t} = \frac{\partial\sigma}{\partial t}\frac{dC_A}{d\sigma} = \frac{-z}{t^2}\frac{dC_A}{d\sigma} = \frac{-\sigma}{t}\frac{dC_A}{d\sigma}$$

Substituting these two expressions into Eq. (3.14), one obtains:

$$\left[u_e - \sigma\left(1 + F_e\bar{f}'_A(C_A)\right)\right]\frac{dC_A}{d\sigma} = 0 \tag{3.16}$$

First, Eq. (3.16) shows that the concentrations depend on σ only, and this was not that obvious! Second, solving Eq. (3.16), one finds that it has two roots, namely:

$$\frac{dC_A}{d\sigma} = 0 \tag{3.17}$$

$$\sigma(C_A) = \frac{z}{t} = \frac{u_e}{1 + F_e\bar{f}'_A(C_A)} \tag{3.18}$$

These two solutions have different physical meanings:

- The first root, given by Eq. (3.17), represents a *concentration plateau*, that is, a constant concentration somewhere within the column or in the chromatogram.[1]
- The second root, given by Eq. (3.18), represents a *concentration wave*. In this wave each concentration value C_A moves at its own specific velocity σ (C_A).[2]

One consequence of these findings is that the solution space of a non-linear chromatographic problem consists of a combination of plateaus and fronts. Another consequence is that in non-linear chromatography, the different concentrations that constitute a front will *migrate at different velocities*, because these velocities are concentration-dependent through the local slope of the adsorption isotherm, $\bar{f}'_A(C_A)$. Consequently, and for purely thermodynamic reasons, we must expect various possible deformations of such fronts during their migration through a column.

Equation (3.16) is thus of primary importance. Notice that it has been established with the lumped model. The porous model would lead (after a few manipulations easily accessible after reading Chapter 1) to the same mass balance as for the lumped model (Eq. (3.13)). As a consequence, the expressions for velocities are not affected by the model selected for representing the system – which is fortunate.

3.3.1.1 Propagation of waves and shocks

Knowledge of the propagation velocities from the adsorption isotherm (given by Eq. (3.18)) allows construction of chromatograms. However, before demonstrating this, we will take a closer look at phenomena that can arise during the propagation of linear and non-linear waves in a chromatographic column.

For a *linear* adsorption isotherm, $\overline{C}_A = \overline{K}_A C_A$, the derivative is $\bar{f}'_A(C_A) = \overline{K}_A =$ const., and from (3.18) we obtain the propagation velocity:

$$\sigma = \frac{u_e}{1 + F_e \overline{K}_A} = \text{const.} \tag{3.19}$$

Accordingly, in a system described by a linear lumped adsorption isotherm all concentrations move with the same velocity through the column.

Figure 3.1 illustrates this relationship. The left side of the figure shows the adsorption isotherm together with the local derivative and the resulting propagation velocity. The right side demonstrates how two different arbitrary initial profiles with corresponding constant feed concentrations C_A^F will propagate further under linear conditions. As indicated by the equal length of the arrows, all concentrations move forward at the same velocity. The logical consequence is that, for linear lumped adsorption isotherms, an initial concentration profile will not change its shape during migration through a column (remember that dispersive and kinetic effects are ignored).

[1] Let us assume that concentration is constant vs. σ in a certain range. For a given time t, different abscissas z related to different velocities thus have the same concentration. This defines a plateau inside the column. Conversely, considering a given abscissa z and different velocities, different times t will define a plateau in the chromatogram.

[2] Note that the velocity σ is lower than that of the (non-adsorbing) mobile phase, u_e, because the denominator of (3.18) is larger than 1, due to $F_e > 0$ and $\bar{f}'_A(C_A) > 0$ for thermodynamically meaningful equilibria (see Chapter 4).

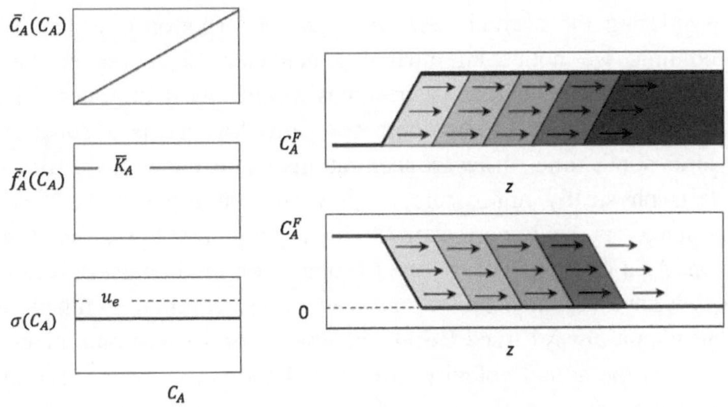

Figure 3.1 Wave propagation for a linear lumped adsorption isotherm. Left (from top): adsorption isotherm, local derivative and propagation velocity. Right: propagation of two arbitrary initial profiles $C_A(t = 0,z)$ (thick lines) through a column at different times (thin lines). The length of the arrows indicates the velocity of corresponding concentration values.

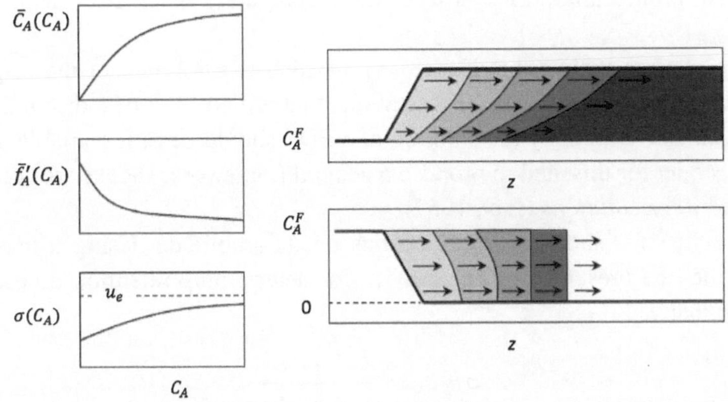

Figure 3.2 Wave propagation for a favorable (convex upward) adsorption isotherm; notation and scenarios as in Figure 3.1.

The situation changes when considering non-linear adsorption. For example, for a *favorable* adsorption isotherm like the well-known Langmuir adsorption isotherm, the derivative $\bar{f}'_A(C_A)$ decreases with increasing concentration C_A. According to Eq. (3.18), the velocity of the concentrations, $\sigma(C_A)$, must then increase with increasing values of C_A, as shown in the left side of Figure 3.2.

Figure 3.2 (right side) shows the impact of this on the initial profiles that we considered already for the linear lumped adsorption isotherm. In the first case (top), which is similar to a regeneration scenario, the faster high concentrations are already ahead of the slower ones. Therefore, the initial profile spreads out the further the wave moves in the column. This well-known phenomenon is called a *simple wave* in mathematical language and a *self-broadening front* in historical chromatographic textbooks. Other common terms for this are *dispersive* or *spreading* wave.

When considering the second case in Figure 3.2 (bottom), which is similar to loading a column, we notice an intriguing behavior. Since higher concentrations move at higher velocities, the wave becomes steeper as it migrates. This effect is known as *self-sharpening*. Another common term for this is a *compressive* wave. Moreover, after some time, faster concentrations catch up with the slower parts of the wave. It is physically impossible for low concentrations to be "overtaken" by higher ones, since this would correspond to having high and low concentrations at the same position. As a consequence, another propagation mode must develop, but what? Up to now, we have assumed that concentrations are continuous functions of space and time, but is that always true? Recall the "bang" of a supersonic plane because it moves faster than the sound velocity: this is a shock wave, i.e. a discontinuity. The high concentration which moves faster than the low concentration is like the plane with respect to sound: a shock or discontinuity develops in the chromatogram. This discontinuity is called a *shock wave* (or *shock* for short). Another term for this is *compressive front*. In our example, the height of the discontinuous part of the front (the middle front in Figure 3.2 bottom right) increases steadily during migration until the complete profile migrates as a true (i.e. rectangular) shock front with a constant velocity σ_{shock}.

It is important to realize that Eq. (3.18) – which is a solution of the original PDE (3.13) – is valid only for those concentrations that are not part of a discontinuity. The reason is that conventionally the solution of a PDE should be differentiable, i.e. continuous. To account for this, and to propose a general framework, shocks are considered to be so-called *weak solutions* (Lax, 1957).

The velocity of a shock, σ_{shock}, depends on its amplitude. Using a mass balance similar to the one presented in Section 1.3 for determining retention times, it can be determined that:

$$\sigma_{shock} = \frac{u_e}{1 + F_e \frac{\Delta \overline{C}_A}{\Delta C_A}} \tag{3.20}$$

where $\Delta \overline{C}_A$ and ΔC_A denote the difference of the lumped solid and fluid-phase concentrations left and right of the discontinuity, respectively; that is, the amplitude of the shock in terms of $\overline{C}_A$ and C_A.

When comparing Eq. (3.18) and Eq. (3.20), it becomes clear that the propagation velocities of waves and shocks differ only in that the former depends on the local slope of the isotherm, $\bar{f}'_A(C_A) = \mathrm{d}\overline{C}_A/\mathrm{d}C_A$, while the latter is described by the slope of the chord of the isotherm, $\Delta \overline{C}_A/\Delta C_A$ involving the initial and feed concentrations.

As a final example, let us consider the case of an *unfavorable* or convex downward adsorption isotherm. For such isotherm models, the derivative $\bar{f}'_A(C_A)$ increases with increasing concentration C_A and, according to Eq. (3.18), the velocity of the concentrations, $\sigma(C_A)$, decreases with increasing values of C_A. As illustrated in Figure 3.3, we observe the opposite behavior to the favorable lumped adsorption isotherm. We notice that during the elution scenario (top) a shock develops, while a simple wave occurs when loading the column (bottom).

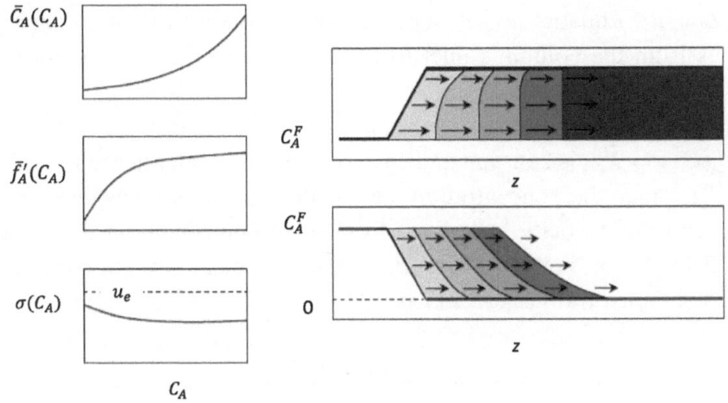

Figure 3.3 Wave propagation for an unfavorable (convex downward) adsorption isotherm; notation and scenarios as in Figure 3.1 and Figure 3.2.

The above results allow us to make the following general rules:

- If velocity decreases when moving from concentration $I(initial)$ to $F(eed)$ the front is dispersive.
- If velocity increases when moving from concentration $I(initial)$ to $F(eed)$ the front is compressive.

Recall that the expressions "simple wave", "self-broadening wave", "dispersive waves" or "fronts" on the one hand, and "shock", "shock waves", "self-sharpening waves", "compressive fronts" on the other hand are synonymous, all being found in the literature. My personal inclination is to the terms "compressive"/"dispersive", but other expressions will be used as well.

3.3.1.2 Construction of breakthrough curves

As demonstrated above, knowledge of the adsorption isotherm $\bar{f}_A(C_A)$ and its derivative $\bar{f}'_A(C_A)$ facilitates determining the velocity of simple waves and shocks. This enables us to construct breakthrough curves and, as shown in Section 3.3.1.3, also complete chromatograms.

As a first step, let us consider introducing a feed stream of concentration C_A^F into a solute-free column, i.e. a column initially equilibrated with pure non-adsorbing solvent. In mathematical terms, this is a so-called Riemann problem, that is, an initial-value problem with uniform initial conditions, $C_A(t = 0,z) = C_A^I$ (here $C_A^I = 0$), and a discontinuity at the boundary, $C_A(t \geq 0,z = 0) = C_A^F$, as shown in Figure 3.4 (left). This type of problem can be solved using the equations given in Section 3.3.1.1.

From the discussion in Section 3.3.1.1 it is clear that the shape of the resulting breakthrough curve will depend strongly on the adsorption isotherm. As for the retention time, it follows from the general solution for the migration velocity Eq. (3.18) that:

$$t_R(C_A) = \frac{L_{col}}{\sigma(C_A)} = t_0\left(1 + F_e\bar{f}'_A(C_A)\right) \tag{3.21}$$

with $t_0 = L_{col}/u_e$. Equation (3.21) specifies the retention time of a concentration C_A that migrates from the column's inlet to its outlet. Note that it does not hold for the special case of a shock wave.

For a linear lumped adsorption isotherm, $\overline{C}_A = \overline{K}_A C_A$, the slope of the adsorption isotherm, $\bar{f}'_A(C_A) = \overline{K}_A$, is constant and any concentration migrates at the velocity given by Eq. (3.19). Thus, the concentration step at the column's inlet given in Figure 3.4 propagates with this velocity and elutes from the column without changing its shape (cf. Figure 3.1). Accordingly, the retention time of a breakthrough curve under linear conditions is obtained from Eq. (3.21) as:

$$t_R = t_0 \left(1 + F_e \overline{K}_A\right) \tag{3.22}$$

It should be noted that Eq. (3.22) also holds for the retention time of a very narrow injection (Dirac pulse), which was derived as Eq. (1.26) using different arguments.

Figure 3.5 (left) shows an example of a breakthrough curve under linear conditions. One recognizes that the elution profile is here nothing else than the rectangular concentration profile introduced at the inlet, but shifted by the retention time. Remember that there is no hydrodynamic dispersion and no mass transfer limitation in this development.

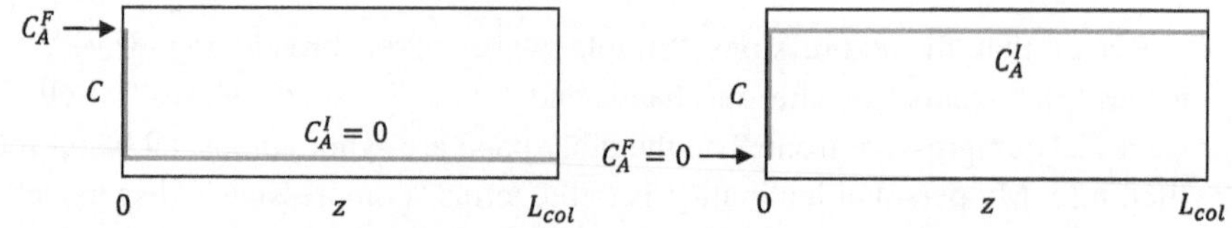

Figure 3.4 Initial and boundary conditions for the loading of a solute-free column ($C_A^I = 0, C_A^F > 0$; left) and regeneration of an equilibrated column ($C_A^I > 0, C_A^F = 0$; right). Note that the feed corresponds to a discontinuity at the inlet boundary.

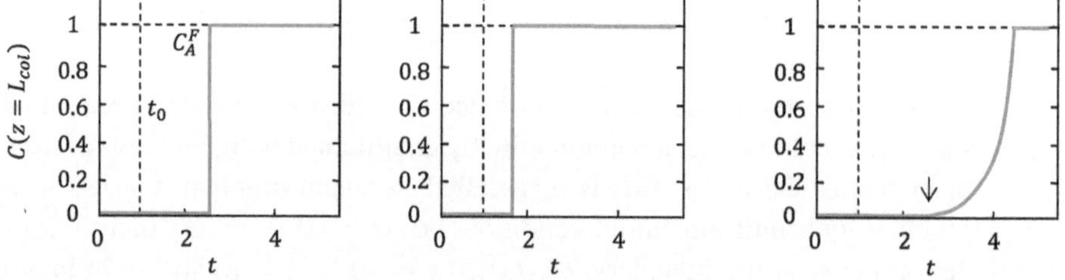

Figure 3.5 Breakthrough curves obtained when loading an initially solute-free column. Left: linear; middle: favorable (Langmuir); right: unfavorable (Freundlich–linear) adsorption isotherms, respectively. The different adsorption isotherm models are presented in Chapter 4. The arrow on the right marks the onset of the simple wave as obtained from $\bar{f}'_A(C_A \to 0)$. Parameters: $C_A^I = 0, C_A^F = 1, t_0 = 1, F_e = 1.5$ (i.e. $\epsilon_e = 0.4$); $\overline{C}_A = C_A$ for linear; $\overline{C}_A = C_A/(1 + C_A)$ for Langmuir; $\overline{C}_A = C_A + C_A^{1.3}$ for Freundlich–linear. Note that the three adsorption isotherms have the same initial slope.

Note that this rectangular profile is not a shock. Shocks occur only due to self-sharpening behavior in non-linear systems. In our case, a rectangular front is obtained since we injected a rectangular profile of which all concentrations move at the same velocity due to linearity. Fronts in which all velocities are equal are also denoted as *contact discontinuities* or *indifferent fronts/waves*.

Next, let us analyze the same case for a favorable lumped adsorption isotherm. Here we find that $\bar{f}'_A(C_A^F) < \bar{f}'_A(C_A^I = 0)$ and from Eq. (3.18) it follows that $\sigma(C_A^F) > \sigma(C_A^I = 0)$. Therefore, the "top" of the boundary discontinuity ($C_A = C_A^F$) in Figure 3.5 moves faster than its foot point ($C_A^I = 0$). As a consequence, the imposed discontinuity will propagate further as a shock. Its velocity is given by Eq. (3.20) and we find for its retention time:

$$t_{R,shock} = t_0 \left(1 + F_e \frac{\Delta \bar{C}_A}{\Delta C_A}\right) \tag{3.23}$$

In our example, the shock amplitude is given by the feed concentration C_A^F and the initial concentration $C_A^I = 0$, so that:

$$t_{R,shock} = t_0 \left(1 + F_e \frac{\bar{C}_A(C_A^F) - \bar{C}_A(C_A^I)}{C_A^F - C_A^I}\right) = t_0 \left(1 + F_e \frac{\bar{C}_A(C_A^F)}{C_A^F}\right) \tag{3.24}$$

An example of a saturation breakthrough curve obtained for a Langmuir adsorption isotherm is shown in Figure 3.5 (middle). In the example, parameters have been adjusted so that the isotherm has the same initial slope as in the linear case above. Comparing Figure 3.5 (left and middle) one observes that the shock front elutes earlier than the rectangular front in the linear case. The practical consequence is that for favorable lumped adsorption isotherms, the higher the feed concentration the quicker a column is loaded completely.

Finally, for an unfavorable lumped adsorption isotherm, $\bar{f}'_A(C_A^F) > \bar{f}'_A(C_A^I = 0)$, and from Eq. (3.18) it is found that $\sigma(C_A^F) > \sigma(C_A^I = 0)$. The top part of the injection profile ($C_A = C_A^F$) here moves more slowly than its lower part ($C_A^I = 0$). Consequently, a simple wave evolves at the inlet and Eq. (3.21) applies directly to the retention times for all concentrations $0 < C_A \leq C_A^F$.

An example is given in Figure 3.5 (right). The same initial isotherm slope has been used as for the linear (left) and the Langmuir (middle) isotherms. As indicated by the arrow, now the onset of the simple wave elutes at the same time as the rectangular front in the linear case (left). Consequently, for unfavorable lumped adsorption isotherms a complete loading of the column will take longer the higher the feed concentration. This behavior is often undesirable in classical adsorption applications, which is why such isotherms are denoted as "unfavorable". Note, however, that this widely used terminology does not make a lot of sense: if an adsorption isotherm gives the desired front shape for saturation, it will give an undesirable front shape for desaturation, and vice versa.

The complementary case to the loading scenario above is the regeneration or elution of a preloaded column, as illustrated in Figure 3.4 (right). Here, the Riemann problem is given by $C_A(t = 0, z) = C_A^I > 0$ and $C_A(t > 0, z = 0) = C_A^F = 0$.

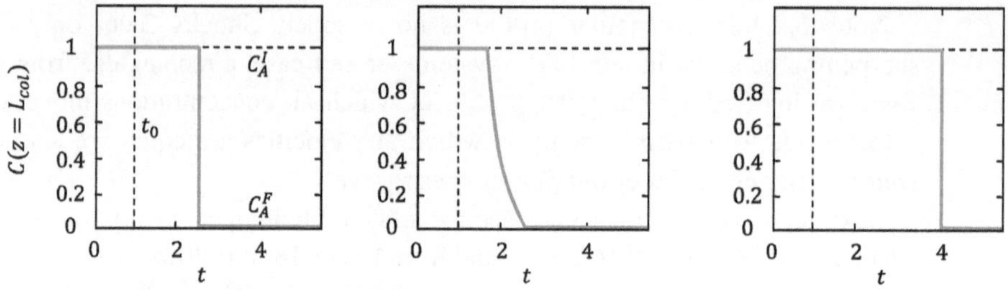

Figure 3.6 Breakthrough curves obtained when regenerating a column with pure solvent. The column is initially equilibrated completely with a concentration C_A^I. Left: linear; middle: favorable; right: unfavorable lumped adsorption isotherms, respectively. Parameters: $C_A^I = 1, C_A^F = 0$; for other parameters see Figure 3.5.

The solution is obtained using the same arguments as above. For a linear lumped adsorption isotherm, the breakthrough curve again corresponds to the rectangular front imposed at the inlet, eluting at a retention time according to Eq. (3.22). For a favorable lumped adsorption isotherm, where we obtained a shock wave in the loading scenario, we now find a simple wave, because the velocity at the boundary $\sigma(C_A^F = 0)$ is lower than that of the initial concentration $\sigma(C_A^I > 0)$. The retention time is given by Eq. (3.21). Finally, and as expected, for an unfavorable lumped adsorption isotherm the situation is also converse. While a simple wave occurred during loading, we find a shock when regenerating, and t_R is given by Eq. (3.24). Figure 3.6 shows regeneration curves for the three cases.

Summarizing, we have seen that the type of breakthrough curve depends on the shape of the adsorption isotherm as well as the combination of initial and boundary conditions. In more general terms, we have seen that an initial discontinuity imposed by a step change of the feed concentration is resolved into either a simple wave, a shock or a contact discontinuity emanating from the inlet of the column.

For linear lumped adsorption isotherms, a step change of the feed concentration propagates further without changing its shape as a rectangular contact discontinuity. In contrast, for favorable lumped adsorption isotherms a shock wave evolves when loading a column, and a simple wave during its regeneration. For unfavorable lumped adsorption isotherms a step change evolves into a dispersed wave when loading, and a shock wave during its regeneration. In all cases the breakthrough curves can be constructed from Eq. (3.21) and Eq. (3.23).

Note that for adsorption isotherms with inflection points (i.e. a non-monotonic derivative) more complex elution profiles may occur. These are discussed in Section 3.3.1.4.

Finally, let us mention that the *average* position of a front is independent of its compressive or dispersive nature. It is always given by:

$$t_{R,avg} = t_0 \left(1 + F_e \frac{\Delta \bar{C}_A}{\Delta C_A}\right)$$

which looks similar to the retention time of a shock. The demonstration of this is the subject of Exercise 3.1.

EXERCISE 3.1 *Determine the average retention time of a front (compressive or dispersive) obtained by feeding a column initially equilibrated with a solution at concentration* C_A^I *with a feed at concentration* C_A^F.

EXERCISE 3.2 *A column is loaded with a feed solution of concentration* C_A^F. *The lumped adsorption isotherm is assumed to be given by* $\overline{C}_A = \frac{1}{\gamma} C_A^2$. *This is an unusual adsorption isotherm (to say the least), but it permits some simple results. Calculate the breakthrough curve if the column is initially solute-free, that is,* $C_A^I = 0$.

EXERCISE 3.3 *A chromatographic column is associated with a dead volume assumed to be equivalent to one mixing cell. Two situations are considered:*

- *situation a: the mixing cell is placed after (downstream of) the column*
- *situation b: the mixing cell is placed before (upstream of) the column.*

Compare the RTDs *associated with the two situations. The column is initially solute-free and is fed with a constant concentration of a solute which is assumed to be adsorbed according to a Langmuir lumped adsorption isotherm (the dead volume does not contain adsorbing material). Characterize the breakthrough curve in both situations (no complex calculations are required).*

3.3.1.3 Construction of chromatograms

Based on the principles presented above, we now consider the complete chromatographic cycle that results from injecting a rectangular pulse of a given volume V_{inj} into a column initially equilibrated with pure solvent.

In general, a pulse injection as shown in Figure 3.7a is simply a sequence of the loading and the elution scenarios discussed in Section 3.3.1.2. Such a pulse corresponds to twice changing the boundary conditions at the inlet. The front side of the injection is a step increase of the inlet concentration at $t = 0$ from the initial concentration, $C_A = C_A^I = 0$, to $C_A = C_A^F$. Following the arguments in the previous section, this will resolve into a right-angled contact discontinuity for a linear, a shock for a favorable, or a simple wave for an unfavorable lumped adsorption isotherm, respectively. The rear of the injection is a step downwards from $C_A = C_A^F$ to $C_A = 0$. This enters the column when the injection is over, after $\Delta t_{inj} = V_{inj}/Q$, and triggers another front that follows the first one. This second front is a contact discontinuity for a linear, a simple wave for a favorable, or a shock for an unfavorable lumped adsorption isotherm, respectively.

For linear systems, the two contact discontinuities migrate at the velocity given by Eq. (3.19). The retention time of the front side is given directly by Eq. (3.22). Since the rear side is triggered later, at $t = \Delta t_{inj}$, its retention time is obtained from the same equation after adding Δt_{inj}. Thus, the chromatogram is characterized by two retention times:

$$t_{R,lead} = t_0 \left(1 + F_e \overline{K}_A\right) \quad (3.25a)$$

$$t_{R,rear} = t_0 \left(1 + F_e \overline{K}_A\right) + \Delta t_{inj} \quad (3.25b)$$

Between the fronts prevails an injection plateau $C_A(t, L_{col}) = C_A^F$, which is always present for linear systems (again assuming a column of infinite efficiency). In other words, in the linear case the chromatogram always corresponds to the injection profile simply shifted by $t_{R,lead}$. An example is shown in Figure 3.7b.

In the non-linear case one has to distinguish between "large" and "small" injections. For sufficiently large injected volumes, the elution profile consists of the two breakthrough curves at the front and rear of the elution profile, connected by the plateau. For smaller injections, the plateau disappears during travel in the column.

For a favorable lumped adsorption isotherm, Eq. (3.24) gives the retention time of the leading shock. The rear side is constructed by adding Δt_{inj} to the retention time of the simple wave given by Eq. (3.21). Thus, the chromatogram is given by:

$$t_{R,lead} = t_0\left(1 + F_e\frac{\bar{f}_A(C_A^F)}{C_A^F}\right) \tag{3.26a}$$

$$t_{R,rear} = t_0\left(1 + F_e\bar{f}'_A(C_A)\right) + \Delta t_{inj} \tag{3.26b}$$

For an unfavorable lumped adsorption isotherm, we have to exchange the positions of the shock and the simple wave, and obtain:

$$t_{R,lead} = t_0\left(1 + F_e\bar{f}'_A(C_A)\right) \tag{3.27a}$$

$$t_{R,rear} = t_0\left(1 + F_e\frac{\overline{C}_A(C_A^F)}{C_A^F}\right) + \Delta t_{inj} \tag{3.27b}$$

Figures 3.7c and 3.7d show corresponding examples for favorable and unfavorable lumped adsorption isotherms.

When injecting "small" volumes under non-linear conditions, the injection plateau vanishes. The reason is that, if triggered too soon after one another, the simple wave will interact with the shock.

Let us elucidate this for a favorable lumped adsorption isotherm. The velocity of the leading shock is determined by the chord of the adsorption isotherm, $\Delta\overline{C}_A/\Delta C_A$, while the velocities in the rear simple wave depend on the local slope, $\overline{C}'_A\,(C_A)$. As illustrated in Figure 3.8, the slope of the chord lies between the local slopes for the initial and inlet concentrations:

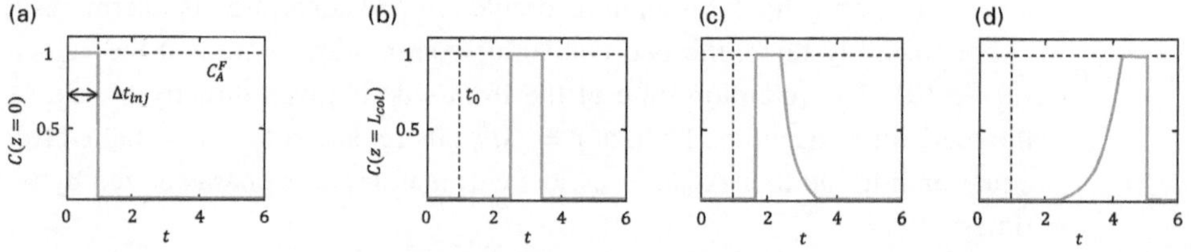

Figure 3.7 Example chromatograms resulting from large injections into an initially solute-free column. Injections are sufficiently large to ensure the existence of a residual plateau. (a) Injection profile; (b)–(d) chromatograms. (b) Linear; (c) favorable (Langmuir); (d) unfavorable (Freundlich–linear) isotherms, respectively. Parameters: $C_A^I = 0, C_A^F = 1, \Delta t_{inj} = 1$; for other parameters see Figure 3.5.

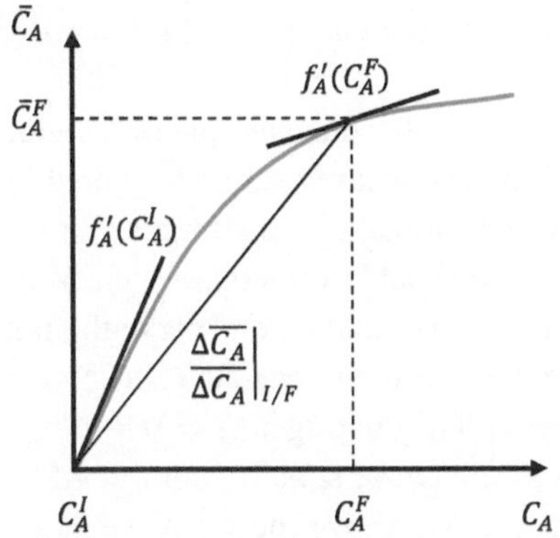

Figure 3.8 Local slope and chord of a favorable lumped adsorption isotherm for injection of a feed at concentration C_A^F into a solute-free column ($C_A^I = 0$).

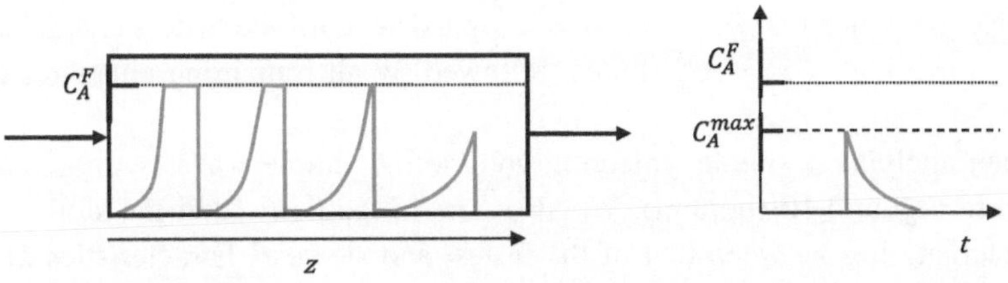

Figure 3.9 Illustration of wave interaction inside the column (left) and the resulting chromatogram (right) for a small injection, in the case of a favorable lumped adsorption isotherm.

$$\bar{f}_A'\left(C_A^I\right) > \left.\frac{\Delta \bar{C}_A}{\Delta C_A}\right|_{I/F} > \bar{f}_A'\left(C_A^F\right) \tag{3.28}$$

The velocity equations (3.18) and (3.20) then tell us that the high concentrations of the simple wave travel faster than the shock, while its lower concentrations must be slower:

$$\sigma\left(C_A^I\right) < \sigma_{shock(I/F)} < \sigma\left(C_A^F\right) \tag{3.29}$$

Consequently, for small injection volumes or long columns, the upper part of the simple wave will eventually catch up with the shock so that the injection plateau vanishes. On their way onward through the column the two fronts will continuously interact with one another. As a result, the shock is "consumed" and its amplitude and velocity decrease steadily. Figure 3.9 illustrates this process and shows a corresponding sample chromatogram.[3]

The process of different concentration fronts interacting dynamically with each other is referred to as *wave interference*. A useful tool for analyzing such phenomena is the representation of the problem in the so-called *physical domain* (*t*,*z*). Here, instead of plotting the concentrations vs. time, the velocities of the concentrations are represented

[3] For unfavorable lumped adsorption isotherms we find "mirrored" chromatograms. In such cases the high concentrations of the leading simple wave are slower than the shock at the rear.

as lines in a distance–time diagram. A more detailed analysis of the wave interference phenomenon is given in Appendix C.

Figure 3.10 (top) illustrates this representation for the example of a "wide" injection and a favorable lumped adsorption isotherm. The leading shock triggered by the beginning of the injection is represented in the (t,z) domain by a straight line starting at the origin, with a slope equal to the inverse of the shock's velocity, σ_{shock}^{-1}, from Eq. (3.20). At positions and times below this line in the (t,z) domain we observe the initial concentration, that is, $C_A\,(t,z) = C_A^I$, while above the line the inlet state $C_A\,(t,z) = C_A^F$ prevails. The rear of the injection causes a simple wave. Originating at $\left(t = 0 + \Delta t_{inj}, z = 0\right)$, the velocities of the different concentrations in this wave span a "family" of straight characteristics with slopes $\sigma^{-1}\,(C_A)$ from Eq. (3.18). Above these wave characteristics we observe the pure solvent introduced after the injection, $C_A\,(t,z) = 0$. The chromatogram on the right is constructed by analyzing the states at the column's outlet, i.e. at $z = L_{col}$. In the example, the first characteristic to reach $z = L_{col}$ belongs to the shock. Afterwards, the injection plateau $C_A(t,z = L_{col}) = C_A^F$ persists, until the fastest characteristic of the simple wave with $C_A = C_A^F$ elutes, followed by all remaining characteristics of this wave.

When applying a smaller injection volume the simple wave is triggered earlier, as shown in Figure 3.10 (bottom). We also note in the figure that the slope of the shock characteristic lies between that of the fastest and slowest characteristics of the simple wave (see Eq. (3.29)). As a consequence, the fastest concentration in the simple wave, C_A^F, reaches the shock within the column at some position $z = z_i$. Afterwards, the two waves start to interact and more and more characteristics of the simple wave catch

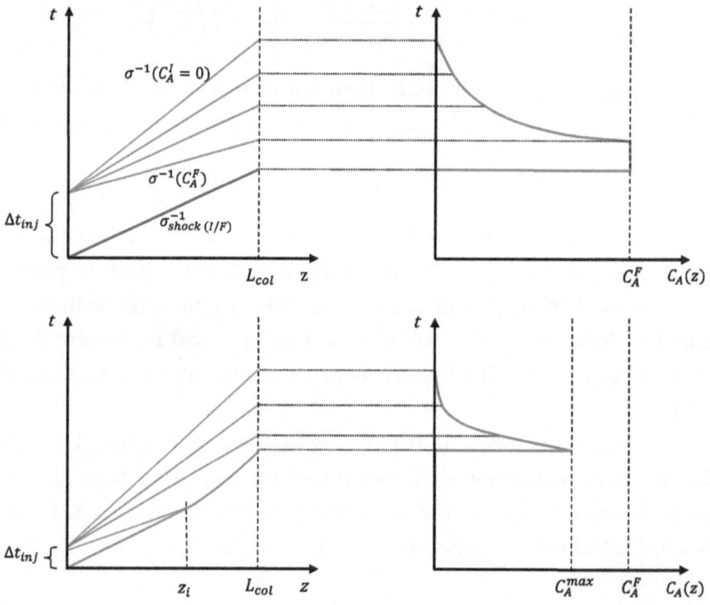

Figure 3.10 Representation of a "wide" (top) and a "narrow" injection (bottom) in the (t,z) domain and resulting chromatograms. Favorable lumped adsorption isotherm.

up with the shock. The shock's height decreases continuously, thus its velocity also decreases perpetually, which renders its characteristic non-linear for $z > z_i$. As illustrated in Figure 3.10 (bottom), the simple wave is adsorbed or blocked by the shock and is not transmitted through the shock. We will see when discussing binary separations that other situations can occur.

Chromatographers are not familiar with the distance–time representation, so it may appear somewhat complicated. In fact, after a short period of adaptation one normally starts to feel comfortable with a tool which has demonstrated its usefulness (see the above and later examples) in analyzing chromatographic processes. A way to become familiar with this distance–time domain and characteristic lines is to consider that we are in a plane flying above the (t,z) space. We are observing the "chromatogram landscape" from above and the characteristics are similar to level lines.

Before constructing chromatograms for some specific conditions, we need to know whether or not they will exhibit an injection plateau. A simple consideration allows a determination of the value of the injection volume, V_{inj}^{plat}, for which the plateau just "vanishes", i.e. shrinks to zero width. For the above example of an initially solute-free column and a favorable lumped adsorption isotherm, this occurs if the shock and the fastest characteristic of the simple wave reach the outlet at exactly the same time. For the duration, $\Delta t_{inj}^{plat} = V_{inj}^{plat}/Q$, of such an injection:

$$\frac{L_{col}}{\sigma\left(C_A^F\right)} + \Delta t_{inj}^{plat} = \frac{L_{col}}{\sigma_{shock(I/F)}} \tag{3.30}$$

From this, we obtain, with the velocity equations Eq. (3.18) and Eq. (3.20):

$$V_{inj}^{plat} = (1 - \epsilon_e)\, V_{col} \left(\left.\frac{\Delta \bar{C}_A}{\Delta C_A}\right|_{I/F} - \bar{f}'_A\left(C_A^F\right) \right) \tag{3.31}$$

Performing an injection of this volume will result in a triangle-like peak with a maximum concentration of C_A^F.

When constructing a chromatogram for a single solute, we need to distinguish two different situations:

- $V_{inj} > V_{inj}^{plat}$: this is what we called above a "wide" injection, for which the elution profile exhibits a plateau. The chromatogram is given by Eq. (3.26a) and Eq. (3.26b) above. (For linear and favorable lumped adsorption isotherms, see Eq. (3.25) and Eq. (3.27).)
- $V_{inj} < V_{inj}^{plat}$: this is what we called above a "narrow" injection, and we will not observe a plateau in the elution profile. The dispersive part of the chromatogram is still given by Eq. (3.26b). However, the outlet concentration does not reach the feed concentration, but we observe a triangle-like peak with a maximum $C_A^{max} < C_A^F$. Considering that C_A^{max} belongs simultaneously to the shock and the simple wave, one can show (Appendix C) that:

$$\bar{f}_A(C_A^{max}) - \bar{f}'_A(C_A^{max})C_A^{max} = \frac{1}{F_e}\frac{\Delta t_{inj}}{t_0}C_A^F = \frac{V_{inj}}{(1-\varepsilon_e)V_{col}}\, C_A^F \tag{3.32}$$

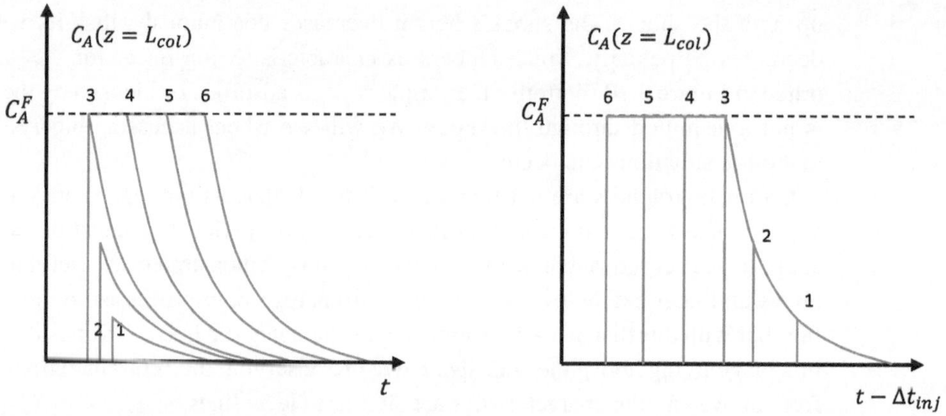

Figure 3.11 Volume overloading series for a favorable lumped adsorption isotherm and a constant feed concentration. The numbers denote increasing injection volumes. Note that case 3 corresponds to V_{inj}^{plat}. Left: overlay of all chromatograms ($t = 0$ corresponds to the onset of the injection). Right: overlay of chromatograms plotted against time corrected by the injection duration, Δt_{inj}.

After determining C_A^{max} from Eq. (3.32), the chromatogram is constructed by calculating the retention times of the simple wave from Eq. (3.26b) for some concentrations between 0 and C_A^{max}. The peak is completed by a shock at $t_{R,lead} = t_{R,shock}\left(0, C_A^{max}\right) = t_{R,rear}\left(C_A^{max}\right)$.

Figure 3.11 (left) shows an overlay of chromatograms obtained from a *volume-overloading series*, i.e. for a series of injections of different volumes. In the example, a system with Langmuir lumped adsorption isotherm was considered. One observes in Figure 3.11 (left) that the peak size increases with increasing volume, and that, once the injection plateau is established, the leading shocks elute at the same time, as given by Eq. (3.26a). In Figure 3.11 (right) the same chromatograms are shown, but the time is corrected for the injection duration. Here it is readily verified that, as stated by Eq. (3.26b), the retention time of the simple wave corrected by the injection duration does not depend on the injection volume but only on the adsorption isotherm parameters.

Note that, for a so-called *mass-overloading series*, where instead of the injected volume the injection concentration is varied, the simple waves coincide without correcting the time by Δt_{inj}.

EXERCISE 3.4 *A column ($V_{col} = 1$ l, $\varepsilon_e = 0.4$) is eluted with an eluent flow rate of $Q = 0.4$ l/min. The* solute A *adsorbs according to the modified Langmuir model given by Eq. (3.2) with $\overline{N}_A = 2$ g/l, $\tilde{K}_A = 0.5$ l/g and $\overline{\lambda}_A = 1$.*

- *Estimate the retention time associated with a Dirac injection of solute A at trace concentration.*
- *Consider two different solutions:* (1) *with $C_A = 0$ and* (2) *with $C_A = 5$ g/l. Draw breakthrough curves resulting from the following experiments:*

- *Case a: the column is initially equilibrated with solution* 1 *and saturated with solution* 2.
- *Case b: the column is initially equilibrated with solution* 2 *and saturated with solution* 1.

EXERCISE 3.5 *Using the data of Exercise 3.4, calculate the column response to an injection of* $V_{inj} = 0.1\ l$ *of solute A at a concentration* $C_A^F = 1$ *g/l. Perform the same task for* $V_{inj} = 1\ l$.

3.3.1.4 Adsorption isotherms with inflection points

The analysis given in the previous sections was restricted to linear, favorable, and unfavorable lumped adsorption equilibria. These are all characterized by a one-signed curvature, i.e. $\bar{f}_A''(C_A) \geq 0$ or $\bar{f}_A''(C_A) \leq 0$. However, different adsorption isotherms exist that can exhibit inflection points at one or more concentrations where $\bar{f}_A''(C_A) = 0$ (see Chapter 4 for additional information). This gives rise to some additional phenomena to be discussed below.

A simple but useful concept for analyzing chromatographic processes for arbitrary adsorption isotherm shapes and combinations of feed and initial concentrations is the so-called *Golden's rule* (after Berkeley doctoral student F. M. Golden, who discovered this rule but never published it). This approach allows one to deduce a chromatogram's shape by analyzing the course of the adsorption isotherm between the initial and boundary conditions. To apply the method, we fix one end of a hypothetical "rubber band" to the adsorption isotherm at the initial condition, I. Then we span the band in a counter-clockwise direction (i.e. we keep the isotherm to the left of the band) and fix the other end at the boundary (feed) condition F, ensuring that the band is pulled tight. If the stretched band forms a chord directly connecting I and F, the corresponding part of the chromatogram is a shock. If the stretched band attaches itself to and takes its course along the curved isotherm, we obtain a simple wave.

Figure 3.12 illustrates the application of Golden's rule for different isotherm types in the case of large injections into preloaded columns ($C_A^I > 0$). For a favorable lumped adsorption isotherm, in Figure 3.12 (top), we see that the band forms a chord from I to F, telling us, as expected, that the leading front of the chromatogram is a shock from C_A^I to C_A^F. In contrast to an initially solute-free column, the shock's migration behavior is determined by both C_A^I and C_A^F. Since we consider a large injection, the shock is followed by an injection plateau. The rear front of the injection corresponds to switching the feed from I to F. Thus we now span the band from F back to I, again keeping the isotherm on the left. Now we observe the band sticking onto the curved adsorption isotherm, and we deduce that the rear front of the chromatogram is a simple wave from C_A^F back to C_A^I. In Figure 3.12 (middle) the same scenario is shown for an unfavorable lumped adsorption isotherm, giving a leading simple wave and a rear shock wave.

A more complex behavior is found for an isotherm with an inflection point, as in Figure 3.12 (bottom). Here the band first forms a chord from C_A^I, before it becomes tangent to the adsorption isotherm at some point a. From there it sticks to the adsorption isotherm until the feed F is reached. According to the rule above, the part from I to

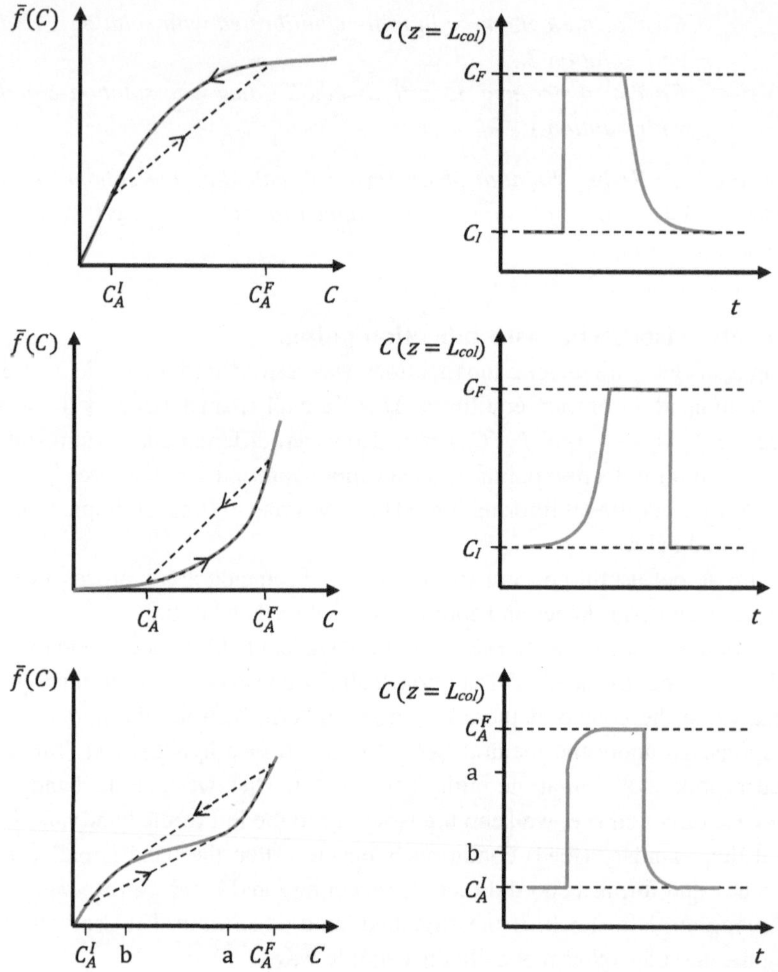

Figure 3.12 Deduction of peak shapes for different isotherm types using Golden's rule for "large" injections into a preloaded column. Solid lines and arrows indicate the front side of the elution profile from the initial condition I to the feed F. Dashed lines and arrows mark the rear side, from F back to I. Adsorption isotherm types: favorable (top), unfavorable (middle), with one inflection point (bottom).

a will form a shock wave, while from a to F a simple wave develops. As shown in the figure, the leading front of the chromatogram is now a so-called *combined wave*, consisting of a shock and a simple wave. Moreover, at point a the band is tangent to both the shock's chord and the isotherm. This means that at this point the shock and the wave have the same velocity, which also means that the simple wave and the shock wave are contiguous (i.e. no plateau in between) in the chromatogram. Furthermore, due to the equal velocities, the self-sharpening tendency of the compressive front vanishes towards its highest concentration. Such a shock, which is tangent to a simple wave, is called a *semi-shock*, as it is only partially self-sharpening. However, for its velocity and retention time Eq. (3.20) and Eq. (3.23) still hold. For the rear side of the chromatogram we observe a similar situation, where again a semi-shock exists from C_A^F down to b, followed by a simple wave. It should be realized that a and b differ from each other, while the inflection point itself is located between these two points.

The presentation can be extended to the very rare situation of an adsorption isotherm with two inflection points, as presented, for example, in (Zhang *et al.*, 2006).

The approach explained above greatly simplifies the analysis of chromatographic processes even under seemingly complex conditions. The method determines the type of waves in a chromatogram together with the relevant concentration values. Based on this information one can construct breakthrough curves and elution profiles using the equations given in Section 3.3.1.3.

3.3.2 More realistic chromatograms: the MC–$\overline{\text{Equil}}$ model

The equilibrium model presented in previous sections allows one to obtain the main features of a chromatogram – the type and retention times of the various possible waves – in a straightforward manner. However, if more realistic chromatograms are desired, hydrodynamic dispersion and mass transfer limitations must be taken into account. A simple model that considers such band-broadening effects on a "rudimentary" level is the MC–$\overline{\text{Equil}}$ model presented in Chapters 1 and 2, which approximates the column to a series of N equilibrium stages. More refined models are discussed in later chapters.

The mass balance for a single solute in the kth equilibrium cell can be written as

$$C_A^{k-1}(t) = C_A^k(t) + \frac{t_0}{N}\frac{dC_A^k(t)}{dt} + F_e\frac{t_0}{N}\frac{d\overline{C}_A^k(t)}{dt} \qquad k \in [1,\ldots,N] \tag{3.33}$$

The plate model assumes that equilibrium is reached, so Eq. (3.33) can be rewritten as:

$$C_A^{k-1}(t) = C_A^k(t) + \frac{t_0}{N}\left[1 + F_e f'_A\left(C_A^k(t)\right)\right]\frac{dC_A^k(t)}{dt} \qquad k \in [1,\ldots,N] \tag{3.34}$$

This is a set of N ordinary differential equations that must generally be solved numerically. The required initial and boundary conditions used in the examples below are expressed as:

$$C_A^k(t=0) = C_A^I \quad k \in [1,\ldots,N] \tag{3.35}$$

$$C_A^0(t \geq 0) = C_A^F \tag{3.36}$$

Figure 3.13 shows breakthrough curves simulated using the MC–$\overline{\text{Equil}}$ model for a system represented by a linearly modified Langmuir lumped adsorption isotherm. Figure 3.13 (left) shows a saturation scenario that leads to a self-sharpening breakthrough curve. It is readily observed that the sharpness of the front increases with increasing column efficiency using higher values of N, until the front almost coincides with the vertical shock obtained from the equilibrium model, Eq. (3.24). It should

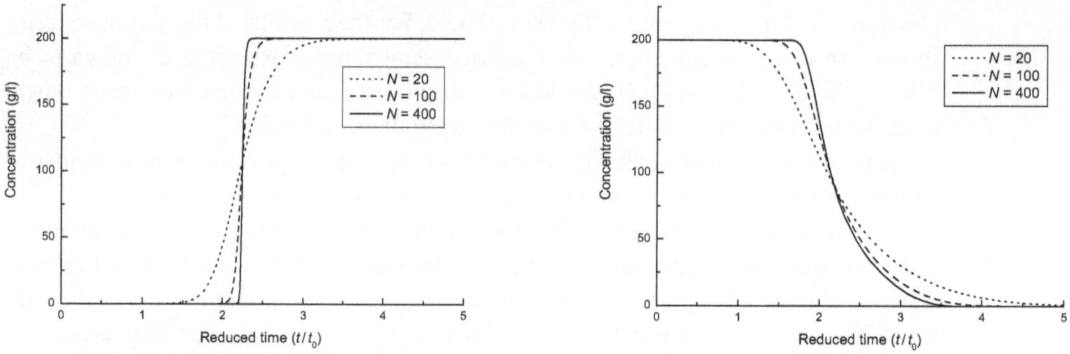

Compressive front simulated with the MC–$\overline{\text{Equil}}$ model.

Dispersive front simulated with the MC–$\overline{\text{Equil}}$ model.

Figure 3.13 Influence of the number of plates on breakthrough curves: linearly modified Langmuir lumped adsorption isotherms. Parameters: $C_A^F = 200$ g/l, $\varepsilon_e = 0.4$ ($F_e = 1.5$), $t_0 = 1$, $\overline{N}_A = 100$ g/l, $\tilde{K}_A = 0.01$ l/g, $\tilde{\lambda}_A = 0.5$.

be noted that the mean position of the shock is independent of N.[4] The dispersive front resulting from regenerating such a column is shown in Figure 3.13 (right). With increasing N, the simple wave again approaches the solution for the equilibrium model given by Eq. (3.21).

It should be noted that, for a given number of plates N, the broadening of the fronts depends strongly on the non-linearity of the adsorption isotherm. This is discussed in more detail in Chapter 7.

3.4 Binary-solute systems: competition between solutes

In order to be suited for practical chromatographic problems, the analysis must be extended to multi-solute systems. To begin with, we will consider binary mixtures that contain two retained solutes, denoted A and B.

As the reader has probably already guessed, the structure of the adsorption isotherms is going to play a key role. Let us first consider the trivial case defined by $\overline{C}_A = \bar{f}_A(C_A)$ and $\overline{C}_B = \bar{f}_B(C_B)$ so that $\bar{f}_A$ depends on A only and $\bar{f}_B$ depends on B only. It is easily guessed that under these circumstances the binary problem reduces to that of two single-solute problems, dealt with in the previous section. The results are thus obtained by superimposition of the solutions for A and B discussed above. However, when non-linear equilibria prevail, this "independence" is rarely observed, due to competition between species. The n-solute lumped adsorption isotherm laws in Eq. (3.6) become, for a binary mixture:

$$\overline{C}_A = \bar{f}_A\left(C_A, C_B\right) \qquad \overline{C}_B = \bar{f}_B\left(C_A, C_B\right) \tag{3.37}$$

[4] It is given by $t_R = t_0 \times \left(1 + F_e \dfrac{\overline{C}_A^F}{C_A^F}\right) = 1.0 \times \left(1 + 1.5 \dfrac{166.66}{200}\right) = 2.25$. The same remark applies to the dispersive front.

We will see that this competition between solutes is responsible for characteristic and singular (if not perplexing) features of non-linear multicomponent phase equilibria in chromatographic processes. Additionally, this will eventually give us powerful tools to characterize multi-component non-linear equilibrium laws, at least on a qualitative basis (i.e. the structure of the laws).

3.4.1 The equilibrium model

As with single-solute systems, we will apply equilibrium theory to derive the main features of chromatograms. If hydrodynamic dispersion and mass transfer limitations can be neglected, the material balances for the two solutes can be written, in analogy to Eq. (3.13), as:

$$
\begin{aligned}
u_e \frac{\partial C_A}{\partial z} + \frac{\partial C_A}{\partial t} + F_e \frac{\partial \overline{C}_A}{\partial t} &= 0 \\
u_e \frac{\partial C_B}{\partial z} + \frac{\partial C_B}{\partial t} + F_e \frac{\partial \overline{C}_B}{\partial t} &= 0
\end{aligned} \tag{3.38}
$$

Again, this mass balance, assuming fluid velocity to be constant, in fact assumes the solutes to be sufficiently diluted in an inert solvent.

Since we assume instantaneous equilibrium, the lumped solid-phase concentrations are related to the fluid-phase concentrations through the lumped adsorption isotherms, Eq. (3.37), and from the chain rule one obtains:

$$
\begin{aligned}
\frac{\partial \overline{C}_A}{\partial t} &= \frac{\partial \overline{C}_A}{\partial C_A}\frac{\partial C_A}{\partial t} + \frac{\partial \overline{C}_A}{\partial C_B}\frac{\partial C_B}{\partial t} \\
\frac{\partial \overline{C}_B}{\partial t} &= \frac{\partial \overline{C}_B}{\partial C_A}\frac{\partial C_A}{\partial t} + \frac{\partial \overline{C}_B}{\partial C_B}\frac{\partial C_B}{\partial t}
\end{aligned}
$$

so that Eq. (3.38) can be rewritten:

$$
\begin{aligned}
u_e \frac{\partial C_A}{\partial z} + \left(1 + F_e \frac{\partial \overline{C}_A}{\partial C_A}\right)\frac{\partial C_A}{\partial t} + F_e \frac{\partial \overline{C}_A}{\partial C_B}\frac{\partial C_B}{\partial t} &= 0 \\
u_e \frac{\partial C_B}{\partial z} + \left(1 + F_e \frac{\partial \overline{C}_B}{\partial C_B}\right)\frac{\partial C_B}{\partial t} + F_e \frac{\partial \overline{C}_B}{\partial C_A}\frac{\partial C_A}{\partial t} &= 0
\end{aligned} \tag{3.39}
$$

Because the partial derivatives of the lumped adsorption isotherms, $\partial \overline{C}_i / \partial C_j$ (i,j = A or B), are assumed to be known, Eq. (3.39) is a set of two equations with two unknowns, C_A and C_B.

If the solutes are not competing, then $\frac{\partial \overline{C}_B}{\partial C_A}$ and $\frac{\partial \overline{C}_A}{\partial C_B}$ vanish and Eq. (3.39) reduces to Eq. (3.14), which describes the single-solute system. This is the trivial case discussed at the beginning of the section and it is ruled out here.

As we did for the single-solute system (cf. p. 67), we now introduce a parameter $\sigma\ (C_A, C_B) = z/t$ analogous to a velocity. Equation (3.39) then becomes:

$$u_e \frac{dC_A}{d\sigma} - \sigma \left(1 + F_e \frac{\partial \bar{C}_A}{\partial C_A}\right) \frac{dC_A}{d\sigma} - \sigma F_e \frac{\partial \bar{C}_A}{\partial C_B} \frac{dC_B}{d\sigma} = 0$$

$$u_e \frac{dC_B}{d\sigma} - \sigma \left(1 + F_e \frac{\partial \bar{C}_B}{\partial C_B}\right) \frac{dC_B}{d\sigma} - \sigma F_e \frac{\partial \bar{C}_B}{\partial C_A} \frac{dC_A}{d\sigma} = 0$$

or, in matrix form:

$$\begin{bmatrix} u_e - \sigma \left(1 + F_e \dfrac{\partial \bar{C}_A}{\partial C_A}\right) & -\sigma F_e \dfrac{\partial \bar{C}_A}{\partial C_B} \\ -\sigma F_e \dfrac{\partial \bar{C}_B}{\partial C_A} & u_e - \sigma \left(1 + F_e \dfrac{\partial \bar{C}_B}{\partial C_B}\right) \end{bmatrix} \begin{bmatrix} \dfrac{dC_A}{d\sigma} \\ \dfrac{dC_B}{d\sigma} \end{bmatrix} = \begin{bmatrix} 0 \\ 0 \end{bmatrix} \tag{3.40}$$

Equation (3.40) shows that the problem depends on σ only and can accept two types of solutions:

- Trivial solutions for which:

$$\frac{dC_A}{d\sigma} = \frac{dC_B}{d\sigma} = 0 \tag{3.41}$$

  As in the single-solute case, these trivial solutions represent *concentration plateaus* in the column or in the chromatogram.
- Non-trivial solutions for which $\sigma\ (C_A, C_B)$ must satisfy:[5]

$$\det \begin{bmatrix} u_e - \sigma \left(1 + F_e \dfrac{\partial \bar{C}_A}{\partial C_A}\right) & -\sigma F_e \dfrac{\partial \bar{C}_A}{\partial C_B} \\ -\sigma F_e \dfrac{\partial \bar{C}_B}{\partial C_A} & u_e - \sigma \left(1 + F_e \dfrac{\partial \bar{C}_B}{\partial C_B}\right) \end{bmatrix} = 0 \tag{3.42}$$

  which are associated with *simple concentration waves*. This is an algebraic equation of the second degree with respect to σ (even though it does not look like one). It has two solutions that must be *real* numbers.[6] These solutions, $\sigma_-(C_A, C_B)$ and $\sigma_+(C_A, C_B)$, define two families of simple waves.

Equation (3.42) is a very important second-order equation in σ known as the *characteristic equation* of the problem. Its two roots in σ correspond to two possible propagation

[5] The non-trivial solution requires us to recall some (unpleasant?) lectures in mathematics. A non-trivial solution is a vector with both non-zero components $\frac{dC_i}{d\sigma}$. This means that the matrix that multiplies the vector in Eq. (3.40) is not invertible, otherwise only the trivial solution exists. However, we all know that an elution curve is not made up of a plateau (constant concentration) only, unless one forgets to inject the solutes. Then, for non-trivial solutions for which the matrix is not invertible, its determinant must vanish.

[6] If not, you made an error in solving the equation or the assumed adsorption isotherm equations $\bar{f}_A$ and/or $\bar{f}_B$ are physically impossible, at least for a restricted composition domain (C_A,C_B).

velocities of the fronts for a given composition (C_A, C_B). A compact solution is obtained by defining:[7]

$$\theta = \frac{u_e/\sigma - 1}{F_e} \quad \text{thus} \quad \sigma = \frac{u_e}{1 + F_e\theta} \tag{3.43}$$

After substituting Eq. (3.43) into Eq. (3.42), one obtains for θ:

$$\theta^2 - \left(\frac{\partial \overline{C}_A}{\partial C_A} + \frac{\partial \overline{C}_B}{\partial C_B}\right)\theta + \frac{\partial \overline{C}_A}{\partial C_A}\frac{\partial \overline{C}_B}{\partial C_B} - \frac{\partial \overline{C}_A}{\partial C_B}\frac{\partial \overline{C}_B}{\partial C_A} = 0 \tag{3.44}$$

which shows that θ depends only on the partial derivatives of the adsorption isotherms at composition (C_A, C_B) and is independent of column void fraction or fluid velocity. The two roots $\theta_\pm$ of the quadratic equation (3.44) are *eigenvalues* of the problem. They allow a determination of corresponding solutions for $\sigma\ (C_A, C_B)$:

$$\sigma_\pm\,(C_A, C_B) = \frac{u_e}{1 + F_e\theta_\pm\,(C_A, C_B)} \tag{3.45}$$

Provided the partial derivatives of the adsorption isotherms are known, Eq. (3.45) allows one to estimate the possible propagation velocities of any given composition (C_A, C_B).

By convention, we will use $\sigma_+ > \sigma_-$, which looks rather classical! Let us, however, notice that this implies $\theta_+ < \theta_-$ and $\left(t_{R+} = \frac{L_{col}}{\sigma_+}\right) < \left(t_{R-} = \frac{L_{col}}{\sigma_-}\right)$, which requires awareness.

As with single-solute systems, the above derivation shows that solution of the chromatographic problem consists of a combination of plateaus (trivial solutions) and waves (non-trivial solutions), where the propagation velocity of a composition within a front is concentration-dependent through the derivatives of the adsorption isotherms. In other words, in a two-solute system, any frontal injection (saturation or regeneration) is resolved into two waves, being either compressive, dispersive or combined.

In addition, since we have seen that the solution of the problem depends on σ only, C_A and C_B are functions of σ only and consequently they must be related. This means that the compositions in a front must follow certain paths in (C_A, C_B) space. The relation describing admissible combinations of the concentrations can be found using the first equation of (3.40), which gives:[8]

$$\frac{dC_B}{dC_A} = \frac{u_e - \sigma_\pm\left(1 + F_e\frac{\partial \overline{C}_A}{\partial C_A}\right)}{\sigma_\pm F_e\frac{\partial \overline{C}_A}{\partial C_B}} \tag{3.46}$$

This expression not only gives a link between the variations of C_A and C_B, but also says that there are two possible directions, related to the two velocities, in which a given composition within a front may evolve. We will see later how to select the correct direction and velocity for a given problem.

[7] θ is defined in this way because we want to find an expression for σ similar to the one derived for single-solute systems.

[8] The reader can check that the same composition path finally given by Eq. (3.48) is obtained by using the second equation.

By introducing $\theta_\pm$, the roots of Eq. (3.44), we find a more compact expression for dC_B/dC_A:

$$\frac{dC_B}{dC_A} = \frac{\theta_\pm - \dfrac{\partial \overline{C}_A}{\partial C_A}}{\dfrac{\partial \overline{C}_A}{\partial C_B}} \tag{3.47}$$

Alternatively, with Eq. (3.44), this also shows that dC_B/dC_A satisfies the quadratic equation:

$$\left(\frac{dC_B}{dC_A}\right)^2 + \frac{\dfrac{\partial \overline{C}_A}{\partial C_A} - \dfrac{\partial \overline{C}_B}{\partial C_B}}{\dfrac{\partial \overline{C}_A}{\partial C_B}} \frac{dC_B}{dC_A} - \frac{\dfrac{\partial \overline{C}_B}{\partial C_A}}{\dfrac{\partial \overline{C}_A}{\partial C_B}} = 0 \tag{3.48}$$

If explicit expressions for the derivatives of the isotherms are available, solving Eq. (3.48) gives the two possible values for dC_B/dC_A.

Some readers may be frightened by Eq. (3.48), if not already: what is this series of ∂ and *d*? Why this glyph difference? Mathematics is a torture! Where are we going?

Where are we going: we are deriving the breakthrough curve by simple (yes!) algebra, i.e. no computer is needed! Perhaps a computer could do the work, but it does not provide understanding, and it provides us with wrong results if the model is wrong. Algebra will provide a lot of understanding!

What about ∂: in Eq. (3.48) $\frac{\partial \overline{C}_i}{\partial C_j}$ are the derivatives (partial) of the equilibrium laws (at constant T). They are defined once and for all, whether we are dealing with thermodynamics, chromatography, distillation, liquid-liquid extraction ...

What about *d*: in Eq. (3.48) $\frac{dC_B}{dC_A}$ is the relative change of C_B with respect to C_A along a wave, as noted above. Let us admit that a one-to-one correspondence between C_A and C_B along a front was not obvious *a priori*! The result is rather rich and surprising; furthermore it can be tested experimentally: when a frontal breakthrough curve is available, one may draw C_B versus C_A and check the dependence predicted by the model. This dependence is a specific feature of the chromatographic process, contrary to the ∂ of the previous item. At this point of the chapter, this relative change, $\frac{dC_B}{dC_A}$, is not known. Equation (3.48) is just the missing link: it is a second-degree equation whose solutions $\frac{dC_B}{dC_A}$ are deduced from knowledge of the partial derivatives of the equilibrium equations. Let us remark incidentally that distillation is not so far from chromatography (especially SMB!). The reader will not be surprised to learn that an equation similar to Eq. (3.48) exists in distillation (Underwood's equation).

Let us stress that, even though mathematics can be cumbersome, it is no more than a toolbox with a very well-defined vocabulary (number, vector, matrix, function, determinant etc.) and alphabet (∂, *d* etc.) which is not simply mechanics: the reader should (must!) wonder (and understand!) the meaning of each term and should not be afraid of symbols.

By analyzing the course of the concentrations in the concentration space (C_A,C_B), the so-called *hodograph plane*, one can simply exchange the role of independent variables (time and space) and dependent variables (concentrations). This is the core of the method of characteristics, a useful mathematical tool for solving hyperbolic PDEs (see for instance (Varma and Morbidelli, 1997)).[9] This is why mentioned paths in the hodograph plane (C_A, C_B) are correspondingly denoted as *characteristic paths* (note: henceforward we reserve the name *characteristics* for the (z,t) plane).

Let us consider some arbitrary composition point within a hypothetical concentration front. The two roots dC_B/dC_A of Eq. (3.48) define two possible directions in which the composition may change within the concentration space (C_A, C_B). When integrating dC_B/dC_A in (C_A,C_B), one obtains two characteristic paths through the given point. The solution of the problem – the change of concentrations that the front will experience during its propagation – must proceed along one of these characteristic paths. Notably, the characteristic paths depend only on the composition and the corresponding partial derivatives; see Eq. (3.47). The velocities of the compositions along a characteristic path are given by Eq. (3.45).

Once we determine the relationship between concentrations C_A and C_B we still need to locate the chromatogram on the time axis. For that, a decision about the nature of the individual fronts has to be made on the basis of Eq. (3.45). As with single-solute systems, the mass balances written up to now assume continuous evolution of the concentrations. That is why the hodograph and the velocity expressions are valid for dispersive fronts only. By analogy with single-solute systems, if σ decreases along a path from the downstream composition I to the upstream composition F, the front is dispersive and thus a simple wave develops. Conversely, if σ increases along a path from the downstream composition I to the upstream composition F, the front is compressive and thus a shock is involved instead of a simple wave.

In the latter case, Eq. (3.45) has to be replaced by:

$$u_{shock} = \frac{u_e}{1 + F_e \dfrac{\Delta \overline{C}}{\Delta C}} \quad \text{applied for species A or B} \tag{3.49}$$

where Δ designates the difference of two states along a path (I–P or P–F), and P designates a composition plateau that will be calculated later.

Appreciating that this approach is pretty remote from typical chromatographers' tools and methodology, it is certainly useful to pause and apply the above theoretical framework to the simple situation of linear systems for which the partial derivatives of the lumped adsorption isotherms are given by:

$$\frac{\partial \overline{C}_A}{\partial C_A} = \overline{K}_A \qquad \frac{\partial \overline{C}_A}{\partial C_B} = 0$$

$$\frac{\partial \overline{C}_B}{\partial C_A} = 0 \qquad \frac{\partial \overline{C}_B}{\partial C_B} = \overline{K}_B$$

9 The reader may find in various articles references to hyperbolic PDEs. Mathematically, *hyperbolicity* means that the eigenvalues of Eq. (3.44) are real and distinct. Physically, the consequence is that solutes will elute in a finite time, as opposed to what would, for instance, be observed when a Fickian diffusion mechanism is involved (elliptic system).

Although it is a non-competing trivial situation, it will illuminate the method.

Under the linearity assumption, the characteristic equation (3.44) has two roots:

$$\theta_- = \overline{K}_B \quad \text{and} \quad \theta_+ = \overline{K}_A$$

so that the velocities obtained from Eq. (3.45) are given by:

$$\sigma_- = \frac{u_e}{1 + F_e\overline{K}_B} \quad \text{and} \quad \sigma_+ = \frac{u_e}{1 + F_e\overline{K}_A} \tag{3.50}$$

These are the classical expressions for the propagation velocities of fronts under linear conditions, the associated retention times being given by $t_{R+} = \dfrac{L_{col}}{\sigma_+}$ and $t_{R-} = \dfrac{L_{col}}{\sigma_-}$. In the linear situation, t_{R+} corresponds to the retention time of the less retained solute A, while t_{R-} corresponds to the retention time of the more retained solute B.

We now have to draw the characteristics in the hodograph plane (C_A,C_B). This can be done by analyzing the system (3.40), which becomes:

$$\left(u_e - \sigma(1 + F_e\overline{K}_A)\right)\frac{dC_A}{d\sigma} = 0 \quad \text{and} \quad \left(u_e - \sigma(1 + F_e\overline{K}_B)\right)\frac{dC_B}{d\sigma} = 0 \tag{3.51}$$

This shows that:

- if $\sigma = \sigma_+$ then dC_A is undefined (first equation of Eq. (3.51)) but dC_B must equal zero (second equation of Eq. (3.51)). This is the definition of a straight line parallel to the C_A axis in the (C_A, C_B) hodograph space.
- if $\sigma = \sigma_-$ then dC_B is undefined (second equation of Eq. (3.51)) but dC_A must equal zero (first equation of Eq. (3.51)). This is the definition of a straight line parallel to the C_B axis in the (C_A, C_B) hodograph space.

Thus the hodograph space is made of vertical and horizontal straight lines.

Knowing that the characteristic lines to be considered must include compositions I and F, one obtains the situation given in Figure 3.14. The solution for a breakthrough curve must follow one of the two characteristics from I, until it intersects the characteristic path that goes through the inlet concentration F, along which it will then proceed into F. Starting from composition I or F, one obtains two different characteristic paths, one vertical, associated with $\sigma = \sigma_-$, and one horizontal, associated with $\sigma = \sigma_+$. The question is which one must be followed to move from I to F.

The problem is elucidated by noticing that in a breakthrough curve, composition I is present earlier than composition F. This means that composition I must be associated with the highest velocities and composition F with the lowest. In other words, *traveling from I to F in the hodograph space must be done with decreasing velocity* σ.

Let us consider the situation given by the saturation proposed in Figure 3.14 (left). As shown in Figure 3.15, starting from composition I and selecting the characteristic associated with the highest velocity (σ_+) requires moving horizontally to point P_2, which represents a plateau of composition (C_A^F,C_B^I). From P_2 one then moves vertically with the slowest front (σ_-) to reach F. This defines the path followed by the concentrations in the hodograph plane.

Similarly, let us consider the regeneration of Figure 3.14 (right). As shown in Figure 3.15, starting from composition I and selecting the characteristic associated with the highest velocity (σ_+) requires moving horizontally to point P_1, which represents a plateau of composition (C_A^F,C_B^I). From P_1 one then moves vertically with the slowest front (σ_-) to reach F. This defines the path followed by the concentrations in the hodograph plane.

The associated breakthrough curves are given in Figure 3.16.

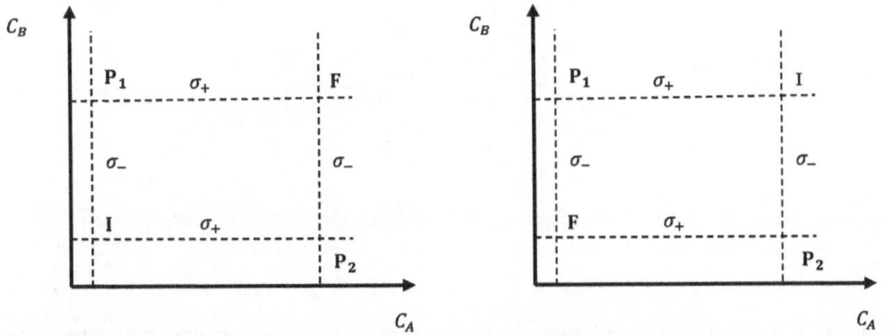

Figure 3.14 Typical hodograph for a linear system. Left (saturation) and right (regeneration) figures are associated with interchanged I and F compositions.

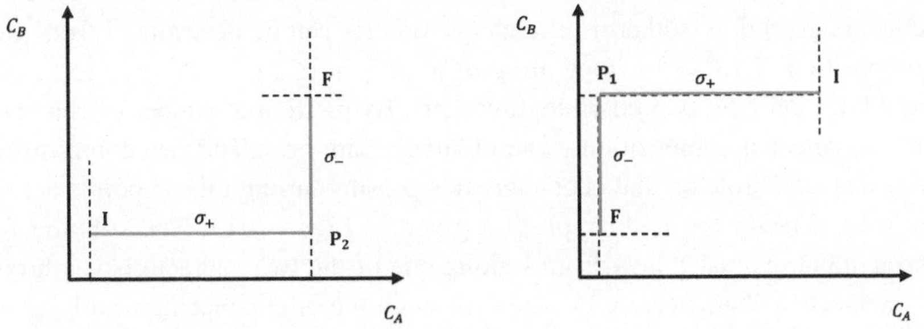

Figure 3.15 Path followed in the hodograph space by the compositions. Left and right figures are associated with interchanged I and F compositions.

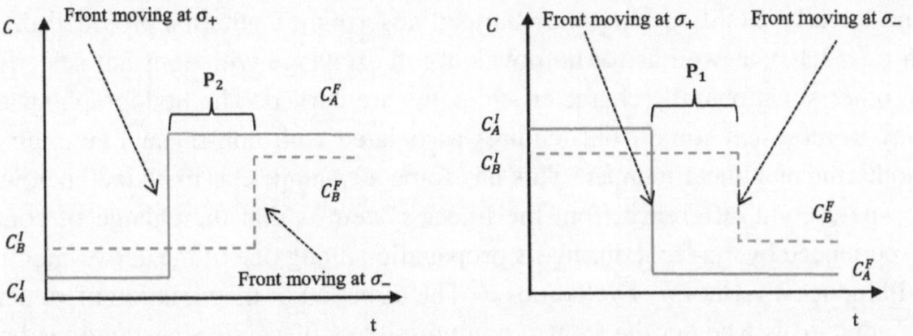

Figure 3.16 Breakthrough curves associated with the hodograph paths given in Figure 3.15.

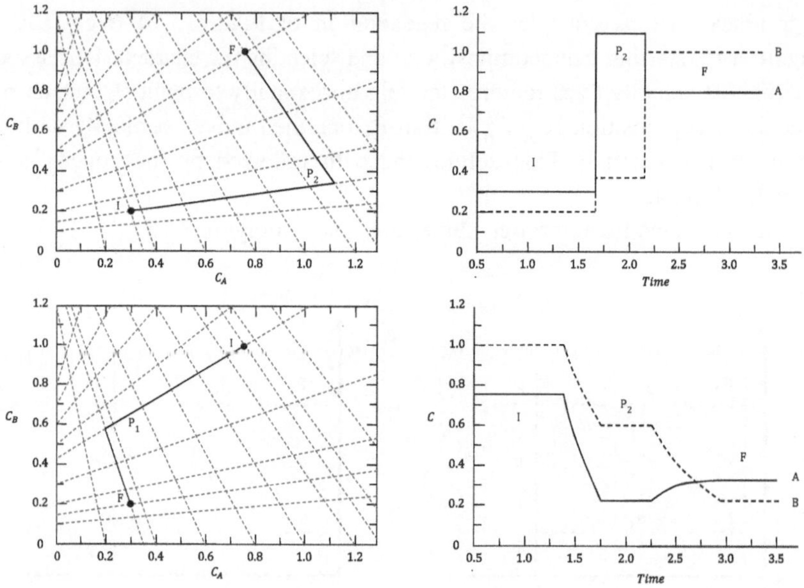

Figure 3.17 Typical hodograph and chromatogram construction as they occur for Langmuir-type isotherms. Top and bottom are associated with opposite I and F compositions.

We now extend this methodology to non-linear systems. In the case of non-linear lumped adsorption isotherms, the characteristics can be determined from any arbitrary composition (C_A,C_B) by integrating dC_B/dC_A in (C_A,C_B), the two roots dC_B/dC_A of Eq. (3.48) defining two possible directions. By plotting a number of characteristics for various selected compositions, one obtains a path grid. The two compositions I and F play a specific role so that characteristics passing through these points are represented by solid lines in the hodograph plot given in Figure 3.17. The solution for a break-through curve must follow from I along one of the two characteristics through I, until it intersects a characteristic that goes through the inlet concentration F, along which it will then proceed into F. Notably, the paths depend only on the composition and the corresponding partial derivatives of equilibrium laws.

The first difference with the linear system is that the characteristic paths are not parallel to the axes. They are straight lines, but not parallel to the axes for Langmuir and linearly modified Langmuir lumped adsorption isotherms provided that $\bar{\lambda}_A = \bar{\lambda}_B$ in Eq. (3.10), as well as for homovalent ion exchange with constant selectivity factors. In other situations, the characteristic paths are curved. The hodograph space is thus a way to represent some front features associated with non-linear Langmuir adsorption isotherms in a linear manner! This has some advantages, as explained below.

The second difference from the linear system is that the change of concentrations experienced by the front during its propagation along one of the above-mentioned paths will induce a variation of velocities σ. The direction of the variation of velocity σ along a path tells us whether the front is compressive or dispersive, as mentioned above.

Let us emphasize the following important practical rules:

- Traveling from I to F in the hodograph space must be done with decreasing velocities σ (for reasons explained in the linear case). As a consequence, the characteristic associated with velocity σ_+ must be selected when departing from I and the characteristic associated with velocity σ_- must be selected when arriving at F.
- When moving along a path (I–P or P–F) in the direction of decreasing velocities, the corresponding front is dispersive, and is otherwise compressive.

The hodograph grid gives us a relatively detailed although concise picture of the solutions: we can calculate plateau compositions, we know in which direction the compositions should move, and we know if the fronts are going to be compressive or dispersive. This is sufficient for a qualitative sketch of the chromatogram along the time axis. Then, calculating the velocities allows the chromatogram to be drawn on a quantitative basis.

The methodology for deriving the complete solution is given below for the case of a binary Langmuir adsorption isotherm (straight paths):

- As with linear systems, from I first take the fast path given by σ_+, until one finds the intersection with the path given by σ_- through the inlet concentration F.
- The intersection between the characteristic paths respectively associated with I and F represents a composition plateau.
- When moving along a path in the direction of decreasing velocities, the corresponding front is dispersive and otherwise compressive. While the hodograph grid is, *a priori*, valid for dispersive fronts only, the extraordinary property of dispersive hodograph paths in the case of Langmuir adsorption isotherms is that they are straight lines, and the same paths are followed by self-sharpening fronts. The composition of plateau P to be used in Eq. (3.49) is thus obtained from the intersection between the straight lines.

The above methodology allows a direct construction of hodographs and then fronts, as shown in Exercise 3.8.

EXERCISE 3.6 *Show that the hodograph plot is made up of straight lines for linearly modified Langmuir lumped adsorption isotherms (Eq. (3.10) with $\overline{\lambda}_A = \overline{\lambda}_B$).*

When characteristic paths are not straight lines, the situation can be more or less complex. If one deals with a situation for which the velocity σ is decreasing along a characteristic path, the methodology presented for the Langmuir situation can be used without difficulty once the hodograph grid has been determined.

The situations associated with increasing velocities along a path leading to shocks require special attention because the hodograph grid is in general valid for dispersive fronts only. For illustration purposes, let us start with composition I. The velocities associated with the shock are determined by Eq. (3.49). Taking into account the fact that the velocity of a front must be independent of the species A or B used in Eq. (3.49), the following equation must hold:

$$\frac{\overline{C}_A^{P*} - \overline{C}_A^I}{C_A^{P*} - C_A^I} = \frac{\overline{C}_B^{P*} - \overline{C}_B^I}{C_B^{P*} - C_B^I} \tag{3.52}$$

where P* is a plateau of a certain composition to be determined. Equation (3.52) illustrates the *coherence* principle proposed by Helfferich and further discussed later.

At first glimpse one may think that different characteristic paths associated with compressive fronts can be determined by using Eq. (3.52), just as we determined characteristic paths associated with dispersive fronts. This is unfortunately not true. The characteristic paths that one can draw for shocks are not universal like the ones derived for dispersive fronts: they depend on the originating point. Let us consider a compressive front and draw the associated path. From I, one increases C_A by a given $\Delta C_A^{'}$ and the problem is to find a point P′ belonging to the path. The coordinates $(C_A^{P'}, C_B^{P'})$ of P′ are obtained so as to satisfy Eq. (3.52) between I and P′.[10] In order to get another point, again starting from I, one increases C_A by a given $\Delta C_A^{''}$. We then determine another point P″ satisfying Eq. (3.52) between I and P″. In case of non-linear paths, the new point does not satisfy Eq. (3.52) between P′ and P″; consequently P″ does not belong to a characteristic path emanating from P′.

The plateau P* composition is determined by identifying an intersection with a characteristic going through F. The paths associated with shocks being in general different from the paths associated with dispersive fronts, plateau composition P^* is, *a priori*, different from the plateaus P_1 or P_2 calculated with the hodograph.

If concentrations are such that the I–P* path is a shock, then the second path P*–F can in principle be a shock or a simple wave.

If the path P*–F is also a shock, the concentrations of this front must satisfy:

$$\frac{\overline{C}_A^{P*} - \overline{C}_A^{F}}{C_A^{P*} - C_A^{F}} = \frac{\overline{C}_B^{P*} - \overline{C}_B^{F}}{C_B^{P*} - C_B^{F}} \tag{3.53}$$

Equations (3.52) and (3.53) allow a determination of the plateau composition P* with an iteration procedure.

If the path P*–F is a simple wave, P* is determined by identifying an intersection with one characteristic of the hodograph going through F.

Finally, thanks to Murphy's law, there is no certainty that the path remains a shock or a simple wave during all its course. In general, one must check whether the shock path is faster or slower than the local, slow, simple wave. If it is slower, things are all right. If it is faster, a combined wave is involved! The latter situation is very intricate and is not discussed further.

One needs to be aware of the difficulties of dealing with complex isotherm shapes. Not only can the details of the mathematics be cumbersome, but in some special (and rare) cases, (Mazzotti, 2009; Mazzotti *et al.*, 2010) have shown theoretically and experimentally that very unexpected results can occur. For instance, using generalized Langmuir adsorption isotherms for associating favorable and unfavorable adsorptions can lead to a new type of solution to the Riemann problem called *delta-shock*. This is neither a shock nor a simple wave, but can be seen as a growing traveling spike superimposed on the discontinuity separating the I and F states. Reading (Mazzotti, 2009) is also a good way to better understand what is allowed and what is not allowed when using characteristics and hodographs to solve a binary problem.

[10] Be careful: Eq. (3.52) gives two roots; the one associated with the adequate velocity must be chosen.

An example of hodograph and chromatogram construction associated with non-linear characteristic paths is given in Exercise 3.9.

The above presentation shows the beauty and the limits of equilibrium theory and of the hodograph technique. With two independent species, it is possible (but not always easy) to find a solution since the progressive construction of the path originating from F will eventually intersect the fast path from I. However, for three independent species or more, there are no longer simple solutions in the general case.

EXERCISE 3.7 *A column is equilibrated with an eluent of a prescribed composition* (C_A^I, C_B^I). *A small injection is performed at the inlet of the column. Estimate the retention time associated with the two peaks obtained at the outlet of the column.*

The previous theory is often presented in a different form, based on the concept of *coherence* (Helfferich and Klein, 1970; Helfferich and Whitley, 1996). It has been shown by (Helfferich and Klein, 1970) that the *propagation velocity of a concentration* C_i is given by:[11]

$$\sigma_i = \frac{u_e}{1 + F_e \left(\dfrac{\partial \bar{C}_i}{\partial C_i}\right)_z} \qquad i = \text{A or B} \tag{3.54}$$

The derivative $\left(\dfrac{\partial \bar{C}_i}{\partial C_i}\right)_z$ is a total derivative, often denoted $\dfrac{D\bar{C}_i}{DC_i}$, which can be expressed as:

$$\begin{aligned} \frac{D\bar{C}_A}{DC_A} &= \frac{\partial \bar{C}_A}{\partial C_A} + \frac{\partial \bar{C}_A}{\partial C_B}\frac{dC_B}{dC_A} \\ \frac{D\bar{C}_B}{DC_B} &= \frac{\partial \bar{C}_B}{\partial C_B} + \frac{\partial \bar{C}_B}{\partial C_A}\frac{dC_A}{dC_B} \end{aligned} \tag{3.55}$$

Since the velocities of concentrations that move together must be identical, one should have $\sigma_A = \sigma_B$ and thus:

$$\begin{aligned} \frac{D\bar{C}_A}{DC_A} &= \frac{D\bar{C}_B}{DC_B} \qquad \text{in the case of a dispersive front} \\ \frac{\Delta\bar{C}_A}{\Delta C_A} &= \frac{\Delta\bar{C}_B}{\Delta C_B} \qquad \text{in the case of a compressive front} \end{aligned} \tag{3.56}$$

Equation (3.56) is the coherence relation which gives a relationship between concentrations (C_A, C_B).[12] A composition is coherent if it moves identically to itself. If all

[11] By using $\left(\frac{\partial C_i}{\partial z}\right)_t = -\left(\frac{\partial C_i}{\partial t}\right)_z \left(\frac{\partial t}{\partial z}\right)_{Ci}$ the mass-balance equation (3.38) gives $-u_e\left(\frac{\partial t}{\partial z}\right)_{Ci} + 1 + F_e \frac{\left(\frac{\partial \bar{C}_i}{\partial t}\right)_z}{\left(\frac{\partial C_i}{\partial t}\right)_z} = 0$. Then, noting that $\left(\frac{\partial t}{\partial z}\right)_{Ci} = \frac{1}{\left(\frac{\partial z}{\partial t}\right)_{Ci}}$ and $\frac{\left(\frac{\partial \bar{C}_i}{\partial t}\right)_z}{\left(\frac{\partial C_i}{\partial t}\right)_z} = \left(\frac{\partial \bar{C}_i}{\partial C_i}\right)_z$, one obtains Eq. (3.54) with $\sigma_i = \left(\frac{\partial z}{\partial t}\right)_{Ci}$.

[12] It is straightforward to verify that $\theta = \frac{D\bar{C}_A}{DC_A} = \frac{D\bar{C}_B}{DC_B}$ and that Eq. (3.56) is equivalent to Eq. (3.44).

the concentrations of a front are coherent, the front is said to be coherent. (Helfferich and Klein, 1970) explained that the postulated coherence property is associated with a "stable" state that is reached irrespective of the operating conditions. If a system is initially non-coherent, it will naturally evolve towards a coherent state.

EXERCISE 3.8 *A column $V_{col} = 1\ l$, $L_{col} = 1\ m$, $\varepsilon_e = 0.4$ is eluted with a 1 l/min flow rate of carrier fluid. The adsorption of the solutes is characterized by a linearly modified Langmuir lumped adsorption isotherm (Eq. 3.10), with $\bar{\lambda}_A = \bar{\lambda}_B = 0$, $\tilde{K}_A = 0.5$ l/g, $\tilde{K}_B = 1$ l/g and $\overline{N}_A = \overline{N}_B = 2$ g/l.*

- *Draw the characteristic paths of the hodograph plane that go through the points defined by ($\alpha : C_A = C_B = 0$) and ($\beta : C_A = C_B = 1.0$) g/l.*
- *Compute the breakthrough curves associated with the following experiments:*
  - *The column is initially equilibrated with composition α and fed with a stream of composition β.*
  - *The column is initially equilibrated with composition β and fed with a stream of composition α.*

EXERCISE 3.9 *Perform the same analysis as in Exercise 3.8 with $\bar{\lambda}_B = 0.25$ instead of $\bar{\lambda}_B = 0$ and for $C_A^F = C_B^F = 5$ g/l.*

The previous theory can be extended to determine the response of the column to a rectangular pulse, i.e. an injection of a given volume of feed mixture at composition (C_A^F,C_B^F) into a column initially equilibrated uniformly with composition (C_A^I,C_B^I). For simplicity, we will assume that the column is initially solute-free and that adsorption isotherms are favorable (consequently, the saturation of the column leads to compressive fronts and its regeneration to dispersive fronts). The method could be extended to other situations.

The injection profile can be considered as the combination of a positive step and of a negative step, as shown in Figure 3.18. The theory given in this chapter allows construction of the response of the system to each of these perturbations, these responses being schematically shown in Figure 3.18.

Provided the adsorption isotherms are favorable, we have shown that the saturation step generates two shocks F_c^1, F_c^2 when the desaturation process generates two

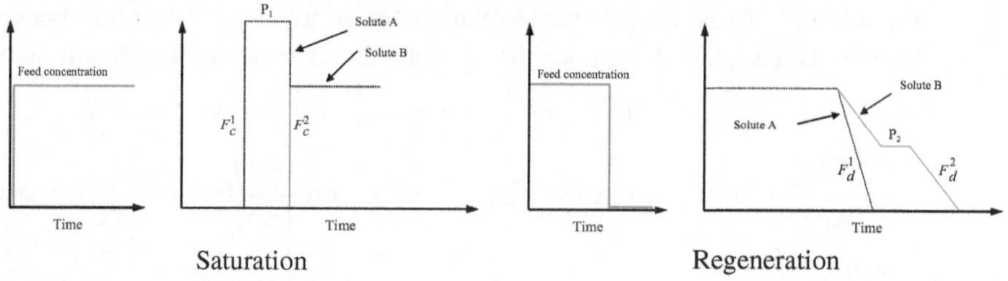

Figure 3.18 Typical normalized breakthrough curves obtained in the case of favorable adsorption isotherms. Left: saturation. Right: regeneration.

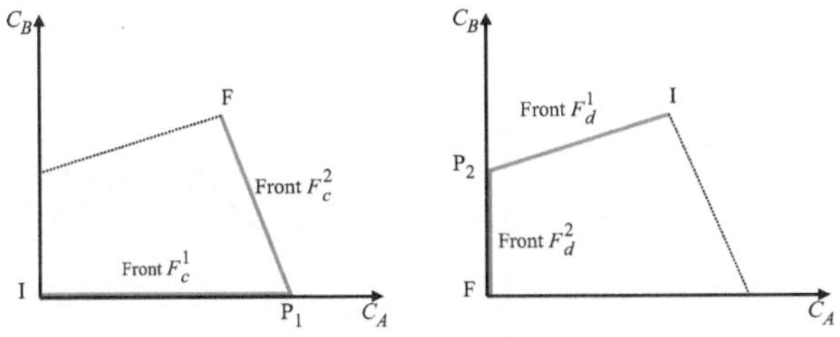

Figure 3.19 Hodographs associated with the saturation and regeneration steps of a column initially free of solutes. Left: saturation step. Right: regeneration step. Favorable (Langmuir-type) adsorption isotherms are assumed.

dispersive fronts F_d^1, F_d^2. We have also shown that the system's key features can be represented using the hodographs given in Figure 3.19. Since the composition of the plateaus can be estimated with the hodograph, the compositions P_1 and P_2 are considered to be known in addition to the compositions I and F.

Let us fix the injection duration Δt_{inj} and discuss how the column length is going to impact the chromatogram on a column eluted with a given fluid velocity u_e.

For very short column lengths, the chromatogram will be almost like the feed rectangular pulse, only the leading and rear parts being impacted by the chromatographic process. More precisely, in the case of favorable lumped adsorption isotherms, the chromatogram will exhibit compressive fronts, followed by a feed plateau zone and then by dispersive fronts. As long as the residual feed plateau exists, the chromatogram thus has four distinct fronts: F_c^1, F_c^2 associated with the saturation, and F_d^1, F_d^2 associated with the regeneration (Figure 3.20, top left). As the column length increases, the relative importance of the residual feed plateau decreases as one of the leading and rear fronts increases.

As the length of the column increases further, the residual feed plateau eventually vanishes. At this point, the compressive slow and downstream front F_c^2 starts interfering with the dispersive fast and upstream front F_d^1 (Figure 3.20, top right). This is the first interference occurring during the propagation of the fronts in the column.

As Z increases, the intermediate zone containing both solutes decreases (Figure 3.20, bottom left) and at a given length, the two solutes A and B are fully separated (Figure 3.20, bottom right). At this level, plateaus P_1 and P_2 of pure solutes may persist or not. Finally, these plateaus will erode due to interferences between fronts F_c^1 and F_d^1 on the one hand and between fronts F_c^2 and F_d^2 on the other hand.

Different abscissas can thus be defined:

- Z_{plat}: abscissa for which the feed plateau disappears
- Z_{resol}: abscissa for which resolution between A and B is just obtained
- Z_{P1}: abscissa for which plateau P_1 disappears
- Z_{P2}: abscissa for which plateau P_2 disappears.

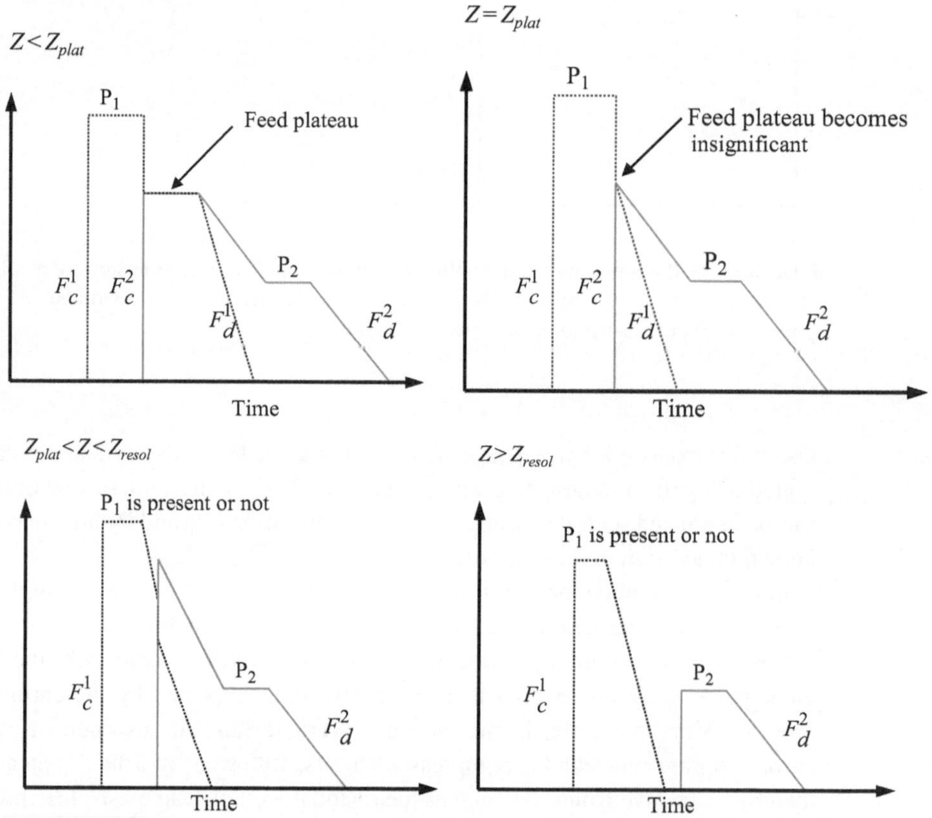

Figure 3.20 Influence of the column length on the qualitative shape of chromatograms for a given injection duration t_{inj}. Favorable lumped adsorption isotherms are assumed.

The (t,z) representation allows visualization of the key information, as shown in Figure 3.21.[13] Notice that, depending on the shape of the adsorption isotherms and of the injected volume, Z_{P1} can be located before or after Z_{resol}.

The main features of the solution of the problem are summarized below, more detailed information being given in Appendix D.

For abscissas smaller than Z_{plat}, the compressive F_c^2 and dispersive F_d^1 fronts move independently, the chromatogram exhibits a feed plateau (Figure 3.20, top left) and the

[13] Instead of fixing Δt_{inj} and discussing the chromatogram shape versus column length, one could have decided to fix the column volume and discuss the chromatogram shape versus injection volume. One would thus define:

- V_{inj}^{plat}: injection volume for which the feed plateau just disappears
- V_{inj}^{resol}: injection volume for which the resolution between solutes A and B is just obtained
- V_{inj}^{P1}: injection volume for which plateau P_1 disappears
- V_{inj}^{P2}: injection volume for which plateau P_2 disappears.

Using the (t,z) representation is, however, more convenient.

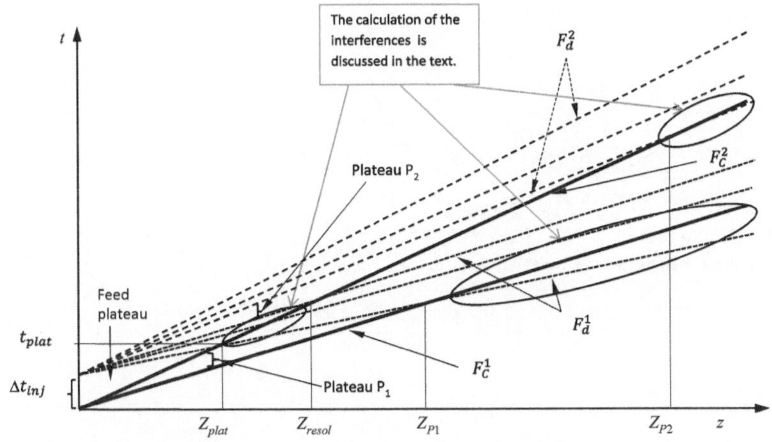

Figure 3.21 Representation of the saturation/regeneration steps in the (t,z) plane. For the sake of clarity, modifications of the lines describing the fronts during interference are not shown.

position of the fronts is given by the methodology presented for the breakthrough curves. The positions of the compressive fronts are given by Eq. (3.49) when the position of the simple waves is given by Eq. (3.45), the time being shifted by Δt_{inj}.

Similarly to the approach followed with single-solute systems, the abscissa Z_{plat} and associated time t_{plat} are calculated by noting that this point is at the intersection between fronts F_d^1 and F_c^2. One obtains:

$$\begin{aligned} Z_{plat} &= \frac{\Delta t_{\text{inj}}}{1/\sigma_{F_c^2(F)} - 1/\sigma_{F_d^1(F)}} \\ t_{plat} &= \frac{Z_{plat}}{\sigma_{F_c^2(F)}} = \Delta t_{inj} + \frac{Z_{plat}}{\sigma_{F_d^1(F)}} \end{aligned} \tag{3.57}$$

The other abscissa that can be calculated relatively simply by using a mass balance on the more strongly retained product is Z_{resol}. We show in Appendix D that:

$$Z_{resol} = \frac{\Delta t_{inj} C_B^F}{C_B^{P2}\left(1/\sigma_{F_d^2(P2)} - 1/\sigma_{F_d^1(P2)}\right) + \int_0^{P_2} \frac{dC}{\sigma_{F_d^2(C)}} - C_B^{P2}/\sigma_{F_d^2(P2)}} \tag{3.58}$$

Equations (3.57) and (3.58) have a practical interest for process design. For sufficiently efficient columns, using a length significantly greater than Z_{resol} is probably not attractive as it would lead to an unnecessary separation of the two solutes. Conversely, using a column length less than Z_{plat} would have little interest as a portion of the chromatogram would have exactly the same composition as the feed. A reasonable column length is certainly greater than Z_{plat} and not too far from Z_{resol}.

An illustration of the use of Eq. (3.57) and Eq. (3.58) for calculating Z_{plat} and Z_{resol} is given in Exercise 3.10.

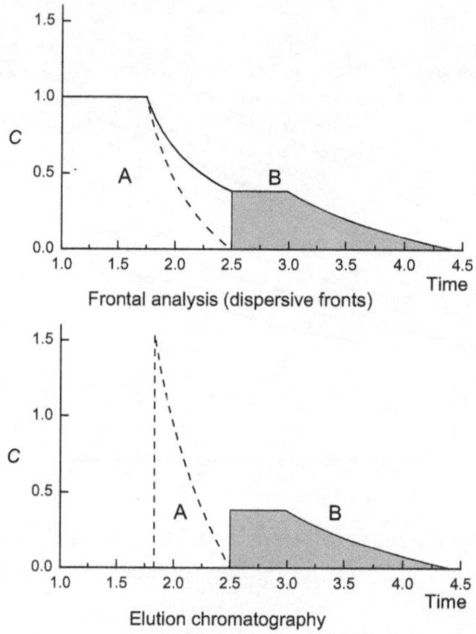

Figure 3.22 Estimate of the maximum injectable amount in elution chromatography.

The determination of Z_{resol} can be related to regeneration front properties. The regeneration fronts of a column initially equilibrated with the feed are qualitatively given in Figure 3.22 (top) for favorable lumped adsorption isotherms. The gray area represents the amount q_B (g or mol) of solute B which is recovered in pure form in a regeneration experiment performed on a column of volume V_{col}.

With relatively simple mass-balance considerations, it can be shown that the maximum injectable volume leading to a complete resolution between the solutes in the same column of volume V_{col} is given by:

$$V_{inj}^{resol} = \frac{q_B}{C_B^F} \quad \text{thus} \quad \Delta t_{inj}^{resol} = \frac{q_B}{QC_B^F} \tag{3.59}$$

The chromatogram obtained with this injected volume is shown qualitatively in Figure 3.22 (bottom).

With one single desaturation experiment or simulation performed on a very efficient column, one thus has access to the maximum injectable volume allowing complete resolution. This gives a way to avoid trial-and-error estimates when no analytical solution is available.

Determination of the other features of the chromatogram requires calculating interferences between fronts. Intuitively, one can easily accept that the interference propagation is affected by the nature of the fronts: two dispersive fronts may not interact as two compressive fronts would. Probably less intuitively, the way the interference develops depends on whether the interfering fronts belong to the same family (like F_d^1 interfering with F_c^1) or not (like F_d^1 interfering with F_c^2).

The main qualitative rules for calculating interferences are:

1 The interaction between shocks is instantaneous, which means that at the very moment the two shocks interfere they are resolved into new shocks.
2 The interaction between simple waves is progressive, which means that some time and distance is required for the waves interfering to be resolved into two new stable simple waves.
3 When a shock and a simple wave of the same family interfere, the simple wave is absorbed (or one could say blocked) by the shock and is not transmitted through the shock. This situation is identical to the one described in the single-solute situation.
4 When a shock and a simple wave of different families interfere, they move across each other (they are transmitted) but their velocities change.
5 When two shocks of the same family interfere, they become superimposed.

We will illustrate qualitatively how these concepts help in determining the chromatograms obtained with favorable adsorption isotherms. For more precise information, the reader is referred to (Rhee *et al.*, 1982, 1989), both being quite mathematically involved, and to the publication by (Rajendran and Mazzotti, 2011) for the so-called generalized Langmuir adsorption isotherms.

For abscissas greater than Z_{plat}, the fronts F_d^1 and F_c^2 start interfering. These fronts belong to two different families, so they are transmitted so that they keep their nature while crossing the interference. Just before the interference starts, front F_c^2 is between the feed composition F and plateau P_1. At the end of the interference, the solutes are fully resolved and the shock separates the plateau P_2 and the zero concentration region. The paths followed by the different fronts are presented in the (t,z) plane as well as in the (C_A,C_B) plane in Figure 3.23. Although compositions at its upstream and downstream sides change continuously as indicated by points K and L, the shock which is at the interface between two dispersive fronts (F_d^1 and F_d^1 transmitted) always remains coherent. The transmitted shock and simple wave are slowed down by crossing the interference, so that at the end of the interference (occurring at Z_{resol}) the fronts are again represented by straight lines in the (t,z) plane, but with a slope which is greater than that before the interference.

For abscissas greater than Z_{P1} (which in general can be greater or smaller than Z_{resol}), fronts F_d^1 and F_c^1 start interfering. Unlike the interference between F_d^1 and F_c^2 which gives rise to the transmission of both fronts, the interference between F_d^1 and F_c^1 involving two fronts belonging to the same family leads to the absorption of the dispersive front F_d^1 by the compressive front F_c^1. This situation is similar to that for single-solute systems. The different representations of this interference are given in Figure 3.24.

For abscissas greater than Z_{P2}, fronts F_d^2 and F_c^2 interfere, and this interference between fronts of the same kind is similar to the one between F_d^1 and F_c^1.

The problem of constructing chromatograms using equilibrium theory has been addressed by (Helfferich and Klein, 1970) and (Rhee *et al.*, 1989). Detailed analytical solutions have been developed by (Golshan-Shirazi and Guiochon, 1989a) and

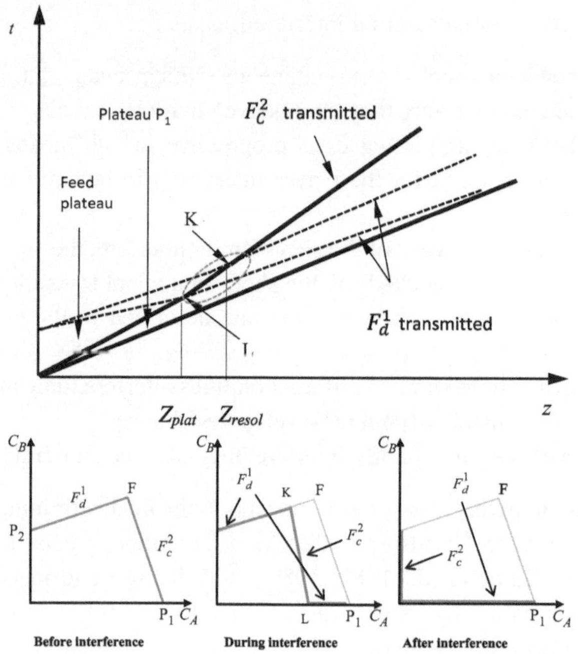

Figure 3.23 Representation of the interference between F_d^1 and F_c^2 in the (t,z) plane (top) and in the hodograph space (bottom). Favorable adsorption isotherms are assumed.

(Golshan-Shirazi and Guiochon, 1989b) for almost the entire chromatographic cycle at column outlet in the case of a binary Langmuir system. The missing part, the solution for the height and retention time of the first shock, can be found in (Siitonen and Sainio, 2011). For a detailed and complete presentation associated with generalized Langmuir adsorption isotherms, the reader is referred to (Rajendran and Mazzotti, 2011). Notice that (Golshan-Shirazi and Guiochon, 1989a), (Siitonen and Sainio, 2011) and (Rajendran and Mazzotti, 2011) use the ω transform presented in Section 3.5.2, which greatly simplifies the resolution in the case of generalized Langmuir adsorption isotherms.

Recall that dealing with complex adsorption isotherm requires a rigorous approach, as it can, in special cases, lead to new sets of solution structures (Mazzotti, 2009; Mazzotti *et al.*, 2010).

EXERCISE 3.10 *Determine Z_{plat} and Z_{resol} for the column described in Exercise 3.8 and $t_{inj} = 0.1$ min.*

3.4.2 More realistic chromatograms: the MC–$\overline{\text{Equil}}$ model

See Section 3.5.4 for the use of the MC–$\overline{\text{Equil}}$ model and qualitative comparison with the prediction of the equilibrium model for multi-solute systems.

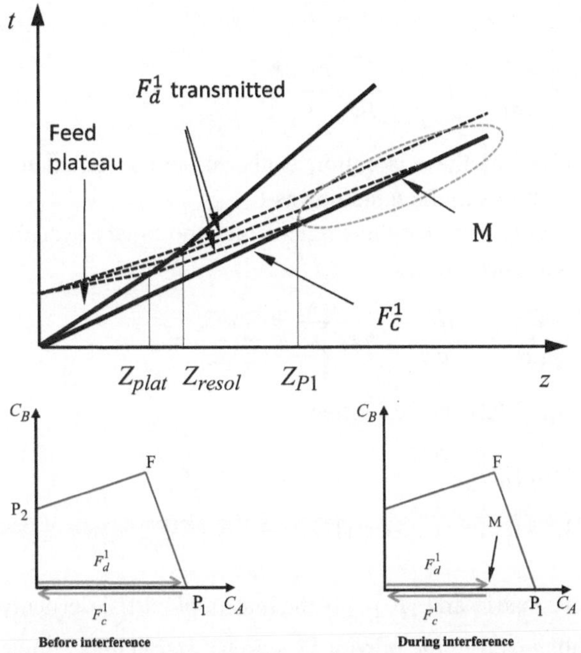

Figure 3.24 Representation of the interference between F_d^1 and F_c^1 in the (t,z) plane (top right) and in the hodograph plane (bottom). Favorable adsorption isotherms are assumed.

3.5 Multi-solute systems: competition between solutes

Let us now consider a column in interaction with a mixture containing n different solutes. The lumped concentrations in the solid phase $\overline{C}_i$ can be related to the fluid-phase concentrations C_i via n adsorption laws given by Eq. (3.6). As with binary systems, the main task of multi-solute system modeling consists in understanding how competition between solutes affects the chromatogram.

3.5.1 The equilibrium model

Assuming no hydrodynamic dispersion and no mass transfer limitation, the mass balances are given by:

$$u\frac{\partial C_i}{\partial z} + \varepsilon_e \frac{\partial C_i}{\partial t} + (1-\varepsilon_e)\frac{\partial \overline{C}_i}{\partial t} = 0 \qquad 1 \le i \le n \tag{3.60}$$

Since the lumped fluid-phase concentrations are related to the fluid-phase concentrations by the equilibrium laws, one can write:

$$\frac{\partial \overline{C}_i}{\partial t} = \sum_{k=1}^{n} \frac{\partial \overline{C}_i}{\partial C_k}\frac{\partial C_k}{\partial t}$$

and Eq. (3.60) becomes:

$$u_e \frac{\partial C_i}{\partial z} + \frac{\partial C_i}{\partial t} + F_e \sum_{k=1}^{n} \frac{\partial \overline{C}_i}{\partial C_k} \frac{\partial C_k}{\partial t} = 0 \qquad 1 \le i \le n \tag{3.61}$$

Since the partial derivatives of the adsorption isotherms are assumed to be known, Eq. (3.61) is a set of n equations in the n unknowns C_i.

Following the methodology used for the single-solute and binary systems, we introduce $\sigma = z/t$, and Eq. (3.61) becomes:

$$u_e \frac{dC_i}{d\sigma} - \sigma \frac{dC_i}{d\sigma} - \sigma F_e \sum_{k=1}^{n} \frac{\partial \overline{C}_i}{\partial C_k} \frac{dC_k}{d\sigma} = 0 \tag{3.62}$$

Using matrix notation, Eq. (3.62) can be written:

$$\begin{aligned} [M(\sigma)]_{n,n} \left[\frac{dC}{d\sigma}\right]_{n,1} &= [0]_{n,1} \\ [M(\sigma)]_{n,n} &= (u_e - \sigma)\,[I]_{n,n} - \sigma F_e \left[\overline{F}'\right]_{n,n} \end{aligned} \tag{3.63}$$

where $[I]_{n,n}$ is the identity matrix and $\left[\overline{F}'\right]_{n,n}$ is the matrix of partial derivatives $\dfrac{\partial \overline{C}_k}{\partial C_i}$.

Generalizing the results given in the previous sections, we obtain an equation which can have two types of roots:

- a trivial solution for which all $\frac{dC_i}{d\sigma}$ equal zero, thus defining plateaus in which concentrations are constant
- a non-trivial solution if the determinant of matrix M equals zero:

$$\det\,[M(\sigma)] = 0 \tag{3.64}$$

Equation (3.64) is the characteristic equation of the problem. This equation has n roots which give n possible propagation velocities and define n fronts.[14]

The method for solving the problem is the same as the one described in Section 3.4.1: the insertion of one root of Eq. (3.64) into Eq. (3.63) gives a set of $n - 1$ independent differential equations which can be solved numerically. The solution defines a family of characteristic paths in the hodograph space. Repeating the procedure with each of the n roots of Eq. (3.64), one obtains a grid of n families of characteristic paths that span the hodograph space, like the two families of characteristic curves in the case of binary mixtures.

As with binary systems, the previous theory can be presented in a different form based on the concept of coherence (Helfferich and Klein, 1970; Helfferich and Whitley, 1996). The propagation velocity of a given concentration is given by:

$$\sigma_i = \frac{u_e}{1 + F_e \left(\dfrac{\partial \overline{C}_i}{\partial C_i}\right)_z} \tag{3.65}$$

[14] The roots must be real and positive. If not, this means that the postulated equations for the equilibrium laws are inconsistent.

The total derivative $\left(\frac{\partial \bar{C}_i}{\partial C_i}\right)_z$ is often called $\frac{D\bar{C}_i}{DC_i}$ and can be expressed as:

$$\frac{D\bar{C}_i}{DC_i} = \sum_{k=1}^{n} \frac{\partial \bar{C}_i}{\partial C_k} \frac{dC_k}{dC_i} \tag{3.66}$$

Since the velocities of the compositions that travel together must be identical, velocity σ_i must be independent of the species and thus the following system must hold:

$$\begin{aligned} &\frac{D\bar{C}_1}{DC_1} = \cdots = \frac{D\bar{C}_i}{DC_i} = \cdots = \frac{D\bar{C}_n}{DC_n} \quad \text{for dispersive fronts} \\ &\frac{\Delta\bar{C}_1}{\Delta C_1} = \cdots = \frac{\Delta\bar{C}_i}{\Delta C_i} = \cdots = \frac{\Delta\bar{C}_n}{\Delta C_n} \quad \text{for compressive fronts} \end{aligned} \tag{3.67}$$

The system of equations (3.67) expresses the coherence condition, which gives $n-1$ relationships between concentrations. This is an alternative way of presenting Eq. (3.64).

The characteristic equation (3.64) has n roots in general, so that n possible velocities are defined. This shows that n fronts are obtained for "n-solute chromatography". As with simple systems containing one or two solutes, front i is a shock (compressive front) if σ_i increases along the characteristic curve and a simple wave (dispersive front) if the reverse holds. In special situations ("S-shaped adsorption isotherms"), the roots σ_i can exhibit a maximum along a path and the corresponding front is combined.

Depending on the initial (I) and final (F) states of the system, n-solute isothermal adsorption consequently leads to n fronts and $n-1$ plateaus, as presented in Figure 3.25. In other words, in an n-solute isothermal system, any initial discontinuity is resolved into at most n simple waves, shocks or combined waves.

More specifically, in the case of a ternary system (Figure 3.25), regardless of the initial and feed compositions, the breakthrough curve exhibits two plateaus (P(1) and P(2)) and three fronts (F_1, F_2 and F_3) which can be compressive or dispersive according to the situation. Don't confuse plateaus P(1) and P(2) describing two plateaus obtained on a ternary system, with P_1 and P_2, the possible plateau compositions defined for a binary-solute system.

The construction of the chromatogram relies on our ability to determine the intersections between the characteristic curves that define the plateau compositions. When $n > 2$, a solution can only be found if the characteristic paths are linear and orthogonal to the axis.[15] These linear/orthogonal paths are encountered in two different situations:

- The adsorption isotherms are linear.
- A suitable orthogonalization procedure is available for special non-linear lumped adsorption isotherms.

[15] For the sake of completeness, the binary system development can be at least partly extended to ternary systems as shown in (Glueckauf, 1949), (Helfferich and Klein, 1970) or (Ruthven, 1984). Instead of being presented in a two-dimensional space, the hodograph can be represented using a triangle representation. A specific composition called the *watershed point* is shown to have important properties. The development being cumbersome, it will not be repeated in this book. Readers interested in knowing more are referred to the above cited articles.

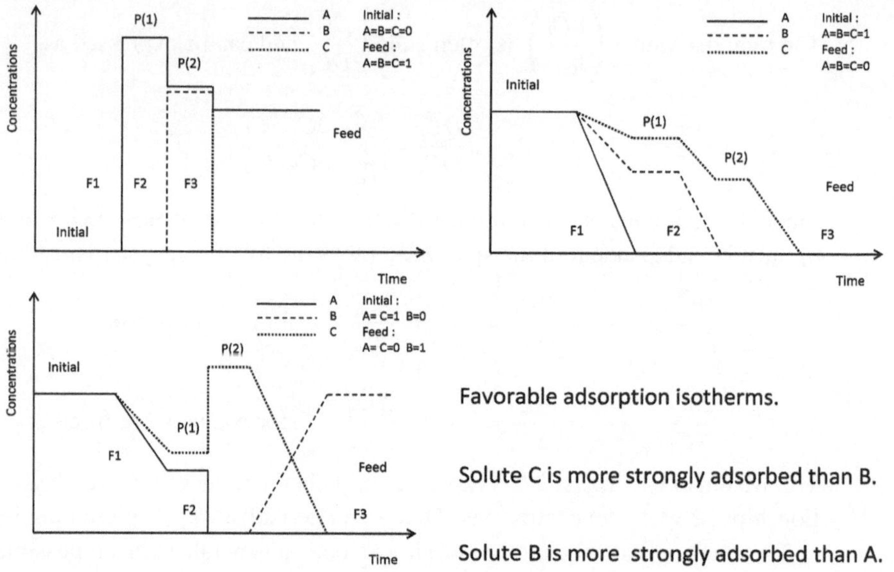

Figure 3.25 Typical chromatograms obtained for ternary systems.

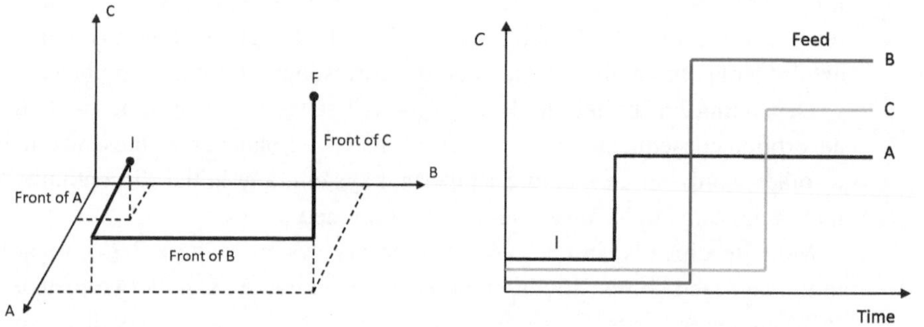

Figure 3.26 Construction of a chromatogram with the hodograph. Linear system with $\overline{K}_A < \overline{K}_B < \overline{K}_C$.

In the case of linear systems, the roots of the characteristic equation, which are given by $\theta_i = \overline{K}_i$, are not concentration-dependent, and the hodograph plot is made of straight lines. The front propagation velocities are thus constant, and the solution can be drawn easily as shown in Figure 3.26.

Fortunately, a suitable orthogonalization procedure is available for the generalized Langmuir lumped adsorption isotherms. Presenting this procedure is the objective of the next section.

3.5.2 An orthogonalization procedure: the ω transform

For non-linear systems, Helfferich described an orthogonalization procedure valid for the case of monovalent ion exchange with constant selectivity coefficient or Langmuir adsorption isotherms (Helfferich and Klein, 1970; Helfferich, 1967). A very similar procedure was given by Rhee and Amundson (Rhee *et al.*, 1970; Rhee, 1981). The concepts and key results are presented below.

The adsorption of the different species on the adsorbent is assumed to follow the Langmuir lumped adsorption isotherm defined by Eq. (3.8) with $\bar{\lambda}_i = 0$. For convenience, we assume that the solutes are ordered in such a way that:

$$(\overline{K}_1 = \overline{N}_1\tilde{K}_1) < (\overline{K}_2 = \overline{N}_2\tilde{K}_2) < \cdots < (\overline{K}_n = \overline{N}_n\tilde{K}_n) \tag{3.68}$$

In order to perform the orthogonalization of the hodograph plane, one defines a set of new variables ω_i, being the roots of the characteristic equation:

$$\sum_{i=1}^{n} \frac{\tilde{K}_i\overline{C}_i}{\overline{N}_i\tilde{K}_i - \omega} = 1 \tag{3.69}$$

For a given set of concentrations C_i in the fluid phase, lumped solid concentrations $\overline{C}_i$ are calculated from the equilibrium laws, and then Eq. (3.69) is solved in ω. It has n real and positive roots such that:

$$0 \le \omega_1 \le \overline{K}_1 \le \cdots \le \omega_n \le \overline{K}_n \tag{3.70}$$

Equation (3.69) allows one to associate n values ω_i to n concentrations C_i.

One can substitute the ω space for the C_i space. The extraordinary property of the ω transform is that the characteristics become straight parallels to the axes (hence the terminology "orthogonalization procedure") in the ω space. Let ω^I and ω^F be the points of the ω space that are associated with the initial and feed compositions. Because of the orthogonality of the characteristic paths in the ω space, along front k only ω_k varies (from ω_k^I to ω_k^F), as for the linear case, where only C_k varied along front k (from C_k^I to C_k^F). This is illustrated in Figure 3.27 (left).

Knowledge of the ω_k^I and ω_k^F roots allows calculation of the compositions of the $n-1$ plateaus P(k) ordered according to their retention order from $k = 1$ to $k = n - 1$:

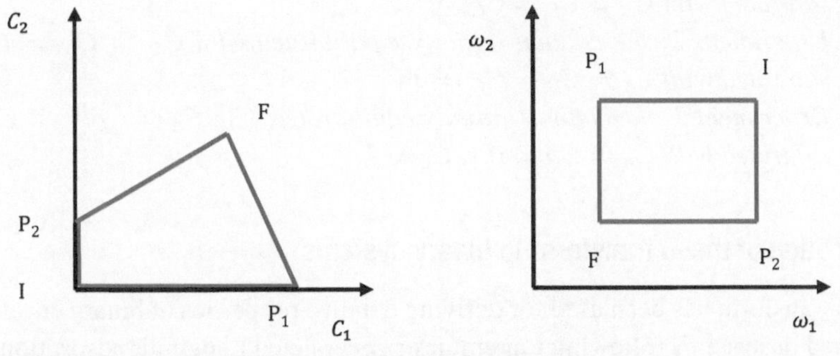

Figure 3.27 Schematic of hodographs in the concentration and ω spaces.

$$C_i^{P(k)} = C_i^I \frac{\prod_{j=1}^{k} \left(\frac{\overline{N}_i \tilde{K}_i}{\omega_j^F} - 1 \right)}{\prod_{j=1}^{k} \left(\frac{\overline{N}_i \tilde{K}_i}{\omega_j^I} - 1 \right)} \tag{3.71}$$

where $C_i^{P(k)}$ is the concentration of species i in plateau k. Moreover, it can be shown that:

- If $\omega_k^I < \omega_k^F$ the front k is dispersive, and its velocity is given by:

$$\sigma_k = \frac{u_e}{1 + F_e \frac{\omega_k}{1 + \sum_i \tilde{K}_i C_i}} \quad \text{for } \omega_k^I \leq \omega_k \leq \omega_k^F \tag{3.72}$$

- If $\omega_k^I = \omega_k^F$ the front k does not appear in the solution.
- If $\omega_k^I > \omega_k^F$ the front k is compressive, and its velocity is given by:

$$\sigma_k = \frac{u_e}{1 + F_e \frac{\omega_k^I}{1 + \sum_i \tilde{K}_i C_i^{P(k)}}} = \frac{u_e}{1 + F_e \frac{\omega_k^F}{1 + \sum_i \tilde{K}_i C_i^{P(k-1)}}} \tag{3.73}$$

These equations allow a complete determination of the breakthrough curves.

Notice that in the case of linearly modified Langmuir lumped adsorption isotherms, provided that $\bar{\lambda} = \bar{\lambda}_1 = \cdots = \bar{\lambda}_n$, the above framework can be used, the fronts being then shifted by $t_0 F_e \bar{\lambda}$ on the time axis.

EXERCISE 3.11 *Apply the ω transform to determine the breakthrough curve of a single-solute system.*

EXERCISE 3.12 *A column $V_{col} = 1\,l$, $L_{col} = 1\,m$, $\varepsilon_e = 0.4$ is eluted with 1 l/min of a prescribed eluent. The adsorption of solutes A, B and C follows Langmuir lumped adsorption isotherms: $\tilde{K}_A = 1$, $\tilde{K}_B = 2$, $\tilde{K}_C = 3$ l/g and $\overline{N}_A = \overline{N}_B = \overline{N}_C = 1$ g/l. Compute the breakthrough curves associated with the following experiments (concentrations in g/l):*

- *Experiment 1: the column initially equilibrated with $C_A = C_B = C_C = 0$ is saturated with $C_A = C_B = C_C = 1$.*
- *Experiment 2: the column initially equilibrated with $C_A = C_B = C_C = 1$ is saturated with $C_A = C_B = C_C = 0$.*
- *Experiment 3: the column initially equilibrated with $C_A = C_C = 1$, $C_B = 0.1$ is saturated with $C_A = C_C = 0.1$, $C_B = 1$.*

3.5.3 Application of the ω transform to binary systems

The ω transform has been used for deriving detailed properties of binary chromatograms obtained on systems following Langmuir or generalized Langmuir adsorption isotherms by (Mazzotti, 2006), (Rajendran and Mazzotti, 2011), (Siitonen and Sainio, 2011). Additionally, let us mention that (Bailly and Tondeur, 1981) published an explicit form of the dispersive fronts in the case of ion exchange. For similarities and differences between ion exchange and adsorption, the reader is referred to Section 3.6.

An illustration of results obtained with the ω transform on binary chromatograms is given below for the determination of the abscissa and times associated with the disappearance of the feed plateau and complete resolution.

We assume that the binary lumped adsorption isotherms can be represented by the generalized expression in Eq. (3.12) with $\overline{\lambda}_A = \overline{\lambda}_B = 0$. The first step consists in solving the characteristic equation derived from Eq. (3.69) for a binary mixture, which becomes:

$$\begin{aligned}&\left(1+p_A\tilde{K}_AC_A^F+p_B\tilde{K}_BC_B^F\right)\omega^2\\&-\left[\overline{K}_B\left(1+p_A\tilde{K}_AC_A^F\right)+\overline{K}_A\left(1+p_B\tilde{K}_BC_B^F\right)\right]\omega+\overline{K}_A\overline{K}_B=0\end{aligned} \tag{3.74}$$

where $\overline{K}_i = \overline{N}_i\tilde{K}_i$ is the lumped Henry's coefficient of solute i and parameters p_i can take the values +1 or −1 (see Section 3.2).

It can be shown that, under conditions of thermodynamic consistency of Eq. (3.12), Eq. (3.74) has two roots $0 < \omega_1^F < \omega_2^F$ satisfying the inequalities:

- Case L: $0 < \omega_1^F \le \overline{K}_A \le \omega_2^F \le \overline{K}_B$
- Case A: $\overline{K}_A < \omega_1^F \le \overline{K}_B \le \omega_2^F < \infty$
- Case M$_A$: $0 < \omega_1^F \le \overline{K}_A < \overline{K}_B \le \omega_2^F < \infty$
- Case M$_B$: $\overline{K}_A \le \omega_1^F \le \omega_2^F \le \overline{K}_B$

The various cases have been described in Section 3.2.

The abscissa and times associated with the disappearance of the feed plateau and complete resolution between the solutes are given in Tables 3.1 and 3.2.

Extension of the results in Tables 3.1 and 3.2 to linearly modified Langmuir lumped adsorption isotherms is trivial if $\overline{\lambda} = \overline{\lambda}_A = \overline{\lambda}_B$: under these circumstances, the times t_{plat} and t_{resol} are shifted by $F_e t_0 \overline{\lambda}$ and the abscissas are unaffected.

EXERCISE 3.13 *Confirm the values of Z_{resol}, Z_{plate} and t_{plate} obtained in Exercise 3.10 using the expressions given in Tables 3.1 and 3.2.*

Table 3.2 gives very important information for elution chromatography design, but at least one piece of information is missing. An important factor for design purposes is associated with the positioning of the times at which the first peak starts, t_{beg}, and at which the last (second) peak ends, t_{end}. Let us consider a favorable adsorption isotherm; in this case, t_{beg} is associated with a shock, while t_{end} represents the end of a dispersive front.

Determination of t_{end} poses no difficulty. One has $t_{end} = \Delta t_{inj} + t_0\left(1 + F_e\overline{f}'_B(0)\right)$.

As long as the first solute plateau remains, thus at distances smaller that Z_{P1}, t_{beg} is associated with a shock between concentrations 0 and P1 so that $t_{beg} = t_0\left(1 + Fe\dfrac{\overline{C}_A^{P1}}{C_A^{P1}}\right)$.

At distances greater than Z_{P1}, the plateau P1 disappears. The expressions allowing us to determine t_{beg} when the plateau is eroded are more complex and have been published by (Siitonen and Sainio, 2011) for Langmuir adsorption isotherms. Abscissa Z_{P1} is given by:

Table 3.1 Determination of abscissas and times associated with the disappearance of the feed plateau for a binary system obeying generalized lumped Langmuir adsorption isotherms (Rajendran and Mazzotti, 2011). $\omega_1^F < \omega_2^F$ are the roots of Eq. (3.74) and $\rho = \frac{1-\varepsilon_e}{\varepsilon_e}\frac{1}{\overline{K}_A\overline{K}_B}$. Generalized lumped Langmuir adsorption isotherms have been defined in Section 3.2.

	Z_{plat}	t_{plat}
Linear case	$\frac{\Delta t_{inj} u_e}{\frac{1-\varepsilon_e}{\varepsilon_e}(\overline{K}_B - \overline{K}_A)}$	$\frac{1+\frac{1-\varepsilon_e}{\varepsilon_e}\overline{K}_B}{\frac{1-\varepsilon_e}{\varepsilon_e}(\overline{K}_B - \overline{K}_A)}\Delta t_{inj}$
Case L	$\frac{t_{inj} u_e}{\rho\omega_1^F\omega_2^F(\overline{K}_B - \omega_1^F)}$	$\frac{1+\rho\omega_1^F\omega_2^F\overline{K}_B}{\rho\omega_1^F\omega_2^F(\overline{K}_B - \omega_1^F)}\Delta t_{inj}$
Case A	$\frac{\Delta t_{inj} u_e}{\rho\omega_1^F\omega_2^F(\omega_2^F - \overline{K}_A)}$	$\frac{1+\rho\omega_1^F\omega_2^F\overline{K}_A}{\rho\omega_1^F\omega_2^F(\omega_2^F - \overline{K}_A)}\Delta t_{inj}$
Case M_A	$\frac{\Delta t_{inj} u_e}{\rho\omega_1^F\omega_2^F(\omega_2^F - \omega_1^F)}$	$\frac{1+\rho\omega_1^F\omega_2^{F2}}{\rho\omega_1^F\omega_2^F(\omega_2^F - \omega_1^F)}\Delta t_{inj}$
Case M_B	$\frac{\Delta t_{inj} u_e}{\rho\omega_1^F\omega_2^F(\overline{K}_B - \overline{K}_A)}$	$\frac{1+\rho\omega_1^F\omega_2^F\overline{K}_B}{\rho\omega_1^F\omega_2^F(\overline{K}_B - \overline{K}_A)}\Delta t_{inj}$

$$Z_{P1} = \frac{u_e\left(1+\tilde{K}_A C_A^{P1}\right)^2 \Delta t_{inj}}{F_e(\overline{N}_A\tilde{K}_A)(\tilde{K}_A C_A^{P1})}\left[1 - \frac{\beta\tilde{K}_B C_B^F}{\alpha\left(\beta+\tilde{K}_A C_A^{P1}\right)^2}\right]$$
$$\alpha = \frac{\overline{N}_B\tilde{K}_B}{\overline{N}_A\tilde{K}_A} \qquad \beta = 1 - 1/\alpha \tag{3.75}$$

When plateau P1 is eroded, the concentration of solute A reaches C_A^{max} at the maximum of the peak, and $t_R^{max}(A)(= t_{beg})$ is the associated retention time. Concentration C_A^{max} is a solution of:

$$L_{fA}\left[1+\frac{1}{\tilde{K}_A C_A^{max}}\right]^2 + L_{fB}\left[1+\frac{1}{\alpha\left(1+\tilde{K}_A C_A^{max}\right)-1}\right]^2 = 1$$
$$L_{fA} = \frac{\tilde{K}_A C_A^F \Delta t_{inj}}{F_e\overline{N}_A\tilde{K}_A t_0} \qquad L_{fB} = \frac{\tilde{K}_B C_B^F \Delta t_{inj}}{F_e\overline{N}_B\tilde{K}_B t_0} \tag{3.76}$$

L_{fA} and L_{fB} are the so-called *loading factors*.

Finally, at abscissas greater than Z_{P1}, $t_R^{max}(A) = t_{beg}$ is given by:

$$t_R^{max}(A) = \Delta t_{inj} + t_0 + F_e t_0\overline{N}_A\tilde{K}_A\left[\frac{1}{\alpha\left(1+\tilde{K}_A C_A^{max}\right)^2} + L_{fA}\frac{\beta}{\left(\tilde{K}_A C_A^{max}\right)^2}\right] \tag{3.77}$$

Table 3.2 Determination of abscissas and times allowing complete resolution for a binary system obeying generalized lumped Langmuir adsorption isotherms (Rajendran and Mazzotti, 2011). $\omega_1^F < \omega_2^F$ are the roots of Eq. (3.74) and $\rho = \frac{1-\varepsilon_e}{\varepsilon_e}\frac{1}{\overline{K}_A\overline{K}_B}$. Generalized lumped Langmuir adsorption isotherms have been defined in Section 3.2.

	z_{resol}	t_{resol}
Linear case	$\frac{\Delta t_{inj} u_e}{\frac{1-\varepsilon_e}{\varepsilon_e}(\overline{K}_B - \overline{K}_A)}$	$\frac{1 + \frac{1-\varepsilon_e}{\varepsilon_e}\overline{K}_B}{\frac{1-\varepsilon_e}{\varepsilon_e}(\overline{K}_B - \overline{K}_A)}\Delta t_{inj}$
Case L	$\frac{\Delta t_{inj} u_e(\overline{K}_B - \omega_1^F)}{\rho\omega_1^F\omega_2^F(\overline{K}_B - \overline{K}_A)^2}$	$\left[1 + \frac{(1+\rho\overline{K}_A^2\omega_2^F)(\overline{K}_B - \omega_1^F)}{\rho\omega_1^F\omega_2^F(\overline{K}_B - \overline{K}_A)^2}\right]\Delta t_{inj}$
Case A	$\frac{\Delta t_{inj} u_e(\omega_2^F - \overline{K}_A)}{\rho\omega_1^F\omega_2^F(\overline{K}_B - \overline{K}_A)^2}$	$\frac{(1+\rho\overline{K}_B^2\omega_1^F)(\omega_2^F - \overline{K}_A)}{\rho\omega_1^F\omega_2^F(\overline{K}_B - \overline{K}_A)^2}\Delta t_{inj}$
Case M_A	$\Delta t_{inj} u_e \frac{X + 2\omega_1^F\omega_2^F - \overline{K}_B(\omega_1^F + \omega_2^F)}{\rho\omega_1^F\omega_2^F(\overline{K}_A - \overline{K}_B)^3}$ $X = \overline{K}_A(2\overline{K}_B - \omega_1^F - \omega_2^F)$	$\frac{X\rho\overline{K}_B^2 + 2(\overline{K}_A\overline{K}_B + \omega_1^F\omega_2^F) - Z}{\rho\omega_1^F\omega_2^F(\overline{K}_A - \overline{K}_B)^3}\Delta t_{inj}$ $X = \overline{K}_A^3 + Y + 3\omega_1^F\omega_2^F\overline{K}_A - \overline{K}_B\omega_1^F\omega_2^F$ $Y = (\overline{K}_B - 2(\omega_1^F + \omega_2^F))\overline{K}_A^2$ $Z = (\overline{K}_A + \overline{K}_B)(\omega_1^F + \omega_2^F)$
Case M_B	$\frac{\Delta t_{inj} u_e}{\rho\omega_1^F\omega_2^F(\overline{K}_B - \overline{K}_A)}$	$\frac{1 + \rho\omega_1^F\omega_2^F\overline{K}_B}{\rho\omega_1^F\omega_2^F(\overline{K}_B - \overline{K}_A)}\Delta t_{inj}$

Equation (3.77) is a generalization to binary mixtures of Eq. (3.32) valid for single-solute systems. The times associated with the beginning and the end of the chromatogram are thus known.

When adsorption isotherms are not Langmuirian, there is in general no analytical solution for t_{beg} and t_{end}. Note, however, that, in general, at least one of these two times is easily accessible. Time t_{end} (*resp.* t_{beg}), being associated with the end (*resp.* beginning) of a dispersive front for a favorable (*resp.* unfavorable) adsorption isotherm, can be determined easily by $t_{end} = t_{inj} + t_0\left(1 + Fe\bar{f}'_B(0)\right)$ (*resp.* $t_{beg} = t_0\left(1 + Fe\bar{f}'_A(0)\right)$).

3.5.4 More realistic chromatograms: the MC–$\overline{\text{Equil}}$ model

As with single-solute systems, equilibrium theory allows one to obtain the main features of a chromatogram, meaning the number and positioning of fronts and their compressive or dispersive character. More realistic curves are obtained with the simple MC–$\overline{\text{Equil}}$ model, which approximates the column by a series of N equilibrium stages (more refined models are given in later chapters).

The mass balance for each cell k is:

$$\begin{aligned} &C_i^{k-1} = C_i^k + \frac{t_0}{N}\frac{dC_i^k}{dt} + \frac{1-\varepsilon_e}{\varepsilon_e}\frac{t_0}{N}\frac{d\bar{C}_i^k}{dt} \qquad k \in [1,N], \quad i \in [1,n] \\ &C_i^0 = C_i^F \end{aligned} \tag{3.78}$$

If adsorption equilibrium is considered to be reached in each cell k, Eq. (3.78) is rewritten as:

$$C_i^{k-1} = C_i^k + \frac{t_0}{N}\frac{dC_i^k}{dt} + \frac{1-\varepsilon_e}{\varepsilon_e}\frac{t_0}{N}\sum_{j=1}^{n}\frac{\partial \bar{C}_i^k}{\partial C_j^k}\frac{dC_j^k}{dt} \qquad k \in [1,N], \quad i \in [1,n] \tag{3.79}$$

The system of equations (3.79) is a set of $n \times N$ ordinary differential equations which must be solved numerically in order to simulate the column behavior. Examples of simulations are given in Figure 3.28 with the following parameters: $\varepsilon_e = 0.4$, $t_0 = 1$, $\bar{\lambda}_A = \bar{\lambda}_B = \bar{\lambda}_C = 0.5$, $\bar{N}_A = \bar{N}_B = \bar{N}_C = 100$ g/l, $\tilde{K}_A = 0.01$, $\tilde{K}_B = 0.015$ and $\tilde{K}_C = 0.025$ l/g.

Compressive fronts (shocks) associated with the saturation step are simulated in Figure 3.28. Notice that their mean positions are independent of N and are given by the equilibrium model. It can be checked that the sharpness of the fronts increases with N, until they become shock-like. However, it is apparent that the column must be equivalent to a significant number of plates in order to approach the "equilibrium model" behavior. Specifically, it is usually difficult to reach the plateau predicted by the equilibrium model unless very efficient columns are used, as a few thousand plates are required.

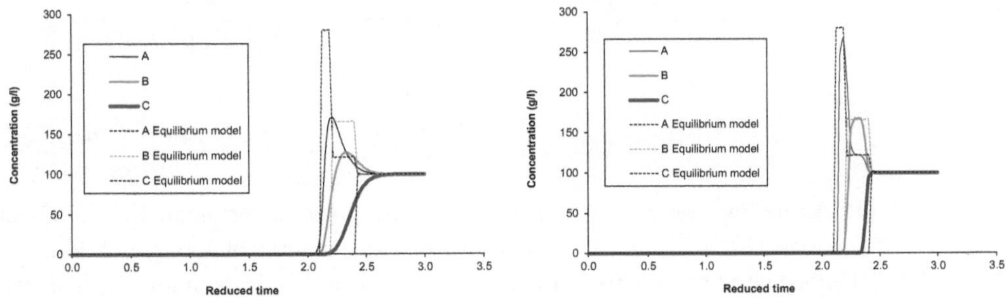

$N = 200$ $N = 2000$

Figure 3.28 Compressive fronts: $C_A^I = C_B^I = C_C^I = 0$; $C_A^F = C_B^F = C_C^F = 200$ g/l. Lumped adsorption isotherm parameters are given in the text.

3.6 Similarities and differences between ion exchange and adsorption

The theory presented so far in this chapter assumes a system in which solutes can reversibly adsorb on a chromatographic medium. It can be adapted to the case of ions interacting with an ion-exchange resin, as explained below.

Ion-exchange equilibria will be presented in detail in Chapter 4, but for the time being, we will use a simple description for illustration purposes. We consider the case of a strong cationic resin, and we assume that a proton H and cations C_i of valences z_i are exchanging between the solution and the resin. All concentrations are expressed in eq/l of phase (fluid or lumped solid).

We assume that the equilibria can be modeled with ideal mass action laws:

$$K_{i/H} = \frac{\overline{C}_i^{z_i+}}{C_i^{z_i+}} \left(\frac{H^+}{\overline{H}^+} \right)^{z_i} \tag{3.80}$$

$K_{i/H}$ are known as the *selectivity coefficients*. They are temperature-dependent and are constant with respect to composition for ideal solutions (both fluid and solid).

Introducing the normality N_0 of the solution and the ionic capacity $\overline{N}^{iex}$ of the ion-exchange resin, one can write the electroneutrality constraints in solution and in the lumped solid:

$$H^+ + \sum_i C_i^{z_i+} = N_0 \quad \text{(eq/l)} \quad \overline{H}^+ + \sum_i \overline{C}_i^{z_i+} = \overline{N}^{iex} \quad \text{(eq/l)} \tag{3.81}$$

When n cations are considered (including H, which is given number n), from Eq. (3.80) and Eq. (3.81) one can show that, modeling *homovalent* ion exchange (thus assuming all $z_i = 1$) with $n - 1$ selectivity coefficients $K_{i/H}$ is equivalent to writing:[16]

[16] More information is given in Section 4.5.2.

$$\overline{C}_i^{z_i+} = \frac{\overline{N}_i\tilde{K}_iC_i^{z_i+}}{1+\sum\limits_i \tilde{K}_iC_i^{z_i+}} \quad i \leqslant \text{n}-1$$
$$\text{with} \quad \tilde{K}_i = \frac{K_{i/H}-1}{N_0} \quad \text{and} \quad \overline{N}_i = \overline{N}^{iex}\frac{K_{i/H}}{K_{i/H}-1} \tag{3.82}$$

Assuming that N_0 is *kept constant* during the experiment, Eq. (3.82) provides relations $\overline{C}_i = \bar{f}_i(C_1,\ldots,C_{n-1})$ allowing calculation of the $n-1$ lumped solid-phase concentrations $\overline{C}_i$ from the $n-1$ concentrations in solution C_i. The relations $\overline{C}_i = \bar{f}_i(C_1,\ldots,C_{n-1})$ are equivalent to lumped adsorption isotherms. Consequently, a homovalent ion-exchange system involving n exchanging ions at constant normality is equivalent to an adsorption system involving $n-1$ solutes.

Additionally, provided that the selectivity coefficients $K_{i/H}$ are constant, one notices that Eq. (3.82) is identical to the Langmuir expression given in Eq. (3.8). Consequently, a homovalent ion-exchange system involving n exchanging ions at constant normality that can be modeled with constant selectivity coefficients is equivalent to an $(n-1)$-solute adsorption system obeying a Langmuir lumped adsorption isotherm model.

The case of *heterovalent* ion exchange leads to more complex and possibly implicit expressions but the main conclusions are the same: an ion-exchange system involving n exchanging ions at constant normality is equivalent to an $(n-1)$-solute adsorption system.

This means that the entire theoretical framework of the equilibrium model developed for adsorption can be used for ion exchange, provided the right definition for the "adsorption isotherms" is used.

While the adsorption theoretical framework can be used for ion exchange, the latter situation is frequently described in terms of selectivity coefficients and ionic fractions, as illustrated, for instance, by (Helfferich and Klein, 1970) or the summary presentation in (Tondeur and Bailly, 1986).

Recall that the assumption of constant normality plays an important role in relating adsorption and ion exchange. There are basically two reasons for the normality to vary during an ion-exchange process: either the normality of the initial (I) and feed (F) compositions differ or some ionic equilibria take place in solution. The latter case will be discussed in the next section.

The case of a normality variation due to differences between initial (I) and feed (F) compositions can significantly impact the behavior of ion-exchange systems. The change in normality is propagated through the bed at a velocity being the one of a tracer (an excluded tracer if conditions of strong Donnan exclusion prevail) and thus different from that of the ion-exchange fronts. The treatment of these interferences between the normality front and the ion-exchange fronts has been proposed by (Grevillot and Tondeur, 1974) for the Na^+/H^+ system, for which normality has little influence on the ion-exchange selectivity, and by (Dodds and Tondeur, 1972) and (Grevillot and Tondeur, 1974) for the Ca^{++}/H^+ system, for which normality has a strong influence on the ion-exchange selectivity.

For a practically oriented document on the use of the equilibrium model and especially on the calculation of interferences occurring in ion-exchange systems, the reader is referred to (Yu and Wang, 1986).

3.7 Qualitative rules and complex multi-solute systems

The equilibrium theory presented above allows a calculation of the column response to step concentration changes for any type of adsorption isotherms in the case of binary systems, and in the case of Langmuir systems (including ion exchange; see Section 3.6), whatever the number of species. As far as the responses to rectangular pulses are concerned, analytical or quasi-analytical solutions are limited to binary Langmuir adsorption isotherms. The methodology can be extended numerically to favorable or unfavorable binary adsorption isotherms, but things become very complex for adsorption isotherms with inflection points and inextricable when the number of species is three or more.

While quantitative solutions are not available in general, the equilibrium model gives some interesting quantitative and fruitful qualitative rules in the case of ordered systems. An ordered system is a system in which all ratios $\overline{C}_i/C_i$ vary in the same direction under a composition change at equilibrium. This means that these ratios follow the same order as the lumped Henry's coefficient $\overline{K}_i$. An immediate consequence is that, for ordered systems, the selectivity coefficients $\alpha_{i/1} = \dfrac{\overline{C}_i}{C_i}\dfrac{C_1}{\overline{C}_1}$ can vary with composition (as opposed to Langmuir adsorption isotherms, for which they are constant) but cannot change their relative order when composition changes.

It can be shown (Klein *et al.*, 1967; Tondeur and Klein, 1967; Tondeur and Bailly, 1986; Helfferich, 1997) that ordered systems present a qualitative pattern of behavior which is similar to the one obtained with constant selectivity factors.

The main qualitative rules are summarized below for a system obeying $\alpha_{n/1} < \alpha_{k/1} < \alpha_{1/1}$. *Warning: the high-affinity solute has the lower number. This is not the convention most chromatographers are used to and not the convention generally used in this book, but we have decided to keep the ordering used in the key publications related to the topic.*

1. The number of fronts generated by a step change in feed composition is equal to the number n of solutes involved in adsorption (resp. $n-1$ in the case of ion exchange at constant normality). Fronts are numbered from 1, which represents the upstream/slow front, to n (resp. $n-1$), which represents the downstream/fast front.
2. In any front of order k, the solutes from 1 to k form the high-sorptivity group and the solutes from $k+1$ to n form the low-sorptivity group.
   - i. If one of the low-sorptivity solute concentrations increases across the front, then all the components of the low-sorptivity group vary similarly.
   - ii. If the low-sorptivity solute concentrations increase across the front, then the high-sorptivity component concentrations decrease.
   - iii. In rules i and ii, "low" and "high" can be interchanged, as can "increase" and "decrease".
3. Solutes k and $k+1$ are called the key solutes in front k.[17] More precisely, solute k is called the low key solute and $k+1$ the high key solute. Their behaviors are

[17] This concept is borrowed from multi-solute distillation.

opposite according to rule 2. Solutes may appear or disappear only in the fronts where they are key solutes. A solute is said to appear when it is present in the effluent curve and absent from the initial state. It is said to disappear when it is absent in the effluent curve and present in the initial state.

4 A front is compressive if the concentrations of the high-sorptivity group are higher on the upstream side than on the downstream side, thus when these solutes are taken up by the solid. The front is dispersive in the opposite situation.

This set of rules can be used to draw qualitative patterns. The reader could, for instance, check their validity by using the breakthrough curves determined in Exercise 3.12.

It must be kept in mind that these rules are valid for ordered systems only. The case of selectivity reversal is relatively common in chromatography and can lead to very different behaviors. A simple example is given by heterovalent ion exchange modeled by an ideal mass action law. As shown in Chapter 4, the fact that valences appear in the selectivity coefficient induces an influence of the solution normality on the affinity of the resin for the different species. This leads to much more complex behaviors which have been studied, for instance, by (Helfferich and Klein, 1970; Tondeur, 1970).

The cases of phase separation (e.g. demixion, precipitation) and partial miscibility (column overloading with partially miscible mixtures) provide an additional level of complexity. Precipitation in chromatographic columns should normally be avoided unless one operates up-flow and in at least partly fluidized beds to prevent clogging. Understanding how solutes can migrate in the simultaneous presence of precipitation and chromatographic interaction can be useful for specific ion-exchange processes, in which calcium sulfate, for instance, can precipitate, or for modeling solute transport in soils. The reader interested in the development of a theory associating ion exchange and precipitation is referred to (Klein, 1986).

The presence of non-miscible fluid phases is generally undesirable in chromatography but should not, *a priori*, be excluded. In different preparative chromatography applications (purification of oils, for instance) the concentration of feed in the injection can be significantly higher than its solubility in the eluent so that a dissolution of the injected fluid phase in the eluent takes place inside the column during the solute migration – an effect never taken into account in preparative chromatography to the best of my knowledge. Fundamental contributions on multi-phase displacement associated with chromatographic interactions have been made for the petroleum industry (Helfferich, 1981).

Much more classical is the presence of reactions in solution, especially in the case of ion exchange. Reactions in solution can be as simple as the ionic dissociation of water. This can have a drastic influence on chromatograms: injecting NaOH or NaCl on a cationic ion-exchange resin in the H form leads to very different breakthrough curve shapes (see Chapter 7). Other important practical situations are associated with weak electrolytes, amino acids, peptides and proteins. A first and "simple" effect of these reactions in solution is to impact *ionic* normality, which may thus no longer be constant. For a presentation of the equilibrium theory of ion exchange in the presence of reactions in solution, with or without precipitation/dissolution, the reader is referred to (Klein, 1981) and (Bryant *et al.*, 1986). An interesting (and simpler) application of the equilibrium model consists in the determination of pH variations that occur in columns that are packed with weak acid ion-exchange resins subject to changes in salt concentrations, as shown in (Carta and Jungbauer, 2011, page 222).

The presence of reactions in solution is not limited to ion exchange, and an example is given below. When injecting either the pure enantiomers of binaphthol or an equimolar racemic mixture of these enantiomers in a non-chiral column eluted with a non-chiral solvent, both enantiomers and racemate elute, as expected, at the same retention time. It has been shown by (Nicoud *et al.*, 1996) that injecting a non-racemic mixture (i.e. a mixture containing both enantiomers, one being in excess) leads to a resolution between the enantiomer in excess and an equimolar racemic mixture. This has proven to be related to the ability of both enantiomers to associate in solution, forming a diastereoisomer which does not have the same properties as the enantiomers. The presence of reactions or equilibria associated with chromatography is probably more frequent than expected, as discussed in Section 7.7; again, understanding the key features of the elementary mechanisms is key for proper modeling.

What type of information can equilibrium theory provide for these complex situations associating chromatographic processes and chemical reactions or equilibria in the fluid or solid phase? Equilibrium theory cannot in general predict the positioning of the fronts or even their compressive or dispersive character, but it will give a simple way to at least estimate the number of fronts.

Very few papers are devoted to the consequences of elementary interactions on the structure and properties of breakthrough curves. Contributions by (Hwang *et al.*, 1988), (Bryant *et al.*, 1986), (Schweich *et al.*, 1993a) and (Schweich *et al.*, 1993b) are counter-examples.

While appreciating that these contributions are qualitative, we think it is worth introducing them because of their potential mid-term value in elucidating mechanisms or predicting breakthrough curves occurring in biochromatography.

We have seen that the response of a column fed with a solution is normally made up of n fronts in the case of n independent solutes (thus $n - 1$ fronts for ion exchange at constant normality involving n solutes). It can be shown that the number of fronts is given by a quantity called *variance* (Helfferich and Whitley, 1996) that can be viewed as the number of degrees of freedom of the system. For instance, the variance of a standard n-solute isothermal adsorption system is n, but it becomes $n + 1$ for non-isothermal systems because temperature is an additional independent variable, and it becomes $n - 1$ for ion exchange because of the electroneutrality constraint. Let us illustrate how this result can be extended to situations where ion exchange and reactions can coexist. Our development is based on the work of (Schweich *et al.*, 1993a). More information on thermodynamic aspects of this approach will be given in Section 4.7.

We consider a system containing n chemical species S_j. These species may be located in the fluid phase or in the chromatographic medium. Some of them may be present in solution only and may not interact with the chromatographic medium. For the sake of clarity, when one component is located in both the fluid and the lumped solid phase, two different species are assumed. We assume that the system can be described by a set of R ($R < n_T$) *independent* stoichiometric reactions having the form:

$$\sum_{j}^{n_T} \nu_{i,j}^{glob} S_j = 0 \quad \text{for reaction } i \tag{3.83}$$

where $\nu_{i,j}^{glob}$ is the (algebraic) stoichiometric coefficient of species S_j in reaction i. Let us stress that we assume all the reactions are at chemical equilibrium.

A stoichiometric reaction is associated with a phenomenological mechanism representing ion exchange, reactions in solution and possibly precipitation/dissolution of solids and gas dissolution. We exclude here mechanisms like precipitation/dissolution of solids and the dissolution of gases that can be associated with a change of the number of phases, and we assume that an appropriate mass action law is associated with Eq. (3.83).

As we assume the existence of R *independent* stoichiometric reactions represented by Eq. (3.83), the rank of the matrix $\left[\nu_{i,j}^{glob}\right]$ is R.

Let us denote species by the concentration symbol, C_j or $\overline{C}_k$ depending on whether they are in the flowing fluid phase or in the chromatographic medium. Then, let us group the n_f species C_j in the fluid phase in one set, and the $n_{\overline{s}}$ species $\overline{C}_k$ in the lumped solid phase in another, contiguously numbered set. One has:

$$n_T = n_f + n_{\overline{s}} \tag{3.84}$$

If coefficients $\nu_{r,j}$ and $\overline{\nu}_{r,k}$ are the respective stoichiometric coefficients in reaction r, reaction i in Eq. (3.83) can be written:

$$\nu_{i,1}C_1 + \cdots + \nu_{i,n_f}C_{n_f} + \overline{\nu}_{i,1}\overline{C}_1 + \cdots + \overline{\nu}_{i,n_{\overline{s}}}\overline{C}_{n_{\overline{s}}} = 0 \tag{3.85}$$

The reactions can consequently be written in matrix form by introducing the stoichiometric matrix of reactions:

$$\begin{bmatrix} \nu_{1,1} & \cdots & \nu_{1,n_f} & \overline{\nu}_{1,1} & \cdots & \overline{\nu}_{1,n_{\overline{s}}} \\ \cdots & \cdots & \cdots & \cdots & \cdots & \cdots \\ \cdots & \cdots & \cdots & \cdots & \cdots & \cdots \\ \cdots & \cdots & \cdots & \cdots & \cdots & \cdots \\ \cdots & \cdots & \cdots & \cdots & \cdots & \cdots \\ \nu_{R1} & \cdots & \nu_{R,n_f} & \overline{\nu}_{R,1} & \cdots & \overline{\nu}_{R,n_{\overline{s}}} \end{bmatrix} \begin{bmatrix} C_1 \\ \cdots \\ C_{n_f} \\ \overline{C}_1 \\ \cdots \\ \overline{C}_{n_{\overline{s}}} \end{bmatrix} = \begin{bmatrix} 0 \\ \cdots \\ 0 \\ 0 \\ \cdots \\ 0 \end{bmatrix} \tag{3.86}$$

The overall stoichiometric matrix is thus composed of two submatrices related to the two phases, and one can write:

$$\left[\nu_{i,j}^{glob}\right] = \left[\left[\nu_{i,j}\right]\left[\overline{\nu}_{i,j}\right]\right] \tag{3.87}$$

Let R, R_f and $R_{\overline{s}}$ be the ranks of matrices $\left[\nu_{ij}^{glob}\right]$, $\left[\nu_{ij}\right]$ and $\left[\overline{\nu}_{ij}\right]$ respectively. It can be shown that the number of retarded fronts is given in general by (Schweich *et al.*, 1993a,b):

$$F = R_f + R_{\overline{s}} - R \tag{3.88}$$

The number of retarded fronts associated with a very complex system can thus be determined almost immediately!

Note that the number of retarded fronts excludes the "non-retarded" front of an unretained tracer (*ERTD*) referred to as a "salinity wave" in oil engineering or a "normality wave" in ion exchange.

Let us illustrate this result with the trivial case of ion exchange between Na and H on a cationic resin. We assume that a solution containing an anion X and sodium ions Na is in contact with a strong cationic resin. The equilibria to be taken into account, with their respective associated mass action laws, are:

$$\begin{aligned} &\mathrm{H^+ + OH^- \rightleftharpoons H_2O} \qquad K_e = H^+OH^- \\ &\mathrm{\overline{H}^+ + Na^+ \rightleftharpoons H^+ + \overline{Na}^+} \qquad K_{Na/H} = \left[\frac{\overline{Na}^+}{Na^+}\right]\left[\frac{H^+}{\overline{H}^+}\right] \end{aligned} \tag{3.89}$$

We have:

- five species in the fluid moving phase, whose concentrations are: H^+, OH^-, Na^+, X^- and H_2O
- two species in the resin phase, whose concentrations are $\overline{H}^+$, $\overline{Na}^+$

Consequently, one obtains $R = 2$, $n_T = 7$, $n_f = 5$ and $n_{\bar{s}} = 2$.

The stoichiometric matrix Eq. (3.86) is written for five species in the fluid phase, two species in the lumped solid phase and two reactions:

$$\left[\nu_{i,j}^{glob}\right] = \begin{bmatrix} \nu_{1,1} & \nu_{1,2} & \nu_{1,3} & \nu_{1,4} & \nu_{1,5} & \bar{\nu}_{1,1} & \bar{\nu}_{1,2} \\ \nu_{2,1} & \nu_{2,2} & \nu_{2,3} & \nu_{2,4} & \nu_{2,5} & \bar{\nu}_{2,1} & \bar{\nu}_{2,2} \end{bmatrix} \tag{3.90}$$

Placing the reactions in the order given in Eq. (3.89), the stoichiometric matrix $\left[\nu_{i,j}^{glob}\right]$ is given by:

$$\begin{matrix} \mathrm{H^+} & \mathrm{OH^-} & \mathrm{Na^+} & \mathrm{X^-} & \mathrm{H_2O} & \mathrm{\overline{H}^+} & \mathrm{\overline{Na}^+} \end{matrix}$$

$$\begin{bmatrix} -1 & -1 & 0 & 0 & 1 & 0 & 0 \\ -1 & 0 & 1 & 0 & 0 & 1 & -1 \end{bmatrix} \tag{3.91}$$

so that:

$$[\nu_{i,j}] = \begin{bmatrix} -1 & -1 & 0 & 0 & 1 \\ -1 & 0 & 1 & 0 & 0 \end{bmatrix} \qquad [\bar{\nu}_{i,j}] = \begin{bmatrix} 0 & 0 \\ 1 & -1 \end{bmatrix} \tag{3.92}$$

One can see that $[\nu_{i,j}]$ has a rank $R_f = 2$ while $[\bar{\nu}_{i,j}]$ has a rank $R_{\bar{s}} = 1$.

Consequently, according to Eq. (3.88), we know that the number of fronts is $F = R_f + R_{\bar{s}} - R = 2 + 1 - 2 = 1$. One thus obtains one retarded front, which is in agreement with the experience of any chromatographer having carried out this simple experiment.

Using stoichiometric matrices for determining the number of fronts associated with the Na/H ion-exchange system is probably making undue use of mathematics (even if the influence of the water dissociation was perhaps not that clear...). We will see in Section 4.7 that this approach can be used with similar ease for systems with a significant degree of complexity.

3.8 Conclusion

Using equilibrium theory, thus by assuming a column of an infinite efficiency in which neither hydrodynamic dispersion nor kinetic limitation occurs, one is able to extract the essence of chromatographic non-linearity. Alternatively, we can consider that the approach gives a kind of asymptotic behavior that chromatographic systems tend to reach should the column be efficient and long enough.

This ideal representation of a complex reality has helped us realize important features. Starting from simple single-solute systems, we have seen that, according to the adsorption isotherm shape, feeding a column at initial composition I with a feed of composition F will lead to either a compressive front (thus a very steep breakthrough curve) or a dispersive front (thus a more or less broad breakthrough curve). The front can even be a combination of the two in the case of S-shaped adsorption isotherms. If a compressive front is obtained when moving from I to F, a dispersive front is obtained if the roles of I and F are interchanged. If a chromatogram is obtained by successively moving the inlet concentration from I to F and then from F to I under non-linear conditions, it must be asymmetrical. The observed tailing or fronting of the peaks is a direct consequence of the adsorption isotherm shape. No reason to blame your column or system supplier (at least not in this case).

Equilibrium theory can be extended to binary systems. While the mathematics is a little bit more complex, in most situations the problem can be solved quasi-analytically (meaning with a limited use of numerical tools). This development has taught us that the breakthrough curves obtained as solutions of the Riemann problem (the response of a column subjected to uniform initial conditions and to a discontinuity at the inlet) contain three composition plateaus (including plateau I and plateau F) separated by two successive fronts. The evolution of the concentrations in the fronts must follow certain trajectories, depending only on the adsorption isotherms and represented in the hodograph plane (C_A,C_B). The evolution of the front velocities along these hodograph paths allows us to decide whether the fronts are compressive or dispersive. The problem can be solved analytically for Langmuir adsorption isotherms, and more generally with limited numerical effort for non-S-shaped adsorption isotherms. S-shaped adsorption isotherms can lead to complex behaviors, and special care must be taken.

Unfortunately, while remaining valid, the methodology based on using specific paths in the hodograph space is generally of limited use for multi-solute systems, as one lacks mathematical resolution techniques. A remarkable breakthrough was made in the 1970s for systems that can be represented by Langmuir adsorption isotherms. Under these circumstances, an astute transformation allows the hodograph space to be orthogonalized, which then allows one to derive analytical solutions for the breakthrough curves.

In general, calculating the response of a system to a rectangular pulse is significantly more complex than calculating the response to a step change, as it requires calculating interferences between fronts, fast-moving compositions belonging to the rear part of the injection eventually catching up with slow-moving compositions belonging to the leading part of the injection. Unfortunately, the extension of the above-mentioned orthogonalization procedures to chromatograms (thus response to a rectangular pulse) is not available for species numbering greater than two.

If the adsorption isotherms are not strictly Langmuirian, and if the selectivity order of the different solutes is not affected by composition, one can show that many qualitative rules associated with Langmuir systems still apply (number of fronts, direction of concentration variations, compressive or dispersive character etc.).

At this point, we start reaching the limits of the equilibrium model: the more complex the system to be described, the more qualitative the results. If one takes one step further by adding equilibria in solution, it is difficult to obtain any information more specific than the number of fronts.

Designing and optimizing chromatographic processes certainly requires much more sophisticated models, as presented in the following chapters. We don't believe, however, that improvements in performance of computer and numerical tools should lead the equilibrium model to be stored in a cupboard containing former inventions that became useless due to technical progress. The straightforward methodology that the equilibrium model offers in calculating true moving beds will be evidence of this, but more generally I believe that mastering simple concepts (even if sometimes associated with complex math), telling us the type of fundamental rules that a system must respect, is of paramount importance to avoid making mistakes or missing something big.

Last but not least, the equilibrium model will give us the spine of the optimization methodology to be presented in Chapter 9.

3.9 Solutions of the exercises

Exercise 3.1

For clarity, and without loss of generality, let us assume that $C_A^I < C_A^F$. The outlet concentration profile is given by a curve, continuous or not, starting at C_A^I and reaching C_A^F at some point in time.

We define the average retention time by:

$$t_R = \frac{\int_0^\infty \left(C_A^F - C_A(t)\right) dt}{\left(C_A^F - C_A^I\right)}$$

A mass balance written at the column level between time 0 and a time t large enough for the column to be saturated with the feed concentration gives:

$$\underbrace{t\,Q\,C_A^F}_{\text{Inlet}} = \underbrace{\varepsilon_e V_{col}(C_A^F - C_A^I) + (1-\varepsilon_e)V_{col}(\overline{C}_A^F - \overline{C}_A^I)}_{\text{Stored}} + \underbrace{Q\int_0^t C_A(t)dt}_{\text{Outlet}}$$

which can be rewritten:

$$Q\int_0^t (C_A^F - C_A(t))dt = \varepsilon_e V_{col}(C_A^F - C_A^I) + (1-\varepsilon_e)V_{col}(\overline{C}_A^F - \overline{C}_A^I)$$

Setting $t = \infty$ and using the definition of t_R, one obtains:

$$Qt_R\left(C_A^F - C_A^I\right) = \varepsilon_e V_{col}(C_A^F - C_A^I) + (1-\varepsilon_e)V_{col}(\overline{C}_A^F - \overline{C}_A^I)$$

so that:

$$t_R = t_0\left(1 + F_e\frac{\overline{C}_A^F - \overline{C}_A^I}{C_A^F - C_A^I}\right)$$

Consequently, the average retention time of the front (compressive or dispersive) depends only on the initial and final concentrations.

Exercise 3.2

One can check that this lumped adsorption isotherm is unfavorable (cf. the definition in Section 3.2). The breakthrough curve obtained is therefore a simple wave, which can be calculated from Eq. (3.21). The derivative of the lumped adsorption isotherm is:

$$\frac{d\overline{C}_A}{dC_A} = \frac{2}{\gamma}C_A$$

After substitution in Eq. (3.21) one obtains:

$$t_R(C_A) = t_0\left(1 + \frac{1-\varepsilon_e}{\varepsilon_e}\frac{2}{\gamma}C_A\right)$$

so that after inversion, we have:

$$C_A\left(t\right) = \frac{t-t_0}{t_0}\frac{\varepsilon_e}{1-\varepsilon_e}\frac{\gamma}{2}$$

According to this result, the breakthrough curve is a straight line beginning at $t = t_0$ with $C_A = 0$, and ending at $C_A = C_A^F$.

Exercise 3.3

As shown in Chapter 2, the RTDs associated with the two situations are identical.

In situation a, the mixing cell is placed after the column. Because of the favorable lumped adsorption isotherm, we know that the column response will be a shock appearing at a certain time $t_{R,shock}$ which is a function of the inlet concentration. This shock will then enter in the mixing cell which will give a response having the shape $C^{out} = C^{inl}(1 - \exp(-t/t_{cell}))$, where t_{cell} is the residence time in the mixing cell. The system response will consequently be a curve remaining at zero concentration until $t_{R,shock}$ and then reaching C_A^F via $C_A^{out} = C_A^F(1 - \exp(-(t - t_{R,shock})/t_{cell}))$.

In situation b, the mixing cell is placed before the column. The response of the mixing cell $C_A^{out} = C^F(1 - \exp(-t/t_{cell}))$ gives the inlet signal of the column. The column inlet will thus see small concentrations at first, and large concentrations afterwards. Because of the non-linear character of the adsorption isotherm, we know that high concentrations travel faster than small concentrations. Consequently the large concentrations will gradually catch up with the small concentrations while traveling inside the column. If the column length is large enough, at the outlet of the column one will obtain a shock at $t_{R,shock} + t_{cell}$.

This illustration shows that systems having two identical RTDs can lead to different responses when non-linearity is involved.

Exercise 3.4

The excluded zero retention time is calculated to be $t_0 = \dfrac{0.4 \times 1}{0.4} = 1$ min. The initial slope of the adsorption isotherm is given by $\overline{K}_A = 1 + 0.5 \times 2 = 2$.

The retention time associated with a Dirac injection is given by:

$$t_R = t_0 \left(1 + \frac{1 - \varepsilon_e}{\varepsilon_e} \overline{K}_A\right) = 4 \text{ min}$$

The lumped adsorption isotherm is favorable, so we know that Case a will give a shock and Case b will give a simple wave.

- Case a: $C_A^I = 0$, $C_A^F = 5$
  The front is a shock and from Eq. (3.2) one can estimate:

$$\overline{C}_A^F = 1 \times 5 + \frac{2 \times 0.5 \times 5}{1 + 0.5 \times 5} = 6.42 \text{ g/l}$$

  and Eq. (3.23) gives: $t_{R,shock} = t_0 \left(1 + \frac{1-\varepsilon_e}{\varepsilon_e} \frac{\overline{C}_A^F}{C_A^F}\right) = 2.92$ min

- Case b: $C_A^I = 5$, $C_A^F = 0$
  The front is dispersive. The derivative of the isotherm is $\bar{f}'(C_A) = 1 + \dfrac{1}{(1 + 0.5 \times C_A)^2}$. Thus, from Eq. (3.21), one can estimate the retention time associated with each given concentration by:

$$t_R(C_A) = 1 \times \left(1 + \frac{1 - 0.4}{0.4} \times \left(1 + \frac{1}{(1 + 0.5 \times C_A)^2}\right)\right)$$

  and one obtains:

C_A (g/l)	0	1	2	3	4	5
$t_R(C_A)$ (min)	4	3.16	2.87	2.74	2.66	2.62

Exercise 3.5

For a linearly modified Langmuir lumped adsorption isotherm, one has:

$$\frac{\bar{f}(C_A^F)}{C_A^F} - \bar{f}'(C_A^F) = \frac{\tilde{K}_A \overline{N}_A \tilde{K}_A C_A^F}{\left(1 + \tilde{K}_A C_A^F\right)^2}$$

Consequently, with the lumped adsorption isotherm parameters of Exercise 3.4 one obtains:

$$\frac{\bar{f}(C_A^F)}{C_A^F} - \bar{f}'(C_A^F) = \frac{1}{(1+0.5)^2} = 0.222 \text{ g/l}$$

From Eq. (3.32), it follows that the maximum injection time which avoids the plateau is $\Delta t_{inj}^{plat} = 0.333$ min, so $V_{inj}^{plat} = Q\Delta t_{inj}^{plat} = 0.132$ l.

- $V_{inj} = 1$ l, thus $\Delta t_{inj} = 2.5$ min: there is a plateau.
  The position of the shock is given by Eq. (3.26):

$$t_{R,shock} = t_0 \left[1 + \frac{1-\varepsilon_e}{\varepsilon_e}\frac{\overline{C}_A^F}{C_A^F}\right] = 1 \times \left[1 + \frac{1-0.4}{0.4} \times \frac{1.66}{1}\right] = 3.5 \text{ min}$$

  The dispersive part of the chromatogram is given by Eq. (3.26):

$$t_{R,rear}(C_A) = 2.5 + 1 \times \left[1 + \frac{1-0.4}{0.4} \times \left(1 + \frac{1}{(1+0.5 \times C_A)^2}\right)\right]$$

  The obtained chromatogram is given in Figure 3.29 (right).
- $V_{inj} = 0.1$ l, thus $\Delta t_{inj} = 0.25$ min: there is no plateau.
  The concentration occurring at the maximum of the peak must satisfy Eq. (3.32):

$$\bar{f}(C_A^{max}) - \bar{f}'(C_A^{max})C_A^{max} = \frac{1}{1-\varepsilon_e}\frac{V_{inj}}{V_{col}}C_A^F$$

  Thus:

$$\frac{\left(\tilde{K}_A \overline{N}_A\right)\left(\tilde{K}_A C_A^{max}\right)}{\left(1 + \tilde{K}_A C_A^{max}\right)^2} C_A^{max} = \frac{1}{1-\varepsilon_e}\frac{V_{inj}}{V_{col}}C_A^F$$

  After introducing the relevant numerical values:

$$\frac{1 \times \left(0.5 \times C_A^{max}\right)}{\left(1 + 0.5 \times C_A^{max}\right)^2} \times C_A^{max} = \frac{1}{1-0.4} \times \frac{0.1}{1} \times 1, \text{ thus } C_A^{max} = 0.81 \text{g/l}$$

  The dispersive part of the front is still given by Eq. (3.26), but the injection time is now 0.25 min instead of 2.5 min. The calculated chromatogram is given in Figure 3.29 (left).

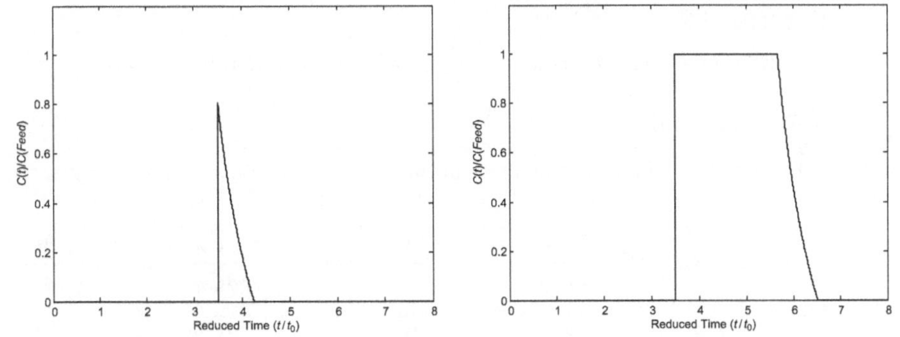

Injected volume of 0.1 l. Injected volume of 1.0 l.

Figure 3.29 For Exercise 3.5.

Exercise 3.6

The partial derivatives of the adsorption isotherms are given by Eq. (3.11) with $\bar{\lambda}_A = \bar{\lambda}_B$. After substitution in Eq. (3.48) one obtains:

$$C_A \left(\frac{dC_B}{dC_A} \right)^2 - \frac{\tilde{K}_A - \tilde{K}_B + \tilde{K}_A \tilde{K}_B (C_B - C_A)}{\tilde{K}_A \tilde{K}_B} \frac{dC_B}{dC_A} - C_B = 0$$

Defining $y = \dfrac{dC_B}{dC_A}$ one obtains: $C_A y^2 - \left[\frac{\tilde{K}_A - \tilde{K}_B}{\tilde{K}_A \tilde{K}_B} + C_B - C_A \right] y - C_B = 0$.

The hodograph is made of straight lines if y is independent of A and B. This is satisfied if $y' = \dfrac{dy}{dC_A} = 0$. The same principle applies for C_B.

Differentiation of the previous equation with respect to C_A gives:

$$C_A \, 2yy' + y^2 - \left[\frac{\tilde{K}_A - \tilde{K}_B}{\tilde{K}_A \tilde{K}_B} + C_B - C_A \right] y' - [y - 1] \, y - y = 0$$

Thus $y' \left[2C_A y - \left[\frac{\tilde{K}_A - \tilde{K}_B}{\tilde{K}_A \tilde{K}_B} + C_B - C_A \right] \right] = 0$ and consequently $y' = 0$. This shows that the hodograph is made up of straight lines.

Exercise 3.7

In Eq. (3.44) the partial derivatives of the adsorption isotherms are estimated at the feed concentration (C_A^F, C_B^F). Some numerical values of θ_+, θ_- are thus obtained, and σ_+, σ_- are calculated from Eq. (3.45).

The retention times are given by: $t_{R+} = \dfrac{L_{col}}{\sigma_+}$ and $t_{R-} = \dfrac{L_{col}}{\sigma_-}$. Velocities will be expressed in m/min.

Exercise 3.8

The zero retention time is $t_0 = \frac{\varepsilon_e V_{col}}{Q} = 0.4$ min. The interstitial fluid velocity is $u_e = \frac{L_{col}}{t_0} = 2.5$ m/min.

Equation (3.11) allows an estimation of the derivatives of the adsorption isotherms:

$$\alpha\colon C_A = C_B = 0 \quad \frac{\partial \overline{C}_A}{\partial C_A} = 1 \quad \frac{\partial \overline{C}_A}{\partial C_B} = 0 \quad \frac{\partial \overline{C}_B}{\partial C_A} = 0 \quad \frac{\partial \overline{C}_B}{\partial C_B} = 2$$

$$\beta\colon C_A = C_B = 1 \quad \frac{\partial \overline{C}_A}{\partial C_A} = 0.32 \quad \frac{\partial \overline{C}_A}{\partial C_B} = -0.16 \quad \frac{\partial \overline{C}_B}{\partial C_A} = -0.16 \quad \frac{\partial \overline{C}_B}{\partial C_B} = 0.48$$

For each composition α or β, Eq. (3.44) has two roots:

$$\alpha\colon \theta_- = 2 \quad \theta_+ = 1 \qquad \beta\colon \theta_- = 0.5788 \quad \theta_+ = 0.2211$$

leading to the front velocities obtained with Eq. (3.45):

$$\alpha\colon \sigma_- = 0.625 \quad \sigma_+ = 1 \qquad \beta\colon \sigma_- = 1.338 \quad \sigma_+ = 1.877 \tag{3.93}$$

The possible slopes $\frac{dC_A}{dC_B}$ at compositions α and β are given by Eq. (3.46):

$$\alpha\colon \quad \left.\frac{dC_B}{dC_A}\right|_- = \infty \qquad \left.\frac{dC_B}{dC_A}\right|_+ = 0$$

$$\beta\colon \quad \left.\frac{dC_B}{dC_A}\right|_- = -1.618 \qquad \left.\frac{dC_B}{dC_A}\right|_+ = 0.618$$

As we have shown in Exercise 3.6 that the characteristic paths (C_B,C_A) are straight lines, they are directly constructed using the slopes at points α and β.

The two possible plateau compositions are then obtained (see Figure 3.30):

$$P_2\colon \quad C_A = 0 \quad C_B = 0.382 \qquad P_1\colon \quad C_A = 1.618 \quad C_B = 0$$

For P_1 and P_2, one can determine:

$$P_1\colon \frac{\partial \overline{C}_A}{\partial C_A} = 0.723 \quad \frac{\partial \overline{C}_A}{\partial C_B} = 0 \quad \frac{\partial \overline{C}_B}{\partial C_A} = -0.2 \quad \frac{\partial \overline{C}_B}{\partial C_B} = 1.05$$

$$P_2\colon \frac{\partial \overline{C}_A}{\partial C_A} = 0.305 \quad \frac{\partial \overline{C}_A}{\partial C_B} = -0.495 \quad \frac{\partial \overline{C}_B}{\partial C_A} = 0 \quad \frac{\partial \overline{C}_B}{\partial C_B} = 1.105$$

which leads to the following velocities:

$$P_1\colon \sigma_- = 0.94 \quad \sigma_+ = 1.714 \quad P_2\colon \sigma_- = 0.972 \quad \sigma_+ = 1.199$$

The hodograph of the system is completely defined.

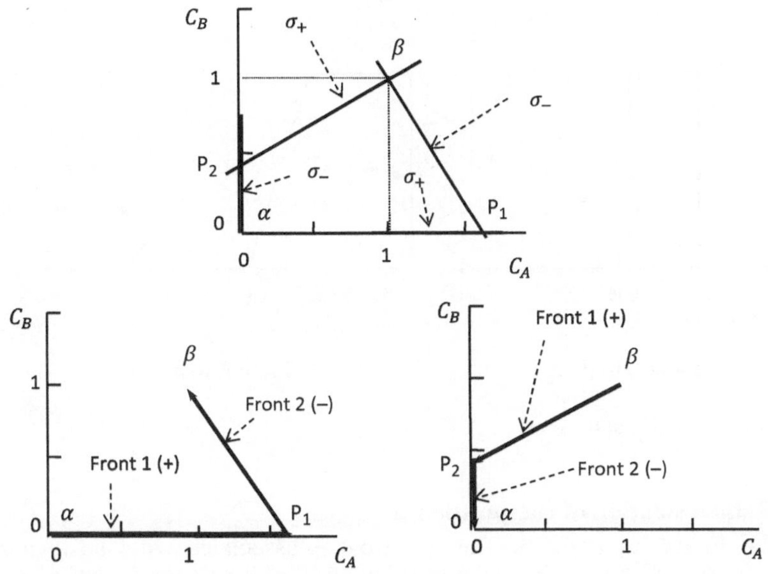

Figure 3.30 Top figure: generic construction. Bottom left: from α to β. Bottom right: from β to α.

Initial state: solution of composition α

One has to start at point α. The first front is associated with the highest velocity, thus σ_+, at point α. One has consequently to select the path on the bottom left of Figure 3.30.

As $\sigma_+(\alpha) < \sigma_+(P_1)$, the front is compressive and its velocity is given by Eq. (3.49):

$$\sigma(F_1) = \frac{2.5}{1 + F_e \dfrac{\bar{C}_A(P_1) - \bar{C}_A(\alpha)}{C_A(P_1) - C_A(\alpha)}} = \frac{2.5}{1 + 1.5\dfrac{0.894}{1.618}} = 1.367$$

As $\sigma_-(P_1) < \sigma_-(\beta)$, the second front is also compressive, its velocity being given by:

$$\sigma(F_2) = \frac{2.5}{1 + F_e \dfrac{\bar{C}(\beta) - \bar{C}(P_1)}{C(\beta) - C(P_1)}}$$

$$= \frac{2.5}{1 + 1.5\dfrac{0.8 - 0}{1 - 0}} = 1.136 \text{ calculated with A}$$

$$= \frac{2.5}{1 + 1.5\dfrac{0.4 - 0.894}{1 - 1.618}} = 1.136 \text{ calculated with B}$$

The retention times associated with the two fronts are:

$$t_R(F_1) = \frac{L_{col}}{\sigma(F_1)} = 0.731 \quad \text{min} \qquad t_R(F_2) = \frac{L_{col}}{\sigma(F_2)} = 0.88 \quad \text{min}$$

This allows construction of the chromatogram qualitatively given in Figure 3.31 (left).

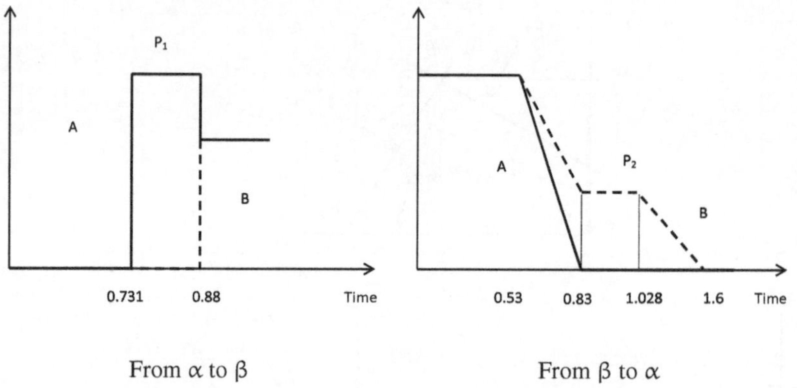

Figure 3.31 Breakthrough curves.

Initial state: solution of composition β

One has to start at point β. The first front is associated with the highest velocity, thus σ_+, at point β. One has consequently to select the path on the bottom right of Figure 3.30.

As $\sigma_+(\beta) > \sigma_+(P_2)$, the front F_1 is a simple wave. It starts at $t_R = \dfrac{L_{col}}{\sigma_+(\beta)} = 0.53$ min and it stops at $t_R = \dfrac{L_{col}}{\sigma_+(P_2)} = 0.83$ min. Front F_2 is also dispersive. It starts at $t_R = \dfrac{L_{col}}{\sigma_-(P_2)} = 1.028$ min and it stops at $t_R = \dfrac{L_{col}}{\sigma_-(\alpha)} = 1.6$ min. This allows construction of the chromatogram qualitatively given in Figure 3.31 (right), curves being replaced by straight lines.

Exercise 3.9

The interstitial fluid velocity is $u_e = \dfrac{L_{col}}{t_0} = 2.5$ m/min.

The numerical resolution of Eq. (3.48) results in the hodograph given in Figure 3.32.

The characteristic paths are no longer straight lines, but the derivation of the plateau composition is not difficult. One obtains $C_A^{P1} = 6.9$ g/l, $C_B^{P1} = 0$, $C_A^{P2} = 0$ g/l, $C_B^{P2} = 2.55$ g/l.

Derivatives of the adsorption isotherms at the key compositions are:

$$\alpha: \quad \frac{\partial \bar{C}_A}{\partial C_A} = 1 \qquad \frac{\partial \bar{C}_A}{\partial C_B} = 0 \qquad \frac{\partial \bar{C}_B}{\partial C_A} = 0 \qquad \frac{\partial \bar{C}_B}{\partial C_B} = 2.25$$

$$\beta: \quad \frac{\partial \bar{C}_A}{\partial C_A} = 0.083 \qquad \frac{\partial \bar{C}_A}{\partial C_B} = -0.069 \qquad \frac{\partial \bar{C}_B}{\partial C_A} = -0.069 \qquad \frac{\partial \bar{C}_B}{\partial C_B} = 0.347$$

$$P_1: \quad \frac{\partial \bar{C}_A}{\partial C_A} = 0.05 \qquad \frac{\partial \bar{C}_A}{\partial C_B} = -0.348 \qquad \frac{\partial \bar{C}_B}{\partial C_A} = -0.0 \qquad \frac{\partial \bar{C}_B}{\partial C_B} = 0.699$$

$$P_2: \quad \frac{\partial \bar{C}_A}{\partial C_A} = 0.2817 \qquad \frac{\partial \bar{C}_A}{\partial C_B} = -0.0 \qquad \frac{\partial \bar{C}_B}{\partial C_A} = -0.2023 \qquad \frac{\partial \bar{C}_B}{\partial C_B} = 0.4087$$

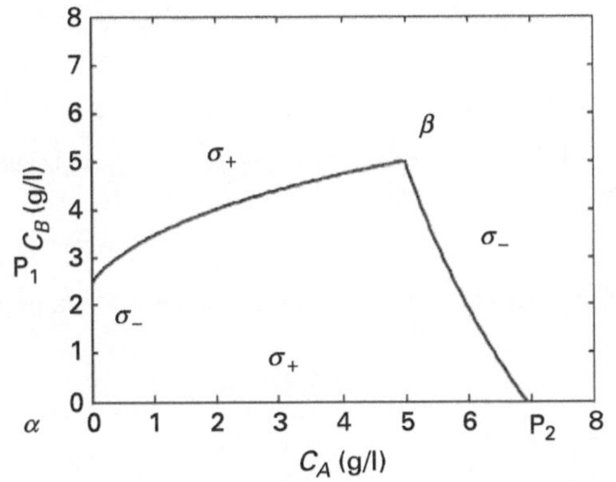

Figure 3.32 Hodograph associated with Exercise 3.9.

Velocities (m/min) determined at the key compositions are:

$$\alpha\text{: } \sigma_- = 0.57 \quad \sigma_+ = 1 \qquad \beta\text{: } \sigma_- = 1.62 \quad \sigma_+ = 2.27$$

$$\mathrm{P}_1\text{: } \sigma_- = 1.22 \quad \sigma_+ = 2.32 \quad \mathrm{P}_2\text{: } \sigma_- = 1.55 \quad \sigma_+ = 1.76$$

Initial state: solution of composition β

This regeneration step is associated with dispersive fronts. The velocities determined above allow positioning of the fronts following exactly the procedure used in Exercise 3.8.

Initial state: solution of composition α

The situation is a bit more complex because two shocks must be considered.

The composition of plateau P* must satisfy at the same time Eq. (3.52) and Eq. (3.53). The initial state being associated with a solute-free fluid, Eq. (3.52) has no solution unless $C_A^{P^*} = C_A^I = 0$ or $C_B^{P^*} = C_B^I = 0$. In our situation it is pretty clear that $C_B^{P^*} = C_B^I = 0$ must be selected, so that $\overline{C}_B^{P^*} = 0$. The determination of $C_A^{P^*} = 0$ is performed so as to satisfy Eq. (3.53), which can be written:

$$\frac{\overline{C}_A^{P^*} - 0.588}{C_A^{P^*} - 5} = \frac{0 - 2.426}{0 - 5} \tag{3.94}$$

One obtains finally $C_A^{P^*} \approx 7$ g/l.

Exercise 3.10

We have $u_e = 2.5$ m/min.

Calculation of plateaus P$_1$ *and* P$_2$*: important velocities*

From the results of Exercise 3.8, we know that the two plateaus have compositions:

$$\mathrm{P}_1 : C_A = 1.618 \quad C_B = 0 \qquad \mathrm{P}_2 : C_A = 0 \quad C_B = 0.382 \quad \text{g/l}$$

The relevant velocities (m/min) are given by:

$$\mathrm{I}: \sigma_- = 0.625 \quad \sigma_+ = 1 \qquad \mathrm{F}: \quad \sigma_- = 1.338 \quad \sigma_+ = 1.877$$
$$\mathrm{P}_1: \sigma_- = 0.94 \quad \sigma_+ = 1.714 \qquad \mathrm{P}_2: \quad \sigma_- = 0.972 \quad \sigma_+ = 1.199$$

Calculation of Z_{plat}

The propagation velocities (m/min) of the relevant compressive and dispersive fronts are given by:

$$\sigma_{F_d^1(F)} = \sigma(F) = 1.877$$

$$\sigma_{F_c^2} = \frac{u_e}{1 + F_e \dfrac{\overline{C}_B^F - \overline{C}_B^{P1}}{C_B^F - C_B^{P1}}} = \frac{2.5}{1 + 1.5\dfrac{0.8 - 0}{1 - 0}} = 1.136$$

Using system (3.57), one obtains:

$$Z_{plat} = \frac{0.1}{1/1.136 - 1/1.877} = 0.288\,\mathrm{m} \qquad t_{plat} = \frac{0.288}{1.136} = 0.254\,\mathrm{min}$$

In other words, for an injection time of 0.1 min, the feed plateau disappears at a column length of 0.288 m.

Calculation of Z_{resol}

This is performed by using Eq. (3.58). Solute B being alone in this part of the chromatogram, the integral contained in Eq. (3.58) is simply calculated by:

$$\int_0^{P_2} \frac{dC}{\sigma_{F_d^2(C)}} = \frac{1}{u_e}\int_0^{P_2}\left(1 + F_e \left.\frac{d\overline{C}_B}{dC_B}\right|_{C_A=0}\right) dC_B$$
$$= \frac{1}{u_e}\int_0^{P_2}\left(1 + F_e \frac{\overline{N}_B\tilde{K}_B}{\left(1 + \tilde{K}_B C_B\right)^2}\right) dC_B$$
$$= \frac{1}{2.5}\left[0.382 + 1.5 \times 2 \times \left(1 - \frac{1}{1 + 1 \times 0.382}\right)\right] = 0.484$$

and from Eq. (3.58), one obtains:

$$Z_{resol} = \frac{t_{inj} C_B^F}{C_B^{P2}\left(\dfrac{1}{\sigma_{F_d^2(\mathrm{P}_2)}} - \dfrac{1}{\sigma_{F_d^1(\mathrm{P}_2)}}\right) + \int_0^{P_2} \dfrac{dC}{\sigma_{F_d^2(C)}} - \dfrac{C_B^{P2}}{\sigma_{F_d^2(\mathrm{P}_2)}}}$$
$$= \frac{0.1 \times 1}{0.382 \times (1/0.972 - 1/1.199) + 0.484 - 0.382/0.972} = 0.604\,\mathrm{m}$$

Exercise 3.11

For a single-solute system, Eq. (3.69) becomes:

$$\frac{\tilde{K}_A \overline{C}_A}{\overline{N}\tilde{K}_A - \omega} = 1$$

thus

$$\omega = \tilde{K}_A \left(\overline{N} - \overline{C}_A\right) = \frac{\overline{N}\tilde{K}_A}{1 + \tilde{K}_A C_A}$$

Consequently, one has:

$$\omega^I = \frac{\overline{N}\tilde{K}_A}{1 + \tilde{K}_A C_A^I} \quad \text{and} \quad \omega^F = \frac{\overline{N}\tilde{K}_A}{1 + \tilde{K}_A C_A^F}$$

The unique plateau concentration is given by Eq. (3.71), which leads to:

$$C_A^P = C_A^I \frac{1 - \dfrac{\overline{N}\tilde{K}_A}{\omega^F}}{1 - \dfrac{\overline{N}\tilde{K}_A}{\omega^I}} = C_A^F$$

We have thus shown that the only possible plateau is the feed composition. Not a big discovery!

- If $\omega^I > \omega^F$ (thus if $C_A^I < C_A^F$), the front is a shock and its velocity is given by Eq. (3.73):

$$\sigma_1 = \frac{u_e}{1 + F_e \dfrac{1}{1 + \tilde{K}_A C_A^F} \dfrac{\overline{N}\tilde{K}_A}{1 + \tilde{K}_A C_A^I}}$$

  which is identical to the general expression for a shock velocity:

$$\sigma_1 = \frac{u_e}{1 + F_e \dfrac{\overline{C}_A^F - \overline{C}_A^I}{C_A^F - C_A^I}}$$

- If $\omega^I < \omega^F$ (thus if $C_A^I > C_A^F$), the front is a simple wave and its velocity is given by Eq. (3.72):

$$\sigma_1 = \frac{u_e}{1 + F_e \dfrac{\omega_k}{1 + \tilde{K}_A C_A}} = \frac{u_e}{1 + F_e \dfrac{\overline{N}\tilde{K}_A}{\left(1 + \tilde{K}_A C_A\right)^2}}$$

  which is identical to the general expression for a simple wave velocity:

$$\sigma_1 = \frac{u_e}{1 + F_e \bar{f}'(C_A)}$$

Exercise 3.12

One calculates $u_e = 2.5$ m/min, $t_0 = 0.4$ min. All concentrations are expressed in g/l.

Experiment 1

The lumped solid-phase concentrations in equilibrium with the initial fluid concentrations are:

$$\overline{C}_A = 0 \qquad \overline{C}_B = 0 \qquad \overline{C}_C = 0$$

whereas the lumped solid-phase concentrations in equilibrium with the feed concentration are:

$$\overline{C}_A = 0.1428 \qquad \overline{C}_B = 0.2857 \qquad \overline{C}_C = 0.4285$$

For the initial conditions, Eq. (3.69) is written:

$$\frac{1 \times 0}{1 \times 1 - \omega} + \frac{2 \times 0}{1 \times 2 - \omega} + \frac{3 \times 0}{1 \times 3 - \omega} = 1$$

The three possible roots are $\omega_1^I = 1$, $\omega_2^I = 2$, $\omega_3^I = 3$.

For the feed composition, Eq. (3.69) becomes:

$$\frac{1 \times 0.1428}{1 \times 1 - \omega} + \frac{2 \times 0.2857}{1 \times 2 - \omega} + \frac{3 \times 0.4285}{1 \times 3 - \omega} = 1$$

The three roots are $\omega_1^F = 0.2748$, $\omega_2^F = 1.2709$, $\omega_3^F = 2.4543$.

One can check that $0 \le \omega_1 \le \overline{N}_1\tilde{K}_1 \le \cdots \le \omega_n \le \overline{N}_n\tilde{K}_n$ for both the I and F compositions.

The concentrations of solute i in the three plateaus P(1), P(2) and P(3) are given by Eq. (3.71):

$$C_i^{P(1)} = C_i^I \frac{1 - \dfrac{\overline{N}_i\tilde{K}_i}{\omega_1^F}}{1 - \dfrac{\overline{N}_i\tilde{K}_i}{\omega_1^I}} \qquad i = \text{A,B,C}$$

$$C_i^{P(2)} = C_i^I \frac{1 - \dfrac{\overline{N}_i\tilde{K}_i}{\omega_1^F}}{1 - \dfrac{\overline{N}_i\tilde{K}_i}{\omega_1^I}} \; \frac{1 - \dfrac{\overline{N}_i\tilde{K}_i}{\omega_2^F}}{1 - \dfrac{\overline{N}_i\tilde{K}_i}{\omega_2^I}} \qquad i = \text{A,B,C}$$

$$C_i^{P(3)} = C_i^I \frac{1 - \dfrac{\overline{N}_i\tilde{K}_i}{\omega_1^F}}{1 - \dfrac{\overline{N}_i\tilde{K}_i}{\omega_1^I}} \; \frac{1 - \dfrac{\overline{N}_i\tilde{K}_i}{\omega_2^F}}{1 - \dfrac{\overline{N}_i\tilde{K}_i}{\omega_2^I}} \; \frac{1 - \dfrac{\overline{N}_i\tilde{K}_i}{\omega_3^F}}{1 - \dfrac{\overline{N}_i\tilde{K}_i}{\omega_3^I}} \qquad i = \text{A,B,C}$$

First plateau One obtains $C_B^{P(1)} = C_C^{P(1)} = 0$ directly, but the expression is undetermined for $C_A^{P(1)}$. In order to determine $C_A^{P(1)}$, let us note that, in order to satisfy Eq. (3.69) when the inlet solution is solute-free, the root ω_1^I must satisfy $\tilde{K}_A\overline{C}_A^I = \tilde{K}_A\overline{N}_A - \omega_1^I$. This leads to $1 - \dfrac{\overline{N}_A\tilde{K}_A}{\omega_1^I} = -\dfrac{\tilde{K}_A\overline{C}_A^I}{\omega_1^I}$ and consequently to:[18]

$$C_A^{P(1)} = C_A^I\omega_1^I \frac{1 - \dfrac{\overline{N}_A\tilde{K}_A}{\omega_1^F}}{-\tilde{K}_A\overline{C}_A^I} = 2.64$$

Second plateau One obtains $C_C^{(P2)} = 0$, but the other expressions are undetermined. Using the same technique as previously, one obtains:

$$C_A^{P(2)} = \frac{\dfrac{\overline{N}_A\tilde{K}_A}{\omega_1^F} - 1}{\tilde{K}_A} \; \frac{1 - \dfrac{\overline{N}_A\tilde{K}_A}{\omega_2^F}}{1 - \dfrac{\overline{N}_B\tilde{K}_B}{\omega_2^I}} = 1.12$$

$$C_B^{P(2)} = \frac{1 - \dfrac{\overline{N}_B\tilde{K}_B}{\omega_1^F}}{1 - \dfrac{\overline{N}_B\tilde{K}_B}{\omega_1^I}} \; \frac{\dfrac{\overline{N}_B\tilde{K}_B}{\omega_2^F} - 1}{\tilde{K}_B} = 1.80$$

Third plateau The three expressions are now undetermined. The same procedure as the one presented above leads to:

$$C_A^{P(3)} = \frac{\dfrac{\overline{N}_A\tilde{K}_A}{\omega_1^F} - 1}{\tilde{K}_A} \; \frac{1 - \dfrac{\overline{N}_A\tilde{K}_A}{\omega_2^F}}{1 - \dfrac{\overline{N}_A\tilde{K}_A}{\omega_2^I}} \; \frac{1 - \dfrac{\overline{N}_A\tilde{K}_A}{\omega_3^F}}{1 - \dfrac{\overline{N}_A\tilde{K}_A}{\omega_3^I}} = 1.00$$

$$C_B^{P(3)} = \frac{1 - \dfrac{\overline{N}_B\tilde{K}_B}{\omega_1^F}}{1 - \dfrac{\overline{N}_B\tilde{K}_B}{\omega_1^I}} \; \frac{\dfrac{\overline{N}_B\tilde{K}_B}{\omega_2^F} - 1}{\tilde{K}_B} \; \frac{1 - \dfrac{\overline{N}_B\tilde{K}_B}{\omega_3^F}}{1 - \dfrac{\overline{N}_B\tilde{K}_B}{\omega_3^I}} = 1.00$$

$$C_C^{P(3)} = \frac{1 - \dfrac{\overline{N}_C\tilde{K}_C}{\omega_1^F}}{1 - \dfrac{\overline{N}_C\tilde{K}_C}{\omega_1^I}} \; \frac{1 - \dfrac{\overline{N}_C\tilde{K}_C}{\omega_2^F}}{1 - \dfrac{\overline{N}_C\tilde{K}_C}{\omega_2^I}} \; \frac{\dfrac{\overline{N}_C\tilde{K}_C}{\omega_3^F} - 1}{\tilde{K}_C} = 1.00$$

Since we have $\omega_k^I > \omega_k^F$, all fronts are shocks and the velocities are given by Eq. (3.73), which leads to:

[18] Similarly, one has $1 - \dfrac{\overline{N}_B\tilde{K}_B}{\omega_2^I} = -\dfrac{\tilde{K}_B\overline{C}_B^I}{\omega_2^I}$ and $1 - \dfrac{\overline{N}_C\tilde{K}_C}{\omega_3^I} = -\dfrac{\tilde{K}_C\overline{C}_C^I}{\omega_3^I}$.

$$\sigma_1 = \frac{2.5}{1 + 1.5\dfrac{1}{1 + 1 \times 2.64 + 2 \times 0 + 3 \times 0}} = 1.77$$

(thus $t_{R1} = L_{col}/\sigma_1 = 1/1.77 = 0.565$)

$$\sigma_2 = \frac{2.5}{1 + 1.5\dfrac{2}{1 + 1 \times 1.12 + 2 \times 1.80 + 3 \times 0}} = 1.628$$

(thus $t_{R2} = L_{col}/\sigma_2 = 1/1.63 = 0.61$)

$$\sigma_3 = \frac{1}{1 + 1.5\dfrac{3}{1 + 1 \times 1.00 + 2 \times 1.00 + 3 \times 1.00}} = 1.52$$

(thus $t_{R3} = L_{col}/\sigma_3 = 1/1.52 = 0.66$)

Calculating the retention times with $t_R = L_{col}/\sigma$, one can draw the chromatogram given in Figure 3.33.

Experiment 2

The same procedure is used and one obtains:

	Front 1	Front 2	Front 3
ω_k^I	0.1428	0.2857	0.4285
ω_k^F	1	2	3
Nature	disp.	disp.	disp.
$C_A^{P(k)}$	0	0	0
$C_B^{P(k)}$	0.16	0	0
$C_C^{P(k)}$	0.2	0.07	0

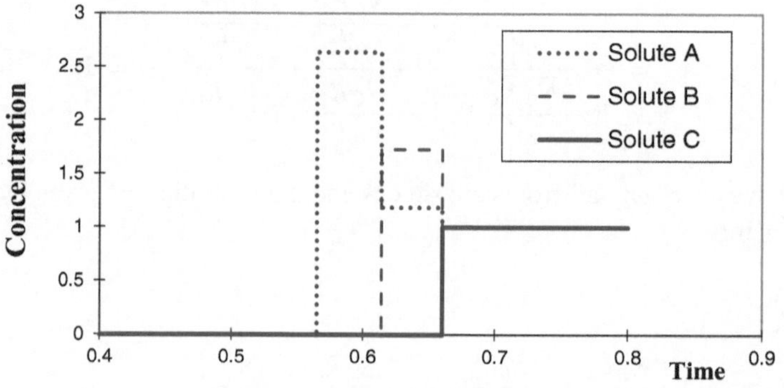

Figure 3.33 Exercise 3.12, experiment 1.

The front velocities are calculated from Eq. (3.72).

Front 1 At the beginning of the front, the concentrations are the initial concentrations, and (3.72) gives:

$$\sigma_1(\text{begin}) = \frac{2.5}{1 + 1.5\dfrac{0.1428}{1 + 1 \times 1 + 2 \times 1 + 3 \times 1}} = 2.06$$

At the end of the front, the concentrations are (0, 0.16, 0.2), leading to another ω value that can be calculated by solving Eq. (3.69); one obtains $\omega = 1$. Equation (3.72) gives:

$$\sigma_1(\text{end}) = \frac{2.5}{1 + 1.5\dfrac{1}{1 + 1 \times 0 + 2 \times 0.16 + 3 \times 0.2}} = 1.40$$

Consequently, front 1 begins at time 0.48 and ends at time 0.71.

Front 2 At the beginning of the front, the concentrations are (0, 0.16, 0.2), and Eq. (3.72) gives:

$$\sigma_2(\text{begin}) = \frac{2.5}{1 + 1.5\dfrac{0.2857}{1 + 1 \times 0 + 2 \times 0.16 + 3 \times 0.2}} = 0.975$$

At the end of the front, the concentrations are (0, 0, 0.07), and the velocity is given by:

$$\sigma_2(\text{end}) = \frac{2.5}{1 + 1.5\dfrac{2}{1 + 1 \times 0 + 2 \times 0 + 3 \times 0.07}} = 0.718$$

Consequently, front 2 begins at time 1.025 and ends at time 1.39.

Front 3 At the beginning of the front, the concentrations are (0, 0, 0.07), and (3.72) gives:

$$\sigma_3(\text{begin}) = \frac{2.5}{1 + 1.5\dfrac{0.4285}{1 + 1 \times 0 + 2 \times 0.16 + 3 \times 0.2}} = 0.53$$

At the end of the front, the concentrations are (0, 0, 0), and the velocity is given by:

$$\sigma_3(\text{end}) = \frac{2.5}{1 + 1.5\dfrac{3}{1 + 1 \times 0 + 2 \times 0 + 3 \times 0.}} = 0.4545$$

Consequently, front 3 begins at time 1.88 and ends at time 2.20.

The chromatogram is given in Figure 3.34 (the dispersive fronts are approximated by straight lines).

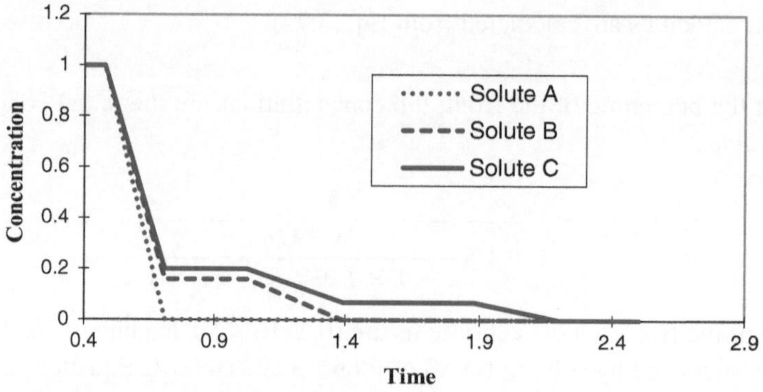

Figure 3.34 Exercise 3.12, experiment 2.

Experiment 3

One obtains:

	Front 1	Front 2	Front 3
ω_k^I	0.35	1.54	2.10
ω_k^F	0.57	1.07	2.88
Nature	disp.	comp.	disp.
C_A^{Pk}	0.417	0.08	0.1
C_B^{Pk}	0.054	0.158	1.0
C_C^{Pk}	0.574	1.09	0.1

Front 1 The velocity of the beginning of the front is:

$$\sigma_1(\text{begin}) = \frac{2.5}{1 + 1.5\dfrac{0.35}{1 + 1 \times 1 + 2 \times 0.1 + 3 \times 1}} = 2.27$$

The velocity of the end of the front is:

$$\sigma_1(\text{end}) = \frac{2.5}{1 + 1.5\dfrac{0.57}{1 + 1 \times 0.417 + 2 \times 0.054 + 3 \times 0.574}} = 1.98$$

Front 2 The velocity of the front is:

$$\sigma_2 = \frac{2.5}{1 + 1.5\dfrac{1.54}{1 + 1 \times 0.08 + 2 \times 0.158 + 3 \times 1.09}} = 1.67$$

Front 3 The velocity of the beginning of the front is:

$$\sigma_3(\text{begin}) = \frac{2.5}{1 + 1.5\dfrac{2.102}{1 + 1 \times 0.08 + 2 \times 0.158 + 3 \times 1.09}} = 1.49$$

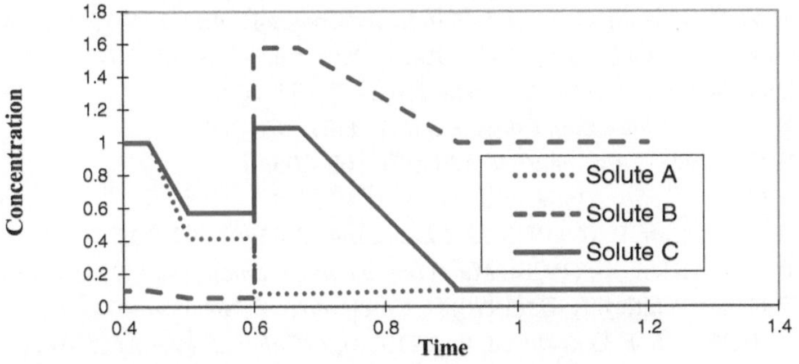

Figure 3.35 Exercise 3.12, experiment 3.

The velocity of the end of the front is:

$$\sigma_3(\text{end}) = \frac{2.5}{1 + 1.5\dfrac{2.88}{1 + 1 \times 0.1 + 2 \times 1 + 3 \times 0.1}} = 1.1$$

The chromatogram is given in Figure 3.35 (the dispersive fronts are approximated by straight lines).

Exercise 3.13

For $C_A^F = C_B^F = 1$ g/l, the roots of the characteristic equation (3.69) are: $\omega_1^F = 0.5528$ and $\omega_2^F = 1.4472$. From the expressions in Tables 3.1 and 3.2, and using $u_e = 2.5$, $\overline{K}_B = 1$, $\overline{K}_B = 2$, $\rho = 1.5/(1 \times 2)$, one obtains:

$$Z_{plat} = \frac{t_{inj} u_e}{\rho \omega_1^F \omega_2^F (\overline{K}_B - \omega_1^F)} = 0.2879 \text{ m}$$

$$t_{plat} = \frac{(1 + \rho \omega_1^F \omega_2^F \overline{K}_B)}{\rho \omega_1^F \omega_2^F (\overline{K}_B - \omega_1^F)} t_{inj} = 0.2534 \text{ min}$$

$$Z_{resol} = \frac{t_{inj} u_e (\overline{K}_B - \omega_1^F)}{\rho \omega_1^F \omega_2^F (\overline{K}_B - \overline{K}_A)^2} = 0.603 \text{ m}$$

confirming the values obtained in Exercise 3.10.

References

Bailly, M. and Tondeur, D. (1981), *Chem. Eng. Sci.* **36**, 455–469.
Bryant, S., Schechter, R. and Lake, L. (1986), *AIChE J.* **32**(5), 751–764.
Carta, G. and Jungbauer, A. (2011), *Protein Chromatography*, Wiley-VCH.
De Vault, D. (1943), *J. Am. Chem. Soc.* **65**(4), 532–540.
Dodds, J. and Tondeur, D. (1972), *Chem. Eng. Sci.* **27**, 2291–2298.
Glueckauf, E. (1949), *Discuss. Faraday Soc.* **7**, 12–25.

Golshan-Shirazi, D. and Guiochon, G. (1989a), *J. Chromatogr.* **484**, 125–151.

Golshan-Shirazi, S. and Guiochon, G. (1989b), *J. Phys. Chem.* **93**(10), 4143 – 4157.

Grevillot, G. and Tondeur, D. (1974), *Chem. Eng. J.* **7**, 187–200.

Helfferich, F. G. (1967), *Ind. Eng. Chem. Fundam.* **6**(3), 362–364.

Helfferich, F. G. (1997), *J. Chromatogr. A* **768**(2), 169–205.

Helfferich, F. G. (1981), *Soc. Pet. Eng. J.* **21**(1), 51–62.

Helfferich, F. G. and Carr, P. W. (1993), *J. Chromatogr. A* **629**(2), 97–122.

Helfferich, F. G. and Klein, G. (1970), *Multicomponent Chromatography*, Marcel Dekker.

Helfferich, F. G. and Whitley, R. D. (1996), *J. Chromatogr. A* **734**(1), 7–47.

Hwang, Y. L., Helfferich, F. G. and Leu, R. J. (1988), *AIChE J.* **34**(10), 1615–1626.

Klein, G. (1981). In A. E. Rodrigues and D. Tondeur, eds., *Percolation Processes: Theory and Applications*, vol. 33 of *NATO ASI Series, Series E: Applied Sciences*, Sijthoff & Noordhoff, pp. 363–423.

Klein, G. (1986). In A. E. Rodrigues, ed., *Ion Exchange: Science and Technology*, vol. 107 of *NATO ASI Series*, Martinus Nijhoff.

Klein, G., Tondeur, D. and Vermeulen, T. (1967), *Ind. Eng. Chem. Fundam.* **6**, 339–351.

Lax, P. D. (1957), *Comm. Pure Appl. Math.* **10**(4), 537–566.

Mazzotti, M. (2006), *Ind. Eng. Chem. Res.* **45**, 5332–5350.

Mazzotti, M. (2009), *Ind. Eng. Chem. Res.* **48**, 7733–7752.

Mazzotti, M., Tarafder, A., Cornel, J., Gritti, F. and Guiochon, G. (2010), *J. Chromatogr. A* **1217**, 2002–2012.

Nicoud, R. M., Jaubert, J.-N., Rupprecht, I. and Kinkel, J. N. (1996), *Chirality* **8**(3), 234–243.

Rajendran, A. and Mazzotti, M. (2011), *Ind. Eng. Chem. Res.* **50**, 352–377.

Rhee, H. K. (1981). In A. E. Rodrigues and D. Tondeur, eds., *Percolation Processes: Theory and Applications*, vol. 33 of *NATO ASI Series, Series E: Applied Sciences*, Sijhoff & Noordhoff, pp. 285–328.

Rhee, H.-K., Aris, R. and Amundson, N. (1982), *AIChE J.* **28**, 423.

Rhee, H.-K., Aris, R. and Amundson, N. R. (1970), *Philos. Trans. R. Soc., A* **267**, 419–455.

Rhee, H.-K., Aris, R. and Amundson, N. R. (1989), *First-Order Partial Differential Equations. Vol. II: Theory and Applications of Hyperbolic Systems of Quasilinear Equations*, Prentice-Hall.

Ruthven, D. M. (1984), *Principles of Adsorption and Adsorption Processes*, John Wiley & Sons.

Schweich, D., Jauzein, M. and Sardin, M. (1993a), *Water Resour. Res.* **29**(3), 723–733.

Schweich, D., Jauzein, M. and Sardin, M. (1993b), *Water Resour. Res.* **29**(3), 735–741.

Siitonen, J. and Sainio, T. (2011), *J. Chromatogr. A* **1218**, 6379–6387.

Tondeur, D. (1970), *Chem. Eng. J.* **1**(4), 337–346.

Tondeur, D. and Bailly, M. (1986). In A. E. Rodrigues, ed., *Ion Exchange: Science and Technology*, vol. 107 of *NATO ASI Series*, Martinus Nijhoff.

Tondeur, D. and Klein, G. (1967), *Ind. Eng. Chem. Fundam.* **6**(3), 351–361.

Varma, A. and Morbidelli, M. (1997), *Mathematical Methods in Chemical Engineering*, Oxford University Press.

Yu, Q. and Wang, N. L. (1986), *Sep. Purif. Methods* **15**, 127.

Zhang, W., Shan, Y. and Seidel-Morgenstern, A. (2006), *J. Chromatogr. A* **1107**, 216–225.

4 Fluid–solid phase equilibria

The roots . . .

4.1 Introduction

With the expression "fluid–solid phase equilibria", we refer to what is often designated in chromatography as "adsorption isotherms". The latter term is so widely used that it can easily be used in an inappropriate way. In a chromatographic process, the interactions of solutes between fluid and solid phases can include mechanisms that can be different from adsorption, including ion exchange, ion or size exclusion, chemical equilibria in the fluid phase and complexation on the surface. In order to avoid confusion, we will use the general term "fluid–solid phase equilibria" and restrict the use of "adsorption isotherm" to situations where an adsorption mechanism is likely to be the predominant factor.

In addition, "adsorption *isotherm*" suggests that a single univariate function relates the fluid-phase concentration to the solid-phase concentration (lumped or not) of a given solute, provided that temperature is kept constant. This can be misleading. Certainly temperature must be kept constant, but pressure must be kept constant as well when working with supercritical fluids. The same applies to normality in ion exchange, as explained in Section 4.5. Additionally, when working with multi-solute mixtures, relating a given solute concentration in the solid phase (lumped or not) to solute concentration in the fluid phase requires that all other concentrations, not only temperature, must be kept constant.

In Chapter 1 we presented basic concepts for describing chromatographic systems as well as simple considerations for relating concentrations of solutes located in the fluid mobile phase and in the particles. Recall that the qualifier *mobile* is important, to make clear that the intragranular fluid is not part of the mobile fluid.

An important assumption that was made in Chapter 1 and that, unless otherwise specified, will be made throughout this book is the rigidity of solute molecules and the absence of conformational changes during their interaction with the chromatographic medium. Similarly, we assume that the chromatographic medium is not subject to structural changes upon interaction with the solutes. These assumptions are fulfilled in the vast majority of situations, but they might, for instance, be questionable in the case of proteins. Additional information is provided in Section 4.7.

We introduced two possible ways to model chromatographic systems:

- The first model (the *porous* model) assumes the particles to consist of a well-defined solid skeleton and intragranular pores. The solutes can be located either on the skeleton or in the intragranular fluid, so two different concentrations are defined inside the particle.
- The second model (the *lumped* model) approximates the particles by a pseudo-homogeneous phase, or lumped solid phase, comprising both the solid skeleton and the intragranular fluid.

The two approaches can be related, provided the intragranular porosity ε_i is well defined and measurable. Under these conditions, the lumped solid concentration is a volume average of the solid skeleton and intragranular fluid concentrations.

The lumped model was used in Chapter 3 as key properties of chromatograms are imposed by the relationship between the concentrations in the fluid mobile phase and the lumped solid concentrations. The porous model will be used in Chapter 5 because a detailed particle description is required to properly model intraparticular mass transfer.

As far as fluid–solid phase equilibria are concerned, we will use one or the other approach, and give specific warnings should one approach be less relevant than the other.

It is clear from Chapter 1 that precise definitions of extragranular and intragranular porosities are of utmost importance for proper description of fluid–solid phase equilibria. As a side comment, knowledge of porosities is also important for subjects not connected to fluid–solid phase equilibria. We will see in subsequent chapters that extragranular porosity influences pressure drop, whereas intragranular porosity impacts mass transfer.

Determination of these porosities is thus a prerequisite, and we now present the different ways for the chromatographer to achieve this task.

4.2 Porosity

The description of the fluid and solid zones contained in a chromatographic column at least requires knowledge of the intraparticular porosity ε_i and the extragranular porosity ε_e, the total porosity ε_T being obtained by a combination of the two:

$$\varepsilon_T = \varepsilon_e + (1 - \varepsilon_e)\varepsilon_i \tag{4.1}$$

This equation, which is a simple consequence of volume definitions given in Chapter 1, is important and is frequently used in this book.

Recall that the porous model assumes an unambiguous difference between extragranular and intragranular volumes. This distinction is only possible provided the intragranular pore diameter is significantly lower than the size of the extragranular pore space. This is generally the case, though this difference is not that big for macroporous particles having pore sizes up to about 500 nm. For these particles it is generally accepted that some (limited) convection occurs in the intragranular pores (see Chapter 5). Throughout this book, however, we will generally admit that the intragranular and extragranular porosities are unambiguously defined.

The fact that porosities are unambiguously defined does not mean that they are constant or independent of experimental conditions. For instance, gels can swell or shrink, particles can be compressed, intraparticular pores can be blocked and so on.

4.2.1 Total porosity

The simplest way to determine the column total porosity ε_T is certainly to determine the retention time (Z_tRT) associated with the injection of a "non-interacting" tracer. By this we mean a solute able to enter the intragranular porosity, but non-adsorbing or more generally non-interacting with the adsorbent. This concept has already been presented in Chapters 1 and 2.

Recall that, assuming incompressibility of the carrier fluid, the Z_tRT obtained with a non-interacting tracer is given by:

$$t_{0R} = \frac{\varepsilon_T V_{col}}{Q} \tag{4.2}$$

so the knowledge of the retention time under given experimental conditions immediately gives access to the total porosity.

The methodology is simple and efficient provided that a "non-interacting" tracer can be identified. This is not a trivial task because of a principal difficulty. We determine total porosity by measuring the retention time of a "non-interacting" tracer, but this can also be defined as a tracer having retention time given by $t_{0R} = \frac{\varepsilon_T V_{col}}{Q}$. We are thus in a catch-22 type of situation! This is why a tracer considered to be non-retained in one publication may be considered to be retained in another if a smaller retention time can be obtained with another (non-excluded!) solute. In this domain, as in many others, we cannot replace the experience and intuition of the chromatographer.

While it is very difficult to be sure, the absence of influence of concentration, eluent composition and temperature on the determined retention time is certainly a good indication of the absence of interaction, so that the determined retention time is in fact a Z_tRT.

Determination of total porosity by means of a "non-interacting tracer" is widely used, and numerous examples are found in the literature. As an illustration, let us mention:

- the use of tetrachloroethylene in normal-phase chromatography
- the frequent use of uracyl or thiourea in reversed-phase chromatography when the eluent is a mixture of water and organic solvent: when the percentage of water in the solvent is higher than, say, 85%, these two tracers may be slightly retained and ammonium nitrate can be used instead
- the possible use of a simple solvent like acetonitrile on polymeric stationary phases (Sajonz *et al.*, 1998) or benzene on reversed-phase silica (Guan and Guiochon, 1996)
- the use of more exotic compounds like cytidine-5′-monophosphate (Miyabe and Guiochon, 2000) on different gels for protein separation, or 1,3,5-tri-tert-butylbenzene on microcrystalline cellulose (Rearden *et al.*, 1998).

The variety of stationary phases and eluents used in chromatography is so wide that a complete list cannot be given. In principle, stationary-phase suppliers and experienced chromatographers always find a decent solution relatively quickly.

A slight adsorption of the tracer will certainly impact the determined value of the total porosity. For modeling purposes, this porosity being essentially used as a reference, *provided one uses a consistent theoretical framework*, it is to be expected that small uncertainties in its exact value are of minor importance, as explained in Appendix E.

The use of deuterated solvents has been proposed for the determination of the total porosity. This technique is certainly a bit expensive and, more importantly, nothing ensures in general that the solvent (deuterated or not) is not adsorbed on the solid. We therefore believe that these ideas are of limited practical value.

Finally, a method that can be used in the laboratory consists in weighing a column after equilibration with two different solvents having different densities. Assuming that the total porosity is independent of the solvent, and neglecting the possible influence of adsorption, the greater the solvent density the greater the column weight. The difference between weights gives the volume accessible to the fluid $\varepsilon_T V_{col}$ times the density difference between the two solvents. Knowing the column geometrical volume V_{col} and the density difference between the two solvents, one has thus access to the total porosity. A similar procedure consists in weighing the column equilibrated with the solvent and then weighing the adsorbent after unpacking and drying. The mass difference between the column and the dried adsorbent gives the total mass and thus the total volume of solvent that was contained in the column. This requires a proper estimate of volumes associated with fittings, frits etc.

4.2.2 Extragranular porosity

Just as total porosity can be determined by the injection of a "non-interacting" tracer, extragranular porosity can be determined by the injection of an "excluded" tracer, meaning a tracer unable to enter the intragranular porosity, because of either size or ion exclusion.

Recall that, assuming incompressibility of the carrier fluid, the retention time Z_eRT obtained with an "excluded" tracer is given by:

$$t_0 = \frac{\varepsilon_e V_{col}}{Q} \tag{4.3}$$

so that knowledge of the Z_eRT under given experimental conditions gives immediate access to the extragranular porosity. Once again, the difficulty lies in the proper selection of the appropriate tracer.

In general, the idea is to select a molecule, normally a polymer, that is large enough that it cannot enter the intragranular pores. An example is the use of polyvinyl pyrrolidone having a molecular mass of 1200 kDa (trade name Kol-90) on gels used for protein separation (Farnan *et al.*, 2002).

The principle can be refined with so-called inversed size exclusion chromatography (ISEC). The qualifier "inverse" expresses the fact that known solutions are used to probe a stationary phase, whereas size-exclusion chromatography uses a defined stationary phase to analyze an unknown solution. The idea is to inject a series of solutes of different sizes (typically polystyrene) and to analyze the relation between the retention volume $V_R = Qt_R$ and the molecular weight of the solute. Plotting the logarithm of the molecular weight versus retention volume gives two approximately linear curves. The first curve corresponds to the behavior of solutes totally excluded from the intragranular pores and the second one to the behavior of solutes able to enter the intragranular pores. The intersection between these two curves gives the extragranular column volume. The ISEC technique has been used successfully to determine extragranular porosity by (Guan and Guiochon, 1996).

In the case of ion exchange, instead of using size exclusion, one may use ion exclusion to select an "excluded" tracer. As an example, if a column is packed with resins in the H form and eluted with water or diluted hydrochloric acid, small injections of hydrochloric acid will lead to retention times that decrease when the concentration of hydrochloric acid decreases. This is due to the Donnan exclusion effect, presented in Section 4.5.4, preventing anions from entering a cationic resin at low concentration. At concentrations lower than about 0.001 eq/l, one can consider the hydrochloric acid to be practically excluded from the resin so that the tracer has access to the extragranular porosity only. At higher concentrations, hydrochloric acid can enter more freely into the intragranular pores, so the associated retention time increases.

Finally, as already mentioned, porosities are mainly used to define references, and a very precise knowledge of extragranular and total porosities is not required for the modeling of chromatographic columns, *provided that one uses a consistent theoretical framework*, as shown in Appendix E. This being said, extragranular porosities outside the range 0.35–0.40 should be viewed with suspicion.

4.2.3 Intragranular porosity

In contrast to extragranular and total porosity determinations, intragranular porosity cannot be determined by direct chromatographic measurements. Assuming that both total and extragranular porosities have been determined, a simple way of determining intragranular porosity is to use the equation $\varepsilon_T = \varepsilon_e + (1 - \varepsilon_e)\varepsilon_i$. This gives a practical way to determine internal porosity, but one must be aware that uncertainties associated with both extragranular and total porosity measurements are cumulative so that the final precision may be a bit questionable.

When either total or extragranular porosity is not available, or simply for double-checking purposes, an independent determination of intragranular porosities can be performed. Classical methods for investigating intragranular porosity (and even pore size and pore size distribution) include nitrogen sorptometry and mercury intrusion porosimetry, as described, for instance, in (Schmidt-Traub *et al.*, 2012) and (Guan-Sajonz *et al.*, 1997). These methods and associated results should in principle be available from chromatographic media suppliers. Obviously, these methods are difficult to apply to gel polymers, not only because the concept of intragranular porosity can be a bit fuzzy but also because these methods normally require drying the solid, while gels require wetness to keep their structure. (Guan-Sajonz *et al.*, 1997) compared nitrogen sorptometry and mercury intrusion porosimetry with ISEC on reversed-phase chromatography silica and concluded that experimentally determined intragranular porosities were in very close agreement.

In addition to possible imprecision associated with the determination of intragranular porosity, an important question is whether or not intragranular porosity is an intrinsic property of the chromatographic medium. The answer is only partially "yes": not only does the ability of some resins to swell or shrink depending on the contacting solvent affect intragranular porosity, but the accessible intragranular porosity can be impacted by solvent composition (Gritti *et al.*, 2003) or by the concentration of adsorbed proteins, as suspected by (Kaczmarski and Bellot, 2003). This means that intragranular porosity may vary according to the physical state of the solid and the accessible porosity may be hindered by the presence of adsorbed solutes.

4.3 Modeling adsorption isotherms

4.3.1 General considerations

The ability of a chromatographic process to separate different solutes is the consequence of the ability of these solutes to distribute themselves more or less selectively between a fluid mobile phase and a chromatographic medium. As already mentioned, this selective distribution can have very different origins including physical adsorption, hydrophobic interactions, affinity, ion exchange etc., acting alone or in combination.

In this chapter we are addressing equilibrium situations, thus situations occurring when the solutes have sufficient time to migrate between the different phases in order to find their thermodynamic equilibrium. We are thus addressing the very fundamental driver for chromatographic separations. The description of how fast a system evolves from an initial situation to the equilibrium situation is detailed in Chapter 5.

Preparative chromatography, being a purification process, is never used with solutions containing a single solute. In addition, the fact that the eluents are often mixtures of different solvents, salts and buffers makes chromatographic systems truly multi-solute systems. It is, however, useful to first study the behavior of ideal hypothetic systems containing a single solute dissolved in a non-adsorbing solvent. The important case of multi-solute systems will be dealt with in Section 4.3.3.

As explained in Chapter 1, the porous model suggests that selective interactions are taking place inside the particle, between the solutes contained in the intragranular fluid and the solutes adsorbed on the skeleton. In order to estimate the lumped solid-phase composition one needs to relate the solute concentration in the intragranular fluid, $\grave{C}_A$, to the solute concentration in the fluid mobile phase C_A. A simple assumption is that these two fluid concentrations are identical. Situations of size-exclusion or ion-exclusion chromatography, for which this assumption is wrong, will be treated in Sections 4.5 and 4.6; but in the present section we will assume that $C_A = \grave{C}_A$.

Let us briefly recall the fundamental definitions given in Chapter 1. When one uses the porous model, the adsorption isotherm is given by the function relating the concentration of solute adsorbed on the skeleton $\bar{\bar{C}}_A$ to the concentration in the intragranular fluid $\grave{C}_A$:

$$\bar{\bar{C}}_A = \bar{\bar{f}}_A(\grave{C}_A) \tag{4.4}$$

$\bar{\bar{f}}_A$ is the "phase equilibrium law" of thermodynamicists.[1]

In the lumped model one can write:

$$\bar{C}_A = \bar{f}_A(C_A) \tag{4.5}$$

so that one relates the lumped solid-phase concentration to the mobile-fluid-phase concentration. Function $\bar{f}_A$, accounting for solutes both adsorbed on the solid skeleton and located in the intragranular fluid, will be called the lumped adsorption isotherm.

Provided that the assumption $C_A = \grave{C}_A$ is satisfied, the two solid-phase concentrations are linearly related by:

$$\bar{C}_A = \varepsilon_i C_A + (1 - \varepsilon_i)\bar{\bar{C}}_A \tag{4.6}$$

Equivalently, the adsorption isotherm and the lumped adsorption isotherm are related by:

$$\bar{f}_A(C_A) = \varepsilon_i C_A + (1 - \varepsilon_i)\bar{\bar{f}}_A(C_A) \tag{4.7}$$

Equation (4.4) is normally considered to be closer to a well-defined thermodynamic adsorption isotherm than Eq. (4.5) is. We don't want to open a philosophical debate, but the assumptions associated with the porous model, including the existence of well-defined pores, are such that we don't believe this model can generally be considered more rigorous than the lumped model. In the absence of solute exclusion, the two adsorption isotherms are in any case linearly related, and one can use the former or the latter (as long as the correct mass balance expressions are used).

We explained in Chapter 3 that adsorption isotherms can be favorable, unfavorable or S-shaped; this characteristic dramatically impacts the chromatogram properties. Because of the linear relation between $\bar{\bar{f}}_A(C_A)$ and $\bar{f}_A(C_A)$, these two adsorption isotherms have the same shape.

Following a remark given in the introduction to this chapter, let us note that $\bar{\bar{C}}_A$ (resp. $\bar{C}_A$) is related by a single curve to C_A via the function $\bar{\bar{f}}_A(C_A)$ (resp. $\bar{f}_A(C_A)$), not only because temperature is fixed but also because we assume a single-solute system.

In order to be thermodynamically acceptable, functions $\bar{f}_A$ and $\bar{\bar{f}}_A$ must be monotonically increasing and obey $\bar{f}_A(0) = \bar{\bar{f}}_A(0) = 0$. As a consequence, rare cases of hysteresis between adsorption and desorption, as reported for instance by (Lin *et al.*, 1991) in the

[1] The definition or measurement of the concentrations on the skeleton is in fact not that trivial: the number of moles of a component in one phase varies depending on the arbitrary location of the dividing surface. Assuming the total number of moles in the system and the number of moles in the fluid phase to be known, the number of moles on the solid is the difference. This can be seen as the *excess* of molecules located in the system compared to a reference in which only the fluid phase is present. The adsorbed molecules being located on a surface, the terminology *surface excess* is often used.

case of a protein adsorption, cannot be representative of adsorption isotherms, i.e. of a thermodynamic equilibrium state: when the composition of one phase is given, thermodynamics stipulates that the compositions of the other phases are uniquely defined (unless metastable equilibrium is considered). Hysteresis, non-monotonic or multi-valued isotherms are the sign either of a hidden multi-solute interaction or of phase separation. As an illustration, hysteresis can be observed with low-boiling-point gases: it is essentially due to capillary condensation, which is responsible for phase separation. The study of the adsorption of a single ionic species subject to ion exchange can be misleading: a second ion (thus a multi-solute system) is necessarily involved! The case of proteins is even more complicated. Despite serious investigations, reasons for reported hysteresis are not clear, and, as noted by (Lin *et al.*, 1991), for example: "although a conformational change of the tetramer on the surface cannot be ruled out ... [it is] more likely that a surface reorientation has taken place".

There are certainly some specific exceptions, but we believe that in most situations apparent irreversibilities (hysteresis) may result from inadequate interpretation of experimental results. Some illustrations are given in (Schweich and Sardin, 1981), who showed that improper modeling of an ion-exchange process by ignoring the multi-solute nature of the process can lead to apparent irreversibilities. Another illustration is given in Section 4.8.4.

Throughout this book, we will consider that adsorption isotherms (lumped or not) are described by monotonically increasing functions so that the possibility of hysteresis is excluded.

At small concentrations, the adsorption isotherms are linear and approximated by:

$$\overline{C}_A = \overline{K}_A C_A \quad \text{with} \quad \overline{K}_A = \overline{f}'_A(0) \tag{4.8}$$

$$\overline{\overline{C}}_A = \overline{\overline{K}}_A C_A \quad \text{with} \quad \overline{\overline{K}}_A = \overline{\overline{f}}'_A(0) \tag{4.9}$$

where $\overline{K}_A$ and $\overline{\overline{K}}_A$ represent the initial slopes of the lumped adsorption isotherm and of the adsorption isotherm. Following the definitions given in Chapter 1, $\overline{K}_A$ and $\overline{\overline{K}}_A$ are called the *lumped Henry's coefficient* and Henry's coefficient, respectively.

We will now detail different models that allow representation of experimental adsorption isotherms. We will use equations that are thermodynamically meaningful (consistent), but we will also keep the thermodynamic aspects as simple as possible. For more general and fundamental considerations on adsorption, the reader is referred to general textbooks by (Ruthven, 1984) and (Toth, 2002), and for a specific focus on chromatography to (Guiochon *et al.*, 2006).

4.3.2 Single-solute systems

A very successful equation for representing non-linear adsorption isotherms was proposed a century ago by (Langmuir, 1918). Assuming that:

- the adsorbent contains a concentration $\overline{\overline{N}}$ of adsorption sites which are all equivalent

- each site can be associated with only one molecule of solute
- there are no interactions between the molecules of the solutes
- the equilibrium between the solutes contained in the fluid, the free sites and the sites associated with solutes follows an ideal mass action law with constant $\tilde{K}_A$
- only one layer of adsorbed molecules can form,

the following equation can be derived:

$$\overline{\overline{C}}_A = \frac{\overline{\overline{N}}\tilde{K}_A C_A}{1 + \tilde{K}_A C_A} \tag{4.10}$$

where $\overline{\overline{N}}$ is normally called the *adsorption capacity* of the adsorbent and $\tilde{K}_A$ the *adsorption constant.*

When C_A is very small (i.e. $\tilde{K}_A C_A \ll 1$), Eq. (4.10) becomes $\overline{\overline{C}}_A = \overline{\overline{N}}\tilde{K}_A C_A$, so that the adsorption isotherm becomes linear and Henry's coefficient is given by $\overline{\overline{K}}_A = \overline{\overline{N}}\tilde{K}_A$. At very high solute concentrations in the fluid, the adsorbed concentration $\overline{\overline{C}}_A$ reaches a plateau given by $\overline{\overline{N}}$.

The Langmuir equation, which generalizes the linear adsorption isotherm and is the simplest model for representing saturation, has proved to adequately represent many experimental situations. It appears that the practical area of application of the Langmuir adsorption isotherm is much wider than what the restricted assumptions would lead one to believe.

If one uses the equivalence given by Eq. (4.7), one obtains the fluid–solid equilibrium based on the lumped solid phase:

$$\overline{C}_A = \varepsilon_i C_A + (1 - \varepsilon_i)\frac{\overline{\overline{N}}\tilde{K}_A C_A}{1 + \tilde{K}_A C_A} = \varepsilon_i C_A + \frac{\overline{N}\tilde{K}_A C_A}{1 + \tilde{K}_A C_A} \tag{4.11}$$

where $\overline{N} = (1 - \varepsilon_i)\overline{\overline{N}}$ is the lumped adsorption capacity of the chromatographic medium.

In our experience, it is often very useful to add a degree of freedom to the Langmuir Equation (4.10) by adding a linear term and thus to define a linearly modified Langmuir adsorption isotherm:

$$\overline{\overline{C}}_A = \overline{\overline{\lambda}}_A C_A + \frac{\overline{\overline{N}}\tilde{K}_A C_A}{1 + \tilde{K}_A C_A} \tag{4.12}$$

The rationale for the added linear term can be the existence of non-selective adsorption sites in large quantities or a simple practical way to get flexibility.

In terms of lumped concentrations, the first linear term added to Eq. (4.12) also gives some flexibility to Eq. (4.11), whose first term is no longer strictly related to intragranular porosity. We thus define a linearly modified Langmuir lumped adsorption isotherm:

$$\overline{C}_A = \overline{\lambda}_A C_A + (1 - \varepsilon_i)\frac{\overline{\overline{N}}\tilde{K}_A C_A}{1 + \tilde{K}_A C_A} = \overline{\lambda}_A C_A + \frac{\overline{N}\tilde{K}_A C_A}{1 + \tilde{K}_A C_A} \tag{4.13}$$

with $\overline{\lambda}_A = \varepsilon_i + \overline{\overline{\lambda}}_A$. Equation (4.13) corresponds to Eq. (3.2) in Chapter 3.

This linear term can be added to any other adsorption isotherm equation, be it of the Langmuir type or not.

Since parameter $\bar{\lambda}_A$ encompasses the solutes contained in the intragranular fluid, it must, *a priori*, be greater than ε_i. This being said, $\bar{\bar{f}}_A$ being possibly an approximate expression for reality, nothing prevents $\bar{\lambda}_A$ from being lower than ε_i and even possibly zero if, at the end, the experimentally obtained lumped concentrations are better represented. We need to be pragmatic and flexible ...

Equation (4.13) has proved useful for describing adsorption on many chiral stationary phases, since these chromatographic media may contain a quantity of non-selective sites (the linear term) and some competitive stereo-selective sites (the second term).

The classical Langmuir (or linearly modified) adsorption isotherm can be extended to the bi-Langmuir adsorption isotherm, accounting, in essence, for the existence of two different types of adsorption sites:

$$\bar{\bar{C}}_A = \frac{\bar{\bar{N}}_1 \tilde{K}_{A1} C_A}{1 + \tilde{K}_{A1} C_A} + \frac{\bar{\bar{N}}_2 \tilde{K}_{A2} C_A}{1 + \tilde{K}_{A2} C_A} \tag{4.14}$$

Langmuir adsorption isotherms, be they strictly Langmuir, linearly modified or bi-Langmuir, allow one to explain, at least qualitatively, many chromatographic behaviors, for instance the asymmetry of overloaded peaks. This broad qualitative success does not mean that this simple model can account for all qualitative or even less quantitative results. A main limitation is that these single-solute Langmuir-type adsorption isotherms are limited to describing favorable behaviors and cannot be used in the case of unfavorable or S-shaped behaviors.

Many non-Langmuir types of adsorption isotherms have been proposed, and the most commonly suggested are presented below. They can be based on purely empirical considerations or can result from more or less complex models that take into account the influence of heterogeneous surfaces, non-ideal adsorbents, multi-layer formation etc. We will not give many details on the origin of these adsorption isotherms, and simply consider them as offering flexible means to represent experimental data.

First, recall the so-called *generalized Langmuir* adsorption isotherm presented in Section 3.2: by changing the sign of the concentration contribution in the denominator, one obtains a simple way to get an unfavorable instead of a favorable adsorption isotherm. This type of adsorption isotherm, sometimes referred to as anti-Langmuir, can only be used in a limited concentration range; otherwise, the solid-phase concentration may become negative.

Often described is the *Freundlich adsorption isotherm*:

$$\bar{\bar{C}}_A = k_F\, C_A^p \text{ usually with } p < 1 \tag{4.15}$$

One must be aware that the first derivative of this expression is infinite when the solute concentration goes to zero, which leads to an unrealistic infinite retention time in analytical chromatography. That is why we do not recommend using this expression except in very special situations, over limited concentration ranges and with great care.

In order to overcome this shortcoming while keeping flexibility, the Langmuir and Freundlich equations can be mixed to give the three-parameter equation suggested by (Peterson and Redlich, 1962):

$$\bar{\bar{C}}_A = \frac{k_1 C_A}{1 + (k_2 C_A)^p} \tag{4.16}$$

This expression reduces to the Langmuir equation when $p = 1$; it is linear at small concentrations, but is not monotonic over all the possible range of fluid concentrations if $p > 1$. It must thus be used with great care for $p > 1$ and, in that case, only over a limited range of concentrations. Additionally, it must be noticed that, unless $p = 1$, Eq. (4.16) does not reach a plateau when the fluid-phase concentration approaches infinity.

A similar modification of the Langmuir adsorption isotherm has been proposed by Toth (Toth, 1971):

$$\bar{\bar{C}}_A = \frac{k_1 C_A}{(1 + k_2 C_A{}^p)^{1/p}} = \frac{\bar{\bar{N}} \tilde{K}_A C_A}{(1 + k_2 C_A{}^p)^{1/p}} \tag{4.17}$$

Compared to Eq. (4.16), the Toth adsorption isotherm (4.17) has the merit of being monotonic irrespective of the value of p. It is qualitatively similar to the Langmuir adsorption isotherm and reaches a plateau when the fluid-phase concentration reaches infinity, but the p parameter allows us to adapt the curvature.

Another extension of the Langmuir adsorption isotherm, given by the BET equation (Guiochon *et al.*, 2006) takes into account the possibility of forming multilayers of solute in the adsorbed phase:

$$\bar{\bar{C}}_A = \bar{\bar{N}} \frac{\tilde{K}_A C_A}{\left(1 - \tilde{K}_{BET} C_A\right)\left(1 - \tilde{K}_{BET} C_A + \tilde{K}_A C_A\right)} \tag{4.18}$$

In addition to the Langmuir parameters $\bar{\bar{N}}$ and $\tilde{K}_A$, this equation involves a parameter $\tilde{K}_{BET}$ which can be seen as the adsorption constant of the solute on a layer of adsorbate. The BET equation reduces to the Langmuir equation when $\tilde{K}_{BET} = 0$; it is linear at small concentrations, but does not tend towards a saturation plateau at large concentrations unless $\tilde{K}_{BET} = 0$.

A powerful and flexible adsorption isotherm equation was derived by (Hill, 1960) on the basis of statistical thermodynamics, showing that an adsorption isotherm can be written as the ratios of two polynomials of the same rank:

$$\bar{\bar{C}}_A = \frac{\bar{\bar{N}}}{p} \frac{C_A(b_1 + 2b_2 C_A + 3b_3 C_A^2 + \cdots + p b_p C_A^{p-1})}{1 + b_1 C_A + b_2 C_A^2 + b_3 C_A^3 + \cdots + b_p C_A^p} \tag{4.19}$$

The part of the numerator within parentheses is the derivative of the denominator. According to this equation, the adsorption isotherm reaches a saturation plateau given by $\bar{\bar{N}}$.

Notice that Eq. (4.19) can also be seen as a Padé approximant, known for its ability to approximate a variety of functions remarkably well. Thus, the ability of Eq. (4.19) to represent experimental data may not only be due to thermodynamic considerations. The second-order truncation of this equation is capable of describing S-shaped adsorption isotherms, thus adsorption isotherms with inflection points:

$$\bar{\bar{C}}_A = \frac{\bar{\bar{N}}}{2}\frac{C_A(\tilde{K}_A + 2b_2C_A)}{1 + \tilde{K}_AC_A + b_2C_A^2} \quad \text{with} \quad b_2 \neq 0 \tag{4.20}$$

All of the above adsorption isotherms $\bar{\bar{f}}_A$ lead to the lumped adsorption isotherms $\bar{f}_A$ using Eq. (4.7). Adding a linear term, as with the Langmuir adsorption isotherm, allows one to take into account the influence of intragranular porosity and offers a simple way to gain flexibility.

A lot of work has been done in order to model adsorption through the so-called Gibbs adsorption isotherm (Myers and Prausnitz, 1965), a concept that will prove its full usefulness when we address multi-solute situations. The basic assumption is that the adsorbate can be treated as a two-dimensional phase for which the fundamental equations of thermodynamics are applied. It is certainly a kind of "overmodeling" for single-solute systems, but understanding the mechanics will be important when extensions to multi-solute systems are considered. The principles are summarized below.

If $\bar{\bar{A}}$ is the surface area (m^2) of the adsorbent, π the so-called spreading pressure (which is in fact the difference between the surface tension of the clean surface and the surface tension of the surface covered by the adsorbed solute – and thus not a pressure) and $\bar{\bar{n}}_A$ the number of moles of solute adsorbed on the surface, the Gibbs–Duhem equation written for an isothermal system gives:

$$-\bar{\bar{A}}d\pi + \bar{\bar{n}}_Ad\bar{\bar{\mu}}_A = 0 \tag{4.21}$$

By dividing both sides of Eq. (4.21) by the volume of the skeleton $\bar{\bar{V}}$ one obtains:

$$-\bar{\bar{A}}_{sp}d\pi + \bar{\bar{C}}_Ad\bar{\bar{\mu}}_A = 0 \tag{4.22}$$

where $\bar{\bar{A}}_{sp} = \bar{\bar{A}}/\bar{\bar{V}}$ is the specific surface area of the adsorbent (m^2/m^3).

At equilibrium, the chemical potential of the adsorbed phase must equal the chemical potential of the fluid phase, so $\bar{\bar{\mu}}_A = \mu_A$. Should the adsorbent be in contact with an ideal gas, the chemical potential can be written:

$$\mu_A = \mu_A^0 + RT\ln(P_A/P_0) \quad \text{thus} \quad d\mu_A = RTd\ln(P_A) \tag{4.23}$$

where P_A is the pressure of solute A in the gas phase and P_0 the standard pressure.

Inserting Eq. (4.22) into Eq. (4.21) one obtains:

$$\frac{d\pi}{d\ln P_A} = RT\frac{\bar{\bar{C}}_A}{\bar{\bar{A}}_{sp}} \tag{4.24}$$

or equivalently:

$$\frac{d\pi}{dP_A} = \frac{RT}{\bar{\bar{A}}_{sp}}\frac{\bar{\bar{C}}_A}{P_A} \tag{4.25}$$

Should the adsorbent be in contact with an ideal liquid solution, the chemical potential can be written:

$$\mu_A = \mu_A^0 + RT\ln(C_A/C_A^{ref}) \quad \text{thus} \quad d\mu_A = RTd\ln(C_A) \tag{4.26}$$

where ^{ref} designates the reference state.
Inserting Eq. (4.26) into Eq. (4.22), one obtains:

$$\frac{d\pi}{dC_A} = \frac{RT}{\overline{\overline{A}}_{sp}} \frac{\overline{\overline{C}}_A}{C_A} \tag{4.27}$$

Equations (4.25) and (4.27) are known as the *Gibbs adsorption isotherm.* They relate the fluid-phase concentration (or pressure) to the adsorbed concentration $\overline{\overline{C}}_A$ provided that the spreading pressure is known.

If an equation of state is postulated to relate the spreading pressure to the number of moles adsorbed on the surface, Eq. (4.25) or Eq. (4.27) allow the derivation of an expression for the adsorption isotherm. Classical expressions are given in (Koter and Terzyk, 2005). For instance, assuming $\pi\overline{\overline{A}} = \overline{\overline{n}}_A RT$ (or $\pi\overline{\overline{A}}_{sp} = \overline{\overline{C}}_A RT$) by analogy with ideal gases, the predicted adsorption isotherm is linear.

The Gibbs adsorption isotherm may be seen as transferring the burden of selecting an adsorption isotherm to the burden of selecting an equation of state. We will see when addressing multi-solute adsorption that the associated merits are in fact significant.

EXERCISE 4.1 *Derive the Gibbs adsorption isotherm associated with the surface equation of state* $\pi = \frac{RT}{\Omega} ln\left(\frac{\overline{\overline{A}}/\overline{\overline{n}}_A}{\overline{\overline{A}}/\overline{\overline{n}}_A - \Omega}\right)$, *where* Ω *represents the surface occupied by one mole of adsorbent.*

The Gibbs approach can be used in a different way. Instead of considering the surface of an adsorbent to be a free space that can be filled by an adsorbate, it can be treated as a "solvent" and the adsorption process viewed as the formation of a "solution" between the adsorbate and the adsorbent. This second approach is called the *vacancy solution model.* With the assumption of an osmotic equilibrium between the two solutions (fluid and adsorbed), (Suwanayuen and Danner, 1980a) derived expressions for the adsorption isotherms for gas-phase adsorption model and (Fukuchi *et al.*, 1982) extended this to liquid systems.

The adsorption isotherm derived from the vacancy solution model has the following structure:

$$C_A = \frac{\overline{\overline{\Theta}}_A}{\tilde{K}_A(1-\overline{\overline{\Theta}}_A)} g(\overline{\overline{\Theta}}_A) \quad \text{with} \quad \overline{\overline{\Theta}}_A = \frac{\overline{\overline{C}}_A}{\overline{\overline{N}}} \tag{4.28}$$

Non-idealities of the adsorbed phase are expressed by the function $g(\overline{\overline{\Theta}}_A)$. For $g = 1$, Eq. (4.28) reduces to the Langmuir adsorption isotherm. The function $g(\overline{\overline{\Theta}}_A)$ depends on the equation chosen for describing the concentration dependence of the adsorbed phase activity coefficients. Classical thermodynamic expressions for activity coefficients proposed by Margules, Van Laar and Wilson have been tested by (Koter and Terzyk, 2005).

This approach has the merit of being consistent and flexible as it allows the use of various known activity coefficient models, but it has the drawback of not being explicit in C_A.[2] The potential of the vacancy solution model was analyzed by (Seidel, 1989) for adsorption on activated carbon.

In summary, the chromatographer has access to many expressions offering enough flexibility to represent different adsorption isotherms. While the classical Langmuir-like expressions allow the representation, at least qualitatively, of a majority of situations, more complex situations, including for instance inflection points, do exist. It would be presumptuous to recommend one specific category of expressions, as the experimental situations can be very diverse. Common sense, however, tells us to keep things as simple as possible and to avoid using expressions carrying some complexity leading to illusory precision.

4.3.3 Multi-solute systems

The main difference between modeling single-solute and multi-solute systems is to be found in the competition between solutes for existing adsorption sites, solute–solute interactions or other effects that can induce a strong influence of one solute on the others. As already mentioned in Chapter 3, should the species behave independently, modeling a multi-solute system would be no more than modeling *n* independent single-solute systems. That is very rarely the case, so that modeling multi-solute systems is roughly synonymous with modeling competition and interactions between species.

While competition between solutes (and carrier fluid constituents) for existing adsorption sites is easily understandable in principle, it is not the only possible source of competition. In reversed-phase chromatography, for instance, it is known that the decrease in retention times observed when alcohol or acetonitrile is added to a water-rich solvent cannot be explained by the ability of these solvents to adsorb, but rather by their ability to increase solute solubility in the carrier fluid.

It is worth mentioning that, in certain conditions, competition can even be "negative": the adsorption of one species is not hindered but favored by the presence of others. In this situation one typically speaks of synergistic adsorption even if adsorption *stricto sensu* is not the key underlying process. As an illustration, the adsorption of fructose on ion-exchange resins is favored by increased quantities of glucose, and similarly the adsorption of glucose on the same ion-exchange resin is favored by increased quantities of fructose (Schmidt-Traub *et al.*, 2012). This situation will be further discussed at the end of this section and in Chapter 10.

It is thus impossible to make general qualitative rules for the influence of composition in adsorption. A first and pragmatic approach consists in generalizing available single-solute adsorption isotherms using common sense or simple arguments – and then to check agreement with reality.

[2] We will see in Section 4.7 that working with non-explicit expressions is in general unavoidable when different interaction mechanisms must be taken into account simultaneously.

A perfect illustration of this approach is given by the Langmuir adsorption isotherm. Using the same simple assumptions as for single-solute systems, one obtains (see Exercise 4.3):

$$\overline{\overline{C}}_i = \frac{\overline{\overline{N}}\tilde{K}_i C_i}{1 + \sum_j \tilde{K}_j C_j} \tag{4.29}$$

As with single-solute systems, experience shows that adding a linear term to Eq. (4.29) provides additional flexibility that allows the representation of many different experimental situations:

$$\overline{\overline{C}}_i = \overline{\overline{\lambda}}_i C_i + \frac{\overline{\overline{N}}\tilde{K}_i C_i}{1 + \sum_j \tilde{K}_j C_j} \tag{4.30}$$

Conversely, using lumped solid concentrations, one obtains a flexible equation representing competitive favorable adsorption isotherms given by:

$$\overline{C}_i = \overline{\lambda}_i C_i + \frac{\overline{N}\tilde{K}_i C_i}{1 + \sum_j \tilde{K}_j C_j} \tag{4.31}$$

The Langmuir (or modified Langmuir) adsorption isotherms are often generalized for multi-solute systems as they are considered to offer a good compromise between simplicity and ability to predict chromatographic column behavior. They often, but certainly not always, provide at least the correct qualitative features.

An important difference between the Langmuir equation and the linearly modified Langmuir equation is that the selectivity coefficient $\alpha_{j/i} = \dfrac{\overline{C}_j C_i}{C_j \overline{C}_i}$ (with $\alpha_{j/i} > 1$) is constant for the former adsorption model, whereas it decreases with concentration for the latter. As decreasing selectivity factors with concentration is quite common, the linearly modified Langmuir equation is often well adapted to represent experimental results.

While the binary Langmuir and linearly modified Langmuir laws are often used for solutes only, they can be used to understand the effect of solvent composition. Let us consider a species s representing an adsorbable constituent contained in the carrier fluid at concentration C_s. Varying C_s influences the adsorption of the solutes i and consequently their retention. As an illustration, by using Eq. (4.31) for i and s, the initial slope of the lumped adsorption isotherm of solute i is given by:

$$\overline{K}_i = \overline{f}'_i(0) = \overline{\lambda}_i C_i + \frac{\overline{N}\tilde{K}_i}{1 + \tilde{K}_s C_s} \tag{4.32}$$

which shows that the lumped Henry's coefficient $\overline{K}_i$ decreases when C_s increases. The multi-solute Langmuir equation thus gives a simple way to explain, at least qualitatively, the evolution of solute retention with the addition of an adsorbed product in the carrier fluid.

All the above Langmuir adsorption isotherms or linearly modified Langmuir adsorption isotherms have assumed the adsorption capacity $\overline{\overline{N}}$ to be a characteristic of the solid, independent of the species. Experience shows, however, that in many situations, the adsorption capacity $\overline{\overline{N}}$ depends on the species and is thus better written $\overline{\overline{N}}_i$. Reasons for this dependence are numerous, the first one associated with the fuzzy definition of adsorption sites in many situations like reversed-phase chromatography.

The fact that adsorption capacities are often expressed in g/l when a more rigorous approach would use mol/l is certainly not a general explanation for the experimental dependence of saturation capacities on species. As an illustration, (Gritti and Guiochon, 2003) obtained on Kromasil-C18 a monolayer saturation capacity of 370 g/l for toluene and 170 g/l for ethylbenzene. Using molar concentrations leads to 4 mol/l for toluene and 1.6 mol/l for ethylbenzene and does not reduce the difference.

Generalizing individual adsorption isotherms having their own saturation capacities $\overline{\overline{N}}_i$ to multi-solute adsorption isotherms poses two problems. The first one is that multi-solute expressions for the Langmuir equation involving different saturation capacities do not lead to a thermodynamically consistent adsorption isotherm as they do not satisfy the Gibbs–Duhem equation. More information on thermodynamic consistency is given at the end of this section. Certainly most chromatographers will not care, although they probably should. The second problem is very practical: how to select a total saturation capacity for the mixture? Should we take some kind of average between the various saturation capacities? Which one?

A simple way to circumvent the problem, which works for binary mixtures, consists in assuming that the adsorbent contains two different types of sites α and β, respectively associated with adsorption capacities $\overline{\overline{N}}^\alpha$ and $\overline{\overline{N}}^\beta$ (Guiochon *et al.*, 2006). We also assume that solute A is able to adsorb on both sites, while solute B is able to adsorb on the first category of sites only, both species following a Langmuirian behavior. Consequently, pure solute A has access to saturation capacity $\overline{\overline{N}}^\alpha + \overline{\overline{N}}^\beta$ whereas pure solute B has access to $\overline{\overline{N}}^\alpha$ only. The adsorption of solute A in the binary mixture is then modeled according to:

$$\overline{\overline{C}}_A = \frac{\overline{\overline{N}}^\alpha \tilde{K}_A^\alpha C_A}{1 + \tilde{K}_A^\alpha C_A + \tilde{K}_B^\alpha C_B} + \frac{\overline{\overline{N}}^\beta \tilde{K}_A^\beta C_A}{1 + \tilde{K}_A^\beta C_A} \tag{4.33}$$

and the adsorption of solute B in the binary mixture is modeled according to:

$$\overline{\overline{C}}_B = \frac{\overline{\overline{N}}^\alpha \tilde{K}_B^\alpha C_B}{1 + \tilde{K}_B^\alpha C_A + \tilde{K}_B^\alpha C_B} \tag{4.34}$$

This concept of species selectively adsorbing on different sites generalizes the linearly modified Langmuir adsorption isotherms and gives a simple way to treat apparent species-dependent adsorption capacities in the case of binary mixtures.

When more flexibility is required, as with single-solute systems, the multi-solute Langmuir expressions can be extended to generalized Langmuir expressions by changing the sign of the concentration contributions in the denominator (see Section 3.2), to Peterson–Redlich or to Toth adsorption isotherms.

The Peterson and Redlich adsorption isotherm (Peterson and Redlich, 1962) is generalized with:

$$\overline{\overline{C}}_i = \frac{k_{1i}C_i}{1+\sum_j (k_{2j}C_j)^p} \tag{4.35}$$

and the Toth adsorption isotherm (Toth, 1971) is generalized with:

$$\overline{\overline{C}}_i = \frac{k_{1i}C_i}{\left(1+\sum_j k_{2j}C_j^p\right)^{1/p}} = \frac{\overline{\overline{N}}\tilde{K}_i C_A}{\left(1+\sum_j k_{2j}C_j^p\right)^{1/p}} \tag{4.36}$$

When the different individual adsorption isotherms are similar in nature (e.g. all Langmuir with identical saturation capacities), extension to multi-solute mixtures is not very difficult *a priori*, and single-solute equations can be generalized with a bit of common sense, at least when solute–solute interactions are not too strong. The situation is more complex when the individual adsorption isotherms are different in nature, e.g. a favorable adsorption isotherm for solute A and an unfavorable adsorption isotherm for solute B.

In order to allow for rational extrapolations, a more rigorous thermodynamic approach must be used. The most successful theory is the ideal adsorbed solution (IAS) theory developed by (Myers and Prausnitz, 1965) for gases and extended to adsorption from dilute liquid solutions by (Radke and Prausnitz, 1972). Using relatively simple assumptions, this theory allows competitive adsorption isotherms to be calculated from single-solute adsorption isotherms without additional parameters. The basic concept, which consists in using the Gibbs adsorption isotherm presented in Section 4.3.2, is summarized below.

For convenience, we define a modified spreading pressures by:

$$\Pi = \frac{\overline{\overline{A}}_{sp}}{RT}\pi \tag{4.37}$$

Let us assume that the individual adsorption isotherms $\overline{C}_i(C_i)$ are known for all the solutes of interest. Modified spreading pressures of pure solutes are calculated by integration of the Gibbs adsorption isotherm 4.27:

$$\Pi_i = \int_0^{C_i^0} \frac{\overline{\overline{C}}_i(u)}{u}\, du \tag{4.38}$$

the superscript 0 characterizing the single-solute state.

In a mixture, equilibrium requires that all solutes exert the same spreading pressure, so:

$$\Pi = \Pi_i \quad \text{independent of solute } i \tag{4.39}$$

The concentrations C_i^0 appearing in Eq. (4.38) are thus fictitious pure-solute concentrations that generate the same modified pressure Π as the mixture. Comparing Eq. (4.38) and Eq. (4.39) shows that these fictitious pure-solute concentrations must satisfy:

$$\Pi_i = \Pi = \int_0^{C_i^0} \frac{\overline{\overline{C}}_i(u)}{u}\, du \quad \text{independent of solute } i \tag{4.40}$$

This gives n equations relating the n unknown fictitious pure-solute concentrations C_i^0 to the modified spreading pressure.

The molar (or mass) fraction of solute i on the adsorbed surface is given by:

$$\bar{\bar{x}}_i = \overline{\overline{C}}_i / \overline{\overline{C}}_{tot} \quad \text{with} \quad \overline{\overline{C}}_{tot} = \sum_i \overline{\overline{C}}_i \tag{4.41}$$

Raoult's law is then used to relate the molar fractions in the adsorbed phase to the concentrations in solution:[3]

$$C_i = \bar{\bar{x}}_i C_i^0(\Pi) \quad \text{for all solutes} \tag{4.42}$$

As the sum of the molar (mass) fractions must equal unity, Eq. (4.42) leads to:

$$\sum_i \frac{C_i}{C_i^0(\Pi)} = 1 \tag{4.43}$$

The last concept of the IAS theory is that the area covered by an ideal solution should be the same whether the solute is pure or in a mixture with other adsorbates. The area occupied by one mole of pure solute i is given by $\bar{\bar{A}}_{sp}/\overline{\overline{C}}_i^0$ (m^2/mol). Component i representing fraction $\bar{\bar{x}}_i$ in the adsorbed phase, the total area occupied by combination of the pure solutes would be $\sum_j \bar{\bar{x}}_i \bar{\bar{A}}_{sp}/\overline{\overline{C}}_i^0$. On the other hand, the total concentration adsorbed is given by $\overline{\overline{C}}_{tot} = \sum_i \overline{\overline{C}}_i$, which occupies an area $\bar{\bar{A}}_{sp}/\overline{\overline{C}}_{tot}$ (m^2/mol). By equating the areas calculated by the two approaches, one obtains:

$$\frac{1}{\overline{\overline{C}}_{tot}} = \sum_i \frac{\bar{\bar{x}}_i}{\overline{\overline{C}}_i^0} \tag{4.44}$$

Equations (4.40) can be integrated analytically for some adsorption isotherms, including Langmuir or Hill types. When the equations (4.40) can be integrated analytically they provide explicit expressions $\Pi(C_i^0)$. Assuming that these expressions can be inversed, one obtains explicit expressions for the fictitious concentrations C_i^0 versus the modified spreading pressure.

For a given set of fluid concentrations C_i, Eq. (4.43) is an equation in Π whose solution subsequently allows calculation of the C_i^0 and then the $\overline{\overline{C}}_i^0$ from the single-solute adsorption isotherms. Raoult's equation (4.42) then allows calculation of all the molar (massic) fractions $\bar{\bar{x}}_i$. This, together with knowledge of the C_i^0, allows calculation of $\overline{\overline{C}}_{tot}$ from Eq. (4.44). Finally, the adsorbed phase concentrations are calculated with $\overline{\overline{C}}_i = \bar{\bar{x}}_i \overline{\overline{C}}_{tot}$.

[3] Consider the analogy with vapor liquid equilibria. If P_A is the partial pressure of A in the vapor phase and y_A the molar fraction of A in the liquid phase, P_A and y_A are related by $P_A = y_A P_{sat}(T)$.

The algorithm for solving the IAS model is thus simple and efficient when analytical expressions are obtained for Eq. (4.40) and when these expressions can be inverted in order to provide explicit $C_i^0(\Pi)$ expressions.

EXERCISE 4.2 *The pure-solute adsorption isotherms of solutes A and B follow a Langmuir expression. Calculate the binary isotherm according to the IAS theory. What happens if one solute follows an anti-Langmuir adsorption isotherm with a negative coefficient in the denominator?*

Unfortunately, analytical expressions are only available for relatively simple adsorption isotherms. In addition to the Langmuir adsorption isotherm, involving a unique saturation capacity for all species, analytical or quasi-analytical expressions have been derived for Langmuir adsorption isotherms with species-dependent saturation capacities, for BET as well as for second-order Hill adsorption isotherms, by (Le Van and Vermeulen, 1981), (Tarafder and Mazzotti, 2012), (Ilic *et al.*, 2010) and (Gritti and Guiochon, 2003). These explicit or quasi-explicit solutions are in general relatively complex and often limited to binary systems, so numerical methods cannot be avoided in general.

Numerical solution of the IAS model is normally relatively difficult, and (Rubiera *et al.*, 2013) have presented different possible approaches as well as their integration in a column model. In addition, (Rubiera *et al.*, 2013) proposed a general approach based on reformulating the equations into a dynamic problem that appears to be robust and stable.

An interesting approximation has been proposed by (Frey and Rodrigues, 1994), who used a Padé approximant to fit the dependence on concentration of the spreading pressure of the single-solute adsorption isotherm. This approximation, which in general works reasonably well for very different single-solute adsorption isotherms, allows us to obtain an explicit formulation of the multi-solute adsorption isotherms, as explained below.

When individual adsorption isotherms are known, an expression for the spreading pressure can be calculated from the Gibbs equation (4.38) so that one has access to $\Pi_i(C_i^0)$ and/or to $C_i^0(\Pi)$. The next step is to fit $C_i^0(\Pi)$ with a three-parameter Padé approximant:

$$C_i^0(\Pi) = \frac{\Pi}{\alpha_i + \beta_i \Pi + \gamma_i \Pi^2} \tag{4.45}$$

It can be shown that an expression for the adsorption isotherm which is compatible with Eq. (4.45) is given by:

$$\bar{\bar{C}}_i^0 = \frac{1}{2\gamma_i C_i^0}\left(\frac{1-\beta_i C_i^0}{\sqrt{(\beta_i^2 - 4\alpha_i\gamma_i)(C_i^0)^2 - 2\beta_i C_i^0 + 1)}} - 1\right) \tag{4.46}$$

While having an uncommon shape for an adsorption isotherm, Eq. (4.46) has proved to be very flexible and capable of representing a variety of experimental data. The parameters $\alpha_i, \beta_i, \gamma_i$ can thus be determined by fitting Eq. (4.46) to the experimental results.

Substituting Eq. (4.45) written for each species into Eq. (4.43) gives:

$$\Pi^2 \sum_j \gamma_j C_j + \Pi \left(\sum_j \beta_j C_j - 1 \right) + \sum_j \alpha_j C_j = 0 \tag{4.47}$$

which is a quadratic equation in Π. Of the two possible solutions, the good one is selected by making sure that at low concentrations the initial slopes of the single-solute and multi-solute adsorption isotherm are identical. Knowledge of Π and of $C_i^0(\Pi)$ then allow us to derive the competitive adsorption isotherms with the frame presented above.

An advantage of the IAS theory is that there are no constraints on the structure of the single-solute isotherms (provided they obey the fundamental rule of being monotonic). It is worth mentioning that this approach has been used successfully even in systems showing a synergistic instead of a competitive effect between species.

A simple example is given by the adsorption of glucose and fructose on resins. (Schmidt-Traub *et al.*, 2012) reported experimental adsorption isotherms for glucose and fructose in water at 65°C that are very well represented by:

$$\begin{aligned} \overline{\overline{C}}_{glu} &= 0.27 C_{glu} + 0.000122 C_{glu}^2 + 0.000103 C_{glu} C_{fru} \quad \text{(g/l)} \\ \overline{\overline{C}}_{fru} &= 0.47 C_{fru} + 0.000119 C_{fru}^2 + 0.000248 C_{glu} C_{fru} \quad \text{(g/l)} \end{aligned} \tag{4.48}$$

The adsorption isotherms are presented in this form to offer a simple visualization of the synergistic effect. Both adsorption isotherms exhibit an anti-Langmuir (unfavorable) behavior, accounted for by the terms C_i^2. When considering mixtures, it can be seen that the adsorption of glucose (resp. fructose) is favored by the presence of fructose (resp. glucose), as expressed by the $C_{glu}C_{fru}$ terms. This synergistic effect is attributed to specific interactions between the hydrated solute molecules and the solvent (Schmidt-Traub *et al.*, 2012). (Nowak *et al.*, 2007) determined mono-solute adsorption isotherms and, using the IAS theory, obtained binary adsorption isotherms in good qualitative agreement with the experimental ones.

In order to refine the precision of the model, (Nowak *et al.*, 2009) then used the NRTL model of activity coefficients in the fluid phase. With this development, they showed that the anti-Langmuirian shapes of the adsorption isotherms were explained by non-idealities in the fluid solution and that the adsorption mechanism follows a Langmuirian mechanism. Non-idealities are not always located where they are expected ...

We have called attention to this fructose/glucose example because it involves a non-classical behavior that can be modeled with thermodynamically consistent tools, but also because this particular behavior has some significant consequences for the important industrial glucose–fructose continuous chromatographic separation processes. Further information is given in Chapter 10.

From a theoretical point of view, the applicability of the IAS model is restricted to dilute solutions, although the notion of "dilute" solutions is somewhat vague. A main limitation comes from the assumption of the absence of solute–solute and adsorbate–adsorbate interactions. Extensions of the IAS theory to higher concentrations and non-ideal situations are based on the introduction of adsorbed phase activity coefficients (real adsorbed solution (RAS) theory (Myers, 1983)). As an illustration, we mention the work of (Heinonen *et al.*, 2012) who have successfully used this approach for modeling competitive and synergistic adsorption on elastic (swelling) ion-exchange resins and integrated it into a column model.

The vacancy solution theory already presented for single-solute adsorption isotherms in Eq. (4.28) can be extended to mixtures (Suwanayuen and Danner, 1980b). Non-idealities are included *a priori* by using concentration-dependent adsorbed phase activity coefficients based on the classical models of Margules, Van Laar and Wilson. To our knowledge, this theory has never been used in chromatographic studies, but its potential has been analyzed by (Seidel, 1989) for adsorption on activated carbons.

It is again worth noting that not all forms of adsorption isotherms can be used, even if they seem, *prima facie*, to exhibit the features required to be thermodynamically acceptable. Using considerations of chemical potentials, (Kvaalen *et al.*, 1985) have shown that the Jacobian matrix of adsorption isotherms has to be diagonalizable to be thermodynamically acceptable. (Flockerzi *et al.*, 2013) have shown that this condition is not satisfied for the bi-Langmuir adsorption isotherms when arbitrary saturation capacities for the different solutes are used. (Jedrzejak *et al.*, 1989) have discussed consistency problems that may arise when modeling multi-solute systems with adsorption isotherms whose Jacobian matrix is not diagonalizable: unexpected and unacceptable results, such as negative front velocities, can be obtained.

The possibility of "bizarre" adsorption isotherms may be due to the fact that the experimentally determined adsorption law lumps a "true" adsorption process (i.e. phase partitioning equilibrium only) together with some other homogeneous equilibria in the fluid or the solid phase (complexation, partial ionization of a weak acid etc.). This shows again (see for instance Section 4.8.4) that one must understand the equilibrium laws, and must not be satisfied only with the goodness of fit of experimental results.

4.4 The case of supercritical fluids

Adsorption isotherms describing the adsorption on solid stationary phases of solutes diluted in supercritical fluids can be defined, interpreted and determined in a way that is similar to that for liquid carrier fluids. For general information on adsorption and supercritical fluid chromatography, the reader is referred to (Guiochon and Tarafder, 2011) and (Rajendran, 2012).

The reader must, however, be aware of a few practical differences:

- Supercritical fluids being compressible, the fluid velocity varies along the axis of the chromatography column. As a consequence, the Z_tRT or Z_eRT cannot be calculated with the simple integrated expressions $t_{0R} = \varepsilon_T V_{col}/Q$ or $t_0 = \varepsilon_e V_{col}/Q$ but result from a more complex integration process over the column. The way to model this effect is presented in Chapter 6. A consequence is that, while perfectly do-able, determining initial slopes of adsorption isotherms from retention times is not as straightforward for supercritical fluids as it is for liquids.

- The expansion of supercritical fluids along the column is associated with endothermic effects that can be significant. A consequence is that operating supercritical fluid columns under isothermal conditions is not as simple as it is for liquid or gas carrier fluids. A systematic check of inlet and outlet temperatures is thus mandatory.

With these two technical specificities in mind, the methods used in liquid chromatography can be used to determine adsorption isotherms in the presence of supercritical fluids. Various examples are reported in the reviews by (Guiochon and Tarafder, 2011) and (Rajendran, 2012). In the same reviews, the reader will find that different experimentally determined adsorption isotherms can be represented by the models proposed in the preceding sections, including Langmuir, second- and third-degree Hill adsorption isotherms, and that the IAS theory has been applied successfully to mixtures.

Of key interest in the case of supercritical fluids is the derivation of equations for Henry's coefficients ($\overline{K}_i$). The first detailed analysis was proposed by (Van Wasen and Schneider, 1975), who showed that Henry's coefficients are affected by pressure and temperature according to:

$$\overline{K}_i = \rho^{-n_i}\varphi_i(T) \tag{4.49}$$

ρ being the average density along the column, estimated for instance by the Peng–Robinson equation of state. More information on this equation of state is given in Appendix F.

Equation (4.49) is useful for representing the observations made by many authors that Henry's coefficients are strongly dependent on the eluent density. The temperature dependence is usually well fit by $\varphi_i(T) = \exp\,(a_i - b_i T)$.

As with gases and liquids, the Langmuir adsorption isotherm has the advantage of generally representing reasonably well the qualitative aspects of non-linear chromatography. In order to represent the influence of pressure and temperature, one can assume the adsorption capacity of the solid stationary phase to be constant and, by analogy with Eq. (4.49), that the coefficient $\tilde{K}_i$ varies as:

$$\tilde{K}_i = \rho^{-n_i}\varphi_i(T) \quad \text{assuming } \overline{N} \text{ constant} \tag{4.50}$$

Langmuir adsorption isotherms involving the coefficients $\tilde{K}_i$ approximated by Eq. (4.50) were successfully used by (Clavier, 1995) to model the behavior of pressure-modulated preparative chromatography purification of fatty acids on reversed-phase silica. This should, however, be seen as a first simple approximation, as the carrier fluid (carbon dioxide or modifier) can also adsorb on the solid and thus compete with solutes for available adsorption capacity. This carrier fluid adsorption has, for instance, been investigated by (Strubinger and Parcher, 1989) on bound silica. If the adsorption of the carrier fluid is not taken into account in the adsorption isotherm model, this may result in an apparent adsorption capacity available to the solutes which varies with experimental conditions.

Additional information on the effect of pressure drop or solvent modifier on adsorption is found in (Rajendran *et al.*, 2005), (Rajendran *et al.*, 2008), (Rajendran, 2012) and (Ottiger *et al.*, 2007).

4.5 The case of ion exchange

The "must-read" on ion exchange is certainly the textbook written 50 years ago by F. Helfferich (Helfferich, 1962).

4.5.1 General considerations

The previous sections were devoted to modeling adsorption chromatography, or more generally interactions between non-charged molecules and a chromatographic medium. This theoretical framework can be adapted to ion exchange with one important modification: the model must recognize that the mechanism is based on an "exchange" between ionic charges, so that if "one charge goes in", "one charge must go out". A second difference with non-charged molecules lies in the electroneutrality constraint: in order to ensure an electrically neutral phase, the number of cationic charges must equal the number of anionic charges. If these constraints are satisfied in both phases initially, they will remain satisfied during the ion-exchange process because of the above-mentioned "exchange" mechanism. This will affect the way we define concentrations. We have seen that in the case of non-charged species, concentrations can be expressed in mol/l or g/l with little practical difference. When ionic species are involved, a choice that makes sense is to use the number of equivalents per liter of phase (fluid or solid), eq/l, because this allows exchange and electroneutrality to be expressed in a straightforward manner.

4.5.2 The lumped model: ideal situations

For illustration purposes, we will mainly consider the case of cation exchange, keeping in mind that a symmetrical treatment can be applied to the case of anion exchange.

We consider a cationic exchange resin in contact with a solution containing protons H at concentration H^+, cations C of valence z at concentration C^{z+} and anions A at concentration A^-, all concentrations being expressed in eq/l. The normality N_0 of the solution is defined by:

$$H^+ + C^{z+} = OH^- + A^- = N_0 \tag{4.51}$$

If the solution is acidic, the hydroxide ions OH^- can be neglected, and Eq. (4.51) becomes:

$$H^+ + C^{z+} = A^- = N_0 \tag{4.52}$$

Unless otherwise specified, we will assume that Eq. (4.52) holds.

The cationic ion exchanger is characterized by a number of fixed anionic sites available for exchanging cations. We will assume that the resin is strong, so that all the sites are ionized, irrespective of the conditions. An illustration of how to model weak resins is given in Section 4.7.1.

An ionic capacity $\overline{N}^{iex}$ is defined by dividing the number of anionic sites by the resin volume $\overline{V}$. The ionic capacities $\overline{N}^{iex}$ are typically in the range of 2–3 eq/l. Similarly, an ionic capacity $\overline{\overline{N}}^{iex}$ is defined by dividing the number of anionic sites by the skeleton volume $\overline{\overline{V}}$. From the above definitions, it follows that $\overline{N}^{iex} = (1 - \varepsilon_i)\overline{\overline{N}}^{iex}$.

The electroneutrality of the resin skeleton is written:

$$\overline{\overline{H}}^{+} + \overline{\overline{C}}^{z+} = \overline{\overline{N}}^{iex} \quad \text{(eq/l)} \tag{4.53}$$

The ion-exchange process between ions H and C can be described by:

$$z\overline{\overline{\mathrm{H}}}^{+} + \mathrm{C}^{z+} \rightleftharpoons z\mathrm{H}^{+} + \overline{\overline{\mathrm{C}}}^{z+} \tag{4.54}$$

which suggests modeling the equilibrium with an ideal mass action law:

$$K_{C/H} = \frac{\overline{\overline{C}}^{z+}}{C^{z+}} \left(\frac{H^{+}}{\overline{\overline{H}}^{+}} \right)^{z} \tag{4.55}$$

$K_{C/H}$ is known as the *selectivity coefficient*. It is temperature-dependent, and can be constant or not with respect to composition, due to non-ideality effects.

If one assumes that the only ions that can enter the resins are those that can exchange on anionic sites, i.e. that anions are excluded from the intragranular pores, the intragranular fluid must be ion-free, and one simply has $\overline{C}^{z+} = (1-\varepsilon_i)\overline{\overline{C}}^{z+}$ and $\overline{H}^{+} = (1-\varepsilon_i)\overline{\overline{H}}^{+}$. Under this assumption, the most practical way to model ion-exchange resins uses the lumped model presented in Chapter 1.

The electroneutrality of the resin lumped particle is:

$$\overline{H}^{+} + \overline{C}^{z+} = \overline{N}^{iex} \quad \text{eq/l} \tag{4.56}$$

The ion-exchange process between ions H and C can be written:

$$z\overline{H}^{+} + C^{z+} \rightleftharpoons zH^{+} + \overline{C}^{z+} \tag{4.57}$$

and the equilibrium law:

$$K_{C/H} = \frac{\overline{C}^{z+}}{C^{z+}} \left(\frac{H^{+}}{\overline{H}^{+}} \right)^{z} \tag{4.58}$$

Unless otherwise specified, we will assume that the anions are excluded from the intragranular pores, so that Eqs. (4.56)–(4.58) hold.

While there is no thermodynamic reason to systematically use H as a reference ion, this choice is meaningful since protons are always present in solution, due at least to the ionic dissociation of water. Using H as a reference thus provides a general scheme for modeling cationic ion exchange.

If one assumes that cation C is *monovalent*, Eq. (4.56) and Eq. (4.58) lead to:

$$\overline{C}^{+} = \frac{\overline{N}^{iex} K_{C/H} C^{+}}{N_0 + \left(K_{C/H} - 1\right) C^{+}} = \overline{N}^{iex} \frac{\frac{K_{C/H}}{N_0} C^{+}}{1 + \frac{K_{C/H}-1}{N_0} C^{+}} \tag{4.59}$$

Consequently, provided that cation C is *monovalent* and that the binary C/H ion exchange is performed at *constant normality*, ion exchange is similar to single-solute adsorption, the "adsorption isotherm" being given by Eq. (4.59). The same result is obtained in the general case of *homovalent* ion exchange, meaning the exchange of two ions having the same valences. Additionally, if the selectivity coefficient $K_{C/H}$ is constant, the "adsorption isotherm" is Langmuir-like (note, however, that $K_{C/H} - 1$ can be either positive or negative). Let us further mention that Eq. (4.59) gives a similarity only and not a full equivalence with adsorption: while concentration C is free to take any value in adsorption, it is capped at N_0 in the case of ion exchange; this forces the denominator of Eq. (4.59) to be positive, fortunately.

The previous result can be extended to multi-solute ion exchange. When n cations ($\neq$ H) are considered, Eqs. (4.52), (4.56) and (4.58) become:

$$H^+ + \sum_i C_i^{z_i+} = A^- = N_0 \tag{4.60}$$

$$\overline{H}^+ + \sum_i \overline{C}_i^{z_i+} = \overline{N}^{iex} \tag{4.61}$$

$$K_{i/H} = \frac{\overline{C}_i^{z_i+}}{C_i^{z_i+}} \left(\frac{H^+}{\overline{H}^+}\right)^{z_i} \tag{4.62}$$

Equation (4.62) can be rewritten:

$$\overline{C}_i^{z_i+} = K_{i/H} \left(\frac{\overline{H}^+}{H^+}\right)^{z_i} C_i^{z_i+} \tag{4.63}$$

Provided that the fluid concentrations are known, after substitution of Eq. (4.63) in Eq. (4.61), one obtains an implicit equation in $\overline{H}^+/H^+$. Solving this implicit equation then allows calculation of all the lumped solid-phase concentrations $\overline{C}_i^{z_i+}$ by using Eq. (4.63).

For the simple case of *monovalent* ions, one obtains:

$$\frac{\overline{H}^+}{H^+} = \frac{\overline{N}^{iex}}{H^+ + \sum_i K_{i/H} C_i^+} \tag{4.64}$$

which leads with Eq. (4.60) and Eq. (4.63) to:

$$\overline{C}_i^+ = \frac{\overline{N}^{iex} K_{i/H} C_i^+}{N_0 + \sum_j \left(K_{j/H} - 1\right) C_j^+} \tag{4.65}$$

Equation (4.65) can be represented by an expression formally identical to the Langmuir adsorption isotherms with species-dependent saturation capacities:

$$\overline{C}_i^+ = \frac{\overline{N}_i \tilde{K}_i C_i^+}{1 + \sum_j \tilde{K}_j C_j^+} \quad \text{with} \quad \tilde{K}_i = \frac{K_{C_i/H} - 1}{N_0} \quad \text{and} \quad \overline{N}_i = \overline{N}^{iex} \frac{K_{C_i/H}}{K_{C_i/H} - 1} \tag{4.66}$$

Consequently, *at constant normality*, an $(n + 1)$-solute ion-exchange system (n cations C_i and H) is *similar* to an n solute adsorption system (plus one inert). Moreover, the apparent "adsorption isotherms" of monovalent ions are Langmuir-like if the selectivity coefficients are constant.

For investigation purposes, it is usually advisable to work at constant fluid normality N_0, and to consider ionic fractions rather than concentrations:

$$X_i = \frac{C_i^{z_i+}}{N_0} \qquad \overline{X}_i = \frac{\overline{C}_i^{z_i+}}{\overline{N}^{iex}} \tag{4.67}$$

With this set of variables, Eq. (4.60) and Eq. (4.61) become:

$$X_H + \sum_i X_i = 1 \qquad \overline{X}_H + \sum_i \overline{X}_i = 1 \tag{4.68}$$

The mass action law (4.62) can be written in terms of ionic fractions:

$$K_{i/H}^x = \frac{\overline{X}_i}{X_i}\left(\frac{X_H}{\overline{X}_H}\right)^{z_i} \quad \text{with} \quad K_{i/H}^x = K_{i/H}\left(\frac{\overline{N}^{iex}}{N_0}\right)^{z_i-1} \tag{4.69}$$

In agreement with the ideal mass action law, experience shows that in general the selectivity coefficient $K_{i/H}$ is *relatively* independent of normality.[4] Consequently, for ideal systems, Eq. (4.69) shows that in the case of a monovalent cation C, the ionic fractions are related by a coefficient $K_{i/H}^x$ which is almost independent of normality and identical to the selectivity coefficient $K_{i/H}$. In case of a multivalent cation C, the ionic fractions are related by a coefficient $K_{i/H}^x$ which increases when normality decreases according to Eq. (4.69).

For binary ion exchange, the so-called *ion-exchange isotherm* is usually represented in an $\left(X_i, \overline{X}_i\right)$ diagram at constant normality N_0. For binary ion exchange, the system of Eq. (4.68) and Eq. (4.69) leads to explicit solutions:

Monovalent cation:

$$\overline{X}_i = \frac{K_{i/H}^x X_i}{1 + \left(K_{i/H}^x - 1\right) X_i}$$

Divalent cation (format adapted to numerical calculations):

$$\overline{X}_i = \frac{K_{i/H}^x X_i}{K_{i/H}^x X_i + \dfrac{(1-X_i)^2}{2} + (1-X_i)\sqrt{K_{i/H}^x X_i + \dfrac{(1-X_i)^2}{4}}} \tag{4.70}$$

As shown in Figure 4.1, it follows from Eq. (4.69) and Eq. (4.70) that the affinity of the resin for monovalent ions towards H is not affected by the solution normality, whereas the affinity of the resin for divalent (or more generally multivalent) ions towards H increases when normality decreases.

Let us now derive Henry's coefficients. It must be pointed out that linear adsorption occurs at moderate concentrations, whereas linear ion exchange occurs at moderate *ionic fractions*. The problem is not to know whether the concentration of a given cation is small, but whether it is significantly smaller than the normality of the solution. More generally, an ion-exchange process can be locally linearized when ionic fraction

[4] In fact, experience shows that this selectivity coefficient is in general slightly decreasing with normality due to non-selective electrolyte uptake.

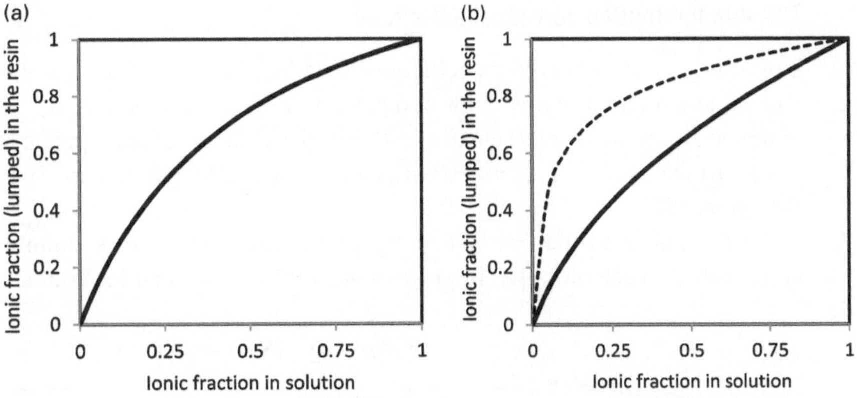

Figure 4.1 Typical ion-exchange isotherms. (a) Monovalent ion against H: $K^x_{i/H} = K_{i/H} = 3$. (b) Divalent ion against H: $K_{i/H} = 3$. Solid line $\overline{N}^{iex}/N_0 = 1$; dashed line $\overline{N}^{iex}/N_0 = 10$.

variations are small, as opposed to an adsorption process, which can be locally linearized when concentration variations are small.

As an illustration, if one considers the situation of binary ion exchange with:

$$X_i \ll 1 \quad \text{and} \quad \overline{X}_i \ll 1 \qquad \text{thus} \quad X_H \approx \overline{X}_H \approx 1 \tag{4.71}$$

the equilibrium relationship Eq. (4.69) then gives:

$$\overline{X}_i \approx K^x_{i/H} X_i \approx K_{i/H} X_i \left(\frac{\overline{N}^{iex}}{N_0}\right)^{z_i - 1} \tag{4.72}$$

or, in the concentration domain:

$$\overline{C}_i \approx \overline{K}_i C_i \qquad \text{with} \qquad \overline{K}_i \approx K_{i/H} \left(\frac{\overline{N}^{iex}}{N_0}\right)^{z_i} \tag{4.73}$$

All expressions given in Chapter 2 for linear chromatography can be used for ion exchange, provided that the initial slope of the adsorption isotherm is estimated with Eq. (4.73). For instance, the retention time associated with a small injection of a cation in a column eluted with hydrochloric acid at *constant normality* is given by:

$$t_R(i) = t_0 \left(1 + \frac{1-\varepsilon_e}{\varepsilon_e} \overline{K}_i\right) = t_0 \left(1 + \frac{1-\varepsilon_e}{\varepsilon_e} K_{i/H} \left(\frac{\overline{N}^{iex}}{N_0}\right)^{z_i}\right) \tag{4.74}$$

In agreement with many experimental results, Eq. (4.74) shows that the retention time of ions in ion-exchange chromatography increases when normality decreases.

The theoretical framework presented in this section is probably a little bit simplistic because it assumes constant selectivity factors and no electrolyte sorption, but it usually gives a good basis for ion-exchange modeling. It will be refined in Sections 4.5.3 and 4.5.4.

4.5.3 The lumped model: non-ideal situations

Unfortunately, the selectivity coefficients used in Section 4.5.2 are not always constant, and some strong variations with composition have been reported, for instance when amino acids are involved (Dye *et al.*, 1990). Some more refined models for describing ion exchange involving activity coefficients are available (Helfferich, 1962; Myers and Byington, 1986).

An alternative to the use of activity coefficient models is the simple and flexible empirical approach proposed by (Dye *et al.*, 1990), which can be written as:

$$K_{i/H} = \langle K_i \rangle \frac{(1/w_i + w_i)X_H + \sum_{k=1}^{n} \langle K_k \rangle (w_k/w_i + w_i/w_k) X_k}{2X_H + \sum_{k=1}^{n} \langle K_k \rangle (w_k/1 + 1/w_k) X_k} \tag{4.75}$$

Each species different from H is characterized by two parameters: $\langle K_i \rangle$, which gives the mean value of the selectivity coefficient over the ionic fraction range, and w_i, which allows the selectivity coefficient to vary with ionic fraction. Parameters $\langle K_i \rangle$ and w_i are determined with the binary mixtures C_i/H.

Notice some interesting properties of Eq. (4.75):

$$K_i = \begin{cases} \langle K_i \rangle & \text{if all } w_k = 1 \\ \langle K_i \rangle \dfrac{(1/w_i + w_i)}{2} & \text{if } X_k \ll X_H \\ \langle K_i \rangle \dfrac{2}{1/w_i + w_i} & \text{if } X_k \gg X_H \end{cases}$$

These allow a simple connection between the average parameters $\langle K_i \rangle$ and the values at the two extremities of the adsorption isotherm.

This approach, with constant or variable selectivity coefficients, assumes that the exchange between ionic solutes is represented by the stoichiometric reaction in Eq. (4.57). Every time a cation of valence z_+ is attached to a binding site, z_+ monovalent ions (H in our example) must leave their binding sites to enter the solution. The sum of the lumped solid concentrations of the different ions expressed in eq/l is constant and equals the ion-exchange capacity $\overline{N}^{iex}$. The situation is more complex when one considers ion-exchange processes involving very large molecules like proteins. The proteins, while interacting via an ion-exchange process, can shield the surface of the ion exchanger, thus blocking more ion-exchange sites than their valence would predict. These blocked sites are not associated with the ionic sites of the protein, but with other counter-ions, and they are unavailable for ion exchange due to steric shielding. This is schematically represented in Figure 4.2.

This effect has been recognized by (Whitley *et al.*, 1989) and modeled by (Brooks and Cramer, 1992) for multi-solute mixtures. The underlying concepts, known as the *steric mass action model*, is briefly summarized below for the case when one single

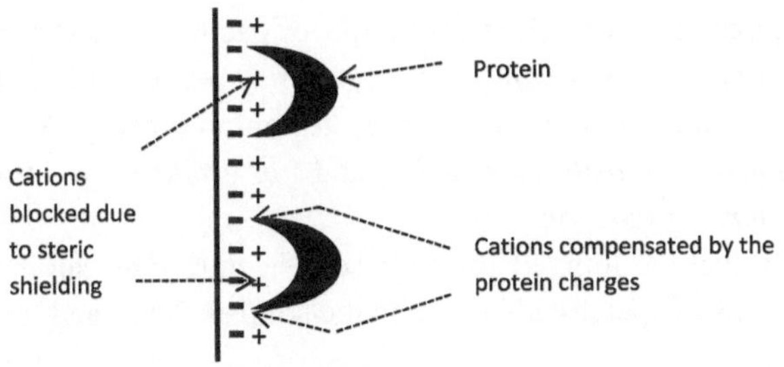

Figure 4.2 Schematic representation of ion-exchange sites shielded by a protein.

protein P of characteristic charge z exchanges with protons H. The electroneutrality Equation (4.56) written for the lumped resin particle is:

$$\overline{H}^{+} + \overline{H}^{+}_{blocked} + \overline{P}^{z+} = \overline{N}^{iex} \quad \text{(eq/l)} \tag{4.76}$$

where $\overline{H}^{+}_{blocked}$ represents the concentration (eq/l) of protons H^+ that are shielded by the protein immobilized and thus unavailable for ion exchange. If one assumes that $\overline{H}_{blocked}$ is proportional to the concentration of fixed protein, Eq. (4.76) can be written:

$$\overline{H}^{+} + (1 + \sigma_P)\overline{P}^{z+} = \overline{N}^{iex} \quad \text{(eq/l)} \tag{4.77}$$

where σ_P is an empirical steric shielding factor. When $\sigma_P = 0$ the protein has no steric shielding effect.

From Eq. (4.77) one sees immediately that the maximum protein concentration in the resin is given by $\overline{N}^{iex}/(1 + \sigma_P)$, expressed in eq/l.

The mass action law is expressed by equations similar to Eq. (4.58):

$$K_{P/H} = \frac{\overline{P}^{z+}}{P^{z+}} \left(\frac{H^{+}}{\overline{H}^{+}} \right)^{z} \tag{4.78}$$

so the introduction of the steric shielding factor does not add any real difficulties to the calculation of the lumped solid concentrations from the above two equations. However, it significantly changes the ion-exchange isotherm shape by preventing the protein from compensating all the sites that valence considerations would predict.

4.5.4 The porous model

A more descriptive presentation of the resin particle, based on the porous model, is useful for a better understanding of the internal mechanisms that can impact not only ion-exchange selectivity but also electrolyte sorption, swelling and intraparticular diffusion.

Before entering into the details, recall that ion-exchange resins are used not only because of their ability to exchange ions but also because of their ability to specifically interact with electrolytes and non-charged species. According to their charge, dipole moment, size etc., species are more or less free to enter the intragranular pores of the resin without being part of an ion-exchange process. Usually, non-ionic species enter freely while anions are excluded from cationic resins due to repulsion by the negatively charged fixed sites. The exclusion of anions also requires exclusion of some associated cations for electroneutrality reasons, so that electrolytes may possibly be excluded from resins. The roles of cation and anion are exchanged for anionic resins. Various terms, including *electrolyte uptake*, *electrolyte exclusion* and *ion exclusion* are used for this effect, also known as *Donnan exclusion*.

In Chapter 1 we noted that the lumped solid-phase concentrations can in general be related to the concentrations of species adsorbed on the solid skeleton and located in the intragranular fluid by:

$$\overline{C} = \varepsilon_i \grave{C} + (1 - \varepsilon_i)\overline{\overline{C}} \quad \text{for ionic or non-ionic solutes} \tag{4.79}$$

When describing adsorption processes in the absence of size exclusion, the assumption $\grave{C} = C$ is normally legitimate, but the situation is certainly different with ion-exchange resins.

In Section 4.5.2, using the lumped model to describe ion-exchange resins, we assumed that the total lumped concentration of cations in the particle $\overline{H}^+ + \overline{C}^{z+}$ equals the lumped ion-exchange capacity of the resin $\overline{N}^{iex}$. A consequence of this assumption is that no anions different from the resin fixed charges can be present in the resin. In other words, we assumed the electrolyte to be totally excluded from the intragranular pores.

In reality, the concentration of ions in the intragranular pores is probably not as high as for adsorption mechanisms ($\grave{C} = C$) but is certainly not zero. Knowing the concentration of ions in the intraparticular porosity is not purely academic as it drives ion exclusion, a technology used at very large scale, for instance, for desalting.

In order to model resin particles more precisely, let us consider a cationic resin in contact with a strong electrolyte C^+A^-. In general, the electroneutrality equation at the lumped resin level can be written:

$$\overline{N}^{iex} + \overline{A}^- = \overline{C}^{z+} + \overline{H}^+ \quad \text{(eq/l)} \tag{4.80}$$

where $\overline{A}^-$ is the lumped concentration of anions A^- in the resin (eq/l).

Note that the use of the lumped concentrations does not assume any specific location for the cations: they can be either on the skeleton or in the intragranular fluid.

Since the anions A^- are assumed to be located in the intragranular pores of the resin only, one can introduce $\grave{A}^- = \overline{A}^-/\varepsilon_i$ as the concentration of anions A^- in the intragranular pores of the resin, so that Eq. (4.80) becomes:

$$\overline{N}^{iex} + \varepsilon_i \grave{A}^- = \overline{C}^{z+} + \overline{H}^+ \quad \text{(eq/l)} \tag{4.81}$$

Equation (4.81) can equivalently be written using the ion-exchange capacity $\overline{\overline{N}}^{iex}$, defined as the number of ion-exchange sites divided by the volume of skeleton, $\overline{\overline{N}}^{iex} = \overline{N}^{iex}/(1 - \varepsilon_i)$. One obtains an expression clearly allocating the fixed ionic charges to the skeleton and the free anions to the intragranular pores when the cations are described using lumped concentrations:

$$(1 - \varepsilon_i)\overline{\overline{N}}^{iex} + \varepsilon_i \grave{A}^- = \overline{C}^{z+} + \overline{H}^+ \quad \text{(eq/l)} \tag{4.82}$$

Following the presentation given in Chapter 1, we define the *partition coefficient* $\grave{K}_A$ of anions A^- as the ratio between their concentration in the intragranular fluid and in the extragranular solution:

$$\grave{K}_{A^-} = \frac{\grave{A}^-}{A^-} = \frac{\grave{A}^-}{N_0} \tag{4.83}$$

Experience shows that the experimental partition coefficient $\grave{K}_{A^-}$ is in general less than 1 and varies with experimental conditions.

In order to understand the key drivers affecting the concentration of ions in the intraparticular porosity, we will start with a simplistic model of a resin that is unable to separate ions of the same valences. Instead of focusing on the "selective adsorption capacity" of the resin, we will focus on the electrical properties of the system. The approach may look a bit strange to chromatographers accustomed to considering that chromatographic media can separate even homovalent ions, but it will bear interesting fruit.

The concept adapts the electrical double layer (EDL) theory presented by (Bolt, 1982) for soil science. Because of electrostatic forces, when an electrolyte is put in contact with a negatively charged surface, the cation concentration increases in moving from the bulk solution to the surface, while the anion concentration moves in the opposite direction. A qualitative picture is given in Figure 4.3a.

A consequence of this description is that electroneutrality is not satisfied over a distance from the surface of the order of the characteristic thickness of the electrical double layer δ_{EDL}.[5] The concentration profiles can be determined by calculating the electrical potential and assuming a Maxwell–Boltzmann distribution for the concentrations. It can be shown that the characteristic thickness of the electrical double layer, δ_{EDL}, is proportional to $1/\sqrt{N_0}$. This thickness is about 3 Å for a normality of 1 eq/l and about 30 Å for a normality of 0.01 eq/l. At distances from the negatively charged surface greater than δ_{EDL}, the influence of the surface vanishes so that the solution is electrically neutral.

The particles of resin are now assimilated into a lattice of electrically charged parallel plates separated by a given distance which reflects the pore size. This distance can be smaller or greater than the thickness of the electrical double layer, depending on the type of resin (gel or macroporous) and the normality of the bulk solution. If the resin pore size is as small as a few ångström, it can be smaller than the characteristic length δ_{EDL} so that the solution contained in these pores is not electrically neutral. In our notation, if the intragranular pores are small enough, intragranular concentrations $\grave{A}^-$ and $\grave{H}^+ + \grave{C}^{z+}$ are not equal.

A schematic representation of the above concepts is presented in Figure 4.3b.

It is apparent that, due to the interference between the concentration profiles associated with the two electrically charged parallel plates, anions can be excluded from the

[5] The first layer is the surface charge (either positive or negative). The second layer is composed of ions attracted to the surface charge via the Coulomb force.

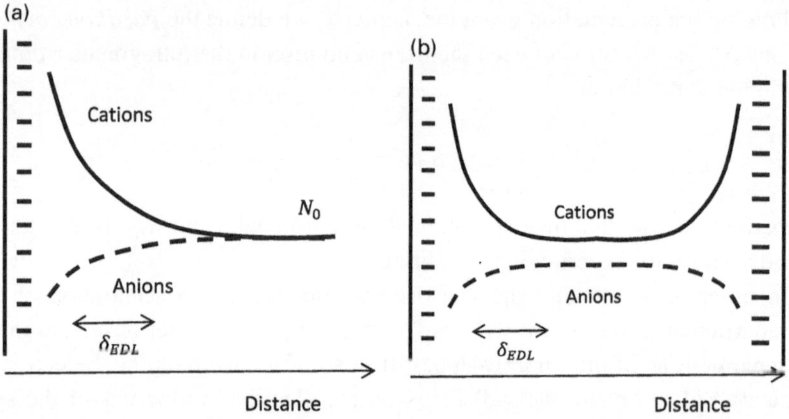

Figure 4.3 Schematic representation of the cation and anion profiles according to EDL theory. (a) Single negatively charged plate in contact with an electrolyte. (b) Two negatively charged plates; distance between the plates is expected to be representative for intragranular pore size.

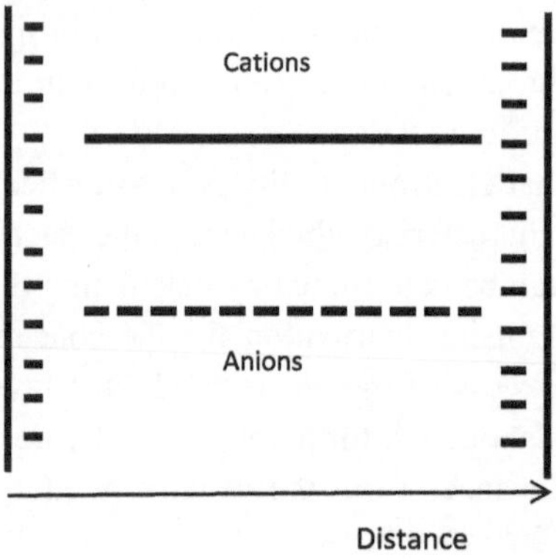

Figure 4.4 Schematic representation of cation and anion profiles according to simplified EDL theory: uniform Maxwell–Boltzmann factor in the pores. The concept is acceptable for gel resins.

interstitial volume when the distance between the plates is smaller than about twice the characteristic length δ_{EDL}.

Rather than a well-identified and positioned boundary between the ions fixed on the skeleton and the ions in the intragranular liquid, the EDL theory considers a continuous variation of the cation and anion concentrations according to their position with respect to the pore walls.

Instead of solving the complex spatially dependent profiles, (Nicoud and Schweich, 1989) proposed a simple model based on averaged Maxwell–Boltzmann factors that are related to the electrical potential between the plates. According to this model, the concentration profiles between the plates are taken to be uniform and approximated by their average values. A schematic is given in Figure 4.4. The free species in the pores at concentrations $\grave{A}^-$, $\grave{H}^+$ and $\grave{C}^{z+}$ are related to the average Maxwell–Boltzmann factor w by:

$$\grave{C}^{z+} = C^{z+}w^z \qquad \grave{H}^+ = H^+w \qquad \grave{A}^- = N_O w^{-1} \tag{4.84}$$

The Maxwell–Boltzmann factor w is a real number that must be greater than or equal to unity. We assume first that there are no specific interactions between the cations and the negative charges fixed on the walls, so that all the cations are contained in the free space between the plates. According to this first simple representation, since the cations are in solution and not specifically bound to the skeleton, the electroneutrality equation holds at the lumped resin level only and not separately in the skeleton and the intragranular pore phases. Additionally, Eq. (4.84) shows that the ions differ only in their valence. In the absence of specific interactions between the cations and the negative charges fixed on the walls, the model is thus not able to predict selectivity between homovalent ions. This assumption will be relaxed later.

We will further assume that the Maxwell–Boltzmann factor is uniform between the "plates" so that concentrations are uniform as well (Figure 4.4). If the Maxwell–Boltzmann factor is known, the concentrations $\grave{C}^{z_c}$, $\grave{H}^+$ and $\grave{A}^-$ of the free species in the resin pores can be calculated by Eq. (4.84). As the Maxwell–Boltzmann factor w is by definition greater than or equal to unity, the anion concentration $\grave{A}^-$ is less than or equal to the cation concentration $\grave{C}^{z+} + \grave{H}$, so the solution contained in the intragranular pores is in general not neutral. Physically one can assume this deviation from electroneutrality to be real for small pores and negligible for wide pores for which $w = 1$.

Under the above assumptions, one has $\overline{\overline{C}}^{z+} = \overline{\overline{H}}^+ = 0$ (all the cations are in the pores) so that $\varepsilon_i\grave{C}^{z+} = \overline{C}^{z+}$ and $\varepsilon_i\grave{H}^+ = \overline{H}^+$, and the electroneutrality equation (4.81) at the resin level is written:

$$\overline{N}^{iex} + \varepsilon_i\grave{A}^- = \varepsilon_i\grave{C}^{z+} + \varepsilon_i\grave{H}^+ \tag{4.85}$$

or, if one prefers, from Eq. (4.82):

$$(1-\varepsilon_i)\overline{\overline{N}}^{iex} + \varepsilon_i\grave{A}^- = \varepsilon_i\grave{C}^{z+} + \varepsilon_i\grave{H}^+ \tag{4.86}$$

The problem is solved by substituting Eq. (4.84) into the electroneutrality equation (4.85), which gives, for a monovalent cation C:

$$w = \frac{1+\sqrt{1+(2\varepsilon_i N_O/\overline{N}^{iex})^2}}{2\varepsilon_i N_O/\overline{N}^{iex}} \tag{4.87}$$

and with Eq. (4.84):

$$\grave{A}^- = N_O \frac{2\varepsilon_i N_O/\overline{N}^{iex}}{1+\sqrt{1+(2\varepsilon_i N_O/\overline{N}^{iex})^2}} \quad \text{with} \quad N_O = A^- \tag{4.88}$$

This equation can be written in a more flexible form, useful for experimental data fitting:

$$\grave{A}^- = \grave{f}_A(A^-) = N_0 \frac{N_0/\overline{N}_D}{1+\sqrt{1+\left(N_0/\overline{N}_D\right)^2}} \quad \text{with} \quad N_0 = A^- \tag{4.89}$$

In principle, the parameter $\overline{N}_D$ equals $\overline{N}^{iex}/(2\varepsilon_i)$, but it can also be seen as an adjustable parameter having the dimensions of concentration (eq/l). The parameter $\overline{N}_D$ should certainly depend on the ionic capacity $\overline{N}^{iex}$ of the resin, but also on the pore geometry, a parameter which is not taken into account in this first simplistic model.

Equation (4.89) is the partition law, $\grave{A}^- = \grave{f}_A(A^-)$, mentioned in Chapter 1, Eq. (1.13). It can be expressed in terms of the partition coefficient $\grave{K}_{A^-} = \grave{A}^-/A^-$:

$$\grave{K}_{A^-} = \frac{N_0/\overline{N}_D}{1+\sqrt{1+\left(N_0/\overline{N}_D\right)^2}} \quad \text{with} \quad N_0 = A^- \tag{4.90}$$

Equation (4.90) shows that the partition coefficient $\grave{K}_{A^-}$ varies with normality. It is proportional to the solution normality for dilute solutions and thus equals zero when normality reaches zero: the ion-exclusion effect is thus predicted. When the normality increases, the partition coefficient $\grave{K}_{A^-}$ increases to reach 1 at large normalities.

The partition law can be investigated thanks to RTD measurements and more specifically by determining the retention time of a small injection of electrolyte (A^-,C^{z+},H^+) in a column equilibrated with the same electrolyte at a very similar concentration to avoid any possible ion-exchange effects. Notice that the peak of anions A^- elutes together with an identical peak of cations in order to ensure electroneutrality of the fluid mobile phase. The theoretical framework developed in Chapters 2 and 3 has shown that this retention time is given by:

$$t_R = t_0\left(1+\frac{1-\varepsilon_e}{\varepsilon_e}\bar{f}'_A\right) \tag{4.91}$$

where $\bar{f}'_A$ represents the derivative of the function $\bar{f}_A(A^-)$ relating the lumped solid-phase concentration $\bar{A}^-$ of anions A to their concentration A^- in the extragranular solution. This function is simply related to the partition law $\grave{f}_A$ by $\bar{f}_A = \varepsilon_i\grave{f}_A$.

Note that Eq. (4.91) uses t_0, the Z_eRT, as reference. When $N_0/N_D \ll 1$, the anion is excluded from the resin, $\grave{f}'_A$ is close to 0 and the retention time is close to t_0. When $N_0/\overline{N}_D \gg 1$, anions are no longer excluded, $\grave{f}'_A$ is close to 1, thus $\bar{f}_A$ is close to ε_i and t_R equals t_{0R}, the Z_tRT. If the extragranular and intragranular porosities are known, a small pulse of electrolyte gives access to $\grave{f}'_A$. If the intragranular porosity is unknown, the same experiment gives access to the product $\varepsilon_i\grave{f}'_A$.

Some typical curves representing the product $\varepsilon_i\grave{f}_A$ versus normality obtained on different resins are given in Figure 4.5. It is apparent that Eq. (4.90) gives enough flexibility to fit experimental curves and that, in agreement with the principles of the EDL model,

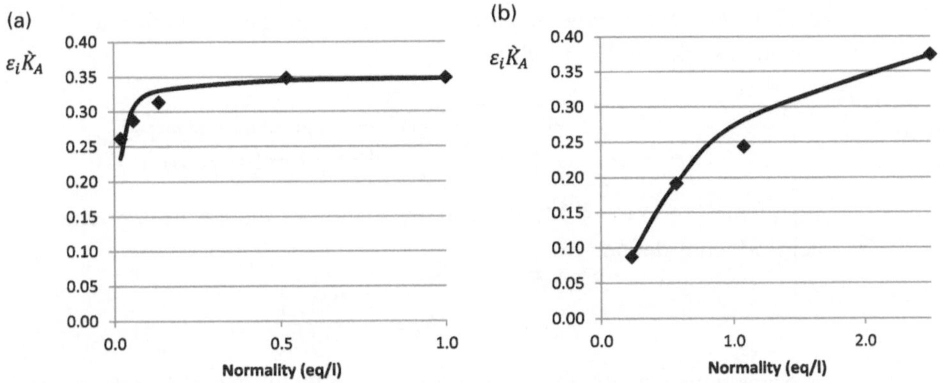

Figure 4.5 Influence of the normality on the partition coefficient of HCl on ion-exchange resins. (a) Macroporous resin (Nicoud, 1987). Continuous curve: Eq. (4.90) with: $\varepsilon_i = 0.35, \overline{N}_D = 0.008$ eq/l. (b) Gel resin (Pepper *et al.*, 1952). Continuous curve: Eq. (4.90) with: $\varepsilon_i = 0.47, \overline{N}_D = 0.58$ eq/l.

ion exclusion is less important for macroporous resins with large pore diameter than for gel resins.

Equation (4.90) has been derived for monovalent ions and there is no difficulty in extending the approach to multivalent ions, at the expense of the resolution of an implicit equation in w.

The previous theory, proposed for strong electrolytes, can be extended to weak electrolytes, as shown below for the case of a weak acid, HA. The ionic dissociation of the acid is modeled by:

$$K_d = \frac{H^+A^-}{(HA)} \tag{4.92}$$

The total concentration of acid in the extragranular solution is given by:

$$(HA)_{tot} = (HA) + A^- \tag{4.93}$$

Taking the electroneutrality constraint $A^- = H^+$ into account, Eq. (4.92) and Eq. (4.93) allow determination of the concentrations in solution. For instance, A^- is given by:

$$A^- = \frac{-K_d + \sqrt{K_d^2 + 4K_d(HA)_{tot}}}{2} \tag{4.94}$$

and concentration HA is then calculated with Eq. (4.93). The concentration of electrolyte in the intragranular pores of the resin is given by:

$$(\grave{H}A)_{tot} = (\grave{H}A) + \grave{A}^- \tag{4.95}$$

Since HA is a non-ionic species, we can assume that:

$$(\grave{H}A) = (HA) \tag{4.96}$$

Additionally, we assume that the concentration of anion A in the particle is given by an equation similar to Eq. (4.90):

$$\grave{A}^- = \frac{\bar{A}^-}{\varepsilon_i} = A^- \frac{A^-/N_D}{1+\sqrt{1+(A^-/N_D)^2}} \tag{4.97}$$

Equations (4.93) to (4.97) allow an evaluation of the partition coefficient $\grave{K}_{HA}$ of the acid, which is defined by:

$$\grave{K}_{HA} = \frac{(\grave{HA})_{tot}}{(HA)_{tot}} \tag{4.98}$$

Some typical qualitative results are given in Figure 4.6. When the ionization constant increases, the acid is more dissociated so the exclusion is more significant. The variation of $\grave{K}_{AH}$ with the acid or experimental conditions gives a basis for separating more or less dissociated acids (or generally electrolytes) with ion-exchange resins. Experimental and simulated results, reporting the influence of experimental conditions on retention times of weak acids on an ion-exchange resin in the H^+ form, are described by (Kaczmarski *et al.*, 2005).

The simple model presented above allows the representation of electrolyte sorption by ion-exchange resins, but it suffers from two important limitations:

- The distance between the electrically charged plates (and thus the pore size) is assumed to be small enough to allow an approximation of the concentration profiles in the intragranular pores by their average values. This approximation is probably valid for gel resins but questionable for macroporous resins. In order to model macroporous resins while keeping the simplicity of the approach, an option is to divide the pores into two zones (Nicoud, 1987). In the first zone, located

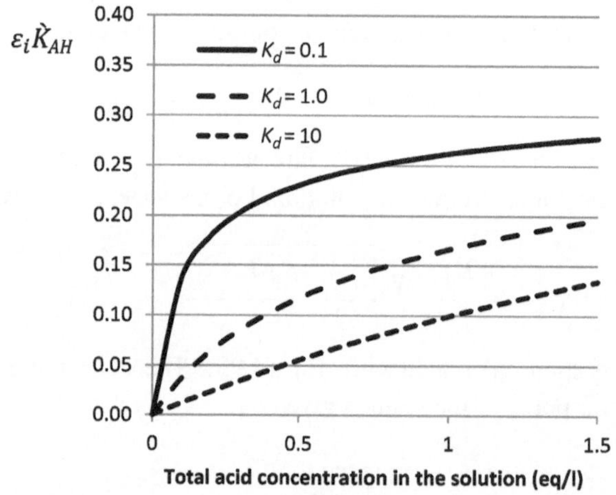

Figure 4.6 Influence of the dissociation constant of a weak acid on its distribution coefficient $\grave{K}_{AH}$. Curves simulated with Eq. (4.98) and $\varepsilon_i = 0.35$, $N_D = 2$ eq/l.

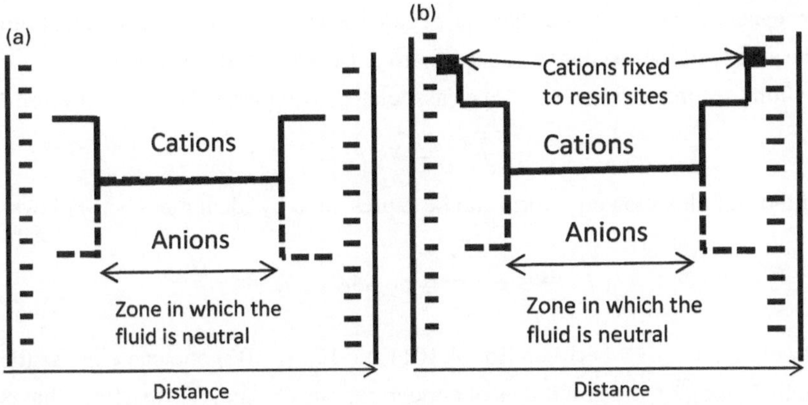

Figure 4.7 Schematic representation of the cation and anion profiles according to the simplified EDL theory. (a) The center of the pores is electrically neutral. The Maxwell–Boltzmann factor equals 1 in the center zone and takes a different but uniform value near the walls; this is an acceptable concept for macroporous resins. (b) In addition to the description in (a), cations can enter into a complexation mechanism with the fixed ionic sites of the resin.

at the center of the pores, the concentrations are assumed to be identical to their values in the extragranular porosity; in the second zone, located near the walls, the concentrations are assumed to obey a Maxwell–Boltzmann distribution similar to the one given by Eq. (4.84). A schematic description is given in Figure 4.7a. The relative thickness of these two zones is an additional parameter of the model. The development does not lead to specific difficulties and is not detailed further. Note that this refinement is still unable to induce a selectivity between homovalent ions.

- No specific interaction is considered between the cations and the negative charges fixed on the pore walls. This absence of specific interaction is the origin of the inability of the model to predict selectivity between ions of the same valence. A possibility of generalizing the model to take into account these selective interactions is presented below and in Figure 4.7b.

We assume that the cations contained in the intragranular pores can enter into a complexation mechanism with the ionic sites N^- of the resin, according to:

$$C^{z_+} + N^- \rightleftharpoons (NC)^{z_+ - 1} \tag{4.99}$$

One cation of valence z_+ is thus assumed to form a complex of valence $z_+ - 1$ with one free site of the resin.

Similarly, H^+ can enter into a complexation mechanism with the free site leading to a charge-free complex.

$$H^+ + N^- \rightleftharpoons (HC)^0 \tag{4.100}$$

In general, one can write that the negative sites of the resin in concentration $\overline{\overline{N}}^{iex}$, i.e. the so-called ion-exchange capacity, are either free (concentration $\overline{\overline{N}}^{-}$), associated with cation C (concentration $\overline{\overline{C}}^{z_+-1}$) or associated with proton H (concentration $\overline{\overline{H}}$):

$$\overline{\overline{N}}^{iex} = \overline{\overline{N}}^{-} + \overline{\overline{C}}^{z_+-1} + \overline{\overline{H}} \tag{4.101}$$

If the complexation equilibria can be represented by ideal mass action laws, one has:

$$\kappa_C = \frac{\overline{\overline{C}}^{z_+-1}}{\grave{C}^{z_+}\overline{\overline{N}}^{-}} \quad \text{and} \quad \kappa_H = \frac{\overline{\overline{H}}}{\grave{H}^{+}\overline{\overline{N}}^{-}} \tag{4.102}$$

By eliminating $\overline{\overline{N}}^{-}$ between Eq. (4.101) and Eq. (4.102) one can express the concentrations $\overline{\overline{H}}$ and $\overline{\overline{C}}^{z_+-1}$ as functions of concentrations $\grave{C}^{z_+}$ and $\grave{H}^{+}$ in a form that is equivalent to a Langmuir adsorption isotherm (should the parameters κ_C and κ_H be constant):

$$\overline{\overline{C}}^{z_+-1} = \frac{\overline{\overline{N}}^{iex}\kappa_C\grave{C}^{z_+}}{1 + \kappa_C\grave{C}^{z_+} + \kappa_H\grave{H}^{+}} \tag{4.103}$$

Notice that Eq. (4.103) is not an experimentally accessible resin–solution adsorption isotherm. It only gives a relationship between internal concentrations of solutes located in zones, being not even electrically neutral.

As an illustration, let us assume that the cation C is monovalent. Under this assumption, the cations $\overline{\overline{C}}$ complexed on the sites are charge-free. The associated concentration is given by Eq. (4.103), and for short we will write it $\overline{\overline{C}}^{z=0}$. As complexed cations mask a fraction of the sites, the electroneutrality Eq. (4.82) can be written:

$$(1 - \varepsilon_i)\overline{\overline{N}}^{-} + \varepsilon_i\grave{A}^{-} = \varepsilon_i\grave{C}^{+} + \varepsilon_i\grave{H}^{+}$$

With the site conservation equation (4.101), one obtains:

$$(1 - \varepsilon_i)\overline{\overline{N}}^{iex} + \varepsilon_i\grave{A}^{-} = (1 - \varepsilon_i)\overline{\overline{C}}^{z=0} + \varepsilon_i\grave{C}^{+} + (1 - \varepsilon_i)\overline{\overline{H}} + \varepsilon_i\grave{H}^{+} \tag{4.104}$$

Substitution of Eqs. (4.103) and (4.84) into the electroneutrality equation (4.104) gives an implicit equation in w.[6] If this equation is solved, all the different concentrations occurring in the particle are known, thanks to Eq. (4.84) and Eq. (4.103).

Finally, the selectivity coefficients $K_{C/H}$ can be calculated from the determined solid-phase concentrations, and their dependence on bulk solution normality assessed. This model, encompassing in a single-frame electrolyte sorption (non-selective) and a selective ion-exchange mechanism, predicts a moderate decrease in the selectivity coefficients $K_{C/H}$ with normality. This is in agreement with many experimental results (Nicoud, 1987).

Not only does the porous model, including the EDL theory, have the advantage of illustrating what happens in the particle, but it also gives a way to describe, in a single frame, electrolyte sorption (Donnan exclusion), ion exchange, swelling (the swelling

[6] For simplicity we assume a uniform w in the pores. There is no difficulty in adding a neutral zone associated with $w = 1$ located at the center of the pores, as qualitatively described in Figure 4.7a. This refinement allows the influence of pore size on resin behavior to be taken into account.

force can be seen as being proportional to the amount of uncompensated fixed negative charge $\overline{\overline{N}}^-$ due to the repulsion forces) and even the variation of diffusion coefficients with various experimental parameters, as presented in Chapter 5. Unfortunately, this approach is seldom used.

4.6 The case of size exclusion

Size exclusion is a technique aimed at separating molecules according to their size. As illustration, many non-ionic compounds can be separated according to their size on polystyrene cross-linked with DVB, while hydrophilic polymers based on methacrylate are useful for separating polar compounds.

Solute molecules can only enter intragranular pores that are large enough to accommodate them. In order to give a simple picture, let us assume that a solute A of size 100 (in arbitrary units) interacts with a particle having pores of size 150, representing a fraction $x\%$ of the total intragranular pore volume, and pores of size 50, representing a fraction $(1 - x)\%$ of the total intragranular pore volume. At equilibrium, one may assume that the intragranular pores of size 150 are filled with the extragranular solution containing concentration C_A of solute A, while the intragranular pores of size 50 are solute-free. The concentration of solute A contained in the intragranular pores per total volume of intragranular pores is thus:

$$\grave{C}_A = xC_A + (1 - x)0 = xC_A \tag{4.105}$$

By analogy with the partition law, Eq. (1.13), given in Chapter 1, which becomes $\grave{C}_A = \grave{K}_A C_A$ under the linearity assumption, this shows that the partition coefficient $\grave{K}_A$ is simply the fraction of intraparticular pores large enough to accommodate solute A. The reality is certainly more complex, as our simple picture does not take into account the pore or solute geometries. It is, however, sufficient to highlight the fact that the partition coefficient $\grave{K}_A$ is more a geometrical factor than a thermodynamic adsorption factor.

The exact relation between molecular size, pore size distribution and exclusion is normally not of primary interest in preparative chromatography, this information being summarized in the partition coefficient $\grave{K}_A$. For further information on the modeling of systems involving distributed-size solutes and/or distributed-size pores, the reader is referred to (Goto and McCoy, 2000) and (Saritha and Madras, 2001).

Because of the very nature of the size-exclusion process, one can expect temperature to have a very limited (if any) impact on the partition. This does not mean that temperature must not be controlled in order to minimize baseline noise and drift for chromatographic determinations. Additionally, temperature has an impact on kinetics and thus on peak shape by increasing the value of diffusion coefficients, as described in Chapter 5.

While in general having little influence, solvent selection can have a complex role. The solvent choice can affect the shape of branched macromolecules, which can be either relatively compact or rather elongated. An example is given by proteins, which can be globular in simple buffers and denatured when a denaturant like sodium dodecyl sulfate is added. By changing the macromolecule shape, the solvent can thus change the partition coefficient $\grave{K}_A$ by affecting the fraction of accessible intragranular pores.

The size-exclusion process is in general considered to be linear, as opposed to adsorption and ion exchange. There are, however, at least two factors that can induce non-linearity:

- Nothing prevents the solutes entering the intragranular pores from subsequently interacting with the solid skeleton via an ion-exchange or adsorption process. This situation is normally not desired, and is avoided by adequate solvent selection when a size-exclusion mechanism is sought for separation purposes.
- The presence of large molecules or adsorbed molecules inside the intragranular pores can block access to other solutes, as mentioned in Section 4.2.3. This mechanism, being probably concentration-dependent, thus non-linear, has not received a lot of modeling attention to the best of our knowledge.

4.7 Phenomenological models

4.7.1 Introductory concepts

Understanding the interactions between solutes and a chromatographic medium can be much more than representing adsorption or ion-exchange mechanisms. This has been briefly discussed in Section 3.7, but recall, for instance, that ionic equilibria in solution play an important role for amino acid, peptide and protein chromatography (see, for instance, (Guélat *et al.*, 2013)). Similarly, when considering weak ion-exchange resins, equilibria occurring in the solid phase are at the center of the chromatographic process. So far we have assumed that molecules are rigid and not subject to conformational changes during interaction with the chromatographic medium. We know this can be wrong, and chromatography can even be a way to promote the refolding of proteins (see, for instance, (Wellhoefera *et al.*, 2013)). For additional information on the possible influence of protein conformation on adsorption, the reader is referred to (Toth, 2002). In recent years, so-called mixed-mode chromatography has appeared, i.e. chromatographic methods that utilize more than one form of interaction between the solid phase and analytes in order to achieve their separation (ion exchange and hydrophobic interactions, for instance). By construction, these chromatographic media involve different mechanisms that must be taken into account simultaneously. The reader will find examples of mixed-mode systems studied in connection with fluid-phase equilibria in (Nfor *et al.*, 2010) and (Bernardi *et al.*, 2013).

The variety of possible situations is such that proposing a general framework is out of the question. This being said, all the situations mentioned above can be described by a set of stoichiometric relationships (or reactions for short) that describe at equilibrium the partition of the solutes between the fluid and the chromatographic medium or the relation between solutes in the different phases. A phenomenological mechanism is this set of stoichiometric relationships.

The development given below is largely based on the work of D. Schweich (Schweich *et al.*, 1993a,b). While initially proposed for describing pollutant migration in soils, the formalism is certainly helpful for structuring the description of complex chromatographic situations.

Let us start with the simple example of the Na/H exchange reaction:

$$\mathrm{Na}^+ + \overline{\overline{\mathrm{H}}}^+ \rightleftharpoons \overline{\overline{\mathrm{Na}}}^+ + \mathrm{H}^+ \qquad K_{Na/H} = \frac{\overline{\overline{Na}}^+\, H^+}{Na^+\, \overline{\overline{H}}^+} \tag{4.106}$$

Thermodynamic equilibrium is described by an ideal mass action law for each stoichiometric relationship. By "ideal mass action law" we mean that activities are identical to concentrations or normality.

Langmuir adsorption can be dealt with similarly. For a single solute S one can write:

$$\mathrm{S} + \overline{\overline{\mathrm{F}}} \rightleftharpoons \overline{\overline{\mathrm{S}}} \qquad \tilde{K}_S = \frac{\overline{\overline{S}}}{S\overline{\overline{F}}} \tag{4.107}$$

where $\overline{\overline{F}}$ is the so-called *free* (or *vacant*) *adsorption site* of the solid phase. The reader may have already guessed that the total concentration of adsorbing material is *conserved* and that:

$$\overline{\overline{F}} + \overline{\overline{S}} = \overline{\overline{N}} \tag{4.108}$$

where $\overline{\overline{N}}$ is the total concentration of adsorbing sites. Combining Eq. (4.107) and Eq. (4.108) gives the usual Langmuir law:

$$\overline{\overline{S}} = \overline{\overline{N}}\frac{\tilde{K}_S S}{1 + \tilde{K}_S S} \tag{4.109}$$

We thus see that this adsorption isotherm equation results from two fundamental and distinct ideas:

- a stoichiometric description of the partition process given by Eq. (4.107)
- supplementary relationships that stipulate how some quantities are conserved, as expressed by Eq. (4.108).

EXERCISE 4.3 *Derive the general Langmuir lumped adsorption isotherm for a multi-component mixture of solutes S_i using the concepts of stoichiometric description (reactions) and conserved total adsorption site concentration.*

EXERCISE 4.4 *A protein may be present in two forms, denoted P (primary) and U (unfolded). The two forms are assumed to be at equilibrium and can adsorb on a chromatographic medium with specific mass action laws. Using a phenomenological description, derive the apparent adsorption isotherm relating the total protein concentration in the solid phase to the total protein concentration in solution.*

In Exercises 4.3 and 4.4, for instance, the mechanism is *phenomenological*. Unlike a *molecular* mechanism, it does not aim to give a fundamental explanation of the interaction process, but simply summarizes some key features, such as:

- the set of species involved and their weighted contribution (stoichiometry)
- electrical charge conservation in ion exchange
- the necessity of a free adsorption site in a Langmuir mechanism
- the existence of an upper bound to the adsorption capacity.

We mentioned at the end of Section 4.3.3 that, in order to be consistent, a thermodynamic model must satisfy certain constraints related to its Jacobian matrix $\left[\partial\overline{\overline{C}}_i/\partial C_k\right]$ (or $\left[\partial\overline{C}_i/\partial C_k\right]$). It is important for equilibrium laws to be thermodynamically consistent, otherwise the chromatographic model may yield wrong, if not nonsensical, predictions. A phenomenological mechanism is thermodynamically consistent because it relies on mass action laws which result from thermodynamics. At least, the author has never met an example of inconsistency when using this approach.

Let us stress that well-known empirical adsorption isotherms do not necessarily obey a phenomenological mechanism. Freundlich adsorption isotherms, Langmuir adsorption isotherms with capacities $\overline{\overline{N}}_i$ that depend on the solute i, and generalized Langmuir adsorption isotherms with a minus sign in front of $\tilde{K}_i$ in the denominator (unfavorable adsorption isotherms) are typical examples. They can lead to some puzzling (if not inconsistent) behaviors, such as the failure of Henry's law for Freundlich (infinite Henry's coefficient) and negative solid-phase concentrations with a minus-signed Langmuir adsorption isotherm at high fluid-phase concentration.

Similarly to the adsorption capacity $\overline{\overline{N}}$ in a Langmuir lumped adsorption isotherm, the total ion exchange capacity $\overline{\overline{N}}^{iex}$ is a *conserved* quantity. Let us stress that "conserved" does not necessarily mean "constant". For instance, the exchange capacity of a weak acid ion exchanger may vary according to the phenomenological mechanistic step:

$$\overline{\overline{\mathrm{NH}}} \rightleftharpoons \overline{\overline{\mathrm{N}}}^- + \overline{\overline{\mathrm{H}}}^+ \qquad K_{ion} = \frac{\overline{\overline{N}}^-\overline{\overline{H}}^+}{\overline{\overline{NH}}} \tag{4.110}$$

where $\overline{\overline{N}}^-$ are the free exchanging sites and $\overline{\overline{NH}}$ are the non-ionized and non-exchanging "potential" sites. Thus the ion-exchange capacity $\overline{\overline{N}}^{iex}$, used for instance in Section 4.5.2, reduces to $\overline{\overline{N}}^-$, which is not constant along the Na/H exchange. Conversely, the total concentration of sites is constant:

$$\overline{\overline{NH}} + \overline{\overline{N}}^- = \overline{\overline{N}}_{tot} \tag{4.111}$$

Combining Eq. (4.106), Eq. (4.110) and Eq. (4.111) gives:

$$\overline{\overline{H}} = \frac{K_{ion}\overline{\overline{N}}_{tot}}{K_{ion} + \overline{\overline{H}}^+} \; \frac{H^+}{H^+ + K_{Na/H}Na^+} \tag{4.112}$$

which is a quadratic equation in $\overline{\overline{H}}$. The positive solution of this equation is rather an unfamiliar isotherm equation; however, it is thermodynamically consistent.

Up to now, the mass action laws associated with the reactions yielded equations that can be analytically solved for the concentrations in the solid phase. However, there is an odd and unpleasant mathematical theorem that stipulates that there is no general closed-form or explicit solution for an algebraic equation of the fifth degree or more: thus, as an illustration, there is no possible explicit "adsorption isotherm" for ion exchange that involves a binary mixture of monovalent and pentavalent ions. One does not need to work with pentavalent ions to find situations preventing the derivation of explicit expressions: one heterovalent ion-exchange mechanism associated with one or two equilibria in solution will usually be sufficient!

The impossibility of obtaining, in general, an explicit expression $\bar{\bar{C}}_i = \bar{\bar{f}}_i(C_1, C_2, \ldots C_n)$ is a general and somewhat unfortunate property of a phenomenological mechanism. This is a consequence of the structure of the well-known equality of chemical potentials at equilibrium:[7] $\mu_i = \bar{\bar{\mu}}_i$ for all species i. Let us remember that the IAS theory does not in general lead to explicit expressions either. Accepting an *implicit* and intricate dependence between the C_i and the $\bar{\bar{C}}_i$, instead of a simple *explicit* expression $\bar{\bar{C}}_i = \bar{\bar{f}}_i(C_1, C_2, \ldots C_n)$, is often the price to pay for thermodynamic consistency.

4.7.2 Some general properties of a phenomenological mechanism

Let S_i and $\bar{\bar{S}}_i$ be a chemical component in the fluid and on the chromatographic solid phase, respectively. These will be considered as two distinct "species", although they have the same chemical formula. The word "species" will be used with this enlarged definition later on. There are n_T such species; n_T is also the number of concentration variables that describe the composition of the fluid–solid mixture. Let n_f and $n_{\bar{\bar{s}}}$ be the number of those species in the fluid and solid phases, respectively. One has:

$$n_T = n_f + n_{\bar{\bar{s}}} \tag{4.113}$$

In the ion-exchange mechanism defined by Eq. (4.106), there are $n_T = 4$ species, with $n_f = n_{\bar{\bar{s}}} = 2$ in each phase.

A phenomenological mechanism can in general be written as a set of R reactions:

$$\sum_{j=1}^{n_f} \nu_{ij} S_j + \sum_{j=1}^{n_{\bar{\bar{s}}}} \bar{\bar{\nu}}_{ij} \bar{\bar{S}}_j = 0 \qquad i \in [1,R] \tag{4.114}$$

where ν_{ij} (resp. $\bar{\bar{\nu}}_{ij}$) is the stoichiometric coefficient (>0 for a product, <0 for a reactant) of S_j (resp. $\bar{\bar{S}}_j$) in the ith reaction. This can be summarized using matrix notation:

$$\left[[\nu]_{R,n_f}, [\bar{\bar{\nu}}]_{R,n_{\bar{\bar{s}}}}\right]_{R,n_T} \begin{bmatrix} [S]_{n_f,1} \\ [\bar{\bar{S}}]_{n_{\bar{\bar{s}}},1} \end{bmatrix} = [0]_{R,1} \tag{4.115}$$

[7] These are chemical potentials that allowed derivation of the IAS theory and of criteria for defining consistent equilibrium laws, and that will allow us to obtain interesting mass-transfer information in Appendix G. Chemical potentials also play a role in chromatography ...

where $[M]_{R,C}$ is a matrix with R rows and C columns. When there will be no ambiguity, the dimensions of the matrix, R and C, will be omitted.

In Eq. (4.115), R is the number of reactions that are assumed to be independent; in other words $\left[[\nu]_{R,n_f},[\bar{\bar{\nu}}]_{R,n_{\bar{\bar{s}}}}\right]_{R,n_T}$ is of rank R. Since there are more species than reactions ($n_T > R$), the row reaction vectors are independent.

Let us use binary Na/H ion exchange on a weak acid exchanger in the presence of chloride as inert anion to show that this *prima facie* complex mathematical formalism is in fact simple. The ion-exchange and site-dissociation reactions are expressed by:

Na^+	H^+	Cl^-	$\overline{\overline{Na}}^+$	$\overline{\overline{H}}^+$	$\overline{\overline{NH}}$	$\overline{\overline{N}}^-$	species
-1	1	0	1	-1	0	0	Na/H ion exchange Eq. (4.106)
0	0	0	0	1	-1	1	Site ionization Eq. (4.110)

(4.116)

It is easily checked that $R = 2$, $n_T = 7$, $n_f = 3$, $n_{\bar{\bar{s}}} = 4$ and that the rows of the stoichiometric matrix in Eq. (4.116) are independent. Additionally, one has:

$$[\nu]_{R,n_f} = \begin{bmatrix} -1 & 1 & 0 \\ 0 & 0 & 0 \end{bmatrix} \qquad [\bar{\bar{\nu}}]_{R,n_{\bar{\bar{s}}}} = \begin{bmatrix} 1 & -1 & 0 & 0 \\ 0 & 1 & -1 & 1 \end{bmatrix}$$

The matrix $\left[[\nu],[\bar{\bar{\nu}}]\right]$ completely defines the problem, and any row implicitly contains the structure of the ideal mass action law:

$$K_i = \prod_{j=1}^{n_f} S_j^{\nu_{ij}} \prod_{j=1}^{n_{\bar{\bar{s}}}} \bar{\bar{S}}_j^{\bar{\bar{\nu}}_{ij}} \tag{4.117}$$

It can be shown that some judicious linear combinations of the species are conserved during the reactions. An intuitive, although partial, proof is given by basic chemistry: atoms are conserved whatever the reactions involved! In the present example of ion exchange, there are three atoms involved, namely Cl, Na and H. One more species is involved: $\overline{\overline{N}}^-$, which can be considered a chemical structure that is conserved. It is a *moiety*, and we will use this word later on. There are thus four conserved moieties involved, i.e. at least four linear combinations of the species that are conserved.

From a mathematical point of view, the following rules can be demonstrated:

1 There are exactly $I = n_T - R$ conserved linear combinations, which are called stoichiometric invariants, moieties being special stoichiometric invariants. The difference $I = n_T - R$ is the so-called *variance* of the problem (at prescribed T and P).

2 The stoichiometric invariants are given by independent vectors that span the null space of matrix $\left[[\nu],[\bar{\bar{\nu}}]\right]$, i.e. vectors $[V]_{n_T,1}$ such that:

$$\left[[\nu],[\bar{\bar{\nu}}]\right]_{R,n_T} [V]_{n_T,1} = [0]_{R,1}$$

The term "invariant" means that the combinations are not affected by the reactions. Stoichiometric invariants are thus constant in a batch system. Conversely, they may vary in an open system such as a chromatographic column owing to a variation in the feed composition only.

One concludes from rule 1 that the Na/H ion-exchange problem on a weak resin involves $I = 7 - 2 = 5$ stoichiometric invariants. As we have already identified four of them (Na, Cl, H, $\overline{\overline{\mathrm{N}}}^-$), one more must be found. The second rule gives the missing culprit. One may check that the following vectors $[V_1]$ to $[V_5]$ are stoichiometric invariants according to rule 2:

$$[V_1] = \begin{bmatrix} 1 \\ 0 \\ 0 \\ 1 \\ 0 \\ 0 \\ 0 \end{bmatrix} \quad [V_2] = \begin{bmatrix} 0 \\ 1 \\ 0 \\ 0 \\ 1 \\ 1 \\ 0 \end{bmatrix} \quad [V_3] = \begin{bmatrix} 0 \\ 0 \\ 1 \\ 0 \\ 0 \\ 0 \\ 0 \end{bmatrix} \quad [V_4] = \begin{bmatrix} 0 \\ 0 \\ 0 \\ 0 \\ 0 \\ 1 \\ 1 \end{bmatrix} \quad [V_5] = \begin{bmatrix} 0 \\ 0 \\ 0 \\ 1 \\ 1 \\ 0 \\ -1 \end{bmatrix} \quad \begin{matrix} Na^+ \\ H^+ \\ Cl^- \\ \overline{\overline{Na}}^+ \\ \overline{\overline{H}}^+ \\ \overline{\overline{NH}} \\ \overline{\overline{N}}^- \end{matrix} \tag{4.118}$$

The meaning of these vectors is found by multiplying each component of [V] by the corresponding chemical symbol: the sum of the resulting components is a conserved quantity. One immediately obtains:

$$\begin{array}{ll} [V_1] : Na^+ + \overline{\overline{Na}}^+ & \text{Sodium is conserved} \\ [V_2] : H^+ + \overline{\overline{H}}^+ + \overline{\overline{NH}} & \text{Hydrogen is conserved} \\ [V_3] : Cl^- & \text{Chlorine is conserved (inert anion)} \\ [V_4] : \overline{\overline{NH}} + \overline{\overline{N}}^- & \text{Exchanging sites (ionized or not) are conserved} \\ [V_5] : \overline{\overline{Na}}^+ + \overline{\overline{H}}^+ - \overline{\overline{N}}^- & \text{Net electric charge on the solid is conserved} \end{array} \tag{4.119}$$

The reader should be convinced that finding the stoichiometric invariants is not a pure math problem of finding the null space of an endomorphism (i.e. a matrix). It is essentially a problem of genuine physical understanding of the chemistry involved, as shown in Eq. (4.119). The most difficult step is to check that you determined independent vectors, knowing that Murphy's law helps in finding non-independent vectors...

Obviously, any linear combination of vectors $[V_1]$ to $[V_5]$ is again a stoichiometric invariant, and some linear combinations can be of great interest. For instance:

$$[U_1] \equiv [V_1] + [V_2] + [V_3] - [V_4] - [V_5] : \; Na^+ + H^+ - Cl^- \quad \text{is conserved} \tag{4.120}$$

This is not a surprise since the sum $Na^+ + H^+ - Cl^-$ reduces to 0 owing to electroneutrality in the fluid phase (OH^- ions are neglected). The same property holds for $[V_5]$ in the last line of Eq. (4.119): the solid phase is electrically neutral.

We have thus obtained a remarkable result: vectors $[U_1]$ and $[V_5]$ show that the *structure* (species involved, + or − signs, valences etc.) of the electroneutrality condition of the phases is *implicitly* contained in the mechanism. Physics and chemistry dictate *only the value* (if any) of the stoichiometric invariant, which is zero in the case of electroneutrality. For an ion-exchange process on a strong acidic exchanger, as depicted by Eq. (4.106), one finds easily that $\overline{\overline{Na}}^{+}+\overline{\overline{H}}^{+}$ is an invariant. Chemistry tells us further that it is constant and equals the exchange capacity; alternatively, it can be viewed as the solid electroneutrality.

The special vectors of the null space, $[U_1]$ and $[V_3]$, have a particular property: they depend on fluid-phase species only. Let R_f and $R_{\bar{\bar{s}}}$ be the ranks of the sub-matrices $[\nu]$ and $[\bar{\bar{\nu}}]$ respectively. It can be shown (Schweich *et al.*, 1993a) that:

3 One may choose $I_f = n_f - R_f$ independent stoichiometric invariants that depend on fluid species exclusively. They are called *fluid stoichiometric invariants*. Fluid electroneutrality and inert ions are examples. In the simple Na/H ion-exchange problem on a weak resin, $n_f - R_f = 3 - 1 = 2$. The corresponding stoichiometric invariants can be $[U_1]$ and $[V_3]$, or any couple of independent linear combinations.

4 One may choose $I_{\bar{\bar{s}}} = n_{\bar{\bar{s}}} - R_{\bar{\bar{s}}}$ independent stoichiometric invariants that depend on solid species exclusively. They are called *solid stoichiometric invariants*. Solid electroneutrality (or exchange capacity) and total adsorption site are examples. In the weak acid ion-exchange problem, $I_{\bar{\bar{s}}} = n_{\bar{\bar{s}}} - R_{\bar{\bar{s}}} = 4 - 2 = 2$. The corresponding stoichiometric invariants can be $[V_4]$ and $[V_5]$, or any couple of independent linear combinations.

5 If the latter fluid and solid invariants are chosen, then the remaining $I_{f\bar{\bar{s}}} = I - I_f - I_{\bar{\bar{s}}} = R_f + R_{\bar{\bar{s}}} - R$ invariants depend on both fluid and solid species. These are called *fluid–solid stoichiometric invariants*. It can be shown that they are responsible for the retained fronts or peaks in the chromatographic process. Applying the result given in Section 3.7, the number of retained fronts obtained on a column subject to a step input change in composition is:

$$F = I_{f\bar{\bar{s}}} = I - I_f - I_{\bar{\bar{s}}} = R_f + R_{\bar{\bar{s}}} - R \tag{4.121}$$

The method used for deriving the stoichiometric invariants is based on some conserved properties (moieties, atoms, electric charge etc.) or on the determination of the null space of the stoichiometric matrix. It follows that these invariants *do not* depend on the exact nature of the mass action law (ideal or not) nor on possible kinetic limitations. This means that:

- Stoichiometric invariants are still conserved and can be used in kinetic studies and all along a batch equilibration process (kinetics is not going to impact electroneutrality or moieties).
- Stoichiometric invariants can be used when some phases appear or disappear (precipitation, dissolution, demixion, degassing etc.), although some species are no longer involved in the mass action laws (pure precipitate, for example).

The concepts presented above are of great help understanding and quantitatively representing the equilibrium laws:

- As already mentioned, the phenomenological mechanism probably ensures thermodynamic consistency.

- Simple reactions (stoichiometries) summarize the heterogeneous and homogeneous interactions among the species in a concise manner.
- Although the set of ideal mass action laws may not be reduced to a set of usual explicit adsorption equilibrium isotherms $\overline{\overline{C}}_i = f_i(C_1, C_2, \ldots C_n)$, they contain, implicitly, all the necessary information required to get the equilibrium composition. Consistency ensures the uniqueness of the equilibrium state, i.e. of the solution to the implicit equations.
- Stoichiometric invariants may be used to support or falsify a postulated mechanism in the early stages of a study. In the determination of the equilibrium laws in a batch system, the invariants are *constant* all along the experiments. If an invariant is found to be experimentally variable, then the mechanism is wrong somewhere. In a chromatographic process, any of the $n_f - R_f$ fluid invariants must behave as a non-retained solute; if not, then the mechanism is wrong somewhere. Note that if the invariants behave as expected, this does not prove that the mechanism is valid: the same invariants can be found for different mechanisms.
- Activity coefficients can in principle be introduced in the mass action laws without any difficulty. Difficulties arise from the requirement to define activity coefficients that are thermodynamically consistent. Recall that stoichiometric invariants are independent of the mass action laws and thus of the activity coefficients.
- The structure and properties of the stoichiometric invariants are valid for a prescribed number of phases in interaction (Schweich *et al.*, 1993a). Usually, the number of phase changes along the process is secondary in chromatography, where precipitation, demixion, degassing are avoided as much as possible – unless it becomes an innovative breakthrough.

4.7.3 Illustrative example: amino acids

We consider here the preliminary study that might precede the separation of a mixture containing an amino acid AAH dissolved in a water solution containing sodium nitrate and nitric acid. The separation is made on a strong acid ion exchanger, and only the positively charged form of the amino acid can enter into an ion-exchange mechanism.

We will:

- propose a simple phenomenological mechanism
- find stoichiometric invariants and moieties
- propose fluid, solid and fluid–solid invariants
- show how a mechanism can be simplified when some species are at known concentrations.

The simplest phenomenological mechanism is given by:

$$H_2O \rightleftharpoons H^+ + OH^- \qquad K_{H20} = \frac{H^+OH^-}{H_2O} \qquad \text{water ionization} \tag{4.122}$$

$$\mathrm{AAH} \rightleftharpoons \mathrm{AA}^- + \mathrm{H}^+ \qquad K_{AA-} = \frac{AA^-H^+}{AAH} \qquad \text{acid ionization} \tag{4.123}$$

$$\mathrm{AAH} + \mathrm{H}^+ \rightleftharpoons \mathrm{AAH}_2^+ \qquad K_{AA+} = \frac{AAH_2^+}{AAHH^+} \qquad \text{amine ionization} \tag{4.124}$$

$$\mathrm{Na}^+ + \bar{\bar{\mathrm{H}}}^+ \rightleftharpoons \bar{\bar{\mathrm{Na}}}^+ + \mathrm{H}^+ \qquad K_{Na/H} = \frac{\bar{\bar{Na}}^+H^+}{Na^+\bar{\bar{H}}^+} \qquad \mathrm{Na^+/H^+}\ \text{exchange} \tag{4.125}$$

$$\mathrm{AAH}_2^+ + \bar{\bar{\mathrm{H}}}^+ \rightleftharpoons \bar{\bar{\mathrm{AAH}_2}}^+ + \mathrm{H}^+ \qquad K_{AA/H} = \frac{\bar{\bar{AAH_2}}^+H^+}{AAH_2^+\bar{\bar{H}}^+} \qquad \mathrm{AAH_2^+/H^+}\ \text{exchange} \tag{4.126}$$

AAH is the non-ionized amino acid; AA^- and AAH_2^+ are the ionized forms. There are $R = 5$ reactions in this simple mechanism. The reader will recognize that these reactions are independent.

The species involved in the system are:

$$\overbrace{\underbrace{\mathrm{H^+, OH^-, Na^+, AA^-, AAH, AAH_2^+, NO_3^-, H_2O}}_{n_f=8},\ \underbrace{\bar{\bar{\mathrm{H}}}^+, \bar{\bar{\mathrm{Na}}}^+, \bar{\bar{\mathrm{AAH}_2}}^+}_{n_{\bar{\bar{s}}}=3}}^{n_T=11} \tag{4.127}$$

The stoichiometric matrix is (reaction order from Eq. (4.122) to Eq. (4.126)):

H^+	OH^-	Na^+	AA^-	AAH	AAH_2^+	NO_3^-	H_2O	$\bar{\bar{H}}^+$	$\bar{\bar{Na}}^+$	$\bar{\bar{AAH_2}}^+$
-1	-1	0	0	0	0	0	1	0	0	0
-1	0	0	-1	1	0	0	0	0	0	0
-1	0	0	0	-1	1	0	0	0	0	0
1	0	-1	0	0	0	0	0	-1	1	0
1	0	0	0	0	-1	0	0	-1	0	1

(4.128)

The submatrix $[\nu]$ is given by the first eight columns, while the submatrix $[\bar{\bar{\nu}}]$ is given by the last three columns, of the stoichiometric matrix.

It is immediately seen that $[\bar{\bar{\nu}}]$ is of rank $R_{\bar{\bar{s}}} = 2$ as it has only two non-zero rows that are not proportional. Alternatively, the $\bar{\bar{\mathrm{Na}}}^+$ and $\bar{\bar{\mathrm{AAH}_2}}^+$ columns are two independent vectors, whereas the $\bar{\bar{\mathrm{H}}}^+$ column is their sum, sign excepted.

The rank of $[\nu]$ is more difficult to guess. Let us first note that the columns OH^-, Na^+ and AA^- are independent column vectors since they have a single non-zero component at different rows. The rank is thus at least 3. Further independent vectors must have third and fifth components that are non-zero: columns AAH and H^+ fulfill this requirement and are independent of each other. We have thus found five independent column vectors, and $[\nu]$ is of rank $R_f = 5$. Readers unfamiliar with this reasoning may also use symbolic calculus software to get the answer.

As a consequence, there are $I = n_T - R = 11 - 5 = 6$ stoichiometric invariants. One may then choose:

- $I_f = n_f - R_f = 8 - 5 = 3$ fluid stoichiometric invariants
- $I_{\bar{\bar{s}}} = n_{\bar{\bar{s}}} - R_{\bar{\bar{s}}} = 3 - 2 = 1$ solid stoichiometric invariant

- $I_{f\bar{\bar{s}}} = I - I_f - I_{\bar{\bar{s}}} = 6 + 3 - 1 = 2$ fluid–solid stoichiometric invariants; there are thus two retarded fronts or peaks in a chromatogram.

Note that there are five moieties, i.e. chemical substructures or atoms that are conserved: NO_3,[8] OH, H, Na and AA. All other chemical species are made of these moieties; for instance, water is made of H and OH, the various forms of the amino acid are made of AA and H, and so on. These five moieties define five stoichiometric invariants:

$$\text{NO}_3 \text{ gives: } NO_3^- \text{ is conserved} \tag{4.129}$$

$$\text{OH gives: } OH^- + H_2O \text{ is conserved} \tag{4.130}$$

$$\text{H gives: } OH^- + H^+ + 2H_2O + AAH + 2AAH_2^+ + \overline{\overline{H}}^+ + 2\overline{\overline{AAH}}_2^+ \text{ is conserved} \tag{4.131}$$

$$\text{Na gives: } Na^+ + \overline{\overline{Na}}^+ \text{ is conserved} \tag{4.132}$$

$$\text{AA gives: } AA^- + AAH + AAH_2^+ + \overline{\overline{AAH}}_2^+ \text{ is conserved} \tag{4.133}$$

Equations (4.129) and (4.130) associated with conservation of moieties give two fluid invariants out of three. The last invariant is immediately given by fluid electroneutrality:

$$H^+ - OH^- + Na^+ - AA^- + AAH_2^+ - NO_3^- = 0 \tag{4.134}$$

The set of three fluid invariants can thus consist of Eq. (4.129), Eq. (4.130) and Eq. (4.134) or any triplet of independent linear combinations of these expressions.

The single solid invariant is immediately given by the solid-phase electroneutrality:

$$\overline{\overline{H}}^+ + \overline{\overline{Na}}^+ + \overline{\overline{AAH}}_2^+ = \overline{\overline{N}} \tag{4.135}$$

Note that Eqs. (4.129)–(4.133), (4.134) and (4.135) are a set of seven stoichiometric invariants: they are thus linearly dependent, as we know that the system has only six stoichiometric invariants. This means that one of the electroneutrality conditions results from the others and from moiety conservation. This should not be a surprise: if one phase is neutral, the charge balance in each of the reactions implies that the second phase is neutral.

Independent fluid–solid invariants must be chosen among H, Na and AA moieties, and one should ensure that they are independent of the previous fluid and solid invariants. Let us try the Na and AA moieties, Eq. (4.132) and Eq. (4.133). By definition these moieties are independent of other moieties like OH and NO_3, which are two fluid invariants. Thus, they could be at most combinations of the electroneutrality conditions of the fluid and solid phases. It is immediately seen that they are not. They are thus independent of the fluid and solid invariants, and Eq. (4.132) and Eq. (4.133) can be chosen as fluid–solid invariants.

[8] Nitrogen could have been chosen instead of the nitrate moiety NO_3. The resulting invariant would have been the sum of Eq. (4.129) and Eq. (4.133), i.e. an expression "heavier" than Eq. (4.129) and which would not remind us that nitrate is an inert species. This shows that some choices of moieties can be more judicious than others, and often more judicious than choosing atoms.

We have already found that there will be $I_{f\bar{\bar{s}}} = R_f + R_{\bar{\bar{s}}} - R = 2$ retarded fronts or peaks in a chromatographic experiment. All the species except NO_3^- may vary in these retarded fronts.

EXERCISE 4.5 *Consider that AAH is alanine ($NH_2 - CH(CH_2) - COOH$). Derive the stoichiometric invariants using atom conservation. Are atoms really convenient invariants?*

The case of species in excess deserves special attention. In the amino acid problem presented above, nobody would consider that water concentration varies significantly. It seems obvious that water concentration may be assumed to be constant and close to 55 mol/l so that H_2O can be eliminated from the set of species of unknown concentration. Assuming the water concentration to be constant allows the mass action law associated with water ionization, Eq. (4.122), to be written in the most classical form:

$$K_e = H^+OH^- \qquad \text{with} \qquad K_e = \frac{K_{H_20}}{H_2O} \tag{4.136}$$

A *degenerate* description of the problem is thus obtained after deletion of water from the species list and from the stoichiometric matrix:

$$\overbrace{\underbrace{H^+, OH^-, Na^+, AA^-, AAH, AAH_2^+, NO_3^-}_{n_f=7}, \underbrace{\bar{\bar{H}}^+, \bar{\bar{Na}}^+, \bar{\bar{AAH_2}}^+}_{n_{\bar{\bar{s}}}=3}}^{n_T=10} \tag{4.137}$$

$$\begin{array}{cccccccccc} H^+ & OH^- & Na^+ & AA^- & AAH & AAH_2^+ & NO_3^- & \bar{\bar{H}}^+ & \bar{\bar{Na}}^+ & \bar{\bar{AAH_2}}^+ \end{array}$$
$$\begin{bmatrix} -1 & -1 & 0 & 0 & 0 & 0 & 0 & 0 & 0 & 0 \\ -1 & 0 & 0 & -1 & 1 & 0 & 0 & 0 & 0 & 0 \\ -1 & 0 & 0 & 0 & -1 & 1 & 0 & 0 & 0 & 0 \\ 1 & 0 & -1 & 0 & 0 & 0 & 0 & -1 & 1 & 0 \\ 1 & 0 & 0 & 0 & 0 & -1 & 0 & -1 & 0 & 1 \end{bmatrix} \tag{4.138}$$

Following the method described above, one may obtain the stoichiometric invariants from the *conserved* moieties. Let us stress that all the moieties are not necessarily conserved in the degenerate mechanism because one species (or several in other problems) is lacking. In the present case, the OH and H moieties are no longer conserved because H_2O is absent.

The mechanistic approach can, however, still be used, as shown below:

- The ranks of the new $[[\nu],[\bar{\bar{\nu}}]]$, $[\nu]$ and $[\bar{\bar{\nu}}]$ matrices are still $R = 5$, $R_f = 5$ and $R_{\bar{\bar{s}}} = 2$ respectively, as in the original problem.
- There are $I = 10 - 5 = 5$ (instead of 6) stoichiometric invariants. They can be given by the two electroneutrality conditions of the fluid and solid phases and the following *conserved* moieties:

$$NO_3 \text{ gives: } NO_3^- \text{ is conserved} \tag{4.139}$$

$$Na \text{ gives: } Na^+ + \overline{\overline{Na}}^+ \text{ is conserved} \tag{4.140}$$

$$AA \text{ gives: } AA^- + AAH + AAH_2^+ + \overline{\overline{AAH}}_2^+ \text{ is conserved} \tag{4.141}$$

Equations (4.139)–(4.141) are invariants of the original problem; they are still valid because they do not involve water. Again, the reader can check that these two invariants together with the two electroneutrality conditions are independent.

We also find that:

- There are now $I_f = n_f - R_f = 2$ fluid stoichiometric invariants. They can be the fluid electroneutrality condition and the NO_3 moiety (given by Eq. (4.139)).
- There is a single solid stoichiometric invariant which is again the exchange capacity.
- There are two fluid–solid stoichiometric invariants, given by Eq. (4.140) and Eq. (4.141).

One may wonder what happened to the invariants deduced from the OH and H moieties as given by Eq. (4.130) and Eq. (4.131). We know that they are no longer invariants; however, Eq. (4.131) minus twice Eq. (4.130), which does not involve H_2O, yields:

$$-OH^- + H^+ + AAH + 2AAH_2^+ + \overline{H}^+ + 2\overline{AAH}_2^+ \quad \text{is conserved} \tag{4.142}$$

This is again a stoichiometric invariant, and it can be shown in general that the invariants of a degenerate problem are suitable independent linear combinations of the invariants of the original problem that do not involve the deleted species.

To conclude this special case, we see that, when deleting some species because their concentrations are presumed to be known:

- The "species" dimension of the problem decreases.
- Some moieties may be non-conserved. They must be discarded or linearly combined so as to make the deleted species disappear from the combination.
- The stoichiometric invariants can be associated with *conserved* moieties or suitable linear combinations of the moieties of the original problem that do not involve the deleted species.

The reader can see that, up to now, the mathematical intricacy has been overcome by physical and chemical arguments for deriving important properties of the system. Quantitative description requires numerical simulation. As an illustration, for a batch experiment, from the initial composition, one may calculate the numerical values of the five invariants which are *constant* along the experiment. This results in five linear equations (two electroneutralities and Eqs. (4.139)–(4.141)). The equilibrium laws in Eqs. (4.122)–(4.126) are the remaining five equations with respect to the ten unknown concentrations (water is excluded). Knowing the equilibrium constants, the unknown species concentrations can then be calculated using standard mathematical software (thermodynamic consistency ensuring the uniqueness of the solution helps a lot).

4.7.4 Illustrative example: induced pH gradients on weak cationic resins

We will now propose another illustration of the methodology with some importance in protein chromatography. (Pabst *et al.*, 2008) studied the possibility of using internally generated pH gradients in response to step changes in buffer concentration on columns packed with weak cationic resins. The authors considered the option of using an acetate buffer, a phosphate buffer or a mixture of the two. Since in this preliminary study the objective is to study possible pH changes as a response to step changes in buffer composition, the system is assumed to be protein-free.

The Na/H ion-exchange reaction, the site dissociation occurring in the resin and the water ionization occurring in solution are described, respectively, by Eq. (4.106), Eq. (4.110) and Eq. (4.122). The acetate dissociation in solution is described by:

$$\mathrm{CH_3COOH} \rightleftharpoons \mathrm{CH_3COO^-} + \mathrm{H^+}$$

$$\text{or } \mathrm{Ac} \rightleftharpoons \mathrm{Ac^-} + \mathrm{H^+} \text{ for short, with } \quad K_{ac} = \frac{Ac^- H^+}{Ac} \tag{4.143}$$

Let us start by considering a system containing sodium, acetate and chloride ions. The species involved in the system are:

$$\overbrace{\underbrace{\mathrm{H^+, OH^-, Na^+, Ac^-, Ac, Cl^-, H_2O}}_{n_f=7},\ \underbrace{\bar{\bar{\mathrm{H}}}^+, \bar{\bar{\mathrm{Na}}}^+, \bar{\bar{\mathrm{NH}}}, \bar{\bar{\mathrm{N}}}^-}_{n_{\bar{\bar{s}}}=4}}^{n_T=11} \tag{4.144}$$

There are $n_T = 11$ species and $R = 4$ reactions, thus $I = 7$. The stoichiometric matrix is (reaction order: Eq. (4.106)-(4.110)-(4.122)-(4.143)):

H^+	OH^-	Na^+	Ac^-	Ac	Cl^-	H_2O	$\bar{\bar{H}}^+$	$\bar{\bar{Na}}^+$	$\bar{\bar{NH}}$	$\bar{\bar{N}}^-$
+1	0	-1	0	0	0	0	-1	1	0	0
0	0	0	0	0	0	0	1	0	-1	1
1	1	0	0	0	0	-1	0	0	0	0
1	0	0	1	-1	0	0	0	0	0	0

(4.145)

The submatrix $[\nu]$ is given by the first seven columns, whereas the submatrix $[\bar{\bar{\nu}}]$ is given by the last four columns, of the stoichiometric matrix.

One immediately sees that $n_f = 7$ and $R_f = 3$, so that $I_f = 4$. Similarly, one sees that $n_{\bar{\bar{s}}} = 4$ and $R_{\bar{\bar{s}}} = 2$ so that $I_{\bar{\bar{s}}} = 2$. One concludes that the number of fluid–solid invariants is $I_{\bar{\bar{s}}} = I - I_f - I_{\bar{\bar{s}}} = 7 - 4 - 2 = 1$. Consequently, a column in equilibrium with the above species and subject to an inlet composition change will generate one retained front. This is in agreement with the results obtained by (Pabst *et al.*, 2008; see their Section 4.1). For the record, the obtained front is dispersive.

Now let us replace the acetate buffer by a phosphate buffer. The phosphate dissociation in solution is described by:

$$\begin{aligned}
&H_3PO_4 \rightleftharpoons H_2PO_4^- + H^+ \\
&\text{or } Ph^0 \rightleftharpoons Ph^- + H^+ \text{ for short, with} \quad K_{ph}^1 = \frac{Ph^- H^+}{Ph} \\
&H_2PO_4^- \rightleftharpoons HPO_4^{2-} + H^+ \\
&\text{or } Ph_- \rightleftharpoons Ph^{2-} + H^+ \text{ for short, with} \quad K_{ph}^2 = \frac{Ph^{2-} H^+}{Ph_-} \\
&HPO_4^{2-} \rightleftharpoons PO_4^{3-} + H^+ \\
&\text{or } Ph^{2-}- \rightleftharpoons Ph^{3-} + H^+ \text{ for short, with} \quad K_{ph}^3 = \frac{Ph^{3-} H^+}{Ph^{2-}}
\end{aligned} \tag{4.146}$$

The species involved in the system are:

$$\overbrace{\underbrace{H^+, OH^-, Na^+, Ph^0, Ph^-, Ph^{2-}, Ph^{3-}, Cl^-, H_2O}_{n_f=9}, \underbrace{\bar{\bar{H}}^+, \bar{\bar{Na}}^+, \bar{\bar{NH}}, \bar{\bar{N}}^-}_{n_{\bar{\bar{s}}}=4}}^{n_T=13} \tag{4.147}$$

We have thus $n_T = 13$ species and $R = 6$ reactions (Eqs. (4.106)-(4.110)-(4.122) and the three phosphate dissociations). Consequently one deduces that $I = 13 - 6 = 7$.

The stoichiometric matrix can be written exactly as we did for the acetate system. One obtains, for the fluid part, $n_f = 9$ and $R_f = 5$ so that $I_f = 4$. The solid stoichiometric matrix is unchanged so $n_{\bar{\bar{s}}} = 4$ and $R_{\bar{\bar{s}}} = 2$, so $I_{\bar{\bar{s}}} = 2$. One concludes that the number of fluid–solid invariants is $I_{\bar{s}} = I - I_f - I_{\bar{\bar{s}}} = 7 - 4 - 2 = 1$. Consequently, a column in equilibrium with the above phosphate buffer and subject to an inlet composition change will generate one retained front. This is again in agreement with the results obtained by (Pabst *et al.*, 2008; see their Section 4.1). For the record, the obtained front is compressive.

Finally, let us mix the acetate buffer with the phosphate buffer. The reader will easily check that there are:

- seven independent stoichiometric reactions: Eqs. (4.106), (4.110) and (4.122) plus the acetate dissociation Eq. (4.143) and the three phosphate dissociations Eq. (4.146)
- 11 species in the fluid phase: H^+, OH^-, Na^+, Ac, Ac^-, Ph^0, Ph^-, Ph^{2-}, Ph^{3-}, Cl^-, H_2O
- 4 species in the solid phase: $\bar{\bar{H}}^+$, $\bar{\bar{Na}}^+$, $\bar{\bar{NH}}$, $\bar{\bar{N}}^-$.

Simple math leads to: $I = 15 - 7 = 8$ invariants. The solid stoichiometric matrix is unchanged, so $n_{\bar{\bar{s}}} = 4$ and $R_{\bar{\bar{s}}} = 2$, so $I_{\bar{\bar{s}}} = 2$. Addressing the fluid phase is slightly more complex, but one quickly finds that $n_f = 11$ and $R_f = 7$, so $I_f = 4$. One concludes that the number of fluid–solid invariants is $I_{\bar{s}} = I - I_f - I_{\bar{\bar{s}}} = 8 - 4 - 2 = 2$. Consequently, a column in equilibrium with the above buffer mixture and subject to an inlet composition change would generate two retained fronts, while only one front is obtained with the individual buffers. This is again in agreement with the results obtained by (Pabst *et al.*, 2008; see their Section 4.1).

The author is not sure whether obtaining a single front for the individual buffers is obvious but, at least for him, obtaining two fronts with the mixture is not!

4.8 Experimental methods

4.8.1 General considerations

Determining adsorption isotherms was certainly not part of the normal practices of chromatographers, but the development of efficient and rather complex continuous chromatographic processes has helped develop awareness of the importance of this information. Would a process engineer seriously consider designing a distillation column without knowing the vapor/liquid equilibrium? Certainly not, and there is no reason for chromatographic process design to be treated differently. The difficulty of the task is neither as high as many chromatographers often think nor as easy as process engineers would dream. This is simply a prerequisite for rational process development that must be accepted.

A review of possible methods was published in 1996 by (Nicoud and Seidel-Morgenstern, 1996); this work was updated and completed by (Seidel-Morgenstern, 2004). Recent information can be found in (Seidel-Morgenstern *et al.*, 2012).

All experimental methods require at least proper temperature and composition control. Other parameters, like fluid flow rate, play a role when specific methods are used.

Prior to detailing the available experimental methods, let us stress the need for systematic consistency checks when adsorption isotherms $\bar{\bar{f}}_i$ (or lumped adsorption isotherms $\bar{f}_i$) are determined. A simple check consists in determining the retention time of a solute injected under very dilute conditions in a carrier fluid that is solute-free.

We have shown that the retention time is related to Henry's coefficient $\bar{\bar{K}}_i$ by:

$$t_R(i) = t_{0R}\left(1 + \frac{1-\varepsilon_T}{\varepsilon_T}\bar{\bar{f}}'_i(0)\right) = t_{0R}\left(1 + \frac{1-\varepsilon_T}{\varepsilon_T}\bar{\bar{K}}_i\right) \tag{4.148}$$

and to the lumped Henry's coefficient $\bar{K}_i$ by:

$$t_R(i) = t_0\left(1 + \frac{1-\varepsilon_e}{\varepsilon_e}\bar{f}'_i(0)\right) = t_0\left(1 + \frac{1-\varepsilon_e}{\varepsilon_e}\bar{K}_i\right) \tag{4.149}$$

The determination of retention times obtained under dilute conditions allows confirmation of the initial slopes $\bar{f}'_i(0)$ (or $\bar{\bar{f}}'_i(0)$) of the experimentally determined adsorption isotherms; this is a simple check that must be carried out systematically.

We will limit our presentation to the determination of the lumped adsorption isotherm $\bar{f}_i$, it being understood that the adsorption isotherm $\bar{\bar{f}}_i$ can be determined with Eq. (4.7).

EXERCISE 4.6 *A column of 1 liter is packed with an adsorbent having an intragranular porosity of 0.5 (supplier data). The total column porosity, determined by injection of a non-retained tracer, is 0.70. A pulse injection of solute A leads to a retention time of 5 min. when the carrier fluid flow rate is 0.4 l/min. Determine the initial slope of the adsorption isotherm using both the porous and the lumped models.*

4.8.2 Single-solute systems

The experimental methods for determining adsorption isotherms $\bar{f}_i$ (or $\bar{\bar{f}}_i$) can be classified into static methods and dynamic methods (Nicoud and Seidel-Morgenstern, 1996); both are presented in the following sections.

4.8.2.1 Static methods

Static methods only use information given by equilibrium states and are not based on the analysis of time-dependent concentration curves.

Batch method

In the classical batch method, known volumes of solute-free adsorbent $\bar{V}$ at incipient wetness (see Exercise 2.7) and of solution V are equilibrated in a stirred vessel. If the initial concentration of solute A in the solution is C_A^I, a mass balance gives:

$$V\, C_A^I = V\, C_A^{Final} + \bar{V}\, \bar{C}_A^{Final} \tag{4.150}$$

where C_A^{Final} and $\bar{C}_A^{Final}$ are the final concentrations of solute A in the solution and in the adsorbent obtained after a sufficient time. Provided that C_A^{Final} is determined experimentally, concentration $\bar{C}_A^{Final}$ can be calculated using Eq. (4.150). The concentrations adsorbed on the skeleton are then obtained from Eq. (4.6) in the absence of solute exclusion.

This methodology was presented in Exercise 2.7, demonstrating the importance of using a solid at incipient wetness.

If different experiments are performed for different C_A^I (or for different adsorbent lumped volume dosages $\bar{V}/V$), the complete adsorption isotherm can be determined. Obviously, C_A^I and the adsorbent lumped volume dosage have to be chosen carefully in order to ensure precision in the determination of C_A^{Final} and thus $\bar{C}_A^{Final}$.

The batch method requires tedious laboratory work and is usually not very accurate. Difficulties arise from inaccurate determination of the adsorbent lumped volume $\bar{V}$ and of the time needed to reach equilibrium, and from possible mixing issues. The precise estimate of $\bar{V}$ is not a trivial task. When particle density is known, a simple weight measurement leads to a volume determination; care must, however, be taken with the exact wetness of the solid, which can impact the result. An alternative, even if a bit primitive, consists in using the material contained in a column of a known volume and extragranular porosity.

This method is certainly well adapted for preliminary investigations and even possibly for high-throughput isotherm determination, as demonstrated by (Nfor *et al.*, 2010), who used 96-well plates to investigate the adsorption of proteins on various media.

The above-mentioned limitations explain why methods using column experiments are normally preferred when higher precision is required.

Adsorption-desorption method

In this technique, a column initially equilibrated with the carrier fluid is fed with a large amount of a solution at concentration C_A^F, large enough that, at some point in time, the outlet concentration reaches the feed concentration. At equilibrium, the quantity of solute A (g or moles) contained in the column is given by:

$$n_A^{tot} = \varepsilon_e\, V_{col}\, C_A^F + (1-\varepsilon_e)\, V_{col}\, \overline{C}_A^F \tag{4.151}$$

The column is then completely desorbed with the eluent. The regeneration solution is subsequently analyzed so that n_A^{tot} is determined. Knowing n_A^{tot}, the concentration $\overline{C}_A^F$ in equilibrium with C_A^F can be calculated with Eq. (4.151).

The determination of the complete adsorption isotherm requires various experiments performed at different concentrations. This method is certainly a bit primitive but it offers a good accuracy.

4.8.2.2 Dynamic methods

The theoretical framework presented in Chapter 3 gives tools for determining adsorption isotherms from dynamic experiments. These methods are based on the mathematical analysis of the response curves obtained from defined variations of the column inlet concentrations.

Analysis of the mean position of fronts (frontal analysis)

This method is based on the determination of breakthrough curves obtained for different feed and initial concentrations. The column is initially equilibrated with a solute concentration in the carrier fluid C_A^I, associated with a lumped solid concentration $\overline{C}_A^I$ which is assumed to be known. At a given time, the inlet concentration is changed from C_A^I to C_A^F. The concentration $C_A^{out}(t)$ exiting the column is recorded or determined with proper sampling and analysis.

The integral mass balance of solute A between the time at which injection of C_A^F starts and a time sufficient to ensure that the entire column is equilibrated with C_A^F is given by:

$$\varepsilon_e V_{col}\left(C_A^F - C_A^I\right) + (1-\varepsilon_e)V_{col}\left(\overline{C}_A^F - \overline{C}_A^I\right) = Q\int_0^\infty \left(C_A^F - C_A^{out}(t)\right) dt \tag{4.152}$$

Knowing the breakthrough curve $C_A^{out}(t)$, Eq. (4.152) can be used to determine the solid-phase concentration $\overline{C}_A^F(C_A^F)$. The main task consists in accurately determining the integral in Eq. (4.152). For single-solute adsorption or desorption, these integrals are marked in Figure 4.8 as shaded areas. The corresponding areas for the adsorption and desorption curves must be identical for reversible physical adsorption processes.

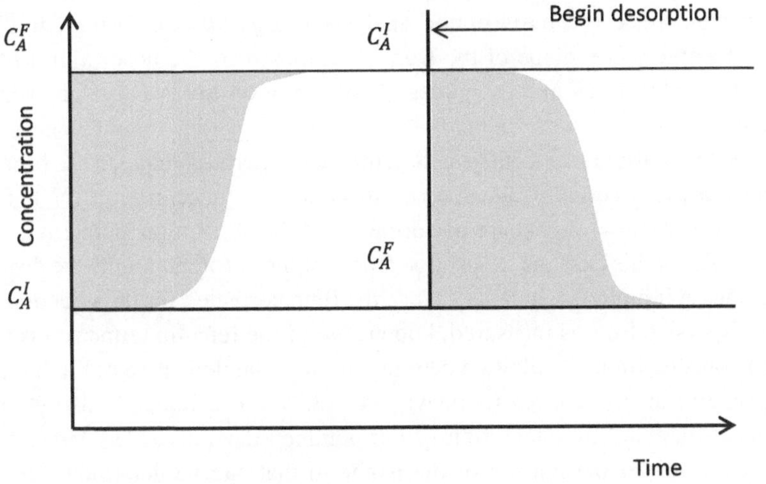

Figure 4.8 Frontal analysis. Left: feed concentration higher than initial concentration. Right: feed concentration lower than initial concentration.

To obtain the complete adsorption isotherm, breakthrough curves must be determined for various feed concentrations. It is expedient to start a series of measurements with a non-preloaded column ($C_A^I = \overline{C}_A^I = 0$) and then to increase the concentration stepwise. It is theoretically sufficient to perform positive concentration steps (adsorption fronts). As a check and to improve the accuracy, the analysis of negative concentration steps (desorption fronts) is recommended however. Obviously, this method works regardless of the dispersive or compressive nature of the fronts. In the case of dispersive fronts, however, a significant imprecision may arise from the integration of the tail.

A simplification is obtained when the measured breakthrough curve is sharp (forms a shock) so that the concentration change occurs abruptly at a time approximated by t_R^{shock}. The integral in Eq. (4.152) is simply given by the product $t_R^{shock}(C_A^F - C_A^I)$, so that one has:

$$\varepsilon_e V_{col}\left(C_A^F - C_A^I\right) + (1-\varepsilon_e)V_{col}\left(\overline{C}_A^F - \overline{C}_A^I\right) = Qt_R^{shock}(C_A^F - C_A^I) \qquad (4.153)$$

This equation, which is similar to Eq. (1.26) derived in Chapter 1, allows one point of the adsorption isotherm $\overline{C}_A^F(C_A^F)$ to be determined from knowledge of t_R^{shock}.

Note that, as shown in Exercise 3.1, Eq. (4.153) can in fact be used irrespective of the nature and shape of the front if t_R^{shock} represents the *mean* position of the front (and not the mid-height as is too often done).

This method is accurate provided that, in addition to proper temperature and composition control, the fluid flow rate is precisely controlled.

Analysis of front shapes

The theoretical framework describing wave propagation presented in Chapter 3 forms the basis of a method allowing us to determine adsorption isotherms from dynamic experiments. Since the basis of the analysis is *Equilibrium Theory*, this method is only applicable if the shape of the fronts is imposed by the adsorption isotherms and not by hydrodynamics or by kinetics. This condition holds only for "very efficient" columns.

The fact that dispersive fronts are determined by thermodynamics has been illustrated in Figure 3.11. A series of injections with identical concentration C_A^F but increasing injection times $t_{inj} = V_{inj}/Q$ are performed, and the adsorption isotherm is assumed to be favorable. In this case the desorption fronts of the chromatograms are dispersive, and they coincide if plotted against $t - \Delta t_{inj}$, the time corrected for the injection time.

If a dispersive front is measured, knowledge of the retention time t_R associated with a given concentration C_A allows a determination of the derivative of the lumped adsorption isotherm at this concentration via Eq. (3.21). The lumped adsorption isotherm $\bar{f}_A$ is itself obtained by integration of the obtained derivative. The larger the injected volume, the longer the portion of the isotherm that can be determined from a single chromatogram.

This method can be extended to unfavorable adsorption isotherms: in that case the dispersive front is obtained in the saturation step. This method is frequently called elution by characteristic points (ECP) or frontal analysis by characteristic points (FACP) (Ma *et al.*, 1990). Only in the latter case does the column outlet concentration reach the injected concentration. This difference does not alter the underlying principle, both ECP and FACP analyzing the diffuse fronts of a chromatogram.

From a theoretical point of view, the method is accurate and a single curve allows the determination of a complete portion of the isotherm. In practice its applicability is restricted, as already mentioned, to columns equivalent to a very high number of plates. In addition to the fact that experimental desorption fronts obtained from different injection times t_{inj} must coincide when plotted against $t - \Delta t_{inj}$, there is no general theory that allows an estimate of the minimum number of plates required for the method to deliver a given precision. This method must consequently be used in association with numerical simulation in order to check that the front shape is only imposed by thermodynamics. Models allowing the performance of these numerical simulations are presented in later chapters.

This method requires access to the concentration dependence on time. This means that either proper sensor calibration or sampling is required. Special care must be given to this aspect as relatively small errors in the concentration/time positioning can detrimentally impact the accuracy of the method. The analysis of dispersive fronts has hitherto been applied only to the determination of single-solute adsorption isotherms. An extension to mixtures is probably too complicated for practical applications.

Perturbation method

Determination of the retention time of a dilute injection of solute into a solute-free carrier fluid allows the initial slope of the adsorption isotherm to be obtained. This information is of key interest for correct positioning of the chromatogram on the time axis.

If one wants to obtain the entire adsorption isotherm, one has to determine the retention times associated with small injections (perturbations) performed after equilibrating the column with different concentrations C_A^I. Assuming that the perturbations are small enough to ensure linearity, knowledge of the retention time allows the local first derivative $\bar{f}'_A(C_A^I)$ of the adsorption isotherm to be determined by:

$$t_R(A) = t_0 \left(1 + \frac{1-\varepsilon_e}{\varepsilon_e} \bar{f}'_A(C_A^I)\right) \tag{4.154}$$

This equation assumes the absence of competition between solutes. Should several peaks be detected, competition between different solutes (cf. the concept of vacancy chromatography described in Section 4.8.3.2) must be suspected.

The adsorption isotherm then is calculated by integration of its first derivative. Another (possibly better) option consists in fitting calculated retention times, using a given adsorption isotherm model, to experimental retention times.

The method does not require the calibration of the detector (provided it is linear), and is not sensitive to the number of plates (provided that mean retention times are determined, not peak maxima). A limitation of accuracy is associated with the loss of information when integrating the derivative to obtain the adsorption isotherm.

Peak maximum method

A simple method, again restricted to columns of high efficiency, investigates the influence of the feed concentration and/or injection volume on the position of the peak maximum. For simplicity, we will restrict our discussion to the case of a favorable adsorption isotherm. If the injected volume is small enough to prevent the existence of the feed plateau, we have shown in Chapter 3 (Eq. (3.32)) that the concentration and retention time of the maximum of the peak are given by:

$$\begin{aligned} \bar{f}_A(C_A^{max}) - \bar{f}'_A(C_A^{max})C_A^{max} &= \frac{\varepsilon_e}{1-\varepsilon_e} \frac{C_A^F \Delta t_{inj}}{t_0} \\ t_R(C_A^{max}) &= t_0 \left(1 + \frac{1-\varepsilon_e}{\varepsilon_e} \bar{f}'_A(C_A^{max})\right) + \Delta t_{inj} \end{aligned} \tag{4.155}$$

Consequently, if an equation describing the adsorption isotherm is postulated, C_A^{max}, then $t_R(C_A^{max})$, can be calculated for any experimental condition. The parameters of the adsorption isotherm can then be determined by fitting the evolution of the retention time of the peak maximum versus the feed concentration.

This method does not require sensor calibration provided that the response curve is linear.

Chromatogram fitting method

This method is often called the "inverse method" as the goal is to determine model parameters from a given chromatogram instead of simulating a chromatogram using known model parameters.

Provided that an adsorption isotherm model and a column model taking into account hydrodynamic and mass-transfer effects are available, one can in principle optimize free parameters describing an adsorption isotherm by minimizing the differences between an experimental overloaded chromatogram and a simulation. Several theoretical models allowing one to simulate chromatograms are presented in this book and can be used for this task. Common sense tells us that non-thermodynamic band spreading must be kept as low as possible in order to maximize the influence of the adsorption isotherm parameters. It is advisable to check the sensitivity of simulated peak shapes to adsorption isotherm parameters in order to have an idea of the expected precision.

This method has been used by (Vajda *et al.*, 2010) to determine the parameters of a BET model. They obtained good agreement with a determination performed by frontal analysis.

This is a method that I have used for many years and I believe it makes good sense: industrial companies are normally reluctant to perform extensive laboratory work, but at the same time have a lot of experimental chromatograms obtained under various conditions associated with trial and error development work. This information is valuable and must be exploited as much as possible.

EXERCISE 4.7 *The adsorption–desorption method was used to determine the adsorption isotherm of solute A on a given adsorbent. Experimental conditions were:* $V_{col} = 4.15$ *ml,* $\varepsilon_e = 0.4$, $Q = 1$ *ml/min. The following results were obtained:*

Feed concentration (g/l)	1	5	10	25	50
Mass of A recovered after desorption (mg)	7.8	32	60	125	209

Propose a possible representation of the adsorption isotherm.

EXERCISE 4.8 *This is a continuation of Exercise 2.7 of Chapter 2. A volume* $\overline{V} = 3$ *ml of porous silica is mixed in a batch vessel with* $V = 10$ *ml of hexane containing a solute at a given initial concentration.* $\overline{V}$ *has been determined by using the silica contained in a column of* $V_{col} = 5$ *ml, having a measured extragranular porosity of* 0.4. *Final concentrations of solute in hexane are determined after 6 hours, a time proven to be sufficient for equilibrium to be reached. Experimental results are given in the table.*

Initial solute concentration (g/l)	1	10	20	50	100	150	200
Final solute concentration (g/l)	0.2	2.15	4.8	16.4	48.8	85.2	127.9

Propose a possible representation for the adsorption isotherm.

4.8.3 Multi-solute systems

As mentioned in Section 4.3.3, the main difference between single-solute and multi-solute systems is to be found in the competition and interactions between species. Should the species behave independently, determining multi-solute adsorption isotherms would be no more than determining *n* independent single-solute adsorption isotherms. Our intent in this section is to address situations where the species do not behave independently.

In an ideal world, knowledge of the single-solute adsorption isotherms should allow us to estimate the multi-solute adsorption isotherms provided that a consistent thermodynamic frame (like IAS, for instance) is available. In that ideal case, no time-consuming multi-solute equilibrium measurements would need to be made.

There is no question that knowledge of single-solute adsorption isotherms is of key interest; however, there are several strong drawbacks to the above approach:

- Determining single-solute adsorption isotherms is not as simple as it may seem *prima facie*, especially because of the possible influence of species in the carrier fluid. As an illustration, studying "single-solute" injections does not ensure the determination of a single-solute adsorption isotherm: due to the coupling between the different species, one can obtain multiple peaks (see the concept of vacancy chromatography explained in Section 4.8.3.2).
- The ability to precisely predict multi-solute adsorption isotherms from single-solute adsorption isotherms is still an open question. Theories like IAS (or even RAS) can help to paint a thermodynamically consistent picture, but they are idealized and their predictions have not yet been compared to a variety of experimental situations. Certainly the available thermodynamic methods for predicting competitive adsorption isotherms are not as reliable and mature as, for example, the ones used for predicting vapor–liquid equilibria.
- An important problem is the precision of the experimentally measured single-solute isotherms. In order to illustrate this problem, let us assume that the solid-phase concentrations are accurate within $\pm 5\%$ so that $\overline{C}_A = 1 \pm 0.05$ for $C_A = 1$ and $\overline{C}_B = 1.1 \pm 0.05$ for $\overline{C}_B = 1$. Consequently the selectivity $\alpha_{B/A}$ is found to be between 1 (impossible separation) and 1.2 (achievable separation). The above example is a bit fallacious but it helps in understanding the difficulty of the task.

For these reasons, it is important to measure competitive adsorption isotherms in addition to single-solute adsorption isotherms and to check consistency by all available experimental and theoretical means. There are different methods available, which are usually just extensions of the methods used to determine single-solute isotherms, so the comments given in Section 4.8.2 about the accuracy of these methods hold for multi-solute systems as well.

4.8.3.1 Static methods

The batch and adsorption–desorption methods are similar to those illustrated in Section 4.8.2.1. It is sufficient to write one mass balance per species, and in any case to make sure that enough time has passed for the system to reach equilibrium. Possible interactions between species can, for instance, be investigated by studying the influence of the concentration of one species in the fluid phase on the solid-phase concentrations of the others.

4.8.3.2 Dynamic methods

The analysis of dynamic experiments for obtaining multi-solute adsorption isotherms is more complicated and is presented below for the case of a binary mixture.

Analysis of the mean position of fronts (frontal analysis)

A column is loaded from an initial state I to a feed state F. After a time sufficient for the fluid concentrations in the column to be identical to those prevailing in the feed state, a mass balance can be written for each solute i:

$$\varepsilon_e V_{col}\left(C_i^F - C_i^I\right) + (1-\varepsilon_e)V_{col}\left(\overline{C}_i^F - \overline{C}_i^I\right) = \int_0^{\infty}\left(C_i^F - C_i^{out}(t)\right)Q dt \qquad (4.156)$$

Figure 4.9 shows adsorption and desorption curves of the two solutes of a binary system showing Langmuirian competitive behavior. As with the single-solute situation, the integrals of Eq. (4.156) that represent the quantity of solute stored in the column (shaded in Figure 4.9) must be estimated. For the first eluted solute (Figure 4.9a) the displaced amount (for $C_A^{out}(t) > C_A^F$) gives a negative contribution to the integral.

One limitation of the method certainly lies in the ability to precisely determine the different integrals of Eq. (4.156). In using this method, it is certainly good practice to check the consistency of some results with the adsorption–desorption method.

Under conditions of high efficiency, for favorable adsorption isotherms the two adsorption fronts are sharp enough to be approximated by shocks appearing at times $t_R^{shock}(1)$ and $t_R^{shock}(2)$. These two retention times are not directly associated with solutes A and B but rather with given compositions. Additionally, as explained in Chapter 3, during the period $\left(t_R^{shock}(1), t_R^{shock}(2)\right)$, the concentration of the less retained solute A exceeds the inlet concentration and reaches an intermediate plateau C_A^{plat} (Figure 3.18). Under these conditions, the mass balance given by Eq. (4.156) can be simplified.

For the adsorption step, one obtains for the second solute:

$$\varepsilon_e V_{col}\left(C_B^F - C_B^I\right) + (1-\varepsilon_e)V_{col}\left(\overline{C}_B^F - \overline{C}_B^I\right) = Q\left(C_B^F - C_B^I\right) t_R^{shock}(2) \qquad (4.157)$$

which is similar to Eq. (4.153) obtained for the single-solute experiments.

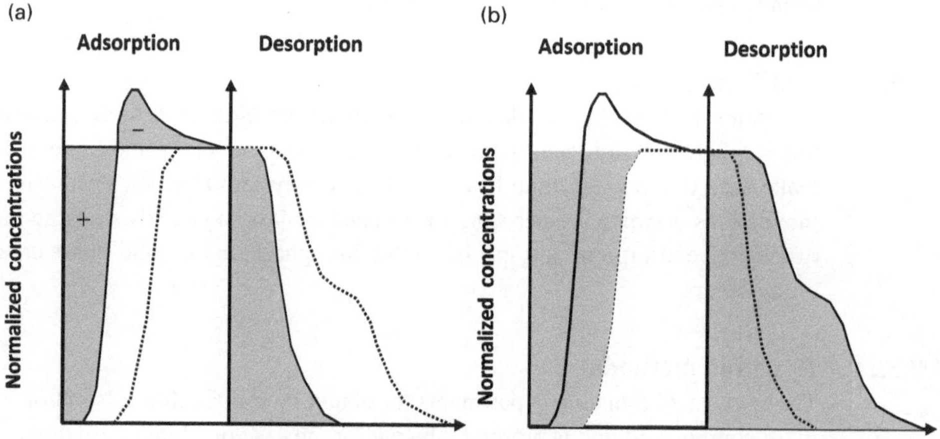

Figure 4.9 Frontal analysis: binary system following a Langmuirian competitive behavior. Solid line: solute A; dashed line: solute B. (a) Integral of Eq. (4.156) shaded for the first solute. (b) Integral of Eq. (4.156) shaded for the second solute.

For the adsorption step, one obtains for the first solute:

$$\varepsilon_e V_{col}\left(C_A^F - C_A^I\right) + (1-\varepsilon_e)V_{col}\left(\overline{C}_A^F - \overline{C}_A^I\right)$$
$$= Q\left(C_A^F - C_A^I\right) t_R^{shock}(1) - Q\left(C_A^{plat} - C_A^F\right)\left(t_R^{shock}(2) - t_R^{shock}(1)\right) \tag{4.158}$$

The term with a negative sign on the right-hand side of Eq. (4.158) accounts for the quantity of solute A displaced during the elution of the plateau at concentration C_A^{plat}.

Equations (4.157) and (4.158) allow determination of $\overline{C}_A^F$ and of $\overline{C}_B^F$ from the positioning of the fronts, and consequently one point of the adsorption isotherm.

As with the static methods, possible competition/interaction between species can be investigated by studying the influence of the concentration of one species on the adsorption of the others.

Perturbation method

The idea is to determine the adsorption isotherm parameters from knowledge of retention times associated with small perturbations, i.e. small injections having a composition similar to the initial composition. We saw in Chapter 3 and especially in Section 3.5 that an n-solute system will respond to the perturbation by giving n peaks or fronts whose velocities are determined by calculating the roots of a given characteristic equation. It is important to understand that the number of peaks obtained at the outlet of the column is not given by the number of solutes contained in the injection, but by the number of adsorbable species, whether present in the feed mixture or in the carrier fluid in equilibrium with the column. For instance, while injecting pure eluent on a column pre-equilibrated with a carrier fluid containing some traces of solutes, one typically obtains as many negative peaks as solutes at the outlet of the system: this is the concept of *vacancy chromatography*. The principle and use of the perturbation method for determining multi-solute adsorption isotherms has been presented in detail by (Tondeur *et al.*, 1996).

Let us illustrate the concept for a binary system. A column is initially equilibrated with a mixture of two solutes A and B at composition I. A small injection of the A+B mixture at a composition very close to I (in order to ensure a linear behavior) is performed at the inlet of the column. Under these circumstances, two peaks are obtained at the outlet of the column with retention times $t_R(1)$ and $t_R(2)$.

From the theoretical framework developed in Section 3.4.1, the retention times of the two peaks are given by:

$$t_R(1) = t_0\left(1 + \frac{1-\varepsilon_e}{\varepsilon_e}\theta_+\right)$$
$$t_R(2) = t_0\left(1 + \frac{1-\varepsilon_e}{\varepsilon_e}\theta_-\right) \tag{4.159}$$

θ_- and θ_+ being the roots of Eq. (3.44). The experimentally determined retention times of the two peaks are thus related to the partial derivatives of the adsorption isotherm at composition I.

By performing several experiments at different initial compositions I, one can determine the parameters of an adsorption isotherm by minimizing the difference between calculated and experimental retention times.

This method is sensitive to various experimental imprecisions, such as the exact initial composition and the complete equilibration of the column. Worth mentioning are the difficulties associated with precise identification of peaks when the baseline is perturbed by significant concentrations of solutes in the carrier fluid.

Hodograph method

An elegant way to determine binary competitive adsorption isotherms from breakthrough curves is given by the hodograph plot introduced in Chapter 3, which allows the information to be represented in a condensed form (Nicoud and Seidel-Morgenstern, 1996). A comparison between theoretical and experimental plots allows the parameters of the model to be determined.

The hodograph method is very straightforward if the system follows the Langmuir model, so that the experimental hodograph plot consists of straight lines. The determination of the adsorption isotherm parameters from the slopes of these lines has been discussed by (Ma *et al.*, 1990).

Since this method relies on the assumption that no dispersion occurs in the column, it is limited to very efficient systems (let us say a few thousand plates); chromatogram simulation is required to check the influence of actual hydrodynamic and/or kinetic dispersion and is practically limited to binary systems.

Peak maximum method

As with single-solute systems, the variation of the positions of the peak maxima versus injected amount can in principle be used to determine the adsorption isotherm parameters. The underlying theoretical tools given in Chapter 3 give some valuable information, but, unfortunately, the existence of an equation similar to Eq. (4.155) even for binary systems is limited to simple situations of Langmuir adsorption isotherms.

The method is limited to very efficient systems (at least equivalent to a few thousand plates, with the minimum requirement to be checked by numerical simulation). It also suffers from an additional shortcoming associated with the requirement to work with relatively well-resolved peaks. This constraint requires that the solutes be relatively dilute at the outlet of the column, so investigating the high concentration range of the adsorption isotherms is difficult.

Instead of focusing on the variation of peak maxima, it is certainly advisable to consider the entire peak, as described in the next method.

Chromatogram fitting method

The inverse method, consisting of determining adsorption isotherms by minimizing the differences between experimental and calculated breakthrough curves or chromatograms, is increasingly popular. This method has been used successfully by different authors, as mentioned in Section 4.8.2.2.

We believe that extracting adsorption isotherm information from overloaded chromatograms, breakthrough curves and generally from available industrial information makes good sense. Methodologies are progressing and recent developments allow the requirements for either sampling or calibration to be minimized (Cornel *et al.*, 2010).

It is worth mentioning the work of (Araujo *et al.*, 2008), who determined adsorption isotherms by a hybrid inverse method using breakthrough curves and periodic steady states of a simulated moving bed, and the work of (Grosfils, 2009), who has made an extensive contribution to the determination of parameters (not limited to adsorption isotherms) on simulated moving beds.

In addition to the need to operate under conditions minimizing the existence of non-thermodynamic sources of dispersion, one must also be aware of a principle limitation occurring when multi-solute systems are considered. While traveling along the column, the concentrations follow specific paths (the hodographs presented in Chapter 3) imposed by the adsorption isotherms. The consequence is that, while inlet compositions are varied from C_i^I to C_i^F, only specific compositions are explored by the system, so one cannot expect meaningful precision on adsorption isotherms in the unexplored regions. Thus it is important to perform different experiments at different feed compositions.

4.8.4 The case of ion exchange

Provided that the total normality is kept constant, all methods described above for adsorption isotherm determination are applicable. The importance of keeping normality constant (and in fact being aware of the influence of normality changes) is illustrated by a simple example involving the batch method.

A volume $\overline{V}$ of resin in the H form is put into contact with a volume V_1 of NaCl 0.1 N in a stirred vessel. Determination of the final concentration C_{Na}^{Final} allows calculation of the concentration $\overline{C}_{Na}^{Final}$ using the mass-balance equation (4.150).

If different experiments are performed for increasing NaCl 0.1 N volumes V_k, increasing concentrations $\overline{Na}^{Final}$ of sodium will be determined so that a complete curve $\overline{C}_{Na}^{up}(C_{Na})$ can be determined.

When the highest desired concentration is reached, a certain volume V_1^{dil} of water is added to the system. The sodium concentration is determined after equilibrium is reached, which then allows determination of a given lumped solid concentration. If different experiments are performed for increasing water volumes V_k^{dil}, decreasing concentrations of Na will be determined so that a complete curve $\overline{C}_{Na}^{down}(C_{Na})$ can be determined. Experience shows that the two curves $\overline{C}_{Na}^{down}(C_{Na})$ and $\overline{C}_{Na}^{up}(C_{Na})$ are different. This is certainly not due to complex hysteresis considerations but to simple ion-exchange properties.

Assuming an ideal ion-exchange situation described by a constant selectivity factor, Eq. (4.59) shows that the lumped solid-phase concentration (eq/l) of sodium ions is related to its liquid-phase concentration by:

$$\overline{C}_{Na} = \frac{\overline{N}^{iex} K_{Na/H} C_{Na}}{N_0 + \left(K_{Na/H} - 1\right) C_{Na}} = \overline{N} \frac{\frac{K_{Na/H}}{N_0} C_{Na}}{1 + \frac{K_{Na/H}-1}{N_0} C_{Na}}$$

$\overline{C}_{Na}$ is not only a function of C_{Na} but is also dependent on N_0. While the "upward" experiments are performed at constant normality 0.1 N by adding volumes of NaCl 0.1 N, the "downward" experiments are performed at decreasing normality by adding water to the system. There is consequently no reason for the $\overline{C}_{Na}^{down}(C_{Na})$ and $\overline{C}_{Na}^{up}(C_{Na})$ curves to be similar. Similar curves would be obtained should the water be replaced by HCl 0.1 N in the "down" experiment so that normality is kept constant.

The impact of normality can be more complex than reported in this simple Na/H exchange. Leaving aside complex molecules, let us mention:

- The case of heterovalent ions: the selectivity factor $K^x_{C/H}$ relating the ionic fractions is strongly normality-dependent for heterovalent ions, whereas it is almost constant for homovalent ions.
- The case of equilibria in solution: it is an interesting exercise to calculate the batch system if one replaces NaCl by NaOH. Because of the dissociation of water, every proton H leaving the resin reacts with a hydroxide ion OH to form water. The ionic fraction of Na tends towards 1 because of the absence of protons in solution, so the Na ions are directed towards the resin just as if the "adsorption isotherm" were irreversible.

The following exercises should make the non-expert reader more familiar with ion exchange.

EXERCISE 4.9 *The desaturation of a cationic column of $V_{col} = 1\ l$ equilibrated with NaCl 0.5 N shows that the total sodium content Na^{tot} in that column is 1.46 eq. A similar experiment performed with the same column equilibrated with (NaCl 0.05 M + HCl 0.05 M) leads to a total sodium content of 0.902 eq. Assuming that the carrier fluid flow rate is 0.4 l/min, estimate:*

- *the retention time associated with a breakthrough curve obtained by feeding a solution containing (NaCl 0.15 M + HCl 0.05 M) on a column initially equilibrated with HCl 0.2 N*
- *the retention time of a "small peak" of sodium on a column initially equilibrated with HCl 0.2 N.*

EXERCISE 4.10 *The column described in Exercise 4.9 is equilibrated with a flow rate of 0.2 l/min of HCl. A small perturbation (Dirac injection) of HCl is injected at the column inlet, leading to a retention time of 1.8 min when HCl concentration (eluent) is 0.001 eq/l and 2.5 min when HCl concentration (eluent) is 0.5 eq/l. Knowing that the intragranular porosity equals 0.5, propose a fitting of the partition coefficient.*

4.8.5 Sensitivity

All the methods presented above have been used for determining adsorption isotherms: a few illustrations can be found in (Nicoud and Seidel-Morgenstern, 1996) and (Seidel-Morgenstern, 2004), and the reader will easily find numerous references in the literature. We stress that a *method* is not a recipe, i.e. it must be understood instead of being

Table 4.1 Influence of the adsorption isotherm on different chromatographic measurements. Reference adsorption isotherm: $\tilde{K}_A = 0.01$, $\tilde{K}_B = 0.015$ g/l and $\tilde{N} = 100$ g/l. Modified adsorption isotherm: $\tilde{K}_A$ and $\tilde{K}_B$ decreased by 5%. FC: frontal chromatography; EC: elution chromatography; TMB: true moving bed. I: initial state; F: feed (50-50 mixture of A and B).

Experiment	Concentration (g/l)	Measured characteristic	Variation (%)
FC	I = 0, F = 1	Retention time of B	3.4
FC	I = 0, F = 100	Retention time of B	0.6
FC, $N = \infty$	I = 0, F = 1	A plateau concentration	0.2
FC, $N = \infty$	I = 0, F = 100	A plateau concentration	0.4
FC, $N = \infty$	I = 1, F = 0	B plateau concentration	0.2
FC, $N = \infty$	I = 100, F = 0	B plateau concentration	5.0
FC	I = 0, F = 1	Amount of A adsorbed	3.0
FC	I = 0, F = 100	Amount of A adsorbed	0.5
EC	I = 0, F = 1	Retention time of A	3.1
TMB	I = 0, F = 100	Extract composition	10.5
TMB	I = 0, F = 100	Raffinate composition	11.5

used blindly. This avoids interpreting the results wrongly and then using "adsorption isotherm curves" that neither predict nor explain breakthrough curves. This is of prime importance when several solutes are competing (possibly with the solvent!), a situation where terms such as "equilibrium curve", "adsorption isotherm" etc. are ambiguous, if ever they make sense.

As adsorption isotherms may subsequently be used for designing chromatographic processes, it is important to know that the sensitivity of a system to adsorption isotherms can vary significantly from one experimental system to other.

In order to demonstrate the influence of the adsorption isotherms on different chromatographic characteristics (retention time, plateau concentration, amount of solute stored in the column etc.), we consider a binary mixture adsorbed on a chromatographic medium according to a Langmuir adsorption isotherm. The influence of a variation of the adsorption constants $\tilde{K}_A$ and $\tilde{K}_B$ (possibly due to temperature or carrier fluid composition fluctuations) on different chromatographic measurements is briefly investigated.

From Table 4.1, it is clear that precise evaluation of the adsorption constants can be a difficult task: a 5% variation in the adsorption constants $\tilde{K}_A$ and $\tilde{K}_B$ may lead to variations lower than 1% in the measured characteristics used for their determination (in the determination of plateau concentration, for instance). Unfortunately, a 5% variation in the adsorption constants $\tilde{K}_A$, $\tilde{K}_B$ leads to a 10% variation in extract and raffinate purities obtained on a true moving bed operated with the flow-rate adequate for the reference adsorption isotherm. Readers not familiar with the true moving bed will find a description of this system in Chapter 8, but the message is that industrial systems may be more sensitive to adsorption isotherms than the laboratory devices used for their determination.

Since each case is unique and each laboratory has its own equipment and expertise, we cannot recommend a specific method, but we stress that one must exercise systematic caution. Because of experimental uncertainties, all methods must be cross-checked, a simple test consisting at least of checking the initial slope of the adsorption isotherms, thanks to linear elution chromatography as already mentioned in the introduction to Section 4.8.1. These sensitivity considerations, and the possible strong dependence of industrial processes on adsorption isotherms, are strong arguments in favor of the inverse method.

4.9 Conclusion

The ability to properly determine adsorption isotherms, or more generally fluid-phase equilibria, is the condition *sine qua non* for the proper modeling of chromatographic processes. As already mentioned, the difficulty of the task is neither as high as many chromatographers think nor as easy as process engineers dream. It is simply a prerequisite for rational process development that must be accepted.

A first important step is the proper definition of the different system phases and associated porosity levels. While porosities can be determined via different established techniques, absolute determination is not an easy task due to the ambiguity of certain definitions. With luck, it can be shown that the exact values of the different porosities are of secondary importance provided that a consistent theoretical scheme is used from the laboratory to the industrial scale.

The single-solute adsorption isotherms can be associated with different shapes and are normally categorized as favorable, unfavorable or S-shaped. The Langmuir-like form (favorable) is certainly the most frequent, but various expressions are available that allow us to represent the various situations that can be met in practice.

The situation of multi-solute systems is more complex, not because of the multiplicity of solutes, but because of the competition – or more generally, interactions – that can occur between the solutes. While in general the multi-solute adsorption is competitive, which means that the presence of one product hinders the adsorption of the others, synergistic effects have also been reported in some situations. When the classical expressions, common sense or intuition are not sufficient to represent multi-solute data, the IAS (possibly RAS) theory is probably the most sensible approach that can be proposed today.

Extension to ion exchange does not bring specific difficulties provided that one systematically draws conclusions associated with the fact that this process requires two species to be exchanged and electroneutrality to be satisfied in the different phases. A consequence is that ion exchange involving n species at *constant normality* is like adsorption associated with $n - 1$ species.

Many situations involve a combination of several mechanisms that can be described by equilibria in the different phases. The stoichiometric equations describing the systems allow derivation of important features of the system, including invariants and the number of fronts.

There are many techniques available for determining adsorption isotherms, be they static or dynamic. For practical reasons, techniques associated with the fitting of experimental chromatograms (the so-called inverse methods) are more and more favored, and we believe that this trend will continue.

4.10 Solutions of the exercises

Exercise 4.1

The equation of state $\pi = \frac{RT}{\Omega} ln\left(\frac{\bar{\bar{A}}/\bar{\bar{n}}_A}{\bar{\bar{A}}/\bar{\bar{n}}_A - \Omega}\right)$ can be rewritten in terms of concentrations:

$$\pi = \frac{RT}{\Omega} ln\left(\frac{\bar{\bar{A}}_{sp}/\bar{\bar{C}}_A}{\bar{\bar{A}}_{sp}/\bar{\bar{C}}_A - \Omega}\right)$$

This gives, after differentiation:

$$d\pi = \frac{RT}{\bar{\bar{A}}_{sp}} \frac{d\bar{\bar{C}}_A}{1 - \bar{\bar{C}}_A \Omega / \bar{\bar{A}}_{sp}}$$

Note that $\frac{\Omega}{\bar{\bar{A}}_{sp}} = \frac{1}{\bar{\bar{N}}}$, so:

$$d\pi = \frac{RT}{\bar{\bar{A}}_{sp}} \frac{d\bar{\bar{C}}_A}{1 - \bar{\bar{C}}_A / \bar{\bar{N}}}$$

Using the Gibbs adsorption isotherm $d\pi = \frac{RT}{\bar{\bar{A}}_{sp}} \bar{\bar{C}}_A \frac{dC_A}{C_A}$, one obtains:

$$\frac{d\bar{\bar{C}}_A}{\bar{\bar{C}}_A\left(1 - \bar{\bar{C}}_A/\bar{\bar{N}}\right)} = \frac{dC_A}{C_A} \tag{4.160}$$

Equation (4.160) can be integrated to give $\frac{\bar{\bar{C}}_A/\bar{\bar{N}}}{1 - \bar{\bar{C}}_A/\bar{\bar{N}}} = kC_A$, which is the Langmuir adsorption isotherm.

Exercise 4.2

The single-solute adsorption isotherms are given by:

$$\bar{C}_A = \frac{\bar{\bar{N}} \tilde{K}_A C_A}{1 + \tilde{K}_A C_A} \quad \bar{C}_B = \frac{\bar{\bar{N}} \tilde{K}_B C_B}{1 + \tilde{K}_B C_B}$$

The modified spreading pressure is calculated from Eq. (4.38):

$$\Pi = \tilde{K}_A \bar{\bar{N}} \int_0^{C_A^0} \frac{1}{1 + \tilde{K}_A u} du = \bar{\bar{N}} \ln\left(1 + \tilde{K}_A C_A^0\right)$$

$$\Pi = \tilde{K}_B \bar{\bar{N}} \int_0^{C_B^0} \frac{1}{1 + \tilde{K}_B u} du = \bar{\bar{N}} \ln\left(1 + \tilde{K}_B C_B^0\right)$$

and consequently one obtains:

$$C_A^0 = \frac{\exp\left(\Pi/\overline{\overline{N}}\right) - 1}{\tilde{K}_A} \qquad C_B^0 = \frac{\exp\left(\Pi/\overline{\overline{N}}\right) - 1}{\tilde{K}_B}$$

Equation (4.43) gives:

$$\frac{\tilde{K}_A C_A}{\exp\left(\Pi/\overline{\overline{N}}\right) - 1} + \frac{\tilde{K}_B C_B}{\exp\left(\Pi/\overline{\overline{N}}\right) - 1} = 1$$

so that the modified spreading pressure is given by:

$$\Pi = \overline{\overline{N}} \ln\left(1 + \tilde{K}_A C_A + \tilde{K}_B C_B\right)$$

One can consequently calculate the fictitious concentrations C_A^0 and C_B^0 as a function of C_A and C_B:

$$C_A^0 = \frac{\exp\left(\Pi/\overline{\overline{N}}\right) - 1}{\tilde{K}_A} = \frac{\tilde{K}_A C_A + \tilde{K}_B C_B}{\tilde{K}_A}$$

$$C_B^0 = \frac{\exp\left(\Pi/\overline{\overline{N}}\right) - 1}{\tilde{K}_B} = \frac{\tilde{K}_A C_A + \tilde{K}_B C_B}{\tilde{K}_B}$$

The fictitious concentrations $\overline{\overline{C}}_A^0$ and $\overline{\overline{C}}_B^0$ of pure solute in equilibrium with C_A^0 and C_B^0 are calculated by:

$$\overline{\overline{C}}_A^0 = \overline{\overline{N}} \frac{\tilde{K}_A C_A + \tilde{K}_B C_B}{1 + \tilde{K}_A C_A + \tilde{K}_B C_B} \qquad \overline{\overline{C}}_B^0 = \overline{\overline{N}} \frac{\tilde{K}_A C_A + \tilde{K}_B C_B}{1 + \tilde{K}_A C_A + \tilde{K}_B C_B}$$

Notice that $\overline{\overline{C}}_A^0 = \overline{\overline{C}}_B^0$.

The molar (massic) fractions are then given by Eq. (4.42):

$$\bar{\bar{x}}_A = \frac{C_A}{C_A^0} = \frac{\tilde{K}_A C_A}{\tilde{K}_A C_A + \tilde{K}_B C_B} \qquad \bar{\bar{x}}_B = \frac{C_B}{C_B^0} = \frac{\tilde{K}_B C_B}{\tilde{K}_A C_A + \tilde{K}_B C_B}$$

The total adsorbed concentration C_{tot} is given by Eq. (4.44):

$$\frac{1}{\overline{\overline{C}}_{tot}} = \frac{\overline{\overline{X}}_A}{\overline{\overline{C}}_A^0} + \frac{\overline{\overline{X}}_A}{\overline{\overline{C}}_A^0} = \overline{\overline{N}} \frac{\tilde{K}_A C_A + \tilde{K}_B C_B}{1 + \tilde{K}_A C_A + \tilde{K}_B C_B} \tag{4.161}$$

The adsorbed phase concentration is finally calculated by $\overline{\overline{C}}_i = \bar{\bar{x}}_i \overline{\overline{C}}_{tot}$, which leads to:

$$\overline{\overline{C}}_A\left(C_A, C_B\right) = \frac{\overline{\overline{N}} \tilde{K}_A C_A}{1 + \tilde{K}_A C_A + \tilde{K}_B C_B} \qquad \overline{\overline{C}}_B\left(C_A, C_B\right) = \frac{\overline{\overline{N}} \tilde{K}_B C_B}{1 + \tilde{K}_A C_A + \tilde{K}_B C_B}$$

Consequently, if the single solute satisfies the Langmuir adsorption isotherms with a unique adsorption capacity $\overline{\overline{N}}$, then, according to the IAS theory, the binary mixture satisfies the binary Langmuir adsorption isotherm.

If one solute follows an anti-Langmuir adsorption isotherm with a negative coefficient in the denominator, the above development is valid in a limited range only; for instance, the spreading pressure must be positive.

Exercise 4.3

For each species j, one writes:

$$S_j + \overline{\overline{F}} = \overline{\overline{S}}_j, \quad K_j = \frac{\overline{\overline{S}}_j}{S_j\overline{\overline{F}}}$$

The conservation of the number of adsorption sites is written:

$$\overline{\overline{N}} = \overline{\overline{F}} + \sum \overline{\overline{S}}_k = \overline{\overline{F}}(1 + \sum K_k S_k)$$

Consequently, one obtains:

$$\overline{\overline{S}}_j = \overline{\overline{N}} K_j S_j / (1 + \sum K_k S_k)$$

which is the classical form of the Langmuir adsorption isotherm when the saturation capacity is independent of the species.

Exercise 4.4

The two forms P and U are assumed to be in equilibrium and their concentrations related by the mass action law $K_f = U/P$.

These two forms can enter into equilibria with the free sites $\overline{\overline{F}}$ that can be described with mass action laws $K_u = \frac{\overline{\overline{U}}}{\overline{\overline{F}}U}$ and $K_p = \frac{\overline{\overline{P}}}{\overline{\overline{F}}P}$.

The conservation of the number of sites gives $\overline{\overline{F}} + \overline{\overline{U}} + \overline{\overline{P}} = \overline{\overline{N}}$.

With the above expressions, one obtains $\overline{\overline{F}} = \dfrac{\overline{\overline{N}}}{1 + (K_u K_f + K_p)P}$.

Consequently:

$$\overline{\overline{U}} = \frac{K_u U \overline{\overline{N}}}{1 + (K_u K_f + K_p)P} \quad \text{and} \quad \overline{\overline{P}} = \frac{K_p P \overline{\overline{N}}}{1 + (K_u K_f + K_p)P}$$

By defining $C = U + P$ and $\overline{\overline{C}} = \overline{\overline{U}} + \overline{\overline{P}}$, one obtains finally:

$$\overline{\overline{C}} = \frac{K_{glob} C \overline{\overline{N}}}{1 + K_{glob} C} \text{ with } K_{glob} = \frac{K_u K_f + K_p}{1 + K_f}$$

The resulting apparent adsorption isotherm thus has a Langmuir form.

Exercise 4.5

The atoms are: H, O, C, N, Na. The five corresponding invariants are:

$$\text{H: } H^+ + \overline{\overline{H}}^+ + 2H_2O + 7AAH + 6AA^- + 8AAH_2^+ + 8\overline{\overline{AAH_2^+}}$$

$$\text{O: } OH^- + H_2O + 2AAH + 2AA^- + 2AAH_2^+ + 2\overline{\overline{AAH}}{}_2^+$$
$$\text{C: } 2AAH + 2AA^- + 2AAH_2^+ + 2\overline{\overline{AAH}}{}_2^+$$
$$\text{N: } AAH + AA^- + AAH_2^+ + \overline{\overline{AAH}}{}_2^+ + NO_3^-$$
$$\text{Na: } Na^+ + \overline{\overline{Na}}{}^+$$

One electroneutrality condition should be added, as already discussed.

Remarks:

- Although the atomic invariants are "obvious" from the point of view of chemistry, they are "heavy expressions" that may involve useless information such as the systematic prefactor 2 in the carbon invariant.
- There are as many independent moieties as atoms in this example. This is not general: in polyethylene synthesis from ethylene, there is a single (and sufficient!) moiety, namely the methylene group, $-CH_2-$, and two atoms, C and H. In polyethylene synthesis, the invariants based on C and H are not independent since they are strictly proportional.

Exercise 4.6

The extragranular porosity is first determined using $\varepsilon_T = 0.7 = \varepsilon_e + 0.5 \times (1 - \varepsilon_e)$, so that $\varepsilon_e = 0.4$.

- **Porous model:** the Z_tRT is used as a reference.
  Under this condition, Eq. (4.148) holds with $t_{0R} = \frac{0.7\times 1}{0.4} = 1.75$ min. This leads to $\overline{\overline{K}}_i = \left(\frac{t_R}{t_{0R}} - 1\right)\frac{\varepsilon_T}{1-\varepsilon_T}$, thus

$$\overline{\overline{K}}_i = \overline{\overline{f}}{}_i'(0) = \left(\frac{5}{1.75} - 1\right)\frac{0.7}{1-0.7} = 4.33$$

- **Lumped model:** the Z_eRT is used as a reference.
  Under this condition, Eq. (4.149) holds with $t_{0R} = \frac{0.4\times 1}{0.4} = 1.0$ min. This leads to $\overline{K}_i = \left(\frac{t_R}{t_0} - 1\right)\frac{\varepsilon_e}{1-\varepsilon_e}$, thus

$$\overline{K}_i = \overline{f}{}_i'(0) = \left(\frac{5}{1.0} - 1\right)\frac{0.4}{1-0.4} = 2.66$$

One can check that $\overline{K}_i = \varepsilon_i + (1-\varepsilon_i)\overline{\overline{K}}_i$.

Exercise 4.7

The recovered mass may be expressed as $m_A = \varepsilon_e V_{col} C_A + (1-\varepsilon_e) V_{col} \overline{C}_A$. Consequently, $\overline{C}_A$ can be estimated for each experiment (note that knowledge of the flow rate is not required). As presented in Table 4.2, the experimental lumped adsorption isotherm $\overline{C}{}_A^F = \overline{f}{}_A^F(C_A^F)$ is correctly represented by a linearly modified Langmuir adsorption isotherm, Eq. (4.13), with $\overline{\lambda}_A = 0.617$, $\overline{N} = 26.3$ mol/l, $\tilde{K}_A = 0.072$ mol/l.

Table 4.2 For Exercise 4.7

C_A (g/l)	1	5	10	25	50
m_A (g)	0.0078	0.032	0.06	0.125	0.209
$\overline{C}_A(exp)$ (g/l)	2.46	9.51	17.4	33.5	50.6
$\overline{C}_A(calc)$ (g/l)	2.39	10.06	17.20	32.35	51.45

Exercise 4.8

The mass balance given by Eq. (4.150) allows a calculation of the concentrations $\overline{C}_A^{Final}$ for the different experiments:

$$\overline{C}_A^{Final} = \frac{V}{\overline{V}}(C_A^I - C_A^{Final})$$

One obtains:

C_A^I (g/l)	1	10	20	50	100	150	200
C_A^{Final} (g/l)	0.2	2.15	4.8	16.4	48.8	85.2	127.9
$\overline{C}_A^{Final}$ (g/l)	2.67	26.17	50.67	112.0	170.7	216.0	240.3

The experimental lumped adsorption isotherm $\overline{C}_A^{Final} = \bar{f}_A(C_A^{Final})$ is correctly represented by a linearly modified Langmuir adsorption isotherm Eq. (4.13) with $\bar{\lambda}_A = 0.5$, $\overline{N} = 200$ mol/l, $\tilde{K}_A = 0.06665$ mol/l. This is confirmed in the following table:

C_A^{Final} (g/l)	0.2	2.15	4.8	16.4	48.8	85.2	127.9
$\overline{C}_A^{Final}$ exp. (g/l)	2.67	26.17	50.67	112.0	170.7	216.0	240.3
$\overline{C}_A^{Final}$ calc. (g/l)	2.73	26.14	50.68	112.65	177.37	212.65	242.95

With knowledge of the lumped adsorption isotherm parameters, one can check that the experiment performed in Exercise 2.7 with an initial concentration of 1 g/l was in the linear part.

Exercise 4.9

The first step consists in determining key thermodynamic parameters from the adsorption–desorption experiment. We will assume the extragranular porosity of the column to be 0.4.

Regeneration of the column equilibrated with NaCl 0.5 N gave $Na^{tot} = 1.46$ eq. Since $Na^{tot} = \left(\varepsilon_e Na + (1-\varepsilon_e)\overline{Na}\right) V_{col}$, one can determine $\overline{Na} = \dfrac{1.46 - 0.4 \times 0.5}{1-0.4} = 2.1$ eq/l.

H and Na being the only two cations present in the system, the resin's total capacity is given by:

$$\overline{H}^{+} + \overline{Na}^{+} = \overline{N}^{iex}$$

When the resin is equilibrated with NaCl, *neglecting possible electrolyte sorption*, we have $\overline{H}^+ = 0$ so that $\overline{N}^{iex} = \overline{Na}^+ = 2.1$ eq/l.

When the column is equilibrated with a mixture of NaCl and HCl, both at 0.05 M, the total sodium content in the column is 0.902 eq. This leads to $\overline{Na} = \dfrac{0.902 - 0.4 \times 0.05}{1 - 0.4} = 1.47$ eq/l. Consequently, one has: $\overline{H} = \overline{N}^{iex} - \overline{Na} = 2.1 - 1.47 = 0.63$ eq/l.

Therefore, *assuming a constant equilibrium constant*, we obtain:

$$K_{Na/H} = \frac{\overline{Na}}{Na}\frac{H}{\overline{H}} = \frac{1.47}{0.05}\frac{0.05}{0.63} = 2.33$$

Retention time of the breakthrough curve

The zero retention time is given by $t_0 = \varepsilon_e V_{col}/Q = 1$ min.

The experiment being performed at constant normality 0.2 N, the ion exchange involving two ions is equivalent to a single-solute adsorption and will thus lead to one front. The equilibrium constant $K_{Na/H}$ being greater than 1, Na is preferred to H (this assertion could be wrong with non-homovalent ions) and the front is thus compressive.

The retention time is given by:

$$t_R = t_0\left(1 + \frac{1-\varepsilon_e}{\varepsilon_e}\frac{\overline{Na}^F - \overline{Na}^I}{Na^F - Na^I}\right) = t_0\left(1 + \frac{1-\varepsilon_e}{\varepsilon_e}\frac{\overline{H}^F - \overline{H}^I}{H^F - H^I}\right)$$

From the ion-exchange capacity and selectivity coefficient determined above, we know that $\left(Na^I = 0, H^I = 0.2\right)$ is in equilibrium with $\left(\overline{Na}^I = 0, \overline{H}^I = 2.1\right)$ and that $\left(Na^F = 0.15, H^F = 0.05\right)$ is in equilibrium with $\left(\overline{Na}^F = 1.84, \overline{H}^F = 0.26\right)$. Consequently:

$$\frac{\overline{Na}^F - \overline{Na}^I}{Na^F - Na^I} = \frac{\overline{H}^F - \overline{H}^I}{H^F - H^I} = 12.66$$

which gives $t_R = 1 \times \left(1 + \dfrac{1-0.4}{0.4} \times 12.66\right) = 20$ min. See Figure 4.10.

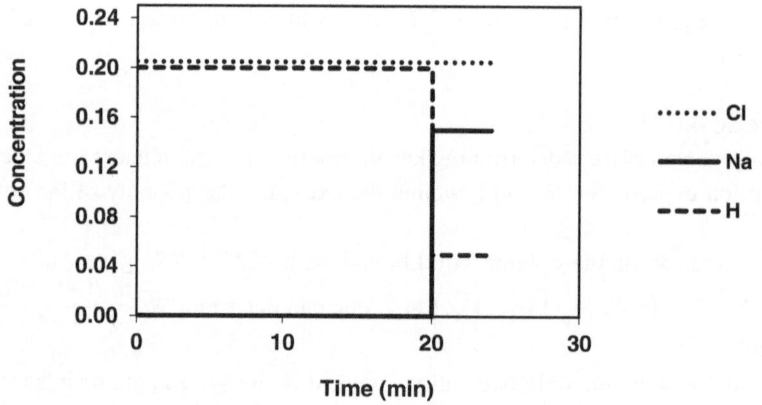

Figure 4.10 Illustration of the profiles obtained for Exercise 4.9.

Retention time of a small peak

Under the linearity assumption, we know from Eq. (4.74) that

$$t_R = t_0 \left(1 + \frac{1-\varepsilon_e}{\varepsilon_e}\overline{K}_{Na}\right)$$

with:

$$\overline{K}_{Na} = K_{Na/H}\,\frac{\overline{N}^{iex}}{N_0} = 2.33 \times \frac{2.1}{0.2} = 24.46$$

One has consequently $t_R = 1 \times \left(1 + \frac{1-0.4}{0.4} \times 24.46\right) = 37.7$ min.

Exercise 4.10

At 0.001 eq/l, the Cl anions can be considered to be completely excluded from the resin beads. The experimental retention time can thus be considered to be the Z_eRT t_0. One can thus calculate the external porosity of the bed:

$$\varepsilon_e = \frac{t_0 Q}{V} = 0.36$$

At 0.5 eq/l, the Cl anions can penetrate the resin particles more freely and the retention time increases. The distribution coefficient is calculated from Eq. (4.91):

$$\varepsilon_i \tilde{K}_A = \frac{\varepsilon_e}{1-\varepsilon_e}\,\frac{t_R - t_0}{t_0} = 0.22 \quad \text{thus} \quad \tilde{K}_A = 0.44$$

This allows an estimate of $N_D = 0.45$ eq/l (Eq. (4.90)).

References

Araujo, J., Rodrigues, R. and Mota, J. (2008), *J. Chromatogr. A* **1189**, 302–313.

Bernardi, S., Getaz, G., Forrer, N. and Morbidelli, M. (2013), *J. Chromatogr. A* **1283**, 46–52.

Bolt, G. H. (1982). In G. H. Bolt, ed., *Soil Chemistry. B: Physico-Chemical Models*, 2nd edn., Elsevier Science.

Brooks, C. and Cramer, S. (1992), *AIChE J.* **38**, 1969.

Clavier, J. Y. (1995), Developpement du procédé de lit mobile simulé avec eluant supercritique, unpublished PhD thesis, INPL, Nancy.

Cornel, J., Tarafder, A., Katuso, S. and Mazzotti, M. (2010), *J. Chromatogr. A* **1217**, 1934–1941.

Dye, S. R., De Carli, J. P. and Carta, G. (1990), *Ind. Eng. Chem. Res.* **29**(5), 849–857.

Farnan, D., Frey, D. and Horvath, C. (2002), *J. Chromatogr. A* **959**, 65–73.

Flockerzi, D., Kaspereit, M. and Kienle, A. (2013), *Chem. Eng. Sci.* **104**, 957–959.

Frey, D. D. and Rodrigues, A. E. (1994), *AIChE J.* **40**(1), 182–186.

Fukuchi, K., Kobuchi, S. and Arai, Y. (1982), *J. Chem. Eng. of Japan* **15**(4), 316–318.

Goto, M. and McCoy, B. (2000), *Chem. Eng. Sci.* **55**, 723–732.

Gritti, F. and Guiochon, G. (2003), *J. Colloid Interface Sci.* **264**(1), 43–59.

Gritti, F., Piatkowski, W. and Guiochon, G. (2003), *J. Chromatogr.* **983**, 51.

Grosfils, V. (2009), Modelling and parametric estimation of simulated moving bed chromatographic processes. PhD thesis, Université Libre de Bruxelles.

Guan, H. and Guiochon, G. (1996), *J. Chromatogr. A* **731**, 27–40.

Guan-Sajonz, H., Guiochon, G., Davis, E., Gulakowski, K. and Smith, D. (1997), *J. Chromatogr. A* **773**, 33–51.

Guiochon, G., Felinger, A., Golshan-Shirazi, S. and Katti, A. M. (2006), *Fundamentals of Preparative and Nonlinear Chromatography*, Elsevier Academic Press.

Guiochon, G. and Tarafder, A. (2011), *J. Chromatogr. A* **1218**, 1037–1114.

Guélat, B., Delegrange, L., Valax, P. and Morbidelli, M. (2013), *J. Chromatogr. A* **1298**, 17–25.

Heinonen, J., Rubiera Landa, H., Sainio, T. and Seidel-Morgenstern, A. (2012), *Sep. Purif. Technol.* **95**, 235–247.

Helfferich, F. G. (1962), *Ion Exchange*, McGraw-Hill.

Hill, T. L. (1960), *Introduction to Statistical Thermodynamics*, Addison-Wesley.

Ilic, M., Flocerzi, D. and Seidel-Morgenstern, A. (2010), *J. Chromatogr. A* **1217**, 2132–2137.

Jedrzejak, A., Gorius, A. and Tondeur, D. (1989), *Chem. Eng. Sci.* **44**, 1315.

Kacmarski, K. and Bellot, J. (2003), *Acta Chromatogr.* **13**, 22–37.

Kaczmarski, K., Mori, M., Glod, B., Kowalska, T. and Tanaka, K. (2005), *Acta Chromatogr.* **15**, 66–81.

Koter, K. and Terzyk, A. (2005), *J. Colloid Interface Sci.* **282**, 335–339.

Kvaalen, E., Neel, L. and Tondeur, D. (1985), *Chem. Eng. Sci.* **40**, 1191–1204.

Langmuir, I. (1918), *J. Am. Chem. Soc.* **40**, 1361–1402.

Le Van, M. D. and Vermeulen, T. (1981), *J. Phys. Chem.* **85**(22), 3247–3250.

Lin, S., Blanco, R. and Karger, B. (1991), *J. Chromatogr.* **557**(1–2), 369–82.

Ma, Z., Katti, A. M., Lin, B. and Guiochon, G. (1990), *J. Phys. Chem.* **94**(17), 6911–6922.

Miyabe, K. and Guiochon, G. (2000), *J. Chromatogr. A* **866**, 147–171.

Myers, A. L. (1983), *AIChE J.* **29**(4), 691–693.

Myers, A. L. and Byington, S. (1986). In A. E. Rodrigues, ed., Ion Exchange: Science and Technology, vol. 107 of *NATO ASI Series, Series E: Applied Sciences*, Martinus Nijhoff, pp. 119–145.

Myers, A. L. and Prausnitz, J. M. (1965), *AIChE J.* **11**(1), 121–127.

Nfor, B., Noverraz, M., Chilamkurthi, S. and Verhaert, P. (2010), *J. Chromatogr.* **1217**, 6829–6850.

Nicoud, R. M. (1987), Influence respective des facteurs thermodynamiques, hydrodynamiques et diffusionnels sur le fonctionnement des echangeurs d'ions Application a un procede d'echange ionique: Le Nymphea PhD thesis, Institut National Polytechnique de Lorraine, Nancy.

Nicoud, R. M. and Schweich, D. (1989), *Water Resour. Res.* **25**(6), 1071–1082.

Nicoud, R. M. and Seidel-Morgenstern, A. (1996), *Isol. Purif.* **2**(3), 165–200.

Nowak, J., Gedicke, K., Antos, D., Piatkowski, W. and Seidel-Morgenstern, A. (2007), *J. Chromatogr. A* **1164**(1–2), 224–234.

Nowak, J., Poplewska, I., Antos, D. and Seidel-Morgenstern, A. (2009), *J. Chromatogr. A* **1216**(50), 8697–8704.

Ottiger, S., Kluge, J., Rajendran, A. and Mazzotti, M. (2007), *J. Chromatogr. A* **1162**, 74–82.

Pabst, T., Antos, D., Carta, G., Ramasubramanyan, N. and Hunter, A. (2008), *J. Chromatogr. A* **1181**, 83–94.

Pepper, K. W., Reichenberg, D. and Hale, D. K. (1952), *J. Chem. Soc.* 3129–3136.

Peterson, D. L. and Redlich, O. (1962), *J. Chem. Eng. Data* **7**(4), 570–574.

Radke, C. J. and Prausnitz, J. M. (1972), *Ind. Eng. Chem. Fundam.* **11**(4), 445–451.

Rajendran, A. (2012), *J. Chromatogr. A* **1250**, 227–249.

Rajendran, A., Gilkison, T. and Mazzotti, M. (2008), *J. Sep. Sci.* **31**(8), 1279–1289.

Rajendran, A., Krauchi, O., Mazzotti, M. and Morbidelli, M. (2005), *J. Chromatogr. A* **1092**, 149–160.

Rearden, P., Sajonz, P. and Guiochon, G. (1998), *J. Chromatogr. A* **813**, 1–9.

Rubiera, L., Flockerzi, D. and Seidel-Morgenstern, A. (2013), *AIChE J.* **59**(4), 1263–1277.

Ruthven, D. M. (1984), *Principles of Adsorption and Adsorption Processes*, John Wiley & Sons.

Sajonz, P., Kele, M., Zhong, G., Sellergen, B. and Guiochon, G. (1998), *J. Chromatogr. A* **810**, 1–17.

Saritha, N. and Madras, G. (2001), *Chem. Eng. Sci.* **56**, 6511–6524.

Schmidt-Traub, H., Kaspereit, M., Engell, S., Susanto, A., Epping, A. and Jupke, A. (2012). In H. Schmidt-Traub, M. Schulte and A. Seidel-Morgenstern, eds., *Preparative Chromatography*, 2nd edn., Wiley-VCH, p. 425.

Schweich, D., Jauzein, M. and Sardin, M. (1993a), *Water Resour. Res.* **29**(3), 723–733.

Schweich, D., Jauzein, M. and Sardin, M. (1993b), *Water Resour. Res.* **29**(3), 735–741.

Schweich, D. and Sardin, M. (1981), *J. Hydr.* **52**, 1–33.

Seidel, A. (1989), *Chem. Tech.* **41**, 525–530.

Seidel-Morgenstern, A. (2004), *J. Chromatogr. A* **1037**, 255–272.

Seidel-Morgenstern, A., Schmidt-Traub, H., Michel, M., Epping, A. and Jupke, A. (2012). In H. Schmidt-Traub, M. Schulte and A. Seidel-Morgenstern, eds., *Preparative Chromatography*, 2nd edn., Wiley-VCH, p. 321.

Strubinger, J. R. and Parcher, J. F. (1989), *Anal. Chem.* **61**(9), 951–955.

Suwanayuen, S. and Danner, R. P. (1980a), *AIChE J.* **26**(1), 68–76.

Suwanayuen, S. and Danner, R. P. (1980b), *AIChE J.* **26**(1), 76–82.

Tarafder, A. and Mazzotti, M. (2012), *Chem. Eng. Technol.* **35**, 102–108.

Tondeur, D., Kabir, H., Luo, A. and Granger, J. (1996), *Chem. Eng. Sci.* **51**(15), 3781–3799.

Toth, J. (1971), *Acta Chim. Acad. Sci. Hung.* **69**(3), 311–328.

Toth, J. (2002), *Adsorption: Theory, Modeling and Analysis*, Marcel Dekker.

Vajda, P., Felinger, A. and Cavazzini, A. (2010), *J. Chromatogr. A* **1217**, 5965–5970.

Van Wasen, U. and Schneider, G. M. (1975), *Chromatographia* **8**(6), 274–276.

Wellhoefera, M., Sprinzla, W., Hahna, R. and Jungbauer, A. (2013), *J. Chromatogr. A* **1319**, 107–117.

Whitley, R., Wachter, R., Liu, F. and Wang, L. (1989), *J. Chromatogr.* **465**, 137–156.

5 Mass transfer

Let's take the time . . .

In the previous chapters, the basics of chromatography modeling were provided under the assumption that kinetic limitations either were non-existent (Chapter 3 for non-linear chromatography) or could be modeled via a first-order approximation (Chapter 2 for linear chromatography). We have shown in Chapter 2 that, as long as kinetic limitations are not extremely strong, a good approximation for modeling chromatographic columns is to lump hydrodynamic dispersion and mass transfer limitations into a single parameter known as the *plate number*. We will show in the following sections and in Chapter 7 that this approach is often sufficient for modeling preparative scale chromatography. In certain situations, however, when kinetic limitations are strong, in the case of strongly non-linear chromatography or when specific precision is required because of challenging purity targets, this approximation may not be accurate enough and more detailed descriptions will be needed.

It is generally accepted that kinetic limitations can be attributed to diffusional transfer occurring either outside the beads, thus referred to as external diffusion, or inside the beads, thus referred to as internal diffusion. Usually, the adsorption step is very fast and is assumed to be kinetically instantaneous. However, because very small particle sizes are now available, the influence of diffusional limitations could in principle decrease to such an extent that, in certain rare cases, the overall kinetics of chromatography may be limited by the kinetics of the adsorption step.

We want to stress the fact that the distinction between external diffusion, internal diffusion and adsorption kinetics is not purely academic, because these three possible processes are impacted differently by operating parameters like particle size and fluid velocity. The ability to identify and properly model these different processes is thus key.

We will use the lumped model and the porous model introduced in Chapter 1 to model mass transfer around and inside particles. The lumped model clearly has the advantage of simplicity, whereas more or less complex descriptions associated with the porous model allow a better understanding of the influence of certain operating parameters on kinetics.

Unless otherwise specified, the presentation in this chapter relies on some important assumptions, already presented in Chapter 1:

- The particles are either "well-formed" particles having a well-identified skeleton and well-defined pores or gel particles.

- The particles are totally rigid so that no swelling occurs.
- The molecules of solute are rigid and are not subject to changes in conformation during the saturation/desaturation process.
- Transport of solutes around the particle is diffusive. The sorption effect that can occur in gas chromatography is neglected.

An additional important assumption is that the transport of solutes other than by convection can be described by Fick's law, which is normally expressed in the following, relatively vague terms: "The diffusion flux is proportional to the concentration gradient via the coefficient of diffusion." In fact, modeling diffusion requires precise definitions to ensure that the right fluxes and gradients are connected by the right diffusion coefficient. The reader can review basic principles by reading Appendix G, where Fick as well as Maxwell–Stefan approaches to diffusion are presented. In particular, the reader may rediscover the trivial concept that diffusion cannot involve a single component! This means that presentations of single-solute systems always imply the existence of a solvent or more generally of a component in large excess compared to the investigated solute. The consequences or even validity of this assumption will be discussed later in this chapter.

In the main body of our presentation, all particles are assumed to have the same size. We will show how particle size distributions may be handled in Section 5.9.

Finally, adsorption-type mechanisms are assumed and used to illustrate the modeling strategy and essential results. The required adaptations to the cases of size exclusion and ion-exchange resins are presented in Sections 5.3.9 and 5.5.

5.1 External transfer

Prior to entering a particle, a solute must be transported from the bulk solution to the particle periphery. A schematic representation of concentration profiles is given in Figure 5.1.

Various models have been proposed to describe the transport of species between a fluid and a particle. Most generally accepted is the *film model*, which assumes that the boundary layer is a motionless fluid skin of uniform thickness through which solutes move by molecular diffusion.

According to this model, the specific flux of species A, denoted J_A, is assumed to be proportional to the difference between the concentration C_A in the bulk solution and the concentration C_A^* in the extragranular fluid at the fluid–solid interface:

$$J_A = k_A \left(C_A - C_A^*\right) \ (\text{mol/m}^2/\text{s}) \tag{5.1}$$

where k_A is the external mass transfer coefficient (m/s) of solute A.

This expression, replacing local derivatives of Fick's law by finite differences, is certainly subject to the validity of the assumptions underlying Fick's law. In particular, the solute is assumed to be immersed in a large excess of solvent.

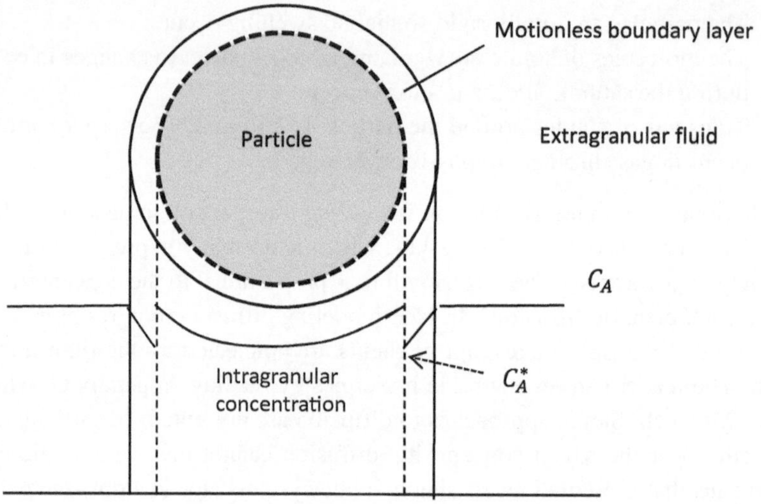

Figure 5.1 Schematic representation of a chromatographic particle (lumped) and surrounding motionless boundary layer.

In order to estimate the mass transfer coefficient k_A, one can assume the particles to be surrounded by a motionless layer of fluid of thickness δ in which diffusion occurs according to Fick's law. Provided that a steady-state regime is reached in the film, and in the absence of chemical reaction and convection, the transfer coefficient is thus simply the ratio of a diffusion coefficient to the boundary layer thickness, so the net flux of A entering the particle is given by:

$$J_A = \frac{D_A}{\delta}(C_A - C_A^*) \quad (\text{mol/m}^2/\text{s}); \quad \text{thus} \quad k_A = \frac{D_A}{\delta} \tag{5.2}$$

By defining $\overline{C}_A$ as the lumped solid concentration of solute A inside the particle, a mass balance on the particle gives:

$$V_p \frac{d\overline{C}_A}{dt} = S_p J_A = S_p \frac{D_A}{\delta}(C_A - C_A^*) \quad (\text{mol/s or g/s}) \tag{5.3}$$

D_A is the diffusion coefficient of solute A in the fluid. As explained in Appendix G, this coefficient is not a specific property of solute A, and is likely to be impacted by the fluid composition.

Equation (5.3) can be rearranged as follows:

$$\begin{aligned} t_A^e \frac{d\overline{C}_A}{dt} &= C_A - C_A^* \\ t_A^e &= \frac{V_p}{S_p}\frac{\delta}{D_A} \\ t_A^e &= \frac{d_p \delta}{6 D_A} \quad \text{for spherical particles} \end{aligned} \tag{5.4}$$

The external diffusion process is thus modeled as a first-order dynamic system given by Eq. (5.4) with a time constant t_A^e, the characteristic time for external diffusion. In order to estimate t_A^e, knowledge of the particle size, the diffusion coefficient and the thickness of the motionless boundary layer is required.

Notice that, if external diffusion is "infinitely fast" so that the characteristic time for external diffusion can be considered to be zero, then $C_A = C_A^*$. The meaning of "infinitely fast" will be discussed in Section 5.6.

Diffusion coefficients in solution can be estimated in various ways, described for instance by (Poling *et al.*, 2001). The main difficulty comes from the estimate of the thickness of the motionless layer δ. We stress that an *a priori* estimate is difficult, that different correlations will typically lead to different results and that expecting more than just the correct order of magnitude is a bit illusory.

Usually, the estimate of δ is performed through correlations among the dimensionless numbers Sherwood (*Sh*), Reynolds (*Re*) and Schmidt (*Sc*), as defined in Eq. (5.5):

$$Re = \frac{d_p u}{\nu} = \frac{\rho d_p u}{\eta} \qquad Sc = \frac{\nu}{D_A} \qquad Sh = \frac{d_p}{\delta} = \frac{k_A d_p}{D_A} \tag{5.5}$$

It can be shown theoretically that the Sherwood number of a sphere placed in an infinite motionless liquid is 2. Similarly, it can be shown that the Sherwood number associated with ideally packed spherical particles in a motionless fluid should be 3.89 (Rexwinkel *et al.*, 1996). As a consequence, various correlations are presented as the sum of two contributions, the first one being a constant equaling about 2, the second one representing the enhancement of the mass transfer coefficient due to the fluid motion. An example is given by the correlation of (Wakao and Funazkri, 1978):

$$Sh = 2 + 1.1 Re^{0.6} Sc^{1/3} \quad 3 < Re < 10\,000 \tag{5.6}$$

Preparative chromatography is typically operated at low *Re* values. Those for HPLC using a particle size of 10–20 μm are typically between 0.005 and 0.05, while they can be about 1 for large-scale, low-pressure SMB packed with 500 μm particles.

Discrepancies between correlations at low particle Reynolds numbers have been reported and analyzed (Rexwinkel *et al.*, 1996; Holger, 1978).

The question of whether the *Sh* number has a limiting value of about 2 for very small *Re* numbers has long been debated and some correlations suggest *Sh* numbers with a limiting value of 0 as *Re* decreases to 0 (Nelson and Galloway, 1975; Kunii and Suzuki, 1967; Cale *et al.*, 1987).

A correlation often used in chromatography is the one given by (Wilson and Geankopolis, 1966):

$$Sh = \frac{1.09}{\varepsilon_e} Re^{1/3} Sc^{1/3} \quad 0.0015 < Re < 55 \tag{5.7}$$

Another correlation is given by (Dwivedi and Upadhyay, 1977):

$$Sh = \frac{Re Sc^{1/3}}{\varepsilon_e} \left(\frac{0.765}{Re^{0.32}} + \frac{0.365}{Re^{0.386}} \right) \quad 0.01 < Re < 15\,000 \tag{5.8}$$

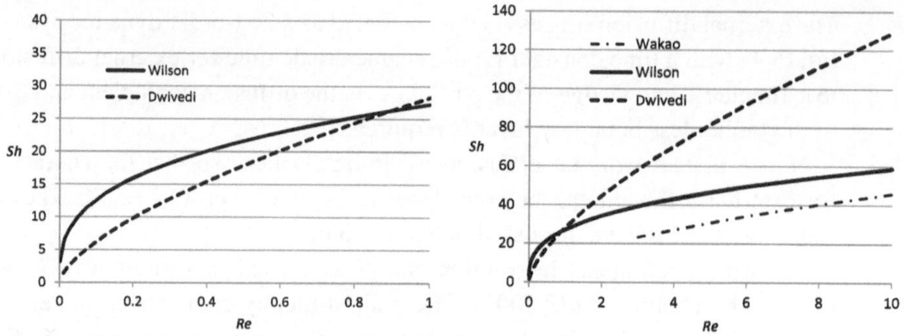

Figure 5.2 Predicted Sherwood numbers according to different correlations, with their validity limits. The Schmidt number *Sc* and the external porosity are set to 100 and 0.4, respectively.

Estimates for the *Sh* number based on the three proposed correlations show significant discrepancies, as shown in Figure 5.2.

Notice that, if one uses correlation (5.7) or (5.8) at *the lowest Re value of its validity range*, and assuming $Sc = 1000$ and $\varepsilon_e = 0.4$, one obtains a minimal *Sh* value of about 3 with the Wilson–Geankopolis correlation and 1.4 with the Dwivedi–Upadhyay correlation. These values are non-zero, and not very far from 2. This shows that correlations like Eq. (5.7) and Eq. (5.8) do not imply that *Sh* goes to zero with *Re*, unless they are used outside their validity range.

As a side comment, should *Sh* equal zero when the *Re* number reaches zero, this would mean that the mass transfer coefficient of a particle in a motionless fluid is zero. This would imply that the particles would be unable to exchange solutes with the fluid and would thus keep their composition indefinitely, regardless of the composition of the fluid they were in contact with. This is clearly physically unrealistic.

(Rexwinkel *et al.*, 1996) and (Holger, 1978) demonstrated that reported values of *Sh* approaching zero are largely due to, according the title of the article, "wrong experiments or wrong experimentations" associated with wall channeling or more generally with improper consideration of axial dispersion.

Even though the correlation of (Wakao and Funazkri, 1978) is probably less pertinent for preparative chromatography because of low values of *Re* typically used in that field, we are not in a position to recommend one correlation over another. We can, however, certainly make the following recommendations:

- Never use a correlation outside the validity domain.
- Read the initial publication and comply with the validity range of *Re* and *Sc* and of basic dimensional parameters like particle size, d_p.
- Do not rely on a flawed predictive character: correlations can provide useful initial orders of magnitude or trends, but experimental determination is mandatory when a minimum level of precision is required.

In typical situations of preparative chromatography, Eqs. (5.6)–(5.8) lead to *Sh* numbers greater than 2 and proportional to $Re^{1/3}$–$Re^{2/3}$. Consequently, as an approximation, we will consider that the characteristic time for external diffusion is proportional to the particle diameter raised to the power 3/2.

EXERCISE 5.1 *A column with $L_{col} = 100$ cm, $D_{col} = 2.5$ cm is packed with microcrystalline cellulose triacetate (25 μm) and eluted at a flow rate of 40 ml/min of ethanol. Estimate the characteristic time for external diffusion.*

Finally, we showed in Chapter 2 that, in the case of linear chromatography, the transfer function $\hat{\mathscr{L}}(s) = \delta\hat{\bar{n}}_A/\delta\hat{n}_A$ plays an important role in column modeling. This transfer function relates the quantities of solute contained in the mobile (δn_A) and stationary ($\delta\bar{n}_A$) phases in a small volume of column where the composition is assumed to be uniform.

Assuming linear equilibrium at the interface between $\overline{C}_A$ and C_A^*, one has $\overline{C}_A = \overline{K}_A C_A^*$, so that the first-order dynamic system given by Eq. (5.4) describing external diffusion gives, in the Laplace domain:

$$\hat{\overline{C}}_A = \frac{\overline{K}_A}{1 + \overline{K}_A t_A^e s}\hat{C}_A$$

This expression allows a determination of the transfer function $\hat{\mathscr{L}}(s)$:

$$\delta\hat{\bar{n}}_A = \hat{\mathscr{L}}(s)\,\delta\hat{n}_A$$
$$\hat{\mathscr{L}}(s) = \frac{1-\varepsilon_e}{\varepsilon_e}\frac{\overline{K}_A}{1 + \overline{K}_A t_A^e s} \tag{5.9}$$

$\hat{\mathscr{L}}(s)$ is in general independent of the position in the column (if operation is isothermal and at uniform fluid velocity) and will be used at various points in this chapter.

5.2 Internal transfer: the lumped Fick model

Modeling transport inside the particle is in principle more difficult than modeling external transport because complex factors like the internal geometry of the particle as well as interactions between solutes and the solid skeleton certainly play a role. We will start our description with the simple lumped model and later build particle descriptions of increasing complexity with the porous model.

5.2.1 Using Fick's law

In previous chapters, solute uptake in chromatographic media was assumed to be instantaneous, i.e. at equilibrium with the bulk solution (Chapter 3), represented by a first-order system (Chapter 2) or governed by external mass transfer (Section 5.1). The particle composition was, furthermore, described by a concentration variable *averaged over the particle volume* and represented by a capital letter $\overline{C}_A$.

Fick's law is expected to account for mass transport within the particle, so we are now looking for a more detailed description of internal concentration profiles. The simplest approach uses the lumped model, assuming the particle to be a pseudo-homogeneous medium and the diffusion process to be characterized by a constant lumped diffusion coefficient $\overline{D}_A$. This model is referred to as the $\overline{\text{Fick}}$ model. Recall that, following the terminology introduced in Chapter 1, a bar over a variable or a model name refers to the fact that lumped solid concentrations are used. (We will see later that a $\overline{\overline{\text{Fick}}}$ approach can be considered if the porous model is used for describing the particles.)

Coming back to the lumped model ($\overline{\text{Fick}}$), the mass balance in spherical geometry and Fick's law are written:

$$\frac{\partial \bar{c}_A(r,t)}{\partial t} = \frac{1}{r^2}\frac{\partial}{\partial r}\left(r^2 J_A\right) \quad \text{with} \quad J_A = \overline{D}_A \frac{\partial \bar{c}_A(r,t)}{\partial r} \tag{5.10}$$

which, assuming a constant diffusion coefficient $\overline{D}_A$ leads to:

$$\frac{\partial \bar{c}_A(r,t)}{\partial t} = \frac{\overline{D}_A}{r^2}\frac{\partial}{\partial r}\left(r^2 \frac{\partial \bar{c}_A(r,t)}{\partial r}\right) = \overline{D}_A \nabla^2 \bar{c}_A(r,t) \tag{5.11}$$

Note the lower-case concentration $\bar{c}_A(r,t)$: this notation, or $\bar{c}_A$ for short, will be used throughout the book for a physical quantity which depends on both time, t, and spatial position, r, inside the particle. The capital letter $\overline{C}_A(t)$ will still designate the concentration averaged over the particle volume:

$$\overline{C}_A(t) = \frac{\int_0^{r_0} r^2 \bar{c}_A(r,t)\,dr}{\int_0^{r_0} r^2 dr} = \frac{3}{r_0^3}\int_0^{r_0} r^2 \bar{c}_A(r,t)\,dr \tag{5.12}$$

Recall that the presentation given in Appendix G shows that the diffusion coefficient $\overline{D}_A$ is the diffusion coefficient of solute A in a given environment. It is thus not a specific property of solute A and it can vary with composition, even in an ideal system. A schematic description of the concentration profiles inside and outside the particle is given in Figure 5.3.

The mass-balance equation (5.11) can be written in a more normalized form:

$$t_A^F \frac{\partial \bar{c}_A(\xi_p,t)}{\partial t} = \frac{1}{\xi_p^2}\frac{\partial}{\partial \xi_p}\left(\xi_p^2 \frac{\partial \bar{c}_A(\xi_p,t)}{\partial \xi_p}\right) \tag{5.13}$$

where $\xi_p = r/r_p$ is the dimensionless particle radius and:

$$t_A^F = r_p^2/\overline{D}_A \tag{5.14}$$

is analogous to a time, and will be called the *Fickian characteristic time for diffusion.*

The mass-balance equation (5.13) shows that the intragranular saturation of the particle is primarily determined by a characteristic time varying as the square of the particle radius while the characteristic time for external diffusion is proportional to the particle radius to the power 3/2.

Since the lumped model ignores the influence of effects like the internal geometry of the particle and fluid–solid interactions taking place inside the particle, the lumped diffusion coefficient $\overline{D}_A$ is expected to be influenced by many factors and thus to be different from the molecular diffusion coefficient of the solute. For the time being we

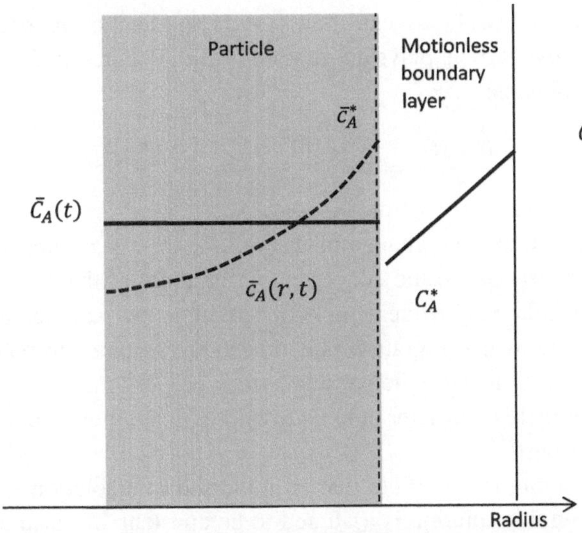

Figure 5.3 Representation of the concentration profiles inside and outside the particle when the lumped model is used for describing intragranular transport.

will consider $\overline{D}_A$ to be a constant phenomenological parameter, and the influence of various factors will be discussed later.

In order to determine the internal concentration profile $\bar{c}_A(r,t)$, the mass-balance equation (5.11) or (5.13) must be associated with proper boundary conditions. The first condition expresses that the derivative at the center of the particle ($r = 0$) must equal zero for symmetry reasons:

$$\left.\frac{\partial \bar{c}_A(r,t)}{\partial r}\right|_{r=0} = \left.\frac{\partial \bar{c}_A(\xi_p,t)}{\partial \xi_p}\right|_{\xi_p=0} = 0 \quad \text{symmetry condition} \tag{5.15}$$

The second condition sets the concentration at the periphery of the particle to a given value:

$$\bar{c}_A(r = r_p, t) = \bar{c}_A(\xi_p = 1, t) \equiv \bar{c}_A^* \tag{5.16}$$

The superscript * refers to the interface between the solid lumped particle and the extragranular fluid.

The lumped interfacial concentration $\bar{c}_A^*$ is assumed to be in equilibrium with the interfacial concentration C_A^* of the solute on the fluid side. The interfacial concentration C_A^* equals the fluid concentration C_A if external mass transfer limitations can be neglected.

It is important to understand that, according to the lumped model, the intragranular saturation process is linear as long as $\overline{D}_A$ is constant: the saturation curves are thus proportional to the interfacial concentration $\bar{c}_A^*$. The possible non-linear adsorption isotherm coupling takes place only at the fluid–particle interface.

Without fully solving Eq. (5.11), we can show that the gradient of lumped concentration at the periphery of the particle plays an important role. Integrating Eq. (5.11) over the entire particle, one obtains:

$$\frac{d\overline{C}_A(t)}{dt} = \frac{3\overline{D}_A}{r_p}\left.\frac{\partial \bar{c}_A(r,t)}{\partial r}\right|_{r_p} \tag{5.17}$$

Equation (5.17) shows that the rate change of the volume-averaged lumped solid concentration $\overline{C}_A(t)$ is proportional to the concentration gradient at the periphery of the particle. This is in fact relatively evident: in order to enter the particle, solutes have to cross its periphery, and as the migration phenomena are assumed to be diffusional, the flux of solute entering or leaving the particle is thus proportional to the gradient at the periphery. Equation (5.17) will later help us to approximate more complex models describing internal transport.

We now consider a solute-free particle that is immersed in a solution. The concentration C_A of solute A in this solution is assumed to be constant. We also assume that external mass transfer limitations can be neglected so that the fluid concentration C_A equals C_A^* and induces a constant peripheral concentration $\bar{c}_A^* = \bar{f}_A(C_A)$.

It can be shown that, under these conditions, the normalized saturation response of the particle is (assuming a constant lumped diffusion coefficient $\overline{D}_A$) given by:

$$U^{Fick}(t) = \frac{\overline{C}_A(t)}{\bar{c}_A^*} = 1 - \sum_{n=1}^{\infty}\frac{6}{\pi n^2}e^{-\pi n^2 t/t_A^F} \tag{5.18}$$

Figure 5.4 illustrates the normalized saturation curve given by Eq. (5.18). It is observed that the initial slope of the curve is very steep, and even theoretically infinite, for reasons explained below. Initially the lumped solid concentration $\bar{c}_A(r,0)$ equals 0 from the center of the particle to its periphery, whereas the external solution requires a non-zero

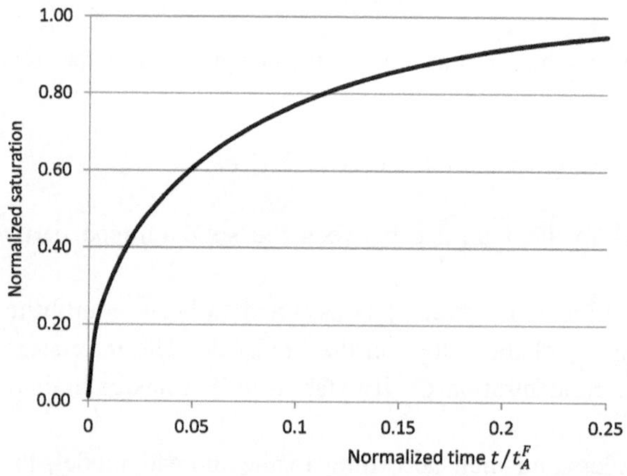

Figure 5.4 Normalized saturation $\overline{C}_A/\bar{c}_A^*$ curve predicted using Fick's law with the lumped model, for a particle immersed in a solution, requiring a constant interfacial concentration $\bar{c}_A^*$. Time is normalized to t_A^F, the Fickian characteristic time for diffusion.

interfacial concentration $\bar{c}_A^*$. This means a discontinuity at the periphery which also means an infinite initial gradient of concentration $\left.\frac{\partial \bar{c}_A(r,t)}{\partial r}\right|_{r_p}$, and eventually an infinite $\left.\frac{d\overline{C}_A(t)}{dt}\right|_{t=0}$ according to Eq. (5.17).

5.2.2 The $\overline{\text{LDF}}$ approximation

While numerically solving Eq. (5.11) with appropriate boundary conditions is perfectly achievable with modern software and computers, its integration in complex column models is time-consuming. It would consequently be interesting to find a simpler formulation, which, while keeping the key characteristics of the Fick model, would allow easier calculation. Even though Chapter 2 is restricted to the linear case, we have seen that the kinetic process can be described by the relationship between the amount of solute in the lumped solid and mobile phases via the transfer function $\hat{\mathscr{L}}(s)$. This suggests that the chromatogram depends primarily on the volume-averaged concentration in the lumped solid phase, and not on the detailed internal concentration profiles. A good part of our modeling efforts will thus use the volume-averaged concentrations only, and, as far as possible, will avoid using more complex internal concentration profiles in the particles.

In order to achieve this, an approximation proposed 65 years ago by (Glueckauf, 1949), called the linear driving force (LDF) approximation, is of key interest. The approximation will be named LDF for short and $\overline{\text{LDF}}$ when used for describing a lumped particle.

The idea consists in replacing Fick's law in spherical geometry by a first-order dynamic system relating the evolution of the volume-averaged lumped solid concentration $\overline{C}_A$ to the difference between this average lumped solid concentration and the lumped concentration at the periphery $\bar{c}_A^*$. With our notation, this approximation is written:

$$t_A^i \frac{d\overline{C}_A}{dt} = \bar{c}_A^* - \overline{C}_A \tag{5.19}$$

where $\bar{c}_A^*$ represents the lumped solid interfacial concentration in equilibrium with C_A^*, the interfacial concentration of the solute in the extragranular fluid.

Parameter t_A^i is the *characteristic time for internal diffusion*, given by:

$$t_A^i = \frac{d_p^2}{60\overline{D}_A} = \frac{r_p^2}{15\overline{D}_A} = \frac{t_A^F}{15} \tag{5.20}$$

With the $\overline{\text{LDF}}$ approximation, internal diffusion is thus modeled like external diffusion as a first-order dynamic system involving a characteristic time which is connected to the particle size and to the lumped internal diffusion coefficient.

In addition, as with the Fick model involving radial concentration profiles, the diffusion process is linear as long as the characteristic time for internal diffusion t_A^i, and consequently the lumped diffusion coefficient $\overline{D}_A$, are constant. The possible non-linear adsorption isotherm coupling takes place at the interface only.

Our knowledge of linear chromatography will help us in justifying the $\overline{\text{LDF}}$ approach and confirming the estimate of t_A^i. In order to do this, we start by considering the transfer function $\hat{\overline{H}}(s)$ relating the volume-averaged lumped solid concentration to the lumped solid concentration at the fluid–solid interface:

$$\hat{\overline{H}}(s) = \frac{\hat{\overline{C}}_A(s)}{\hat{\overline{c}}_A^*(s)} \tag{5.21}$$

$\hat{\overline{C}}_A(s)$ is the Laplace transform of the volume-averaged lumped concentration; $\hat{\overline{c}}_A^*(s)$ is the Laplace transform of the lumped solid concentration at the particle's periphery.

The transfer function $\hat{\overline{H}}(s)$ is determined from the kinetic model used to describe the solute diffusion in the particle. It can be shown (Sardin *et al.*, 1991) that, for spherical particles, the mass-balance equation (5.11) leads to:

$$\hat{\overline{H}}^{Fick}(s) = \frac{3\,\text{cotanh}\left(\sqrt{st_A^F}\right)}{\sqrt{st_A^F}} - \frac{3}{st_A^F} \tag{5.22}$$

On the other hand, one immediately obtains from the mass-balance equation (5.19):

$$\hat{\overline{H}}^{LDF}(s) = \frac{1}{1+st_A^i} \tag{5.23}$$

With either $\hat{\overline{H}}^{Fick}$ or $\hat{\overline{H}}^{LDF}$, one can determine the transfer function $\hat{\mathscr{L}}(s) = \delta\hat{\overline{n}}_A/\delta\hat{n}_A$ introduced in Chapter 2, relating the quantities of solute in the mobile (δn_A) and in the lumped solid ($\delta\overline{n}_A$) phases in a small volume of column where the composition is assumed to be uniform in each phase.

In the absence of external mass transfer limitations, taking into account the equilibrium at the interface one has $\hat{\overline{c}}_A^* = \overline{K}_A\hat{C}_A$, so:

$$\hat{\mathscr{L}}^{Fick}(s) = \frac{1-\varepsilon_e}{\varepsilon_e}\overline{K}_A\hat{\overline{H}}^{Fick}(s) \text{ and } \hat{\mathscr{L}}^{LDF}(s) = \frac{1-\varepsilon_e}{\varepsilon_e}\overline{K}_A\hat{\overline{H}}^{LDF}(s) \tag{5.24}$$

We have shown in Chapter 2 that two models associated with transfer functions $\hat{\mathscr{L}}(s)$ having similar zero- and first-order moments will lead to chromatograms having the same mean retention times and variances.

The zero- and first-order moments of the transfer functions $\hat{\mathscr{L}}^{Fick}(s)$ and $\hat{\mathscr{L}}^{LDF}(s)$ are calculated by Van der Laan's theorem, presented in Chapter 2. The calculation shows that the zero-order moments are $\hat{\mathscr{L}}^{Fick}(0) = \hat{\mathscr{L}}^{LDF}(0) = \frac{1-\varepsilon_e}{\varepsilon_e}\overline{K}_A$ for both models and that identity of the first moments is obtained if the characteristic times are related by $t_A^i = t_A^F/15$, which is Eq. (5.20).

We have thus shown that, under the linearity assumption, the $\overline{\text{Fick}}$ and $\overline{\text{LDF}}$ models will lead to chromatograms having similar mean retention times and variances, provided that the equivalence given by Eq. (5.20) is used.

Identity of mean retention times and variances doesn't imply an identity between the peaks. Some common features and differences between the $\overline{\text{Fick}}$ and $\overline{\text{LDF}}$ approaches can be visualized on normalized response curves $U(t)$ of particles suddenly immersed in a solution of concentration C_A under conditions inducing a constant peripheral concentration $\bar{c}_A^*$.

The normalized response associated with the $\overline{\text{LDF}}$ approximation is given by:

$$U^{LDF}(t) = \frac{\overline{C}_A(t)}{\bar{c}_A^*} = 1 - e^{-t/t_A^i} \tag{5.25}$$

which is to be compared with expression (5.18) when Fick's law is used.

The shapes of the normalized saturation curves $U(t)$ are given in Figure 5.5. The two simulated uptake curves exhibit similar features and the overall saturation process is well represented by the simple $\overline{\text{LDF}}$ approximation. However, one can see that the fractional uptake curve obtained with the $\overline{\text{LDF}}$ approximation is smoother than the one obtained with Fick's law, which starts with an infinite slope.

The main difference between Fick's law and the $\overline{\text{LDF}}$ approximation on uptake curves occurs at small normalized times. By comparing the initial derivative $d\overline{C}_A(t)/dt\big|_{t=0}$ given by Eq. (5.17) and Eq. (5.19), we can see that the LDF approximation has the effect of replacing a gradient at the periphery of the particle, which can in principle reach infinity, by a finite concentration difference.

An example of chromatogram simulation is given in Section 5.11. It shows little difference between the $\overline{\text{Fick}}$ and $\overline{\text{LDF}}$ models, even with significant internal diffusion limitations. This is a direct consequence of the properties of the function $\hat{\mathscr{L}}(s)$.

Being an approximation of Fick's law, the $\overline{\text{LDF}}$ approximation may appear to be less rigorous. While the argument makes sense, the reality is more complex, for several reasons. As shown by (Sircar and Hufton, 2000), the $\overline{\text{LDF}}$ approximation works very well for column modeling because detailed kinetic characteristics at the particle level are lost

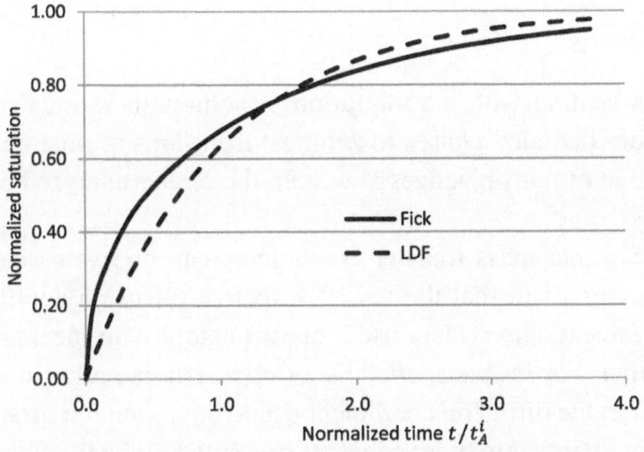

Figure 5.5 Normalized saturation $\overline{C}_A/\bar{c}_A^*$ curve predicted Fick's law and the $\overline{\text{LDF}}$ approximation, for a particle immersed in a solution, requiring a constant interfacial concentration $\bar{c}_A^*$. Time is normalized to the characteristic time for diffusion t_A^i.

at the column level due to various integrations and averaging. Additionally, taking into account particle heterogeneity, thanks to a distribution of apparent diffusion coefficients due to variable pore size or connectivity, drastically minimizes the differences between Fick's law and $\overline{\text{LDF}}$ calculated curves.

Instead of describing internal mass transfer by cumbersome models of ever-increasing complexity (various particle shapes, pore size distribution, multi-level porosity, adsorbed-phase diffusivity etc.) and finally using the dynamics of the volume-averaged concentration, it seems wise to describe the complexity with an "averaged" dynamics, i.e. the simplest description with a suitable internal diffusion time. This is the aim of the $\overline{\text{LDF}}$ model.

An exception that could lead to visible differences between Fick's law and $\overline{\text{LDF}}$ may be found in RPSA (rapid pressure swing adsorption), for which rapid pressure swings could push the system to work at "low times" of the saturation curve (Todd and Webley, 2002). These situations are outside the scope of this book.

Finally, the $\overline{\text{LDF}}$ approximation can be extended to multiple-solute systems without major difficulties by writing a mass-balance equation (5.19) for each solute:

$$\begin{aligned} t_k^i \frac{d\overline{C}_k}{dt} &= \left[\bar{c}_k^* - \overline{C}_k\right] \\ t_k^i &= \frac{r_p^2}{15\overline{D}_k} \end{aligned} \tag{5.26}$$

We should not forget that this formulation implies the existence of a component (solvent) in large excess. For a better understanding of multi-component diffusion, the reader is referred to Appendix G. Additionally, recall that the $\overline{\text{LDF}}$ approximation implicitly assumes constant lumped diffusion coefficients.

This $\overline{\text{LDF}}$ approximation is extremely convenient, and we strongly recommend it when internal limitations are to be taken into account. It normally gives a good compromise between simplicity and precision.

5.2.3 Typical experimental results

There are many articles dealing with the migration of solutes in various solid media, from silica to resins, from bounded phases to zeolites, from ions to organic molecules in supercritical fluids. One of our objectives is to help the reader analyze these diverse presentations.

By "diverse", we mean that mass transfer coefficients can be expressed in m/s or s^{-1} according to the authors, and that the terms "effective diffusion coefficient" and "apparent diffusion coefficient" are widely used, unfortunately with meanings varying from one article to another. For instance, *effective* or *apparent* can refer to the lumped diffusion coefficient $\overline{D}_A$, to the diffusion coefficient occurring in the intragranular pores corrected for the porosity effect, or to other concepts presented in the following sections. There are certainly some good reasons for this variety of terms, but interpreting the literature results requires great attention; better to have clear and well-defined concepts in mind!

Table 5.1 Typical orders of magnitude for lumped diffusion coefficients.

Chromatographic medium	$\overline{D}_A$ m^2/s	Solutes
Silica or reversed-phase	10^{-10}–10^{-11}	Small organic molecules (MW below 1000)
	10^{-11}–10^{-13}	Proteins according to particle type and protein size
Modified silica or gels for optical isomer separations	10^{-11}	Various organic molecules
Ion-exchange resins	10^{-9}	Hydrogen ion
	10^{-10}	Monovalent ions
	10^{-11}	Divalent ions

As already mentioned, the lumped diffusion coefficients are expected to be influenced by many factors that are not explicitly taken into account in the lumped model. Their numerical values can vary considerably from one situation to another, and orders of magnitude are given in Table 5.1. Table 5.1 shows that $\overline{D}_A$ may span decades according to the situation. Conversely, D_A, the molecular diffusion coefficient, is specific to the solute–solvent system, irrespective of the adsorbent. In order to normalize things, it is convenient to define the ratio $\overline{g}_A$ between the molecular diffusion coefficient of the solute and the lumped diffusion coefficient:

$$\overline{g}_A = \frac{D_A}{\overline{D}_A} \tag{5.27}$$

A high value of $\overline{g}_A$ means that diffusion inside the particle is hindered, for whatever reason.

It is apparent that, for the conditions typically selected for preparative chromatography, typical orders of magnitudes of the ratio $\overline{g}_A$ are between 10 and 100. However, ratios as high as 700 have been reported for proteins, as discussed later in this section.

In order to demonstrate the influence of operating parameters on lumped diffusion coefficients, we will now discuss the influence of Henry's coefficient and of solute concentration.

5.2.3.1 Influence of Henry's coefficient

(Farnan *et al.*, 2002) investigated the influence of Henry's coefficients for certain proteins (bovine serum albumin: BSA; and α-lactalbumin: α-L) on lumped diffusion coefficients in macroporous and gel particles. Retention time measurements of excluded and non-retained tracer and proteins allowed determination of porosities, as well as Henry's and lumped Henry's coefficients, $\overline{\overline{K}}_A$ and $\overline{K}_A$. The authors investigated the influence of Henry's coefficients in a domain varying from about 0 to 10, corresponding to typical retentions in preparative chromatography. Experimental conditions associated with the work of (Farnan *et al.*, 2002) are summarized in Table 5.2, which also includes some terminology bridges.

Table 5.2 Summary table showing the influence of Henry's coefficient on diffusion coefficients. Results from Farnan *et al.*, 2002.

Chromatographic media	• Macroporous anion exchanger having a pore size of 1000 Å • Gel-filled particle anion exchanger
Solutes	BSA and α-L
Porosity measurement	By injection of NaCl and polyvinyl pyrrolidone (MW 1200 kD)
Adsorption isotherms	Linearity assumed
Method for varying $\overline{K}_{\mathbf{A}}$	Addition of NaCl in the eluent (50 nm Tris-HCl, pH 8.5).
Method for determining diffusion coefficient (cf. experimental methods in Section 5.10)	Measurement of peak asymmetry vs. fluid velocity.
Terminology bridge	The equilibrium constant K_{eq} corresponds to $\overline{\overline{K}}_A$ in our terminology. Apparent diffusion coefficient corresponds to $\overline{D}_A\overline{K}_A$ in our terminology. Effective diffusion coefficient corresponds to $\varepsilon_i\grave{D}_A$ in our terminology. Surface diffusion coefficient D_s corresponds to $(1-\varepsilon_i)\overline{\overline{D}}_A$ in our terminology. The factor $\left(\frac{\varepsilon_i\lambda}{\theta}\right)_{app}$ relating the apparent diffusion coefficient to the molecular diffusion coefficient D_A corresponds to $\left(\frac{\varepsilon_i\lambda}{\theta}\right)_{app}=\frac{\overline{K}_A}{\bar{g}_A}$ in our terminology.

(Farnan *et al.*, 2002) determined their so-called *apparent diffusion coefficient*, corresponding to the product $\overline{D}_A\overline{K}_A$ in our terminology. Understanding this choice requires taking a more precise look at intraparticular mass balances, which will be done later.

On macroporous particles, the authors obtained apparent diffusion coefficients, which are independent of Henry's coefficient, for both BSA and α-L. In other words, they obtained lumped diffusion coefficients $\overline{D}_A$, proportional to the reciprocal of the lumped Henry's coefficients.

On gel particles, however, they obtained a distinct linear increase of apparent diffusion coefficients versus Henry's coefficient $\overline{\overline{K}}_A$ for both BSA and α-L. Translated into our terminology, this means that lumped diffusion coefficients $\overline{D}_A$ are much less dependent on Henry's coefficients for these gel particles than they are for macroporous particles.

The results expressed in terms of variation of the ratio $\bar{g}_A$ versus Henry's coefficient $\overline{\overline{K}}_A$ for both proteins on both resins are presented in Figure 5.6. The difference in behavior between the gel and the macroporous resins appears very clearly: while the ratios $\bar{g}_A$ are about 3–5 at zero Henry's coefficient for both resins, they can reach about 10 on gel resins and about 80 on macroporous resins, while Henry's coefficient reaches

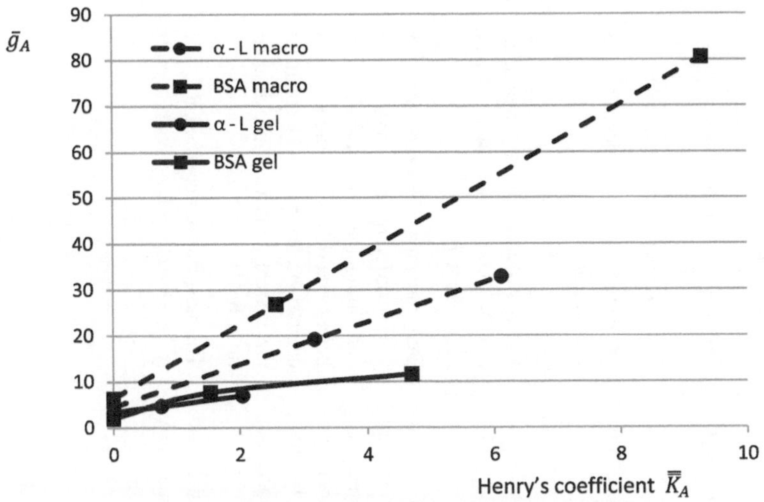

Figure 5.6 Influence of Henry's coefficient on coefficient $\bar{g}_A$. Based on results from (Farnan *et al.*, 2002).

about 5–10. Additionally it seems that the curve $\bar{g}_A(\overline{\overline{K}}_A)$ is linear for macroporous resins whereas it reaches a plateau for the porous gel.

Generally, there is little doubt that the ratio $\bar{g}_A$ increases (thus lumped diffusion coefficients normally decrease) with increasing Henry's coefficients, to an extent that depends on the exact particle structure and solutes.

Similar conclusions regarding the connection between lumped diffusion coefficients and retention have been drawn by (Nicoud and Schweich, 1989) for ion exchange, and will be discussed in Section 5.5.

Using the porous model, we will show in Section 5.3 that this effect can be explained both by adsorption equilibrium taking place inside the particle and by the possibility of adsorbed particles diffusing along the pore walls.

5.2.3.2 Influence of solute concentration

Many authors have reported the influence of solute concentration on lumped diffusion coefficients, and some key results are summarized below. Key experimental information, as well as a short terminology bridge, are given in Table 5.3.

(Yoshida *et al.*, 1994) investigated normalized saturation curves for BSA on different stationary phases with batch and shallow-bed methods, both methods being described in Section 5.10. Operating conditions were such that the liquid concentration during the experiment can be considered constant for the shallow-bed technique and constant within 10–15% for the batch method.

Adsorption isotherms were determined prior to the kinetic studies and were represented by a Langmuir equation with the following parameters:

- Adsorbent 1: $\overline{\overline{N}} = 1645$ kg/m^3, $\tilde{K}_A = 9.23$ m^3/kg, $\varepsilon_i = 0.896$
- Adsorbent 2: $\overline{\overline{N}} = 1727$ kg/m^3, $\tilde{K}_A = 11.8$ m^3/kg, $\varepsilon_i = 0.891$.

Table 5.3 Summary of results on the influence of solute concentration on diffusion coefficients.

	Yoshida *et al.*, 1994	**Rearden *et al.*, 1998**	**Sajonz *et al.*, 1998**
Chromatographic media	Two different strongly basic chitosan ion exchangers	Microcrystalline cellulose triacetate	Imprinted stationary phases
Solutes	Bovine serum albumine	S-Tröger's base	D and L phenylalanine anilide
Porosity measurement (total porosity)	Internal porosity reported to be 0.9; external porosity not reported (and not required by the methodology)	Measured via injection of 1,3,5-tri-butyl-benzene	Measured via injection of acetonitrile
Adsorption isotherms	Langmuir	Langmuir	Either bi-Langmuir or Freundlich; the D-isomer is less retained than the L-isomer
Method for determining diffusion coefficients (cf. experimental methods in Section 5.10)	Quick saturation and desaturation of a shallow bed. Batch method limiting fluid concentration variation to 10–15%.	Determination of breakthrough curves for different concentration steps at different fluid velocities, and determination of plate number. Concentration steps are relatively small, so the system behaves approximately linearly.	Determination of breakthrough curves for different concentration steps and fitting the mass transfer coefficient via PD–LDF model. Concentrations are relatively small, so the system behaves approximately linearly.
Terminology bridge	The effective diffusion coefficient corresponds to an averaged $\overline{D}_A$ over the explored range of concentrations. The adsorbed concentration q equals $(1-\varepsilon_i)\,\overline{\overline{C}}_A$. The saturation capacity q_∞ equals $(1-\varepsilon_i)\,\overline{\overline{N}}$. The adsorption constant K equals our constant $\tilde{K}_A$.	The mass transfer coefficient k_f corresponds to $k_f = 1/t_A^i$, consequently $k_f \sim \overline{D}_A$.	The mass transfer coefficient k_f corresponds to $k_f = 1/t_A^i$, consequently $k_f \sim \overline{D}_A$.

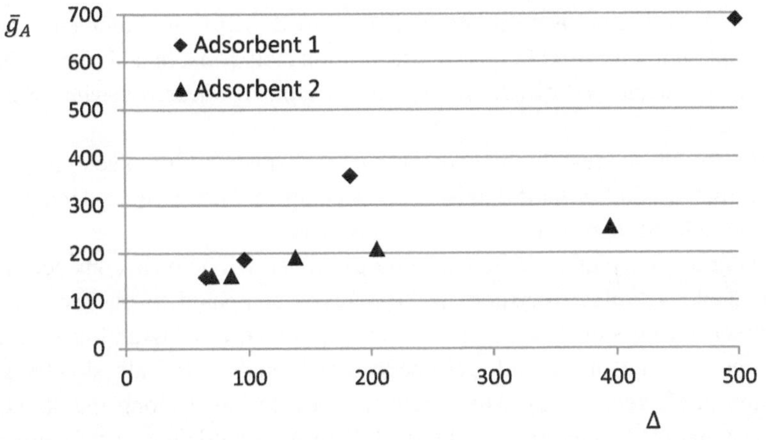

Figure 5.7 Presentation of results obtained by (Yoshida *et al.*, 1994), showing the influence of solution concentration on the parameter $\bar{g}_A$. See text for parameter Δ definition and experimental description.

Kinetic studies were performed with liquid concentrations in the range 0.3 to 3 kg/m^3, thus exploring the non-linear part of the adsorption isotherms.

(Yoshida *et al.*, 1994) determined the so-called *effective diffusion coefficient* (lumped diffusion coefficient in our terminology) by analyzing the initial part of the normalized saturation curves. From their reported value of molecular diffusion for BSA in solution (6.84 10^{-11} m^2/s), one can readily determine the ratio $\bar{g}_A$. Their results are reported in Figure 5.7 versus the parameter $\Delta = \frac{1-\varepsilon_i}{\varepsilon_i}\frac{\bar{\bar{C}}_A}{C_A}$, which is proportional to the chord of the adsorption isotherm.[1] Because of the Langmuir adsorption isotherms, the parameter Δ decreases when concentration increases.

As shown in Figure 5.7 the ratio $\bar{g}_A$ varies dramatically with the parameter Δ. One can also notice that for high values of Δ, thus working at low concentrations, the ratio $\bar{g}_A$ can be as high as 700, meaning that the lumped coefficient of the protein is 700 times smaller than it is in solution!

(Rearden *et al.*, 1998) investigated the mass transfer of S-Tröger's base on microcrystalline cellulose triacetate by frontal analysis. Information on their methodology is given in Table 5.3. The authors obtained a distinct increase of the lumped diffusion coefficient $\overline{D}_A$ (proportional to the mass transfer coefficient in their terminology) with concentration.

(Sajonz *et al.*, 1998) studied mass transfer of D and L phenylalanine anilide on imprinted stationary phases with a methodology similar to that of (Rearden *et al.*, 1998). They reported that the lumped diffusion coefficient $\overline{D}_A$ (thus mass transfer coefficient in their terminology) increases in concentration, the dependence on concentration being stronger for the most retained enantiomer and at low temperature.

[1] (Yoshida *et al.*, 1994) reported their results in the form $\overline{D}_A\left(1+\frac{1}{\Delta}\right)$ vs. $1/\Delta$ and obtained straight lines. A justification for this format will be presented in Section 5.3.

In conclusion, there is a consensus that lumped diffusion coefficients increase with concentration in the case of favorable adsorption isotherms. In addition we have already mentioned that lumped diffusion coefficients typically decrease when Henry's coefficients increase.

These results suggest that the adsorption mechanism taking place inside the particle plays a role. Taking this mechanism into account requires a more detailed description of the particle, to be provided by the porous model.

In order to get a picture of how the porous model will help us, consider the following simple analogy. Solute transport in the pores is compared to the transport of people in a street having sidewalks. In the street, cars move fast (assuming no traffic jam in the pores ...) and on the sidewalk pedestrians (adsorbed) walk slowly. The lumped diffusion coefficient $\overline{D}_A$ measures how fast people travel along the street, by car or by foot without distinction. The lumped diffusion coefficient is thus affected by the proportion of pedestrians and people traveling in cars. It is also affected by the velocity of the cars (we assume they move more or less freely) and by the velocity of the pedestrians. If the pedestrians stop in bars (adsorption sites) beside the sidewalks, the overall (lumped) velocity is decreased.

With this analogy in mind, we can now move forward.

5.3 Internal transfer: the $\overline{\overline{\text{Fick}}}$ model

The porous model will help to clarify the situation and the experimental results, but, to look ahead, we want to stress that the influence of porosity, or more generally of particle structure, Henry's coefficients or concentration, on lumped diffusion coefficients is not fully understood or predictable. This is clearly an area of research, and, in this area as in many others, there is no substitute for experimentation, at least for the foreseeable future.

Prior to starting our modeling presentation, let us summarize the key assumptions that will be used in this section:

1. Particles are well formed, having a solid internal structure characterized by a constant value of intragranular porosity.
2. There is no possibility of folding/unfolding for the solutes.
3. The physico-chemical properties of the particles are homogeneous.
4. There is no convection in the pores.
5. Equilibrium between solutes contained in the intragranular fluid and adsorbed on the skeleton is instantaneous.
6. Solutes are not excluded from the particle; they can enter freely into the intragranular pores and thus have access to the entire intragranular porosity.
7. The Fickian diffusivities introduced below are constant, and in particular independent of concentration.

Assumptions 4, 5 and 6 will be relaxed in Sections 5.3.7, 5.3.8 and 5.3.9.

5.3.1 General description

The porous model presented in Chapter 1 makes a distinction between the concentration $\grave{C}_A$ of solutes in the intragranular pores of the particles and the concentration $\bar{\bar{C}}_A$ of solutes adsorbed on the skeleton. The lumped solid-phase concentration is then related to these concentrations by:

$$\overline{C}_A = \varepsilon_i \grave{C}_A + (1 - \varepsilon_i)\,\bar{\bar{C}}_A \tag{5.28}$$

While the lumped model does not allow us to differentiate between the solutes in the intragranular liquid of the pores and adsorbed on the solid skeleton, our aim, using the porous model, is precisely to investigate the influence of this difference on the mass transfer.

During the saturation/desaturation process, the intragranular fluid and adsorbed concentrations vary with time and position so that local concentrations $\bar{\bar{c}}_A\,(r,t)$ and $\grave{c}_A\,(r,t)$ must be considered. Equation (5.28), valid for volume-averaged concentrations, can also be written locally:

$$\bar{c}_A(r,t) = \varepsilon_i \grave{c}_A(r,t) + (1 - \varepsilon_i)\,\bar{\bar{c}}_A(r,t) \tag{5.29}$$

or for short:

$$\bar{c}_A = \varepsilon_i \grave{c}_A + (1 - \varepsilon_i)\,\bar{\bar{c}}_A$$

As with the lumped model, note the use of lower-case notation for the local concentrations and capital letters to designate concentrations averaged over the particle volume.

Let us mention that, if volume-averaged concentrations $\grave{C}_A$ and $\bar{\bar{C}}_A$ are defined by an equation similar to Eq. (5.12) for the lumped volume-averaged concentration $\overline{C}_A$, then these three concentrations do not satisfy the equilibrium law in the case of non-linear adsorption. In order to get a set of concentrations satisfying the equilibrium law from a known lumped volume-averaged concentration $\overline{C}_A$, one has to simultaneously solve Eq. (5.28) and $\bar{\bar{C}}_A = \bar{\bar{f}}_A(\grave{C}_A)$.

If one considers that the molecules adsorbed on the skeleton are *immobilized* and that only the molecules located in the intragranular pore fluid are *able to diffuse*, the mass-balance equation (5.10) must be rewritten as:

$$\frac{\partial \bar{c}_A\,(r,t)}{\partial t} = \frac{1}{r^2}\frac{\partial}{\partial r}(r^2 J_A) \quad \text{with} \quad J_A = \varepsilon_i \grave{D}_A \frac{\partial \grave{c}_A\,(r,t)}{\partial r} \tag{5.30}$$

which gives, assuming a constant diffusion coefficient:

$$\frac{\partial \bar{c}_A\,(r,t)}{\partial t} = (1 - \varepsilon_i)\,\frac{\partial \bar{\bar{c}}_A\,(r,t)}{\partial t} + \varepsilon_i \frac{\partial \grave{c}_A\,(r,t)}{\partial t} = \varepsilon_i \grave{D}_A \nabla^2 \grave{c}_A\,(r,t) \tag{5.31}$$

where $\grave{D}_A$ represents the diffusion coefficient of solute A in the intragranular pores. $\grave{D}_A$ is expected to be more or less simply related to the molecular diffusion coefficient, whereas many effects are likely to be lumped into the coefficient $\overline{D}_A$.

The mass-balance equation (5.31) expresses that, while a global mass balance involves the evolution of the lumped solid concentration, diffusion can only occur into the intragranular pores, thus in proportion to concentration $\grave{c}_A$.

The mass-balance equation (5.31) must be associated with two boundary conditions:

- The first one requires that the derivative at the center of the particle ($r = 0$) must equal zero for symmetry reasons:

$$\left.\frac{\partial \bar{\bar{c}}_A(r,t)}{\partial r}\right|_{r=0} = \left.\frac{\partial \grave{c}_A(r,t)}{\partial r}\right|_{r=0} \tag{5.32}$$

- The second one sets the concentration at the periphery of the particle to a given concentration:

$$\grave{c}_A\left(r = r_p, t\right) \equiv \grave{c}_A^* \tag{5.33}$$

As with the lumped model, superscript * is used to refer to the peripheral fluid–particle interface.

In the absence of solute exclusion (one of the key assumptions of this section), the concentration $\grave{c}_A^*$ in the intragranular pores at the periphery of the particle is identical to the concentration C_A^* in the extragranular fluid, so one has $\grave{c}_A^* = C_A^*$.

A schematic description of the system is given in Figure 5.8, which shows continuity of the fluid concentrations at the periphery. The fluxes being identical on both sides of the periphery, the slopes of the extragranular and intragranular fluid concentration profiles are identical on both sides of the periphery as well, assuming diffusion coefficients in the extragranular fluid and in the intragranular fluid are identical.

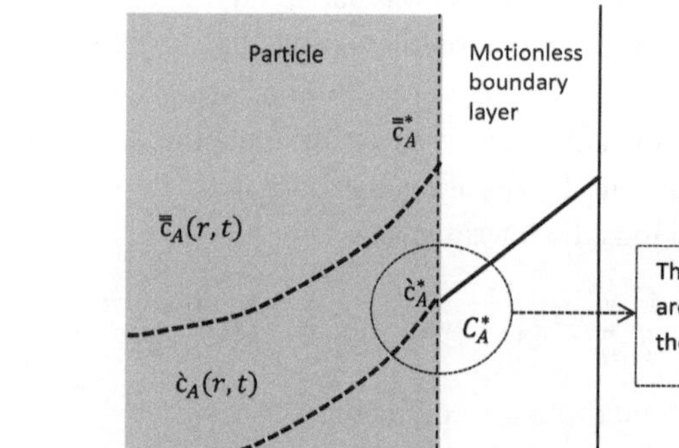

Figure 5.8 Representation of the concentration profiles inside and outside the particle when the porous model is used for describing intragranular transport. Lower-case quantities refer to local concentration.

We will discuss later how to determine c_A^* but, for the time being, we simply consider it as a variable of the problem.

The diffusion coefficient $\grave{D}_A$ can in principle be related to the molecular diffusion coefficient, assuming that a proper description of the pore structure of the solid is available. In other words, the fact that the pores are not straight, have variable size, can be interconnected, or could even in principle allow mass transfer through their walls are factors impacting $\grave{D}_A$.

The fact that the pores are not straight can be characterized by a *tortuosity factor* τ, and $\grave{D}_A$ is usually related to the bulk diffusion coefficient (excluding the situation of Knudsen diffusion) by (Jackson, 1977):

$$\grave{D}_A = D_A/\tau \tag{5.34}$$

Assuming no steric hindrance, the tortuosity factor is expected to be independent of the solute, and is usually between 2 and 5. For a presentation of tortuosity and possible confusion around it, the reader is referred to (Epstein, 1989).

Some attempts have been made to derive the tortuosity from simple knowledge of intragranular porosity. A frequently used relation is (Guiochon *et al.*, 2006):[2]

$$\grave{D}_A = \varepsilon_i \left(\frac{1}{2-\varepsilon_i}\right)^2 D_A \tag{5.35}$$

which is equivalent to $\tau = \dfrac{1}{\varepsilon_i}\left(\dfrac{2-\varepsilon_i}{1}\right)^2$. Intragranular porosities being typically between 0.3 and 0.7, Eq. (5.35) forecasts tortuosities τ between 2.5 and 10.

Equations (5.34) and (5.35) should not give an illusion of precision as they are based on very simple descriptions of the particle. Experience shows that even a parameter as simple as the intragranular porosity can be impacted by the concentration of solute in the particle, as reported and studied for instance by (Piatkowski *et al.*, 2003) and (Coquebert de Neuville, Thomas and Morbidelli, 2013). The reality is that estimating $\grave{D}_A$ from D_A is a complex task and the best course is to admit some ignorance (renamed tortuosity . . .) and an inability to predict precisely. Once again, experimentation cannot be replaced.

$\grave{D}_A$ is the diffusion coefficient that would be observed in the absence of adsorption inside the particle. Only a few studies have investigated internal diffusion of solutes in such a way that diffusion can be decoupled from adsorption. One example is given by (Schröder *et al.*, 2006), who studied the diffusion coefficients $\grave{D}_A$ of proteins in underivatized media, so that adsorption effects were eliminated. According to the protein size, they obtained $\varepsilon_i\grave{D}_A$ values of 1.5–5×10^{-11} m^2/s on agarose, leading to tortuosity values between 1.7 and 3.0. These values are at the low end of what was expected, and the fact that tortuosity varies with protein size shows that steric hindrance certainly plays a role.

[2] In certain references, the diffusion coefficient in the pores is considered to be the product $\breve{D}_A = \varepsilon_i\grave{D}_A$ so that intragranular porosity and diffusion coefficient in solution are lumped in a single parameter $\breve{D}_A$. In this case, Eq. (5.35) is written $\breve{D}_A = \left(\frac{\varepsilon_i}{2-\varepsilon_i}\right)^2 D_A$.

Results presented in Section 5.2.3, describing lumped diffusion coefficients sometimes 100 to 700 times lower than molecular diffusion coefficients, show that, if tortuosity is one factor, others may even be more important.

Mass transfer in particles is certainly affected by geometry, but the mass-balance equation (5.31) forces us to realize that it may be even more sensitive to thermodynamics. Modeling kinetics without taking into account adsorption effects occurring in the particle is rather like modeling columns without taking into account adsorption isotherms.

The mass-balance equation (5.31) has been established with the assumption that adsorbed solutes are immobile. Whether the adsorbed solutes are motionless is debatable, but assuming that the adsorbed solute can move at the surface of the pore walls according to a "surface diffusion" mechanism characterized by a diffusion coefficient $\overline{\overline{D}}_A$ has proven to be fruitful for explaining various experimental results.

As mentioned by (Ruthven, 1984), "surface diffusion" is the generally accepted terminology, but the implication of a two-dimensional surface is unnecessarily restrictive. *Surface diffusion*, or preferably *adsorbed-phase diffusivity*, is our term to designate the diffusion coefficient of the solute when Fick's law describes its transport in the adsorbed phase. This can be:

- surface diffusivity on a solid surface
- molecular diffusivity of the solute in a suitable immiscible liquid coated on the solid support; it can differ from the bulk diffusivity if the coated film is a few molecules thick and/or if the solid support modifies the properties of the immiscible liquid
- the "electrolyte diffusivity" of charged solute in an intense electric field in solution close to the charged structure of an ion-exchanging material
- "microdiffusivity" in confined micropores.

In each of the cases cited above, diffusivity has its own properties which depend as usual on the solute and on its environment.

Under the assumption of the existence of adsorbed-phase diffusivity, the mass-balance equation (5.31) must be rewritten as:

$$\begin{aligned}\frac{\partial \bar{c}_A(r,t)}{\partial t} &= \varepsilon_i \frac{\partial \grave{c}_A(r,t)}{\partial t} + (1-\varepsilon_i)\frac{\partial \bar{\bar{c}}_A(r,t)}{\partial t} \\ &= \varepsilon_i \grave{D}_A \nabla^2 \grave{c}_A(r,t) + (1-\varepsilon_i)\overline{\overline{D}}_A \nabla^2 \bar{\bar{c}}_A(r,t)\end{aligned} \tag{5.36}$$

Equation (5.36) expresses the contributions of both the intragranular and adsorbed solutes to the overall diffusion process. It assumes that the diffusion coefficients $\grave{D}_A$ and $\overline{\overline{D}}_A$ are constant. The boundary conditions associated with Eq. (5.36) are those given in Eqs. (5.32) and (5.33).

Assuming linearity of the adsorption isotherm will allow us to quickly derive some interesting characteristics of the lumped diffusion coefficients and to explain some experimental results.

5.3.2 Linear adsorption

Recalling that Henry's coefficient and the lumped Henry's coefficient are related by $\bar{K}_A = \varepsilon_i + (1-\varepsilon_i)\,\bar{\bar{K}}_A$, when both linear adsorption and local equilibrium are assumed one has, everywhere in the particle:

$$\bar{\bar{c}}_A = \bar{\bar{K}}_A \grave{c}_A \quad \text{and} \quad \bar{c}_A = \bar{K}_A \grave{c}_A \tag{5.37}$$

Using the volume-averaged concentrations, one obtains similarly:

$$\bar{\bar{C}}_A = \bar{\bar{K}}_A \grave{C}_A \quad \text{and} \quad \bar{C}_A = \bar{K}_A \grave{C}_A \tag{5.38}$$

The linear relations (5.37) allow us to write the mass-balance equation (5.36) in a more compact form:

$$\frac{\partial \grave{c}_A(r,t)}{\partial t} = \frac{\varepsilon_i \grave{D}_A + (1-\varepsilon_i)\,\bar{\bar{K}}_A \bar{\bar{D}}_A}{\varepsilon_i + (1-\varepsilon_i)\bar{\bar{K}}_A} \nabla^2 \grave{c}_A\,(r,t) \tag{5.39}$$

or, equivalently because $\bar{c}_A = \bar{K}_A \grave{c}_A$:

$$\frac{\partial \bar{c}_A(r,t)}{\partial t} = \frac{\varepsilon_i \grave{D}_A + (1-\varepsilon_i)\,\bar{\bar{K}}_A \bar{\bar{D}}_A}{\varepsilon_i + (1-\varepsilon_i)\bar{\bar{K}}_A} \nabla^2 \bar{c}_A\,(r,t) \tag{5.40}$$

Equation (5.40) is identical to the mass-balance equation (5.11) associated with the lumped model, provided one uses:

$$\bar{D}_A = \frac{\varepsilon_i \grave{D}_A + (1-\varepsilon_i)\,\bar{\bar{K}}_A \bar{\bar{D}}_A}{\varepsilon_i + (1-\varepsilon_i)\bar{\bar{K}}_A} = \frac{\varepsilon_i \grave{D}_A + \left(\bar{K}_A - \varepsilon_i\right) \bar{\bar{D}}_A}{\bar{K}_A} \tag{5.41}$$

This shows that under *linear* conditions, modeling the transfer using the porous model is strictly equivalent to modeling the transfer using the lumped model provided that one uses a lumped diffusion coefficient defined by Eq. (5.41). In particular, the transfer functions and the $\overline{\text{LDF}}$ approximation in Section 5.2.2 remain valid.

Additionally, this shows that, everything else being constant, the lumped diffusion coefficient should show a dependence on Henry's coefficient $\bar{\bar{K}}_A$ (or the lumped Henry's coefficient $\bar{K}_A$) and that the respective contributions of the pore and adsorbed-phase diffusivity processes are given by:

- Intragranular fluid diffusion: its influence decreases when Henry's coefficient increases (because the fraction of solute contained in the internal liquid decreases). At small Henry's coefficient, pore diffusion is predominant and $\bar{D}_A$ is proportional to the reciprocal of $\bar{K}_A$. In the absence of adsorbed-phase diffusivity, $\bar{D}_A$ is proportional to the fraction of mobile solutes in the particle.
- Adsorbed-phase diffusivity: its relative influence increases with Henry's coefficient and possibly becomes the predominant factor. Henry's coefficient being positive (and lumped Henry's coefficient greater than the intragranular porosity), the contribution of the adsorbed-phase diffusivity to the lumped diffusion coefficient

is always positive. At significant values of Henry's coefficient, the lumped diffusion coefficient $\bar{D}_A$ equals the adsorbed-phase diffusivity coefficient $\bar{\bar{D}}_A$.

Equation (5.41) can be presented in a more normalized form by introducing the ratio of the surface to pore diffusion coefficients:

$$\bar{\bar{\theta}}_A = \bar{\bar{D}}_A / \grave{D}_A \tag{5.42}$$

If one relates the pore diffusion coefficient to the molecular diffusion coefficient via the tortuosity factor τ defined in Eq. (5.34), one obtains a normalized expression for the ratio $\bar{g}_A$:

$$\begin{aligned} \bar{g}_A &= \frac{\tau \bar{K}_A}{\varepsilon_i + \left(\bar{K}_A - \varepsilon_i\right) \bar{\bar{\theta}}_A} = \frac{\tau \left(\varepsilon_i + (1-\varepsilon_i)\bar{\bar{K}}_A\right)}{\varepsilon_i + (1-\varepsilon_i)\,\bar{\bar{K}}_A \bar{\bar{\theta}}_A} \\ \bar{D}_A &= D_A / \bar{g}_A \end{aligned} \tag{5.43}$$

Assuming that the molecular diffusivity is known, Eq. (5.43) allows the lumped diffusion coefficient to be related to defined geometrical and adsorption parameters.

Typical Henry's coefficients are in the range of 1 to 10 in preparative chromatography, and if one assumes an internal porosity of 0.5 and a tortuosity between 2 and 5, Eq. (5.43) gives $\bar{g}_A$ ratios in the range of 4 to 100 in the absence of adsorbed-phase diffusivity. Assuming $\bar{\bar{\theta}}_A = 0.1$ limits the range from 4 to 35. This confirms the orders of magnitude given in Section 5.2.3.

More precisely, Eq. (5.43) allows a representation of the results from (Farnan *et al.*, 2002) presented in Section 5.2.3 and in Figure 5.6:

- First, the increase of the coefficient $\bar{g}_A$ with Henry's coefficients is predicted!
- At low values of Henry's coefficient $\bar{\bar{K}}_A$, the ratio $\bar{g}_A$ tends towards $\bar{g}_A = \tau$, a geometrical factor expected to be relatively independent of the solute and about 2–5. This is confirmed in Figure 5.6.
- When Henry's coefficient $\bar{\bar{K}}_A$ increases, the ratio $\bar{g}_A$ is expected to increase indefinitely in the absence of adsorbed-phase diffusivity ($\bar{\bar{\theta}}_A = 0$) or to reach a plateau given by $\bar{g}_A = \tau / \bar{\bar{\theta}}_A$ in the presence of adsorbed-phase diffusivity.
- The straight lines obtained for the macroporous resin tend to indicate negligible adsorbed-phase diffusivity in this chromatographic medium. In macroporous media, the model confirms the presence of two very distinct categories of solute, the first one diffusing inside the pores, the second being immobile so that $\bar{\bar{\theta}}_A = 0$.
- The curves obtained with the gel suggest the existence of an adsorbed-phase diffusivity mechanism associated with $\bar{\bar{\theta}}_A \approx 1/5$, as determined from the height of the plateau obtained at significant $\bar{\bar{K}}_A$ values. In a gel, the distinction between fixed and immobile solutes is not as clear as it is for a macroporous particle, so it is not really a surprise to find some differences between the two categories of solutes, but not as clearly as for macroporous particles.

- Finally, in the case of macroporous resins, while remaining in the same order of magnitude, the initial slopes of the $\bar{g}_A\left(\bar{\bar{K}}_A\right)$ curves are greater for BSA (large molecule) than they are for α-lactalbumin (smaller molecule). The initial slope being given by the tortuosity according to Eq. (5.43), this result is at least compatible with the observation of (Schröder *et al.*, 2006) that tortuosity factors are larger for larger molecules.

The $\overline{\overline{\text{Fick}}}$ model presented in this section is thus able to represent and explain the dependence of lumped diffusion coefficients on certain parameters such as Henry's coefficient. This success should not, however, lead the reader to believe that the approach is fully general and gives a precise description of what occurs inside the particle. As an illustration, the concept of adsorbed-phase diffusivity is relatively vague and the concept of tortuosity somehow related to the solute is not fully consistent with a geometrical interpretation of this parameter. Further information on the ability of the $\overline{\overline{\text{Fick}}}$ model to represent and explain experimental results will be given in Section 5.3.5.

EXERCISE 5.2 *A solute diffuses in a chromatographic medium consisting of 20 μm particles. Some RTD measurements lead to estimates of the intragranular porosity of about 0.6 and Henry's coefficient of* $\overline{K}_A = 3$. *The molecular diffusion coefficient of the solute is about* 10^{-9} *m²/s. Can we estimate the lumped average diffusion coefficient?*

5.3.3 Non-linear adsorption

The non-linearity of the adsorption isotherm can play a role at two different levels in mass transfer:

- at the particle and extragranular fluid interface via the relation $\bar{c}_A^* = \bar{f}_A(C_A^*)$
- inside the particle, in association with the diffusion process itself.

The coupling at the interface is relatively easy to understand and will be discussed in Section 5.6. We will now generalize the presentation given in Section 5.3.2 in order to understand how adsorption non-linearity can affect the diffusion process.

If one assumes that the intragranular concentration and the solid skeleton concentration are related through the equilibrium law $\bar{\bar{f}}_A$, one has, everywhere in the particle:

$$\bar{\bar{c}}_A(r,t) = \bar{\bar{f}}_A\left(\grave{c}_A(r,t)\right) \tag{5.44}$$

This equation should be associated with the mass-balance equation (5.36) in order to calculate the internal profiles. Equation (5.36) is not very handy as it involves simultaneously two different concentrations being non-linearly related. It would be convenient to modify this equation to involve only one concentration (preferably the lumped solid concentration) in a mass balance similar to the one for the $\overline{\text{Fick}}$ model, possibly at the expense of a concentration-dependent apparent diffusion coefficient.

Unfortunately, this is not feasible in general, and in order to understand the matter we will start with a simple situation assuming a quasi-linear local equilibrium. By this we mean that the particles are initially in equilibrium with a solution of concentration

C_A^I so that internal concentration profiles are initially uniform and given by $\grave{C}_A^I = C_A^I$ and $\bar{\bar{C}}_A^I = \bar{\bar{f}}_A\left(C_A^I\right)$. We further assume that the particles are then put in contact with a solution at concentration C_A not very different from C_A^I. We will express C_A as a variation from the initial concentration so that one has: $C_A = C_A^I + \delta C_A$.

We are looking for an expression involving the lumped solid-phase concentration, so that we express the local lumped solid concentration as:

$$\bar{c}_A(r,t) = \bar{C}_A^I + \delta\bar{c}_A(r,t) \quad \text{or} \quad \bar{c}_A = \bar{C}_A^I + \delta\bar{c}_A \text{ for short} \tag{5.45}$$

The perturbation $\delta\bar{c}_A$ is assumed to be small enough so that one can use first-order approximations:

$$\begin{aligned} \grave{c}_A^I &= C_A^I + \left.\frac{d\grave{C}_A}{d\bar{C}_A}\right|_{C_A^I} \delta\bar{c}_A \\ \bar{\bar{c}}_A &= \bar{\bar{C}}_A^I + \left.\frac{d\bar{\bar{C}}_A}{d\bar{C}_A}\right|_{C_A^I} \delta\bar{c}_A \end{aligned} \tag{5.46}$$

Notice that, in Eq. (5.46), we derive the fluid concentration with respect to the lumped solid-phase concentration, whereas the adsorption isotherm is normally expressed in the form $\bar{C}_A = \bar{f}_A(C_A)$. Once more, this proscribes adsorption isotherms with hysteresis.

Introducing the first-order approximations into Eq. (5.36) yields:

$$\frac{\partial \delta\bar{c}_A(r,t)}{\partial t} = \left[\varepsilon_i \grave{D}_A \left.\frac{d\grave{C}_A}{d\bar{C}_A}\right|_{C_A^I} + (1-\varepsilon_i)\,\bar{\bar{D}}_A \left.\frac{d\bar{\bar{C}}_A}{d\bar{C}_A}\right|_{C_A^I}\right] \nabla^2 \delta\bar{c}_A(r,t)$$

so that after rearrangement one obtains:

$$\frac{\partial \delta\bar{c}_A(r,t)}{\partial t} = \bar{D}_A(C_A^I)\nabla^2\delta\bar{c}_A(r,t) \tag{5.47}$$

The lumped diffusion coefficient $\bar{D}_A(C_A^I)$ is given by:

$$\bar{D}_A(C_A^I) = \frac{\varepsilon_i \grave{D}_A + (1-\varepsilon_i)\bar{\bar{f}}'_A(C_A^I)\bar{\bar{D}}_A}{\varepsilon_i + (1-\varepsilon_i)\bar{\bar{f}}'_A(C_A^I)} \tag{5.48}$$

Alternatively one can express the ratio $\bar{g}_A$:

$$\bar{g}_A(C_A^I) = \frac{\tau \bar{f}'_A(C_A^I)}{\varepsilon_i + \left(\bar{f}'_A(C_A^I) - \varepsilon_i\right)\bar{\bar{\theta}}_A} = \frac{\tau\left(\varepsilon_i + (1-\varepsilon_i)\bar{\bar{f}}'_A(C_A^I)\right)}{\varepsilon_i + (1-\varepsilon_i)\bar{\bar{f}}'_A(C_A^I)\bar{\bar{\theta}}_A} \tag{5.49}$$

This shows that, with the assumption of a quasi-linear equilibrium, the diffusion process still obeys a Fickian mechanism, but with a lumped diffusion coefficient directly impacted by the local slope of the adsorption isotherm. The diffusional transport of the solutes is influenced by the derivative of the adsorption isotherm, thus by the concentrations, just like the convectional transport in a column (see Chapter 3). This cannot be a surprise!

Incidentally, notice that the expression for the lumped diffusion coefficient (5.48) is a generalization of Eq. (5.41), which was valid for $C_A^I = 0$. The extension of this development to complete saturation, for which quasi-linear local equilibrium cannot be assumed, is not an easy task, as explained below.

After using derivation properties of the adsorption equilibrium law (5.44) and rearranging the mass-balance equation (5.36), one can show that:

$$\frac{\partial \bar{c}_A(r,t)}{\partial t} = \bar{D}_A^{loc} \nabla^2 \bar{c}_A(r,t) - \varepsilon_i \left(\grave{D}_A - \bar{\bar{D}}_A \right) \frac{\bar{f}''_A(\grave{c}_A)}{\left(\bar{f}'_A(\grave{c}_A)\right)^3} \left(\frac{\partial \bar{c}_A(r,t)}{\partial r} \right)^2 \tag{5.50}$$

or:

$$\frac{\partial \grave{c}_A(r,t)}{\partial t} = \bar{D}_A^{loc} \nabla^2 \grave{c}_A(r,t) + (1-\varepsilon_i) \bar{\bar{D}}_A \frac{\bar{\bar{f}}''_A(\grave{c}_A)}{\bar{f}'_A(\grave{c}_A)} \left(\frac{\partial \grave{c}_A(r,t)}{\partial r} \right)^2 \tag{5.51}$$

with:

$$\bar{D}_A^{loc} = \frac{\varepsilon_i \grave{D}_A + (1-\varepsilon_i) \bar{\bar{f}}'_A(\grave{c}_A) \bar{\bar{D}}_A}{\varepsilon_i + (1-\varepsilon_i) \bar{\bar{f}}'_A(\grave{c}_A)} = \frac{\varepsilon_i \grave{D}_A + \left(\bar{f}'_A(\grave{c}_A) - \varepsilon_i \right) \bar{\bar{D}}_A}{\bar{f}'_A(\grave{c}_A)} \tag{5.52}$$

Equations (5.50), (5.51) and (5.52) are a generalization to non-linear equilibria of Eq. (5.40), which is valid for linear equilibrium only.

The right-hand side of the mass-balance equation (5.50) or (5.51) is the sum of two terms:

- The first one is the classical term obtained with Fick's law, but involving a diffusion coefficient $\bar{D}_A^{loc}$ varying with concentration.
- A second term, which is not present in the classical Fickian formulation, is a kind of correction term to be applied when one uses either lumped or pore concentrations.
  - When lumped solid concentrations $\bar{c}_A$ are used, this correction term is a consequence of the non-linearity of the adsorption isotherm and of the fact that intragranular and adsorbed solutes diffuse at different speeds. One can check that this correction term vanishes either if the adsorption isotherm is linear ($\bar{f}''_A(\grave{c}_A) = 0$) or if the diffusion coefficient in the intragranular pores and on the surface are identical ($\grave{D}_A = \bar{\bar{D}}_A$).
  - When pore concentrations $\grave{c}_A$ are used, this correction term is a consequence of the non-linearity of the adsorption isotherm and of the existence of adsorbed-phase diffusivity. One can check that this correction term vanishes either if the adsorption isotherm is linear ($\bar{f}''_A(\grave{c}_A) = 0$) or if the adsorbed-phase diffusivity coefficient equals zero.

When one uses the lumped model associated with a constant lumped diffusion $\bar{D}_A$, one not only neglects the concentration dependence of the local diffusion coefficient, but, in addition, one assumes that the lumped concentrations obey a Fickian mass balance, which is not true in general.

In order to investigate the influence of non-linearity on the diffusion process, let us normalize the local diffusion coefficient $\overline{D}_A^{loc}$ by $\overline{D}_A(0)$, the lumped diffusion coefficient given by Eq. (5.41) observed at a zero solute concentration. We then define:

$$\overline{D}_A^{loc} = \overline{D}_A(0)\,\psi(c_A) \tag{5.53}$$

By comparing Eqs. (5.41) and (5.52), we see that $\psi(c_A)$ is a dimensionless number given by:

$$\psi(c_A) = \frac{\left[1 + \frac{(1-\varepsilon_i)}{\varepsilon_i}\overline{\overline{f}}'_A(c_A)\,\overline{\overline{\theta}}_A\right] \Big/ \left[1 + \frac{(1-\varepsilon_i)}{\varepsilon_i}\overline{\overline{K}}_A\overline{\overline{\theta}}_A\right]}{\left[1 + \frac{(1-\varepsilon_i)}{\varepsilon_i}\overline{\overline{f}}'_A(c_A)\right] \Big/ \left[1 + \frac{(1-\varepsilon_i)}{\varepsilon_i}\overline{\overline{K}}_A\right]} \tag{5.54}$$

While relatively unfriendly at first glimpse, the parameter $\psi(c_A)$ in fact has simple features allowing quantification of the deviations from linearity and differences between the lumped model and the porous model:

- Deviation from linearity: for linear adsorption isotherms, one has $\overline{\overline{f}}'_A(c_A) = \overline{\overline{K}}_A$, so $\psi(c_A) = 1$.
- Difference between models: the lumped model and the porous model become physically identical if solutes contained in the pores or adsorbed on the skeleton can diffuse with the same diffusion coefficient, thus if the ratio $\overline{\overline{\theta}}_A = 1$. From Eq. (5.54), it appears that $\psi(c_A) = 1$ when $\overline{\overline{\theta}}_A = 1$. Under this assumption one has $\overline{D}_A^{loc} = \overline{D}_A(0) = \grave{D}_A = \overline{\overline{D}}_A$.

Consequently, when the adsorption isotherm is linear, or when the porous model and the lumped model are indistinguishable, the parameter $\psi(c_A) = 1$. The impact of non-linearity and the deviation between the porous model and the lumped model are thus expected to increase as $\psi(c_A)$ increasingly differs from 1.

In general, the adsorbed-phase diffusivity coefficient is expected to be lower than pore diffusion so that the ratio $\overline{\overline{\theta}}_A$ is less than 1. Under these conditions, the coefficient $\psi(c_A)$ increases with concentration in the case of a favorable adsorption isotherm and decreases in the opposite situation as shown in Figure 5.9. As a direct consequence, according to Eq. (5.53) the local diffusion coefficient $\overline{D}_A^{loc}$ increases with concentration in the case of a favorable adsorption isotherm and decreases in the opposite situation.

EXERCISE 5.3 *Estimate the lumped diffusion coefficient $\overline{D}_A$ when the solute concentration inside the particle is sufficiently high for the derivative $\overline{\overline{f}}'_A(c_A)$ to be practically zero.*

In order to illustrate the impact of non-linearity on uptake curves, we will start neglecting surface diffusivity (thus assuming $\overline{\overline{\theta}}_A = 0$). This assumption, which maximizes the difference in mobility between the solutes, maximizes the impact of non-linearity.

Two different situations are now considered, representing reasonable preparative chromatography conditions. The adsorption isotherm is taken to be Langmuirian and the relevant parameters are given in Table 5.4. Cases A and B are associated with adsorption

Table 5.4 Parameters of adsorption isotherms used to simulate the influence of non-linearity on uptake curves. $\varepsilon_i = 0.5$, $\bar{\bar{\theta}}_A = 0$, ψ calculated with Eq. (5.54).

	Case A	Case B
$\bar{\bar{N}}$ (g/l)	20	100
$\tilde{K}_A$ (l/g)	0.05	0.2
$\bar{\bar{K}}_A = \bar{\bar{N}}\tilde{K}_A$	1	20
$\overline{K}_A = \varepsilon_i + (1-\varepsilon_i)\,\bar{\bar{K}}_A$	1	10.5
$\bar{g}_A(0) = \frac{D_A}{\overline{D}_A(0)}$ Calculated with Eq. (5.43).	2τ	21τ
C_A (g/l) to reach 25% of capacity $\bar{\bar{N}}$	6.7 ($\psi = 1.28$)	1.7 ($\psi = 1.73$)
C_A (g/l) to reach 50% of capacity $\bar{\bar{N}}$	20 ($\psi = 1.60$)	5.0 ($\psi = 3.50$)
C_A (g/l) to reach 95% of capacity $\bar{\bar{N}}$	400 ($\psi = 2.0$)	95.0 ($\psi = 20.0$)

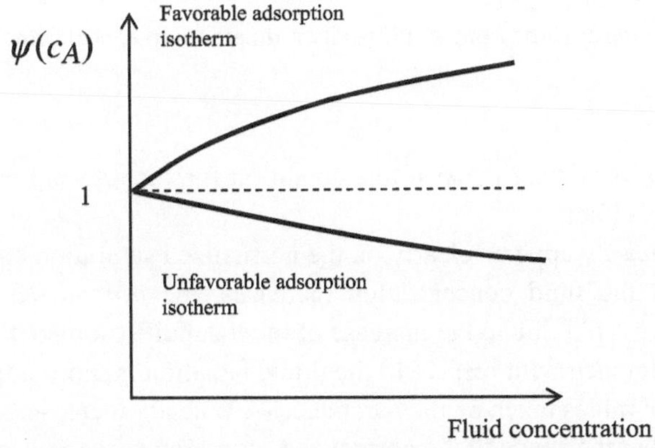

Figure 5.9 Qualitative evolution of the coefficient ψ (c_A) for different types of adsorption isotherms.

capacities $\bar{\bar{N}}$ of 20 and 100 g/l, respectively. These values are classical for optical isomer separations and for separations on reversed-phase or silica. Parameter $\tilde{K}_A$ has been selected in order to provide reasonable values for the lumped Henry's coefficient $\overline{K}_A$. Case A (resp. B) is associated with a Henry's coefficient that can be considered relatively small (resp. high) for many preparative chromatography applications.

One can see from Table 5.4 that, under these conditions, the lumped diffusion coefficients $\overline{D}_A(0)$ obtained for very dilute solutes are respectively about 2τ and 21τ times lower than the molecular diffusion coefficients.

The concentrations in the mobile phase required for $\bar{\bar{C}}_A$ to reach 25%, 50% and 95% of the capacity $\bar{\bar{N}}$ are calculated and given for each case in Table 5.4. The associated coefficients ψ are then calculated for these different concentrations with Eq. (5.54). For concentrations associated with a 95% capacity usage, the coefficient ψ is about 2 (resp. 20) for case A (resp. case B). This shows that, during the saturation/desaturation process of the particle, the local diffusion coefficient increases by 2 times compared to

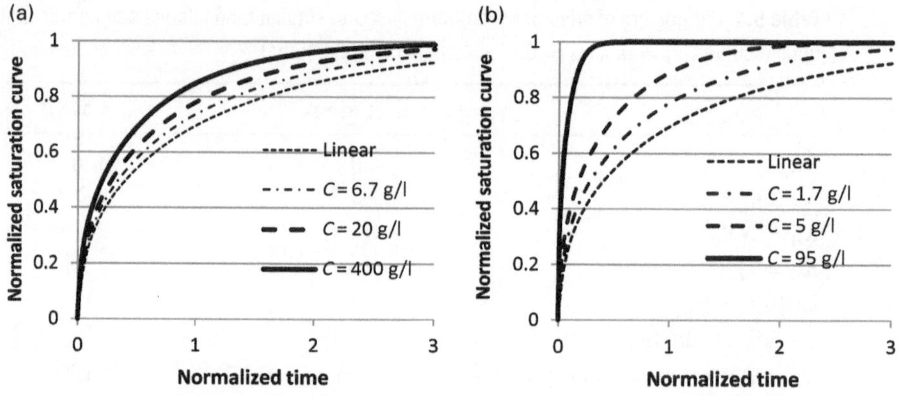

Figure 5.10 Normalized saturation curves (lumped concentration) simulated with the porous model: (a) case A; (b) case B. Parameters are given in Table 5.4. Concentrations are those which allow 0%, 25%, 50% and 95% of saturation capacity. The time is normalized to $t_A^i = \frac{r_p^2}{15\bar{D}_A(0)}$. Intragranular porosity and the adsorbed-phase diffusivity ratio are set to $\varepsilon_i = 0.5$ and $\bar{\bar{\theta}}_A = 0$.

the diffusion coefficient occurring under dilute condition for case A, and by 20 times for case B. That is a big effect!

The effect of non-linearity appears clearly on the normalized saturation curves given in Figure 5.10. When the fluid concentration increases, the *normalized* saturation becomes more and more rapid due to the increase of the local diffusion coefficients.

As expected, the difference with respect to the dilute situation is more important for case B due to the higher values taken by the parameter ψ. When ψ increases, not only is the saturation quicker, but the shape of the normalized saturation curve is also impacted.

When the influence of surface diffusivity, measured by the ratio $\bar{\bar{\theta}}_A$, increases, the coefficient ψ decreases, so the influence of concentration on saturation curves is expected to decrease. As an illustration, for a concentration C_A associated with 95% of $\bar{\bar{N}}$, $\psi = 2$ for case A (resp. 20 for case B) for $\bar{\bar{\theta}}_A = 0.0$ and 1.81 (resp. 6.70 for case B) for $\bar{\bar{\theta}}_A = 0.1$. When $\bar{\bar{\theta}}_A$ approaches 1, the normalized saturation curves become independent of concentration.

We will see in Section 5.3.4 that this relatively complex behavior, requiring a solution of Fick's law in spherical geometry with non-linear constraints, can in general be represented by a modified LDF approximation.

Finally, let us qualitatively connect the impact of the non-linear diffusion process to chromatograms. We have seen in Chapter 3 that the non-linearity of the adsorption isotherm imposes key front characteristics, including their mean positioning, sharpening and broadening behaviors, thus peak asymmetry. The non-linearity of the diffusion process, thus the variation of the lumped diffusion coefficient with concentration, will add another source of asymmetry. The fact that either pore or adsorbed-phase diffusivity is predominant is going to impact the shape of breakthrough curves, as explained by (Ma *et al.*, 1996). This will be discussed further in Chapter 7.

5.3.4 $\overline{\overline{\text{LDF}}}$ approximations for the non-linear porous model

Solving the mass-balance equation (5.36) together with the adsorption law equation (5.44) and then calculating lumped averaged concentrations is numerically convoluted, so adapting the LDF approximation presented in Section 5.2.2 for representing the key features of the porous model under non-linear conditions is of key interest. A possible path forward is presented below.

In order to extend the LDF approximation to non-linear situations involving variable diffusion coefficients, one can consider using the structure of Eq. (5.19):

$$t_A^{i\,var}\frac{d\overline{C}_A}{dt} = \left(\bar{c}_A^* - \overline{C}_A\right) \tag{5.55}$$

but using a concentration-dependent characteristic time for diffusion, $t_A^{i\,var}$, given by:

$$t_A^{i\,var} = \frac{d_p^2}{60\overline{D}_A^{var}} \tag{5.56}$$

where $\overline{D}_A^{var}$ is a lumped diffusion coefficient varying with concentration. We will name this model $\overline{\overline{\text{LDF}}}^{\text{var}}$. The terminology $\overline{\overline{\text{LDF}}}$ reminds us that this model is based on the porous model.

An expression for $t_A^{i\,var}$ based only on the volume-averaged lumped concentration is unlikely to be adequate for covering a variety of situations: due to the non-linear variation of the local diffusion coefficient $\overline{D}_A^{loc}$ with concentration, two different profiles associated with the same volume-averaged lumped concentration can be associated with different saturation/desaturation dynamics. We are thus going to relate $t_A^{i\,var}$ to both $\overline{C}_A$ and $\bar{c}_A^*$, which represent the only concentrations appearing in our model, in order to take into account the direction of the internal concentration profile.

The role of the interfacial composition can be seen by integrating Eq. (5.36) over the entire particle, which leads to:

$$\frac{d\overline{C}_A}{dt} = \left(\frac{3\varepsilon_i\grave{D}_A}{r_p} + \frac{3(1-\varepsilon_i)\overline{\overline{D}}_A}{r_p}\overline{\overline{f}}'_A(\grave{c}_A^*)\right)\left.\frac{\partial\grave{c}_A(r,t)}{\partial r}\right|_{r_p}$$

Using the properties of the derivation and of the adsorption isotherm, one obtains:

$$\frac{d\overline{C}_A(t)}{dt} = \frac{3\overline{D}_A^{loc*}}{r_p}\left.\frac{\partial\bar{c}_A(r,t)}{\partial r}\right|_{r_p} \tag{5.57}$$

$$\overline{D}_A^{loc*} = \frac{\varepsilon_i\grave{D}_A + (1-\varepsilon_i)\overline{\overline{f}}'_A(\grave{c}_A^*)\overline{\overline{D}}_A}{\varepsilon_i + (1-\varepsilon_i)\overline{\overline{f}}'_A(\grave{c}_A^*)} \tag{5.58}$$

where $\overline{\overline{f}}'_A(\grave{c}_A^*)$ represents the first derivative of the adsorption isotherm $\overline{\overline{f}}_A$ calculated with the interfacial concentration $\grave{c}_A^*$ assumed to be identical to C_A^*.

Equation (5.57) is similar to Eq. (5.17) obtained for the lumped model and suggests that the variation of the lumped volume-averaged concentration is strongly influenced by the value of the local diffusion coefficient at the periphery of the particle.

Let us further notice that the lumped diffusion coefficient at the fluid–solid interface, $\overline{D}_A^{loc*}$, and the local lumped diffusion coefficient $\overline{D}_A^{loc}$ are calculated by similar expressions (5.52) and (5.58).

Based on different simulations, it appears that a good option is to choose for $\overline{D}_A^{var}$ the average value of $\overline{D}_A^{loc*}$ calculated with the fluid concentration occurring at the particle–fluid interface and of $\overline{D}_A^{loc}$ calculated with the average intragranular concentration $\grave{C}_A$ in equilibrium with the volume-averaged concentration $\overline{C}_A$:

$$\overline{D}_A^{var} = \frac{\overline{D}_A^{loc*} + \overline{D}_A^{loc}(\grave{C}_A)}{2} \tag{5.59}$$

One can use a more normalized form by expressing the lumped coefficient $\overline{D}_A^{var}$ as the product of the lumped diffusion coefficient occurring at zero concentration $\overline{D}_A\,(0)$ by a correction factor ψ^{var} having a structure similar to the one in Eq. (5.54):

$$\overline{D}_A^{var} = \overline{D}_A\,(0)\,\psi^{var} \tag{5.60}$$

$$\psi^{var} = \frac{\psi\left(c_A^*\right) + \psi\left(\grave{C}_A\right)}{2}$$

$$\psi\,(x) = \frac{\left[1 + \frac{(1-\varepsilon_i)}{\varepsilon_i}\bar{\bar{f}}'_A\,(x)\,\bar{\bar{\theta}}_A\right] \Big/ \left[1 + \frac{(1-\varepsilon_i)}{\varepsilon_i}\bar{\bar{K}}_A\bar{\bar{\theta}}_A\right]}{\left[1 + \frac{(1-\varepsilon_i)}{\varepsilon_i}\bar{\bar{f}}'_A\,(x)\right] \Big/ \left[1 + \frac{(1-\varepsilon_i)}{\varepsilon_i}\bar{\bar{K}}_A\right]}$$

The mass-balance equation (5.55), together with Eq. (5.59) or (5.60), provides a simple way to simulate the particle saturation with the $\overline{\overline{\text{LDF}}}^{\text{var}}$ model. For a given set of concentrations $\overline{C}_A$ and $\bar{c}_A^*$, pore concentrations $\grave{C}_A$ and $\grave{c}_A^*$ are calculated from the equilibrium law, and the diffusion coefficient $\overline{D}_A^{var}$ is directly calculated from Eq. (5.59) or (5.60). The time derivative of the volume-averaged lumped concentration $\overline{C}_A(t)$ is calculated with Eq. (5.55). Iterating the procedure allows a simulation of the evolution of $\overline{C}_A(t)$.

We will now check the accuracy of the $\overline{\overline{\text{LDF}}}^{\text{var}}$ approximation by comparing saturation/desaturation curves obtained for case B in Table 5.4. Recall that the non-linearity impact is very strong for this case, and that, at the highest concentration, the local diffusion coefficient is 20 times higher than its value under dilute conditions.

In order to investigate both saturation and desaturation, the following experiment is simulated: initially solute-free particles are put in contact with a solution containing a prescribed solute concentration for a certain duration, the particles are then abruptly put in contact with a solute-free solution. In order to allow for the saturation step to be largely completed, its duration is fixed at three times $t_A^i = \frac{r_p^2}{15\overline{D}_A(0)}$, the characteristic time for internal diffusion at zero concentration.

The simulated curves obtained with the porous model, the linear LDF model and the $\overline{\overline{\text{LDF}}}^{\text{var}}$ model are given in Figure 5.11 for different solute concentrations and for $\bar{\bar{\theta}}_A$ ratios equaling either 0 or 0.25. One can see in Figure 5.11 (bottom) that, at low concentrations, so that when the ψ factor is about one, the $\overline{\text{LDF}}$ and $\overline{\overline{\text{LDF}}}^{\text{var}}$ curves are almost identical, the two models becoming strictly identical at zero solute concentration. The difference between the $\overline{\text{LDF}}$ and $\overline{\text{Fick}}$ models has already been noted in Section 5.2.2.

More interesting are the figures at the top of Figure 5.11, for which non-linearity is very significant. One can see that the $\overline{\overline{\text{LDF}}}^{\text{var}}$ approximation is remarkably well able to predict the influence of concentration on saturation/desaturation curves, even when the local diffusion coefficient varies by a factor as high as 20 during the experiment. The improvement compared to the classical $\overline{\text{LDF}}$ model is very significant.

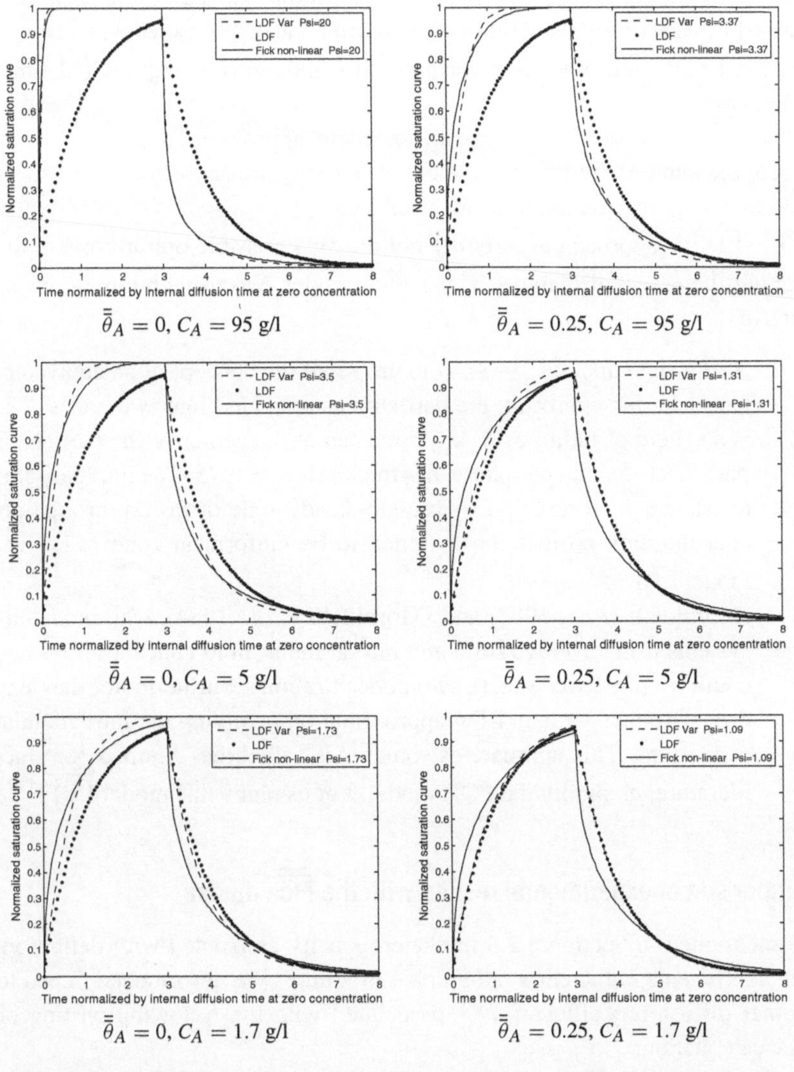

Figure 5.11 Normalized saturation curves (lumped concentration) simulated with the $\overline{\overline{\text{Fick}}}$ model, the $\overline{\text{LDF}}$ model and the $\overline{\overline{\text{LDF}}}^{\text{var}}$ model. The time is normalized to $t_A^i = \frac{r_p^2}{15\overline{D}_A(0)}$. Intragranular porosity is set to $\varepsilon_i = 0.5$. Other parameters are given in Table 5.4. The parameter ψ in the legends is calculated from Eq. (5.60) applied to the extragranular fluid concentration.

Comparing simulations on the left ($\overline{\overline{\theta}}_A = 0$) and on the right ($\overline{\overline{\theta}}_A = 0.25$) in Figure 5.11 shows that, the higher $\overline{\overline{\theta}}_A$, the closer the $\overline{\overline{\text{LDF}}}$ and $\overline{\overline{\text{LDF}}}^{\text{var}}$ models. This was fully expected as we showed that increasing $\overline{\overline{\theta}}_A$ decreases the influence of non-linearity.

The saturation/desaturation curves presented in Figure 5.11 tend to magnify the differences between models because the particles are submitted to a step change in interfacial concentration. In the more realistic situation of a chromatographic column simulation, the influence of external diffusion as well as hydrodynamic dispersion will require smoother variations of the interfacial concentration. The $\overline{\overline{\text{LDF}}}^{\text{var}}$ approximation, having "passed the test" of the saturation/desaturation curves, is expected to be even closer to the non-linear $\overline{\overline{\text{Fick}}}$ prediction for column simulations – with a CPU time about 100 times lower.

The $\overline{\overline{\text{LDF}}}^{\text{var}}$ approach is thus an excellent approximation, which we recommend should the internal limitation and the non-linearity impact be too strong for the classical $\overline{\overline{\text{LDF}}}$ linear approximation to be acceptable.

The $\overline{\overline{\text{LDF}}}^{\text{var}}$ approach is certainly not the only possible option for simplifying while keeping the essential features of the $\overline{\overline{\text{Fick}}}$ model. Worth mentioning are the contributions of:

- (Carta and Cincotti, 1998), who proposed the concept of an equivalent film resistance. In this approach, the particle is subdivided into two zones. The first zone is a sphere of radius $\delta = 4r_p/5$ located at the center of the particle. The second one, located at the periphery, has thickness $\delta = r_p/5$. The intraparticle diffusional resistance is represented by pseudo-steady-state diffusion through zone 2 when concentration profiles are assumed to be uniform in zone 1. Let us name this model $\overline{\overline{\text{LDF}}}^{1/5}$.
- (Morbidelli *et al.*, 1982) and (Morbidelli *et al.*, 1984), who modeled adsorption systems using average solid and intragranular fluid concentrations and assuming equilibrium between these two concentrations. Furthermore, they expressed the flux of solutes with an LDF approximation involving the intragranular fluid concentrations. This approach is sometimes called the "lumped pore model" in the literature, or simply the POR model. Let us name this model $\overline{\overline{\text{LDF}}}^{\text{POR}}$.

5.3.5 Comparison of experimental results with the $\overline{\overline{\text{Fick}}}$ model

We mentioned in Section 5.2.3 the heterogeneity associated with definitions for diffusion coefficients in the chromatographic literature. We unfortunately need to introduce another diffusion coefficient, $\overline{D}_A^{ps}$, associated with the following writing of the mass-balance equation:

$$\frac{\partial \bar{c}_A(r,t)}{\partial t} = \overline{D}_A^{ps} \nabla^2 \grave{c}_A(r,t) \qquad (5.61)$$

We will call $\overline{D}_A^{ps}$ a pseudo-lumped diffusion coefficient.

The origin of Eq. (5.61) is clear when adsorbed-phase diffusivity equals zero so that only solutes contained in the intragranular pores can diffuse. Under this assumption, one has simply $\overline{D}_A^{ps} = \varepsilon_i \grave{D}_A$. For some reason, however, Eq. (5.61) is also frequently used when adsorbed-phase diffusivity is considered, which is a bit confusing. Under these conditions, one has, under linear adsorption conditions:

$$\overline{D}_A^{ps} = \varepsilon_i \grave{D}_A + (1 - \varepsilon_i)\,\overline{\overline{K}}_A \overline{\overline{D}}_A = \overline{K}_A \overline{D}_A \qquad (5.62)$$

The coefficient $\overline{D}_A^{ps}$ is often mentioned in the literature either as *effective* or *apparent* diffusion coefficient.

We showed in Section 5.3.2 that the dependence of lumped diffusion coefficients $\overline{D}_A$ on Henry's coefficient, as published by (Farnan *et al.*, 2002), is well given by the $\overline{\overline{\text{Fick}}}$ model. For readers willing to study the original paper, we warn that the reported "apparent diffusion coefficient" is in fact $\overline{D}_A^{ps}$ and that the "adsorbed-phase diffusivity coefficient" defined by the authors corresponds to $(1 - \varepsilon_i)\,\overline{\overline{D}}_A$ as mentioned in Table 5.2.

The non-linear $\overline{\overline{\text{Fick}}}$ model also explains the concentration dependence of lumped diffusion coefficients on concentration mentioned in Section 5.2.3. This is discussed below.

Using the shallow-bed and batch methods, (Yoshida *et al.*, 1994) obtained a strong dependence of the lumped diffusion coefficient on concentration. They obtained fairly linear plots of $\overline{D}_A\left(1 + \frac{1}{\Delta}\right)$ versus Δ, defined as $\Delta = \frac{1-\varepsilon_i}{\varepsilon_i}\frac{\overline{\overline{C}}_A}{C_A}$.

(Yoshida *et al.*, 1994) determined lumped diffusion coefficients by analyzing the initial part of the saturation curve calculated from the Fick model. Concentration variations being small, one can assume the saturation of the particles to be described by the quasi-linear model presented in Section 5.3.3. The determined lumped diffusion coefficient $\overline{D}_A$ is thus expected to be related to the extragranular fluid concentration by Eq. (5.48). Knowledge of the adsorption isotherms in Section 5.2.3.2 allows a calculation of the influence of concentration on $\overline{D}_A$ and the parameter Δ.

As shown in Figure 5.12, results obtained by (Yoshida *et al.*, 1994) are well represented with two adjustable parameters only: the lumped diffusion coefficient at zero concentration and the ratio $\overline{\overline{\theta}}_A$.

Additionally, recall that, as presented in Section 5.2.3:

- (Rearden *et al.*, 1998) obtained a significant increase of the lumped diffusion coefficient $\overline{D}_A$ with concentration.
- (Sajonz *et al.*, 1998) reported lumped diffusion coefficients $\overline{D}_A$ increasing with concentration, the dependence on concentration being stronger:
  - for the most retained enantiomer
  - at low temperature where retention is stronger.

These results, obtained under quasi-linear situations, are qualitatively compatible with the $\overline{\overline{\text{Fick}}}$ theoretical framework presented in Section 5.3.3. The increasing dependence of lumped diffusion coefficients on concentration when retention increases is a direct consequence of the influence of the derivatives of the adsorption isotherms.

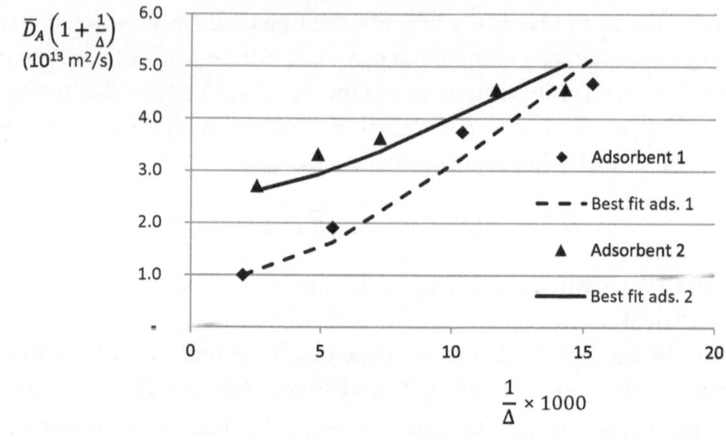

Figure 5.12 Representation of the results obtained by (Yoshida *et al.*, 1994) assuming that the lumped diffusion coefficient is determined using the quasi-linear approximation. Best fit: $\bar{\bar{\theta}}_A = 0.06$, $\bar{D}_A\ (0) = 0.9 \times 10^{-13}$ m²/s for adsorbent 1 and $\bar{\bar{\theta}}_A = 0.25$, $\bar{D}_A\ (0) = 2.5 \times 10^{-13}$ m²/s for adsorbent 2. See details in the text and in Table 5.3.

The ability of the $\overline{\overline{\text{Fick}}}$ model to represent the influence of both Henry's coefficient and concentration on lumped diffusion coefficients should not, however, lead the reader to believe that everything is perfectly understood. As an illustration, (Miyabe and Guiochon, 2002) studied internal diffusion coefficients of various small molecules on different reversed-phase silica. By adjusting the MeOH–water composition of the eluent the authors showed that a single decreasing curve is obtained when the ratio $\bar{\bar{D}}_A/D_A$ is plotted versus Henry's coefficient $\bar{K}_A$ for various solutes and adsorbents. Moreover, this curve suggests that molecular diffusion and adsorbed-phase diffusivity coefficients are intimately related. As mentioned by the authors, these "unreasonable" results suggest that new models and formulations must be derived for adsorbed-phase diffusivity.

More generally, it seems that adsorbed-phase diffusivity coefficients vary with concentration and with Henry's coefficient (Miyabe and Guiochon, 2000a,b, 2002).

When using a $\overline{\overline{\text{Fick}}}$ model neglecting adsorbed-phase diffusivity, different authors reported a decrease of pore diffusivities with increasing protein concentration (Melter *et al.*, 2008; Chang and Lenhoff, 1998; Conder and Hayek, 2000; Johnston and Hearn, 1991). Reasons for this are not obvious (pore constriction effect? inadequate description of the solute's interaction with the skeleton?).

Let us mention the difficulty of investigating the influence of concentration on adsorbed-phase diffusivity coefficients as it consists in analyzing a "fourth-order" phenomenon. The chromatographic peaks or breakthrough curves are primarily positioned according to the adsorption isotherm, then additional broadening can be approximated by a plate number (first-order), the variation of plate number with velocity can be related to diffusion coefficients (second-order) and, finally, investigation of the variation of this diffusion coefficient with concentration leads to adsorbed-phase diffusivity (third-order). Studying the variation of adsorbed-phase diffusivity coefficient with experimental conditions (fourth-order) possibly exceeds the precision of available models.

We believe that we reach here the limits of the simplistic model whose key assumptions were presented in the introduction of Section 5.3. When protein diffusion in chromatographic media is considered, the fact that solute size and pore size are of the same order of magnitude may play a significant role. As an illustration, (Coquebert de Neuville, Tarafder and Morbidelli, 2013) and (Coquebert de Neuville, Thomas and Morbidelli, 2013) studied the influence of pore size distribution and protein adsorption on internal mass transfer. Another investigation on hindered protein diffusion is found in (Gutenwik *et al.*, 2004). A less frequently studied but possibility important factor to be taken into account is the elasticity of some chromatographic media particles and thus the impact of swelling/shrinking on intragranular diffusion (Heinonen *et al.*, 2012).

Finally, only very limited results are available on multiple-solute systems. For the very important field of proteins, the available contributions are normally limited to binary systems (Martin *et al.*, 2005; Lewus and Carta, 1999; Russell and Carta, 2005). (Lewus and Carta, 1999) confirmed the importance of taking non-linear coupling into account in representing binary experimental results.

5.3.6 Special situation: the biporous model

Some chromatographic media can consist of microparticles having their own porosity, these microparticles being agglomerated into the larger particles used for packing the columns. Under these circumstances, it can be appropriate to describe the particles by taking into account two different levels of porosity. A simple approach consists in considering a particle made of:

- macroporous intragranular pores associated with porosity ε_i^M in which solutes can diffuse with a diffusion coefficient $\grave{D}_A^M$
- microporous pores representing fraction ε_i^μ of the volume of particle different from the macropores, in which solutes can diffuse with a diffusion coefficient $\grave{D}_A^\mu$.

This model, which will be called the *biporous model*, is an extension of the porous model: the adsorbed phase is now considered to be a two-dimensional phase (microparticles) in which the solutes can diffuse. This is a further illustration of the Lego approach:

- The *biporous model* involves macroporous intragranular pores and microporous particles containing the adsorbed solutes.
- When the microparticles are considered to be a one-dimensional homogeneous phase, one obtains the porous model.
- When the adsorbed phase and the macroporous intragranular pores are lumped together, one obtains the lumped model.

Simple volumic considerations lead to:

$$\varepsilon_i = \varepsilon_i^M + \left(1 - \varepsilon_i^M\right) \varepsilon_i^\mu \tag{5.63}$$

If one calls $\grave{c}_A^M$ and $\grave{c}_A^\mu$ the concentrations of solute in respectively the macropores and in the micropores, the average solute concentration contained in the total intragranular porosity is given by:

$$\varepsilon_i \grave{c}_A = \varepsilon_i^M \grave{c}_A^M + \left(1 - \varepsilon_i^M\right) \varepsilon_i^\mu \grave{c}_A^\mu \tag{5.64}$$

The lumped average solid concentration is still given by $\bar{c}_A = \varepsilon_i \grave{c}_A + (1 - \varepsilon_i)\,\bar{\bar{c}}_A$.

One can decide to approximate the microparticles by a pseudo-homogeneous medium, thus to model them with a lumped model and to write:

$$\bar{c}_A^\mu = \varepsilon_i^\mu \grave{c}_A^\mu + (1 - \varepsilon_i^\mu)\bar{\bar{c}}_A \tag{5.65}$$

The concentration $\bar{c}_A^\mu$ is the lumped solid-phase concentration in the microparticles: it agglomerates solutes contained in the micropores and solutes adsorbed on the solid skeleton. This concentration can be related to the concentration in the intragranular macropores by an equilibrium law:

$$\bar{c}_A^\mu = \bar{f}_A^\mu(\grave{c}_A^M) \tag{5.66}$$

The lumped solid concentration $\bar{c}_A$ at the particle level is related to the lumped solid concentration $\bar{c}_A^\mu$ at the microparticle level by:

$$\bar{c}_A = \varepsilon_i^M \grave{c}_A^M + \left(1 - \varepsilon_i^M\right) \bar{c}_A^\mu \tag{5.67}$$

The complete set of equations associated with Fick's law to describe diffusion in such a system is pretty complex and is not reported here. (Leitão and Rodrigues, 1999) proposed a simplification based on two LDF approximations involving volume-averaged concentrations. The first LDF approximation gives the mass balance at the particle level and expresses the contribution of the solutes contained in the macropores to the particle saturation:

$$\left(1 - \varepsilon_i^M\right) \frac{d\bar{C}_A^\mu}{dt} + \varepsilon_i^M \frac{d\grave{C}_A^M}{dt} = \frac{C_A^* - \grave{C}_A^M}{t_A^M} \tag{5.68}$$

$t_A^M = \frac{d_p^2}{60(\varepsilon_i^M \grave{D}_A^M)}$ is the characteristic time for diffusion associated with the macropore diffusion process.[3] $\grave{D}_A^M$ is the diffusion coefficient of the solutes in the macropores.

The second LDF approximation characterizes the mass balance for the solutes contained in the lumped microparticles:

$$\frac{d\bar{C}_A^\mu}{dt} = \frac{\bar{f}_A^\mu(\grave{C}_A^M) - \bar{C}_A^\mu}{t_A^\mu} \tag{5.69}$$

$t_A^\mu = \frac{d_{p\mu}^2}{60\bar{D}_A^\mu}$ is the characteristic time for diffusion associated with the microparticles. $\bar{D}_A^\mu$ is the lumped diffusion coefficient of the solute in the microparticles.

This two-LDF approximation, which can be justified by the transfer function approach already used in Section 5.2.2, allows a relatively complex diffusion process to be modeled with only two ODEs.

[3] Don't confuse this with t_A^m, the characteristic time for overall mass transfer.

In the case of a linear adsorption isotherm, the adsorption law Eq. (5.66) can be written:

$$\overline{C}_A^{\mu} = \overline{K}_A^{\mu} \grave{C}_A^{M} \tag{5.70}$$

Using Eq. (5.65) and Eq. (5.67), one shows that coefficient $\overline{K}_A^{\mu}$ is related to the particle average Henry's coefficients by :

$$\begin{aligned} \overline{K}_A^{\mu} &= \varepsilon_i^{\mu} + (1-\varepsilon_i^{\mu})\overline{\overline{K}}_A \\ \overline{K}_A &= \varepsilon_i^{M} + \left(1-\varepsilon_i^{M}\right)\overline{K}_A^{\mu} \end{aligned} \tag{5.71}$$

Using the Laplace transform to derive the transfer function $\hat{\overline{H}}^{Bipor}$ connecting the lumped average solid concentration to the lumped solid concentration at the fluid–particle interface presents no difficulty. From Eq. (5.68) and (5.69), one obtains:

$$\hat{\overline{H}}^{Bipor}(s) = \frac{\hat{\overline{C}}_A}{\hat{\overline{c}}_A^*} = \frac{1}{\overline{K}_A}\frac{\varepsilon_i^M + \frac{\overline{K}_A - \varepsilon_i^M}{1+st_A^{\mu}}}{1 + st_A^M\left(\varepsilon_i^M + \frac{\overline{K}_A - \varepsilon_i^M}{1+st_A^{\mu}}\right)} \tag{5.72}$$

Following our classical methodology (see Chapter 2 and Section 5.2.2), equating the first-order derivatives of the functions $\hat{\overline{H}}$ allows us to show that the two-LDF approximation modeled by Eqs. (5.68) and (5.69) can be approximated by a first-order dynamic system:

$$\frac{d\overline{C}_A}{dt} = \frac{c_A^* - \overline{C}_A}{t_A^{\mu/M}} \tag{5.73}$$

with time constant:

$$t_A^{\mu/M} = \overline{K}_A t_A^M + \frac{\overline{K}_A - \varepsilon_i^M}{\overline{K}_A} t_A^{\mu} \tag{5.74}$$

When the characteristic time for diffusion in the microparticles is very small, the kinetic process is limited by the diffusion in the macropores. The characteristic time for the overall process is given by $t_A^{\mu/M} = \overline{K}_A t_A^M = \overline{K}_A \frac{d_p^2}{60(\varepsilon_i^M \grave{D}_A^M)}$. Recall that, using the porous model, we showed that $\overline{D}_A = \frac{\varepsilon_i^M \grave{D}_A}{\overline{K}_A}$, so $t_A^{\mu/M}$ is no more than t_A^i calculated with $\overline{D}_A$. As expected under these circumstances, with the microparticles acting as a solid phase at equilibrium with the solute contained in the macropores, the biporous model is equivalent to the porous model and thus to the lumped model associated with characteristic time t_A^i.

When the characteristic time for diffusion in the macropores is very small, the kinetic process is limited by the diffusion in the microparticles. The characteristic time for the overall process is $t_A^{\mu/M} = \frac{\overline{K}_A - \varepsilon_i^M}{\overline{K}_A} t_A^{\mu}$. Notice that this time becomes zero for $\overline{K}_A = \varepsilon_i^M$, which means that solute does not enter and is not adsorbed in the microparticles.

If the kinetic process is limited by the diffusion in the microparticles, the overall characteristic time for diffusion $t_A^{\mu/M}$ is independent of particle size (but dependent on microparticle size). This shows that observing kinetics independent of particle size does not systematically mean that adsorption kinetics is limiting.

5.3.7 Special situation: internal convection

The previous developments rely on the assumption that no convection induced by the extragranular fluid flow occurs in the particles. Even though this is usually the case, it has been shown that convection may occur in strongly porous particles with 1000–7000 Å pore size, leading to a significant enhancement of internal transfer (Rodrigues *et al.*, 1982; Afeyan *et al.*, 1990; Rodrigues, 1993). This effect is often called *perfusion chromatography.*

Taking into account internal convection adds a considerable amount of complexity to the diffusion model because the spherical symmetry of the particles is broken due to the existence of an intraparticular flow in the direction of the longitudinal axis of the column.

(Rodrigues *et al.*, 1982), (Carta *et al.*, 1992) and (Leitão and Rodrigues, 1995) proposed a simplification, approximating the internal transfer by an "augmented" lumped diffusion coefficient:

$$\overline{D}_A^{Augm} = \overline{D}_A/h(\nu) \tag{5.75}$$

where $1/h(\nu)$ is the enhancement factor for diffusivity, given by:

$$\begin{aligned} h(\nu) &= \frac{3}{\nu}\left(\frac{1}{\tanh \nu} - \frac{1}{\nu}\right) \approx \left(1+\frac{\nu}{3}\right)^{-1} \\ \nu &= \frac{\xi u d_p}{6\grave{D}_A} \quad \text{intraparticle Péclet number} \end{aligned} \tag{5.76}$$

The term ξu represents the intraparticular convective velocity, which is assumed to be proportional to the extragranular fluid velocity via Darcy's law (Afeyan *et al.*, 1990). The proportionality factor ξ is an adjustable parameter of the model. For the HPLC separation of proteins on large pore media, ξ values up to 0.01 have been reported, leading to ν values up to 10. Parameters ξ and ν are obviously dramatically lower for conventional chromatographic media.

At low fluid velocities, ξu is close to zero, therefore ν is close to zero and $h(\nu) = 1$. There is no transport enhancement due to internal convection and $\overline{D}_A^{Augm} = \overline{D}_A$. If the fluid velocity u increases, then ν increases and $h(\nu)$ asymptotically reaches $3/\nu$. Under this asymptotic condition, one has $\overline{D}_A^{Augm} = \overline{D}_A \nu/3 = \frac{\xi u d_p}{18} \frac{\tau}{g_A}$ so that the "augmented" lumped diffusion coefficient no longer depends on the true diffusion coefficient.

The modeling of the particle's saturation/desaturation can be performed with the LDF approximation exactly as in the case of pure internal diffusion, but with a modified characteristic time accounting for internal convection/diffusion (Leitão and Rodrigues, 1995):

$$h(\nu)\, t_A^i \frac{d\overline{C}_A}{dt} = \bar{c}_A^* - \overline{C}_A \tag{5.77}$$

(Leitão and Rodrigues, 1999) extended this concept to the biporous model and showed that Eq. (5.68) can be used provided t_A^M is calculated using $t_A^M = \frac{d_p^2}{60\bar{D}_A} h(\nu)$.

The authors then used the PD model for simulating column hydrodynamic dispersion and showed that breakthrough curves obtained with the two-LDF approximation are in good agreement with the complete model involving Fick's law. The CPU time is a few orders of magnitude lower for the two-LDF model – which again proves the value of the LDF approach.

5.3.8 Special situation: adsorption kinetics

We have assumed so far that the adsorption step of the solutes on the adsorption sites was infinitely fast with respect to the diffusion process. This assumption is acceptable in the vast majority of situations, although some exceptions can probably be found (Liberti and Passino, 1985). Recall that experience showing that chromatography kinetics is in general affected by particle size gives evidence that adsorption kinetics is normally fast compared to diffusion. Conversely, the absence of influence of particle size on kinetics does not mean that adsorption kinetics is limiting, as explained in Section 5.3.6.

We will use the $\overline{\overline{\text{Fick}}}$ model and assume that the adsorbed solutes are immobile. The solute transport in the intragranular pores is thus modeled using Fick's law, Eq. (5.31), adsorbed-phase diffusivity is neglected and adsorption kinetics is modeled using a chemical reaction kinetics formalism.[4]

In the case of a linear adsorption isotherm, it is expedient to model the adsorption kinetics by:

$$\frac{\partial \bar{\bar{c}}_A}{\partial t} = k_{ads}\left[\overline{\overline{K}}_A \grave{c}_A - \bar{\bar{c}}_A\right] \tag{5.78}$$

where k_{ads} is the adsorption rate constant (s^{-1}), assumed to be independent of particle size. When the adsorption rate constant reaches infinity, equilibrium is reached. We define $t^{ads} = 1/k_{ads}$, the characteristic time for adsorption.

More sophisticated expressions for Eq. (5.78), and possible extension to non-linear situations, are discussed in (Ma *et al.*, 1996). We will limit ourselves to the simple assumption of Eq. (5.78), our goal being essentially limited to helping the reader to recognize adsorption kinetics limitations, should they occur.

Under the assumption that adsorption kinetics can be modeled with Eq. (5.78), the saturation process of the particle can be modeled with a first-order dynamic system given by:

$$t_A^{i/ads} \frac{d\overline{C}_A}{dt} = \bar{c}_A^* - \overline{C}_A \tag{5.79}$$

[4] In case surface diffusion is taken into account together with adsorption kinetics, mass balance and associated boundary conditions should be written differently, as shown by (Ma *et al.*, 1996). This situation is not covered in this book.

$$t_A^{i/ads} = \frac{\overline{K}_A - \varepsilon_i}{\overline{K}_A} t_A^{ads} + t_A^i \tag{5.80}$$

A particle in which both diffusion and adsorption kinetics must be taken into account can thus be modeled using the LDF approximation, the global characteristic time for diffusion being the sum of the characteristic time for diffusion and the characteristic time for adsorption. This result, simply obtained by our now classical method based on transfer functions, is derived in Exercise 5.4.

EXERCISE 5.4 *Using the theoretical framework used to introduce the LDF approximation in Section 5.2.2, demonstrate Eq. (5.80).*

5.3.9 Special situation: size exclusion

The development presented in Sections 5.2 and 5.3 assumed the ability of the solutes to freely enter the intragranular pores of the particles. The aim of this section is to investigate the consequences of relaxing this assumption on the saturation kinetics of the particles.

In Chapters 1 and 4, the ability of a chromatographic medium to exclude solutes for whatever reason was expressed in terms of a partition law relating the average solute concentration in the intragranular pores to the solute concentration in the extragranular fluid. Under the linearity assumption, this partition law is written $\grave{C}_A = \grave{K}_A C_A$. The partition constant $\grave{K}_A$ was interpreted in Section 4.6 as the fraction of intragranular pores large enough to accommodate the solute.

If the pore size distribution is assumed to be homogeneous over the entire particle, then the fraction of pores large enough to allow solutes to enter the pores is the same at the periphery of the particle and internally. Consequently, the periphery of the particle acts as an entry barrier: the solutes able to cross this barrier are then able to diffuse freely in the particle.

Under the assumption of homogeneous pore size distribution discussed above, interfacial concentrations $\grave{c}_A^*$ (particle side) and C_A^* (extragranular fluid side) are related by the partition law $\grave{c}_A^* = \grave{f}_A(C_A^*)$ so that, under the linearity assumption:

$$\grave{c}_A^* = \grave{K}_A C_A^* \tag{5.81}$$

A schematic representation of the concentration profiles is given in Figure 5.13.

According to this representation, the possible selectivity between molecules of different sizes takes place at the periphery of the particles. After having crossed the peripheral barrier, the solute diffuses inside the particle and is possibly allowed to adsorb on the pore walls, the adsorbed solutes themselves diffusing should an adsorbed-phase diffusivity mechanism take place. Assuming a linear adsorption interaction, this intragranular diffusion process is described by Eq. (5.40) and can be characterized by a lumped diffusion coefficient $\overline{D}_A$.

The difference with the situation of non-exclusion is that the interfacial concentration $\bar{c}_A^*$ is now given by $\bar{c}_A^* = \overline{K}_A \grave{c}_A^* = \overline{K}_A \grave{K}_A C_A^*$ instead of $\bar{c}_A^* = \overline{K}_A C_A^*$.

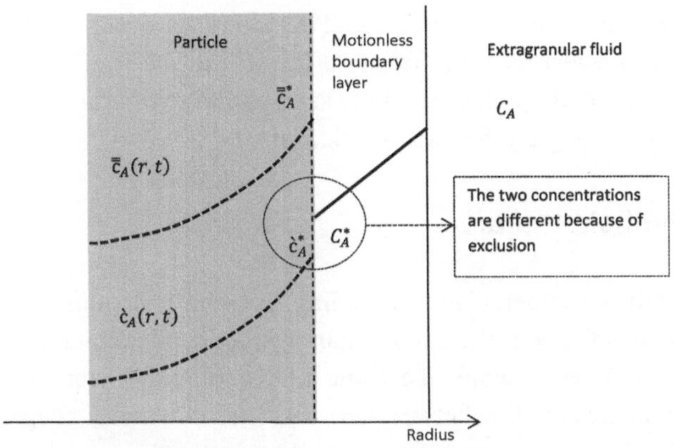

Figure 5.13 Representation of the concentration profiles inside and outside the particle when the porous model is used for describing intragranular transport in connection with solute exclusion.

Under the assumption of a homogeneous pore size distribution, the dynamics of saturation in the presence of size exclusion is thus exactly the dynamics of saturation in the absence of size exclusion, except that the lumped averaged solid concentration reaches a plateau $\overline{K}_A \grave{K}_A C_A^*$ instead of a plateau $\overline{K}_A C_A^*$. When size-exclusion chromatography is used, it is often the case, or at least desired, that the solutes be non-interacting with the solid skeleton. Under this condition, one has $\overline{K}_A = \varepsilon_i$ so that the lumped solid-phase concentration reaches a plateau given by $\overline{C}_A = \varepsilon_i \grave{K}_A C_A$.

As a consequence of the above, the expressions given in Section 5.3.2 still hold, and the LDF approximation in Section 5.2.2 can still be used, provided that the mass balance equation (5.19) is written with the appropriate interfacial boundary condition.

The general methodology involving the transfer function $\hat{\mathscr{L}}(s) = \delta\hat{\bar{n}}_A / \delta\hat{n}_A$ for modeling chromatography columns presented in Chapter 2 and demonstrated, for instance, in Sections 5.2.2, 5.3.6 and 5.3.8 applies with:

$$\hat{\mathscr{L}}_{exc}^{LDF}(s) = \frac{1-\varepsilon_e}{\varepsilon_e} \overline{K}_A \grave{K}_A \hat{\overline{H}}^{LDF}(s) \tag{5.82}$$

The geometrical description of the particle can be refined to describe more precisely the pore structure. As an illustration, one can modify the biporous model presented in Section 5.3.7 and consider a particle made of:

- macroporous intragranular pores associated with a porosity ε_i^M in which solutes can diffuse freely
- microporous pores representing a fraction ε_i^{μ} of the particle volume different from macroporous pores and in which solutes are partly prevented from entering.

The framework presented in Section 5.3.6 still applies provided that a partition law is introduced to relate the solutes in the microparticles and in the macropores.

In contrast to the standard porous model, in which exclusion takes place at the periphery of the particle, the biporous model allows exclusion to take place inside the particles, at the microporosity level. After a mathematical development of no difficulty, one can express the overall characteristic time for mass transfer as the sum of a contribution from the macropores (depending on the particle size) and a contribution associated with the microparticles (depending on the microparticle size).

5.4 Mass transfer: multiple solutes

Assuming constant diffusion coefficients, modeling a system containing n species interacting linearly with the adsorbent is no more than using n similar equations. Assuming constant diffusion coefficients again, the same holds for non-linear systems in the absence of thermodynamic coupling between species, such as when n independent Langmuir adsorption isotherms are involved. These situations are rare, so non-linear multiple-solute adsorption is typically associated with non-linear coupling between species.

In case of diffusion coefficients varying with composition, non-linear coupling between species impacts mass transfer, irrespective of the nature or even the existence of adsorption: diffusion can become non-linear even in a homogeneous phase (fluid or lumped solid).

By "multiple-solute systems" we thus refer to situations involving a coupling between solutes, and presenting the influence of this coupling is the task of this section. In order to do so, we need to start with the very basics of Fick's law.

5.4.1 General considerations on Fick's law

In Sections 5.1 and 5.2, when we presented single-solute systems, we wrote Fick's law as:

$$J_A = -D_A \nabla C_A \tag{5.83}$$

This equation is generally written so automatically and systematically that one easily overlooks the underlying assumptions and definitions. A diffusion process refers to the motion of one species relative to others, so it involves at least a binary mixture. Especially important when using Fick's law, one needs to keep in mind that the diffusion coefficient D_A in fact represents the diffusion coefficient $D_{A,S}$ of solute A in "some mixture", possibly the solvent. The fact that D_A is the diffusion coefficient of solute A in a given surrounding media, and not an intrinsic property of the solute, may look a little bit semantic, but this will have important consequences when we extend our description to multi-component (multi-solute) systems. The reader will find more information on approaches to diffusion modeling in Appendix G.

Fick's law is often extended to multiple-solute systems by postulating that the diffusional flux of species k is proportional to the concentration gradient of that species:

$$J_k = -D_k \nabla C_k \tag{5.84}$$

D_k is the diffusion coefficient of solute k in the surrounding fluid or medium.[5]

[5] To be precise, we should write D_k in the form $D_{k/surroundingfluid}$: not handy at all, but at least it would remind us of the basic principles.

The system of equations (5.84) is normally written for $n - 1$ species (normally the solutes). The flux of the last species (normally the solvent) can be obtained by equating the sum of the diffusional fluxes to zero.

D_k representing the diffusion coefficient of k in "some mixture", we show in Appendix G that this diffusion coefficient has in general very good reason to vary with composition (which is unfortunate) and that Eq. (5.84) is a significant simplification of reality.

Where multiple-solute systems are concerned, Fick's law is often written in a generalized matrix form:

$$\begin{bmatrix} J_1 \\ \\ J_{n-1} \end{bmatrix} = -\left[D_{k,j}^G\right] \begin{bmatrix} \nabla C_1 \\ \\ \nabla C_{n-1} \end{bmatrix} \tag{5.85}$$

The generalized Fickian diffusion coefficient $D_{k,j}^G$ relates the flux of species k to the concentration gradient of species j. Note that $D_{k,j}^G$ is, *a priori*, different from $D_{k,j} = D_{j,k}$, the binary diffusion coefficient of j in k (identical to the binary diffusion coefficient of k in j) determined for the binary mixture (j,k).

By adding free parameters, the generalized Fick's law as written in Eq. (5.85) certainly helps to better represent experimental data than Fick's law in Eq. (5.84).

However, we show in Appendix G that this generalization does not really help. It does not address fundamental issues of Fick's law with multiple-solute systems. The consequence can be unpredictable variability of the generalized diffusion coefficients, whose physical meaning is unclear.

We further show in Appendix G that the Fickian diffusivities as defined in Eq. (5.85) for a fluid mixture of more than two species are in general dependent on composition. They are constant when solutes are diluted in an excess of solvent, be it a pure species or a mixture, or when all the *binary* diffusion coefficients $D_{k,j}$ (again, not to be confused with $D_{k,j}^G$) are identical.

The situation of dilute species (solutes diluted enough to evolve in a surrounding medium with constant properties) or of similar solutes having similar diffusion coefficients is not uncommon in chromatography. This suggests that assuming constant Fickian diffusivities as defined in Eq. (5.84) in multi-solute systems is an acceptable starting point. However, one must keep in mind that there are exceptions when high concentrations, species different in nature or conformational changes are involved.

In this section, we will assume that the diffusion flux can be represented by Eq. (5.84) with a constant diffusion coefficient. Information on the limits of this assumption and more consistent ways of addressing diffusion are discussed in Section 5.13 and Appendix G.

5.4.2 External transfer

With the assumption that Eq. (5.84) with constant diffusion coefficient holds, generalization of the presentation in Section 5.1 presents no difficulty: one has simply to write

Eq. (5.1), thus Eq. (5.4), for each *solute* taken into consideration. That is what is done in many situations.

Again, the above approximation is true in situations where some species (solutes) are diluted in a solvent in large excess. In all generality, Eq. (5.1) must, *a priori*, be written for each component (including solvent), and the sum of the fluxes must equal zero. This imposes one constraint between the concentrations or between the characteristic times for mass transfer, or the existence of convection in the film. In practice, one solves Eq. (5.1) for the solutes only, and "forgets" the solvent. The solvent concentration is thus implicitly imposed, so that the sum of the fluxes equals zero. In general, the solvent is in large excess so that its concentration can be approximately constant, which at the same time ensures that the sum of the fluxes equals zero. In case of concentrated solutions, the need to use more rigorous approaches based on Section 5.12 should be considered.

Additionally, let us note that Eq. (5.1), thus Eq. (5.4), implicitly assumes that the concentration profiles in the film are linear and at steady state. This assumption is normally reasonable for diffusion of non-charged species in liquid phase, but we will see in Section 5.5 that using the film model for ion exchange requires care. Additionally, the assumption of linear concentration profiles is normally not compatible with reactions occurring in solution.

5.4.3 Internal transfer

Using the lumped model with the assumption that Eq. (5.84) with constant diffusion coefficient holds, generalizing the presentation in Section 5.2 presents no difficulty: one has simply to write Eq. (5.11) for each solute. The remarks in Section 5.4.2 on modeling external transfer hold: we implicitly assume the existence of a solvent in large excess.

The situation is more complex with the porous model, as detailed below.

Assuming that the diffusion process can be modeled with constant diffusion coefficients, the mass-balance equation (5.36) remains valid and can be written for all solutes $i \in [1,n]$:

$$\varepsilon_i \frac{\partial \grave{c}_k(r,t)}{\partial t} + (1-\varepsilon_i)\frac{\partial \bar{\bar{c}}_k(r,t)}{\partial t} = \varepsilon_i \grave{D}_k \nabla^2 \grave{c}_k(r,t) + (1-\varepsilon_i)\bar{\bar{D}}_k \nabla^2 \bar{\bar{c}}_k(r,t) \tag{5.86}$$

Equation (5.86) must be associated with classical boundary conditions, expressing symmetry at the center of the particle and concentration at the periphery.

With the assumption that adsorption equilibrium is reached between the solutes contained in the intragranular pores and those adsorbed on the skeleton, one has, everywhere in the particle:

$$\bar{\bar{c}}_k = \bar{\bar{f}}_k\left(\grave{c}_1, \ldots, \grave{c}_n\right) \quad k \in [1,n] \tag{5.87}$$

We will show that this thermodynamic equilibrium will induce a coupling between the different species.

As with single-solute systems, the mass-balance equation (5.86) can be rearranged to introduce the influence of the derivatives of the adsorption isotherms in the diffusion process. Because of the complexity of the associated expressions, we will limit our presentation to the situation of quasi-linear equilibrium, extending the presentation given for single-solute systems.

We assume that the particles are initially in equilibrium with a solution of concentrations C_k^I so that the internal concentration profiles are initially uniform and given by concentrations $\grave{C}_k^I = C_k^I$ and $\bar{\bar{C}}_k^I = \bar{\bar{f}}_k\left(C_1^I,\ldots,C_1^I\right)$. We are looking for an expression involving lumped solid concentrations, so we will express the local lumped concentrations as a perturbation of the initial uniform concentrations:

$$\bar{c}_k(r,t) = \bar{C}_k^I + \delta\bar{c}_k(r,t) \quad k \in [1,n] \quad \text{or} \quad \bar{c}_k = \bar{C}_k^I + \delta\bar{c}_k \quad \text{for short} \tag{5.88}$$

Assuming that the variations are small enough to allow the use of first-order approximations:

$$\begin{aligned} \grave{c}_k^I &= C_k^I + \sum_j \left.\frac{\partial \grave{C}_k}{\partial \bar{C}_j}\right|_{C^I} \delta\bar{c}_j \\ \bar{\bar{c}}_A &= \bar{\bar{C}}_k^I + \sum_j \left.\frac{\partial \bar{\bar{C}}_k}{\partial \bar{C}_j}\right|_{C^I} \delta\bar{c}_j \end{aligned} \tag{5.89}$$

By injecting the first-order approximations (5.89) into the mass-balance equation (5.86), one obtains a mass balance describing the evolution of the lumped variations:

$$\frac{\partial \delta\bar{C}_k}{\partial t} = \sum_j \bar{D}_{k,j} \nabla^2 \delta\bar{C}_j \tag{5.90}$$

$$\bar{D}_{k,j} = \varepsilon_i \grave{D}_k \left.\frac{\partial \grave{C}_k}{\partial \bar{C}_j}\right|_{C^I} + (1-\varepsilon_i)\bar{\bar{D}}_k \left.\frac{\partial \bar{\bar{C}}_k}{\partial \bar{C}_j}\right|_{C^I} \tag{5.91}$$

Note that $\bar{D}_{k,j}$ is not a binary diffusion coefficient; it is a phenomenological diffusion coefficient lumping molecular diffusion coefficients, possible influence of adsorption etc. into one parameter.

Equations (5.90) and (5.91) show that the lumped concentration variation obeys a multi-solute Fick's law. The variation of concentration of one solute inside the particle is in general influenced by the gradient of the other solutes via the cross-diffusion coefficients $\bar{D}_{k,j}$. This coupling via cross-diffusion coefficients, given by Eq. (5.90) and (5.91), is due to the non-linearity of the adsorption isotherms and not to the mechanism of diffusion, the $\grave{D}_k$ and $\bar{\bar{D}}_k$ being assumed constant.

Using the expression $\bar{c}_k = \varepsilon_i \grave{c}_k + (1-\varepsilon_i)\,\bar{\bar{c}}_k$, one can then express the lumped diffusion coefficients as functions of the derivative $\left.\frac{\partial \grave{C}_k}{\partial \bar{C}_j}\right|_{C^I}$, and one obtains:

$$\bar{D}_{k,j} = \varepsilon_i \left(\grave{D}_k - \bar{\bar{D}}_k\right) \left.\frac{\partial \grave{C}_k}{\partial \bar{C}_j}\right|_{C^I} \quad \text{if } j \neq k \tag{5.92}$$

$$\bar{D}_{k,k} = \varepsilon_i \grave{D}_k \left.\frac{\partial \grave{C}_k}{\partial \bar{C}_k}\right|_{C^I} + \left(1 - \varepsilon_i \left.\frac{\partial \grave{C}_k}{\partial \bar{C}_k}\right|_{C^I}\right) \bar{\bar{D}}_k \tag{5.93}$$

Estimation of the derivatives in Eq. (5.92) and (5.93) requires access to the inverse of the adsorption isotherm law, or numerical tools.

One can see from Eq. (5.92) that the lumped cross-diffusion coefficients equal zero in the absence of thermodynamic coupling between the species $\left(\left.\frac{\partial \grave{C}_k}{\partial \overline{C}_j}\right|_{C^I} = 0, j \neq k \right)$ or in case pore and adsorbed-phase diffusivity coefficients are identical.[6] In particular, in the specific situation of linear adsorption isotherms, one has $\left.\frac{d\grave{C}_k}{d\overline{C}_i}\right|_{C^I=0} = 0$ if $j \neq k$, so Eqs. (5.92) give:

$$\overline{D}_{k,j} = 0 \quad \text{if} \quad j \neq k$$

Additionally, if one considers small concentration so that $C^I \approx 0$, one has $\left.\frac{d\grave{C}_k}{d\overline{C}_k}\right|_{C^I} = \frac{1}{\overline{K}_A}$, so Eq. (5.93) gives:

$$\overline{D}_{k,k} = \varepsilon_i \grave{D}_k \frac{1}{\overline{K}_A} + \left(1 - \varepsilon_i \frac{1}{\overline{K}_A}\right) \overline{\overline{D}}_k = \frac{\varepsilon_i \grave{D}_k + \left(\overline{K}_A - \varepsilon_i\right) \overline{\overline{D}}_k}{\overline{K}_A}$$

which is Eq. (5.41).

The expressions (5.90), (5.92) and (5.93) show the complex dependence of the lumped diffusion coefficients on concentration. Not only do the different solutes impact the adsorption of solute k, but they also impact its ability to diffuse in the particle.

Relaxing the quasi-linearity assumption and addressing non-linear systems in general poses a difficulty already mentioned for the single-solute systems: the mass balance is in general not compatible with a Fick's law type of expression, even with concentration-dependent diffusion coefficients. The expressions obtained are complex and give only qualitative information. Should one decide to model non-linear multiple-solute systems diffusion with lumped diffusion coefficients, one should expect:

- the saturation process of species k to be influenced by all other species via the lumped auto-diffusion coefficients $\overline{D}_{k,k}$ and lumped cross-diffusion coefficients $\overline{D}_{k,j}$
- the lumped auto-diffusion coefficients $\overline{D}_{k,k}$ to increase with concentration C_k for favorable adsorption isotherms, and the opposite for unfavorable adsorption isotherms
- the lumped cross-diffusion coefficients $\overline{D}_{k,j}$ to exhibit complex variations with concentration associated with the local derivatives of the adsorption isotherms
- cross-diffusion between species to occur due to:
    - the coupling induced by the adsorption isotherms and differences between pore and surface diffusions
    - the existence of molecular cross-diffusion coefficients.

6 $\left.\frac{\partial \grave{C}_k}{\partial \overline{C}_j}\right|_{C^I} = 0, j \neq k$, does not imply $\left.\frac{\partial \overline{C}_j}{\partial \grave{C}_k}\right|_{C^I} = \infty, j \neq k$. One has $\left.\frac{\partial \grave{C}_k}{\partial \overline{C}_j}\right|_{C^I} = \left.\frac{\partial \overline{C}_j}{\partial \grave{C}_k}\right|_{C^I} = 0, j \neq k$, which means that species k does not influence the adsorption of species j. In mathematical terms, the Jacobian is diagonal, its inverse as well.

The above items assume that diffusion coefficients $\grave{D}_k$ and $\overline{\overline{\overline{D}}}_k$ are constant with composition; unfortunately, this is not guaranteed. The dependence of $\grave{D}_k$ and $\overline{\overline{\overline{D}}}_k$ on concentration can obviously be an additional source of non-linearity. Going beyond these qualitative considerations requires numerically solving the set of equations (5.86) and (5.87).

It is worth mentioning that, as with column modeling, the average concentrations of least retained species can reach a maximum during the saturation process of a particle; see for instance (Martin *et al.*, 2005).

In order to give a flavor of the possible influence of non-linear diffusion on chromatography, we briefly refer to the work of (Ma *et al.*, 1996), who included the non-linear $\overline{\overline{\overline{\text{Fick}}}}$ model taking into account pore diffusion and adsorbed-phase diffusivity in a column model and solved the associated equations numerically. They provided some simulations and comparison with experimental data for a ternary ion-exchange system on zeolites and for a binary-solute adsorption system on activated carbon. While the mean positioning of breakthrough curves is imposed by thermodynamics only, kinetics can significantly affect breakthrough curve shapes. As an illustration, when the less retained species diffuses in the particle significantly faster than the more retained species, front dispersion can be such that the initial fraction of the breakthrough curve associated with the more retained species can appear before the breakthrough curve associated with the less retained species. The authors described this phenomenon as "elution order change" or "kinetic separation".[7] This is certainly a bit of an extreme case, but it illustrates how non-linear diffusion can impact chromatograms.

5.4.4 The $\overline{\text{LDF}}$ and $\overline{\overline{\overline{\text{LDF}}}}^{\text{var}}$ approximations

Using the lumped model approach, extending the $\overline{\text{LDF}}$ approximation to multiple-solute systems presents no difficulty when constant diffusion coefficients (or constant characteristic times for internal transfer) are considered: one has simply to write the mass-balance equation (5.26) for each solute, the solvent being considered to be in large excess. The non-linear coupling takes place at the particle–fluid interface only.

The situation is much more complex if one wants to account for variable internal diffusion coefficients due to non-linear interactions occurring inside the particle, and thus to extend the $\overline{\overline{\overline{\text{LDF}}}}^{\text{var}}$ approximation to multiple-solute systems. We propose an approach that we believe makes sense in Appendix H. This approach generalizes the framework for linear systems and the $\overline{\overline{\overline{\text{LDF}}}}^{\text{var}}$ methodology developed successfully for non-linear single-solute systems. However, we lack numerical simulations to broadly compare this approach with the $\overline{\overline{\overline{\text{Fick}}}}$ model.

Finally, as already mentioned in Section 5.3.4, the contributions of (Carta and Cincotti, 1998), (Martin *et al.*, 2005), (Morbidelli *et al.*, 1982) and (Morbidelli *et al.*, 1984) may provide a way to capture some features of the non-linearity of intragranular

[7] "Elution order change" is probably not a truly appropriate terminology because mean elution order is not impacted by mass transfer.

mass transfer while avoiding most of the difficulties associated with the general $\overline{\overline{\text{Fick}}}$ model.

5.5 The case of ion exchange

As with non-ionic species, the kinetics of ion-exchange processes is normally limited by diffusion. While the general modeling principles remain the same, modeling ionic species diffusion requires two important adaptations:

- The first is associated with the electroneutrality constraint. We will see that this constraint will restrict our freedom to select independent characteristic times for diffusion.
- The second is due to the appearance of an electrical potential gradient during ion diffusion. The consequence is that Fick's law must be replaced by the Nernst–Planck equations (Helfferich, 1962):

$$N_k = -D_i\left(\nabla C_k + \frac{F}{RT}C_k z_k \nabla\varphi\right) \quad (\text{eq}/(\text{m}^2\,\text{s})) \tag{5.94}$$

where F is the Faraday constant, R the ideal gas constant and φ the electrical potential.

Using the Nernst–Planck equations for modeling chromatographic columns is a source of significant complexity, so our objective will be to adapt as much as possible the LDF approximations proposed for adsorption. We will illustrate the methodology for cation exchange, knowing that anion exchange can be treated in exactly the same way.

Prior to starting our presentation, recall that, according to the convention presented in Chapter 4, all ionic concentrations are expressed in eq/m^3.

5.5.1 Internal transfer

We first consider the binary exchange between two cations A and B modeled with the lumped model. Solutions are assumed to be ideal and electrolyte sorption is neglected. Under these conditions, it can be shown (Helfferich and Plesset, 1958; Helfferich, 1962) that the Nernst–Planck equations can be replaced by the Fick equations:

$$\frac{\partial \overline{C}_A}{\partial t} = \overline{D}_{AB}^{var}\nabla^2\overline{C}_A \qquad \frac{\partial \overline{C}_B}{\partial t} = \overline{D}_{AB}^{var}\nabla^2\overline{C}_B \tag{5.95}$$

provided that a variable *lumped inter-diffusion coefficient* is used:[8]

$$\overline{D}_{AB}^{var} = \frac{\overline{D}_A\overline{D}_B\left(z_A\overline{C}_A + z_B\overline{C}_B\right)}{z_A\overline{C}_A\overline{D}_A + z_A\overline{C}_A\overline{D}_B} \tag{5.96}$$

Equations (5.95) and (5.96) show that, even in the simple case of the lumped model and ideal solutions, the ionic diffusion process is non-linear if the two exchanging species have different diffusion coefficients.

[8] If Eq. (5.96) is expressed using molar concentrations, valences would appear via their square.

Note: following the discussion of Section 5.4.1, coefficients $\overline{D}_A$ and $\overline{D}_B$ are binary diffusivities in the solution, and, as such, can also vary with concentration. This impact is not taken into account.

From Eq. (5.96), one can see that the lumped inter-diffusion coefficient is dominated by the species at the smallest concentration: $\overline{D}_{AB}^{var} = \overline{D}_A$ if $\overline{C}_A << \overline{C}_B$ and $\overline{D}_{AB}^{var} = \overline{D}_B$ if $\overline{C}_A >> \overline{C}_B$. Assuming that $\overline{D}_A \neq \overline{D}_B$, a practical consequence is that the saturation kinetics depends on the direction of the exchange.

Ion exchange involving by definition at least two ions, the individual lumped diffusion coefficients $\overline{D}_A$ and $\overline{D}_B$ cannot be determined independently and are normally determined by fitting with a series of experimental saturation curves.

A method allowing investigation of the individual lumped diffusion coefficients consists in using isotopes. If a resin in equilibrium with cation A (no B) is put in contact with a solution at the same normality containing an isotope A*, the two isotopes will diffuse until the concentrations are equilibrated. This isotopic equilibration only involves an ion-exchange mechanism between cation A and isotope A*, this ion-exchange mechanism being associated with a selectivity of one. Determination of the saturation curves allows a determination of the lumped isotopic diffusion coefficient $\overline{D}_A = \overline{D}_{Aiso}$, both isotopes being assumed to have the same diffusion coefficient. In the following, we will speak equivalently of lumped self-diffusion coefficients or lumped isotopic diffusion coefficients.

This methodology was already used 60 years ago in the pioneering work of Soldano and coworkers (Boyd and Soldano, 1953; Boyd *et al.*, 1954; Boyd and Soldano, 1954), who investigated different cation and anion lumped self-diffusion coefficients in various resins. Their results show that:

- There is a strong influence of the ion valences on the lumped isotopic diffusion coefficients: based on results obtained with Na^+, Zn^{2+}, Y^{3+} and Th^{4+} ions on the classical Dowex 50 resin, it seems that the lumped self-diffusion coefficients decrease by about one order of magnitude when the valences increase by one unit. This is contrary to observations made in solution. As an illustration, the Na^+ ion diffuses 1.7 times faster than the Ca^{2+} ion in solution but typically 10–15 times faster in cationic gel resins.
- It seems that, the stronger the affinity of the ion for the resin, the higher the ratio $\overline{g}_A$ of the lumped self-diffusion coefficients to the diffusion coefficient in solution. As an illustration, the ratio $\overline{g}_A$ equals about 12 for Na^+, 15 for Rb^+, Cs^+ and K^+ and about 25 for Ag^+. The selectivity coefficients versus potassium $K_{C/K}$ ($K_{C/K} = 1$ for K=C by definition) are respectively 0.8 for Na^+, about 1 for Cs^+ and Rb^+ and about 3 for Ag^+

Isotopic measurement can also be used to assess the influence of composition on the isotopic diffusion coefficient. Not surprisingly, $\overline{D}_A$ is also influenced by composition.

The dependence of lumped self-diffusion coefficients on selectivity, composition and ion valences cannot be explained as an effect of the Nernst–Planck equations as these simply reduce to Fick's law for isotopic experiments. These results suggest that ion diffusion in resins is impacted by the interactions of the ions with the matrix. The situation is thus similar to the situation met in adsorption, and the porous model will help us understand the impact of experimental conditions on lumped diffusion coefficients.

In Section 4.5.4, we used the theory of the electrical double layer (EDL) associated with complexation on the fixed ionic sites of the resin to give a description of the positioning of the ions in the particles. They can be located either in the intragranular pores in an electroneutral zone, in a zone relatively close to the pore walls in which electroneutrality is not satisfied, or associated with the fixed ionic sites of the resin. The ions located in the first two zones can certainly be considered to be mobile whereas the ions entering into a complexation mechanism with the fixed ionic sites are likely to be immobile (or at least less mobile).

Connecting this particle description with the isotopic diffusion which is assumed to follow Fick's law:

$$\frac{\partial \bar{c}_A}{\partial t} = \varepsilon_i \grave{D}_A \nabla^2 \grave{c}_A \tag{5.97}$$

requires recognizing that isotopes are distributed between the pores and the wall according to:

$$\frac{\grave{c}_A}{\bar{c}_A} = \frac{\grave{c}_{Aiso}}{\bar{c}_{Aiso}} = \text{constant during the isotopic exchange} \tag{5.98}$$

Provided that relevant model parameters are known, the ratio $\frac{\grave{c}_A}{\bar{c}_A}$ can be calculated with the theoretical framework given in Section 4.5.4.

From Eqs. (5.97) and (5.98) one obtains:

$$\begin{gathered} \frac{\partial \bar{c}_A}{\partial t} = \bar{D}_A \nabla^2 \bar{c}_A \\ \bar{D}_A = \varepsilon_i \grave{D}_A \frac{\grave{c}_A}{\bar{c}_A} = \text{constant} \end{gathered} \tag{5.99}$$

so that the lumped isotopic diffusion coefficient can be related to the diffusion coefficient in solution and to the fraction of mobile species, which is given by the thermodynamic description of the particles.

(Nicoud and Schweich, 1989) and (Nicoud, 1987) showed that the porous model associated with the EDL theory and Fick's law allows at least a qualitative explanation of the dependence of lumped isotopic diffusion coefficients on ion valences, selectivity coefficients and resin composition. This model has been extended by (Hasnat and Jvekar, 1996) to the more complex ion-exchange situations involving the use of the Nernst–Planck equations instead of the Fick equations.

For practical results, see the work of (Rodriguez *et al.*, 1998) on the determination of H^+, Na^+, K^+ diffusion coefficients and the attempt by (Rendueles de la Vega *et al.*, 1996) to define a corrected diffusivity that allows the simulation of binary ion-exchange processes with Fick's approach. Both works are based on a porous model, assuming the solution in the pores to be neutral.

In terms of the lumped model, we can say that, due to the electrical potential induced by ion diffusion, a binary ion-exchange process is characterized by a lumped inter-diffusion coefficient $\bar{D}_{AB}^{var}$ given by Eq. (5.96). This lumped inter-diffusion coefficient depends on composition and self-diffusion coefficients $\overline{D}_A$, these self-diffusion coefficients being themselves impacted by different factors including thermodynamic interactions.

Rather than solving Fick's law in spherical geometry, we know from Section 5.2 that a simple model for describing mass transfer is the LDF approximation with constant or possibly variable characteristic times for diffusion. The mass balance equations describing the exchange between protons H and a cation A can be written at the particle level:

$$\begin{aligned} t_A^i \frac{d\overline{C}_A}{dt} &= \left[\bar{c}_A^* - \overline{C}_A\right] \\ t_H^i \frac{d\overline{H}}{dt} &= \left[\bar{h}^* - \overline{H}\right] \end{aligned} \tag{5.100}$$

where t_A^i and t_H^i are the characteristic times for diffusion of the cation A and proton H. If one neglects the possible electrolyte sorption, the electroneutrality in the resin requires that:

$$\overline{H} + \overline{C}_A = \bar{h}^* + \bar{c}_A^* = \overline{N}^{iex} \tag{5.101}$$

and consequently:

$$\frac{d\overline{C}_A}{dt} = -\frac{d\overline{H}}{dt} \tag{5.102}$$

From the electroneutrality equation (5.101), one has $\overline{H} - \bar{h}^* = -\left(\overline{C}_A - \bar{c}_A^*\right)$, so Eq. (5.100) and (5.102) can only be satisfied if the characteristic times for diffusion t_A^i and t_H^i are identical. Consequently, in the case of binary ion exchange, only one characteristic time for diffusion can be freely selected. Similar conclusions are derived for multiple-solute ion exchange showing that, if ion exchange involves n cations, only $n-1$ mass transfer times can be chosen independently.

The simplest way to handle this constraint is to write $n-1$ mass balances and to determine the concentration of the last solute from the electroneutrality equation. For instance, if the solution contains n solutes including proton H one can write:

$$t_k^i \frac{d\overline{C}_k}{dt} = \left[\overline{C}_k - \bar{c}_k^*\right] \text{ for all the species } k \text{ except proton H} \tag{5.103}$$

$$\overline{H} + \sum_k \overline{C}_k = \bar{h}^* + \sum_k \bar{c}_k^* = \overline{N}^{iex} \tag{5.104}$$

Equations (5.103) and (5.104) allow calculation of the evolution of the lumped solid concentrations versus time.

The easiest option consists in working with constant characteristic times t_k^i. This approximation is certainly good enough in a variety of situations, but does not allow us, for instance, to represent the influence of the ion-exchange direction on saturation kinetics.

The theoretical framework presented at the beginning of this section gives some information on how to estimate characteristic transfer times, at least in the case of binary systems.

We know from Eq. (5.96) that the characteristic time for diffusion should be connected with a variable diffusion $\bar{D}^{var}_{AB}$. This situation is similar to the one met in Section 5.3.4 where we extended the $\overline{\text{LDF}}$ approximation to the non-linear porous model. Consequently, the $\overline{\overline{\text{LDF}}}^{\text{var}}$ approach can be used for representing binary ion exchange associated with variable inter-diffusion coefficients. A characteristic time $t_A^{i\ var}$ can be based on a diffusion coefficient $\bar{D}^{var}_{AB}$, being the averaged value obtained by applying Eq. (5.96) on the volume-averaged lumped solid concentration and at the interface.

Using a different approach, (Carta *et al.*, 1999; Carta and Lewus, 1999) proposed an LDF approximation of the Nernst–Planck diffusion equations used in the lumped model that is explicit for homovalent and quasi-explicit for heterovalent ion-exchange systems. Agreement between approximate solutions and resolution of Nernst–Planck diffusion equations in spherical geometry is very good. Another approach, proposed by (Melis *et al.*, 1996), also allows the complex Nernst–Planck equations in a spherical geometry to be replaced by a sound first-order approximation. Let us call these approaches $\overline{\overline{\text{LDF}}}^{\text{NP}}$.

5.5.2 External transfer

In this section, as we use the film model, we assume linear concentration profiles, and thus the absence of reactions in solution. Rigorously, the approach should consequently not be used (at least without adaptation) for amino acids, peptides or proteins, for instance. Practice is something else, but awareness is required.

Diffusion of ions in the film surrounding particles should also in principle be described by the Nernst–Planck equation, instead of Fick's law. As with intragranular mass transfer, this requires the introduction of variable diffusion coefficients as described by (Van Brocklin and David, 1972).

A difference between intragranular and extragranular diffusion is that the former is performed at constant normality imposed by the resin saturation capacity (neglecting electrolyte uptake), whereas the normality in the film is not, *a priori*, constant according to the Nernst–Planck equations (Van Brocklin and David, 1972).

A first option for using the LDF approximation is to define a mass transfer time for each species (including H), and to allow the solution normality at the interface to vary (Nicoud and Schweich, 1988). The principle is demonstrated below.

By analogy with the presentation given in Section 5.1, the external transport of ions can be described by:

$$\frac{d\bar{C}_k}{dt} = \frac{1}{t_k^e}\left(C_k - c_k^*\right) \text{ for all the species} \tag{5.105}$$

with the electroneutrality constraints:

$$H + \sum_k C_k = N_0 \qquad H^* + \sum_k c_k^* = N_0^* \tag{5.106}$$

Neglecting electrolyte sorption, the electroneutrality on the resin side requires $\sum_k \frac{d\overline{C}_k}{dt} = 0$ so that, with Eq. (5.105), one should have:

$$\sum_k \frac{C_k - C_k^*}{t_k^e} = 0 \tag{5.107}$$

In association with the equilibrium laws, Eq. (5.107) allows the normality N_0^* at the interface to be calculated.

An alternative consists in writing the mass balance for all species except proton H:

$$t_k^e \frac{d\overline{C}_k}{dt} = \left[C_k - c_k^*\right] \text{ for all the species except proton H} \tag{5.108}$$

and requiring a constant normality in the film:

$$N_0 = N_0^* = H + \sum_j C_j = H^* + \sum_k c_k^* \tag{5.109}$$

As with internal transport, we model the particle saturation in contact with n different cations by writing $n - 1$ mass balances, and determine the concentration of the last species from the electroneutrality equation.

This alternative is a bit simpler to program than the first one and does not lead to significant differences in the situations of practical interest: characteristic times for diffusion can normally be determined to adequately represent breakthrough curves.

5.6 The rate-controlling step

In the previous sections, we showed how to model external and internal diffusion with more or less refined descriptions. These processes are coupled via the assumption of equilibrium at the particle periphery. We also showed how the methodology can be extended to take into account a possible kinetics of adsorption.

The different kinetic processes being influenced differently by operating parameters like particle size, porosity and superficial velocity, it is of interest to know if a specific step is limiting the entire process. It is relatively common for one step to be dominant over the others and thus to be referred to as the rate-controlling (or rate-limiting) step. When one step is controlling (or limiting), improving the other steps does not change the overall saturation/desaturation kinetics. This is certainly important information: if external diffusion is controlling, improving the intragranular porosity of the particles is useless but increasing fluid velocity can help. If the diffusion in microparticles is controlling, decreasing the particle size is not going to help. If internal diffusion controls, options include both intragranular porosity and particle size.

As already mentioned, it is very rare to see the adsorption step playing a visible role; if it did, decreasing the particle size would not improve the kinetics. We will consequently assume in this section that the adsorption step is instantaneous, so that only external and internal diffusion have to be considered.

We will use LDF approximations to analyze the behavior of chromatographic systems when both internal and external mass transfer limitations are taken into account. Our analysis will be limited to single-solute systems.

The mass transfer rates in the fluid and solid phases must be identical because the interface, being two-dimensional, has no capacity. Consequently one must have:

$$\frac{d\overline{C}_A}{dt} = \frac{(C_A - C_A^*)}{t_A^e} = \frac{(\overline{c}_A^* - \overline{C}_A)}{t_A^i} \tag{5.110}$$

The second equality of Eq. (5.110) can be written in terms of an "operating line" relating the interfacial concentrations:

$$\overline{c}_A^* = -\frac{t_A^i}{t_A^e} C_A^* + \overline{C}_A + \frac{t_A^i}{t_A^e} C_A \tag{5.111}$$

Equation (5.111) expresses the fact that, for a constant fluid–solid composition $(C_A, \overline{C}_A)$, the peripheral concentrations are related via a straight line of slope $-\frac{t_A^i}{t_A^e}$ crossing the *y*-axis at $\overline{C}_A + \frac{t_A^i}{t_A^e} C_A$. The interfacial concentrations are also related by the adsorption isotherm, as shown in Figure 5.14.

From any given pair of concentrations $(C_A, \overline{C}_A)$ existing at a given time and position in the column, Eq. (5.111), together with the equilibrium relation, allows a determination of the interfacial composition $(C_A^*, \overline{c}_A^*)$. The driving force can be represented as an arrow from $(C_A, \overline{C}_A)$ to $(C_A^*, \overline{c}_A^*)$. The arrow has projections $\Delta C = C_A - C_A^*$ and $\Delta\overline{C} = \overline{C}_A - \overline{c}_A^*$ on the axes, and its length is a measure of the magnitude of the driving force.

According to the mass-balance equation (5.4) describing external transfer, increasing the flux requires either $\Delta C = C_A - C_A^*$ to increase or t_A^e to decrease. From a practical point of view, one has no way to act on ΔC, so only t_A^e can be tuned (by decreasing particle size or increasing the *Sh* number). Similar reasoning prevails for internal diffusion.

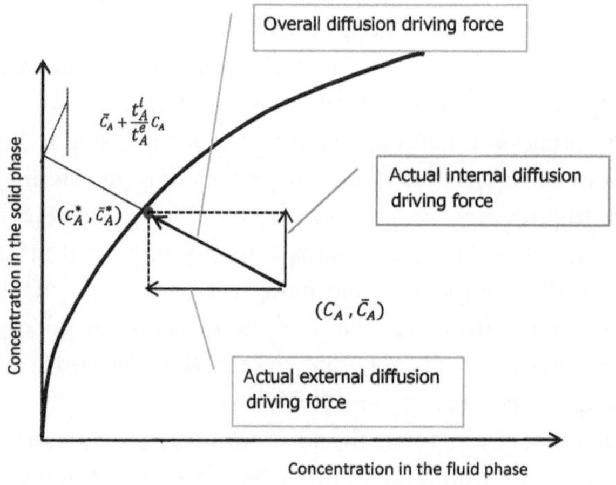

Figure 5.14 Schematic construction of the interfacial concentrations when both internal and external diffusion are considered.

In summary, one can say that:

- When the external diffusion step is controlling (qualitatively when $t_A^e \gg t_A^i$) one has $\overline{C}_A \approx \overline{c}_A^*$, which means that internal diffusion has time enough to flatten the internal profiles. Under these conditions, the driving force arrow is practically horizontal.
- When the internal diffusion step is controlling (qualitatively when $t_A^e \ll t_A^i$) one has $C_A \approx C_A^*$, which means that external diffusion has time enough to flatten the external profiles. Under these conditions, the driving force arrow is practically vertical.

We will now derive a criterion allowing us to identify the rate-limiting step. In case of a linear adsorption isotherm, Eq. (5.111) becomes:

$$C_A^* = \frac{C_A + \frac{t_A^e}{t_A^i}\overline{C}_A}{1 + \frac{t_A^e}{t_A^i}\overline{K}_A} \qquad \overline{c}_A^* = \overline{K}_A C_A^* \tag{5.112}$$

From Eq. (5.112) one concludes that:

- if $T' = \frac{t_A^e}{t_A^i}\overline{K}_A \ll 1$ one has $C_A = C_A^*$, there is no gradient outside the particle, so decreasing the time for external diffusion would have no effect; the process is limited by internal diffusion
- if $T' = \frac{t_A^e}{t_A^i}\overline{K}_A \gg 1$ one has $\overline{C}_A = \overline{c}_A^*$, there is no gradient inside the particle, so decreasing the time for internal diffusion would have no effect; the process is limited by external diffusion.

Let us note that the T' criterion will be confirmed in Section 5.7 for linear systems when we derive general transfer functions taking into account all the different kinetic processes.

According to the T' criterion, for a given ratio $\frac{t_A^e}{t_A^i}$ of the characteristic times for diffusion, a strong lumped Henry's coefficient $\overline{K}_A$ favors limitation by external diffusion. The reality is, however, a bit more complex, as we have shown in Section 5.3.3 that the characteristic time for internal diffusion t_A^i is in general impacted by the lumped Henry's coefficient $\overline{K}_A$. We address this matter in the following lines.

In order to relate this criterion to more operational parameters, we need to introduce the respective expressions for the characteristic times for internal and external diffusion:

$$T' = \frac{t_A^e}{t_A^i}\overline{K}_A = \frac{\dfrac{d_p\delta}{6D_A}}{\dfrac{d_p^2}{60\overline{D}_A}}\overline{K}_A \tag{5.113}$$

so that, introducing the Sh number and $\overline{g}_A = D_A/\overline{D}_A$ one obtains:

$$T' = \frac{10\overline{K}_A}{\overline{g}_A Sh} \tag{5.114}$$

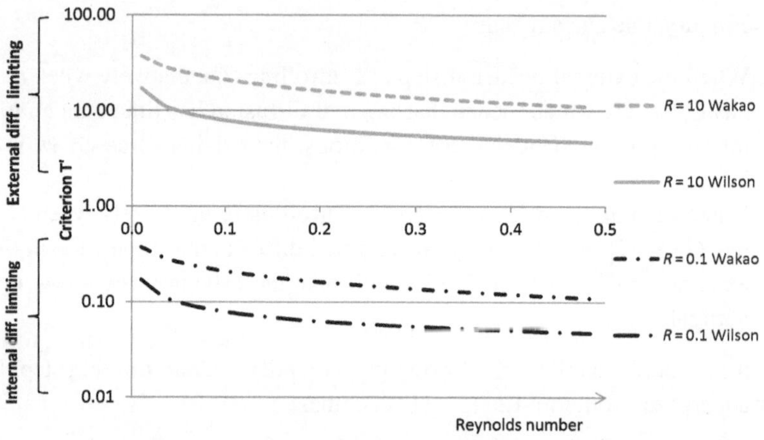

Figure 5.15 Estimate of the T′ criterion for a linear adsorption isotherm and different values of the ratio $R = \overline{K}_A/\overline{g}_A$. The *Sh* numbers are calculated with the correlations of (Wilson and Geankopolis, 1966), (Wakao and Funazkri, 1978) in Section 5.1, with $Sc = 1000$.

The T' criterion thus depends on the ratio $\overline{g}_A$ between molecular and lumped diffusion coefficient, the *Sh* number and the local derivative of the adsorption isotherm. The T' criterion calculated for linear systems is given in Figure 5.15 versus the Reynolds number at different values of the ratio $R = \overline{K}_A/\overline{g}_A$. Figure 5.15 shows that different correlations will lead to different detailed conclusions, but qualitatively the message is clear: the influence of the Reynolds number is relatively secondary and external diffusion is controlling when the ratio R is greater than about 1.

From the presentation in Section 5.3.2, we can make some general comments on the ratio R. Using Eq. (5.43) for $\overline{g}_A$, one obtains immediately:

$$R = \overline{K}_A/\overline{g}_A = \frac{\varepsilon_i + \left(\overline{K}_A - \varepsilon_i\right)\overline{\overline{\theta}}_A}{\tau} \tag{5.115}$$

If the adsorbed-phase diffusivity contribution can be neglected, either because $\overline{\overline{\theta}}_A = 0$ or because the solute is not retained $\left(\overline{K}_A = \varepsilon_i\right)$, then Eq. (5.115) shows that $R = \frac{\varepsilon_i}{\tau}$. With typical intragranular porosities and tortuosities, R is about 0.1 or lower. Consequently, internal diffusion is likely to be the limiting step. This represents the vast majority of situations for adsorption chromatography.

If adsorbed-phase diffusivity cannot be neglected or is dominant, a situation more likely to occur in gels, Eq. (5.115) shows that $R = \overline{K}_A/\overline{g}_A = \frac{\left(\overline{K}_A - \varepsilon_i\right)\overline{\overline{\theta}}_A}{\tau}$. The value of the ratio $\overline{\overline{\theta}}_A$ is, *a priori*, not known precisely but we know that it must be less than 1. For typical preparative chromatography applications, $\overline{K}_A$ is less than about 10 so that, according to the value of $\overline{\overline{\theta}}_A$, R can possibly be on the order of 1. Consequently, situations for which both internal and external diffusion can play a role are possible.

The case of ion exchange is an exception: as explained in Chapter 4, ion exchange at low normality is associated with very significant lumped Henry's coefficient $\overline{K}_A$ values

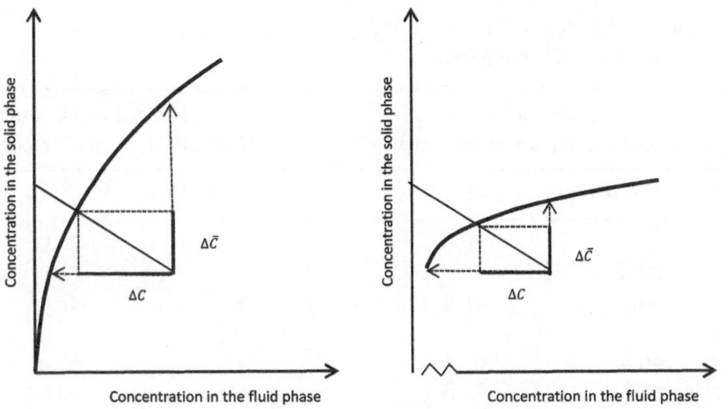

Figure 5.16 Influence of the slope of the adsorption isotherm on the concentration gradient.

so that the ratio $R = \overline{K}_A/\overline{g}_A$ can easily be significantly above 10. It is consequently frequent to have external diffusion controlling in ion-exchange processes.

How can these results be generalized to non-linear systems? The relative position of $(C_A, \overline{C}_A)$ with respect to the adsorption isotherm and the adsorption isotherm shape play important roles, as shown in Figure 5.16: for a given slope of the operating line $\frac{t_A^i}{t_A^e}$, the relative values of $\Delta\overline{C}$ and ΔC can be significantly impacted.

In order to extend the T' criterion to non-linear adsorption isotherms, we will use the methodology proposed by (Nicoud and Schweich, 1989) for ion exchange. Assuming that the particles are in contact with a solution of constant concentration, by differentiation of the two last terms of Eq. (5.110) with respect to time, one obtains:

$$\frac{d\overline{C}_A}{dt} = \left(1 + \frac{t_A^i}{t_A^e}\frac{dc_A^*}{d\overline{C}_A^*}\right)\frac{d\overline{c}_A^*}{dt} \tag{5.116}$$

Equation (5.116) relates the evolution of $\overline{C}_A$ and $\overline{c}_A^*$ versus time. If $\frac{d\overline{C}_A}{dt} = \frac{d\overline{c}_A^*}{dt}$, then $\overline{C}_A$ varies like $\overline{c}_A^*$ over time, and one can conclude that internal diffusion is not limiting the overall process and consequently that external diffusion is limiting.

Consequently, a simple and general criterion T' can be proposed from Eq. (5.116):

$$T' = \frac{t_A^e}{t_A^i}\frac{d\overline{c}_A^*}{dC_A^*}\left(= \frac{t_A^e}{t_A^i}\overline{K}_A \quad \text{for linear systems}\right) \tag{5.117}$$

- If $T' \gg 1$, external mass transfer is controlling.
- If $T' \ll 1$, internal mass transfer is controlling.

The T' criterion shows that the rate-limiting step can, *a priori*, change during a saturation or desaturation process because of the variation of the slope of the adsorption isotherm with concentration. This shows that, in the case of favorable adsorption isotherms, the influence of external mass transfer limitations is expected to be strongest at low loading and the influence of internal limitations is expected to be strongest at

Table 5.5 For Exercise 5.5. Na → Ca: Na^+ replacing Ca^{++} in the resin.
Ca → Na: Ca^{++} replacing Na^+ in the resin.

	Normality: 1 eq/l Normalized saturation curve (%)		Normality: 0.1 eq/l Normalized saturation curve (%)	
Time (s)	Na → Ca	Ca → Na	Na → Ca	Ca → Na
8		20.0		20.0
18	20.0			
50	34.7	46.9	184	46.9
70			20.0	
100	46.9	63.3	245	61.2
150	55.1	73.5	306	69.4
200	61.2	79.6	347	75.5
250	65.3	83.7	388	81.6
300	69.4	87.8	408	85.7
350	73.5	91.8	429	87.8
400	77.6	93.9	449	89.8

significant loadings. In other words, the top of a chromatographic peak can be limited by internal diffusion, and the base by external diffusion.

EXERCISE 5.5 *(Dodds and Tondeur, 1974) studied Na^+/Ca^{++} ion-exchange kinetics on 800 μm resin particles initially in the Na^+ form (resp. Ca^{++}) put in contact with solutions containing a Ca^{++} salt (resp. Na^+) either at 1 eq/l or at 0.1 eq/l. This shallow-bed technique (see Section 5.10.2 for details) allows the particles to be in contact with a solution of constant composition and containing the entering ion. The normalized saturation curves are given in Table 5.5. Additional experiments have shown that saturation curves obtained at 1 and at 2 eq/l are almost identical. Using the lumped model, estimate the internal lumped diffusion coefficients. Comment on the respective influence of internal and external diffusion on the saturation curves.*

5.7 Expression of the column transfer function (linear adsorption isotherms)

We can now model the saturation of chromatographic particles taking simultaneously into account external mass transfer, internal mass transfer and the adsorption step. The next step is to associate this kinetic description with a column model. In order to achieve this task, we will make full use of the methodology presented in Chapter 2, which will again prove its power. In order to simplify the presentation, we will neglect the kinetics associated with the adsorption step.

The methodology is straightforward in the Laplace domain and illustrates the Lego approach mentioned before.

1. We know from Section 5.2.2 how to relate the volume-averaged lumped concentration in the stationary phase $\hat{\bar{C}}_A(s)$ to the solid concentration at the periphery of

the particle $\hat{\bar{c}}_A^*$ by the transfer function $\hat{\bar{H}}(s)$:

$$\hat{\bar{C}}_A(s) = \hat{\bar{H}}(s)\,\hat{\bar{c}}_A^* \qquad (5.118)$$

This transfer function will be denoted $\hat{\bar{H}}^{Fick}, \hat{\bar{H}}^{LDF}$ or $\hat{\bar{H}}^{modelx}$ according to the particle model selected.

2. We can relate the solid concentration $\hat{\bar{c}}_A^*$ at the periphery of the particle to the concentration in the fluid at the periphery of the particle via the equilibrium constant:

$$\hat{\bar{c}}_A^* = \overline{K}_A \hat{\grave{c}}_A^* \qquad (5.119)$$

3. Following the approach suggested in Chapter 1 and in Section 5.3.9, if one suspects an exclusion (size or ion) process behaving linearly, one can connect the concentration in the intragranular pores at the interface to the concentration in the extragranular liquid at the interface by:

$$\hat{\grave{c}}_A^* = \grave{K}_A \hat{C}_A^* \qquad (5.120)$$

In the absence of exclusion, one has $\grave{K}_A = 1$.

4. We should not forget that solutes must cross the motionless boundary layer in order to penetrate the particle. Equation (5.4) is written in the Laplace domain:

$$st_A^e \hat{\bar{C}}_A(s) = \hat{C}_A - \hat{C}_A^* \qquad (5.121)$$

5. With the four preceding relations, one can relate the lumped solid concentration to the concentration in the extragranular fluid:

$$\hat{\bar{C}}_A(s) = \frac{\hat{\bar{H}}(s)\,\overline{K}_A \grave{K}_A}{1 + st_A^e \hat{\bar{H}}(s)\,\overline{K}_A \grave{K}_A} \hat{C}_A \qquad (5.122)$$

6. From Eq. (5.122), we then derive the transfer function relating the quantity of solute contained in the mobile phase (δn_A) and in the lumped stationary phase ($\delta \bar{n}_A$) in a small volume of column where the composition is assumed to be uniform:

$$\hat{\mathscr{L}}(s) = \frac{1-\varepsilon_e}{\varepsilon_e} \frac{\hat{\bar{H}}(s)\,\overline{K}_A \grave{K}_A}{1 + st_A^e \hat{\bar{H}}(s)\,\overline{K}_A \grave{K}_A} = \frac{\delta \hat{\bar{n}}_A}{\delta \hat{n}_A} \qquad (5.123)$$

7. We know from Eq. (2.61) that, if $\hat{E}_e$ is the transfer function of the column for a solute excluded from the intragranular porosity (Z_eRTD), the transfer function of the retained solute is given by:

$$\hat{E}(s) = \hat{E}_e\left[s\left(1 + \hat{\mathscr{L}}(s)\right)\right] \qquad (5.124)$$

8. The problem is solved! The response of the column to any input signal is obtained in a few milliseconds of CPU time via the inversion of eq. (5.124), as explained in Chapter 2 and Appendix A.

As we anticipate some skepticism, let us focus on known specific situations and derive some simple characteristic features of our system.

We will assume that no exclusion occurs, so $\grave{K}_A = 1$. Additionally, we model the internal diffusion process with the LDF approximation so that $\hat{\overline{H}} = \hat{\overline{H}}^{LDF}(s) = \frac{1}{1+st_A^i}$ as given by Eq. (5.23). Under these conditions, Eq. (5.123) gives:

$$\hat{\mathscr{L}}^{LDF}(s) = \frac{1-\varepsilon_e}{\varepsilon_e}\frac{\overline{K}_A}{1+s\left(t_A^i + t_A^e\overline{K}_A\right)} \tag{5.125}$$

Equation (5.125) is identical to Eq. (2.64) with:

$$t_A^m = t_A^i + t_A^e\overline{K}_A \tag{5.126}$$

For those who feel more comfortable in the time domain, Eq. (5.125) and Eq. (5.126) are equivalent to:

$$t_A^m\frac{d\overline{C}_A}{dt} = \overline{K}_A C_A - \overline{C}_A \tag{5.127}$$

which is exactly Eq. (2.44) given in Chapter 2, the structure of the characteristic time for mass transfer t_A^m now being explicit.

Equations (5.125) and (5.127) prove that, for linear systems modeled with LDF approximations, the complete kinetic process can be modeled with a global mass transfer time constant given by Eq. (5.126). The global mass transfer time appears thus as the sum of two kinetic contributions.

The most important contribution to t_m^A is associated with the *rate-limiting step*. Equation (5.126) thus gives an additional proof of the criterion in Section 5.6 for determining the rate-limiting step: external diffusion prevails if $T' = \frac{t_e(A)\overline{K}_A}{t_i(A)} \gg 1$, internal diffusion prevails if $T' = \frac{t_e(A)\overline{K}_A}{t_i(A)} \ll 1$.

EXERCISE 5.6 *Equation* (5.127) *has been derived using the LDF approximation. Explain why using the Fick model would also lead to Eq.* (5.127) *should one be willing to model the entire kinetic process with a first-order system.*

The above illustration can be extended without difficulty to the case of adsorption kinetics. In this situation, following the development presented in Section 5.3.8, one can show that:

$$t_A^m = \frac{\overline{K}_A - \varepsilon_i}{\overline{K}_A}t_A^{ads} + t_A^i + t_A^e\overline{K}_A \tag{5.128}$$

According to the development presented in Section 5.3.7, if one wants to take into account possible internal convection, one can simply use:

$$t_A^m = \frac{\overline{K}_A - \varepsilon_i}{\overline{K}_A}t_A^{ads} + h(\nu)\,t_A^i + t_A^e\overline{K}_A \tag{5.129}$$

Dealing with transfer functions associated with Fick's law instead of LDF approximations is certainly less handy for deriving key features. We learned from Chapter 2 that LDF approximations being determined by identification of the first derivative of the function $\hat{\mathscr{L}}(s)$ guarantees that chromatograms simulated by Fick's law and the LDF approximation have the same mean retention time and variances. This is already a lot, but it does not guarantee identity of the higher moments of the peaks (skewness, for instance).

Assessing the possible differences between $\overline{\text{Fick}}$ and $\overline{\text{LDF}}$ models requires using numerical tools to simulate chromatograms via inversion of the transfer function, Eq. (5.124). Chromatograms simulated with weakly dispersive flow ($J = 2000$ mixing cells) but strong internal diffusion limitation and infinitely fast external transfer and adsorption are given in Figure 5.17. These conditions explore areas in which differences between the $\overline{\text{Fick}}$ and $\overline{\text{LDF}}$ models are expected to be maximized. Comparison with the MC–$\overline{\text{Equil}}$ model involving a single parameter N is also provided in Figure 5.17. Note that the same set of simulation parameters was also used in Figure 2.8 of Chapter 2.

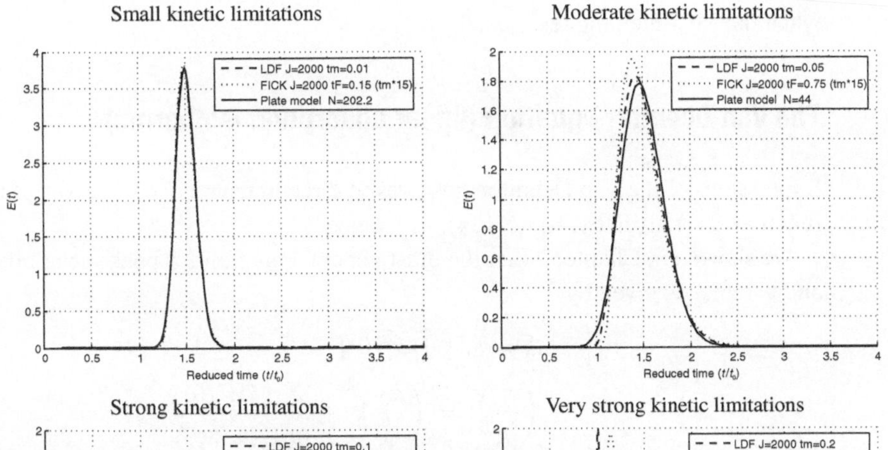

Figure 5.17 Comparison of chromatograms predicted by the MC–$\overline{\text{Fick}}$, MC–$\overline{\text{LDF}}$ and MC–$\overline{\text{Equil}}$ plate model. Simulation performed by numerical inversion of Eq. (5.124), with $t_0 = 1, t_A^e = t_A^{ads} = 0$ $\overline{K}'_A = 0.5$. Fickian times for diffusion t_A^F and t_A^i are related by Eq. (5.20). N is calculated with Eq. (2.49).

The two top figures are associated with kinetic limitations which are small to moderate. It can be seen that the three models provide very similar chromatograms. The qualifier "moderate" is relative: on the top left figure, the number of plates is $N = 202$, thus about ten times less than J, the number of mixing cells, this difference being due to kinetic limitations. When kinetic limitations are very strong (on the last figure the apparent plate number $N = 11$ when $J = 2000$), the models taking into account mass transfer differ significantly from the plate model.

It is interesting to note that, in the case of very strong kinetic limitations, the chromatograms simulated with the MC–$\overline{\text{Fick}}$ and MC–$\overline{\text{LDF}}$ models appear to be thinner than the one using the MC–$\overline{\text{Equil}}$ model. This is pure illusion because the peaks have exactly the same variance, though they do not have the same base width or mid-height width owing to peak asymmetry. The illusion comes from the fact that the maxima of the peaks obtained with the MC–$\overline{\text{Fick}}$ and MC–$\overline{\text{LDF}}$ models are positioned earlier than the mean retention times, and a significant tailing takes place.

The chromatograms simulated using the MC–$\overline{\text{Fick}}$ and MC–$\overline{\text{LDF}}$ models are reasonably close, and we consider the LDF approximation to be good enough for column modeling. One needs, however, to be aware that under extremely strong kinetic limitations the MC–$\overline{\text{LDF}}$ model forecasts a residual peak at zero retention time that is absent when the MC–$\overline{\text{Fick}}$ model is used.

5.8 The Van Deemter equation (linear adsorption isotherms)

Expressions of the Van Deemter curve result directly from the derivatives of the transfer function $\hat{E}(s)$ given by Eq. (5.124).

We showed in Chapter 2 that the most general equation for peak mean retention time and variance is given by:

$$t_R = t_0[1 + \hat{\mathscr{L}}(0)]$$
$$\left(\frac{\sigma}{t_R}\right)^2 = \left(\frac{\sigma_0}{t_0}\right)^2 + \frac{2}{t_R}\frac{-\hat{\mathscr{L}}'(0)}{1+\hat{\mathscr{L}}(0)} \tag{5.130}$$

where t_0 and σ_0 are the retention time and the variance associated with the Z_eRTD.

As mentioned in Chapter 2, Eq. (5.130) proves in general that, for a given hydrodynamic model, two transfer models for solute partition between the fluid and solid phase having the same $\hat{\mathscr{L}}(0)$ and $\hat{\mathscr{L}}'(0)$ will lead to the same variance and thus HETP. Equation (5.130) also shows that, under the assumption of linear equilibrium, the variances associated with hydrodynamic dispersion and mass transfer are additive, irrespective of the hydrodynamic and kinetic models.

Let us consider a column in which external diffusion and internal diffusion must be taken into account but adsorption kinetics can be neglected. Injecting the zero- and first-order moments of the transfer function $\hat{\mathscr{L}}^{LDF}(s)$ in Eq. (5.125) into Eq. (5.130) gives:

$$\left(\frac{\sigma}{t_R}\right)^2 = \left(\frac{\sigma_0}{t_0}\right)^2 + \frac{2\overline{K}'_A}{1+\overline{K}'_A}\frac{t_A^m}{t_R} \tag{5.131}$$

When the model used for characterizing the Z$_e$RTD is the MC model, one has $\left(\frac{\sigma_0}{t_0}\right)^2 = \frac{1}{J}$; when the model used for characterizing the Z$_e$RTD is the PD model, one has $\left(\frac{\sigma_0}{t_0}\right)^2 = \frac{2}{Pe}$. Consequently, from Eq. (5.131) one obtains:

$$\begin{aligned}\left(\frac{\sigma}{t_R}\right)^2 = \frac{1}{N} = \frac{1}{J} + \frac{2\overline{K}'_A}{1+\overline{K}'_A}\frac{t_A^m}{t_R} \\ = \frac{2}{Pe} + \frac{2\overline{K}'_A}{1+\overline{K}'_A}\frac{t_A^m}{t_R}\end{aligned} \tag{5.132}$$

The HETP is directly obtained from Eq. (5.132), and, introducing the interstitial fluid velocity u_e, one has:

$$\begin{aligned}HETP = \frac{L_{col}}{N} = L_{col}\left(\frac{\sigma}{t_R}\right)^2 = \frac{L_{col}}{J} + \frac{2\overline{K}'_A}{\left(1+\overline{K}'_A\right)^2}t_A^m u_e \\ = \frac{2L_{col}}{Pe} + \frac{2\overline{K}'_A}{\left(1+\overline{K}'_A\right)^2}t_A^m u_e\end{aligned} \tag{5.133}$$

Equation (5.133), which is valid for linear systems only, gives a complete generalization and justification of Eqs. (2.49), (2.50) and (2.51) in Chapter 2. Irrespective of the kinetic model used, if one defines $\overline{K}'_A t_A^m$ as the negative of the first derivative of $\hat{\mathscr{L}}(s)$ according to Eq. (2.68), the column will be equivalent to a number of plates given by Eq. (5.133).

Let us consider the situation where external and internal mass transfer are modeled with LDF approximations, and possible internal convection and kinetics of adsorption are taken into account. By inserting the expressions for the different characteristic times into Eq. (5.129), one obtains:

$$t_A^m = \frac{d_p^2}{60\overline{D}_A}h(\nu) + \overline{K}_A\frac{d_p\delta}{6D_A} + \frac{\overline{K}_A - \varepsilon_i}{\overline{K}_A}\frac{1}{k_{ads}} \tag{5.134}$$

$h(\nu)$ usually close to 1

We will explain in Chapter 6 that the axial dispersion coefficient can be expressed as the sum of two contributions. The first one is related to the molecular diffusion coefficient of the solute. The second one is related to random directional changes of the fluid near the particles and is proportional to the fluid velocity and to the particle diameter:

$$D_{ax} = \vartheta_{diff}D_A + \vartheta_{eddy}u_e d_p \tag{5.135}$$

The parameter ϑ_{diff} is typically close to 0.7; the parameter ϑ_{eddy} is normally close to 0.5.

Using Eq. (5.135) to express the Péclet number, and taking into account the expressions for the individual transfer times contributing to t_A^m, Eqs. (5.134) and (5.133) can be written:

$$HETP = \frac{2\vartheta_{diff} D_A}{u_e} + 2\vartheta_{eddy} d_p + \frac{2\overline{K}'_A}{\left(1+\overline{K}'_A\right)^2}\left(\frac{d_p^2}{60\overline{D}_A} h(\nu) + \overline{K}_A \frac{d_p \delta}{6D_A} + \frac{\overline{K}_A - \varepsilon_i}{\overline{K}_A}\frac{1}{k_{ads}}\right) u_e \quad (5.136)$$

$h(\nu)$ usually close to 1

The HETP is now related to defined physico-chemical parameters.

Equation (5.136) is equivalent to expressions obtained by (Kucera, 1965) and presented, for instance, by (Guiochon *et al.*, 2006) and (Schmidt-Traub *et al.*, 2012). Let us stress, however, that we derived this equation with a general methodology based on system dynamics: readers should now be able to derive their own model, build a transfer function like assembling Lego and then obtain the HETP expression after a few derivative calculations ... thanks to symbolic calculus software or to much scratch paper, several pencils and many erasers.

It should be noticed that Eq. (5.136) is often presented in a slightly different form:

$$HETP = \frac{2\vartheta_{diff} D_A}{u_e} + 2\vartheta_{eddy} d_p + 2\left(\frac{\overline{K}'_A}{1+\overline{K}'_A}\right)^2 \frac{\varepsilon_e}{1-\varepsilon_e}\left(\frac{d_p^2}{60\overline{D}_A^{ps}} h(\nu) + \frac{d_p \delta}{6D_A} + \frac{\overline{K}_A - \varepsilon_i}{\overline{K}_A^2}\frac{1}{k_{ads}}\right) u_e \quad (5.137)$$

$h(\nu)$ usually close to 1

Equations (5.136) and (5.137) are equivalent, but in the former case the internal diffusion coefficient to be used is the lumped diffusion coefficient, while in the second case the pseudo-lumped diffusion coefficient $\overline{D}_A^{ps} = \overline{D}_A \overline{K}_A$ (see Eq. (5.62)) must be used. This may be a source of confusion.

Additionally, when considering the influence of the lumped Henry's coefficient $\overline{K}_A$ on HETP, one must keep in mind that, according to the model presented in Section 5.3, $\overline{D}_A$ is, *a priori*, a function of this coefficient according to Eq. (5.41).

For flexibility in representing experimental results, the above HETP expressions are normally used in the form:

$$HETP = \frac{L_{col}}{N} = \frac{A_e}{u_e} + B_e + C_e u_e \quad (5.138)$$

with

$$A_e \approx 2D_A$$
$$B_e \approx d_p \text{ to } 4d_p$$
$$C_e = \frac{2\overline{K}'_A}{\left(1+\overline{K}'_A\right)^2} t_A^m$$

Equation (5.138), known as the Van Deemter equation (Van Deemter *et al.*, 1956), is a useful mathematical representation of the influence of the fluid velocity on the column efficiency.

This curve exhibits a minimum, so that we can define an "optimal" velocity, $u_m = \sqrt{A/C}$, which maximizes the efficiency of the column. Velocities higher than u_m are generally (not to say systematically) used, the drawback of losing some efficiency being more than compensated by the benefits of decreased cycle times.

In liquid chromatography, the contribution of molecular diffusion to axial dispersion is usually negligible, so the A_e term is close to 0 and Eq. (5.138) can be simplified:

$$HETP = \frac{L_{col}}{N} = B_e + C_e u_e \tag{5.139}$$

Equation (5.139) is a *first approximation* which gives a good tool for estimating column efficiencies in liquid chromatography (Ruthven and Ching, 1993). In the case of gas chromatography, molecular diffusion is usually not negligible, and Eq. (5.138) has to be used without simplification.

The B_e coefficient is expected to be independent of the solutes and to qualify the quality of the packing only. In fact things are probably a bit more complex, as explained below. (Miyabe and Guiochon, 2000b) studied mass transfer of p-tert-butylphenol (PTNP) on C18 by the perturbation method and frontal analysis (see Section 5.10) at different concentrations. As expected, they obtained a linear dependence of the HETP on fluid velocity for the various concentrations, but the extrapolated HETP values at zero velocity vary with concentration. This effect is attributed by the authors to the influence of concentration on viscosity, thus diffusion coefficients, thus axial dispersion, as expected from Eq. (5.135).

The Van Deemter curve is often presented in reduced units $\tilde{h}(= HETP/d_p)$ and $\tilde{u}$. The reduced velocity $\tilde{u} = \frac{u_e d_p}{D_A}$, which is known as the particle Péclet number in chemical engineering, is extremely convenient for deriving normalized efficiency plots. Reduced velocities are typically in the range 10–100 in preparative HPLC (particle size 10–40 μm) and 100–1000 in conventional "low-pressure" chromatography (particle size 200–1000 μm).

From (5.136), the HETP can be written in reduced units:

$$\tilde{h} = \frac{HETP}{d_p} = \frac{\alpha}{\tilde{u}} + \beta + \gamma\tilde{u} \tag{5.140}$$

with $\alpha \approx 2$ and $\beta \approx 2$ to 4 for adequately packed columns, and:

$$\gamma = \frac{2\overline{K}'_A}{\left(1+\overline{K}'_A\right)^2}\left(\frac{\bar{g}_A}{60}h(\nu) + \frac{\overline{K}_A}{6Sh} + \frac{1}{k_{ads}}\frac{\overline{K}_A - \varepsilon_i}{\overline{K}_A}\frac{D_m}{d_p^2}\right) \tag{5.141}$$

In the common situation of internal diffusion without internal convection being the limiting step, Eq. (5.141) reduces to:

$$\gamma = \frac{2\overline{K}'_A}{\left(1+\overline{K}'_A\right)^2}\frac{\bar{g}_A}{60} \tag{5.142}$$

and, using Eq. (5.43), one obtains:

$$\gamma = \frac{2\overline{K}'_A}{\left(1+\overline{K}'_A\right)^2} \frac{\tau \overline{K}_A}{\varepsilon_i + \left(\overline{K}_A - \varepsilon_i\right)\overline{\overline{\theta}}_A} \tag{5.143}$$

When the adsorbed solutes are motionless ($\overline{\overline{\theta}}_A = 0$), using $\tau = 3$, $\varepsilon_i = 0.5$ and $\overline{K}_A$ ranging from 0.5 to 10, one obtains γ in the range of 2×10^{-2} to 10^{-2}. Similar calculations show that γ is about 10^{-2} in the case of an adsorbed solute that diffuses freely ($\overline{\overline{\theta}}_A = 1$). In summary, γ is normally in the range of 10^{-2} to 5.10^{-2} for classical preparative chromatography conditions.

Equations (5.140) and (5.142) show that, in reduced units, the Van Deemter equation is expected to be practically independent of particle diameter, molecular diffusion coefficient and, to a large extent, eluent characteristics.

This normalization is extremely interesting since it allows results obtained with very different experimental conditions to be represented on a single curve. In earlier work (Nicoud and Perrut, 1993), Van Deemter curves were obtained on similar axial compression columns packed with the same stationary phase (irregular C18 silica 15 to 35 μm from Biorad RSL) for liquid and supercritical CO_2 eluents. As shown in Figure 5.18 even though experimental conditions are extremely different both results can be presented on the same curve. Moreover, this curve can be fitted with $\beta \approx 2$ and $\gamma = 0.45$, in good agreement with the expected orders of magnitude. Considering that diffusion coefficients are approximately 50 times greater in supercritical fluids than in liquids, the superficial velocity corresponding to the same reduced velocity, thus to the same reduced plate height, is 50 times greater in supercritical chromatography than in liquid chromatography.

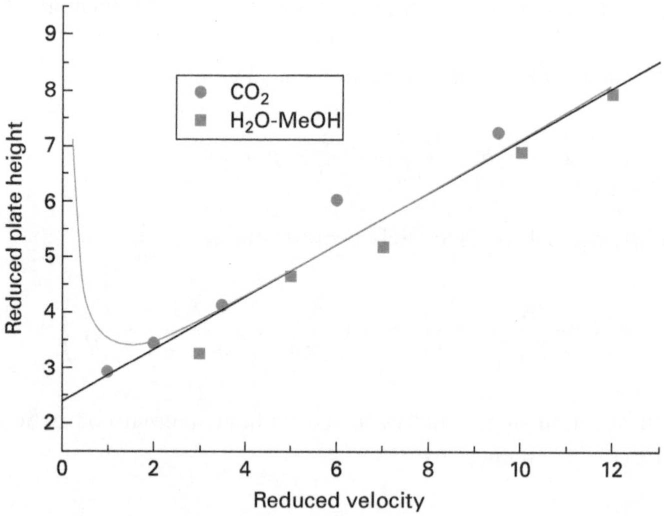

Figure 5.18 Van Deemter curves (in reduced units) obtained on irregular C18 silica with liquid and supercritical fluids. Experimental conditions are given in (Nicoud and Perrut, 1993).

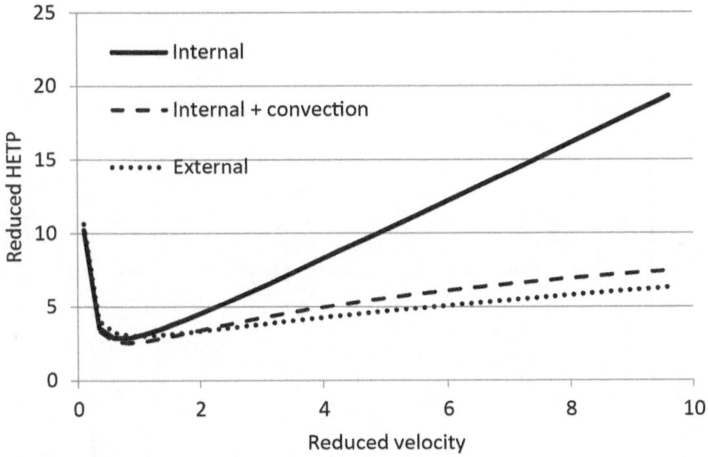

Figure 5.19 Comparison of qualitative Van Deemter curves when internal diffusion, internal diffusion + convection or external diffusion is limiting.

We summarize in Table 5.6 the different Van Deemter equations that can be obtained from Eq. (5.136) according to the rate-limiting step. Van Deemter curves calculated with expressions given in Table 5.6 are presented in Figure 5.19.

It should be noted that the linear dependence of the HETP on velocity as suggested by Eq. (5.139) only holds when internal diffusion is the limiting step. When internal convection or external mass transfer limitations occur, the influence of fluid velocity on the HETP is not as important.

EXERCISE 5.7 *A column (L_{col} = 100 cm, D_{col} = 2.5 cm) is packed with cellulose triacetate CTA (25 μm). The flow rate is set to 40 ml/min. The lumped internal diffusion coefficient has been determined to be 10^{-11} m²/s. Estimate the global diffusion time for a linearly retained solute ($\overline{K}_A = 1$). Estimate the apparent column efficiency (number of plates).*

EXERCISE 5.8 *A column (L_{col} = 25 cm, D_{col} = 30 cm) is packed with silica (15 μm). The silica has relatively wide pores and an intragranular porosity of 0.5. The number of plates required to achieve separation of small molecules having retention factors close to 1 has been determined to be 5000. Knowing that the solutes have a molecular diffusion coefficient of 10^{-5} cm²/s, give a rough estimate of the adequate eluent flow rate.*

EXERCISE 5.9 *A column (L = 100 cm, D_{col} = 2.5 cm) is packed with resin particles (350 μm). The flow rate is set to 40 ml/min. Estimate the global diffusion time for a linearly retained divalent ion ($\overline{K}_A = 50$). Estimate the apparent column efficiency.*

EXERCISE 5.10 *Some experiments performed on a column packed with 20 μm particles of intragranular porosity 0.5 have led to the following reduced HETP ($\tilde{h}$)*

Table 5.6 Some possible expressions for the Van Deemter equation. Expected parameters are: $\alpha \approx 2$, $\beta \approx 2$ *to* 4 and $A_e \approx 2D_A$, $B_e \approx 2$ *to* $4d_p$.

Rate-limiting step	Normalized HETP	Comments	HETP	Comments
Internal diffusion (no convection) FREQUENT	$\frac{\alpha}{\tilde{u}} + \beta + \gamma\tilde{u}$	$\gamma = \frac{2\overline{K}'_A}{\left(1+\overline{K}'_A\right)^2}\frac{\overline{g}_A}{60}$ is independent of $\tilde{u}$	$\frac{A_e}{u_e} + B_e + C_e u_e$	$C_e = \frac{2\overline{K}'_A}{\left(1+\overline{K}'_A\right)^2}\frac{d_p^2}{60\overline{D}_A}$
Internal diffusion and convection SELDOM	$\frac{\alpha}{\tilde{u}} + \beta + \gamma\tilde{u}$	$\gamma = \frac{2\overline{K}'_A}{\left(1+\overline{K}'_A\right)^2}\frac{\overline{g}_A}{60}h(\nu)$; $h(\nu)$ varies with $\tilde{u}$	$\frac{A_e}{u_e} + B_e + C_e u_e$	$C_e = \frac{2\overline{K}'_A}{\left(1+\overline{K}'_A\right)^2}\frac{d_p^2}{60\overline{D}_A}h(\nu)$
External diffusion ($Sh \gg 2$) FREQUENT FOR ION EXCHANGE	$\frac{\alpha}{\tilde{u}} + \beta + \gamma\tilde{u}$	$\gamma = \frac{\overline{K}'_A}{\left(1+\overline{K}'_A\right)^2}\frac{\overline{K}_A}{6Sh}\gamma'$ Assuming $Sh \sim Re^{1/2}$, γ varies like $\tilde{u}^{-1/2}$	$\frac{A_e}{u_e} + B_e + Cu_e$	$C_e = \frac{2\overline{K}'_A}{\left(1+\overline{K}'_A\right)^2}\overline{K}_A\frac{d_p\delta}{6D_A}$
Adsorption VERY SELDOM	$\frac{\alpha}{\tilde{u}} + \beta + \gamma\tilde{u}$	$\gamma = \frac{2\overline{K}'_A}{\left(1+\overline{K}'_A\right)^2}\frac{\overline{k}_A - \varepsilon_i}{\overline{K}_A}\frac{1}{k_{ads}}\frac{D_A}{d_p^2}$ is proportional to d_p^{-2}	$\frac{A_e}{u_e} + B_e + C_e u_e$	$C_e = \frac{2\overline{K}'_A}{\left(1+\overline{K}'_A\right)^2}\frac{\overline{K}_A - \varepsilon_i}{\overline{K}_A}\frac{1}{k_{ads}}$

$$h(\nu) = \frac{3}{\nu}\left(\frac{1}{\tanh\nu} - \frac{1}{\nu}\right) \approx \left(1 + \frac{\nu}{3}\right)^{-1} \text{ with } \nu = \frac{\xi u d_p}{6\dot{D}_A} = \xi\tau/6\tilde{u} \text{ and } \tilde{u} = \frac{u_e d_p}{D_A}$$

for different lumped Henry's coefficients (varied by slightly modifying the eluent composition). Estimate the pore and adsorbed-phase diffusivity coefficients.

$u_e(m/s)$	3×10^{-4}	6×10^{-4}	1.2×10^{-3}	3×10^{-3}
$\overline{K}_A = 1.0$	2.9	3.8	5.7	11.1
$\overline{K}_A = 2.0$	3.3	4.6	7.2	15.0
$\overline{K}_A = 5.0$	3.4	4.9	7.7	16.3

5.9 Influence of particle size distributions

Classical chromatographic media are often not made of monodisperse particles but rather exhibit a particle size distribution, which can be wide for less expensive materials. The theory presented in the previous sections assumes knowledge of a particle diameter characterizing the medium. The ability to define a "mean particle diameter" from a particle size distribution is thus of key interest for mass transfer, but also for pressure-drop and hydrodynamic dispersion estimates, as discussed in Chapter 6.

As non-spherical particles are probably less and less common and can be modeled like spherical particles with characteristic times but different shape factors (Sardin *et al.*, 1991; Villermaux, 1981), we will concentrate on spherical particles.

In practice, particle size distribution in *number* is expressed as a discrete distribution obtained from sieving or particle-size analysis, $n_{(k)}$ particles having a diameter between $d_{(k)}$ and $d_{(k)} + \delta d_{(k)}$.

The particle size distribution is fully characterized if one knows the *fraction in number* $\varphi_{(k)}^{num}$ of particles having a diameter between $d_{(k)}$ and $d_{(k)} + \delta d_{(k)}$, which is given by:

$$\varphi_{(k)}^{num} = \frac{n_{(k)}}{\sum_k n_{(k)}} \tag{5.144}$$

Particle size distributions are often given in terms of mass (or of volume) by suppliers (with the risk of masking the presence of fine particles). The *mass* (or *volume*) *fraction* of particles having a diameter between $d_{(k)}$ and $d_{(k)} + \delta d_{(k)}$ is given by:

$$\varphi_{(k)}^{vol} = \frac{n_{(k)} d_{(k)}^3}{\sum_k n_{(k)} d_{(k)}^3} \tag{5.145}$$

From knowledge of the number distribution $\varphi_{(k)}^{num}$ or the volumic distribution $\varphi_{(k)}^{vol}$, one can define different mean diameters. The mean diameter in number is defined as:

$$d_p^{num} = d_p^{1;0} = \frac{\sum_k n_{(k)} d_{(k)}}{\sum_k n_{(k)}} = \sum_k \varphi_{(k)}^{num} d_{(k)} \tag{5.146}$$

The mean diameter in volume is defined as:

$$d_p^{vol} = d_p^{3;0} = \left(\frac{\sum_k n_{(k)} d_{(k)}^3}{\sum_k n_{(k)}} \right)^{1/3} = \left(\sum_k \varphi_{(k)}^{num} d_{(k)}^3 \right)^{1/3} \tag{5.147}$$

The mean diameter in volume is obviously larger than the mean diameter in number as the influence of large particles is greater in the former definition.

More generally, we can extend definitions (5.146) and (5.147) by defining:

$$d_p^{i;j} = \left(\frac{\sum_k \varphi_{(k)}^{num} d_{(k)}^i}{\sum_k \varphi_{(k)}^{num} d_{(k)}^j} \right)^{1/(i-j)} \tag{5.148}$$

Parameters i and j can be integers or not.

For the flexibility required to derive key results, we will restrict our analysis to linear adsorption isotherms. Under the linearity assumption, the saturation process of a population of particles having a unique diameter $\langle d_p \rangle$ can be modeled by a first-order dynamic system (LDF approximation), which is written in the Laplace domain as:

$$\hat{\overline{C}}_A = \frac{\overline{K}_A \hat{C}_A}{1 + t_A^m s} = \hat{M}(s) \hat{C}_A \tag{5.149}$$

The characteristic time t_A^m for mass transfer depends on the particle diameter $\langle d_p \rangle$ according to the kinetic process taken into account.

Now, let us assume that the saturation processes of the individual particles can be modeled by a first-order dynamic system depending on their size. We will also assume that the physico-chemical characteristics of the particles (e.g. porosity, adsorption isotherm, diffusion coefficients) are independent of particle size.

For particles having a diameter between $d_{(k)}$ and $d_{(k)} + \delta d_{(k)}$, one can thus write:

$$t_{A(k)}^m \frac{d\overline{C}_{A(k)}}{dt} = \left(\overline{K}_A C_A - \overline{C}_{A(k)} \right) \tag{5.150}$$

where $\overline{C}_{A(k)}$ refers to the lumped solid concentration of solute in particles of size (k) and C_A the concentration in the extragranular fluid.

This gives, in the Laplace domain:

$$\hat{\overline{C}}_{A(k)} = \frac{\overline{K}_A \hat{C}_A}{1 + t_{A(k)}^m s} = \hat{m}_k(s) \hat{C}_A \tag{5.151}$$

where $\hat{m}_k(s)$ is the transfer function of the class k of particles connecting the lumped solid concentration to the extragranular fluid concentration.

The Laplace transform of the mean lumped solid concentration of the entire distribution is given by:

$$\hat{\bar{C}}_A = \frac{\text{quantity of A in all the particles}}{\text{volume of all the particles}} = \frac{\sum_k \varphi_{(k)}^{num} d_{(k)}^3 \hat{\bar{C}}_{A(k)}}{\sum_k \varphi_{(k)}^{num} d_{(k)}^3} \tag{5.152}$$

Equations (5.151) and (5.152) can be rewritten:

$$\hat{\bar{C}}_A = \hat{M}_\Sigma(s)\,\hat{C}_A$$
$$\hat{M}_\Sigma(s) = \frac{\sum_k \varphi_{(k)}^{num} d_{(k)}^3 \hat{m}_k(s)}{\sum_k \varphi_{(k)}^{num} d_{(k)}^3} \tag{5.153}$$

The transfer function associated with the particle size distribution is thus a linear combination of the transfer functions of the different classes.

Modeling the system either using an average particle size $\langle d_p \rangle$ or by taking into account the particle size distribution will lead to the same result if the transfer functions are identical, so $\hat{M} = \hat{M}_\Sigma$. Introducing the characteristic times for mass transfer, this would require:

$$\frac{1}{1 + t_A^m s} = \frac{\sum_k \varphi_{(k)}^{num} d_{(k)}^3 \dfrac{1}{1 + t_{A(k)}^m s}}{\sum_k \varphi_{(k)}^{num} d_{(k)}^3} \quad \text{which is impossible in general} \tag{5.154}$$

The right-hand and left-hand sides of the above equation being of different orders in s, this equation is impossible in general (no t_A^m can meet such a challenge!). Consequently, there is no mass transfer time t_A^m that can deliver a transfer function identical to the one associated with a distribution of particle sizes. Modifying the complex Eq. (5.153) by introducing a suitable "average" diameter $\langle d_p \rangle$ to be used in the definition of t_A^m requires accepting approximations.

A possible approximation is to equate the derivatives at $s = 0$ of $\hat{M}_\Sigma(s)$ and $\hat{M}(s)$ so that these two transfer functions have the same characteristic times. This leads to:

$$t_A^m = \frac{\sum_k \varphi_{(k)}^{num} d_{(k)}^3 t_{A(k)}^m}{\sum_k \varphi_{(k)}^{num} d_{(k)}^3} \tag{5.155}$$

Using Eq. (5.155), we get a mass transfer time t_A^m which will best represent the overall saturation curve of the distribution.

Another option is to emphasize the initial slope of the particle saturation, which is the reciprocal of the mass transfer time. Under this option, by equating the initial slopes one obtains:

$$\frac{1}{t_A^m} = \frac{\sum_k \varphi_{(k)}^{num} d_{(k)}^3 \frac{1}{t_{A(k)}^m}}{\sum_k \varphi_{(k)}^{num} d_{(k)}^3} \tag{5.156}$$

Equation (5.155) is certainly the approach that must be used for simulating columns. Equation (5.156) can be adequate for specific kinetic studies, focusing on the initial part of saturation curves.

Let us consider the unlikely case of a predominant adsorption kinetics. In this case, $t_{A(k)}^m$ is independent of the particle size, so that both Eq. (5.155) and (5.156) give $t_A^m = t_{A(k)}^m$ independent of particle size.

More interesting is the case of internal limitation. Under these circumstances, we have shown that the characteristic times for diffusion are proportional to $d_{(k)}^2$ so that, assuming diffusion coefficients to be independent of (k), Eq. (5.155) gives finally:

$$\langle d_p \rangle = d_p^{5;3} = \sqrt{\frac{\sum_k \varphi_{(k)}^{num} d_{(k)}^5}{\sum_k \varphi_{(k)}^{num} d_{(k)}^3}} \tag{5.157}$$

Equation (5.157) gives the average particle diameter representing the best overall particle saturation/desaturation curve when internal diffusion is limiting.

Alternatively, Eq. (5.156) leads to:

$$\langle d_p \rangle = d_p^{3;1} = \sqrt{\frac{\sum_k \varphi_{(k)}^{num} d_{(k)}^3}{\sum_k \varphi_{(k)}^{num} d_{(k)}^1}} \tag{5.158}$$

Equation (5.158) gives the average particle diameter representing the best initial saturation/desaturation curve when internal diffusion is limiting.

Similar calculations can be made when external diffusion or adsorption limitations prevail, and Table 5.7 summarizes the results.

In order to fix these ideas, let us use these expressions to estimate the different average particle sizes to be used for approximating the distribution in Figure 5.20. We assume

Table 5.7 Definition of average particle size according to the rate-limiting step.

Limiting step	$\langle d_p \rangle$ matching initial slope of saturation curves	$\langle d_p \rangle$ matching overall saturation curve
Adsorption kinetics ($t_A^m \sim d_p^0$)	No influence of diameter	No influence of diameter
Internal diffusion ($t_A^m \sim d_p^2$)	$d_p^{3;1}$	$d_p^{5;3}$
External diffusion ($Sh > 2 \rightarrow t_A^m \sim d_p^{3/2}$)	$d_p^{3;\left(\frac{3}{2}\right)}$	$d_p^{\left(\frac{9}{2}\right);3}$

Table 5.8 Calculation of average particle size according to the rate-limiting step; size distribution given in Figure 5.20.

Mean diameter	i	j	$d_p^{i;j}$ (Eq. (5.148))
In number	1	0	44.2
In volume	3	0	49.4
To match initial saturation slope in the case of internal diffusion limitation	3	1	52.1
To match entire saturation process in the case of internal diffusion limitation	5	3	60.0
To match initial saturation slope in the case of external diffusion limitation	3	3/2	53.5
To match entire saturation process in the case of external diffusion limitation	9/2	3	59.2

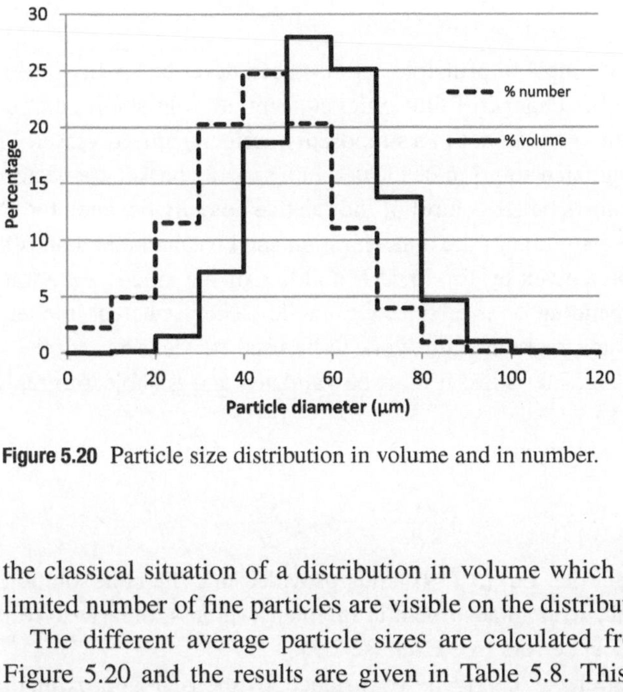

Figure 5.20 Particle size distribution in volume and in number.

the classical situation of a distribution in volume which is visually Gaussian, while a limited number of fine particles are visible on the distribution in number.

The different average particle sizes are calculated from the distribution given in Figure 5.20 and the results are given in Table 5.8. This simple example shows that the mean diameter to be considered varies considerably with the process to be modeled. When the average diameter in number is 44 μm, the best choice for modeling internal diffusion is 60 μm, and this value is significantly greater than the average diameter in volume. Notice that, should the mean diameter in number be used instead of the best mean diameter for internal diffusion, this would impact the internal time for diffusion by about $(60/44)^2$ or about 100%.

The above equations are, *a priori*, valid for linear chromatography only. In non-linear ion exchange, assuming either strong external mass transfer limitations or strong internal limitations at initial saturation stages, (Nicoud and Schweich, 1989) obtained expressions identical to those in Table 5.7 for matching initial slopes.

EXERCISE 5.11 *An adsorbent is composed primarily of 100 μm particles but contains 10% in number of 20 μm particles and 10% in number of 200 μm particles. Estimate the mean particle diameters in number and in mass, and the best-suited diameters for modeling situations of internal and external mass transfer limitations.*

5.10 Experimental methods

Many different methodologies for determining mass transfer coefficients have been described in the literature, and five important methods are discussed in this section. We do not believe that one method can in general be recommended over others, each having advantages and drawbacks.

5.10.1 Batch methods

This method is very simple in principle and has been described many times. It consists in monitoring the time change of the solute concentration in a solution put in contact with a given amount of adsorbent in a supposedly perfectly stirred vessel. The particles can be free in the agitated solution or located in a specific basket connected to a stirrer. The latter option allows better control of the relative velocity between the fluid and the particles. The time evolution of the concentration can be calculated thanks to two equations: a mass balance given by Eq. (5.159) and one of the kinetic equations presented in this chapter, depending on the "selected" model. The "selected" model is no more than an assumed interpretation framework to be used for determining the equilibrium and mass transfer characteristics: it must be validated and is subject to replacement by another one!

The mass balance is given by:

$$\overline{V}\frac{d\overline{C}_A}{dt} = -V_{sol}\frac{dC_A}{dt} \tag{5.159}$$

Let us stress the fact that Eq. (5.159) holds provided that the bulk solution of volume V_{sol} is put in contact with an adsorbent at incipient wetness, otherwise results may be impacted by significant errors (see Exercise 2.7).

The kinetic equation can be given, for instance, by the $\overline{\text{LDF}}$ approximation for internal mass transfer:

$$t_A^i\frac{d\overline{C}_A}{dt} = \left(\bar{c}_A^* - \overline{C}_A\right) \tag{5.160}$$

The interfacial concentration $\bar{c}_A^*$ is assumed to be in equilibrium with the fluid concentration C_A if external mass transfer limitations can be neglected.

The system of Eq. (5.159) and Eq. (5.160) allows simulation of the evolution of the concentration in the extragranular fluid over time. The characteristic time t_A^i is determined by minimizing the differences between the calculated and experimental concentration curves $C_A(t)$. Reaching a minimum does not guarantee that the selected model is the "best" one. It only means that your computer gave you the best result it has found ...

This method is adequate for investigating internal diffusion (again depending on the quality of the mixing) but cannot be recommended when external mass transfer limitations are expected to play a role. Using external mass transfer coefficients determined in a batch system to simulate a column system is probably a guarantee of headaches. A possibility for checking that external mass transfer limitations can be neglected consists in making sure that the stirring speed (in fact, the relative velocity of the particles with respect to the fluid) is not impacting the results (provided that the cell is perfectly mixed, independently of the stirring speed).

This batch method has been used, for instance, by (Yoshida *et al.*, 1994).

There are technically two ways to follow the system evolution: one can follow variations of either the fluid concentrations or the lumped solid-phase concentrations.

Variation of fluid concentrations:

- Advantages: relatively simple; allows the use of a continuous record of fluid concentration when proper calibration is available; can be extended to multi-solute systems
- Drawbacks: cannot give access to meaningful external mass transfer coefficients; fluid concentrations are intrinsically time-variable so studying the influence of concentration on transport is very difficult; not well adapted for determining small time constants because of imprecisions associated with the initial part of the experiment information.

An alternative consists in determining lumped solid concentrations. The particles are quickly withdrawn from the device and the fluid attached to the particles quickly removed, by quick washing or by vacuum sucking.[9] (How fast is "quick"? This again shows that one must assess the results as well as their interpretation.) The particles are then desorbed and the quantity of solute on the adsorbent determined. Repeating the experiment at different times allows one to obtain the saturation curve $\overline{C}_A(t)$. The characteristic time for diffusion is then determined by minimizing the differences between the calculated and experimental concentration $\overline{C}_A(t)$ curves. This alternative allows one to work at almost constant fluid concentration C_A if small volumes of adsorbent are used compared to the fluid volume.

Variation of lumped solid-phase concentrations:

- Advantages: direct measurement of the saturation curve; possibility of working at constant fluid concentration; can be extended to multiple-solute systems
- Drawbacks: cannot give access to meaningful external mass transfer coefficients; possible imprecision due to the desaturation requirement; not well adapted for

[9] This must be done without removing the solutes contained in the intragranular pores if $\overline{C}_A$ is the monitored concentration.

determining small time constants because of imprecisions associated with the initial part of the experiments and with particle removal from the device.

5.10.2 Shallow-bed method (1)

Another method involving saturation/desaturation consists in saturating and then desorbing a shallow-bed column. If the shallow-bed length is small enough (let us say a few particle diameters) and the mean residence time small enough (can be less than 1 s), the concentrations in the column can be considered homogeneous and identical to the inlet concentrations, so that the particles are saturating in contact with a fluid of constant concentration. After a given time, the fluid contained in the extragranular liquid is quickly removed, either by washing or by vacuum sucking.[10] The particles are then desorbed and the quantity of solute determined.

Saturation and desorption performed at different times allow determination of the saturation curve of the particles. Kinetic parameters are then determined by comparison of the model and experimental results. It is wise to check that the characteristic time of the saturation curve is at least one or two orders of magnitude greater than t_0, and that the concentrations exiting the column are effectively identical to the inlet concentrations.

This method has been used, for instance, by (Yoshida *et al.*, 1994).

Shallow-bed method with saturation/desaturation measurements:

- Advantages: possibility of working at constant fluid concentration; can be extended to multiple-solute systems
- Drawbacks: requires repeating similar experiments several times; possible imprecision due to the desaturation requirement; not well adapted for determining small time constants because of the imprecisions associated with the initial part of the experiments and with the constraint of flushing the shallow bed.

5.10.3 Shallow-bed method (2)

The shallow bed presented in Section 5.10.2 can be used in a very different manner. The bed, while remaining shallow, is now long enough to be equivalent to a certain number J of mixing cells, preferably at least about 5. It is fed with a solution containing a known concentration C_A^F of solute. If the residence time of the fluid in the column is small enough compared to the characteristic times for mass transfer, under certain conditions breakthrough curves can show a very specific pattern: first, the solute concentration quickly (at times about t_0) reaches a plateau at a given concentration; then it rises again to the feed concentration. An illustration of this behavior for a very favorable adsorption isotherm is given in Figure 5.21.

The height of the plateau concentration is linked to the ratio of the Z_eRT to the mass transfer time under certain conditions, as detailed below.

[10] Again, this must be done without removing the solutes contained in the intragranular pores if $\overline{C}_A$ is the monitored concentration.

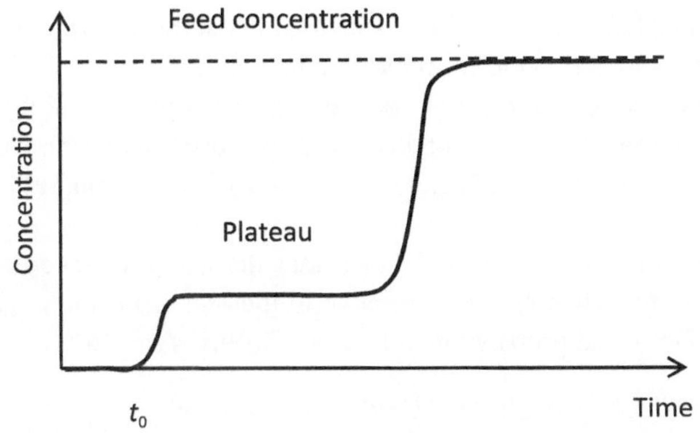

Figure 5.21 Possible breakthrough curve obtained on a shallow bed.

According to the MC–$\overline{\text{LDF}}_{\text{E+I}}$ model, the mass balance of solute A in cell k is written:

$$C_A^{k-1} = C_A^k + \frac{t_0}{J}\frac{dC_A^k}{dt} + \frac{t_0}{J}\frac{1-\varepsilon_e}{\varepsilon_e}\frac{d\overline{C}_A^k}{dt} \qquad k \in [1, J]$$

$$\frac{d\overline{C}_A^k}{dt} = \frac{\overline{C}_A^{k*} - \overline{C}_A^k}{t_A^i} = \frac{C_A^k - C_A^{k*}}{t_A^e} \qquad \text{with } \overline{C}_A^* = f_A\left(C_A^*\right) \tag{5.161}$$

When external diffusion is limiting, one has $C_A^{k*} \ll C_A^k$ as explained in Section 5.6. Under these conditions, $\frac{d\overline{C}_A^k}{dt} = \frac{C_A^k}{t_A^e}$ and Eq. (5.161) can be written:

$$C_A^{k-1} = C_A^k + \frac{t_0}{J}\frac{dC_A^k}{dt} + \frac{t_0}{J}\frac{1-\varepsilon_e}{\varepsilon_e}\frac{C_A^k}{t_A^e}$$

which shows that a plateau ($\frac{dC_A^k}{dt} = 0$) can be reached if:

$$C_A^{k-1} = C_A^k + \frac{1}{J}\frac{t_0}{t_A^e}\frac{1-\varepsilon_e}{\varepsilon_e}C_A^k$$

Finally, reproducing the calculation on the J cells one obtains:

$$\frac{C_A^{Plat}}{C_A^F} = \left(1 + \frac{1-\varepsilon_e}{\varepsilon_e}\frac{1}{J}\frac{t_0}{t_A^e}\right)^{-J} \quad \text{thus} \quad \frac{C_A^{Plat}}{C_A^F} = \exp\left(-\frac{t_0}{t_A^e}\right) \text{ if } J = \infty \tag{5.162}$$

By determining the height of the plateau, Eq. (5.162) allows one to obtain an estimate of the characteristic time for external diffusion. Knowledge of the number of mixing cells J is required, but luckily Eq. (5.162) is relatively independent of J provided that this value is about 5 or more.

One key assumption in deriving Eq. (5.162) is that the system is strongly limited by external diffusion. The reader found in Section 5.6 a means for assessing this assumption. Another implicit assumption is that the saturation time of the bed is long enough compared to t_0 for the plateau to exist. This is favored when the adsorption capacity of the particles is large compared to the concentration in solution.

Consequently, this method can certainly be recommended in external mass transfer limitations and favorable adsorption isotherms. It has been used successfully by (Nicoud and Schweich, 1988) and (Nicoud and Schweich, 1989) for ion exchange at low concentration.

Using perturbation methods, it can be shown that the above methodology can be extended to less favorable adsorption isotherms or to internal diffusion limitations, but it then loses its simplicity and probably its precision (Gorius *et al.*, 1991).

Shallow-bed method with plateau measurement:

- Advantages: easy to implement; can in principle be extended to multiple-solute situations
- Drawbacks: requires external diffusion to be limiting; requires favorable adsorption isotherms and dilute conditions; application to other situations is much more difficult; assessing the real number of mixing cells is difficult as it requires Z_eRTD measurements associated with very small Z_eRT.

5.10.4 The Van Deemter method

A very simple method for investigating kinetics consists in studying the Van Deemter curve. If the variance of peaks resulting from small injections is determined at various fluid velocities, the mass transfer time can be estimated from the slope of the Van Deemter curve (see Eq. (5.138)). It is then possible to investigate the influence of parameters such as particle diameter, retention factor, fluid velocity and temperature on the mass transfer time. This method is precise provided that the contribution of axial dispersion to the peak variance is small with respect to the mass-transfer contribution. This is normally not an issue.

This method has been successfully used, for instance, by (Farnan *et al.*, 2002) to investigate the effect of the lumped Henry's coefficient on internal diffusion coefficients and by (Melter *et al.*, 2008) to study protein intraparticular diffusion.

It must be noted that nothing prevents this method from being used at different steady-state concentrations provided that the perturbation injections are small enough to ensure linearity: instead of using a solute-free solvent, one can use a solvent containing a prescribed solute concentration and perform a small perturbation. Instead of using the initial slope of the isotherm, one then has to use its local slope in the HETP equations. This is detailed in Exercise 5.12.

As shown in Chapter 2, proper determination of the HETP requires determining variances, a task that can be handled without major difficulty with a little bit of experience. Alternative methods for extracting the influence of kinetics from a chromatographic peak may be of value. For instance, (Carta and Bauer, 1990; Carta, 1998) proposed taking advantage of the shift of the peak maximum towards short times when the kinetic contribution increases. On the basis of this concept, they proposed a correlation leading to the mass transfer coefficient from the variation of the position of the peak maximum.

Finally, instead of injecting peaks one can elect to analyze fronts, as demonstrated by (Rearden *et al.*, 1998; Miyabe and Guiochon, 2000a,b), who used this method to investigate the influence of concentration on diffusion coefficients.

EXERCISE 5.12 *Derive the Van Deemter curve when the eluent contains a prescribed and constant concentration C_A^I of a solute. Hint: reread the note devoted to deviation variables in Section 2.1, and make a suitable change of concentration scale.*

Van Deemter method:

- Advantages: easy to implement; can be used for multiple-solute situations; experimental conditions close to those in preparative implementations; allows one in principle to study the influence of concentration via the "locally linear" method. Works irrespective of the rate-limiting step
- Drawbacks: difficulty (not to say impossibility) of being used outside linear conditions.

5.10.5 Curve fitting (the inverse method)

A potentially robust method consists in fitting mass transfer coefficients on peaks or fronts obtained on a chromatographic column, under either linear or non-linear conditions. Common sense tells us that adsorption isotherms must be known precisely, otherwise uncertainties associated with peak positioning and broadening may ruin the efficiency of the method. It is advisable to start fitting adsorption isotherms with an experiment at low velocity so that kinetic limitations are minimized, and then to increase velocity to investigate mass transfer.

This method, used by (Sajonz *et al.*, 1998) for assessing internal diffusion, gave results similar to those obtained with the Van Deemter method.

Interpretation of results is more complex under non-linear conditions. Because of experimental imprecision on peak tailing, we don't believe the method will allow, in a single experiment performed at high concentration, the determination of the concentration influence on lumped diffusion coefficients. It is rather a method allowing one to determine an average coefficient over the concentration range explored by the peak or front. If one wants to investigate concentration dependence more precisely, one must perform different experiments at different concentrations.

Finally, the method can be extended to multiple-solute systems. However, investigating the possible concentration dependence of mass transfer becomes even more difficult in this case, owing to non-linearity and competition between solutes. For a theoretical and practical presentation of the possibility of multi-parameter determination in chromatography by curve fitting, the reader is referred to (Grosfils, 2009).

As already mentioned for adsorption isotherm measurement, we believe that this curve fitting method (the so-called inverse method) makes good sense because it makes use of information (chromatograms) normally available in any development laboratory.

Curve fitting:

- Advantages: easy to implement; can be used for multiple-solute situations; experimental conditions close to those used in real implementation
- Drawbacks: needs a reliable model, neither oversophisticated (too many adjustable parameters) nor oversimplified (poor ability to fit the experimental curve); non-kinetic parameters (equilibrium, hydrodynamics) should be determined independently and kept constant along with curve fitting.

5.11 Influence of mass transfer on chromatograms

A detailed discussion of column modeling for single- and multiple-solute systems is provided in Chapter 7. This section is aimed at describing the qualitative influence of mass transfer limitations on single-solute breakthrough curves. This will allow us to emphasize again the possible limits associated with the common (and very convenient) practice of using a single parameter, the plate number N, to characterize peak broadening irrespective of its origin (hydrodynamics or kinetics).

In order to illustrate the problem, we will use the MC–$\overline{\text{LDF}}_{\text{E+I}}$ model; the corresponding mass balances are given in Eq. (5.161). In the case of linear adsorption isotherms, we have shown in Chapter 2 and in this chapter that this kinetic model may be approximated by the simpler MC–$\overline{\text{Equil}}$ model:

$$C_A^{k-1} = C_A^k + \frac{t_0}{N}\frac{dC_A^k}{dt} + \frac{t_0}{N}\frac{1-\varepsilon_e}{\varepsilon_e}\frac{d\overline{C}_A^k}{dt} \qquad k \in [1,N]$$
$$\overline{C}_A^k = f_A\left(C_A^k\right) \tag{5.163}$$

with:

$$\frac{1}{N} = \frac{1}{J} + \frac{2\overline{K}_A'}{1+\overline{K}_A'}\frac{t_A^m}{t_R} \qquad t_A^m = t_A^i + \overline{K}_A t_A^e \tag{5.164}$$

Because the equivalence given in Eq. (5.164) is based on identical variances only and valid only for *linear systems*, it does not ensure that chromatograms obtained with the MC–$\overline{\text{Equil}}$ and MC–$\overline{\text{LDF}}_{\text{E+I}}$ models are similar. This has already been presented in Chapter 2 (Figure 2.8) and Section 5.7 (Figure 5.17) in the case of linear systems associated with very strong kinetic limitations. We will show that non-linearity can exacerbate the differences between models.

In order to illustrate the matter, we will use a simple example. A single solute is assumed to adsorb according to a Langmuir adsorption isotherm with $\overline{N} = 100$ g/l, $\tilde{K}_A = 0.1$ l/g, so the lumped Henry's coefficient is $\overline{K}_A = 10$. Dilute solute injections are performed on a column at different velocities so that a Van Deemter curve is determined. The results are assumed to be compatible with Eq. (5.164) using: $\overline{K}_A = 10$, $\varepsilon_e = 0.4$, $J = 100$, $t_A^m = 1$ min. The characteristic time for mass transfer t_A^m is assumed to be independent of concentration and fluid velocity.

We now analyze the response of the column to a front of concentration C_A^F for t_0 values of 5, 2, 0.1 and 0.05 min. When t_0 decreases, t_R decreases, so the column is equivalent to a lower number of plates N, as provided by Eq. (5.164).

As we do not know if mass transfer time t_A^m is to be attributed to external or internal diffusion limitations, we will simulate both options, and models MC–$\overline{\text{LDF}}_\text{E}$, MC–$\overline{\text{LDF}}_\text{I}$ and MC–$\overline{\text{Equil}}$ are thus compared. Notice that in case external limitation only is considered, one has $t_A^e = t_A^m / \overline{K}_A = 0.1$ min, and in case internal limitation only is considered, one has $t_A^i = t_A^m = 1$ min.

We start by considering a feed concentration of $C_A^F = 0.01$ g/l, ensuring linear conditions. The results of simulations are given in Figure 5.22. Even when the column is equivalent to $N = 4$ plates only, to be compared to a number of mixing cells of $J = 100$, models MC–$\overline{\text{LDF}}_\text{E}$, MC–$\overline{\text{LDF}}_\text{I}$ and MC–$\overline{\text{Equil}}$ give almost identical breakthrough curves.

At extremely low t_0 only, the $\overline{\text{LDF}}$ models differ from the MC–$\overline{\text{Equil}}$ model. Notice in the bottom right figure the "plateau" obtained with the $\overline{\text{LDF}}$ models when $t_0 =$ 0.05 min: this is the basis for the shallow-bed method described in Section 5.10.3. The kinetic limitations are so extreme for this situation that the equivalent number of

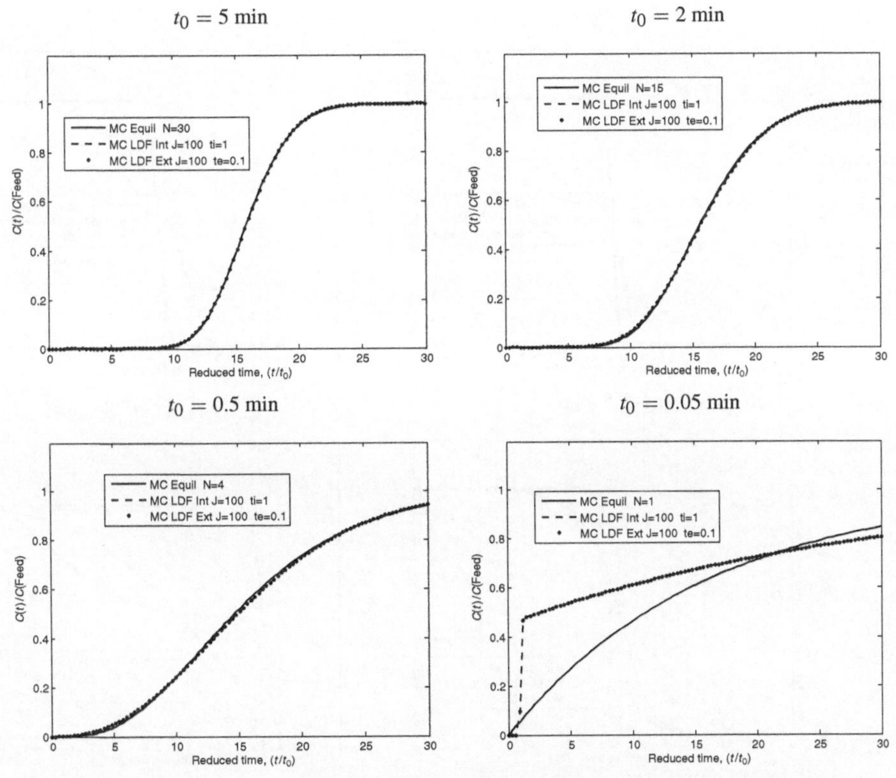

Figure 5.22 Simulation of breakthrough curves under linear conditions. $\overline{N} = 100$ g/l, $\tilde{K}_A = 0.1$ l/g, $t_A^m = 1$ min, $C_A^{inl} = 0.01$ g/l. See details in the text.

plates is below 1 according to Eq. (5.164), but rounded to 1 for the breakthrough curve simulations.

We now conduct the same numerical experiment except that the feed concentration is increased from 0.01 to 100 g/l so that we enter the non-linear region. The number of plates is still calculated with Eq. (5.164). The simulated curves are given in Figure 5.23.

The results are very different: even at $t_0 = 5$ min, differences between the simulated curves are visible, and these differences increase dramatically when t_0 decreases. It is apparent from Figure 5.23 that the MC–$\overline{\text{Equil}}$ model is relatively close to the MC–$\overline{\text{LDF}_\text{E}}$ model and pretty far from the MC–$\overline{\text{LDF}_\text{I}}$ model: this is purely a consequence of the set of parameters chosen for this simulation, and there is no degree of generality associated with this statement.

The message is clear: with non-linear chromatography, approximating the column by a number of plates determined under dilute conditions may not be sufficient.

Finally, note that the plateau announced in Section 5.10.3 when very strong external kinetic limitations prevail is clearly visible for both linear and non-linear simulations (bottom right simulations in Figures 5.22 and 5.23). The plateau determined from

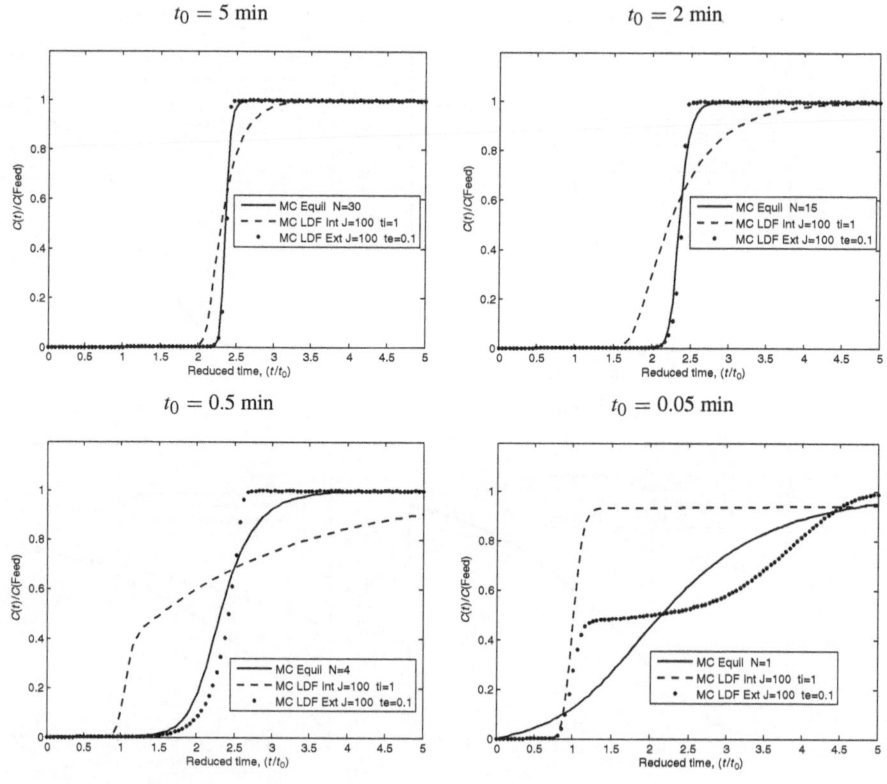

Figure 5.23 Simulation of breakthrough curves under non-linear conditions. $\overline{N} = 100$ g/l, $\tilde{K}_A = 0.1$ l/g, $t_A^m = 1$ min, $C_A^F = 100$ g/l. See details in the text.

Eq. (5.162) is in good agreement with the plateaus obtained with the MC–$\overline{\text{LDF}}_E$ model (bottom right simulations in Figures 5.22 and 5.23):

$$\frac{C_A^{Plateau}}{C_A^{inl}} = \left(1 + \frac{1-\varepsilon_e}{\varepsilon_e}\frac{1}{J}\frac{t_0}{t_A^e}\right)^{-J} = \left(1 + \frac{1-0.4}{0.4}\frac{1}{100}\frac{0.05}{(\frac{1}{10})}\right)^{-100} = 0.47$$

The situations associated with $t_0 = 0.05$ min certainly appear a bit extreme. For some reason, however, it is similar to the operating conditions prevailing for ion exchangers installed in nuclear spent pools for removing radioactive contaminants, presented in Section 10.3.3.

Care must certainly be taken when using the plate model for simulating non-linear conditions associated with strong kinetic limitations. Common sense also tells us that even greater care must be taken for multiple-solute systems: a single plate number cannot represent band spreading for different solutes having different mass transfer characteristics.

5.12 Perspectives: from Fick to Maxwell–Stefan?

The models presented for mass transfer of solutes around and inside chromatographic particles are based on important assumptions, listed at the beginning of this chapter. Recall that a very important assumption is that Fick's law (Nernst–Planck in the case of ions) applies. Introducing the concept of local equilibrium and of adsorbed-phase diffusivity, we had interesting successes and some important experimental observations were explained.

This being said, detailed investigation shows that parameters like porosity, tortuosity, pore diffusion coefficient and adsorbed-phase diffusivity sometimes vary significantly with concentration and other parameters, in an unpredictable manner.

One can argue that, should the concentrations used in Fick's law be replaced by activity coefficients, the model would gain in rigor. This is possibly true, but the price for that is a significant increase in complexity and the introduction of extra parameters.

Addressing multiple-solute systems adds further difficulty. Predicting multiple-solute system behavior, knowing that we are struggling to account for concentration dependence of single-solute lumped diffusion coefficients, appears to be a task of tremendous difficulty. It would be great finding a way, a model, a theoretical framework allowing us to deal with more constant diffusion coefficients.

When Daniel Schweich told me, after a discussion on mass transfer, "you should consider moving from Fick's law to the Maxwell–Stefan approach", I imagined being forced to work on complex, "super-Fickian-like" equations, a bit like moving from matrices to tensors. In fact, I discovered a fascinating and rejuvenating approach, allowing us to connect our chromatographic world to other parts of science.

For the chromatographer, the must-read contributions are certainly the book and review articles written by Wesselingh and Krishna (Wesselingh and Krishna, 2006; Krishna and Wesselingh, 1997) and, in case you become Maxwell–Stefan–dependent, you may even read *Multicomponent Mass Transfer* (Taylor and Krishna, 1993).

In addition, I certainly recommend the article written by (Mota, 2004), which gives an excellent description and perspective on how to use the Maxwell–Stefan theory in adsorption. Our goal here is not to give a detailed presentation, but to share information that should be simple and convincing enough to encourage chromatographers to move, or at least explore, in that direction.

The underlying concepts of the Maxwell–Stefan approach are briefly presented in Appendix G and even more briefly summarized below. The idea consists in writing a force balance, with frictional forces exerted on species i equilibrating with driving forces (deriving from a potential) exerted on the same species. Friction forces exerted by solute j on solute i are assumed to be proportional to the molar fraction of solute j and to the difference in diffusional velocities between the two species via coefficients of friction $\zeta_{i,j}$. The driving force is expressed as the gradient of the chemical potential (and electrical potential as the case may be).

The so-called *Maxwell–Stefan diffusivities* $Đ_{i,j}$ are related to the coefficients of friction by $Đ_{i,j} = \frac{RT}{\zeta_{i,j}}$, and the above-mentioned force balance leads to:

$$C_{tot}\nabla x_i = \sum_{j \neq i} \frac{x_j J_i - x_i J_j}{Đ_{i,j}} \tag{5.165}$$

at constant T,P and in the absence of external force which is the classical Maxwell–Stefan equation.

This does not look very handy at first glance: instead of expressing flux versus gradient, it expresses gradient versus flux. This little inconvenience is largely compensated by important advantages related to the fact that diffusion coefficients are now related to well-defined physical parameters.

Fickian diffusivities can be related to Maxwell–Stefan diffusivities $Đ_{i,j}$ and one can show that these two diffusivities are identical in the case of binary-component systems.[11] In the more general case of multi-component systems, these diffusivities are different and the Maxwell–Stefan approach shows that Fickian diffusivities cannot, in general, be constant with composition, even for ideal solutions. Taking the risk of oversimplifying, moving from the Fick to the Maxwell–Stefan approach offers a chance of trading variable diffusivities for constant diffusivities. At least in a majority of situations, experience shows that Maxwell–Stefan diffusivities are much more constant than Fickian diffusivities.

Let us consider the lumped model for describing the particles, in order to illustrate how the use of the Maxwell–Stefan equations can help in modeling mass transfer in chromatography. The Maxwell–Stefan approach, including chemical potentials in the driving force, gives a simple way to take the effects of non-linearity of the adsorption isotherm into account. We show in Appendix G that $\overline{D}_{1,2} = \overline{Đ}_{1,2}\frac{\overline{C}_1}{C_1}\frac{dC_1}{d\overline{C}_1}$, which relates Fickian lumped diffusivities to Maxwell–Stefan diffusivities in the lumped particle.

This result shows that, under linear equilibrium conditions, the two diffusivities $\overline{D}_{1,2}$ and $\overline{Đ}_{1,2}$ are identical. Assuming $\overline{Đ}_{1,2}$ constant, the above equation forecasts a Fickian lumped diffusion coefficient increasing with concentration in the case of a favorable

[11] This means, in chromatography, one solute and only one solvent!

adsorption isotherm. That is qualitatively in agreement with many experiments discussed in this chapter. When we had to use a complex porous model to explain this variation, using the Maxwell–Stefan equations with the simple lumped model immediately gives, at least qualitatively, the same behavior.

This approach has been used successfully by (Lewus and Carta, 1999) to model the diffusion of lysozyme and cytochrome-C in highly porous polystyrene-coated silica particles. The authors showed that taking into account the coupling between diffusion fluxes (the gradient of i influencing the flux of j) was needed to obtain a reasonable prediction of binary-solute (three-component with the solvent) mass transfer from single-solute experiments.

Nothing prevents us from using the Maxwell–Stefan approach in more sophisticated models like the porous model associated with adsorbed-phase diffusivity. It is quite the opposite, as the Maxwell–Stefan approach allows greater flexibility than Fick's law. As an illustration, one can consider the matrix as an additional component of a system that will contribute to frictional forces.

This type of model has been used by (Krishna, 1990) and (Krishna, 1993) to describe adsorbed-phase diffusivity. The influence of surface coverage on the lumped Fickian diffusion coefficients has been determined, and it has been shown that different literature results can be interpreted with relatively constant Maxwell–Stefan diffusivities. In addition, (Kaczmarski *et al.*, 2002) and (Kaczmarski *et al.*, 2003) used the Maxwell–Stefan approach to model mass transfer of enantiomers using the porous model. They confirmed the role of surface diffusion and the importance of the coupling between the species for mass transfer.

As a second illustration of the flexibility permitted by the Maxwell–Stefan approach, let us consider the existence of an electrical potential gradient in addition to chemical gradients. Differences in ion mobilities create an electrical gradient, to be added to the driving force. Taking into account electroneutrality and no-current conditions, one can eliminate the electrical potential and relate fluxes to concentration gradients.

This approach has the merit of directly explaining various phenomena that cannot be explained with Fick's law (Wesselingh and Krishna, 2006) and that are very important for properly modeling ion exchange:

- The diffusivities in water of H and Cl ions are very different (let us say a factor of 5). When considering a solution of HCl in water, despite their different diffusivities these two ions must travel together in order to ensure electroneutrality and zero current. This is possible due to the electrical potential appearing as soon as these two ions separate.
- If one adds NaCl traces to the HCl solution, it can be shown that, due to the electrical potential caused by the H ion, the Na ion can diffuse against its composition gradient!
- The diffusivity of lysozyme in water is about five times higher than its diffusivity in an electrolyte like KCl. This is explained by the fact that, in the absence of an electrolyte, the 13 chloride ions surrounding the protein to ensure electroneutrality "pull" the protein so that a diffusion coefficient, being a kind of average of the diffusivities of the 14 ions, is measured.

Accounting for these effects becomes relatively straightforward with the Maxwell–Stefan approach. (Krishna and Wesselingh, 1997) showed how the Maxwell–Stefan approach generalizes the Nernst–Planck equations for modeling the transport of ions in solution or in resins. (Silva and Lito, 2007) compared the Maxwell–Stefan and Nernst–Planck approaches in representing saturation curves for removal of mercury from aqueous solution using microporous ion-exchange particles.

Finally, when included in a simple model for describing external diffusion, the Maxwell–Stefan equations simply show that the normality cannot be constant in the film surrounding a particle, and that the sense of normality variation depends on the ion-exchange direction (Wesselingh and Krishna, 2006). This solves in an elegant way the difficulties mentioned in Section 5.5.

What could be a future evolution of modeling that would do better than the $\overline{\text{Fick}}$ and $\overline{\overline{\text{Fick}}}$ models? Probably, not to say certainly, the use of the Maxwell–Stefan approach would greatly help, but solving these equations in spherical geometry is probably not what we want for efficient modeling. A finite-difference approximation (LDF type) of the Maxwell–Stefan equations may offer an excellent direction for modeling kinetics with relatively simple and physically meaningful tools. The winners could be $\overline{\overline{\text{LDF}}}^{\text{MS}}$- and $\overline{\text{LDF}}^{\text{MS}}$-type models.

5.13 Conclusion

In the vast majority of situations, chromatographic processes are limited by the requirement for the solutes to diffuse from the bulk solution to the chromatographic medium (external transfer) and then inside the chromatographic medium itself (internal transfer).

External transport is conveniently modeled with an LDF approximation, each solute being associated with a given characteristic time for external diffusion. The characteristic times for external diffusion are related to the molecular diffusion coefficients, to particle size and to the thickness of a presumably motionless layer of fluid surrounding the particle. The thickness of this motionless layer can be correlated to the Reynolds number. In the case of ion exchange, due to the electroneutrality constraint it is convenient to define a mass transfer time for all ionic species except for one, the mass transfer times being based on inter-diffusion coefficients. The inter-diffusion coefficients are in general concentration-dependent, as accounted for by the Nernst–Planck equations.

It should be noted that chemical reactions and equilibria in solution that are frequent in ion-exchange processes can significantly impact external mass transfer.

A simple way to model internal diffusion is to use the lumped model associated with constant lumped diffusion coefficients. This model can then be approximated by the $\overline{\text{LDF}}$ approach, each solute being associated with a characteristic time for internal diffusion.

For most situations of practical interest, this approximation is sufficient for modeling chromatographic processes. The characteristic times for internal diffusion are related to the lumped diffusion coefficients and the particle diameter. There is a body of evidence that the lumped diffusion coefficients are impacted not only by the molecular diffusion and by particle geometric characteristics but also by Henry's coefficient, and more generally by fluid-phase equilibria. Investigating the dependence of lumped diffusion coefficients on experimental conditions including concentration requires developing a porous model that takes into account the chromatographic medium structure and the thermodynamic equilibria occurring inside the particles. Porous models can be more or less sophisticated, but their common ground is to make a distinction between solutes free to diffuse in the intragranular pore and solutes adsorbed on pore walls. Explaining experimental variations in lumped diffusion coefficients generally requires assuming that the adsorbed solutes are not totally immobile but can diffuse via an adsorbed-phase diffusivity mechanism.

Size exclusion and ion exchange have their own specificities but these different chromatographic processes can be described with the theoretical principles outlined above.

Moving to multi-solute systems presents no difficulty as long as one uses the LDF approximation with constant mass transfer times. Should the variations in lumped diffusion coefficients be such that the above approach is not adequate, modeling the system can then be associated with a significant degree of complexity. Using Fick's law certainly has significant limitations, and descriptions based on Maxwell–Stefan approaches are probably the most promising.

In the case of linear systems, the methodology presented in Chapter 2 based on transfer functions allows the integration of any kinetic model into any hydrodynamic model and then either the simulation of chromatograms or derivation of key characteristics like variance and skewness. The above approach shows that the contribution of external diffusion, internal diffusion and hydrodynamic dispersion to peak variance are additive for linear systems. This allows one to immediately recognize whether one diffusion step is controlling the kinetic process.

Identifying the rate-controlling step in non-linear situations does not pose specific difficulties as long as the LDF approximation with constant characteristic times is used. The rate-controlling step can change with concentration so that the top of the peaks can in principle be limited by one diffusion process and the bottom by another

Finally, recall that breakthrough curves and chromatograms are first positioned by thermodynamic constraints as explained in Chapter 3. Using a plate model (MC–$\overline{\text{Equil}}$) can generally account for additional broadening with a precision compatible with design purposes. Introducing diffusion with constant lumped diffusion coefficients is a refinement (MC$-\overline{\text{LDF}^{\text{E}}}$, MC$-\overline{\text{LDF}^{\text{I}}}$, PD$-\overline{\text{LDF}^{\text{E}}}$, PD$-\overline{\text{LDF}^{\text{I}}}$, MC–$\overline{\text{Fick}}$, PD–$\overline{\text{Fick}}$ and so on), and introducing variations in lumped diffusion coefficients is generally considered an unnecessary refinement, right or wrong. The key is to appreciate the level of precision that is required for a given process design: reaching 85% purity with good selectivity and reaching 99.9% purity with moderate selectivity are very different subjects.

As far as precision is concerned, I am not convinced that the significant complexity of the $\overline{\overline{\text{Fick}}}$ type of approach can be justified for simulation/design purposes by the gains it brings in rigor or precision. These models are helpful for investigation purposes, but are based on a simplified description of particle geometry, the vague concept of adsorbed-phase diffusivity, the internal homogeneity of parameters such as pore size, constant individual diffusion coefficients and the use of Fick's law, while the Maxwell–Stefan equations should be more widely adopted.

The art of modeling requires recognizing the processes most affecting chromatograms. By over-modeling a process like kinetics, by focusing on differences between second- and third-order approximations, one may forget that other effects such as hydrodynamics may dominate at this level of precision. This is the subject of the next chapter.

5.14 Solutions of the exercises

Exercise 5.1

The superficial velocity is $u = Q/\Omega_{col} = 1.36 \times 10^{-3}$ m/s. At 20°C, the viscosity of ethanol is close to 10^{-3} Pa s, and its density is 791 kg/m^3. A typical value for the diffusion coefficient in a liquid phase is $D_A = 10^{-9}$ m^2/s. One can thus estimate:

$$Re = \frac{\rho u d}{\eta} = \frac{791 \times 1.36\ 10^{-3} \times 25\ 10^{-6}}{0.001} = 2.7\ 10^{-2}$$

$$Sc = \frac{\nu}{D_A} = \frac{\eta}{\rho D_A} = \frac{0.001}{791 \times 10^{-9}} = 1264$$

Consequently, using the Wilson–Geankopolis or the Dwivendi–Upadhyay correlation, one obtains:

	Wilson–Geankopolis	Dwivendi–Upadhyay
Sh	8.84 (with $\varepsilon_e = 0.4$)	2.85
$\delta = \frac{d_p}{Sh}$ (µm)	2.83	8.78
$t_A^e = \frac{d_p \delta}{6D}$ (s)	0.012	0.037

We can thus consider that the characteristic time for external diffusion is about 0.02 s.

Exercise 5.2

The estimate can only be very rough. According to the porous model, the lumped diffusion coefficient is related to the molecular diffusion coefficient via the ratio $\bar{g}_A$:

$$\bar{g}_A = \frac{\tau \bar{K}_A}{\varepsilon_i + (\bar{K}_A - \varepsilon_i)\,\bar{\bar{\theta}}_A}$$

We don't take a big risk postulating a tortuosity of about 2–3. The intragranular porosity and Henry's coefficient being known, only the ratio $\bar{\bar{\theta}}_A$ is unknown. If one assumes $\bar{\bar{\theta}}_A = 0$, one obtains:

$$\bar{g}_A = \frac{3 \times 3}{0.5 + 0} = 18$$

We have no information allowing us to estimate $\bar{\bar{\theta}}_A$ but we can probably assume that a value of 0.25 is relatively high. To put things in perspective, the results of (Farnan *et al.*, 2002) are compatible with $\bar{\bar{\theta}}_A = 0.2$. If one assumes $\bar{\bar{\theta}}_A = 0.25$, one obtains:

$$\bar{g}_A = \frac{3 \times 3}{0.5 + (3 - 0.5) \times 0.25} = 8$$

From the above calculations, one can reasonably guess (rather than calculate) the lumped average diffusion coefficient to be 10–20 times lower than the molecular diffusion coefficient. Finally, we obtain $\bar{D}_A = 0.5 - 1 \times 10^{-10}$ m²/s.

Exercise 5.3

Equations (5.53) and (5.54) give $\bar{D}_A^{loc} = \bar{D}_A(0) \frac{\bar{K}_A}{\varepsilon_i + (\bar{K}_A - \varepsilon_i).\bar{\bar{\theta}}_A}$, independent of concentration, so $\bar{D}_A = \bar{D}_A^{loc}$.

$\bar{D}_A(0)$ is given by Eq. (5.43): $\bar{D}_A(0) = D_A/\bar{g}_A$, $\bar{g}_A = \frac{\tau.\bar{K}_A}{\varepsilon_i + (\bar{K}_A - \varepsilon_i)\bar{\bar{\theta}}_A}$. Consequently, one has:

$$\bar{D}_A = \bar{D}_A^{loc} = D_A \frac{\varepsilon_i + (\bar{K}_A - \varepsilon_i)\,\bar{\bar{\theta}}_A}{\tau \bar{K}_A} \frac{\bar{K}_A}{\varepsilon_i + (\bar{K}_A - \varepsilon_i)\,\bar{\bar{\theta}}_A} = \frac{D_A}{\tau}$$

The lumped local diffusion coefficient $\bar{D}_A$ is constant and given by the ratio of the diffusion coefficient in solution to the tortuosity. It is independent of the intragranular porosity.

Exercise 5.4

Under the condition of linear adsorption, and assuming a motionless adsorbed solute, Eq. (5.78) gives, in the Laplace domain:

$$\hat{\bar{\bar{c}}}_A = \frac{\bar{\bar{K}}_A \hat{c}_A}{1 + s t_A^{ads}} \tag{5.166}$$

Consequently, noting that $\bar{K}_A = \varepsilon_i + (1 - \varepsilon_i)\,\bar{\bar{K}}_A$ and that $\hat{\bar{c}}_A = \varepsilon_i \hat{c}_A + (1 - \varepsilon_i)\,\hat{\bar{\bar{c}}}_A$, one has, everywhere in the particle:

$$\hat{\bar{c}}_A = \hat{\bar{h}}^{ads}(s)\,\hat{c}_A$$
$$\hat{\bar{h}}^{ads}(s) = \frac{\bar{K}_A + \varepsilon_i t_A^{ads} s}{1 + t_A^{ads} s} \tag{5.167}$$

The function $\hat{\bar{h}}^{ads}(s)$ is the transfer function connecting the lumped concentrations to the concentrations in the intragranular pores everywhere in the particle. Notice that in the absence of adsorption kinetics limitation, one has $\hat{\bar{h}}^{ads}(s) = \bar{K}_A$.

The mass-balance equation (5.31), describing the solute transfer inside the particle, is thus written in the Laplace domain:

$$\hat{\bar{h}}^{ads}(s)\, s\hat{\bar{c}}_A(r,s) = \varepsilon_i \grave{D}_A \nabla^2 \hat{\bar{c}}_A(r,s) \tag{5.168}$$

Equation (5.168) is an ODE in r, whose solution gives the lumped concentration profile inside the particle. By integration over r, one obtains the average lumped concentration. The transfer function connecting the average lumped solid concentration to the lumped concentration at the periphery of the particle is thus obtained.

It is easy to see that Eq. (5.168) is exactly the Laplace transform of the mass balance obtained for a lumped particle, Eq. (5.11), if one replaces $\overline{D}_A$ by $\varepsilon_i \grave{D}_A / \bar{h}^{ads}(s)$. This means that, assuming the diffusion coefficient, internal porosity and time for adsorption are independent of the position in the particle, the transfer function $\overline{H}(s)$ given in Eq. (5.22) still applies provided that $\overline{D}_A$ is replaced by $\varepsilon_i \grave{D}_A / \bar{h}^{ads}(s)$. Consequently, one obtains:

$$\hat{\overline{H}}^{Fick/ads}(s) = \frac{\hat{\overline{C}}_A(s)}{\hat{\bar{c}}_A^*} = \frac{3\ \mathrm{Cotanh}(\sqrt{s\vartheta(s)})}{\sqrt{s\vartheta(s)}} - \frac{3}{s\vartheta(s)} \tag{5.169}$$

with $\vartheta(s) = \bar{h}^{ads}(s) r_0^2 / \varepsilon_i \grave{D}_A$.

At the fluid–solid interface, one has $\hat{\bar{c}}_A^* = \hat{\bar{h}}^{ads}(s)\, \hat{c}_A^*$. Neglecting external transfer limitations, one has $\hat{C}_A = \hat{c}_A^*$ so that one obtains:

$$\hat{\overline{C}}_A(s) = \hat{\overline{H}}^{Fick/ads}(s)\, \hat{\bar{h}}^{ads}(s)\, \hat{C}_A(s)$$

One can thus determine the transfer function $\hat{\mathscr{L}}^{Fick/ads}$ relating the quantity of solute contained in the mobile (δn_A) and stationary ($\delta \bar{n}_A$) phases in a small volume of column where the composition is assumed to be uniform:

$$\hat{\mathscr{L}}^{Fick/ads}(s) = \frac{1-\varepsilon_e}{\varepsilon_e} \hat{\overline{H}}^{Fick/ads}(s)\, \hat{\bar{h}}^{ads}(s)$$

By following the methodology presented in Section 5.2.2, we will approximate $\hat{\mathscr{L}}^{Fick/ads}$ by a first-order transfer function $\hat{\mathscr{L}}^{LDF/ads}$ having the same zero- and first-order moments. After some (painful if done by hand) derivations, one obtains:

$$\hat{\mathscr{L}}^{LDF/ads}(s) = \frac{1-\varepsilon_e}{\varepsilon_e} \left(\frac{\overline{K}_A}{1 + \left(\frac{\overline{K}_A - \varepsilon_i}{\overline{K}_A} t_A^{ads} + t_A^i \right) s} \right) \tag{5.170}$$

Equation (5.170) allows the approximation of a system limited by internal diffusion and adsorption with a first-order system having a time constant $t_A^{i/ads} = \frac{\overline{K}_A - \varepsilon_i}{\overline{K}_A} t_A^{ads} + t_A^i$.

In other words, we have proven that chromatograms simulated by taking into account internal diffusion and adsorption kinetics will have the same variance as chromatograms simulated by taking into account only diffusion, provided that one uses a global mass transfer time given by $t_A^{i/ads}$.

Exercise 5.5

The saturation data are presented graphically in Figure 5.24. It is clear that the saturation process is quicker when Ca^{++} replaces Na^{+} and favored by a high normality. Should internal diffusion be the limiting step, the mass balance at the particle level has the form (Eq. (5.108)):

$$\frac{d\overline{C}_A}{dt} = \frac{(\overline{c}_A^* - \overline{C}_A)}{t_A^i}$$

where A refers to the entering solute.

The particles being immersed in a solution containing only the entering solute, the interfacial concentration $\overline{c}_A^*$ equals the ion-exchange capacity of the resin, $\overline{N}^{iex}$. The equations describing the particle saturation are thus independent of the solution normality.

Should external diffusion be the limiting step, the mass balance at the particle level has the form (Eq. (5.105)):

$$\frac{d\overline{C}_A}{dt} = \frac{(C_A - C_A^*)}{t_A^e}$$

where A refers to the entering solute.

The particles being immersed in a solution containing only the entering solute, the concentration C_A equals the normality of the solution, N_0. The equations describing the particles saturation are thus directly impacted by the solution normality.

The fact that saturation curves at 1 and at 2 eq/l are almost identical suggests that internal diffusion is the controlling step under these conditions.

The lumped model and the Nernst–Planck equations leading to a variable diffusion coefficient given by Eq. (5.96) explain the fact that saturation kinetics is affected by the direction of the exchange.

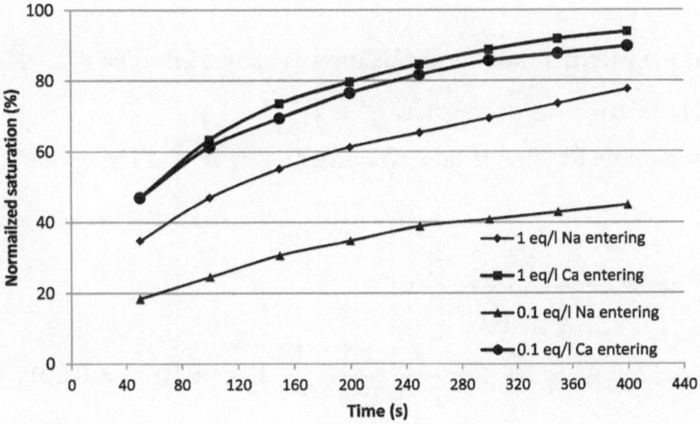

Figure 5.24 For solution of Exercise 5.5.

We are looking for a rough estimate of diffusion coefficients, thus a back-of-the-envelope calculation. From Figure 5.4, we know that 20% of saturation is reached at normalized time $\frac{t}{t_A^F} = 4 \times 10^{-3}$.

The 20% saturation is obtained at 8 s when Ca^{++} is entering and at 18 s when Na^+ is entering the resin. This leads to:

$$t_A^F\left(\mathrm{Ca^{++}entering}\right) = \frac{8}{4 \times 10^{-3}} = 2000\ \mathrm{s}$$
$$t_A^F\left(\mathrm{Na^{+}entering}\right) = \frac{18}{4 \times 10^{-3}} = 4500\ \mathrm{s}$$

We know from Eq. (5.96) that the variable lumped inter-diffusion coefficient varies with ion concentration. This is confirmed by the different values of t_A^F obtained.

With knowledge of the particle radius (400 μm), one obtains:

$$\bar{D}\left(\mathrm{Ca^{++}entering}\right) \approx 8 \times 10^{-11}\ \mathrm{m^2/s} \qquad \bar{D}\left(\mathrm{Na^{+}entering}\right) \approx 3.5 \times 10^{-11}\ \mathrm{m^2/s}$$

Determining Ca^{++} and Na^+ diffusion coefficients would require solving the Nernst–Planck equations.

Interpreting the results obtained at lower normality requires taking into account both internal and external diffusional limitations. Because of the non-linear coupling induced at the lumped solid/fluid interface by the non-linear ion-exchange isotherm, no simple formula can be used. Diffusion coefficients in solution can only be determined by fitting the numerical solution of the model, taking into account both internal and external diffusion, to the experimental saturation curves.

Exercise 5.6

The transfer functions $\hat{\mathscr{L}}^{LDF}(s)$ and $\hat{\mathscr{L}}^{Fick}(s)$ have the same zero- and first-order moments as explained in Section 5.2.2. Equation (5.127) being a first-order approximation obtained with $\hat{\mathscr{L}}^{LDF}$, one would thus obtain the same equation with $\hat{\mathscr{L}}^{Fick}$.

Exercise 5.7

In Exercise 5.1, the external time for diffusion was estimated to be $t_A^e = \frac{d_p\delta}{6D_A} = 0.02$ s. Knowing that $\bar{D}_A = 10^{-11}\ \mathrm{m^2/s}$, one has $t_A^i = \frac{d_p^2}{60\bar{D}_A} = 1$ s.

The global mass transfer time is thus $t_A^m = \bar{K}_A t_A^e + t_A^i \approx 1$ s. Using the Van Deemter equation, one has:

$$HETP = 2d_p + \frac{2\bar{K}'_A}{\left(1+\bar{K}'_A\right)^2}\frac{t_A^m}{\varepsilon_e}u$$
$$= 2 \times 25 \times 10^{-6} + \frac{2 \times 1.5}{(1+1.5)^2}\frac{1}{0.4}1.36 \times 10^{-3} = 0.0017\ \mathrm{m}$$

thus $N = L_{col}/HETP = 588$ plates.

Exercise 5.8

For this type of situation, one can guess that internal diffusion is the limiting step. One can consequently write the HETP as:

$$HETP \approx 2d_p + \frac{2\overline{K}'_A}{\left(1+\overline{K}'_A\right)^2} t_A^m \frac{\mathrm{u}}{\varepsilon_e} \quad \text{with } t_A^m = t_A^i$$

The molecules being small, Henry's coefficient moderate and silica having wide pores, one can assume that adsorbed-phase diffusivity plays a negligible role. Under this assumption, one has $\overline{g}_A = \frac{\tau \overline{K}_A}{\varepsilon_i}$, thus about 10. Assuming $\overline{g}_A = 10$, the characteristic time for internal diffusion is calculated to be:

$$t_A^i = \frac{\left(15 \times 10^{-6}\right)^2}{60 \times 10^{-10}} = 3.75 \times 10^{-2}\ s$$

With a retention factor $\overline{K}'_A = 1$ and an extragranular porosity close to 0.4, one has:

$$HETP = 2 \times 15 \times 10^{-6} + \frac{2}{(1+1)^2} 3.75 \times 10^{-2} \frac{u}{0.4}$$
$$= 3 \times 10^{-5} + 4.68 \times 10^{-2} u$$

Obtaining 5000 plates in 25 cm requires an HETP of 5×10^{-5} m. With the above equation, one obtains a superficial fluid velocity $u = 0.43 \times 10^{-3}$ m/s.

This is the type of calculation that one may typically do if an estimate is absolutely required on very short notice with no other information available – to be absolutely avoided otherwise.

Exercise 5.9

The superficial fluid velocity is $u = Q/\Omega_{col} = 1.35\ 10^{-3}$ m/s. At 20°C, the viscosity of water is close to 10^{-3} Pa s and its density close to 1000 kg/m^3. Typical molecular diffusion coefficients are $D_A = 10^{-5}$ cm^2/s, and one can estimate:

$$Re = \frac{\rho u d}{\eta} = \frac{1000 \times 1.35\ 10^{-3} \times 350\ 10^{-6}}{0.001} = 0.47$$

$$Sc = \frac{\nu}{D_A} = \frac{\eta}{\rho D} = \frac{0.001}{1000 \times 10^{-9}} = 1000$$

If one uses the Wakao correlation, the Sherwood number is given by:

$$Sh = 2 + 1.8 Re^{1/2} Sc^{1/3} = 14.2$$

and the thickness of the liquid layer is $\delta = \frac{d}{Sh} = \frac{25 \times 10^{-6}}{14.2} = 1.76 \times 10^{-6}$ m.

The characteristic time for external diffusion is thus $t_A^e = \frac{d_p \delta}{6 D_A} = 0.1$ s. According to Table 5.1, if one guesses $\overline{D}_A = 10^{-11}$ m²/s, one has:

$$t_A^i = \frac{d_p^2}{60 \overline{D}_A} = 204 \text{ s}$$

The global diffusion time is thus:

$$t_A^m = \overline{K}_A t_A^e + t_A^i \approx 209 \text{ s}$$

Using the Van Deemter equation, one has:

$$HETP = 2d_p + \frac{2\overline{K}_A'}{\left(1+\overline{K}_A'\right)^2} t_A^m u_e$$
$$= 2 \times 350 \times 10^{-6} + \frac{2 \times 75}{(1+75)^2} \times 209 \times \frac{1.35 \times 10^{-3}}{0.4} = 0.018 \text{ m}$$

Thus $N = \dfrac{L_{col}}{HETP} = 55$ plates.

Exercise 5.10

Plots of the reduced HETP ($\tilde{h}$) values versus interstitial velocity give straight lines, in agreement with Eq. (5.140), which can be written:

$$\tilde{h} = cst + \frac{C_e}{d_p} u_e \quad \text{with} \quad \frac{C_e}{d_p} = \frac{2\overline{K}_A'}{\left(1+\overline{K}_A'\right)^2} \frac{1}{d_p} t_A^m$$

Assuming an extragranular porosity of 0.4, one can calculate the different lumped retention factor $\overline{K}_A'$ values from the respective lumped Henry's coefficients $\overline{K}_A$. The slope of $\tilde{h}$ versus fluid velocity then leads directly to t_A^m.

Assuming that internal diffusion prevails, knowledge of $t_A^m = t_A^i$ allows the lumped diffusion coefficient $\overline{D}_A$ to be estimated. The results are summarized in the following table.

Experiment	$\overline{K}_A'$	$\frac{C_e}{d_p}$ (s/m)	t_A^m (s)	$\overline{D}_A$ (m²/s)	$1/\overline{K}_A$
$\overline{K}_A = 1.0$	1.5	3.05×10^3	1.27×10^{-1}	5.25×10^{-11}	1.0
$\overline{K}_A = 2.0$	3.0	4.35×10^3	2.32×10^{-1}	2.88×10^{-11}	0.5
$\overline{K}_A = 5.0$	7.5	4.77×10^3	4.60×10^{-1}	1.45×10^{-11}	0.2

The fact that Eq. (5.41) can be written $\overline{D}_A = \overline{\overline{D}}_A + \frac{\varepsilon_i\left(\grave{D}_A - \overline{\overline{D}}\right)}{\overline{K}_A}$ suggests plotting $\overline{D}_A$ vs. $\frac{1}{\overline{K}_A}$. In doing so, we get a straight line intercepting the y-axis at $\overline{\overline{D}}_A = 5 \times 10^{-12}$ m²/s and having a slope $\varepsilon_i\left(\grave{D}_A - \overline{\overline{D}}\right) = 4.75 \times 10^{-11}$ m²/s, allowing the determination of $\grave{D}_A = 10^{-10}$ m²/s.

Exercise 5.11

The calculations are straightforward and give:

Mean diameter	i	j	$d_p^{i;j}$
In number	1	0	102.0
In volume	3	0	117.0
Internal diffusion: matching initial slope	3	1	125.3
Internal diffusion: matching entire saturation	5	3	158.1
External diffusion: matching initial slope	3	3/2	129.1
Internal diffusion: matching entire saturation	9/2	3	154.1

While this distribution looks like it should be centered at 100 μm, the equivalent diameter for internal diffusion exceeds 154 μm.

Exercise 5.12

In Exercise 2.8, we established the transfer function of the MC–$\overline{\text{LDF}}$ model by using deviation variables and assuming that the adsorption isotherm was linear over the entire concentration range. We now relax the linearity assumption by using a local linearization of the adsorption isotherm and generalizing this approach irrespective of the hydrodynamic model.

The two mass-balance equations describing the solute transfer at the particle level are:

$$t_A^i \frac{d\overline{C}_A}{dt} = \overline{c}_A^* - \overline{C}_A \tag{5.171}$$

$$t_A^e \frac{d\overline{C}_A}{dt} = C_A - C_A^* \tag{5.172}$$

Using the same approach as in Exercise 2.8, we define deviation variables:

$$\Delta C_A = C_A - C_A\,(t=0) = C_A - C_A^I$$

$$\Delta \overline{C}_A = \overline{C}_A - \overline{C}_A\,(t=0) = \overline{C}_A - \overline{C}_A^I$$

where $\overline{C}_A^I$ represents the lumped solid-phase concentration in equilibrium with C_A^I. The same definitions are used for the interfacial concentrations:

$$\Delta C_A^* = C_A^* - C_A^I$$

$$\Delta \overline{c}_A^* = \overline{c}_A^* - \overline{C}_A^I$$

Our objective being to determine the response of the column to a small perturbation, one can linearize the adsorption isotherm around $(C_A^I, \overline{C}_A^I)$:

$$\overline{c}_A^* = \overline{C}_A^I + \overline{f}_A'\left(C_A^I\right)(C_A^* - C_A^I)$$

so $\Delta \overline{c}_A^* = \overline{f}_A'\,(C_A^I)\,\Delta C_A^*$.

After reinjecting the expressions for the deviation variables into Eq. (5.171) and Eq. (5.172), and adding the two equations, one obtains:

$$t_A^m \frac{d\Delta\overline{C}_A}{dt} = \bar{f}'_A\left(C_A^I\right)\Delta C_A - \Delta\overline{C}_A$$

$$t_A^m = t_A^i + \bar{f}'_A\left(C_A^I\right)t_A^e$$

This shows that Eq. (2.45), relating the Laplace transforms of the fluid and of the lumped solid-phase concentrations, holds for the deviation variables, provided that the lumped Henry's coefficient $\overline{K}_A$ is replaced by the local slope $\bar{f}'_A\left(C_A^I\right)$ of the adsorption isotherm. Consequently, the same applies to the transfer function $\hat{\mathscr{L}}(s)$ defined in Eq. (2.64).

The transfer function $\hat{E}_e(s)$ relates the deviation variables of the outlet and inlet concentration of an excluded solute. Consequently, the transfer function in Eq. (2.47) relates the deviation variables of the outlet and inlet concentrations of our solute, so that Eq. (2.67) expressing the variance of the chromatographic peak holds for the MC model.

We immediately conclude that the Van Deemter equations derived for a solute-free column apply to a column initially equilibrated with a solution at concentration C_A^I, provided that the lumped Henry's coefficient $\overline{K}_A$ is replaced by the local slope $\bar{f}'_A\left(C_A^I\right)$ of the adsorption isotherm.

References

Afeyan, N. B., Gordon, N. F., Mazsaroff, I., Varady, L., Fulton, S. P., Yang, Y. B. and Regnier, F. E. (1990), *J. Chromatogr. A* **519**(1), 1–29.

Boyd, G. and Soldano, B. (1953), *J. Am. Chem. Soc.* **75**, 6091–6099.

Boyd, G. and Soldano, B. (1954), *J. Am. Chem. Soc.* **75**, 6105.

Boyd, G., Soldano, B. and Bonner, O. (1954), *J. Phys. Chem.* **58**, 456–459.

Cale, T. S., Lawson, J. M. and Ludlow, D. K. (1987), *Chem. Eng. Commun.* **56**(1–6), 169–181.

Carta, G. (1998), *Chem. Eng. Sci.* **43**, 2877.

Carta, G. and Bauer, J. (1990), *AIChE J.* **36**, 147.

Carta, G. and Cincotti, A. (1998), *Chem. Eng. Sci.* **53**, 3483.

Carta, G., Cincotti, A. and Gao, G. (1999), *Sep. Sci. Technol.* **34**(1), 1–16.

Carta, G., Gregory, M. E., Kirwan, D. J. and Massaldi, H. A. (1992), *Sep. Technol.* **2**(2), 62–72.

Carta, G. and Lewus, R. (1999), *Sep. Sci. Technol.* **34**, 2685–2697.

Chang, C. and Lenhoff, A. (1998), *J. Chromatogr. A* **827**, 281–293.

Conder, J. and Hayek, B. (2000), *Biochem. Eng. J.* **6**, 15–223.

Coquebert de Neuville, B., Tarafder, A. and Morbidelli, M. (2013), *J. Chromatogr. A* **1298**, 26–34.

Coquebert de Neuville, B., Thomas, H. and Morbidelli, M. (2013), *J. Chromatogr. A* **1314**, 77–85.

Dodds, J. and Tondeur, D. (1974), *J. Chim. Phys.* **71**, 238.

Dwivedi, P. N. and Upadhyay, S. N. (1977), *Ind. Eng. Chem. Process Des. Dev.* **16**(2), 157–165.

Epstein, N. (1989), *Chem. Eng. Sci.* **44**, 777–779.

Farnan, D., Frey, D. and Horvath, C. (2002), *J. Chromatogr. A* **959**, 65–73.

Glueckauf, E. (1949), *Discuss. Faraday Soc.* **7**, 12–25.

Gorius, A., Bailly, M. and Tondeur, D. (1991), *Chem. Eng. Sci.* **46**(2), 685–692.

Grosfils, V. (2009), Modelling and parametric estimation of simulated moving bed chromatographic processes. PhD thesis, Université Libre de Bruxelles.

Guiochon, G., Felinger, A., Golshan-Shirazi, S. and Katti, A. M. (2006), *Fundamentals of Preparative and Nonlinear Chromatography*, Elsevier Academic Press.

Gutenwik, J., Nilsson, B. and Axelsson, A. (2004), *Biochem. Eng. J.* **19**, 1–7.

Hasnat, A. and Jvekar, V. (1996), *AIChE J.* **42**(1), 161–175.

Heinonen, J., Rubiera Landa, H., Sainio, T. and Seidel-Morgenstern, A. (2012), *Sep. Purif. Technol.* **95**, 235–247.

Helfferich, F. G. (1962), *Ion Exchange*, McGraw-Hill.

Helfferich, F. G. and Plesset, M. S. (1958), *J. Chem. Phys.* **28**, 418–424.

Holger, M. (1978), *Chem. Eng. Sci.* **33**, 913–919.

Jackson, R. (1977), *Transport in Porous Catalysts*, Elsevier.

Johnston, A. and Hearn, M. (1991), *J. Chromatogr.* **557**, 335–358.

Kaczmarski, K., Cavazzini, A., Szabelski, P., Zhou, D., Liu, X. and Guiochon, G. (2002), *J. Chromatogr. A* **962**, 57–67.

Kaczmarski, K., Gubernak, M., Zhou, D. and Guiochon, G. (2003), *Chem. Eng. Sci.* **58**, 2325–2338.

Krishna, R. (1990), *Chem. Eng. Sci.* **45**(7), 1779–1791.

Krishna, R. (1993), *Gas Sep. Purif.* **7**(2), 91–104.

Krishna, R. and Wesselingh, J. (1997), *Chem. Eng. Sci.* **52**(6), 861–911.

Kucera, E. (1965), *J. Chromatogr.* **19**(2), 237–248.

Kunii, D. and Suzuki, M. (1967), *Int. J. Heat Mass Transfer* **10**(7), 845–852.

Leitão, A. and Rodrigues, A. E. (1995), *Chem. Eng. J.* **60**, 81–87.

Leitão, A. and Rodrigues, A. E. (1999), *Biochem. Eng. J.* **3**, 131–139.

Lewus, R. and Carta, G. (1999), *AIChE J.* **45**(3), 512–522.

Liberti, L. and Passino, R. (1985). In J. A. Marinski and Y. Marcus, eds., *Ion Exchange and Solvent Extraction, Vol. 9*, Marcel Dekker, pp. 175–210.

Ma, Z., Whitley, R. and Wang, N. (1996), *AIChE J.* **42**(5), 1244–1262.

Martin, C., Iberer, G., Ubiera, A. and Carta, G. (2005), *J. Chromatogr. A* **1079**, 105–115.

Melis, S., Markos, J., Cao, G. and Morbidelli, M. (1996), *Ind. Eng. Chem. Res.* **35**, 3629–3636.

Melter, L., Butté, A. and Morbidelli, M. (2008), *J. Chromatogr. A* **1200**, 156–65.

Miyabe, K. and Guiochon, G. (2000a), *J. Chromatogr. A* **866**, 147–171.

Miyabe, K. and Guiochon, G. (2000b), *J. Chromatogr. A* **890**, 211–223.

Miyabe, K. and Guiochon, G. (2002), *J. Chromatogr. A* **961**, 23–33.

Morbidelli, M., Servida, A., Storti, G. and Carra, S. (1982), *Ind. Eng. Chem. Fundam.* **21**(2), 123–131.

Morbidelli, M., Storti, G., Carra, S., Niederjaufner, G. and Pontoglio, A. (1984), *Chem. Eng. Sci.* **39**, 384.

Mota, J. (2004). In G. Lu and X. Zhao, eds., *Nanoporous Materials: Science and Engineering*, Imperial College Press, p. 694.

Nelson, P. A. and Galloway, T. R. (1975), *Chem. Eng. Sci.* **30**(1), 1–6.

Nicoud, R. M. (1987), Influence respective des facteurs thermodynamiques, hydrodynamiques et diffusionnels sur le fonctionnement des echangeurs d'ions. Application a un procede d'echange ionique: Le Nymphea PhD thesis, Institut National Polytechnique de Lorraine, Nancy.

Nicoud, R. M. and Perrut, M. (1993). In G. Ganetsos and P. E. Barker, eds., *Preparative and Production Scale Chromatography*, vol. 61 of *Chromatographic Science Series*, Marcel Dekker, pp. 47–77.

Nicoud, R. M. and Schweich, D. (1988), *J. Chim. Phys.* **85**(2), 239–246.

Nicoud, R. M. and Schweich, D. (1989), *Water Resour. Res.* **25**(6), 1071–1082.

Piatkowski, W., Gritti, F., Kaczmarski, K. and Guiochon, G. (2003), *J. Chromatogr. A* **989**, 207–219.

Poling, B., Prausnitz, J. and O'Connell, J. (2001), *The Properties of Gases and Liquids*, 5th edn., McGraw-Hill.

Rearden, P., Sajonz, P. and Guiochon, G. (1998), *J. Chromatogr. A* **813**, 1–9.

Rendueles de la Vega, M., Loureiro, J. and Rodrigues, A. E. (1996), *Chem. Eng. J.* **61**, 123–126.

Rexwinkel, G., Heesink, A. and Van Swaaij, W. (1996), *Chem. Eng. Sci.* **52**(21–22), 3995–4003.

Rodrigues, A. E. (1993), *LC-GC International* **6**(1), 20–29.

Rodrigues, A. E., Ahn, B. and Zoulalian, A. (1982), *AIChE J.* **28**, 541–546.

Rodriguez, J., Valverde, J. and Rodrigues, A. E. (1998), *Ind. Eng. Chem. Res.* **37**, 2020–2028.

Russell, A. and Carta, G. (2005), *AIChE J.* **51**, 2469–2480.

Ruthven, D. M. (1984), *Principles of Adsorption and Adsorption Processes*, John Wiley & Sons.

Ruthven, D. M. and Ching, C. B. (1993). In G. Ganetsos and P. E. Barker, eds., *Preparative and Production Scale Chromatography*, vol. 61 of *Chromatographic Science Series*, Marcel Dekker, pp. 629–671.

Sajonz, P., Kele, M., Zhong, G., Sellergen, B. and Guiochon, G. (1998), *J. Chromatogr. A* **810**, 1–17.

Sardin, M., Schweich, D., Leij, F. and Van Genuchten, M. V. (1991), *Water Resour. Res.* **27**(9), 2287–2307.

Schmidt-Traub, H., Kaspereit, M., Engell, S., Susanto, A., Epping, A. and Jupke, A. (2012). In H. Schmidt-Traub, M. Schulte and A. Seidel-Morgenstern, eds., *Preparative Chromatography*, 2nd edn., Wiley-VCH, p. 425.

Schröder, M., Von Lieres, E. and Hubbuch, J. (2006), *J. Phys. Chem. B* **110**, 1420–1436.

Silva, C. and Lito, P. (2007), *Chem. Eng. Sci.* **62**, 6939–6946.

Sircar, S. and Hufton, J. (2000), *Adsorption* **6**, 137–147.

Taylor, R. and Krishna, R. (1993), *Multicomponent Mass Transfer*, John Wiley & Sons.

Van Brocklin, L. P. and David, M. M. (1972), *Ind. Eng. Chem. Fundam.* **11**(1), 91–99.

Van Deemter, J. J., Zuiderweg, F. J. and Klinkenberg, A. (1956), *Chem. Eng. Sci.* **5**(6), 271–289.

Villermaux, J. (1981). In A. E. Rodrigues and D. Tondeur, eds., *Percolation Processes: Theory and Applications*, vol. 33 of *NATO ASI Series, Series E: Applied Sciences*, Sijthoff & Noordhof, pp. 83–140.

Wakao, N. and Funazkri, T. (1978), *Chem. Eng. Sci.* **33**(10), 1375–1384.

Wesselingh, J. and Krishna, R. (2006), *Mass Transfer in Multicomponent Mixtures*, VSSD.

Wilson, E. J. and Geankopolis, C. J. (1966), *Ind. Eng. Chem. Fundam.* **5**(1), 9–14.

Yoshida, H., Yoshikawa, M. and Katoaka, T. (1994), *AIChE J.* **40**(12), 2035–2044.

6 Hydrodynamics of chromatographic columns

Delivering unique performance or . . . seriously struggling.

In previous chapters the hydrodynamics of chromatographic columns has been described with simplified assumptions: fluids and chromatographic beds were assumed to be incompressible, possible temperature effects were neglected, velocity profiles were taken to be radially uniform, axial dispersion presumed to obey a Fickian mechanism, and so on. These assumptions allowed us to model the columns either with the mixing cells in series (MC) model or with the plug flow plus axial dispersion (PD) model. These models have the merits of simplicity and flexibility as they allow a large range of dispersion situations, from perfect mixing to plug flow with only one parameter, to be represented.

The real world can be more complex, because chromatographic beds are not indefinitely stable and can plug, distributors are imperfect, velocities are not always radially uniform, temperature effects can impact dispersion, large differences in density or viscosity between the feed and the eluent can induce instabilities etc. Addressing these matters and understanding their impact on column performance is the subject of this chapter.

Modeling hydrodynamics requires addressing three subjects of specific relevance to the design of chromatographic processes:

- Pressure drop: a key parameter for hardware design (pumps, columns, etc.) and sometimes for protecting the adsorbent
- Zero (total or excluded) retention time: a key reference for chromatogram positioning that can be affected by velocity heterogeneity and fluid compressibility
- Hydrodynamic dispersion: to ensure that the separation ability of the chromatographic medium is not spoiled by undesirable contributions to band broadening.

After presenting the modeling of "ideal" situations, we will relax some assumptions and discuss their influence on the scale-up of real systems.

6.1 Ideal systems

As explained in Chapter 1, we define "ideal" systems as chromatographic systems in which:

- the properties of the chromatographic bed (porosity and permeability) are uniform along the axial and radial directions
- the chromatographic bed is incompressible and stable
- the fluid velocity and temperature are radially uniform.

We consider this situation to be "ideal" because it allows a straightforward scale-up of a chromatographic process: for given bed length and fluid velocity, the chromatograms are independent of the column diameter. Scale-up becomes as simple as increasing the diameter of the column to accommodate higher quantities of material while maintaining the same bed length and superficial flow velocity.

Assuming this ideal situation, we will successively model the pressure drop and the hydrodynamic dispersion, as this information is mandatory for the scaling-up of chromatographic processes. In subsequent sections we will address more complex situations, including the effect of column diameter on chromatographic performance.

6.1.1 Pressure-drop modeling

Since the pioneering work of Darcy, the description of fluid flow through porous media has been based essentially on empirical correlations obtained from numerous experimental results. This is attributed to the fact that the solution of rigorous fluid mechanics equations in such complicated geometries is impossible except for very simplified situations.

For laminar flow, an accurate pressure-drop correlation has been derived with the Kozeny–Kármán model, which is based on the representation of a porous bed as a bundle of unconnected cylindrical and tortuous pores with identical diameters. The bed length, as seen by the fluid particles, is $l = L_{col}\theta$, where L_{col} is the column length and θ the tortuosity factor (Figure 6.1). For more on tortuosity and possible confusion about it, the reader is referred to (Epstein, 1989). In such pores, under laminar conditions the fluid motion is described by Poiseuille's law. This leads to the well-known Kozeny–Kármán equation for the pressure drop in a packed bed:

$$\frac{\Delta P}{L} = h_k \frac{\mu}{d_p^2} \frac{(1-\varepsilon_e)^2}{\varepsilon_e^3} u \qquad \text{validity range: } Re = \frac{\rho u d_p}{\mu} < 1 \tag{6.1}$$

For most granular beds, it has been shown that $h_k \approx 150 - 180$ (in MKSA units).

The Kozeny–Kármán equation can be rewritten in order to introduce the so-called bed permeability κ (m^2):

$$\frac{\Delta P}{L} = \frac{1}{\kappa}\mu u \qquad \kappa = \frac{d_p^2}{h_k} \frac{\varepsilon_e^3}{(1-\varepsilon_e)^2} \tag{6.2}$$

The concept of bed permeability is often used as a convenient parameter for modeling pressure drop: when the permeability is known, the pressure drop is then simply related to fluid viscosity and velocity by Eq. (6.2).

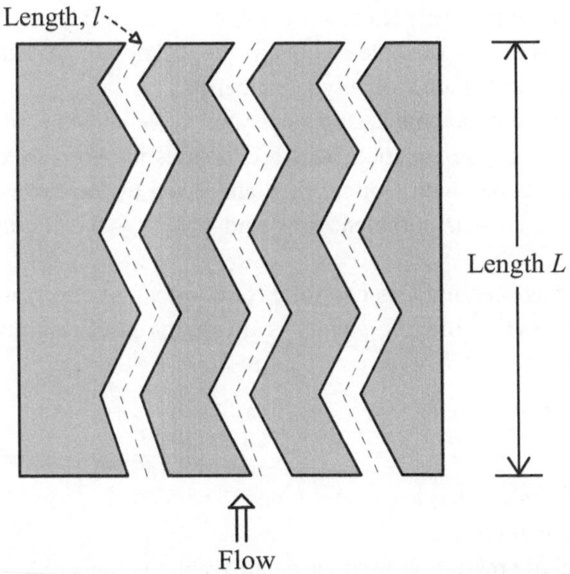

Figure 6.1 Illustration of the Kozeny–Kármán model. The chromatographic bed is represented as a bundle of unconnected tubes with identical diameters and tortuous pathways.

For design purposes, we will often use the proportionality between the pressure drop per unit length and the fluid velocity (either superficial or interstitial) and write:

$$\frac{\Delta P}{L} = \delta_s u \qquad \frac{\Delta P}{L} = \delta_e u_e \tag{6.3}$$

The Kozeny–Kármán equation predicts a permeability proportional to the square of the particle diameter. It is also extremely sensitive to the extragranular porosity. An order of magnitude to have in mind is that a change of 0.01 in extragranular porosity around the classical value of 0.35 leads to a 10% change in bed permeability.

In HPLC, with particle size in the range 10–100 μm, typical superficial velocities are about 10^{-3} m/s. Considering a density of about 1000 kg/m^3 and a viscosity of about 10^{-3} Pa s, Reynolds numbers are typically in the range 10^{-2} to 10^{-1}. Pressure drops are typically in the range of 50–200 bar/m.

In large-scale, low-pressure chromatography (normally adsorption or ion exchange), with particle sizes of about 500 μm, superficial fluid velocities of 1–2 × 10^{-3} m/s are typically used. This leads to Reynolds numbers in the range 0.5–1.0. Pressure drops are typically a few bar/m to a few tens of bar/m.

Because $Re < 1$ for liquid chromatography, the flow is laminar and the Kozeny–Kármán equation is well adapted for estimating the pressure drop.

In the case of supercritical fluid chromatography (SFC), particle sizes in the range 10–50 μm together with superficial fluid velocities in the range 5×10^{-3} to 1×10^{-2} m/s are typical. The fluid velocities are much higher than in liquid chromatography because lower viscosity allows a lower pressure drop and higher diffusivities allow more efficient mass transfer. Typical viscosities are in the range 0.02×10^{-3} to 0.05×10^{-3}

Pa s and density 500–800 kg/m^3. This leads to Re values between 0.1 and possibly 10, so experimental conditions can in principle be outside the validity range of the Kozeny–Kármán equation. Pressure drops are typically a few bar/m.

In the rare cases of chromatographic systems operated at $Re > 1$, the pressure drop can be estimated using the Ergun equation, which expresses the pressure drop as the sum of a contribution associated with laminar flow and given by the Kozeny–Kármán equation, and one associated with turbulent flow and given by the Burke–Plummer equation.

The Ergun equation[1] has been used successfully for convex particles (particles without holes, as for instance in a Raschig ring), which is always the case in chromatographic applications:

$$\frac{\Delta P}{L} = h_k \frac{\mu}{d_p^2} \frac{(1-\varepsilon_e)^2}{\varepsilon_e^3} u + h_B \frac{\rho}{d_p} \frac{1-\varepsilon_e}{\varepsilon_e^3} u^2 \tag{6.4}$$

The Burke–Plummer constant h_B has been determined to be $h_B \approx 1.75$ (in MKSA units) from numerous experimental results.

A first factor influencing pressure drop in chromatography is the solid-phase granulometry. Smaller particles have a positive impact on mass transfer kinetics, as shown in Chapter 5, but an adverse impact on pressure drop.

A second factor influencing pressure drop is fluid viscosity. Injection of a viscous feed mixture, prediluted in the eluent or not, can have a significant impact on temporary increases in pressure drop, possibly causing the flow rate to decrease and thus significantly limiting the productivity of the system. In addition, one must be aware of the non-linear viscosity behavior of mixtures. For instance, at 20°C, while the viscosity of pure water is about 10^{-3} Pa s and the viscosity of pure ethanol is 1.2×10^{-3} Pa s, a 50–50 mixture has a viscosity of about 3×10^{-3} Pa s. This is why there is some reluctance to use ethanol–water mixtures despite their possible interest for toxicological reasons.

EXERCISE 6.1 *An axial compression column of 0.4 m length packed with 10 μm particles is eluted with an ethanol–water mixture ($\rho \approx 1000$ kg/m^3, $\mu = 0.002$ Pa s). The bed extragranular porosity is 0.4 and superficial velocity 0.001 m/s. Estimate the pressure drop. Perform the same calculation using a porosity of 0.35.*

While the Kozeny–Kármán and Ergun expressions can be considered relatively accurate, it is highly recommended that the constants h_k (and h_B) be determined experimentally for a given application. For instance, any deviation from particle sphericity, or a heterogeneous particle size distribution, will significantly impact the value of the constants h_k and h_B.

If the experimental constants are significantly different from the expected ones, then technical problems might account for this. For instance, the outlet frit could be plugged, especially when the particle diameter distribution is wide. Small particles, coming either from the solid phase or from the fluid mobile phase, tend to migrate within the bed and

[1] Also called the Darcy–Brinkmann–Forchheimer equation by physicists.

to accumulate near or inside the frits. Frit plugging may drastically alter (not to say totally ruin) the column efficiency. Let us note that we use the term "frit" to refer to any type of multi-layer grid.

This leads us to propose the first fundamental rule for chromatography: "In the case of strange column performance, be it relative to pressure drop or to the number of plates, clean the frits!" A corollary is: "If you want to keep your column working for a decent period of time, don't remove the filters on the fluid lines!"

When I say that non-respect of the above rules is the origin of about 50% of the problems that I have had to address, I am normally not totally believed. This is, however, true.

6.1.2 Hydrodynamic dispersion

In Chapter 2 we introduced the MC and PD models to characterize the hydrodynamics of chromatographic columns. We showed that the dispersion of the Z_eRTD was related to the number of mixing cells (J) or, equivalently, to the Péclet number ($Pe = \frac{u_e L_{col}}{D_{ax}}$), according to the model selected. Equating variances in the two models allowed us to show that $Pe = 2J$. We will now relate the J and Pe parameters to physical processes occurring in the chromatographic bed.

For solutes unable to enter inside the particles (or for non-porous particles having no physical interaction with the solutes), there are at least two possible causes for dispersion:

- Molecular diffusion: each solute can diffuse in the extragranular fluid with its own molecular diffusion coefficient D_m. In the absence of flow and of interaction with the particles, a band of solute positioned in the extragranular volume would spread according to a Fickian mechanism, characterized by a diffusion coefficient close to the molecular diffusion coefficient D_m. We say "close" because the presence of the particles affects this diffusion process via a tortuosity factor.
- Eddy diffusion (dispersion): the second reason for dispersion is directly related to fluid motion in porous media. The chromatographic bed, consisting of particles and voids, causes deviations from plug flow. The stream lines can branch for a given distance and then rejoin. Because these branches are likely to be associated with different lengths, travel times along them are not identical. These differences in travel times induce some dispersion, which is called eddy diffusion (dispersion).

With arguments based on a random walk, similar to those for the kinetic theory of gases, one can show to a good approximation that the second process is diffusional in nature, which means that dispersion can be characterized by Fick's law associated with a given dispersion coefficient.[2] This also means that the eddy diffusivity reduces to the product of a mean velocity times a *mean free path*. The former is the interstitial fluid velocity, the latter is taken to be the particle diameter, so that $D_{eddy} = u_e d_p$.

[2] For a more complete analysis, the reader is referred to (Koch and Brady, 1985).

Additionally, as a first approximation, one can consider that the contributions of these two sources of dispersion are additive so that the axial dispersion coefficient is given by:

$$D_{ax} = \vartheta_{diff} D_m + \vartheta_{eddy} u_e d_p \tag{6.5}$$

The parameter ϑ_{diff} is the reciprocal of a bed tortuosity factor, and is typically close to $\frac{1}{\sqrt{2}} \simeq 0.7$, as mentioned in Section 5.8. Experimental evidence shows that the parameter ϑ_{eddy} is normally close to 0.5.

Equation (6.5) is certainly a useful simplification, but it does not fully represent experimental results, and more precise correlations are available. Readers interested in knowing more on the subject, and especially on the influence of column length, particle size, particle size distribution, particle shape and correlations relating D_{ax} to different physico-chemical parameters, are encouraged to read the review by (Delgado, 2006).

Equation (6.5) is sometimes used in a dimensionless form:

$$\frac{1}{Pe_{ax}} = \vartheta_{diff} \frac{\varepsilon_e}{\tilde{u}} + \vartheta_{eddy} \approx \vartheta_{diff} \frac{\varepsilon_e}{\tilde{u}} + \frac{1}{2} \tag{6.6}$$

Pe_{ax} is the Péclet number for axial dispersion based on the particle diameter: $Pe_{ax} = \frac{u_e d_p}{D_{ax}}$. The reduced fluid velocity $\tilde{u} = u d_p / D_m$, introduced in Section 5.8, is often called the particle Péclet number by physicists, and it also equals the product $Re \times Sc$. Thus the reader should not be surprised to find similar correlations but involving different dimensionless numbers in the literature, and should not be confused by the various definitions of the Péclet number. Reduced velocities are typically in the range 10–100 in preparative HPLC (particle size 10–40 μm) and 100–1000 in conventional "low-pressure" chromatography (particle size 200–1000 μm).

Equation (6.6) shows that, at high enough reduced velocities, Pe_{ax} should reach approximately 2.

Equations (6.5) and (6.6) allow an evaluation of the column Péclet number, $Pe = \frac{u_e L_{col}}{D_{ax}}$, for subsequent use in the mass balance equations. Let us stress the fact that, for physical as well as historical reasons, the fluid velocity to be used in the different definitions can be the superficial velocity (*Re* number) or the interstitial velocity (*Pe* number). This is a frequent source of mistakes, and special attention is required.

At this point, a distinction must be made between gases, liquids and supercritical eluents.

If the eluent is a liquid, typical molecular diffusion coefficients are about 10^{-9} m^2/s. Consequently, the first term on the right-hand side of Eq. (6.5) is expected to be slightly above 10^{-10} m^2/s. Using typical fluid velocities in HPLC systems of 10^{-3} m/s, the second term Eq. (6.5) is in the range 0.5×10^{-8} m^2/s, for a particle size of 10 μm, to about 0.5×10^{-7} m^2/s, for particle sizes of about 100 μm. The axial dispersion coefficient is thus dominated by the eddy dispersion term and increases with the fluid velocity. Under these conditions the *Pe* number is given by $Pe = 2J = \dfrac{u_e L_{col}}{\vartheta_{eddy} u_e d_p} = \frac{1}{\vartheta_{eddy}} \dfrac{L_{col}}{d_p}$. Consequently, *Pe* and the number of mixing cells are simply proportional to the ratio of the column length to the particle diameter. Increasing the column length or decreasing the particle size will thus increase *Pe* and *J* proportionally, independently of fluid velocity.

If the eluent is a gas, molecular diffusion coefficients being about 10 000 times greater than liquid diffusion coefficients, the first term on the right-hand side of Eq. (6.5) can be important or even dominant. At low fluid velocities, when the first term is dominant, the axial dispersion coefficient is thus independent of fluid velocity. The *Pe* number is given by $Pe = 2J = \frac{u_e L_{col}}{\vartheta_{diff} D_m}$ and becomes independent of particle size. At low fluid velocities, a good way to reduce hydrodynamic dispersion is to *increase* fluid velocity. In other words, we are located on the left side of the minimum of the Van Deemter curve (see Section 5.8), a zone to be avoided for production purposes.

If the eluent is a supercritical fluid, the order of magnitude of diffusion coefficients is 10^{-7} m/s, between that of liquids (10^{-9} m^2/s) and that of gases (10^{-5} m^2/s). The first term on the right-hand side of Eq. (6.5) is expected to be slightly greater than 10^{-8} m^2/s. The fluid velocity is normally in the range 10^{-3} to 10^{-2} m/s, so for particle sizes of 10 to 100 μm the second term is in the range 0.5×10^{-8} to 0.5×10^{-6} m^2/s. Consequently, while eddy dispersion is likely to play the biggest role in a majority of situations, the predominance of molecular diffusion cannot be ruled out. This behavior is intermediate between those of gases and liquids but is in general closer to liquids.

The above considerations on the influence of fluid velocity are limited to hydrodynamics, and do not take into account possible mass transfer limitations. When kinetic limitations and axial dispersion are taken into account together, we showed in Chapter 2 and Section 5.8 that the column can be approximated by a certain number of plates *N* in which equilibrium is assumed to have been reached. The dependence of *N* on fluid velocity is given by the Van Deemter curve (see Eq. (5.138)), sometimes expressed in reduced units (see Eq. (5.140)).

If one feels more comfortable using lumped axial dispersion coefficients than plate numbers, one can use the PD–$\overline{\text{Equil}}$ model, characterized by an *apparent axial dispersion coefficient* D_{ax}^{app} or equivalently by an *apparent Péclet number* $Pe_{ax}^{app} = \frac{u_e L_{col}}{D_{ax}^{app}}$. Just as we wrote the analogy $Pe = 2J$ in Chapter 2, we can here write $Pe_{ax}^{app} = 2N$, so:

$$HETP = \frac{L_{col}}{N} = \frac{2L_{col}}{Pe_{ax}^{app}} = 2\frac{D_{ax}^{app}}{u_e} = \frac{A_e}{u_e} + B_e + C_e u_e \tag{6.7}$$

where A_e, B_e and C_e are the coefficients of the Van Deemter equation (5.138).

EXERCISE 6.2 *To be done adequately, a given separation requires a column equivalent to about 500 plates. The Van Deemter curve can be represented in reduced units (Eq. (5.140)) with* $\alpha = 2$, $\beta = 3$ *and* $\gamma = 2$. *The fluid has a viscosity of* 10^{-3} *Pa s, the diffusion coefficient is about* 10^{-9} *m*^{2}*/s. Consider using a system filled with 20 μm particles operated either at 20 bar or at 50 bar pressure drop. The extragranular porosity is taken to be 0.35. Compare the fluid velocities and bed length to be used in each situation.*

Operating a chromatographic process consists in successively injecting solutions of different compositions, so axial gradients are always present in the system. These gradients are then smoothed during the transport in the column due to different processes, including hydrodynamic axial dispersion. The axial dispersion coefficient D_{ax} is thus of primary importance in designing chromatographic processes.

In an ideal system, in which concentration profiles are radially uniform, the radial dispersion coefficient D_{rad} would have no effect. Chromatographic systems being designed so that radial non-uniformities are minimized, knowledge of the radial dispersion coefficient D_{rad} is normally considered of secondary importance to knowledge of the axial dispersion coefficient. This is certainly a good first approximation, but, real systems being imperfect, some knowledge of the radial dispersion coefficient D_{rad} is of interest as it can help in reducing radial non-uniformities. Additionally, we will see in Section 6.5 that radial dispersion limits the effects of viscous fingering.

Ideally, we would like to have axial dispersion coefficients as low as possible and radial dispersion coefficients as high as possible. Unfortunately, the opposite holds! As with Eq. (6.6), it has been proposed to express Pe_{rad}, a particle Péclet number for radial dispersion $\left(Pe_{rad} = \frac{u_e d_p}{D_{rad}}\right)$, as a function of the reduced fluid velocity. A simplified expression is given by:

$$\frac{1}{Pe_{rad}} \approx \vartheta_{diff} \frac{\varepsilon_e}{\tilde{u}} + \frac{1}{12} \tag{6.8}$$

One cannot expect more than orders of magnitude from this expression, but it shows that, at significantly reduced fluid velocity, Pe_{rad} is about six times higher than Pe_{ax}. In other words, at significant velocities, the radial dispersion coefficient D_{rad} is about six times smaller than the axial dispersion coefficient D_{ax}.

For a more comprehensive presentation of radial dispersion coefficients, the reader is again referred to (Delgado, 2006). Readers interested in a more precise description of hydrodynamic dispersion, based on more detailed models, possibly involving CFD, are referred to (Augier *et al.*, 2008), (Daneyko *et al.*, 2011), (Desmet, 2013) and (Koch and Brady, 1985).

6.1.3 Fluid compressibility

One important assumption in the preceding section is that both fluid and bed are incompressible.

When the carrier fluid is compressible (a gas or a supercritical fluid), a change in pressure along the column will eventually lead to a change in density ρ, in fluid velocity u and possibly in viscosity μ, should this parameter be pressure-dependent. In general, we thus have to consider profiles $P(z)$, $u(z)$, $\rho(z)$ and $\mu(z)$, where z is the abscissa in the column, equal to zero at the inlet and L_{col} at the outlet. Fluid compressibility has a practical consequence for chromatographers: the volumetric flow rate and fluid velocities, are no longer constant along the column axis.

Modeling the flow of compressible fluids requires knowledge of an equation of state allowing one to calculate the fluid density under given pressure and temperature conditions:

$$\rho = f(P,T) \qquad \text{equation of state} \tag{6.9}$$

This equation of state can take very simple forms, such as $\rho = cst$ for incompressible fluids or $\rho = MP/(RT)$ for ideal gases. It can be highly sophisticated for complex compressible mixtures. The relatively simple and flexible Peng–Robinson equation of state is given in Appendix F; for more detailed information, the reader is referred to thermodynamics textbooks including (Poling *et al.*, 2001).

It is convenient to use a normalized quantity for fluid density, known as the *compressibility factor Z*:

$$Z = \frac{MP}{\rho RT} \tag{6.10}$$

Z is normally obtained by resolving an implicit equation derived from the equation of state (see Appendix F).

The compressibility factor is 1 for an ideal gas, and is normally less than 1 for compressible fluids under conditions of practical interest for chromatography. For an incompressible liquid, density is independent of pressure so $Z \sim P$ at constant temperature.

A conservation balance allows the fluid velocity at position z to be related to the mass flow rate Q^{mass}, the local compressibility factor and pressure:

$$u(z) = \frac{Q_{mass}}{\rho \Omega_{col}} = \frac{Z(z)RT}{MP(z)} \frac{Q_{mass}}{\Omega_{col}} \tag{6.11}$$

Equation (6.11), expressing fluid velocity $u(z)$ versus mass flow rate, can be modified in order to relate $u(z)$ to the pressure P^{out}, temperature T^{out} and compressibility factor Z^{out} at the outlet of the column:[3]

$$u\,(z) = u\,(out)\,\frac{P^{out}}{P(z)}\frac{T(z)}{T^{out}}\frac{Z(z)}{Z^{out}} \tag{6.12}$$

Notice that, in the case of an incompressible fluid, P being proportional to Z at constant temperature, the above equation becomes $u\,(z) = cst$ under isothermal conditions.

In all generality, viscosity is pressure- and temperature-dependent, so one needs to know:

$$\mu = g\,(P,T) \quad \text{viscosity dependence} \tag{6.13}$$

Some expressions allowing one to estimate viscosity for supercritical fluids are given in Appendix F. Much more information will be found in (Poling *et al.*, 2001).

The influence of pressure on both fluid velocity and viscosity does not allow the Kozeny–Kármán and Ergun equations to be used in the integrated form presented in Section 6.1. Instead, these equations must be used locally, in a differential form:

$$\frac{dP}{dz} = h_k \frac{\mu\,(z)}{d_p^2}\frac{(1-\varepsilon_e)^2}{\varepsilon_e^3}u(z) + h_B \frac{\rho(z)}{d_p}\frac{1-\varepsilon_e}{\varepsilon_e^3}u(z)^2 \tag{6.14}$$

We assume that h_k and h_B are independent of P and T.

[3] It is normally appropriate to use the conditions at the column outlet (and not the inlet) as a reference. The reason is that columns are normally controlled by outlet pressure and outlet flow rate control devices, so the outlet fluid velocity is normally the most easily accessible.

In order to highlight the position-dependent variables we can write:

$$\frac{dP(z)}{dz} = -\breve{A}\mu(z)u(z) - \breve{B}\rho(z)u(z)^2 \tag{6.15}$$

where $\breve{A} = h_k \frac{1}{d_p^2} \frac{(1-\varepsilon_e)^2}{\varepsilon_e^3}$ and $\breve{B} = h_B \frac{1}{d_p} \frac{1-\varepsilon_e}{\varepsilon_e^3}$ are constants, independent of position in the column. The coefficient $\breve{A}$ is the reciprocal of the permeability when laminar flow conditions prevail.

The endothermicity of the fluid's expansion during its travel through the column, as well as the heat generated by viscous flow, are neglected in this section, so the entire process is assumed to be isothermal. Possible reasons for various temperature profiles are discussed in Section 6.3.

Pressure and axial velocity profiles are determined by the simultaneous solution of Eqs. (6.10), (6.13) and (6.15). There is no specific difficulty, but numerical tools are now required even for the simple question of estimating pressure drop.

One important consequence of the variation of the volumetric flow rate (and fluid velocity) inside the column is that the Z_eRT cannot be calculated by the classical equation $t_0 = \varepsilon_e V_{col}/Q$. Knowing the fluid velocity profile along the column axis, the zero retention time must be determined by integration of "local retention times" over the entire column:

$$t_0 = \int_0^{L_{col}} \frac{\varepsilon_e}{u(z)} dz \qquad t_{0R} = \int_0^{L_{col}} \frac{\varepsilon_T}{u(z)} dz \tag{6.16}$$

When $u(z)$ is constant, Eq. (6.16) reduces to the classical integrated expressions.

Using Eq. (6.12) expressing the local velocity as a function of the outlet velocity, one can express the retention time as a function of the zero retention time calculated from the outlet parameters and a correction factor j. This correction factor depends on the pressure and compressibility factor over the column. Under isothermal conditions, one obtains:

$$\begin{aligned} t_0 &= j t_0^{out} \qquad t_0^{out} = \frac{\varepsilon_e L_{col}}{u^{out}} \\ t_{0R} &= j t_{0R}^{out} \qquad t_{0R}^{out} = \frac{\varepsilon_T L_{col}}{u^{out}} \\ j &= \frac{1}{L_{col}} \frac{Z^{out}}{P^{out}} \int_0^{L_{col}} \frac{P(z)}{Z(z)} dz \end{aligned} \tag{6.17}$$

Notice that, in the case of incompressible fluids, $j = 1$ and $u = u^{out}$ so that Eq. (6.17) reduces to $t_0 = \varepsilon_e L_{col}/u$ ($t_{0R} = \varepsilon_T L_{col}/u$).

6.1.3.1 Gas chromatography

As an illustration, we now address the case of a low-pressure gas eluent considered to be an ideal gas.[4] Under this condition, the compressibility factor Z equals 1. Assuming

[4] When preparative gas chromatography (which has led to very few applications) is used for purifying liquid mixtures, the ability to vaporize the liquid flow rate at the column inlet is essential. The lower the pressure, the easier the vaporization. The maximum injection flow rate is thus directly related to the inlet pressure.

laminar flow, the solution of Eqs. (6.10), (6.12), (6.13) and (6.15) is easy because gas viscosity is nearly independent of pressure. Under isothermal operation, one obtains:

$$u(z) = u^{out} \frac{P^{out}}{P(z)} \tag{6.18}$$

$$\mu = \text{constant} \tag{6.19}$$

$$\frac{dP(z)}{dz} = -\check{A}\mu u(z) \tag{6.20}$$

This leads to the classical solution:

$$P^{inl2} - P^{out2} = 2\check{A}\mu u^{out} P^{out} L_{col} \tag{6.21}$$

It is noteworthy that, unlike for liquids, the difference between the squares of the inlet and outlet pressures is proportional to the column length.

The correction factor j defined in Eq. (6.17) is called the James–Martin factor in the case of an ideal gas, and one obtains:

$$j = \frac{3}{2} \frac{(P^{inl}/P^{out})^2 - 1}{(P^{inl}/P^{out})^3 - 1} \tag{6.22}$$

Since P^{inl}/P^{out} depends on the flow rate, j also depends on the flow rate, and the retention time is not linearly related to the outlet flow rate Q^{out}.

It can also be shown that the pressure drop influences dispersion in the column (Dixon and Ma, 1988; Pazdernik and Schneider, 1981). For an ideal gas, the actual plate number is equal to the plate number that would have been obtained without pressure drop times the James–Martin factor:

$$N = jN_{\text{isobaric}} \tag{6.23}$$

Consequently, since the James–Martin factor j is less than 1, the pressure drop reduces to the equivalent number of plates of the column.

6.1.3.2 Supercritical fluid chromatography

The hydrodynamic behavior of a supercritical eluent is expected to be intermediate between that of a liquid (fluid velocity constant, pressure drop varying linearly with bed length, retention times proportional to bed length) and that of an ideal gas (fluid velocity increasing along the column axis, pressure drop varying with the square root of the bed length, retention times not proportional to bed length). At this stage, recall that typical orders of magnitude for viscosities are 10^{-4} to 10^{-5} Pa s for supercritical fluids versus 10^{-3} Pa s for liquids, and about 10^{-5} Pa s for gases, and that viscosities increase with density for supercritical fluids.

Pressure-drop estimates are thus of primary importance for the economic evaluation of gas chromatography processes: the higher the inlet pressure, the lower the possible injection flow rate; that is why hydrogen is normally preferred as a vector gas compared to nitrogen.

The resolution of the system of equations (6.10), (6.12), (6.13) and (6.15) is not associated with specific numerical difficulties. The real difficulty lies in the selection of an adequate equation of state, and in the selection of the $\mu(T,P)$ expression.

In the case of carbon dioxide eluent, (Perrut, 1987) showed that experimental pressure drop and zero retention time results can be represented to an acceptable accuracy by choosing the Peng–Robinson equation of state and the viscosity correlations recommended by Stiel–Thodos which are given in Appendix F. Contributions associated with different equations of state and viscosity correlations are found in (Rajendran *et al.*, 2005), (Rajendran *et al.*, 2008) and (Rajendran, 2012).

In order to illustrate the impact of fluid compressibility and viscosity dependence on pressure, equations (6.10), (6.12), (6.13) and (6.15) have been solved for pure carbon dioxide (T_c = 304.2 K, P_c = 74 bar), using parameters that can be considered typical for supercritical fluid chromatography (SFC) applications. The results are given in Figure 6.2.

At 325 K, two sets of outlet pressure have been used: 100 bar, at which the fluid is relatively expanded (ρ = 352 kg/m^3, Z = 0.462), and 250 bar, at which the fluid is fairly dense (ρ = 818 kg/m^3, Z = 0.497). The associated viscosities are respectively 3.2×10^{-5} Pa s at 100 bar and 7.2×10^{-5} Pa s at 250 bar.

In Figure 6.2, pressure, velocity, density and viscosity are expressed as the ratios of these parameters to their values at the outlet of the column. These curves show that relative density and velocity exhibit variations of about 30% within the bed for the less dense fluid (100 bar) and almost 10% for the more dense fluid (250 bar). The impact of fluid compressibility on velocity variation can thus be significant and must be taken into account. It is interesting to note that, given that the fluid velocity and viscosity vary in opposite directions, these effects in a way compensate one another when the

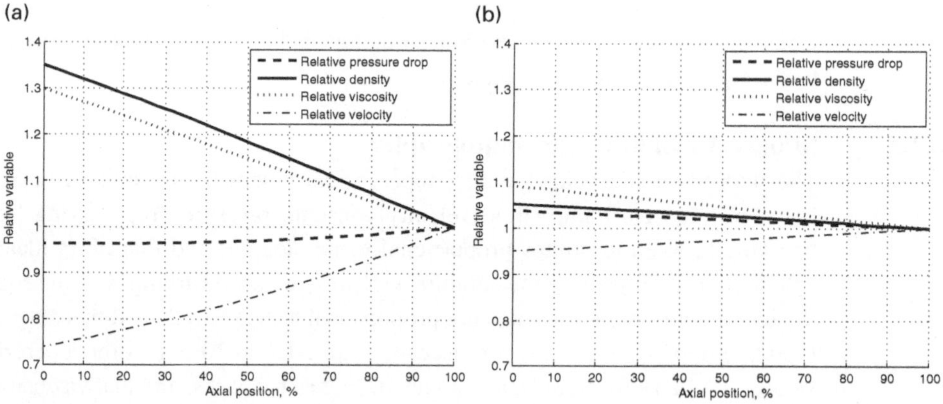

Figure 6.2 Axial profiles of normalized pressure drop, density and viscosity in SFC. Variables are normalized to their values at outlet conditions. Peng–Robinson and Stiel–Thodos correlations are used to estimate density and viscosity (Appendix F). Parameters used in the Kozeny-Kármán equation are h_k = 150 (MKSA), extragranular porosity ε_e = 0.4 and L_{col} = 0.5 m. The fluid velocity at the outlet of the column is $u = 10^{-2}$ m/s. (a) Pure carbon dioxide at T = 325 K and outlet pressure 100 bar. (b) Pure carbon dioxide at T = 325 K and outlet pressure 250 bar.

Kozeny–Kármán equation is used, so the pressure drop per unit length is more constant than expected. For a bed length of 0.5 m, one obtains about a 15 bar pressure drop when the outlet pressure is 100 bar, and 40 bar pressure drop when the outlet pressure is 250 bar. This is consistent with the fact that the viscosity at 250 bar is approximately 2.25 times larger than it is at 100 bar. Based on this simulation, one can conclude that, while viscosity, density and velocity vary significantly over the column (possibly by 30% or more), pressure drop is *approximately* linear with column length, as observed with liquids. This is due to the above-mentioned compensation between viscosity and velocity when the Kozeny–Kármán equation is used.

The variation of fluid density and velocity with axial position given in Figure 6.2 can be used to calculate the zero retention time and pressure drop versus column length. The results are given in Figure 6.3 and show that the deviation from linearity of the variation of zero retention time with column length is more pronounced at moderate pressure (100 bar) than it is at higher pressure (250 bar). Deviations from zero retention time calculated on the basis of the outlet velocity can be 10–15% at moderate pressures.

While real, the evolution of fluid velocities and viscosities along the column are moderate (at least compared to the behavior of ideal gases) so that, in order to avoid the resolution of the differential equation (6.15), one can use a back-of-the-envelope calculation. Usually, the pressure at the outlet of the column is known; fluid velocity, specific gravity and viscosity can thus be estimated at this pressure. A first pressure drop can be guessed by using the integrated form of the Kozeny–Kármán equation (6.1) with the fluid viscosity and velocity values at the outlet of the column. This leads to an estimate of the pressure at the inlet of the column. The same procedure (estimate of fluid velocity, specific gravity, viscosity, global pressure drop) is iterated, replacing the outlet pressure by the mean pressure in the column. One iteration is normally enough to get a good approximation of pressure drop, average fluid velocity and thus zero retention time.

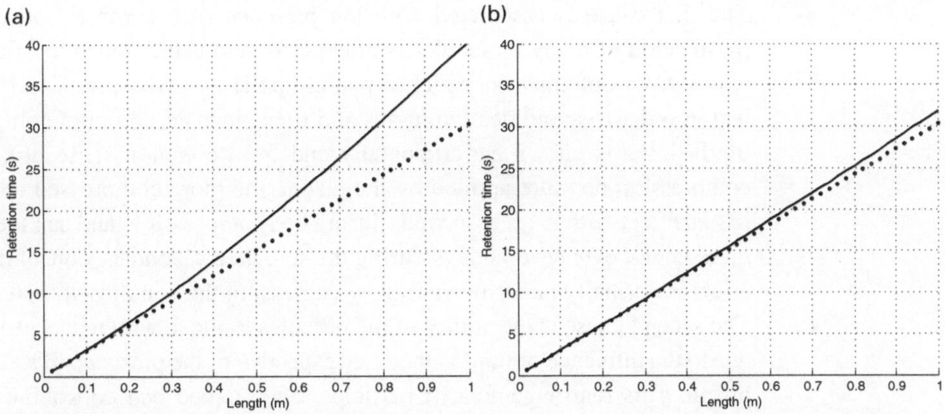

Figure 6.3 Influence of column length on zero retention time in SFC. Conditions are as given in Figure 6.2. Dotted line: zero retention time calculated using outlet velocity. (a) Pure carbon dioxide at $T = 325\ K$ and outlet pressure 100 bar. (b) Pure carbon dioxide at $T = 325\ K$ and outlet pressure 250 bar.

EXERCISE 6.3 *An SFC column of 45 cm length and 5 cm diameter is packed with particles of 10 μm diameter ($\varepsilon_e = 0.4$) and is eluted with 50 kg/h of CO_2 at 325 K. The pressure at the outlet of the column is set to 125 bar. Using the Peng–Robinson equation of state as well as Thodos correlations for non-polar fluids given in Appendix F, estimate the pressure drop in the column and the Z_eRT.*

Finally, we should not forget that fluid density affects the adsorption isotherms of the solutes in supercritical fluid chromatography. This means that the pressure profile in the column not only affects the zero retention time but also the retention. This can only be accounted for by solving detailed mass-balance equations, as briefly described in Chapter 7.

Additional information on the effect of pressure drop on the hydrodynamics of SFC is found in (Rajendran *et al.*, 2005), (Rajendran *et al.*, 2008) and (Rajendran, 2012).

6.2 Bed compressibility

6.2.1 General considerations

Though it is often neglected for modeling purposes, experience shows that the compressibility of chromatographic beds can have a significant impact on process performance. By "compressibility" we mean the ability of a bed to shrink or to expand, as well as the ability of the solid phase to deform under the influence of physical forces. Compressibility associated with a loss or a gain in bed length and porosity can be reversible or not.

We will first describe the main reasons for a chromatographic bed to shrink or possibly expand over time:

- The first cause is associated with the presence of a compression piston. The purification of many molecules is now performed using dynamic axial compression (DAC) columns, in which a piston applies a given pressure P_{mech} to the bed in order to avoid dead volumes and to achieve an optimum compactness of the bed. As is known experimentally and from common sense, the higher the compression pressure applied by the piston, the more compressed the bed. This has an effect within a few seconds during bed compression, and can then continue more gently over several days during the so-called secondary consolidation. This effect is normally partly reversible by decreasing the compression pressure.
- The second cause is the ability of the particles to move with respect to each other due to the influence of fluid velocity, irrespective of the presence of a compression piston. This rearrangement of particles, often called bed consolidation, allows some holes to be filled and increases the compactness of the bed. It can show effects over hours or days, depending on parameters that are not totally understood. It is irreversible, can be magnified if broken particles are formed, and can even plug the chromatographic bed.

- The third cause comes from the tendency of many ion-exchange resins to shrink and swell when changing ionic form or when put into contact with solutions of different concentrations. Changes in bed length as high as 25% can be experienced with moderately reticulated resins. This induces high bed stresses and significantly impacts extragranular porosity. This has an effect in a few minutes, the time required for the resins to change form, and is normally reversible.
- Finally, in addition to a pure repositioning of particles among themselves, piston compression or eluent velocity can induce particle deformation. This is the case when soft gels are used, typically for protein purification. The consequence can be as extreme as a plugged column, so that it becomes virtually impermeable to the fluid. This can have an effect within a few seconds if fluid velocity is too high, and is generally partly irreversible.

These four factors affecting bed length also affect extragranular porosity and possibly intragranular porosity, as well as particle shape. This can impact permeability more or less severely.

Notice that the first two factors can affect bed compression even with incompressible media (by which we mean that the particles themselves are incompressible): the bed itself can possibly shrink or expand.

The column length and total porosity are easily related if one assumes that the volume $\overline{\overline{V}}$ of the skeleton of the particles is constant. Under this assumption, the conservation of the volume of skeleton leads to:

$$(1-\varepsilon_{T1})\,L_{col1}=(1-\varepsilon_{T2})\,L_{col2} \tag{6.24}$$

where ε_{T1} and L_{col1} are the total porosity and bed length in state 1 and ε_{T2} and L_{col2} in state 2.

Introducing the intragranular and extragranular porosities, Eq. (6.24) can be written:

$$(1-\varepsilon_{e1})\,(1-\varepsilon_{i1})\,L_{col1}=(1-\varepsilon_{e2})\,(1-\varepsilon_{i2})\,L_{col2} \tag{6.25}$$

Equations (6.24) and (6.25) only assume conservation of the skeleton volume, and are thus not dependent on any assumption regarding particle deformation.

We are especially interested in extragranular porosity variation, which directly impacts the pressure drop. In order to derive the extragranular porosity variation from the column length variation, one needs to know the intragranular porosity.

The total and intraparticular porosities can in principle be determined by measuring zero and excluded retention times. From their difference, one has access to the extraparticular porosity, as explained in Chapter 4. When one works with solid particles like silica, which break without deforming, the intraparticular porosity is constant. The situation is more complex with soft particles, but for simplicity we will assume that all bed length variation is attributable to extragranular porosity variation.

If one assumes that the particles do not change shape, so that their intragranular porosity is constant, Eq. (6.25) leads to:

$$(1-\varepsilon_{e1})\,L_{col1}=(1-\varepsilon_{e2})\,L_{col2} \tag{6.26}$$

Later on in this section, we will assume that Eq. (6.26) holds.

The possible decrease in extragranular porosity yields to an increase in pressure drop, as forecast by the Kozeny–Kármán equation. Recall that, for porosities around 0.35, a change of 0.01 in extragranular porosity leads to a 10% change in bed permeability. Small variations in bed length are thus likely to influence permeability significantly.

Experience shows that compressibility/expansion effects are in general dependent on column geometry: for a given bed length, larger diameter columns are normally more sensitive to compression effects, so the scale-up is not as straightforward as normally anticipated.

Using the detailed mechanics of porous media to precisely model the bed evolution under stress probably exceeds the scope of this book.[5] We will thus limit our presentation to simple considerations, which will prove sufficient to explain experimental results at least semi-quantitatively. Let us call this the *simplistic model.*

6.2.2 Simplistic model

The key to understanding how column geometry can influence bed compressibility, and thus permeability, is to understand how the column wall can influence the bed structure. Let us start with purely qualitative considerations. In a column of a significant length (let us say 100 cm) and very small diameter (let us say 1 cm), the wall of the column restricts the freedom of the particles to move and thus to rearrange under the influence of viscous forces. This is often mentioned as a "wall support" effect. When the column diameter increases, the influence of this effect decreases and the particles have more freedom to rearrange. The effect is likely to vanish when the column diameter is significantly larger than its length.

Attempts to model the behavior of the chromatographic bed under stress have been described. The simplest models treat the bed as a one-dimensional elastic continuum with pore space, and describe the change in solid stress, strain and pressure drop along the axis of the vertical bed (Verhoff and Furjanic, 1983; Shadday, 2006; Keener *et al.*, 2008).

This approach has been extended to more precise two-dimensional descriptions of the bed, allowing simulation of axial and radial pressure gradients, porosity and velocity profiles (Östergren *et al.*, 1998; Östergren and Trägardh, 1999). These latter models will not be discussed further in this book.

The chromatographic bed is kept in place as a result of the action of various forces with different origins. The pressure drop in the column is influenced by the flow direction. While working upwards in columns having no upper part to maintain the top of the bed in a fixed position, the bed can expand up to fluidization. This so-called expanded-bed chromatography mode is not covered in our presentation.

[5] These are very interesting contributions, coming especially from so-called reservoir engineering, a branch of petroleum engineering, and could be adapted to chromatography. See, for instance, (Wang *et al.*, 2009; Kováčik, 1999; Wang, 1984).

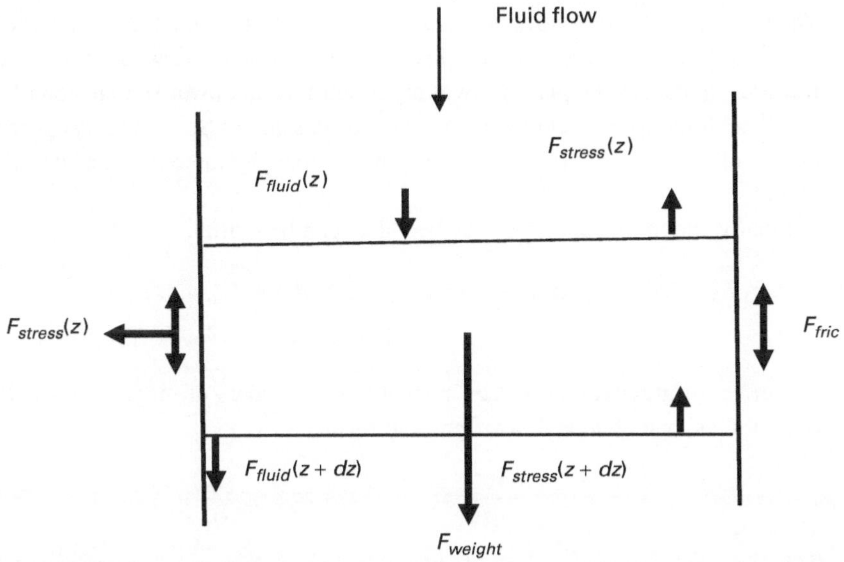

Figure 6.4 Schematic representation of the forces acting on a slice of a chromatographic bed. Downwards flow.

A schematic description of the forces acting on a slice of a chromatographic bed operated downwards is presented in Figure 6.4. The pressure of the fluid entering the column section results in a seepage force $F_{fluid}(z) = \Omega_{col}P_{fluid}(z)$ acting in the direction of the flow. At the outlet of the slice, the fluid pressure action results in a seepage force $F_{fluid}(z + dz) = \Omega_{col}P_{fluid}(z + dz)$, also acting in the direction of the flow. Acting downward at the column centroid is the force associated with the weight of solid-phase material, $F_{weight}(z) = g\left(\rho_{bed} - \rho_{fluid}\right)\Omega_{col}dz$. The density ρ_{bed} is the ratio of the mass of a slice of bed divided by the volume of the slice.

The forces acting downwards must be balanced by an upward force carrying the bed. This force, F_{stress}, is expressed as the product of the column cross section Ω_{col} and a pressure referred to as the stress, $S(z)$. While the stress can have an axial and a radial contribution, only the axial contribution is taken into account in this presentation. The forces associated with this stress are respectively $F_{stress}(z) = \Omega_{col}S(z)$ and $F_{stress}(z + dz) = \Omega_{col}S(z + dz)$ at levels z and $z+dz$.

F_{stress} is the force likely to be the source of bed or particle compression, the forces associated with fluid pressure being unlikely to have any effect on the particles as the intragranular pores are filled by the fluid.

In addition, a frictional force F_{fric} is assumed to exist at the wall of the column, opposed to any motion of the bed. It is probably rather intuitive to say that the larger the peripheral surface of the slice and the higher the stress, the higher the frictional force. A simple model assumes this friction force to be proportional to both the stress and the peripheral surface of the slice:

$$F_{fric}(z) = \beta S(z) D_{col}dz \quad \beta \text{ is dimensionless} \tag{6.27}$$

The sign of the friction force may vary according to the situation: when the bed moves downwards due to the action of a piston or of the viscous flow, the frictional force acts upwards. In the case of particle swelling promoting an upward expansion of the bed, the frictional force would act downwards (see discussion in Section 6.2.5). The proportionality coefficient β is considered to be an adjustable phenomenological coefficient of our model.

Finally, the force balance on the bed slice can be written:

$$\begin{aligned} F_{fluid}(z) + F_{stress}(z+dz) &= F_{fluid}(z+dz) + F_{stress}(z) \\ &\quad + g(\rho_{bed} - \rho_{fluid})\,\Omega_{col} dz \mp F_{fric}(z) \end{aligned} \tag{6.28}$$

The effect of gravitational forces normally being orders of magnitude smaller than the other forces, one obtains in a differential form:

$$\frac{dS(z)}{dz} = \frac{dP_{fluid}(z)}{dz} \mp S(z)\frac{\beta}{D_{col}} \tag{6.29}$$

This force balance must be satisfied for the abscissa varying between 0 and L_{col}.

The bed is assumed to be compressible, so the determination of the column length is not a trivial question. The volume of particles contained in a slice of volume $\Omega_{col}dz$ is $d\bar{V} = (1-\varepsilon_e)\Omega_{col}dz$. By integration, one obtains $\int_0^{L_{col}}(1-\varepsilon_e)\,dz = \frac{\bar{V}}{\Omega_{col}} = cst$, where $\bar{V}$ is the total volume of beads contained in the column, which is assumed to be constant (see Eq. (6.26)). This means that, if one knows that a bed length L_{col0} and an extragranular porosity ε_{e0} are obtained under no stress, one has:

$$\int_0^{L_{col}}(1-\varepsilon_e)\,dz = (1-\varepsilon_{e0})\,L_{col0} \tag{6.30}$$

Introducing a normalized abscissa $\xi = z/L_{col}$, Eq. (6.29) can be written:

$$\begin{aligned} &\frac{dS(z)}{d\xi} + \breve{\beta}S(z) = \frac{dP_{fluid}(z)}{d\xi} \\ &\xi = \frac{z}{L_{col}} \\ &\breve{\beta} = \frac{\beta L_{col}}{D_{col}} \end{aligned} \tag{6.31}$$

The meaning of the differential force balance can be discussed in simple limiting terms. In the absence of frictional force at the wall, the viscous force acting in the fluid direction (downwards in our example) must be balanced by forces in the opposite direction. The gradient of stress and fluid pressure are thus identical.

When the wall frictional force increases, a portion of the fluid force is compensated by the wall, so that the stress in the bed is lower than in the absence of wall friction. The bed is thus expected to be less compact due to the influence of wall friction.

Fluid drag, as a body force, is proportional to the square of the column diameter, whereas the wall frictional force is proportional to the column diameter. Fluid drag can thus collapse the bed in large-diameter columns packed with compressible gels. This simple model clearly shows that an effect of the L_{col}/D_{col} ratio can be expected on bed compactness, making scale-up not as easy as normally desired.

At the inlet of the column, the bed can either be free under no mechanical stress, or under a mechanical stress imposed by a compression piston. If one denotes by P_{mech} the mechanical pressure exerted by the piston on the solid particles, one has:

$$S\,(\xi = 0) = P_{mech} \tag{6.32}$$

In the absence of mechanical pressure, the bed inlet is assumed to be under no stress, so $S\,(\xi = 0) = 0$.

Note: the pressure P_{mech} should not be confused with the compression target $P_{compress}$ set on the hydraulic jack of the DAC equipment. Because of the fluid pressure at the inlet of the column one has $P_{mech} = P_{compress} - P_{fluid}(inlet)$.

Integration of Eq. (6.31), together with boundary conditions in Eq. (6.32), will give the axial distribution of the stress $S\,(z)$. This is important information, as it will tell us, for example, where the bed is prone to contract (or expand).

In order to better understand the key features of the model, we will start with the simple situation of an incompressible bed packed with an incompressible medium, so that the bed properties are uniform in the axial direction. With the assumptions of constant fluid velocity and axial uniformity, the pressure drop per unit length is constant, so $\dfrac{dP_{Fluid}\,(z)}{d\xi} = \Delta P_0$. ΔP_0 is the pressure drop occurring in the bed in the absence of deformation.

Under the non-compressibility (thus axial uniformity) assumption, the force-balance equation (6.31) is finally written:

$$\frac{dS(z)}{d\xi} + \breve{\beta} S\,(z) = \Delta P_0 \tag{6.33}$$

with the boundary condition $S\,(\xi = 0) = P_{mech}$.

The fluid pressure relative to atmospheric pressure varies between $P\,(\xi = 1) = 0$ at the column outlet (assumed to be at atmospheric pressure) and $P\,(\xi = 0) = \Delta P_0$ at the column inlet, by the definition of ΔP_0.

By defining a normalized stress $\sigma\,(\xi) = \frac{S}{\Delta P_0}$, the force balance equation (6.33) can be written in a normalized form:

$$\frac{d\sigma\,(z)}{d\xi} + \breve{\beta}\sigma\,(z) = 1 \tag{6.34}$$

the boundary condition being $\sigma\,(0) = \frac{P_{mech}}{\Delta P_0}$.

The evolution of the normalized stress along the column axis is obtained by solving Eq. (6.34):

$$\sigma\,(\xi) = \frac{1}{\breve{\beta}}\left(1 - e^{-\breve{\beta}\xi}\right) + \frac{P_{mech}}{\Delta P_0} e^{-\breve{\beta}\xi} = \frac{1}{\breve{\beta}} + \left(\frac{P_{mech}}{\Delta P_0} - \frac{1}{\breve{\beta}}\right) e^{-\breve{\beta}\xi} \tag{6.35}$$

We now consider a few limiting situations that will help understanding the key features of Eq. (6.35).

When the $\breve{\beta}$ factor equals zero (due either to a negligible frictional factor β or to a very small L_{col}/D_{col} ratio), the expression for the normalized stress becomes:

$$\sigma(\xi) = \xi + \frac{P_{mech}}{\Delta P_0} \tag{6.36}$$

so that:

$$S(\xi) = \Delta P_0 \xi + P_{mech} \tag{6.37}$$

This shows that, in the absence of frictional force, the stress in the incompressible bed *increases* linearly with the normalized abscissa, from P_{mech} to $P_{mech} + \Delta P_0$.

Another limiting situation is obtained in the absence of piston compression, so that $P_{mech} = 0$. Under this condition, one obtains for the incompressible bed:

$$\sigma(\xi) = \frac{1}{\breve{\beta}}\left(1 - e^{-\breve{\beta}\xi}\right) \tag{6.38}$$

so that the stress equals zero at the column inlet and *increases* gradually with the abscissa.

In both cases, there is a non-zero mechanical stress at the column outlet, even in the absence of mechanical compression: chromatographers know that outlet frits must be kept in place by appropriate mechanical devices . . .

Alternatively, if the pressure drop ΔP_0 in the bed is negligible with respect to the mechanical pressure P_{mech} applied by the piston on the bed, one obtains:

$$\sigma(\xi) = \frac{P_{mech}}{\Delta P_0} e^{-\breve{\beta}\xi} \tag{6.39}$$

so that the normalized stress is maximum at the column inlet and *decreases* afterwards.

Consequently, whereas the stress increases versus the abscissa due to viscous forces in the absence of mechanical compression, it is expected to decrease with the abscissa in the case of a significant mechanical compression. When both viscous forces and mechanical compression are involved, the variation of the stress with the abscissa can be one way or the other, depending on the sign of $\left(\frac{P_{mech}}{\Delta P_0} - \frac{1}{\breve{\beta}}\right)$, as shown by Eq. (6.35).

The situation is more complex with compressible beds or particles: when the bed or possibly the particles are compressible, the stress impacts the local porosity, which in return impacts pressure drop and thus the stress. One thus has to solve an implicit set of equations.

In order to model compressible beds, a constitutive equation relating strain (deformation of the bed or particles) to stress must be postulated. For instance, one can relate the bed length variation to the stress via the Young modulus E. Under the elasticity assumption (reversibility of deformation), Hooke's law allows the Young modulus to be defined as the ratio of stress to strain, $\frac{\Delta L}{L}$:

$$\frac{\Delta L}{L} = \frac{S(z)}{E} \tag{6.40}$$

The Young modulus E thus has units of pressure.

If this local equation is written over different axial slices of the column, after summation and averaging, one can show that the total bed deformation is the ratio of the average value of the stress $\langle S \rangle$ in the column to the Young modulus:

$$\frac{\Delta L_{col}}{L_{col}} = \frac{\langle S \rangle}{E} \tag{6.41}$$

Alternatively, instead of using a Young modulus, one can pragmatically postulate a phenomenological relation between bed porosity, bed length and stress.

As an illustration, (Verhoff and Furjanic, 1983) and (Shadday, 2006) postulated a simple empirical relation of the form $\varepsilon_e(S) = \dfrac{\varepsilon_{e0}}{1 + S/E_\varepsilon}$, where ε_{e0} is the bed extragranular porosity in the absence of stress and E_ε is an adjustable parameter. Under simple assumptions, the coefficient E_ε can be related to the Young modulus E,[6] so we finally have an approximate expression:

$$\begin{aligned} &\varepsilon_e(S) = \frac{\varepsilon_{e0}}{1 + S/E_\varepsilon} \quad \text{or} \quad \varepsilon_e(\sigma) = \frac{\varepsilon_{e0}}{1 + \sigma \Delta P_0 / E_\varepsilon} \\ &E_\varepsilon = E\frac{\varepsilon_{e0}}{1 - \varepsilon_{e0}} \end{aligned} \tag{6.42}$$

As a reminder of the connection between the parameter E_ε and the Young modulus E, we will call E_ε the "Young modulus E_ε". The quotes remind us that this expression is not totally correct, first because E and E_ε have different numerical values, but also because Eq. (6.42) is a phenomenological expression that can represent experimental results even if not strictly justified by Hooke's law.

Equation (6.42) is a constitutive equation that allows our system of equations to be closed. Other constitutive equations could in principle be used, for instance to take into account the deformation of particles which acts not only on the extragranular porosity but also on particle shape and consequently on the constant in the Kozeny–Kármán equation. We will assume that Eq. (6.42) is good enough for our purposes and will be used as a primary equation, allowing us to relate permeability to stress.

The pressure drop ΔP_0 under no stress is calculated by the Kozeny–Kármán equation:

$$\frac{\Delta P_0}{L_{col0}} = h_k \frac{\mu}{d_p^2} \frac{(1 - \varepsilon_{e0})^2}{\varepsilon_{e0}^3} u \tag{6.43}$$

which can be written in terms of permeability:

$$\Delta P_0 = \frac{1}{\kappa_0} \mu u L_{col0} \tag{6.44}$$

[6] A force F applied to surface area A of an elastic solid of length L should induce a length variation δL such that $\frac{\delta L}{L} = \frac{F}{AE} = \frac{S(z)}{E} = \frac{\sigma \Delta P_0}{E}$. Assuming that the particles are not deformed, using Eq. (6.26) to relate bed length and extragranular porosity for the unstressed and stressed beds one obtains

$$\varepsilon_e(\sigma) = \frac{\varepsilon_{e0} - \frac{\sigma(z)\Delta P_0}{E}}{1 - \frac{\sigma(z)\Delta P_0}{E}}$$

Equating first-order terms in $\frac{\sigma(z)\Delta P_0}{E}$ of the above equation and of Eq. (6.42) leads to $E_\varepsilon = E\frac{\varepsilon_{e0}}{1-\varepsilon_{e0}}$.

where κ_0 is the permeability obtained in the absence of mechanical stress:

$$\kappa_0 = \frac{d_p^2}{h_k} \frac{\varepsilon_{e0}^3}{(1-\varepsilon_{e0})^2} \tag{6.45}$$

Using the constitutive equation (6.42) and the Kozeny–Kármán equation (6.44), the local pressure drop can be written, after a few mathematical manipulations:

$$\frac{dP_{Fluid}(z)}{d\xi} = \Delta P_0 \frac{L_{col}}{L_{col0}} f(\sigma) \approx \Delta P_0 f(\sigma) \tag{6.46}$$

with:

$$f(\sigma) = \left(1 + \frac{\sigma \Delta P_0}{E_\varepsilon}\right) \left(\frac{1 + \frac{1}{1-\varepsilon_{e0}} \sigma \Delta P_0}{E_\varepsilon}\right)^2 \tag{6.47}$$

$f(\sigma)$ expresses the variation of the $\frac{(1-\varepsilon_e)^2}{\varepsilon_e^3}$ term of the Kozeny–Kármán equation. It reduces to 1 in the absence of stress, and increases when the normalized stress increases.

To derive Eq. (6.46), we assumed that $\frac{L_{col}}{L_{col0}} \approx 1$. This is a legitimate assumption because the impact of bed compressibility on pressure-drop variation is mainly due to $f(\sigma)$ and only marginally due to the variation in bed length.

The pressure drop $\frac{dP_{fluid}(z)}{d\xi}$ is thus presented as the pressure drop that would occur in the absence of compressibility ΔP_0, multiplied by the correction factor $f(\sigma)$.

EXERCISE 6.4 *When the compression pressure P_{mech} on a 10 cm diameter DAC column is increased from 20 bar to 50 bar, the bed length varies from 25 cm to 24 cm. Calculate the impact on permeability assuming an external porosity of about 0.35.*

With Eq. (6.46), the force-balance equation (6.31) can thus be written in the normalized form:

$$\begin{aligned} \frac{d\sigma(z)}{d\xi} + \breve{\beta}\sigma(z) &= f(\sigma) \\ \sigma(0) &= \frac{P_{mech}}{\Delta P_0} \end{aligned} \tag{6.48}$$

where the correction factor $f(\sigma)$ is given by Eq. (6.47). This generalizes Eq. (6.34), which was valid under the non-compressibility assumption.

Solving the normalized force-balance equation (6.48) allows a determination of the normalized stress profile, thus the stress profile, inside the column. One can then determine the porosity profile thanks to Eq. (6.42).

Knowledge of the stress and porosity profiles finally allows a determination of the pressure at the column inlet by integration of the pressure gradient (6.46) along the column axis:

$$P(\xi = 0) = \Delta P_0 \int_0^1 f(\sigma(\xi)) d\xi \tag{6.49}$$

One has $P(\xi = 1) = 0$, as the column outlet is normally under atmospheric pressure.

The procedure for calculating the various profiles in the bed is thus relatively straightforward. Assuming that the bed porosity under no stress is known, the column permeability is calculated via Eq. (6.45). The pressure drop ΔP_0 in the absence of stress is then calculated for any given fluid velocity or column length by Eq. (6.44). Equations (6.47) and (6.48) are solved in order to get the normalized stress profile. Finally, pressure drop is determined via Eq. (6.49).

Some numerical solutions of the above-mentioned system are given in Figure 6.5. The curves have been simulated using a "Young modulus E_ε" of 100 bar, which represents a compressibility intermediate between non-compressible silica beds and highly compressible soft gel beds. The friction factor $\breve{\beta} = 0$ is set to zero. A higher friction factor would not affect the qualitative results of this discussion.

In Figure 6.5a, the profiles are given for a moderate pressure drop ΔP_0. The term "moderate" should be considered with respect to the "Young modulus E_ε". The axial stress and pressure profiles are nearly linear, and porosity is almost constant.

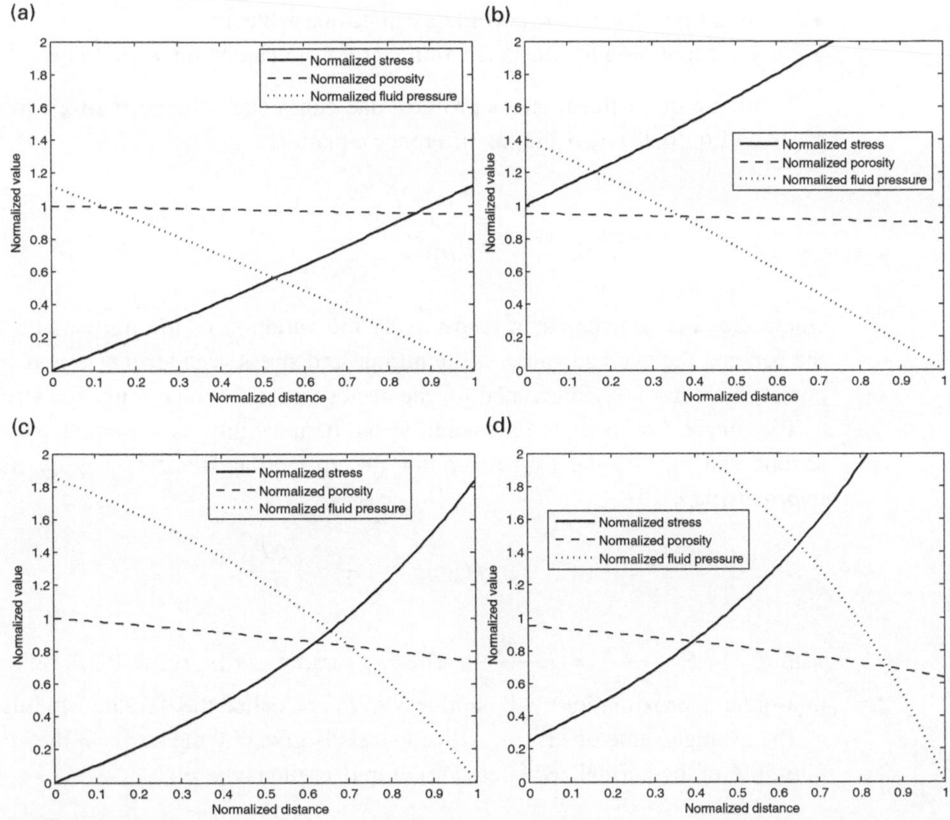

Figure 6.5 Simulated stress, pressure-drop and porosity profiles in a chromatographic bed using the simplistic model. $E_\varepsilon = 100$ bar, $\breve{\beta} = 0$, $\varepsilon_{e0} = 0.4$ for all figures. (a) $\Delta P_0 = 5$ bar, $P_{mech} = 0$ bar. (b) $\Delta P_0 = 5$ bar, $P_{mech} = 5$ bar. (c) $\Delta P_0 = 20$ bar, $P_{mech} = 0$ bar. (d) $\Delta P_0 = 20$ bar, $P_{mech} = 5$ bar.

In Figure 6.5c, the flow rate is increased so that ΔP_0 represents about 20% of the "Young modulus E_ε". The non-linearity of the profiles is becoming significant and, as expected from the structure of the Kozeny–Kármán equation, a relatively limited variation of 15% in extragranular porosity leads to a significant non-linearity of the pressure profile.

In Figures 6.5b and 6.5d, a relatively modest 5 bar mechanical pressure is applied to the inlet of the bed. Comparing the left- and right-hand figures, one notices that the additional mechanical pressure has a significant impact on both pressure and porosity profiles.

Should the non-dimensional friction parameter be different from zero, part of the stress would be absorbed by the wall, so that the strain (bed compression) would be lower. The profiles would thus show less non-linearity.

In order to avoid numerical integration of Eq. (6.48), one can further simplify the simplistic model using averaged values and deriving a *first-order approximation.* This provides simple analytical solutions in the case of beds showing *limited compressibility*, so that:

- stress profiles can be considered to be quasi-linear
- the expression for the correction factor $f(\sigma)$ can be linearized in σ.

Assuming quasi-linear stress profiles, one can replace the normalized force balance given by Eq. (6.48) by a simple difference equation:

$$\begin{aligned} \Delta\sigma + \breve{\beta}\langle\sigma\rangle &= f(\langle\sigma\rangle) \\ \sigma(0) &= \frac{P_{mech}}{\Delta P_0} \end{aligned} \tag{6.50}$$

where $\Delta\sigma$ and $\langle\sigma\rangle$ represent respectively the variation of the normalized stress over the bed and the average value of the normalized stress. The term $f(\langle\sigma\rangle)$ represents the correction factor $f(\sigma)$ calculated for the average value of the normalized stress.

The impact of bed compressibility on permeability is assumed to be limited, so one can replace the expression for the correction factor $f(\langle\sigma\rangle)$ by a *first-order approximation*:

$$f(\langle\sigma\rangle) \approx 1 + \langle\sigma\rangle\frac{\Delta P_0}{E_{fo}} \tag{6.51}$$

with $E_{fo} = E_\varepsilon \dfrac{1-\varepsilon_e}{3-\varepsilon_e} = E\dfrac{\varepsilon_e}{3-\varepsilon_e}$. The E_{fo} parameter (the subscript *fo* referring to the first-order approximation) will, similarly to E_ε, be called the "Young Modulus E_{fo}".

The average value of the normalized stress is given by the stress at the column inlet plus 50% of the normalized stress variation over the bed:

$$\langle\sigma\rangle = \sigma\,(\xi = 0) + \frac{1}{2}\Delta\sigma = \frac{P_{mech}}{\Delta P_0} + \frac{1}{2}\Delta\sigma \tag{6.52}$$

or, equivalently, $\Delta\sigma = 2\left(\langle\sigma\rangle - \frac{P_{mech}}{\Delta P_0}\right)$.

Solving Eqs. (6.50), (6.51) and (6.52) leads to:

$$\langle\sigma\rangle = \frac{1 + 2\dfrac{P_{mech}}{\Delta P_0}}{2 + \breve{\beta} - \dfrac{\Delta P_0}{E_{fo}}} \tag{6.53}$$

By definition of the normalized stress, one obtains the average stress:

$$\langle S\rangle = \Delta P_0 \langle\sigma\rangle = \frac{\Delta P_0 + 2P_{mech}}{2 + \breve{\beta} - \dfrac{\Delta P_0}{E_{fo}}} \tag{6.54}$$

Recall that, due to this stress, the bed is assumed to have shrunk by $\frac{\Delta L_{col}}{L_{col0}} = \frac{\langle S\rangle}{E}$ as provided by Eq. (6.41).

Knowledge of the average stress finally allows a determination of the pressure drop in the column, and Eq. (6.49) becomes:

$$\begin{aligned} \Delta P &= \Delta P_0 f(\langle\sigma\rangle) \\ &= \Delta P_0 \left[1 + \frac{\Delta P_0 + P_{mech}}{E_{fo}\left(2 + \breve{\beta} - \frac{\Delta P_0}{E_{fo}}\right)} \right] \end{aligned} \tag{6.55}$$

where ΔP_0, representing the pressure drop under no stress, is calculated by Eq. (6.44).

Equation (6.55) gives a very simple and flexible relationship representing the influence of the operating parameters on pressure drop with two parameters $\breve{\beta}$ and E_{fo}. Comparison with numerical simulations shows that Eq. (6.55) can predict pressure drop to within 20% of the value obtained with the resolution of the differential system, provided that $\Delta P/\Delta P_0$ is no more than about 2. This is thus well adapted to the modeling of incompressible chromatographic media leading to limited bed compressibility.

Finally, Eq. (6.55) giving the pressure drop in the column can be expressed in terms of relative permeability:

$$\frac{\kappa_0}{\kappa} = 1 + \frac{\Delta P_0 + P_{mech}}{E_{fo}\left(2 + \breve{\beta} - \frac{\Delta P_0}{E_{fo}}\right)} \tag{6.56}$$

where κ and κ_0 are the permeabilities of the bed obtained respectively in the presence and in the absence of stress.

We will call the set of Eqs. (6.50)–(6.56) the *first-order approximation* of the simplistic model.

Prior to facing our simplistic model with experimental results, let us again discuss the key assumptions:

- The bed is seen as a *one-dimensional* elastic continuous medium where any radial distribution of stress is neglected. It is difficult to "guesstimate" how important this radial distribution might be, but we believe it makes sense to focus first on the axial distribution of properties. The column axis, being the direction of both flow and piston compression, is clearly the primary direction. Radial heterogeneity can intuitively be seen as a "second-order" phenomenon.

- A very important assumption is the elasticity assumption and thus the reversible effects of flow rate and compression pressure. This is certainly questionable, and, as mentioned in the introduction, some effects are irreversible (especially those due to particle motion) and thus cannot be accounted for by this model. Chromatographers know that void volumes which may appear after a few hours of elution do not disappear when flow is stopped... The elasticity of beds packed with soft particles is probably more intuitively acceptable than the elasticity of beds packed with incompressible particles. Some experiments show, however, that beds have some elasticity even when packed with incompressible media (Section 6.2.3).
- Another key assumption is that the coefficient of friction β is constant. Variation of this parameter is certainly possible, and friction can even be non-linear. We don't have physical evidence allowing us to further discuss this matter.

This simplistic model is certainly based on a highly idealistic description of the chromatographic bed. It will, however, demonstrate the great merit, with simple arguments and equations, of offering at least a semi-quantitative description of how the L_{col}/D_{col} ratio can impact the very important process parameter which is pressure drop.

6.2.3 Incompressible chromatographic media

Incompressible media, which are typically silica-based materials, are normally used in DAC columns. The bed length compressibility can, because of piston compression, reach a few percent when one increases the compression pressure from a few bar to the usual 50–100 bar compression range. Experimental Young moduli E are about 1000 bar; the large value of E shows that the chromatographic beds have a very small but measurable compressibility. As mentioned previously, chromatographic beds packed with incompressible particles are in general not totally elastic, so the Young moduli measured during compression and during relaxation are normally a bit different.

Bed compressibilities are limited for incompressible media, so we will compare experimental results to the first-order approximation of the simplistic model. To do so, we will use:

- the results of (Cherrak and Guiochon, 2001), who packed a column of 50 mm diameter with different weights of reversed-phase silica (Kromasil C8, 10 μm) so that bed lengths from about 25 mm to about 200 mm were obtained. Their results are summarized in Table 6.1.
- the results of (Cherrak *et al.*, 2002), who investigated further the influence of bed length and mechanical pressure P_{mech} on permeability. These authors studied the influence of the compression pressure on bed permeability of a 50 mm internal diameter DAC column packed with Luna C18 bound silica with 10 μm particle diameter. Compression pressure varied between 20 and 86 bar.
- unpublished experience of the author in large-scale HPLC.

Table 6.1 Results obtained by (Cherrak and Guiochon, 2001) on a 50 mm internal diameter column. Italic numbers are the results of our own calculations and assumptions; see details in the text. Note: the longest column results for permeability/porosity are disregarded (as mentioned by the authors, they lead to somewhat abnormal results); these numbers are indicated with a solid line through them.

M (g)	Settled L_{col0} (mm)	41 bar L_{col} (mm)	Relax L_{colR} (mm)	E_{bed} comp (bar)	41 bar ε_T	$\kappa \times 10^{10}$ (cm^2)	Settled ε_{T0}	Settled ε_{e0}
30	28	26.6	27.4	888	0.57	3.69	*0.592*	*0.285*
60	53	50.5	52.3	815	0.600	5.05	*0.619*	*0.332*
120	109	105.0	107.0	1246	0.620	6.41	*0.632*	*0.355*
180	165	159.0	161.0	1273	0.616	6.78	*0.628*	*0.348*
240	207	200	203.0	1691	~~0.64~~	~~6.66~~	~~*0.652*~~	~~*0.389*~~

Settled bed characteristics

We start by discussing the results obtained by (Cherrak and Guiochon, 2001) for settled beds. It is clear from their results that column length (the second column of Table 6.1) does not vary proportionally to the mass of silica, but that shorter beds seem to be more compact than longer beds. In order to relate this to a porosity change, we will use the results associated with beds submitted to compression. After filling the column tube with a slurry containing a known weight of silica in methanol, a compression $P_{mech} =$ 41 bar was applied to the bed. The bed length was recorded prior (L_{col0}) and after (L_{col}) application of the compression pressure. The authors used tracer injections to measure the total porosity of the bed submitted to compression (no similar measurements were made on the settled bed). From knowledge of the total porosity after compression and of the bed lengths before and after compression, we estimated the porosity of the gravity-settled bed, ε_{T0}, with Eq. (6.24). The total porosity of the gravity-settled bed, ε_{T0}, increased with bed length (column 8 in Table 6.1). This result is certainly impacted by the exact procedure used for settling the bed, and shows the need to use compression devices to get compact beds.

We have estimated the value of the intragranular porosity ($\varepsilon_i = 0.43$) from results obtained on analytical columns. Knowledge of the intragranular porosity, assumed to be uniform, allows calculation of the extragranular porosity for the gravity-settled bed from knowledge of its total porosity. The resulting extragranular porosity ε_{e0} looks a bit smaller than the values usually accepted, and increases with the initial bed length (column 9 in Table 6.1). The extragranular porosity ε_{e0} must be considered as entry data of the model as it does not take into account the influence of gravity during initial filling.

Influence of P_{mech} and L_{col} on bed compressibility

As shown in Table 6.1 (column 3), beds shrank by a few percent under the influence of mechanical compression. The compression pressure was then released and a new bed length L_{colR} was measured after this relaxation. This was systematically higher than L_{col} but lower than the initial bed length L_{col0}. This shows that the elasticity assumption is not fully verified, so the compression effect is not fully reversible. This non-reversibility is known to be reduced after two or three compression/decompression steps so that columns gain elasticity after their first compression (Cherrak *et al.*, 2002). This is most probably because a few compression/decompression steps give some particles the opportunity to move and fill holes, this motion being irreversible. We will now focus on results obtained for the compression step.

(Cherrak and Guiochon, 2001) defined a "Young modulus E_{bed}" by $\frac{\Delta L_{col}}{L_{col}} = \frac{P_{mech}}{E_{bed}}$, where ΔL_{col} represents the difference in lengths of the gravity-settled bed and the compressed bed in the absence of fluid flow. The "Young modulus E_{bed}" has the merit of being directly obtained from experimental data, but it is related to P_{mech}, differently from the Young modulus E, which is related to the average stress $\langle S \rangle$ as provided by Eq. (6.41).

In order to obtain the same variation $\frac{\Delta L_{col}}{L_{col}}$ with the two definitions E_{bed} and E, one has to use $E_{bed} = E\frac{P_{mech}}{\langle S \rangle}$. Taking into account the expression for the average stress given by Eq. (6.54) in the absence of fluid flow ($\Delta P_0 = 0$), one obtains:

$$E_{bed} = E\frac{\left(2+\breve{\beta}\right)}{2} = E\left(1+\beta\frac{L_{col}}{D_{col}}\right) \tag{6.57}$$

It is apparent (column 5 Table 6.1) that the experimental "Young modulus E_{bed}" varies linearly with the bed length, in agreement with Eq. (6.57). This result can be considered a first "sanity check" of our model. A good fit is obtained with $E = 700$ bar and $\beta = 0.6$, parameters that we will use later.

Influence of fluid velocity on permeability

(Cherrak and Guiochon, 2001) and (Cherrak *et al.*, 2002) reported that bed permeability was independent of fluid velocity, this result probably being systematically obtained when working with that type of rigid particles.

Our model predicting a bed compressibility, it predicts a variation of permeability with fluid velocity via the term ΔP_0 in Eq. (6.56). This influence is in fact totally negligible within the normal range of parameters. For a porosity of about 0.35, one obtains $E_{fo} = 700\frac{0.35}{3-0.35} = 92$ bar, and for a L_{col}/D_{col} ratio of 1.5 the $\breve{\beta}$ parameters is about 1. At a very small flow rate, so that $\Delta P_0 = 0$, by using Eq. (6.56) one obtains $\frac{\kappa}{\kappa_0} = 1.15$. At a significant flow rate, so that $\Delta P_0 = 10$ bar, one obtains $\frac{\kappa}{\kappa_0} = 1.19$. This 3–4% variation is not visible experimentally.

While the first-order approximation of our simplistic model can predict the influence of pressure on bed compactness, thus on permeability, using the experimentally determined E_{fo} and β parameters, it predicts permeabilities almost independent of fluid velocity, in agreement with the experimental results. This can be considered a second "sanity check".

Influence of P_{mech} and L_{col} on permeability

(Cherrak *et al.*, 2002) investigated the influence of compression/relaxation on the "Young modulus E_{bed}" at different mechanical pressures. They showed that after three compression/relaxation steps the bed becomes elastic, so that compression/relaxation

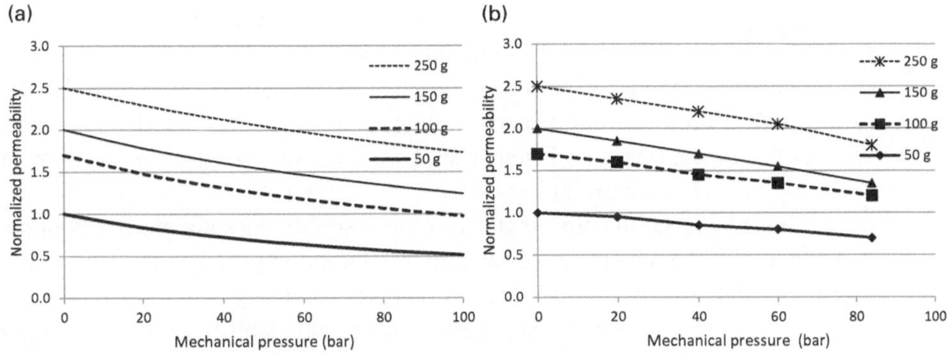

Figure 6.6 Influence of compression (mechanical) pressure and solid quantity on column normalized permeability. (a) Simulated: simplistic model, $E = 600$ bar, $\beta = 0.9$, $\varepsilon_{e0} = 0.35$. (b) Experimental: (Cherrak *et al.*, 2002).

effects are reversible. We will focus on the results obtained with the first compression, for which the most detailed information has been disclosed.

As with the results of (Cherrak and Guiochon, 2001), we use the variation of the "Young modulus E_{bed}" versus bed length to determine our model parameters. A good fit is obtained with $E = 600$ bar and $\beta = 0.9$. The fact that these parameters are very similar to those obtained in the previous study is not surprising because the two chromatographic media are very similar.

Knowledge of the Young modulus E and the assumption that external porosity is about $\varepsilon_{e0} = 0.35$ allows a determination that $E_{fo} = 79$ bar. Additionally, knowledge of the friction parameter β allows the determination of the parameter $\breve{\beta}$ for the different column geometries.

With this set of parameters, we have estimated the influence of the mass of silica (thus bed length) and mechanical pressure on normalized column permeabilities with Eq. (6.56). Normalized permeability is defined as the ratio between permeability under given conditions and permeability obtained for the longest bed (250 g of silica) under no mechanical pressure. Results are given in Figure 6.6. Notice that the simulated results obtained at zero mechanical pressure are identical to the experimental results by construction, permeability κ_0 being entry data of the model.

One sees that the model predicts fairly well the evolution of the permeability of the bed at constant length and varying mechanical pressures, as well as its evolution at constant mechanical pressure and varying lengths.

Influence of bed diameter on permeability

Finally, few if any results have been published on the influence of column diameter at given column length on bed permeability. While our simplistic model predicts that compressibility should be impacted by column geometry only via the ratio L_{col}/D_{col}, it is probably worth sharing a bit of qualitative experience.

Using E and β parameters having orders of magnitude compatible with those determined above, one can predict the influence of the mechanical pressure and of L_{col}/D_{col} on bed permeability. Simulations and experience (for DAC column diameters ranging from 5–10 cm to about 100 cm) are in qualitative agreement: there is an influence, though relatively moderate, of the mechanical pressure and column diameter (at constant length) on permeability. The bed permeability is expected to decrease with L_{col}/D_{col} until this ratio reaches about 2–5, depending on the mechanical pressure applied on the bed. This is important information for scale-up, as it can impact pressure drop by up to perhaps 20–30%.

It is important to notice that bed compressibility does not materially improve column efficiency: the adverse effect of compressibility on productivity due to the decrease in permeability is not compensated by a higher efficiency.

While the influence of compressibility on permeability for beds packed with incompressible media is relatively limited, let us again say about 20–30% maximum under practical conditions, we will see that the situation is more dramatic when compressible chromatographic media are used.

EXERCISE 6.5 *The permeability of a 50 mm DAC column packed with 10 μm particles was determined at different bed lengths under mechanical pressures of 25 and 50 bar. The resulting permeabilities, given in the table, were independent of fluid velocity in the range* 10^{-4} *to* 10^{-3} *m/s.*

L_{col} (m)	0.05	0.1	0.2	0.3
		$P_{mech} = 25$ bar		
κ (m^2)	1.05×10^{-13}	1.08×10^{-13}	1.11×10^{-13}	1.13×10^{-13}
		$P_{mech} = 50$ bar		
κ (m^2)	9.6×10^{-14}	1.00×10^{-13}	1.07×10^{-13}	1.08×10^{-13}

Using the first-order approximation of the simplistic model, determine the influence of the fluid velocity on pressure drop for columns of 20 cm bed length and of 10 and 100 cm diameter packed with the same material under a mechanical pressure of 50 bar and eluted with a carrier fluid having a viscosity of 1 cP.

6.2.4 Compressible chromatographic media

By compressible media or soft gels we mean particles that are subject to physical deformation due to the impact of viscous drag or possibly piston compression. The situation of particles which can swell or shrink as a consequence of a chemical change is addressed in Section 6.2.5.

In contrast to what has been seen for incompressible particles, the relation between pressure drop and superficial velocity normally becomes non-linear for compressible particles. There is a lot of experimental evidence (Keener *et al.*, 2008; Stickel and Fotopoulos, 2001; Verhoff and Furjanic, 1983) that, if the fluid velocity exceeds a given value, the pressure drop increases exponentially, and that this value increases when the ratio L_{col}/D_{col} increases. A convincing illustration of this phenomenon for protein separation media is given in (Stickel and Fotopoulos, 2001). The shapes of the ΔP vs. u curves are given in Figures 6.7 and 6.8.

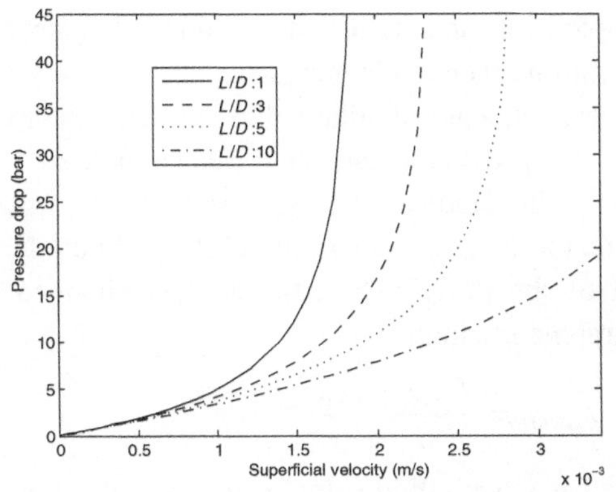

Figure 6.7 Simulated influence of the fluid velocity on pressure drop at different L_{col}/D_{col} ratio obtained by solving of the differential equations (6.47), (6.48), (6.49) of the simplistic model. $P_{mech} = 0$, $L_{col0} = 0.5$ m, $\beta = 0.7$, $E = 13$ bar, $\varepsilon_0 = 0.35$, $\mu = 1$ cP, $d_p = 50\mu$m.

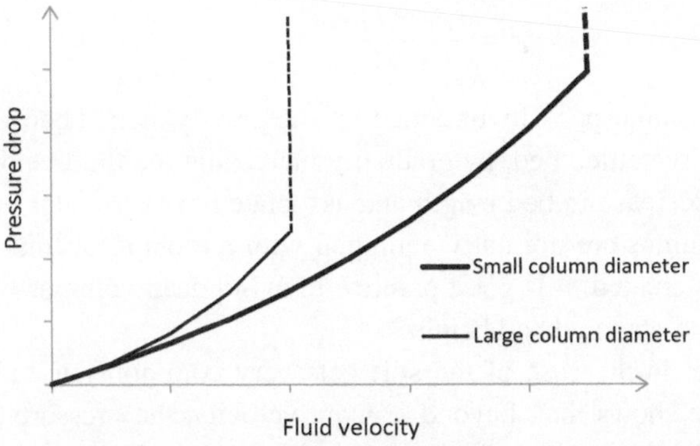

Figure 6.8 Qualitative influence of the fluid velocity on pressure drop for columns of a given length packed with compressible particles. Model of (Stickel and Fotopoulos, 2001).

There is little that can be done about the phenomenon of abrupt pressure drop increase: the problem is not to model in detail what happens at very high velocities (we know the conclusion perfectly well: the column is blocked!) but to stay away from this range.

The simplistic model presented above (without the first-order approximation) can be used to predict, at least qualitatively, these experimental results. Note, however, that the assumption of non-deformable particles in the constitutive equation (6.42), as well as the constancy of the proportionality coefficient of the Kozeny–Kármán equation, are highly questionable in the case of strong deformation. In Figure 6.7, some simulations associated with a soft material ($E = 13$ bar) clearly show the extreme non-linear influence of fluid velocity on the pressure drop, beyond a given velocity, as well as the influence of the L_{col}/D_{col} ratio. This behavior must absolutely be understood and controlled for scale-up, otherwise the larger column (lower L_{col}/D_{col} ratio) may be totally blocked at velocities adapted for the smaller one. In this case, chromatography is not ideal according to the definition given in Chapter 1.

While our simplistic model can at least qualitatively represent the experimental results, it is worth mentioning the pragmatic and empirical approach followed by (Stickel and Fotopoulos, 2001), who studied agarose-based media. These authors defined the bed compression factor λ_{bed} as the ratio of the column length under given conditions to the length of the gravity-settled bed L_{col0} obtained by simple sedimentation of the chromatographic medium:

$$\lambda_{comp} = \frac{L_{col0} - L_{col}}{L_{col0}} \tag{6.58}$$

L_{col} being the bed length observed at a given fluid velocity and associated pressure drop. Note: according to our simplistic model, $\lambda_{comp} = \frac{\langle \sigma \rangle}{E}$.

Derived on the assumption that the volume of skeleton $\overline{\overline{V}}$ and the intragranular pore volume of the particles are constant, Eq. (6.26) relating porosity and bed length can be written:[7]

$$\varepsilon_e = \frac{\varepsilon_{e0} - \lambda_{comp}}{1 - \lambda_{comp}} \tag{6.59}$$

where ε_{e0} is the extragranular porosity obtained for the gravity-settled bed.

Starting from a gravity-settled bed, by gradually increasing the fluid velocity (down-flow), one observes a decrease in bed length and associated extragranular porosity. Let us mention that, as columns are normally equipped with a moving piston, either manually or automatically actuated, it is good practice to avoid dead volumes by adjusting the piston position to the obtained bed height.

As mentioned above, in the case of the soft gels very commonly used for protein purification, experience shows that, beyond a given velocity, the pressure drop in the column increases sharply, just as if the bed were totally blocked. The bed compressibility associated with this situation is called the *critical bed compression factor* λ_{comp}^{crit}, and is associated with a critical fluid velocity u^{crit}. For rigid particles, the situation of critical velocity is never met in practice.

Three very important results have been reported by (Stickel and Fotopoulos, 2001):

- From different experiments, it is clear that λ_{comp}^{crit} characterizes the chromatographic medium itself and is largely independent of the column geometry.
- The bed compression factor plotted against fluid velocity is approximately linear up to the critical velocity:

$$\lambda_{comp} = \lambda_{comp}^{crit} \frac{u}{u^{crit}} \tag{6.60}$$

  This equation expresses that the bed shrinks constantly with velocity, until the critical velocity is reached. Note: this result is fully compatible with our simplistic model. Equations (6.41) and (6.54) show that the bed compressibility is

[7] The second condition is in principle not satisfied with deformable particles.

proportional to ΔP_0, thus to the fluid velocity, when ΔP_0 is small enough to ensure that $\langle \sigma \rangle \approx 1$.

- The critical velocity can be related to the geometric characteristics of the column via a simple empirical linear equation:

$$u^{crit} L_{col0} = p \frac{L_{col0}}{D_{col}} + q \tag{6.61}$$

where p and q are empirical constants characterizing the chromatographic medium itself and independent of column geometry.

The critical velocity is thus directly affected by the ratio L_{col}/D_{col} of the column.

The way to use correlation (6.61) to calculate the pressure drop under specific situations is as follows:

- The length of the gravity-settled bed L_{col0} is determined; ε_{e0} is thus determined.
- Two or three critical velocities must be determined in order to identify parameters p and q and λ_{comp}^{crit}. Note that these experiments can, *a priori*, be performed on the same column at different bed lengths.

This information can then be used to forecast the behavior of different column geometries:

- For a given column geometry, the critical velocity is determined from Eq. (6.61).
- For $u > u^{crit}$, the pressure drop reaches infinity.
- For $u < u^{crit}$, Eq. (6.60) allows determination of the bed compression factor λ_{bed}.
  - The bed length and porosity are determined from Eqs. (6.58) and (6.59).
  - The Kozeny–Kármán equation (6.1) is used to evaluate pressure drop.

Qualitatively, the influence of fluid velocity on pressure drop for columns of given length but different diameters is given in Figure 6.8. At small velocities, the bed compression is moderate, and the pressure drop varies linearly with fluid velocity, independently of column diameter. When the fluid velocity increases, the impact of bed compressibility on the extragranular porosity starts to be visible, and the pressure drop becomes an increasingly non-linear function of fluid velocity. The critical velocity decreases when diameter increases. When the critical velocity is reached, the pressure drop goes quickly towards infinity.

The approach of (Stickel and Fotopoulos, 2001) has the merit of being simple and practical. As highlighted at different occasions in the text, it is at least qualitatively compatible with our simplistic model, which gives a deeper scientific understanding.

EXERCISE 6.6 *A compressible medium made of 100 μm particles is considered for protein purification. A column of 5 cm diameter is used for a gravity-settled bed. For a given quantity of the medium, a column of 25 cm length is obtained, and tracer injection allows determination of an extragranular porosity of 0.45. By gradually increasing the flow rate (downwards), one obtains a very sharp pressure drop increase at a fluid velocity of* $u = 1.3 \times 10^{-3}$ *m/s when the bed length is 21.5 cm.*

A similar experiment was performed with an initial bed length of 10 cm (resp. 50 cm), and a sharp pressure drop increase was obtained for a fluid velocity of $u = 2 \times 10^{-3}$ *m/s (resp.* 9×10^{-4} *m/s) when the bed length was 8.5 cm (resp. 42 cm).*

A purification is planned using a 50 cm diameter column of 40 cm bed length. Estimate the anticipated influence of fluid velocity on pressure drop in the preparative column.

6.2.5 Swelling and shrinking media

Some particles give rise to bed compression or expansion due to their ability to swell or shrink depending on the fluid composition. For instance, ion-exchange resin particle diameters can vary by up to about 30% when moving from one ionic form to another. This diameter is expected to influence pressure drop according to the Kozeny–Kármán equation.

The reality is more complex because bed expansion is laterally and axially constrained by the column wall. Wall friction prevents the resins from moving freely along the longitudinal axis, so the relative volume variation of the bed length is lower than the volume variation of the beads (Shadday, 2006). The existence of additional stress due to constrained swelling by the column wall has long been recognized, and some authors have reported the bursting of glass ion-exchange resin columns due to swelling.

Briefly mentioned in connection with Eq. (6.27), the modeling of bed shrinking/swelling can be performed with the simplistic model, with one important adaptation: the propensity of the bed to expand or shrink is counterbalanced by frictional forces, whose sign can change during a particular saturation/desaturation process. Let us consider the force balance presented in Figure 6.4. In the absence of swelling/shrinking, fluid flow being oriented downwards, the bed tends to move downwards, so frictional forces are directed upwards. Assume now that the resin is swelling, so that it expands upwards. The frictional forces must then be directed downwards to oppose the bed motion.

The swelling/shrinking process is strongly impacted by the column geometry. With particles gaining some freedom to expand in large-diameter columns, one can expect less compact beds at lower L_{col}/D_{col} ratios, and more compact beds at higher L_{col}/D_{col} ratios. This result has been confirmed by (Shadday, 2006), who studied the expansion of cationic resins during a regeneration step, replacing hydrogen ions with sodium ions. This ionic form change is associated with about 30% bead volume change in a non-constrained environment. When performing this regeneration in columns, they obtained bed expansions varying from 30% for $L_{col}/D_{col} = 2$ to 15% for $L_{col}/D_{col} = 6$. This stressed expansion impacts bed porosity, thus pressure drop, which is about four times greater for $L_{col}/D_{col} = 6$ than for $L_{col}/D_{col} = 2$. A decrease in permeability when the L_{col}/D_{col} ratio increases is observed when significant swelling occurs. This result is the opposite of what is observed when compressibility is due to viscous forces or piston compression, but it is explained using the same theoretical framework.

This stressed expansion can lead to another counterintuitive result: when the resin swells, its diameter increases, so one can expect a decrease in pressure drop. On the other hand, the induced stress leads to a decrease in porosity, thus permeability; as a result, the overall pressure drop can increase. This result has been confirmed by (Shadday, 2006), who recorded a pressure increase when resins were regenerated from the hydrogen form to the sodium form: while the beads are 30% larger after regeneration, the pressure drop increased about three-fold.

Finally, in contrast to the effect of viscous forces and piston compression, which are steady-state, swelling and shrinking phenomena are associated with transitory behavior related to the saturation/desaturation process. The transient swelling/shrinking is associated with a fluid flow rate variation due to carrier fluid accumulation in the particles: the fluid flow rate decreases locally as the resin swells, while it increases as the resin shrinks. This can be modeled by a continuity equation (Shadday, 2006), and coupling between this continuity equation, the ion-exchange mass balance and the force balance allows the complete saturation/regeneration steps to be modeled. This makes modeling rather complex; the reader is again referred to (Shadday, 2006) for more detailed information.

6.2.6 Conclusion

Chromatographic beds can be compressible under the influence of viscous flow, piston compression or the swelling/shrinking of some ion-exchange resins. The resulting stress being more or less adsorbed by frictional forces occurring at the column wall, the column geometry represented by the L_{col}/D_{col} ratio can impact the permeability.

Permeability can decrease with scale-up in diameter:

1. Using DAC technology with incompressible media, at a given compression pressure and fluid velocity, permeability can be about 20–30% lower at small L_{col}/D_{col} ratio than at large L_{col}/D_{col} ratio.
2. When soft particles are used (with or without compression), not only is the permeability impacted by the L_{col}/D_{col} ratio, but pressure drop increases non-linearly with fluid velocity, and can even increase exponentially when velocities exceed a given limit. This fluid velocity limit decreases when the L_{col}/D_{col} ratio decreases. When the fluid velocity limit is exceeded, the result can be as bad as a totally blocked column.

Permeability can also increase with scale-up in diameter:

3. The situation associated with saturation/desaturation of some ion-exchange resins can be very complex, as the compressibility due to viscous flow can act in synergy with or in opposition to the tendency of the particles to swell or shrink when changing forms. When the swelling is significant, bed permeability can decrease even if the particles become larger, this effect being especially important at significant L_{col}/D_{col}, thus for small columns.

The simplistic model presented in this section can normally represent key features reasonably well, and can thus give valuable information on when the L_{col}/D_{col} ratio should be taken into consideration. It is less practical and pertinent for representing swelling situations. As mentioned in point 3 above, the negative consequences of swelling on permeability are expected to decrease with scale-up, so that a direct scale-up of a laboratory column can lead to a loss of opportunity, rather than to a catastrophic result.

6.3 Radial profiles

Chromatographic columns are almost systematically considered to be radially uniform; in other words, the fluid velocity is independent of the radial coordinate.

Even if it is adequately designed, no column (except the "yet-to-be-invented" perfect column) can deliver perfect radially uniform profiles. The problem is thus to know whether or not radial profiles can materially impact the performance of the system. The answer is: "It's a possibility." Certainly, the lower the selectivity between the solutes, the higher the target purity, the greater the possible influence of radial profiles on performance.

Prior to explaining how these velocity profiles can impact column efficiency, thus performance, let us discuss the possible reasons for their existence.

The first reason relates to the fact that no distributor is perfect.[8] Consider that, in the case of large-scale HPLC systems, a distributor should deliver, within a distance of a few millimeters, a plug flow across a diameter exceeding 1 m from an inlet pipe with a diameter of about 1 cm. Under these conditions, obtaining a fluid velocity constant within a few percent between the center and the periphery of the column can be considered an excellent technical performance. A few percent deviation will in some situations be meaningless, although it can ruin the efficiency expected for very small particles.

Secondly, a distributor, even if "perfectly" designed, can see its performance dramatically reduced by fine particles from the adsorbent or from an imperfectly filtered carrier fluid. This can be largely avoided by properly cleaning the fluids and using spherical solid particles, but one should be prepared to face this issue sooner or later. In the case of efficiency deterioration, possibly associated with a small pressure-drop increase, remember rule number one: "Clean the frits!"

The third reason, which is too often underestimated or even forgotten, is related to temperature gradients. If the carrier fluid is not at the same temperature as the column wall, radial and longitudinal temperature profiles will be established in the column. The viscosity of the fluid being temperature-dependent, this will induce radial velocity profiles. Illustrations will be given in subsequent sections. If the first precept is to clean the frits, the second is certainly to "Jacket your column and control temperature carefully!"

The final reason is also associated with temperature. However, instead of arising from improper temperature control, this effect comes from the heat dissipation associated with pressure drop. Heat dissipation is totally negligible when low-pressure chromatography is used, but can be relevant in HPLC systems, associated with a pressure drop

[8] The reader will find some contributions from computational fluid dynamics on designing distributors, for instance in (Toumi *et al.*, 2012; Yuan *et al.*, 1999; Pathak *et al.*, 2008).

exceeding 50 bar and requiring very tight velocity profile control. Note further that the enthalpy of adsorption/desorption is normally totally negligible in liquid and supercritical fluid chromatography.

Numerous articles describe porosity and thus velocity variation at the very periphery of the column. By "very periphery" or "vicinity" we mean a few particle diameters. While accepting that this effect may exist in principle, we seriously doubt whether it can materially impact the efficiency of preparative columns. This effect is likely to develop over a distance of a few particle diameters, thus about 50–100 μm for high-efficiency columns. The impacts associated with fluid distributor design and temperature effects take place over distances of a few centimeters and thus totally override the "near-to-wall effect".

Finally, note that a radial velocity profile has no impact on pressure drop or on mean retention times, which are imposed by the average velocity. Radial velocity profiles can, however, significantly impact the peak variance and asymmetry.

We will now discuss how temperature profiles induce velocity profiles, and then how these radial velocity profiles affect the RTD.

6.3.1 Temperature profiles

Prior to discussing the equations describing heat transfer in chromatographic columns, let us analyze in simple terms the impact of the heat generated by energy dissipation due to viscous fluids percolating through the bed.

The amount of heat generated per unit volume of bed by viscous flow is given (Poppe and Kraak, 1983) by:

$$h_u = u\frac{dP}{dz} \tag{6.62}$$

where h_u is expressed in W/m^3 of column.

Neglecting possible longitudinal velocity, porosity and viscosity variations, the pressure drop per unit length is constant, so Eq. (6.62) can be approximated by a simple difference:

$$h_u = u\frac{\Delta P}{L} \tag{6.63}$$

Now, assuming that the column is adiabatic and has reached a temperature steady state, the heat balance says that the heat exiting the column equals the heat entering the column plus the heat generated by viscous flow:

$$Q\rho C_p T^{inl} = Q\rho C_p T^{out} + V_{col}u\frac{\Delta P}{L} \tag{6.64}$$

The simple heat balance of Eq. (6.64) shows that the adiabatic temperature elevation is simply the ratio of the pressure drop (Pa) to the product of heat capacity of the eluent

(J/(kg K)) and the fluid density:

$$\Delta T_{adiab} = T^{inl} - T^{out} = \frac{\Delta P}{\rho C_p} = \frac{\Delta P}{L_{col}} \frac{L_{col}}{\rho C_p} \tag{6.65}$$

This adiabatic temperature elevation ΔT_{adiab} gives the maximum observable temperature increase due to viscous dissipation; real systems being non-adiabatic, the actual temperature rise will be lower than this value. Notice that ΔT_{adiab} increases linearly with pressure drop, thus with column length.

For illustration purposes, consider a column eluted with methanol ($\rho = 800$ kg/m^3, $C_p = 2500$ J/(kg K)) subject to a pressure drop of 100 bar. According to Eq. (6.65), the eluent temperature increases by about 4.8 K due to the heat dissipated by viscous flow. If the inlet eluent temperature is 20°C, the outlet temperature should thus be about 25°C. This temperature increase will induce a viscosity and thus a velocity profile. Correlations show that the viscosity of methanol is about 0.60 cP at 20°C and 0.56 cP at 25°C, a difference of 7%. Should the temperature elevation result in a radial temperature gradient of the same magnitude, this could lead to a radial velocity difference of 7% between the center and the wall of the column. Again, this is a maximum estimate for a relatively extreme case, involving a 100 bar pressure drop taking place in an adiabatic column.

The model can be refined by writing more or less sophisticated heat balances taking into account heat accumulation in the solid and fluid phases, axial and radial heat conduction in the bed, heat transfer and heat accumulation in the column mechanical parts, possible axial dispersion and so on. For additional information, the reader is referred to (Kaczmarski *et al.*, 2008; Ching *et al.*, 2002).

Deriving precise temperature profiles is difficult because various complex factors can play a role, e.g. the precise geometry and thermal inertia of the column (distributors, for instance), the exact values of axial and radial heat transfer coefficients, non-ideal properties of concentrated solutions.

Deriving precise velocity profiles in connection with temperature profiles is a second significant difficulty which requires the simultaneous solution of momentum, heat and mass balances, as done, for instance, by (Ching *et al.*, 2002). As minute variations in velocity can significantly impact a chromatogram's peak shape in the case of highly efficient columns, accurate prediction of dispersion is extremely difficult. Instead of dealing with complex models giving possibly illusory prediction ability, we believe that only first approximations are really required for modeling temperature profiles.

Modeling transitory thermal behavior can only be justified when the column environment or eluent/feed temperature are not properly controlled. This is simply poor practice, which must be banished so that the system is in thermal steady state when operated for purification (unless new, creative and perfectly controlled processes are considered). We will thus limit our discussion to steady-state situations.

The thermal conductivity of particle beds (including intergranular fluid) is typically in the order of 0.5–1.0 W/(m K), while that of stainless steel is typically 90 W/(m K). This means that the temperature of the column at its inner diameter is likely to be very similar to the temperature at the tube's outer diameter, itself assumed to be identical to the constant external temperature.

The influence of the radial velocity profile as well as axial dispersion on the heat balance is probably secondary and will be neglected. Finally, we assume that the fluid velocity is constant along the column axis so that $\frac{dP}{dz} = \frac{\Delta P}{L_{col}}$.

With these assumptions, the relevant simplified heat-balance equation at the column level can be written at any axial (z) and radial (r) position:[9]

$$\rho C_p u \frac{\partial T}{\partial z} = \lambda_{bed} \left[\frac{1}{r} \frac{\partial T}{\partial r} + \frac{\partial^2 T}{\partial r^2} \right] + u \frac{\Delta P}{L_{col}} \tag{6.66}$$

where C_p is the fluid-phase heat capacity (J/(kg K)) and λ_{bed} the bed radial conductivity (W/(m K)).

The boundary conditions are given by:

$$T(z = 0) = T^{inl} \text{ constant}$$
$$T(z, r = R_{col}) = T^{ext} \text{ constant}$$

T^{ext} is either the temperature of the controlled room or the temperature of the fluid circulating in the jacket.

Defining $\xi_c = r/R_{col}$ and rearranging the heat balance, one obtains:

$$\begin{aligned} \breve{L} \frac{\partial T}{\partial z} &= \frac{1}{\xi_c} \frac{\partial T}{\partial \xi_c} + \frac{\partial^2 T}{\partial \xi_c^2} + \Delta \breve{T} \\ \breve{L} &= \frac{\rho C_p u}{\lambda_{bed}} R_{col}^2 \\ \Delta \breve{T} &= u \frac{dP}{dz} \frac{R_{col}^2}{\lambda_{bed}} \approx u \frac{\Delta P}{L_{col}} \frac{R_{col}^2}{\lambda_{bed}} \end{aligned} \tag{6.67}$$

Two important parameters are introduced:

- $\breve{L}$ has the dimension of a length. We will see that internal temperature profiles tend to stabilize at distances on the order of $\breve{L}$.
- $\Delta \breve{T}$ has the dimension of a temperature. In physical terms, $\Delta \breve{T}$ characterizes the possible temperature increase in the column due to the balance between energy generation by viscous flow and the ability of the bed to dissipate this energy.

Solving Eq. (6.67), one obtains the temperature of the carrier fluid at different axial and radial positions in the column. An analytical solution is available (Poppe and Kraak, 1983) and involves *prima facie* inconvenient Bessel functions, but modern software makes using Bessel functions almost as convenient as using logarithms. The solution is:

$$T(\xi_c, z) = T^{ext} + \frac{\Delta \breve{T}}{4} \left\{ 1 - (\xi_c)^2 + 8 \sum_{i=1}^{\infty} \left(-\frac{1}{\alpha_i^3} + \frac{T^{inl} - T^{ext}}{\alpha_i \Delta \breve{T}} \right) \frac{J_0(\alpha_i \xi_c)}{J_1(\alpha_i)} \exp\left(-\alpha_i^2 \left(\frac{z}{\breve{L}} \right) \right) \right\} \tag{6.68}$$

where the coefficients α_i are the roots of $J_0(\alpha_i) = 0$. The first 20 α_i are given by:

[9] As there is no risk of confusion, we use the same letter to denote radial position in the column and in the particle (Chapter 5).

i	1	2	3	4	5
α_i	2.4048	5.5201	8.6537	11.7915	14.9309
i	6	7	8	9	10
α_i	18.0711	21.2116	24.3525	27.4935	30.6346
i	11	12	13	14	15
α_i	33.7758	36.9171	40.0558	43.1998	46.3412
i	16	17	18	19	20
α_i	49.4826	52.6241	55.7655	58.9069	62.0485

and the subsequent ones by $\alpha_{i+1} = \alpha_i + \pi$. Note: the convergence of Eq. (6.68) is not very quick. Be prepared to consider about a hundred terms for a decent approximation.

Let us discuss different limiting cases in order to extract the main properties of this solution. First, as a sanity check, recalling that $J_0(\alpha_i) = 0$, one can verify that, for $\xi_c = 1$, Eq. (6.68) becomes $T(\xi_c = 1, z) = T^{ext}$. When $z = \infty$ (in practice, at lengths about equal to or greater than $\breve{L}$), Eq. (6.68) becomes:

$$T(\xi_c, z = \infty) = T^{ext} + \frac{\Delta\breve{T}}{4}\left\{1 - (\xi_c)^2\right\} \tag{6.69}$$

so the temperature profiles tend to stabilize towards a parabolic profile, becoming independent of the axial position and of the inlet temperature. The temperature at the center of the column exceeds the external temperature by $\Delta\breve{T}/4$.

By integrating Eq. (6.69) over the column radius, one obtains the radially averaged fluid temperature at very long lengths:

$$\langle T(z = \infty)\rangle = T^{ext} + \frac{\Delta\breve{T}}{8} = T^{ext} + \frac{1}{8}u\frac{\Delta P}{L_{col}}\frac{R_{col}^2}{\lambda_{bed}} \tag{6.70}$$

While the temperature was assumed to increase linearly with the column length for adiabatic systems, the ability of the system to exchange heat limits the asymptotic temperature increase to $\Delta\breve{T}/8$.

At small abscissa, thus near the column inlet, Eq. (6.68) does not lead to simple and friendly expressions. We know, however, that $T(\xi_c = 1, z) = T^{ext}$ at the column periphery and that the temperature inside the bed must be very similar to the inlet temperature.

We have thus shown that the axial temperature profile in the column evolves between two limiting situations along the axis:

- At the inlet, the fluid is at a temperature very close to the inlet temperature T^{inl}, only a small fraction of the fluid near the column wall being at temperature T^{ext}.
- At large distances from the inlet, the temperature profile reaches a constant parabolic pattern, independent of the inlet temperature, dictated by the external temperature and the energy produced by viscous flow.

The transition between the two situations occurs over lengths of the order of $\breve{L}$, as suggested by Eq. (6.68). This characteristic length $\breve{L}$ is proportional to the fluid velocity and to the square of the column radius. This means that the higher the fluid velocity or the larger the column radius, the greater the length required to reach a constant pattern.

As an illustration, we consider a column of 1 m diameter in which the pressure drop is 100 bar/m for a fluid velocity of 10^{-3} m/s. For a fluid similar to methanol (ρ = 800 kg/m^3, C_p = 2500 J/(kg K)) and assuming $\lambda_{bed} \approx 1$ W/(m K), one has:

$$\breve{L} = \frac{\rho C_p u}{\lambda_{bed}} R_{col}^2 = \frac{800 \times 2500 \times 0.001}{1.0} 1^2 = 2000 \text{ m!}$$

The huge value of $\breve{L}$ means that, with reasonable parameters, it is impossible for large-scale columns to reach the constant pattern resulting from the equilibrium between heat generated by viscous flow and heat transferred at the periphery of the column. This means that, for a column of a typical length (significantly below 2000 m!), the main fraction of fluid inside the column is at a temperature very similar to the inlet temperature; only a small fraction of the fluid at the very periphery is at a temperature close to the external temperature.

Because of the heat generated by viscous flow, the fluid temperature is expected to increase gradually with column length and, at a distance of about $\breve{L}$, to reach a temperature elevation of:

$$\frac{\Delta \breve{T}}{8} = \frac{1}{8} u \frac{\Delta P}{L_{col}} \frac{R_{col}^2}{\lambda_{bed}} = \frac{1}{8} 0.001 \times 100 \times 10^5 \frac{1^2}{1.0} = 1250 \text{ K!}$$

which is generally above the boiling point of the solvent and of some parts of the chromatographic equipment.

Let us now consider the situation of a laboratory column of 0.01 m diameter. One obtains:

$$\breve{L} = \frac{800 \times 2500 \times 0.001}{1.0} 0.01^2 = 0.2 \text{ m}$$

a characteristic length that is on the order of a typical column length. In contrast to the large-scale column, the laboratory column is likely to reach the asymptotic state, which means that the outlet temperature is relatively disconnected from the inlet temperature and basically imposed by the external temperature and the heat generated by viscous flow.

Additionally, one obtains for the laboratory column:

$$\frac{\Delta \breve{T}}{8} = \frac{1}{8} u \frac{\Delta P}{L_{col}} \frac{R_{col}^2}{\lambda_{bed}} = \frac{1}{8} 0.001 \times 10 \times 10^5 \frac{0.01^2}{1.0} = 0.125 \text{ K}$$

which is relatively negligible.

The radial temperature profiles are thus likely to be relatively uniform in laboratory columns and significantly more pronounced in large-scale columns. This certainly requires awareness in scale-up. Note that the simple measurement of T^{out} and comparison with T^{inl} and T^{ext} gives valuable information on what is going on in the column.

For estimating (or "guesstimating") radial velocity profiles in the column, important information is given by the averaged temperature profile *along the z axis*. From

Eq. (6.68) one obtains:

$$< T(r) >= T^{ext} + \frac{\Delta\breve{T}}{4}\Bigg\{1 - (\xi_c)^2 + 8\sum_{i=1}^{\infty}\left(-\frac{1}{\alpha_i^3} + \frac{T^{inl} - T^{ext}}{\alpha_i \Delta\breve{T}}\right)\frac{J_0(\alpha_i\xi_c)}{J_1(\alpha_i)}\frac{1 - \exp\left(-\alpha_i^2\left(\frac{L_{col}}{L}\right)\right)}{\alpha_i^2\left(\frac{L_{col}}{L}\right)}\Bigg\} \quad (6.71)$$

Equation (6.71) gives an *axially averaged* temperature profile, consequently, knowing the temperature dependence of viscosity, an *axially averaged* viscosity profile. The axially averaged velocity profile can, at least qualitatively, be estimated by assuming $u(r) \sim 1/\mu(r)$.

For illustration purposes, various simulations are presented in Figures 6.9 and 6.10 of the example considered above. In Figure 6.9, axially averaged temperature profiles are presented. Assuming $T^{ext} = T^{inl} = 295$ K, it is clear that the temperature profiles are already stabilized at short bed length for the laboratory column, while internal temperature increases continuously with bed length in the case of the large-scale column.

Profiles obtained for $T^{ext} = 295$ K on columns of 0.5 m length and different diameters are given in Figure 6.10. The influence of the difference between T^{ext} and T^{inl} is also investigated. When $T^{ext} = T^{inl} = 295$ K (Figure 6.10a), the temperature rise at the center of the column is 1 K, irrespective of the column diameter. For a given set of physico-chemical parameters, this temperature rise is proportional to the pressure drop per unit length, according to Eq. (6.69). This temperature increase due to viscous forces, which can be a few degrees for HPLC, can have a significant effect on the performance of HPLC systems, for which the influence of any deviation from plug flow can be drastic (Dapremont *et al.*, 1998; Welsch *et al.*, 1996). For columns packed with large particles, not only is the temperature rise small, but hydrodynamics as well as kinetic dispersion easily override the influence of a moderate velocity profile. As shown in Figure 6.10b,

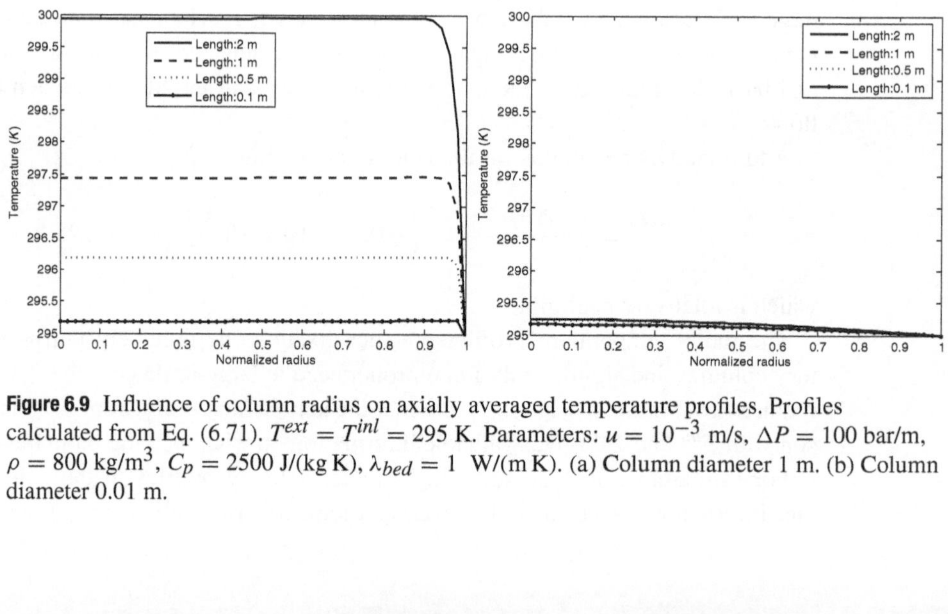

Figure 6.9 Influence of column radius on axially averaged temperature profiles. Profiles calculated from Eq. (6.71). $T^{ext} = T^{inl} = 295$ K. Parameters: $u = 10^{-3}$ m/s, $\Delta P = 100$ bar/m, $\rho = 800$ kg/m^3, $C_p = 2500$ J/(kg K), $\lambda_{bed} = 1$ W/(m K). (a) Column diameter 1 m. (b) Column diameter 0.01 m.

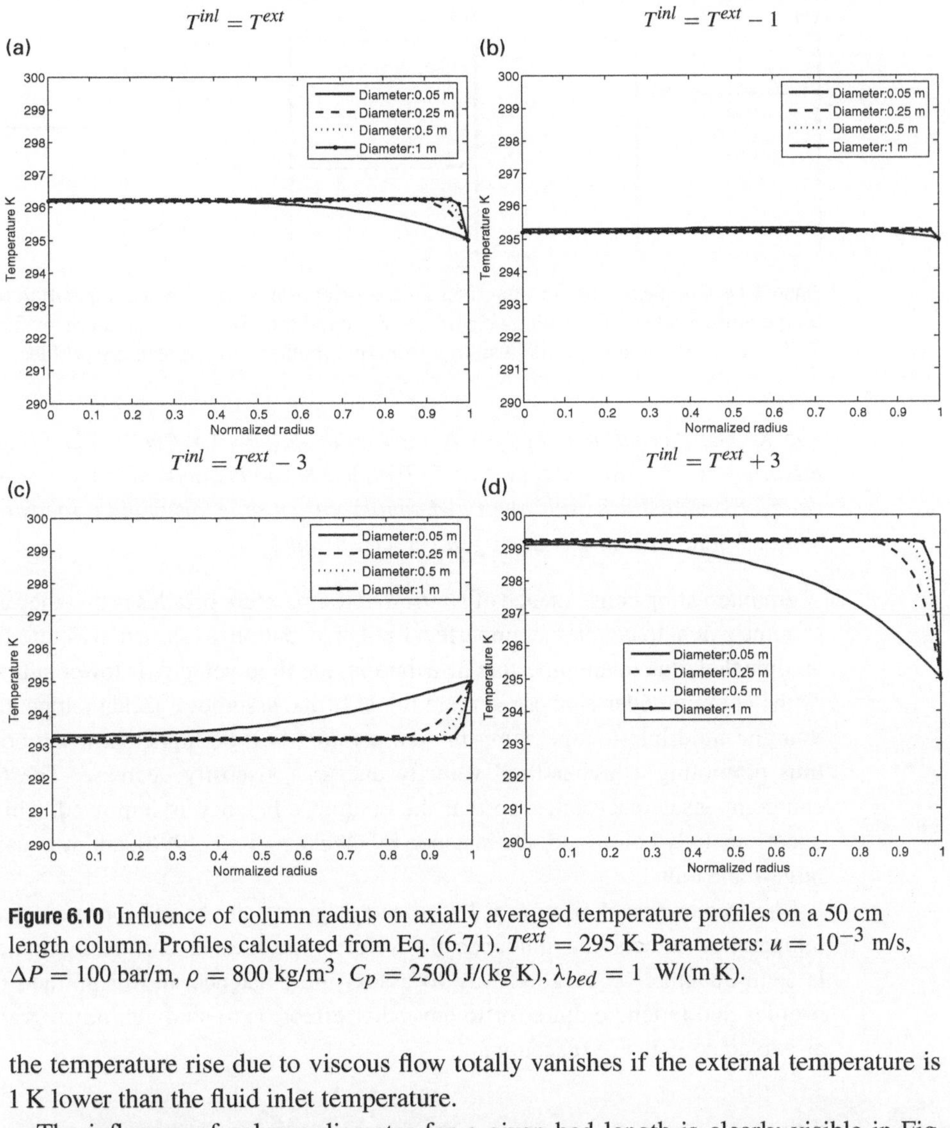

Figure 6.10 Influence of column radius on axially averaged temperature profiles on a 50 cm length column. Profiles calculated from Eq. (6.71). $T^{ext} = 295$ K. Parameters: $u = 10^{-3}$ m/s, $\Delta P = 100$ bar/m, $\rho = 800$ kg/m^3, $C_p = 2500$ J/(kg K), $\lambda_{bed} = 1$ W/(m K).

the temperature rise due to viscous flow totally vanishes if the external temperature is 1 K lower than the fluid inlet temperature.

The influence of column diameter for a given bed length is clearly visible in Figures 6.10c and 6.10d. While the temperature at the center (very close to T^{inl}) and at the wall of the column (very close to T^{ext}) are independent of column diameter, profiles tend to be more uniform in small-diameter columns, approaching the asymptotic parabolic shape.

Finally, one can see, by comparing Figures 6.10c and 6.10d, that the temperature gradient between column center and wall is higher when the column is heated than when it is cooled. This is easily understood because a temperature rise due to frictional forces adds to heating and compensates cooling.

EXERCISE 6.7 *A column of 60 cm diameter and 40 cm length packed with 10 μm particles is eluted with a mixture containing mainly water/ethanol so that* $\rho = 900$ *kg/m*^{3}*,* $C_p = 3000$ *J/(kg K). The fluid inlet temperature as well as room temperature are set at*

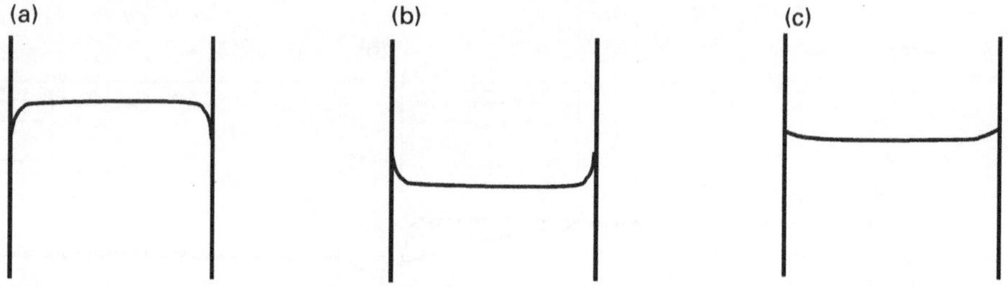

Figure 6.11 Compensation for imperfect fluid distribution profiles by the imposition of a temperature gradient. (a) Radial velocity profile (isothermal). (b) Temperature profile due to $T^{ext} > T^{in}$. (c) Velocity profile resulting from distribution and temperature profiles.

295 K. The thermal conductivity of the bed is $\lambda_{bed} = 1$ W/(m K). The viscosity of the mixture is 2.25 cP at 290 K and 1.95 cP at 300 K and is assumed to vary linearly in that range. Estimate the axially averaged temperature profile for a fluid velocity of 10^{-3} m/s. "Guesstimate" the axially averaged velocity profile.

An interesting consequence of *controlled* temperature profiles is that they can be used to some extent to counteract imperfect fluid distribution, as shown in Figure 6.11. Let us assume that, due to an imperfect distribution, the fluid velocity is lower at the periphery of the column under isothermal conditions. If one assumes a jacket temperature higher than the fluid inlet temperature, the temperature will be higher near the column wall, thus promoting a higher fluid velocity due to a viscosity decrease. The two effects can compensate each other so that the column efficiency is improved. This has been experimentally confirmed, for instance by (Welsch *et al.*, 1996), and it is used in some large-scale units.

The framework of Section 6.3.1, aimed at giving the basics for understanding how temperature effects can influence the velocity profiles in chromatographic columns, is semi-quantitative. A good way to assess the existence and magnitude of velocity profiles due to temperature (or to any other effect) is to study chromatographic peaks, as explained in the next section.

6.3.2 RTD in the presence of radial profiles

The simplicity and flexibility of the unidirectional PD and MC models, as well as their ability to represent most situations, has possibly led the chromatographic community to underestimate the influence of radially non-uniform velocity profiles. As mentioned in Chapter 2, even if different physical models can represent linear chromatograms equally well, one must try to derive models as close as possible to physical reality if one wants to properly address non-linear behaviors.

An example of the influence of a non-uniform radial velocity profile on a chromatogram is given in Figure 6.12. This example will be further analyzed later in the section. The chromatogram, obtained on a DAC column, does not look bad, but a detailed observation reveals peak tailing. The tailing is relatively moderate for this column, but in any case visible. The influence of this tailing can be more or less

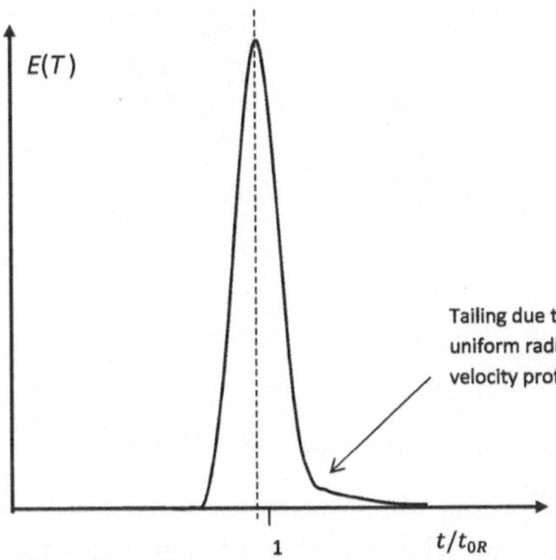

Figure 6.12 Experimental chromatogram (Nicoud and Perrut, 1993; Etienne *et al.*, 1989). Eluent CH_2Cl_2. DAC column of 80 mm id. Tracer: Acetophenone. Silica: 20–40 μm.

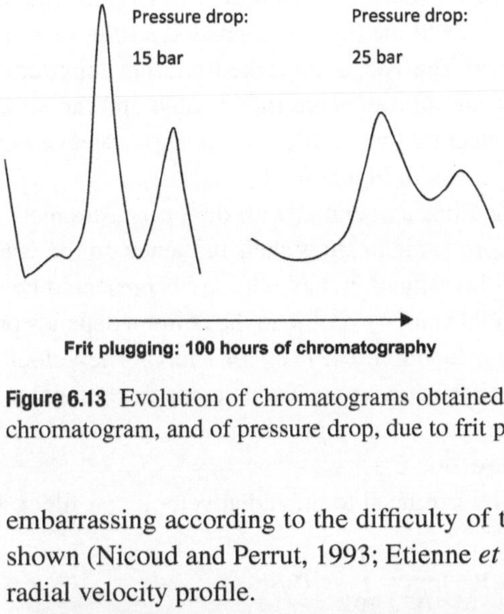

Figure 6.13 Evolution of chromatograms obtained on a 300 mm id DAC column preparative chromatogram, and of pressure drop, due to frit plugging (Nicoud and Perrut, 1993).

embarrassing according to the difficulty of the separation to be performed. It has been shown (Nicoud and Perrut, 1993; Etienne *et al.*, 1989) that it arises from a non-uniform radial velocity profile.

The consequences can be more dramatic when uneven radial velocity profiles appear, owing for instance to "frit plugging" as shown in Figure 6.13. The 300 mm id DAC column has been progressively plugged due to particles moving inside the bed and then partly plugging the distributors. This phenomenon is normally associated with pressure-drop increase and can seriously ruin the column efficiency, as shown in the right in Figure 6.13.

The experimental determination of radially non-uniform velocity profiles is not an easy task. Some work involving NMR and computed tomography, electrochemical or fluorescence detectors has been reported (Astrath *et al.*, 2007). These technologies have only been used for small-diameter columns, in the range of 5–20 cm. An interesting and simple visual method is available when DAC technology is used with small particles. Under these conditions, the bed is normally compact enough that it can be removed

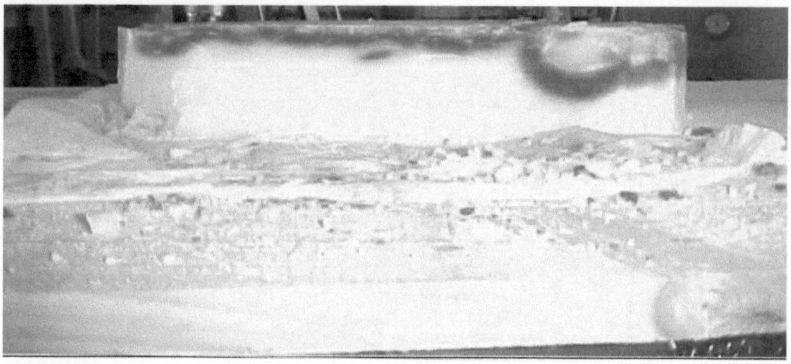

Figure 6.14 Frit plugging: experimental verification with dye concentration profile determination; see methodology in the text. The zone of altered velocity is clearly visible in the right-hand part of the picture.

from the column in its entirety, thanks to the piston motion. Dyes (for instance Cibacron Blue) are injected into the column, and the flow is stopped at a time such that dyes are still contained inside the column. The bed is unpacked and cut longitudinally along the axis. Internal radial concentration profiles are thus visible and the velocity profile is then determined from the concentration profile. An example of dye concentration profiles showing frit plugging is given in Figure 6.14.

Because of the technical difficulties associated with direct measurement of velocity profiles, the most convenient approach is to study their influence on the characteristics of the chromatogram. A rational investigation methodology is presented below.

Let us assume that the superficial velocity profile in the column depends on the radius only and can be represented by a function $u_r(r)$ (u_r for short). The velocity profile is thus assumed to be constant along the column axis and versus time; in other words, it is approximated by the axially averaged velocity profile, just as we considered an axially averaged temperature profile in Section 6.3.1.

The average superficial velocity is related to the radial velocity profile u_r by:

$$u = \frac{2}{R_{col}^2} \int_0^{R_{col}} r u_r dr \tag{6.72}$$

As a consequence of the radial velocity profile, a radial concentration profile is established in the column, so concentrations $c_A(r,z,t)$ depend on radial position, abscissa and time. Like the convention used in Chapter 5 for addressing intraparticular concentration profiles, we will use capital letters $C_A(z,t)$ to represent the average of the local concentrations $c_A(r,z,t)$ over the column cross section.

The column is mentally divided into successive concentric rings, and for each of these rings we define:

- the inlet radial concentration in the carrier fluid, which is assumed to be identical to the inlet concentration of the column:

$$c_A^{inl}(r,t) = c_A(r,z=0,t) = C_A^{inl} \tag{6.73}$$

This assumes a perfect fluid distribution at the column inlet. Should an imperfect distribution be suspected, this can be modeled by introducing pure delays between the inlet pipe and the different rings (Nicoud and Perrut, 1993).

- the outlet radial concentration:

$$c_A^{out}(r,t) = c_A(r,z = L_{col},t) \tag{6.74}$$

At the outlet of the column, all the streams associated with the different rings merge, so that a mass balance allows a determination of the outlet concentration from the different radial concentrations. Noting that $2\pi r u_r dr$ represents the fluid flow rate in the ring between r and $r + dr$, one obtains:

$$QC_A^{out}(t) = \int_0^{R_{col}} c_A^{out}(r,t)\, 2\pi r u_r dr$$

so that:

$$C_A^{out}(t) = \frac{2}{uR_{col}^2} \int_0^{R_{col}} c_A^{out}(r,t)\, r u_r dr \tag{6.75}$$

Modeling the column behavior, and especially the RTD, is easily performed using the tools presented in Chapter 2. If one assumes a linear behavior and the absence of radial transfer between the rings, the Laplace transform of the outlet radial concentration $\hat{c}_A^{out}(r,s)$ can be related to the Laplace transform of the inlet radial concentration $\hat{c}_A^{inl}(r,s) = \hat{C}_A^{inl}(s)$ by a certain transfer function $\hat{e}_r(s)$:

$$\hat{c}_A^{out}(r,s) = \hat{e}_r(s)\, \hat{C}_A^{inl}(s) \tag{6.76}$$

Because of the linearity of the Laplace transform, from Eq. (6.75) and Eq. (6.76) one can relate the transfer function $\hat{E}(s)$ of the column to the local transfer functions $\hat{e}_r(s)$:

$$\hat{E}(s) = \frac{\hat{C}_A^{out}(s)}{\hat{C}_A^{inl}(s)} = \frac{2}{uR_{col}^2} \int_0^{R_{col}} \hat{e}_r(s)\, r u_r dr \tag{6.77}$$

Assuming that the individual transfer functions and velocity profiles are known, one can then calculate any linear chromatogram by inversion of the transfer function $\hat{E}(s)$.

For the time being, no assumption has been made about the nature of the local transfer functions $\hat{e}_r(s)$. They can for instance be associated with identical or different physicochemical parameters such as axial dispersion coefficient, Henry's coefficient or mass transfer time.

The moments $\mu_n(r)$ (μ_{nr} for short) of the local transfer functions $\hat{e}_r(s)$ can be calculated via Van der Laan's theorem. Because of the linearity of the derivation, one obtains the moments associated with the column transfer function $\hat{E}(s)$ from Eq. (6.77):

$$\mu_n = \frac{2}{uR_{col}^2} \int_0^{R_{col}} \mu_{nr} r u_r dr \tag{6.78}$$

Equation (6.78) gives a general relation allowing us to determine the moments of the column transfer function, provided that the moments of the local transfer functions and velocity profiles are known.

In order to illustrate the methodology and to derive key results, we will now assume that each ring can be modeled with the MC–$\overline{\text{Equil}}$ model presented in Chapter 2. We will further assume that the number of plates characterizing each ring is independent of the radius r. This simplification is justified by our intention to focus on the dominant phenomenon, which is the radial velocity profile.[10] We will also assume that Henry's coefficients are independent of the radial position. This assumption is certainly questionable if there is a radial temperature profile.

Adapting the methodology to other models and/or relaxing some of the above assumptions is again a matter of using the Russian Lego concept presented in Chapter 2.

The local transfer functions associated with the MC–$\overline{\text{Equil}}$ model are, assuming radial uniformity of number of plates and Henry's coefficients, given by:

$$\hat{e}_r(s) = \left[1 + \frac{st_R}{N}\frac{u}{u_r}\right]^{-1} \tag{6.79}$$

where t_R is the retention time obtained with the average superficial velocity u.

The expressions for the moments associated with the local transfer functions are those given in Chapter 2 for the MC–$\overline{\text{Equil}}$ model, after replacing t_R with $t_R\frac{u}{u_r}$. One obtains:

$$\begin{aligned} \mu_{1r} &= t_R\frac{u}{u_r} \\ \mu_{2r} &= \frac{N+1}{N}t_R^2\left(\frac{u}{u_r}\right)^2 \\ \mu_{3r} &= \frac{(N+1)(N+2)}{N^2}t_R^3\left(\frac{u}{u_r}\right)^3 \end{aligned} \tag{6.80}$$

The moments of the column transfer function are related to the local moments by the integration provided by Eq. (6.78):

$$\begin{aligned} \mu_1 &= t_R h_1^{corr} & h_1^{corr} &= \frac{2}{R_{col}^2}\int_0^{R_{col}} rdr = 1 \\ \mu_2 &= \frac{N+1}{N}t_R^2 h_2^{corr} & h_2^{corr} &= \frac{2}{R_{col}^2}\int_0^{R_{col}} \frac{u}{u_r}rdr \\ \mu_3 &= \frac{(N+1)(N+2)}{N^2}t_R^3 h_3^{corr} & h_3^{corr} &= \frac{2}{R_{col}^2}\int_0^{R_{col}} \left(\frac{u}{u_r}\right)^2 rdr \end{aligned} \tag{6.81}$$

The nth order moment of the column transfer function appears as a product of the nth order moment that would be obtained with a uniform radial velocity profile with a correction factor h_n^{corr}, depending on the velocity profile only. Irrespective of the radial velocity profile, one has $h_1^{corr} = 1$, so the first-order moment is not affected by the velocity profile. Higher-order correction factors h_n^{corr} are greater than 1, and equal to 1 only for a uniform radial velocity profile.

[10] We are aware of results indicating an influence of the radial position on the axial dispersion coefficient. The experimental difficulties and underlying assumptions are such that we believe that these results must be viewed with great care.

A similar presentation can be given with a discrete, instead of a continuous, velocity profile (Nicoud and Perrut, 1993; Etienne *et al.*, 1989). The column is approximated by n_R concentric rings. Ring i represents fraction ω_i of the column section Ω_{col}, and is eluted by fraction α_i of the total flow rate Q. Fluid velocity u_i is given by $u_i = u\alpha_i/\omega_i$.

The following relations follow from the above definitions:

$$\sum_{i=1}^{n_R} \alpha_i = \sum_{i=1}^{n_R} \omega_i = \frac{1}{u}\sum_{i=1}^{n_R} \omega_i u_i = 1 \qquad u = \sum_{i=1}^{n_R} \omega_i u_i \tag{6.82}$$

The global transfer function of the chromatographic column is:

$$\hat{E}(s) = \sum_{i=1}^{n_R} \alpha_i \hat{e}_i(s) \tag{6.83}$$

where $\hat{e}_i(s)$ represents the transfer function associated with ring i.

The retention time t_R in each ring is given by $t_{Ri} = t_R \frac{u}{u_i}$, so the transfer function associated with the MC–Equil model for ring i is:

$$\hat{e}_i(s) = \left[1 + \frac{st_R}{N}\frac{u}{u_i}\right]^{-1} \tag{6.84}$$

One immediately concludes that Eq. (6.81) still holds, provided that the correction factors are calculated with:

$$\begin{aligned}
\mu_1 &= t_R h_1^{corr} & h_1^{corr} &= 1 \\
\mu_2 &= \frac{N+1}{N} t_R^2 h_2^{corr} & h_2^{corr} &= u^2 \sum_{i=1}^{n_R} \frac{\alpha_i}{u_i^2} = \sum_{i=1}^{n_R} \frac{\omega_i^2}{\alpha_i} \\
\mu_3 &= \frac{(N+1)(N+2)}{N^2} t_R^3 h_3^{corr} & h_3^{corr} &= u^3 \sum_{i=1}^{n_R} \frac{\alpha_i}{u_i^3} = \sum_{i=1}^{n_R} \frac{\omega_i^3}{\alpha_i^2}
\end{aligned} \tag{6.85}$$

Either a continuous or discrete approach can thus be selected, and they are equivalent provided that Eqs. (6.78), (6.79) and (6.80), valid for the continuous approach, are replaced by Eqs. (6.83), (6.84) and (6.85) when the discrete approach is used.

With knowledge of the moments given by Eq. (6.81) or Eq. (6.85), one can then calculate the variance and skewness of the RTD, as presented in Chapter 2:

$$\begin{aligned}
\sigma^2 &= \mu_{2c} = \mu_2 - \mu_1^2 = \frac{t_R^2}{N}\left(1 + (N+1)\left(h_2^{corr} - 1\right)\right) \\
S_k &= \frac{\mu_{3c}}{\mu_{2c}^{3/2}} = \frac{2}{\sqrt{N}} \frac{1 + \frac{(N+1)(N+2)}{2}\left(h_3^{corr} - 1\right) - \frac{3N(N+1)}{2}\left(h_2^{corr} - 1\right)}{\left(1 + (N+1)\left(h_2^{corr} - 1\right)\right)^{3/2}}
\end{aligned} \tag{6.86}$$

One can check that these expressions reduce to those for a uniform velocity profile when both correction factors h_2^{corr} and h_3^{corr} equal 1.

The HETP of the column is deduced from the variance given by Eq. (6.86), so one obtains:

$$HETP = L_{col}\frac{\sigma^2}{t_R^2} = \frac{L_{col}}{N} + L_{col}\left(1 + \frac{1}{N}\right)\left(h_2^{corr} - 1\right)$$

If one defines $HETP_{unif} = \frac{L_{col}}{N}$, the HETP that would be obtained with a uniform radial concentration profile, one has:

$$HETP = HETP_{unif} + L_{col}\left(1 + \frac{1}{N}\right)\left(h_2^{corr} - 1\right) \tag{6.87}$$

The HETP is thus expressed as the sum of $HETP_{unif}$, which would be obtained for a uniform radial velocity profile, plus a correction term depending on the shape of the velocity profile. The correction term is proportional to the column length as long as one assumes $N \gg 1$. This is a very important result: in contrast to the classical and desired situation obtained with uniform velocity profiles, in the presence of non-uniform radial velocity profiles the HETP *varies linearly with column length*. Alternatively, experimental evidence of a linear or quasi-linear variation of the HETP with column length should give a strong warning of the possible existence of non-uniform radial velocity profiles.

There is another way to assess the existence of non-uniform velocity profiles. Let us assume the column to be axially symmetrical, so that its behavior is independent of the flow direction. Reversing the flow at some point in time, and detecting chromatograms at the initial inlet of the column, allows cancellation of the irreversible influence of velocity profiles. Peak efficiency and skewness should be improved when determined by the flow-reversal technique. This has been confirmed experimentally, for instance by (Mosciarello *et al.*, 2001).

From the HETP expression (6.87), one can relate the apparent number of plates of the column, N_{app} (resulting from both axial hydrodynamic dispersion and radial velocity profiles), to the number of plates, N_{unif} (resulting from the axial hydrodynamic dispersion only, thus obtained with a uniform radial velocity profile), by:

$$\frac{1}{N_{app}} = \frac{1}{N_{unif}} + \left(1 + \frac{1}{N_{unif}}\right)\left(h_2^{corr} - 1\right) \tag{6.88}$$

Let us further illustrate the possible linear variation of the HETP with column length using the results of (Nicoud and Perrut, 1993).

In order to investigate the influence of non-uniform radial velocity profiles in an 80 mm diameter DAC pilot scale column, the authors used a mask to partly obstruct the periphery of the inlet frit. Internal velocity profiles were assessed by the visual method using dyes, presented at the beginning of this section. The dye profiles are radially symmetrical, and a typical result obtained with a mask of about 5 mm width is given in Figure 6.15. This dye profile was approximated by seven rings, and associated parameters are given in the table in Figure 6.15. From the set of parameters α_i and ω_i, one can determine h_2^{corr} =1.001 017 from Eq. (6.85).

RTD measurements were performed for different column lengths and associated peak variances calculated. The key results are given in Figure 6.16. The HETP curve exhibits a linear variation with column length, in good agreement with the prediction of Eq. (6.87).

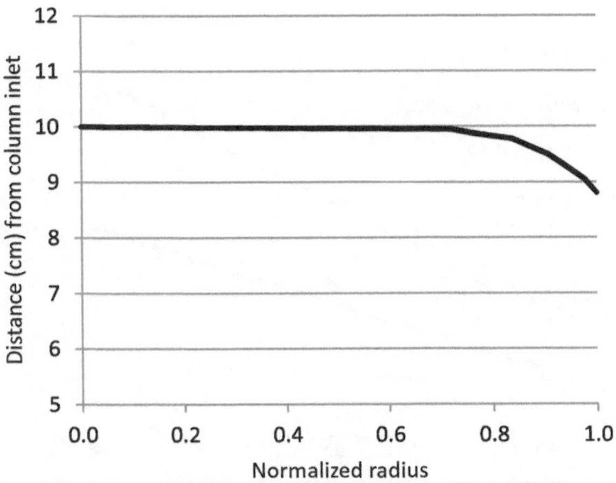

Ring i	R_{min}/R_{col}	R_{max}/R_{col}	L_i (cm)	u_i (m/s)	ω_i	α_i
1	0	0.721	9.975	7.76E-03	0.520	0.531
2	0.721	0.767	9.910	7.71E-03	0.069	0.070
3	0.767	0.837	9.825	7.64E-03	0.112	0.113
4	0.837	0.907	9.640	7.50E-03	0.122	0.120
5	0.907	0.953	9.350	7.27E-03	0.087	0.083
6	0.953	0.977	9.125	7.10E-03	0.045	0.042
7	0.977	1.000	8.925	6.94E-03	0.046	0.042

Figure 6.15 Dye and consequent velocity profiles obtained in an 80 mm diameter DAC column (Nicoud and Perrut, 1993; Etienne *et al.*, 1989). A mask of about 5 mm width was inserted to obstruct the flow at the periphery of the column. Top: the dye profile obtained after $t_{elu} = 1.5$ min of elution. Bottom: approximate representation of the velocity profile. Dyes were not adsorbed, so the velocity profile is determined with $u_i = \varepsilon_T L_i / t_{elu}$, $\varepsilon_T = 0.7$.

The HETP dependence on bed length is best represented by $h_2^{corr} = 1.0006$ (solid line), whereas our estimate of $h_2^{corr} = 1.001\,017$ from the velocity profiles overestimates the effect a bit (solid line of Figure 6.16). The two values are, however, pretty close, so that there is a real consistency between the shape of the velocity profile and the variation of the HETP with bed length.

It is worth mentioning that the HETPs estimated via peak mid-width measurements are independent of column length, as shown in Figure 6.16. This is further evidence that, as mentioned in Chapter 2, determining chromatogram variances, thus the number of plates, by approximate methods can be dangerous and can lead to wrong conclusions. This is because peak tailing is not accounted for by these approximate methods.

This is further illustrated in Figure 6.17, which shows chromatograms of a 20 cm length column simulated using the MC–$\overline{\text{Equil}}$ model with and without taking into account the discrete velocity profile given in Figure 6.15. Taking into account the non-uniform velocity profile requires inverting the transfer function defined by Eqs. (6.83) and (6.84). Under the flow-rate conditions prevailing in Figure 6.16, one can see that the HETP at zero bed length equals about 100 μm; as a consequence, in the absence

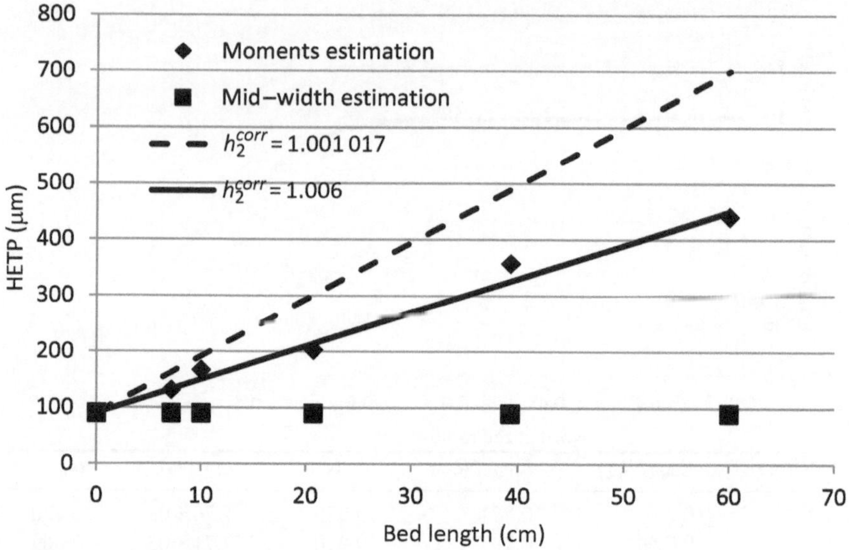

Figure 6.16 Influence of bed length on the HETP for a DAC column of 80 mm internal diameter (Nicoud and Perrut, 1993; Etienne *et al.*, 1989). A mask of about 5 mm width has been inserted to partly obstruct the flow at the periphery of the column (see Figure 6.15). Silica: C18 Alltech 12–45 μm. Eluent: Methanol. Tracer: Acetophenone. Average fluid velocity is similar to the one in Figure 6.15.

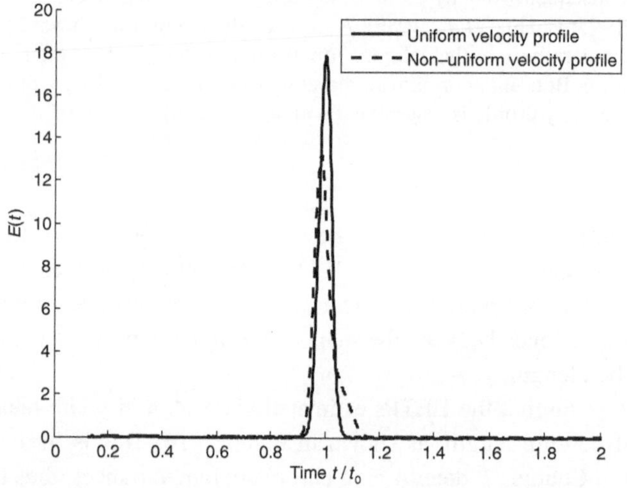

Figure 6.17 Simulated chromatograms (MC–$\overline{\text{Equil}}$ model) of a column of 80 mm internal diameter and 200 mm length in which the velocity profile presented in Figure 6.15 prevails.

of non-uniform velocity profiles, a column of 20 cm length would be characterized by $N_{unif} = 2000$. We used this parameter and the discrete velocity profiles of Figure 6.15 to simulate the RTD of a 20 cm length column. We can see in Figure 6.17 that the mid width of the peaks simulated with or without taking into account the velocity profiles are very similar, even if the peaks themselves are different. The impact of the velocity profile is especially impacting the tail of the peaks.

In order to further illustrate the influence of velocity profiles, let us postulate a simple and flexible parabolic velocity profile by:

$$u_r(r)/u = \gamma + 2(1-\gamma)\left(1-\xi_c{}^2\right) \tag{6.89}$$

where γ is the ratio of the fluid velocity at the periphery $u_{min} = u(\xi_c = 1)$ of the column to the average fluid velocity ($\gamma = \frac{u_{min}}{u} \leq 1$). With the parameter γ, one can simply adapt the span of the radial velocity profile. For instance, if one selects $\gamma = 0.95$, the radial velocities $u_r(r)$ are within 5% of the average velocity u.

The correction factors are calculated from the radial velocity profile expression (6.89):

$$h_2^{corr} = \frac{1}{2}\frac{1}{\gamma - 1}Ln\left(\frac{\gamma}{2-\gamma}\right) \tag{6.90}$$

$$h_3^{corr} = \frac{1}{\gamma(2-\gamma)} \tag{6.91}$$

With the expressions for the correction factors, one can determine the influence of the radial velocity profile on the moments of the chromatograms, thus on variance and skewness.

For illustration purposes, the influence of a parabolic velocity profile on the ratio N_{app}/N_{unif} (see Eq. (6.88)) is presented in Figure 6.18. For columns of a limited efficiency (for instance $N_{unif} = 100$, which is typical for low-pressure ion-exchange chromatography), one sees in Figure 6.18a that the ratio $N_{app}/N_{unif} = 0.5$ for $\gamma = 0.85$. In other words, should the velocity profile exhibit variations of about 15% with respect to the average velocity, the apparent number of plates of the column would be half of what it would be in the absence of the velocity profile.

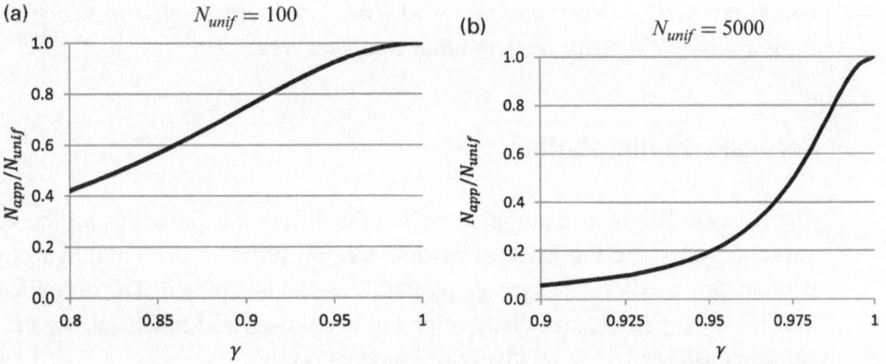

Figure 6.18 Influence of the span parameter γ of the parabolic velocity profile on the ratio N_{app}/N_{unif}. See text for details.

For columns of high efficiency ($N_{unif} = 5000$, which is typical for preparative HPLC), one sees in Figure 6.18b that the ratio $N_{app}/N_{unif} = 0.5$ for $\gamma = 0.98$. In other words, should the velocity profile exhibit variations of about 2% with respect to the average velocity, the apparent number of plates of the column would be half of what it would be with a uniform velocity profile. This means that HPLC chromatography columns must be designed with extreme care in order to avoid ruining the potential efficiency of the particles. This shows that the temperature profile determined in Exercise 6.7, even if of limited magnitude, can certainly disrupt the good performance of an HPLC column.

The reader can easily check with Eq. (6.91) that, in addition to the loss in apparent number of plates, the existence of non-uniform radial velocity profiles induces an increased skewness.

Expressions similar to (6.90) and (6.91) can be derived analytically or numerically for velocity profiles that might be more representative of specific situations. For instance, the parabolic profile can be replaced by a higher-order polynomial that could better represent relative uniformity of the velocity profile at column center and significant variation near the column wall. Analytical expressions with polynomials are available (especially with the assistance of modern symbolic calculators ...). Note that if numerical integration is required to calculate correction factors, special care must be taken as to the precision. The terms $(h_n^{corr} - 1)$ are multiplied in Eq. (6.86) by factors proportional to N or N^2, which can be very big numbers, so a minute error in h_n^{corr} can have a significant impact on variance and skewness.

EXERCISE 6.8 *The inlet distributor of a chromatography column is partly plugged so that 10% of the column surface is eluted at a fluid velocity 10% lower than the fluid velocity in the non-plugged part. Estimate the influence of this partial plugging on the apparent number of plates of the chromatogram.*

The following two exercises are only for the readers willing to further exercise their skills in handling RTD and transfer functions.

EXERCISE 6.9 *Show how any given RTD can be represented by a model that approximates the system by a pipe with a radial velocity profile.*

EXERCISE 6.10 *Determine the RTD associated with a pipe in which laminar flow occurs. Calculate mean retention time and variances.*

6.4 Particle size distribution

The Kozeny–Kármán equation, as well as the different expressions for the Van Deemter curve, describe the influence of particle size on pressure drop and hydrodynamic dispersion, provided that an average particle size can be defined. The dependence of mass transfer upon particle size distribution has been discussed in Chapter 4, and this matter can be considered to be relatively well understood.

The situation for pressure drop is much more complex and less understood. The reason for this complexity is mainly the ability of particles of different sizes to move

Table 6.2 Particle sizes. From (Dewaele and Verzele, 1983).

Rounded particle diameter (μm) (Coulter)	3	4	5	6	8	10
"Effective" particle diameter (μm)	2.89	3.50	5	5.77	7.63	10.8

with respect to each other, fine particles being able to fill some holes, thus reducing the overall bed extragranular porosity. Mixing particles of different sizes not only affects the "average" particle size, but it also impacts porosity. Unfortunately, little information and few tools are available in the literature.

In order to illustrate the matter, we refer to the work by (Dewaele and Verzele, 1983), who measured pressure drop and the efficiencies of chromatographic columns packed with C18 silica bound spherical particles. Different samples of a practically mono-dispersed size distribution ranging from 3 to 10 μm were available. Different mixtures have been prepared by mixing known weight amounts of the mono-dispersed particle samples.

For each mono-dispersed particle sample, the authors determined particle size by Coulter counting and reported rounded values. For each sample, they also defined and determined an "effective" particle size diameter, being the particle diameter to be used in the Kozeny–Kármán equation to represent the experimental pressure-drop measurements. As shown in Table 6.2, there is good agreement between the "effective" particle size diameter obtained from the Kozeny–Kármán equation and the rounded value obtained by Coulter counting.

The authors then prepared three different mixtures (a,b,c) by mixing equal weights of samples of the two particle sizes, as described in Table 6.3. The weight-averaged diameter of the three mixtures is given in Table 6.3. Pressure-drop measurement on these three mixtures allowed determination of an effective particle diameter using the Kozeny–Kármán equation. While the weight-averaged diameters of the three mixtures a,b,c are in the order $a < b < c$, the "effective" particle diameters are in the order $a > b > c$. The biggest distortion between weight-averaged and effective diameters is obtained for mixture c, obtained by mixing the two most widely different particle sizes: the weight-averaged diameter is 6.5 μm while the effective diameter is only 3.3 μm. It seems that, when two samples containing very different particle sizes are mixed, the effective pressure-drop diameter is imposed by the smaller particles. This result, obtained with simple binary mixtures, demonstrates that the influence of small particles is significantly higher than what a weight-averaged estimate would lead one to believe. If you mix 1 cm stones with 1 mm sand, the pressure drop will be determined by the sand ...

(Dewaele and Verzele, 1983) then studied 10 different weight composition mixtures of the 3 μm and 8 μm particles: the pressure drop exhibits a non-linear behavior with respect to the weight proportion of fine particles in the mixture. Once again, the small particles tend to determine the pressure drop. The results can be expressed in terms of an excess pressure drop, calculated as the difference between the actual pressure drop and what would be obtained using the weight-averaged particle size in the Kozeny–Kármán

Table 6.3 Pressure drop and number of plates determined for mixtures of equal weight of two different particle sizes. Column: 100 mm × 4.6 mm. From (Dewaele and Verzele, 1983).

Mixture	Rounded particle diameter (μm)	Pressure drop (bar)	Weight-averaged diameter (μm)	Effective particle diameter (μm)
a	4+6	74	5.0	4.35
b	3+8	130	5.5	3.28
c	3+10	128	6.5	3.31

equation. By definition, the excess pressure drop equals zero for the mono-dispersed particle samples. The pressure drop across a mixed bed is always larger than the pressure drop predicted with the weight-averaged diameter, the excess increasing rapidly when the fraction of small particles is low, and reaches a maximum when the two particle sizes are mixed in equal proportions.

Although the work is limited to mixtures of two particle sizes only, it clearly emphasizes the strong and non-linear influence of small particle sizes on pressure drop. This means that the pressure drop is larger when the particle size distribution is wide.

This not only shows that for a given pressure drop one needs to favor homogeneous particle size distribution, but it also orients towards the use of spherical particles because this geometry is less favorable to the formation of finer particles from breakage.

In practice, a precise monitoring of pressure drop is very important as it is a precursor of the appearance of fine particles. This should give a warning that column efficiency could then rapidly deteriorate due to frit plugging, as shown in Figure 6.13. Recall that frit plugging consequences can be as bad as forcing an immediate stop of the column. Once again, don't forget rule 1: "Clean the frits."

To the best of our knowledge, particle size distribution has a relatively predictable influence on column efficiency: the Van Deemter curve is influenced largely by mass transfer and can be predicted for a given particle size distribution, thanks to the methodology presented in Chapter 5. The influence of particle size distribution on hydrodynamic dispersion is discussed in (Delgado, 2006), and there is evidence that the larger the particle size distribution, the larger the axial dispersion. However, we consider this influence to be secondary, dispersion in chromatographic columns being mainly imposed by kinetics and possibly by velocity profiles.

6.5 Viscosity and gravity effects

The preceding theoretical framework has assumed that the fluid flowing in the column is monophasic. In fact, due to productivity requirements, the feed material is sometimes injected in a very concentrated form, so solubilization can be tedious. Consequently, at least during some period the fluid in the column can be biphasic, and this can lead to very significant alteration of column behavior.

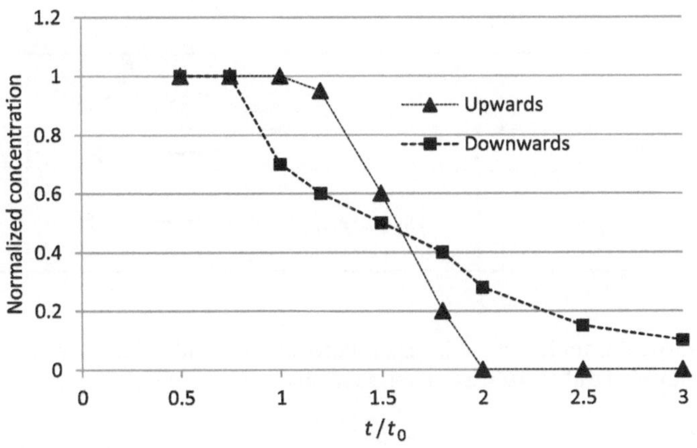

Figure 6.19 Regeneration curves of glycerol by water on ion-exchange resin particles of about 500 μm, initially equilibrated with glycerol at 500 g/l in water, showing the influence of the flow direction. Room temperature. (R.M. Nicoud, unpublished results.)

In order to demonstrate possible difficulties, we consider the regeneration by water of an ion-exchange column equilibrated with glycerol. The chromatographic interactions between glycerol and the resin beads are minimal, and one can assume that the glycerol has access to the intragranular porosity but is not adsorbed.

When glycerol is fed or regenerated at concentrations up to a few tens of grams per liter, breakthrough curves exhibit a classical behavior, characteristic of the injection of a linearly retained compound on a low-pressure, low-efficiency column equivalent to a few tens of plates per meter. When the column initially contains glycerol at significant concentrations (500 g/l) and is regenerated by water, breakthrough curves tend to deform, and are dependent on the flow direction (Figure 6.19). Additionally, results are less reproducible at high concentrations than at low concentrations. The influence of flow direction as well as questionable reproducibility are certainly embarrassing, and not appreciated by chromatographers.

If one could look inside the column, one would see that the glycerol/water interface does move as a plug flow, but that instabilities around the initially uniform surface gradually build up and may amplify, so that this interface does not remain stable. Finally, various "spikes", "channels" or "fingers" appear. The word "fingering" is now generally used to describe this phenomenon, and, as differences in viscosities are normally the driver, one normally speaks of "viscous fingering".

This phenomenon has long been recognized during displacement of highly concentrated sugars by water (Hill, 1952), and was initially called "channeling". Viscous fingering is often met when size-exclusion chromatography is used for purifying proteins, and was reported as early as 1961 by (Flodin, 1961). Later on, (Czok *et al.*, 1991) investigated this phenomenon further, on size-exclusion chromatography of proteins, modifying the viscosity of the carrier or of the feed by adding various concentrations of glycerol; they reported irreproducibility. (Cherrak *et al.*, 1996) reported irreproducibility of toluene peak shapes when the carrier fluid and the sample solvent had different concentrations of isopropanol in methanol. (Castells and Castells, 1998) observed

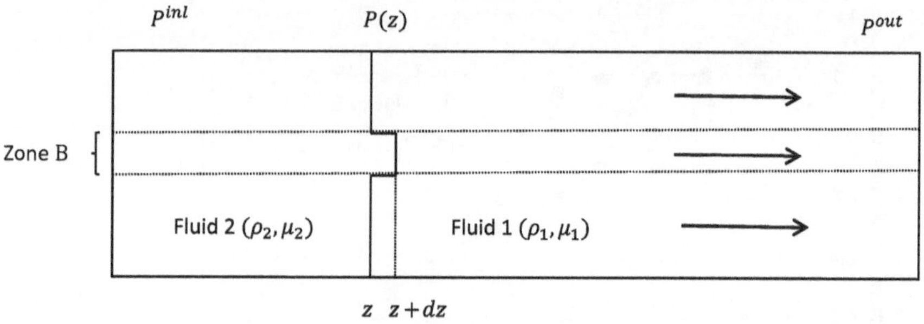

Figure 6.20 Schematic illustration of the interface between two fluids, 1 and 2. A small part B of interface A has moved ahead of the mean interfacial position.

erratic peak maxima and shoulders in reversed-phase chromatography with water–acetonitrile and water–alcohol mixtures. We have reason to believe that this list is not complete and that similar situations occur when, for instance, natural oils are separated by reversed-phase chromatography.

Various work aimed at visualizing the development of viscous fingering in columns has been published. One technique consists in matching the refractive index of the mobile phase, glass beads particles and column. This allows a visualization of the development of colored component in the column, with sometimes very impressive and surprising pictures (Shalliker *et al.*, 1999, 2007; Catchpoole *et al.*, 2006). The consequences of viscous fingering can be as bad as totally destroying the separation ability of a chromatographic column: not only are the peaks distorted but they can exhibit erratic patterns, the retention times can be severely impacted and a single solute can appear at two or three different retention times depending on the "fingers" it has traveled with.

There is no question that modeling viscous fingering is a difficult task, and that chromatographers are not used to addressing unstable and irreproducible phenomena. Incidentally, let us mention that viscous fingering is the only unstable and irreproducible phenomenon that will be discussed in this book.

A simple model of biphasic flow will help us understand the origin of fingering. A column in which fluid 2 having density and viscosity (ρ_2, μ_2) displaces a downstream fluid 1 with density and viscosity (ρ_1, μ_1) is schematically presented in Figure 6.20.

We assume that, for whatever reasons, a small portion B of interface A between the two fluids is displaced by a distance *dz* along the column axis. Even with perfect distributors, the interface between two fluids is not totally planar for various reasons, such as:

- non-uniformity of the chromatographic beds or the particles themselves introducing a perturbation at the d_p scale (predictable and reproducible perturbations)
- local concentration fluctuations due to thermal motion of the molecules, accidental effects such as slight pressure or temperature change, vibrations etc. (unpredictable and irreproducible perturbations).

These fluctuations can be infinitesimal, so one may think they cannot materially influence column behavior. This is true... unless something amplifies the initial perturbations to such an extent that they can become visible.

This important concept of stability can be understood with a relatively simple consideration of the relation between pressure and fluid velocity. Work presented by (Rousseaux *et al.*, 2007) is briefly summarized below.

Including the hydrostatic contribution to pressure, the pressure drop over a given distance L in a bed of permeability κ eluted with a fluid of viscosity μ can be related to the fluid velocity u by:

$$u = \frac{\kappa}{\mu}\left(\frac{p^{inl} - p^{out}}{L} + \rho\delta g\right) \tag{6.92}$$

where g is the gravitational acceleration and δ is a parameter which is $+1$ for downward vertical flow, -1 for upward vertical flow or 0 for horizontal flow. Intermediate situations are not of practical interest.

In each area, A or B, the velocity of fluid 1 is identical to that of fluid 2 in the column in Figure 6.20, and if one assumes that there is no momentum exchange between zones A and B, one can write:

For area A:

$$u^A = \frac{\kappa}{\mu_2}\left(\frac{p^{inl} - P(z)}{z} + \rho_2\delta g\right) = \frac{\kappa}{\mu_1}\left(\frac{P(z) - p^{out}}{L_{col} - z} + \rho_1\delta g\right) \tag{6.93}$$

For area B:

$$u^B = \frac{\kappa}{\mu_2}\left(\frac{p^{inl} - P(z+dz)}{z+dz} + \rho_2\delta g\right) = \frac{\kappa}{\mu_1}\left(\frac{P(z+dz) - p^{out}}{L_{col} - z - dz} + \rho_1\delta g\right) \tag{6.94}$$

After a few mathematical manipulations, one obtains for the velocity difference between A and B:

$$u^A - u^B = \frac{\kappa}{\mu_2 z/L_{col} + \mu_1\left(1 - z/L_{col}\right)}\left[\frac{u^B}{\kappa}\left(\mu_2 - \mu_1\right) - \delta g\left(\rho_2 - \rho_1\right)\right]\frac{dz}{L_{col}} \tag{6.95}$$

From Figure 6.20 it is apparent that the system is stable if $u^A > u^B$, as this would mean that the main interface A would catch up with the perturbation. If $u^A < u^B$, the length of the perturbation would grow and thus amplify. The stability condition $u^A > u^B$ requires $\left[\frac{u^B}{\kappa}\left(\mu_2 - \mu_1\right) - \delta g\left(\rho_2 - \rho_1\right)\right] > 0$ according to Eq. (6.95). Since u^B differs only marginally from the average flow velocity u, one can state that the system is stable if:

$$\frac{u}{\kappa}\left(\mu_2 - \mu_1\right) - \delta g\left(\rho_2 - \rho_1\right) > 0 \qquad \text{stability condition} \tag{6.96}$$

This stability criterion contains two terms, one related to the viscosity difference and one to the density difference.

Let us first assume that the densities are identical ($\rho_2 = \rho_1$) but that the viscosities are different. To satisfy the stability condition (6.96), one should have $\mu_2 > \mu_1$. The fact that $\mu_2 > \mu_1$ requires that the perturbation move less rapidly than the main interface, so that the main section is catching up with the perturbation. If $\mu_1 > \mu_2$, fluid 2 is protruding through the interface in fluid 1, forming a kind of finger.

In the absence of gravitational effects (either because the column is horizontal or because the fluids are of identical density) the interface is therefore unstable when the displacing fluid is less viscous than the displaced fluid. For a stable flow, the displacing fluid must act like a piston that is more "rigid" than the displaced fluid. This type of instability is called a *Saffman–Taylor instability*.

Let us assume instead that $\mu_2 = \mu_1$, or that the second term of the stability criterion in Eq. (6.96) is dominant. The stability criterion (6.96) becomes $\delta\,(\rho_2 - \rho_1) < 0$, which says that, irrespective of whether the flow direction is upwards or downwards, the interface is stable when the lower fluid is denser. If the upper fluid is denser, a so-called *Rayleigh–Taylor instability* is observed.

Notice that, for the glycerol regeneration curve in Figure 6.19, the less viscous fluid (water) displaces the more viscous fluid (glycerol), and in addition the more dense fluid is above for upwards flow. Upwards flow thus combines all the ingredients for fingering.

A simple order-of-magnitude estimate shows that the first term of criterion (6.96) is normally much larger than the second one unless the pressure drop is very low or the permeability very high. This shows that, in general, viscosity differences rather than density differences are responsible for the instability, thus for fingering.

In preparative chromatography, the injection profile is normally a rectangular pulse, so a lower-viscosity fluid will at some time displace a higher-viscosity fluid. One can conclude that instability will always occur, either at the front or at the tail of a chromatogram. Feed mixtures being normally more viscous than carrier fluids, viscous fingering is generally expected to take place at the tail of the peaks.

This instability may not be visible because destabilizing effects due to the viscosity difference can be compensated by a stabilization mechanism. This mechanism is driven by the dispersion phenomena occurring in the column, which will eventually lead to dilution and thus miscibility of the two fluids. For the sake of completeness, one can notice that, even in the case of immiscible fluids, superficial tension will stabilize the process.

Neglecting the influence of density and gravity, it can be shown (Tan and Homsy, 1986) that for miscible fluids the stability criterion (6.96) must be replaced by:

$$\left.\frac{d\mu}{dx}\right|_{x=0} + \left.\frac{d\mu}{dx}\right|_{x=1} < 0 \qquad \text{stability criterion for miscible fluids} \tag{6.97}$$

where x is the molar fraction of the downstream fluid 1 in the binary mixture ($x = 0$ for fluid 2; $x = 1$ for fluid 1).

If the fluid viscosity decreases monotonically with composition x, criterion (6.97) is satisfied and the system is stable. The system is unstable if the fluid viscosity increases with x. Additionally, in the case of a monotonic variation of viscosity with concentration, criteria (6.97) and (6.96) are equivalent.

The situation is more complex for mixtures like alcohol–water or acetonitrile–water, in which viscosity can exhibit a local maximum versus concentration. The derivatives may be of opposite signs and the interface may be stable even if (6.96) is not true, and vice versa.

The modeling of the instability can be extended to describe what is expected to occur at the onset of the perturbation using linear stability analysis (LSA). LSA analyzes the early stages of the development of an instability from a base steady state. Without introducing quantitative expressions, we can note that LSA shows that viscous fingering can be characterized by three main parameters (Tan and Homsy, 1986; Rousseaux *et al.*, 2007):

- the "logviscosity" ratio $R = \ln\left(\frac{\mu_2}{\mu_1}\right)$, which looks rather intuitive from the above presentation
- the ratio $\check{D} = \frac{D_{rad}}{D_{ax}^{app}}$ between radial and axial dispersion coefficients; the higher this ratio, the higher the radial dispersion coefficient, the higher the ability of the dispersion process to erase perturbations of a given width
- the fluid velocity: the growth rate of the instability at onset increases with fluid velocity.

This analysis then allows one to estimate the size and growth rate of the most probable (dangerous and unstable) fingers.

This analysis has been extended to linearly retained solutes by (Mishra *et al.*, 2007). They have shown that the above results can still be used but that the "logviscosity" ratio must be divided by $1 + K'_A$. Fingering is thus reduced by adsorption as a consequence of the reduction of the propagation of the concentration front.

Viscous fingering is, *a priori*, not impacted by column diameter as long as column diameter is significantly higher than, let us say, 10 particle diameters, which is always the case in practice. This is why (Fernandez *et al.*, 1996) proposed the use of capillary columns to avoid the viscous fingering phenomena. Practical implementation is certainly questionable.

The use of LSA is limited to the *onset* of development of viscous fingering. After this initial stage, the non-linearity of fluid dynamics (no relation to the non-linearity of adsorption isotherms) must be taken into account as it can, for instance, lead fingers to merge or to split. The extension of the model to this non-linear range has been presented by (Rousseaux *et al.*, 2011) for solutes not interacting with the stationary phase. Equations and numerical efforts are too complex to be reported here. Let us, however, mention that this approach allows the first moments of a chromatogram to be related to viscous fingering. Because of the irreproducibility of viscous fingering, these moments must be calculated on a statistical basis.

6.6 Conclusion

Chromatography is normally considered to be an ideal process that can be easily scaled up using constant column length and fluid velocity. This condition is, at least approximately, often met in practice so that scaling up (provided that the right technology is used) is in general not an issue.

In ideal systems:

- radial velocity, temperature and concentration profiles are uniform.
- column dispersion under linear conditions, possibly characterized by the number of plates, is well described by the classical framework in Chapters 2, 4 and 5. A practical example is the Van Deemter curve, which predicts an HETP independent of column length, generally varying linearly with fluid velocity and with the square of particle diameter.

An ideal system does not exclude the use of compressible fluids such as supercritical fluids or gases. Under these conditions, the modeling of fluid velocity, pressure drop and dispersion requires taking axial profiles into account. This adds a bit of complexity but does not lead to real difficulty provided that an adequate equation of state and a correlation expressing viscosity versus pressure and temperature are available.

Finally, let us mention that characteristics of ideal systems such as pressure drop and plate number can evolve over time due to the existence or appearance of fine particles that can migrate and plug the bed or distributors. Particle size distribution and particle shape must be considered with special care, not only because they may induce an initial pressure drop that is higher than anticipated, but also because the motion of fine particles or the production of fines may impact column stability. Narrow particle size distributions of spherical particles are normally preferred to ensure better stability.

Chromatographic systems are considered non-ideal according to our terminology when the ratio L_{col}/D_{col} influences performance characteristics such as permeability, velocity profile and peak shape. In addition to "suboptimal" distributor design, an important factor in non-ideality can be temperature, because of either improper control of inlet and external temperature or the heat generated by viscous frictional forces in the case of HPLC columns operated at a 50–100 bar pressure drop.

Another cause of non-ideality is bed or particle compressibility under the stress imposed by piston compression or viscous flow. Under these conditions, when soft gels are used the variation of pressure drop versus fluid velocity and column length can exhibit very strong non-linearity. Alternatively, at constant length and fluid velocity, permeability can vary with column diameter.

Finally, a word on the not-very-well-known viscous fingering effect, which represents a source (the only one?) of instability and non-reproducibility in production-scale chromatography. In short, one must be very careful when a fluid of low viscosity pushes a fluid of high viscosity, as this can totally ruin column efficiency.

6.7 Solutions of the exercises

Exercise 6.1

The Re number is given by $Re = \frac{1000\times10^{-3}\times10\times10^{-6}}{2\times10^{-3}} = 0.005$. This value is well below 1, so the Kozeny–Kármán equation (6.1) can be used. One obtains:

$$\frac{\Delta P}{L} = h_k \frac{\mu}{d_p^2} \frac{(1-\varepsilon_e)^2}{\varepsilon_e^3} u$$
$$= 150 \frac{0.002}{\left(10 \times 10^{-6}\right)^2} \times \frac{(1-0.4)^2}{0.4^3} \times 0.001 = 16875000 \text{ Pa/m}$$

The pressure drop is thus 169 bar/m, so finally 169 × 0.4 = 67 bar on the 0.4 m-long column.

If the extragranular porosity is 0.35 instead of 0.4, the calculation leads to a pressure drop of 295 bar/m, representing a 75% increase.

Exercise 6.2

The required number of plates ($N = 500$) is given by:

$$N = \frac{L_{col}}{HETP} = \frac{L_{col}}{d_p} \frac{1}{\tilde{h}}$$

Consequently, one has:

$$N = \frac{L_{col}}{d_p} \frac{1}{\frac{\alpha}{\tilde{u}} + \beta + \gamma \tilde{u}}$$

Using the numerical values:

$$L_{col} = N d_p \left(\frac{\alpha}{\tilde{u}} + \beta + \gamma \tilde{u} \right) = 0.01 \times \left(\frac{2}{\tilde{u}} + 3 + 2\tilde{u} \right)$$

The pressure drop is given by:

$$\frac{\Delta P}{L_{col}} = h_k \frac{\mu}{d_p^2} \frac{(1-\varepsilon_e)^2}{\varepsilon_e^3} u = h_k \frac{\mu}{d_p^2} \frac{(1-\varepsilon_e)^2}{\varepsilon_e^3} \frac{D_m}{d_p} \tilde{u}$$

so with a 20 bar pressure-drop target one has:

$$\frac{20 \times 10^5}{L_{col}} = 150 \frac{10^{-3}}{\left(20 \times 10^{-6}\right)^2} \frac{(1-0.35)^2}{0.35^3} \frac{10^{-9}}{20 \times 10^{-6}} \tilde{u} = 1.85 \times 10^5 \tilde{u}$$

and with a 50 bar pressure-drop target one has:

$$\frac{50 \times 10^5}{L_{col}} = 150 \frac{10^{-3}}{\left(20 \times 10^{-6}\right)^2} \frac{(1-0.35)^2}{0.35^3} \frac{10^{-9}}{20 \times 10^{-6}} \tilde{u} = 1.85 \times 10^5 \tilde{u}$$

The number of plates and pressure-drop constraints give two equations that can be solved for L_{col} and $\tilde{u}$. For the 20 bar pressure-drop target one obtains $L_{col} = 0.8$ m and $\tilde{u} = 13.4$, so $u = 6.7 \times 10^{-4}$ m/s. For the 50 bar pressure, drop target, $L_{col} = 1.27$ m and $\tilde{u} = 21.2$, so $u = 1.06 \times 10^{-3}$ m/s.

Working at 50 bar instead of 20 bar allows an increased fluid velocity (thus a reduced column diameter).

Exercise 6.3

The simplified method is used. A first computation at 125 bar and 325 K gives a density of 535 kg/m^3 (Peng–Robinson equation of state) and a viscosity of 4.14×10^{-5} Pa s (equation of Thodos for non-polar compounds). This leads to a fluid velocity of $u^{out} = \frac{50/3600}{\frac{3.14}{4} \times 0.05^2} \frac{1}{535} = 0.0132$ m/s, which, using the Kozeny–Kármán Eq. (6.1), leads to a pressure drop of:

$$\Delta P = L_{col} h_k \frac{\mu}{d_p^2} \frac{(1-\varepsilon_e)^2}{\varepsilon_e^3} u$$
$$= 0.45 \times 150 \frac{4.14 \times 10^{-5}}{\left(10^{-5}\right)^2} \times \frac{(1-0.4)^2}{0.4^3} 0.0132 = 2074916 \text{ Pa}$$

thus about 21 bar.

A second computation carried out at 135 bar (mean pressure in the column assuming pressure drop of 21 bar) and 325 K gives a density of 583 kg/m^3 and a viscosity of 4.56×10^{-5} Pa s. This leads to a fluid velocity of $u^{avg} = \frac{50/3600}{\frac{3.14}{4} \times 0.05^2} \frac{1}{583} = 0.012$ m/s. Using the Kozeny–Kármán equation, Eq. (6.1), leads to a pressure drop of:

$$\Delta P = L_{col} h_k \frac{\mu}{d_p^2} \frac{(1-\varepsilon_e)^2}{\varepsilon_e^3} u$$
$$= 0.45 \times 150 \frac{4.56\ 10^{-5}}{\left(10^{-5}\right)^2} \times \frac{(1-0.4)^2}{0.4^3} 012 = 2285415 \text{ Pa}$$

thus about 23 bar.

The approximate mean retention time is $t_0^{avg} = \frac{0.4 \times 0.45}{0.012} = 15$ s.

Exercise 6.4

Let us assume that the extragranular porosity ε_{e1} equals 0.35 when the bed length equals 25 cm. From Eq. (6.26), the extragranular porosity associated with a bed length of 24 cm is given by $(1-\varepsilon_{e2}) = (1-0.35)\frac{25}{24}$, so $\varepsilon_{e2} = 0.323$. From the Kozeny–Kármán equation, we know that everything but porosity being constant, pressure drop varies as $\frac{(1-\varepsilon_e)^2}{\varepsilon_e^3}$.

Consequently, the ratio of the pressure drop in condition (1) to that in condition (2) is given by:

$$\frac{\Delta P(2)}{\Delta P(1)} = \frac{\frac{(1-\varepsilon_{e2})^2}{\varepsilon_{e2}^3}}{\frac{(1-\varepsilon_{e1})^2}{\varepsilon_{e1}^3}} \frac{24}{25} = \frac{13.6}{9.85} \frac{24}{25} = 1.32$$

thus a 32% pressure-drop increase.

Notice that, if one had neglected the influence of the column length ratio $\frac{24}{25}$, one would have obtained a 38% pressure-drop increase. This justifies the approximation used in Eq. (6.46).

Exercise 6.5

The first step consists in determining the permeability under no stress, κ_0, the "Young modulus E_{fo}" and the friction factor by minimizing differences between the experimental and calculated (Eq. (6.56)) permeability values.

Because the permeability is reported to be independent of fluid velocity in the investigated range, Eq. (6.56) is simplified to:

$$\frac{\kappa_0}{\kappa} = 1 + \frac{P_{mech}}{E_{fo}\left(2+\breve{\beta}\right)} = 1 + \frac{P_{mech}}{E_{fo}\left(2+\beta\frac{L_{col}}{D_{col}}\right)}$$

A decent agreement with experimental data is obtained with $\kappa_0 = 1.185 \times 10^{-13}$, $E_{fo} =$ 75 bar and $\beta = 0.8$.

The value of "Young modulus E_{fo}" is compatible with $E = 575$ bar if one assumes an external porosity of 0.35. This value is in reasonable agreement with experimental results reported in Section 6.2.3. It would certainly be good practice to confirm this value with experimental values of bed compressibility, but these data are not reported.

The pressure drop in a given column is calculated from Eq. (6.55):

$$\Delta P = \Delta P_0 \left[1 + \frac{\Delta P_0 + P_{mech}}{E_{fo}\left(2+\beta\frac{L_{col}}{D_{col}} - \frac{\Delta P_0}{E_{fo}}\right)}\right]$$

where the pressure drop under no stress is calculated from Eq. (6.44), $\Delta P_0 = \dfrac{1}{\kappa_0}\mu u L_{col0}$. Using the numerical values of permeability κ_0 and viscosity μ one obtains, in MKSA units:

$$\Delta P_0 = 8.44 \times 10^9\, uL_{col0}$$

After substitution in Eq. (6.55) and using the numerical values of P_{mech}, E_{fo} and β, one obtains:

$$\Delta P = 8.44 \times 10^9\, uL_{col0}\left[1 + \frac{8.44 \times 10^9 uL_{col0} + 50 \times 10^5}{75 \times 10^5\left(2 + 0.7\dfrac{L_{col}}{D_{col}} - \dfrac{8.44 \times 10^9\, uL_{col0}}{75 \times 10^5}\right)}\right]$$

in MKSA units.

Numerical application shows that the pressure-drop curves are substantially linear versus flow rate, but at the typical 10^{-3} m/s velocity the pressure drop in a 100 cm diameter column is expected to be about 15% higher than its value in a 10 cm diameter pilot-scale column.

Exercise 6.6

The results allow a determination of critical velocity u^{crit} and critical compression factor λ_{bed}^{crit} for the different initial bed lengths L_{col0}:

L_{col0} (cm)	10	25	50
L_{col} (cm) at critical velocity	8.5	21.5	42
λ_{bed}^{crit}	0.15	0.14	0.16
u^{crit} (m/s)	2 10^{-3}	1.3 10^{-3}	9 10^{4}

The evolution of critical fluid velocity u^{crit} versus column length L_{col0} can be represented with Eq. (6.61). The experimental results are well represented with $p = 3\ 10^{-5}$ m^2/s and $q = 1.5\ 10^{-4}$ m^2/s.

One notices that the critical bed compression factor is about 0.15, independently of column length.

With knowledge of the p and q parameters, one can calculate the critical velocity u^{crit} associated with the preparative column:

$$u^{crit} = \frac{3 \times 10^{-5}\frac{0.5}{0.4} + 1.5 \times 10^{-4}}{0.4} = 4.7 \times 10^{-4} \text{ m/s}$$

The critical compression factor taken to be 0.15, one obtains from Eq. (6.60): $\lambda_{comp} = \frac{0.15}{4.7\times 10^{-4}}u = 320u$ with u in m/s.

By injecting the expression for λ_{comp} in Eq. (6.59), one obtains the influence of fluid velocity on porosity: $\varepsilon_e = \dfrac{0.45 - 320u}{1 - 320u}$ (u in m/s), this being only valid for $u < u^{crit}$.

One can then use the expression for porosity in the Kozeny–Kármán equation for estimating the pressure drop: $\dfrac{\Delta P}{0.5} = 150\dfrac{10^{-3}}{\left(100 \times 10^{-6}\right)^2}\dfrac{(1 - \varepsilon_e)^2}{\varepsilon_e^3}u.$

The previous expression shows that the bed can shrink even at very low velocities and pressure drop (Figure 6.21). It is certainly wise to operate at fluid velocities lower than about 4×10^{-4} m/s.

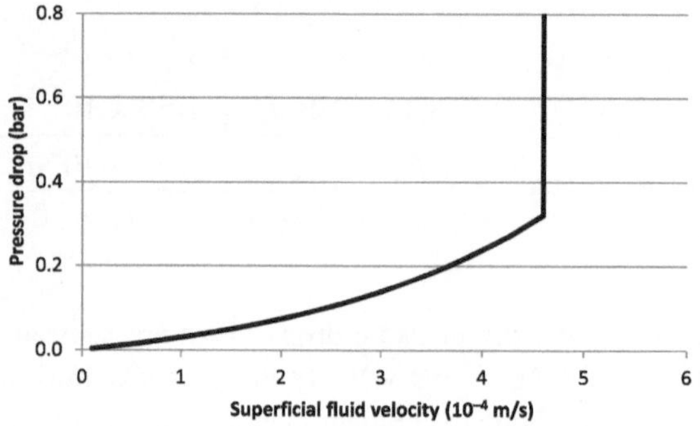

Figure 6.21 For solution of Exercise 6.6.

Exercise 6.7

The first step consists in estimating the pressure drop in the column. The viscosity at 295 K is estimated to be 2.1 cP. We assume an external porosity of 0.4. With the Kozeny–Kármán equation one obtains:

$$\frac{\Delta P}{L} = 150 \frac{2.1 \times 10^{-3}}{\left(10 \times 10^{-6}\right)^2} \frac{(0.6)^2}{0.4^3} 0.001 = 17\,718\,750 \text{ Pa/m} = 177 \text{ bar/m}$$

The second step consists in estimating the $\breve{L}$ and $\Delta \breve{T}$ parameters:

$$\breve{L} = \frac{\rho C_p u}{\lambda_{bed}} R_{col}^2 = \frac{900 \times 3000 \times 0.001}{1} 0.3^2 = 243 \text{ m}$$

$$\Delta \breve{T} = u \frac{\Delta P}{L_{col}} \frac{R_{col}^2}{\lambda_{bed}} = 0.001 \times 177 \times 10^5 \times \frac{0.3^2}{1} = 1593\ K$$

As $T^{inl} = T^{ext}$, Eq. (6.71) becomes:

$$\langle T(r) \rangle = T^{ext} + \frac{\Delta \breve{T}}{4} \left\{ 1 - (\xi_c)^2 + 8 \sum_{i=1}^{\infty} \left(-\frac{1}{\alpha_i^3} \right) \frac{J_0(\alpha_i \xi_c)}{J_1(\alpha_i)} \frac{1 - \exp\left(-\alpha_i^2 \left(\frac{L_{col}}{\breve{L}}\right)\right)}{\alpha_i^2 \left(\frac{L_{col}}{\breve{L}}\right)} \right\}$$

so the axially averaged radial temperature profile is known.

By adding the first 200 terms, one obtains the temperature profile. This profile is uniform and equals 296.3 K from the center of the column to a normalized radius of about 0.9. It then decreases sharply to 295 K at the column wall.

We know that the viscosity is 2.1 cP at 295 K and about 2.14 cP at 296.3 K. Consequently, the fluid velocity at the wall is "guesstimated" to be about 2.1/2.14 = 98% of its value at the center of the column. We will see in Section 6.3.2 whether this difference is likely to impact column performance.

Exercise 6.8

Call 1 the non-plugged zone and 2 the plugged zone. We have $\omega_1 = 0.9$ and $\omega_2 = 0.1$. Knowing that the fluid velocity in the plugged part is 10% lower than the fluid velocity in the non-plugged part, one can write, in arbitrary units, $u_1 = 1, u_2 = 0.9$.

The average velocity u is given by $u = u_1\omega_1 + u_2\omega_2 = 1 \times 0.9 + 0.9 \times 0.1 = 0.99$.

The fractions of the flow rate associated with areas 1 and 2 are given by:

$$\alpha_1 = \frac{u_1\omega_1}{u} = \frac{0.9}{0.99} = 0.909\,090\,91 \quad \alpha_2 = \frac{u_2\omega_2}{u} = \frac{0.9 \times 0.1}{0.99} = 0.090\,909\,091$$

Finally, from Eq. (6.85) one obtains:

$$h_2^{corr} = \sum_{i=1}^{n} \frac{\omega_i^2}{\alpha_i} = \frac{0.9^2}{0.909\,090\,91} + \frac{0.1^2}{0.090\,909\,091} = 1.001$$

The apparent number of plates N_{app} resulting from the heterogeneity is related to the number of plates N_{unif} that prevails in the absence of heterogeneity by Eq. (6.88):

$$\frac{1}{N_{app}} = \frac{1}{N_{unif}} + \left(1 + \frac{1}{N_{unif}}\right)\left(h_2^{corr} - 1\right)$$

so:

$$\frac{1}{N_{app}} = \frac{1}{N_{unif}} + \left(1 + \frac{1}{N_{unif}}\right) \times 0.001$$

If $N_{unif} = 10$, one obtains $N_{app} = 9.9$, which is very close. If $N_{unif} = 1000$, one obtains $N_{app} = 500$, which is pretty different.

Exercise 6.9

We consider a pipe of length L_{col} and radius R_{col} in which a fluid flows with a radial velocity profile $u_r(r)$ (u_r for short). For simplicity, and without loss of generality, we assume that u_r is maximum at the center of the column.

Neglecting axial dispersion, the molecules that will exit the column at a time between t and $t + dt$ are located in a ring associated with velocity u_r such that:

$$t(r) = L_{col}/u_r$$

From the properties of the RTD (Chapter 2), $E(t)dt$ gives the fraction of the outlet stream which has been in the column for a time between t and $t + dt$. The fraction of the stream passing between r and $r + dr$ is given by $\frac{dQ}{Q} = \frac{2\pi r u_r dr}{\pi R_{col}^2 u}$, where u is the average fluid velocity.

Consequently, one has:

$$E(t)dt = \frac{2ru_r dr}{R_{col}^2 u} \tag{6.98}$$

After differentiation of the relation $t(r) = L_{col}/u_r$ one obtains $dt = -L_{col}\frac{-2}{u_r^2}\frac{du_r}{dr}dr$. After injection of dt in the expression for the RTD one obtains:

$$\tau E(t) = -2\xi_c \left(u_r^{norm}\right)^3 \frac{d\xi_c}{du_r^{norm}} \tag{6.99}$$

where τ is the mean residence time L_{col}/u, $\xi_c = r/R_{col}$ and u_r^{norm} is the normalized velocity profile u_r/u.

Noting that $t = \tau/u_r^{norm}$, Eq. (6.99) becomes an ODE in u_r^{norm}. Its resolution (with proper integration constant) answers the question: starting from any experimental RTD, one can thus determine a velocity profile perfectly representing this RTD.

Exercise 6.10

The methodology is the same as for Exercise 6.9. Laminar flow in a pipe is associated with a parabolic velocity profile $u_r(\xi_c) = u_{max}(1 - \xi_c^2)$, with $\xi_c = r/R_{col}$. The velocity u_{max} is the velocity at the center of the pipe when the mean velocity is $u = u_{max}/2$.

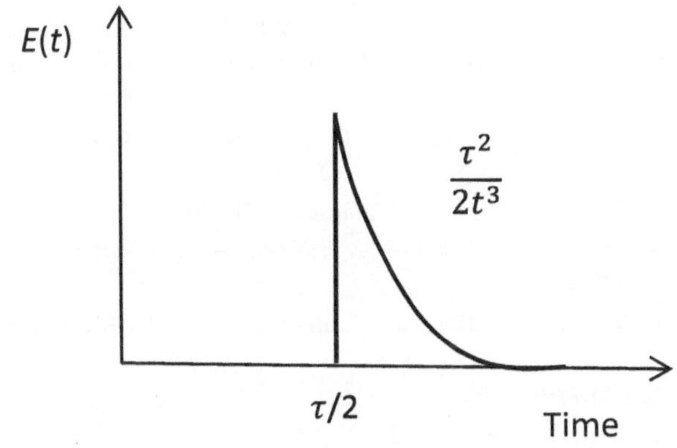

Figure 6.22 For solution of Exercise 6.10.

Inserting $t = \tau/u_r^{norm}$ and differentiating the expression for the fluid velocity in Eq. (6.99), one obtains:

$$E(t) = -2\xi_c \frac{\tau^2}{t^3} \frac{d\xi_c}{du_r^{norm}} = -2\xi_c \frac{\tau^2}{t^3} \frac{1}{-4\xi_c} = \frac{1}{2} \frac{\tau^2}{t^3} \tag{6.100}$$

Noting that the smallest possible retention time is $\tau/2$ (central stream), one obtains:

$$E(t) = \frac{\tau^2}{2t^3} H(t - \tau/2)$$

Consequently, the RTD has the shape given in Figure 6.22.

The moments of the distribution are then calculated:

$$\mu_1 = \int_0^\infty tE(t)dt = \frac{\tau^2}{2} \int_{\tau/2}^\infty \frac{1}{t^2} dt = \frac{\tau^2}{2} \left[\frac{-1}{t} \right]_{\tau/2}^\infty = \tau$$

$$\mu_2 = \int_0^\infty t^2 E(t)dt = \frac{\tau^2}{2} \int_{\tau/2}^\infty \frac{1}{t} dt = \frac{\tau^2}{2} [\ln(t)]_{\tau/2}^\infty = \infty$$

The second moment and thus the variance associated with this simple model are thus infinite. In order to get a finite variance, molecular diffusion must be taken into account; see the famous work by (Taylor, 1953) and (Aris, 1956).

References

Aris, R. (1956), *Proc. Roy. Soc. Lond. A* **235**, 67–77.

Astrath, D.-U., Lottes, F., Vu, D. T., Arlt, W. and Stenby, E. H. (2007), *Adsorption* **13**(1), 9–19.

Augier, F., Laroche, C. and Brehon, E. (2008), *Sep. Pur. Technol.* **63**, 466–474.

Castells, C. and Castells, R. (1998), *J. Chromatogr. A* **805**, 55.

Catchpoole, H., Shalliker, R., Dennis, G. R. and Guiochon, G. (2006), *J. Chromatogr. A* **1117**, 137.

Cherrak, D. E., Al-Bokari, M., Drumm, E. C. and Guiochon, G. (2002), *J. Chromatogr. A* **943**(1), 15–31.

Cherrak, D. E. and Guiochon, G. (2001), *J. Chromatogr. A* **911**(2), 147–166.

Cherrak, D., Guernet, E., Cardot, P., Herrenknecht, C. and Czok, M. (1996), *Chromatographia* **46**, 647.

Ching, C., Wu, Y., Liusso, M., Wozny, G., Laiblin, T. and Arlt, W. (2002), *J. Chromatogr. A* **945**, 117–131.

Czok, M., Katti, A. and Guiochon, G. (1991), *J. Chromatogr.* **550**, 705.

Daneyko, A., Khirevich, S., Holtzel, A., Seidel-Morgenstern, A. and Tallarek, U. (2011), *J. Chromatogr. A* **1218**, 8231–8248.

Dapremont, O., Cox, G. B., Martin, M., Hilaireau, P. and Colin, H. (1998), *J. Chromatogr. A* **796**(1), 81–99.

Delgado, J. (2006), *Heat Mass Transfer* **42**(4), 279–310.

Desmet, G. (2013), *J. Chromatogr. A* **1314**, 124–137.

Dewaele, C. and Verzele, M. (1983), *J. Chromatogr.* **260**, 13–21.

Dixon, A. G. and Ma, Y. H. (1988), *Chem. Eng. Sci.* **43**(6), 1297–1302.

Epstein, N. (1989), *Chem. Eng. Sci.* **44**, 777–779.

Etienne, C., Nicoud, R. M. and Sardin, M. (1989). In *2ème Congrès Français de Génie des Procédés*, vol. 9 of *Recents Progrès en Génie des Procédés*, Lavoisier Technique et Documentation, p. 417.

Fernandez, E., Norton, T., Jung, W. and Tsavalas, J. (1996), *Biotechnol. Prog.* **12**(4), 480–487.

Flodin, P. (1961), *J. Chromatogr.* **5**, 103.

Hill, S. (1952), *Chem. Eng. Sci.* **1**, 247.

Kaczmarski, K., Gritti, F. and Guiochon, G. (2008), *J. Chromatogr. A* **1177**, 92–104.

Keener, R., Fernandez, E., Maneval, J. and Hart, R. (2008), *J. Chromatogr. A* **1190**, 127–140.

Koch, D. and Brady, J. (1985), *J. Fluid. Mech.* **154**, 399–427.

Kováčik, J. (1999), *J. Mater. Sci. Lett.* **18**(13), 1007–1010.

Mishra, M., Martin, M. and De Wit, A. (2007), *Phys. Fluids* **19**, 073101.

Mosciarello, J., Purdon, G., Coffman, I., Root, T. and Lightfoot, E. (2001), *J. Chromatogr. A* **908**, 131–141.

Nicoud, R. M. and Perrut, M. (1993). In G. Ganetsos and P. E. Barker, eds., *Preparative and Production Scale Chromatography*, vol. 61 of *Chromatographic Science Series*, Marcel Dekker, pp. 47–77.

Östergren, K. and Trägardh, C. (1999), *Chem. Eng. J.* **72**, 153–161.

Östergren, K., Trägardh, C., Enstad, G. and Mosby, J. (1998), *AIChE J.* **44**(1), 2–12.

Pathak, N., Norman, C., Kundu, S., Nulu, S., Durst, M. and Fang, Z. (2008), *BioProcess Int* **October**, 72–81.

Pazdernik, O. and Schneider, P. (1981), *J. Chromatogr. A* **207**(2), 181–191.

Perrut, M. (1987), *J. Chromatogr. A* **396**, 1–17.

Poling, B., Prausnitz, J. and O'Connell, J. (2001), *The Properties of Gases and Liquids*, 5th edn., McGraw-Hill.

Poppe, H. and Kraak, H. (1983), *J. Chromatogr.* **282**, 399–412.

Rajendran, A. (2012), *J. Chromatogr. A* **1250**, 227–249.

Rajendran, A., Gilkison, T. and Mazzotti, M. (2008), *J. Sep. Sci.* **31**(8), 1279–1289.

Rajendran, A., Krauchi, O., Mazzotti, M. and Morbidelli, M. (2005), *J. Chromatogr. A* **1092**, 149–160.

Rousseaux, G., De Wit, G. and Martin, M. (2007), *J. Chromatogr. A* **1149**, 254–273.

Rousseaux, G., Martin, M. and De Wit, A. (2011), *J. Chromatogr. A* **1218**, 8353–8361.

Shadday, M. A. (2006), *Chem. Eng. Sci.* **61**(8), 2688–2700.

Shalliker, R., Broyles, B. and Guiochon, G. (1999), *J. Chromatogr. A* **865**, 73.

Shalliker, R., Catchpoole, H., Dennis, G. and Guiochon, G. (2007), *BioProcess Int.* **January**, 32–37.

Stickel, J. and Fotopoulos, A. (2001), *Biotechnol. Prog.* **17**(4), 744–751.

Tan, C. and Homsy, G. (1986), *Phys. Fluids* **29**, 3549.

Taylor, G. I. (1953), *Proc. Roy. Soc. Lond. A* **219**, 186–203.

Toumi, A., Dingenen, J., Genolet, J., *et al.* (2012). In H. Schmidt-Traub, M. Schulte and A. Seidel-Morgenstern, eds., *Preparative Chromatography*, Wiley-VCH.

Verhoff, F. and Furjanic, J. J. (1983), *Ind. Eng. Chem. Process Des. Dev.* **22**(2), 192–198.

Wang, J. C. (1984), *J. Mater. Sci.* **19**(3), 801–808.

Wang, X.-S., Jiang, X.-W., Wan, L., Song, G. and Xi, Q. (2009), *Int. J. Rock Mech. Min. Sci.* **46**, 1175–1181.

Welsch, T., Schmid, M., Kutter, J. and Kalman, A. (1996), *J. Chromatogr. A* **728**, 299–306.

Yuan, Q., Rosenfeld, A., Root, T., Klingenberg, D. and Lightfoot, E. (1999), *J. Chromatogr. A* **831**, 149–165.

7 Simulating chromatographic columns

The art of adapting complexity to needs, or the art of modeling the elephant.

7.1 Introduction

As stated in previous chapters, the behavior of chromatographic columns is governed by:

- thermodynamic factors, describing the interactions of the solutes and carrier fluid components with the chromatographic medium at equilibrium, as well as possible equilibria in the fluid phase or solid phase
- kinetic factors, describing how fast solutes move in the fluid and solid phases to reach an equilibrium state from a non-equilibrium state
- hydrodynamic factors, describing how the fluid flows into the column.

We generally assume that these factors can be modeled independently of each other, with coupling occurring only at interfaces. This is a very useful postulate, which is the basis of classical chromatography modeling. From a theoretical standpoint, however, it might be subject to discussion: for highly porous particles, the distinction between hydrodynamics and internal transport occurring partially through convection is somewhat vague. In the case of gas adsorption, the adsorption of solutes on the solid may lead to hydrodynamic perturbations. This effect, called the sorption effect, has been described by (Valentin, 1981) and (Le Van *et al.*, 1988). We will consider that these refinements are not required for our purposes, and, except as specified in Section 7.8, we will assume that the *decoupling postulate* applies between the above factors.[1]

In the past decades, a lot of different more or less refined models have been proposed, and it is out of the scope of this book to review them all. Let us express that the different possible models differ by their representation of hydrodynamic, kinetic or thermodynamic contributions.

The simplest model[2] is the equilibrium model (PF–$\overline{\text{Equil}}$ in our terminology), which assumes plug flow without hydrodynamic dispersion and infinitely fast mass transfer,

[1] Let us further state that thermodynamic factors are never influenced by kinetics or hydrodynamics.

[2] "Simple" refers to the system description; the simplicity of the associated mathematics is another story.

so that the moving fluid and lumped solid concentrations are in equilibrium. Presented in Chapter 3, this model can be used as a reference as it gives the average position of the fronts, which is independent of hydrodynamic dispersion and mass transfer limitations. All other models should asymptotically approach the equilibrium model as the influences of hydrodynamic dispersion and mass transfer limitations vanish. As a side comment, one could have a model assuming equilibrium between the *fluid* (moving or not) concentrations and the *solid skeleton* concentrations on the one hand, and plug flow on the other hand. This model, called PF–$\overline{\overline{\text{Equil}}}$ in our terminology, is strictly equivalent to PF–$\overline{\text{Equil}}$.

By proposing more sophisticated models, we will illustrate the Lego or Russian-doll approach already mentioned. We will see that different bricks can be added, subdivided or changed for modeling purposes. Starting from a given representation, when further details or precision is required, a given model (brick) will typically be subdivided into sub-models (smaller bricks) offering a more detailed description of the system. If a house represents a model, then sub-models will represent rooms, roofs or walls. Sub-sub-models represent cables, cement, plaster. Sub-sub-sub-models represent water, sand, clay. If you want to conceive a town, thinking at the house level is sufficient; if you are an architect, you need to go to the room level; if you are a construction engineer you need to consider cement; and so on. For chromatography, averaging concentrations associated with the sub-models then allows a connection between more and less detailed models: modeling is also the art of averaging.

Let us illustrate the concept with the elementary models (bricks) that have been presented in previous chapters.

When the equilibrium model is used, plug flow (elementary brick) is assumed in the column. The model can be refined by replacing this brick by several equilibrium plates (sub-bricks) and thus by considering the MC–$\overline{\text{Equil}}$ model. In this case, one requires an infinity of elementary bricks for the two models to be identical.

The MC–$\overline{\text{Equil}}$ model assumes equilibrium to have been reached between the lumped solid phase and the fluid mobile phase in each plate. When moving from the $\overline{\text{Equil}}$ model to the more detailed $\overline{\text{LDF}}$ models (see Chapter 5), one allows the fluid and lumped solid phases to explore non-equilibrium regions, and to evolve towards equilibrium with a certain dynamics. The equilibrium brick is replaced by first-order dynamic systems.

One can further refine the dynamic description by recognizing the existence of radial concentration profiles inside the particles, and by solving Fick's law in spherical geometry. The $\overline{\text{LDF}}$ model is based on the averaging of the concentration profiles as predicted from the $\overline{\text{Fick}}$ model, which consequently appears as sub-model of the $\overline{\text{LDF}}$ approximation.

We can go a step further by recognizing the porous structure of the particles (see Chapters 1 and 5). The lumped concentrations used in the lumped models are the result of an averaging of the intragranular and skeleton concentrations of the porous models. The lumped $\overline{\text{Fick}}$ model is thus the result of an averaging of the porous $\overline{\overline{\text{Fick}}}$ sub-model.

One can go another step further and define a sub-sub-model by acknowledging a difference between micro- and macroporosity. The adsorbed-phase composition becomes an average between solutes contained in the micropores and adsorbed on the skeleton (Section 5.3.6).

Building models of different complexity by introducing sub-models is not limited to mass transfer modeling. Recall the modeling of column hydrodynamics in the case of a non-uniform radial velocity profile (Section 6.3.2). Modeling the column with a plate model is no more than averaging the radial concentration profile obtained from more detailed sub-models. The same can also be done with axial temperature or pressure profiles, as shown by the James–Martin relation for gases or the more complex supercritical fluid chromatography (SFC) modeling, when fluid compressibility must be taken into account (Section 6.1.3).

Even thermodynamics is not kept aside. Recall the size-exclusion modeling with the biporous model (Section 4.6 and Section 5.3.9) and the electrical double layer sub-model for describing ion exchange (Section 4.5.4). In the latter example, some fraction of the intragranular fluid can not even satisfy local electroneutrality, while lumped concentrations certainly do after proper averaging.

Thanks to Theorem 2.61, the theory of linear chromatography gives an extremely powerful tool for comparing this cascade of more or less refined models. Basically, the mean retention time of peaks or fronts is independent of the mass transfer sub-models. The contribution of mass transfer to the variance of peaks or fronts depends only on the first-order moment of the transfer function describing the mass transfer sub-models (Eq. (2.67)). This allows us to show that many *prima facie* different sub-models will eventually lead to the same expression for the variance of chromatographic peaks, and thus to similar Van Deemter curves.

Unfortunately, no equivalent theorem exists for non-linear systems, and numerical tools must be used for comparing different models. The need for numerical simulation should not, however, preclude the use of experience and common sense.

Comparing in detail all the models that can be built with the bricks proposed in the previous chapters would probably require writing a useless encyclopedia. We will simply focus on the main categories of models, and let readers adapt the methodology and models to their own problems.

Table 7.1 proposes a simple classification of the different models according to the terminology in Chapter 1 and according to their complexity. They will be detailed further in the following sections.

In order to compare models, we need to quantify the differences between simulated chromatograms, and consequently to define a simple *dissimilarity criterion*.[3]

In addition to visual inspection and thus qualitative comparison between peaks, one can compare variances, skewness or any other geometrical parameter that may appear relevant in a given situation. Proposing a simple dissimilarity criterion that could be pertinent in all situations is not an easy task (if it is possible at all). Following is the best compromise between simplicity and generality that I can suggest.

Consider two chromatographic peaks $C^1(t)$ and $C^2(t)$. These can be associated with one species and two different models, with two different species, or with two different

[3] Breakthrough curves being normally of less interest for preparative chromatography purposes.

Table 7.1 Possible chromatographic column models, first qualitative comments. See detailed presentation in the text.

Model	Description	Comments	Where in this book and elsewhere?
PF–$\overline{\text{Equil}}$ (equilibrium model, sometimes called ideal model)	Hydrodynamic dispersion and mass transfer limitations are neglected; only the influence of the adsorption isotherms is taken into account.	This model must be considered as a reference: all the other models must lead to identical front positioning.	Presented in detail in Chapter 3.
MC–$\overline{\text{Equil}}$ (sometimes called equilibrium stage model)	The column is approximated by a series of equilibrium plates. Hydrodynamic dispersion and mass transfer limitations are modeled with a single parameter: the number of plates N. Lumped solid concentrations are used.	This is a very simple and powerful extension of the PF–$\overline{\text{Equil}}$ model. In most situations, it is sufficient for process development needs.	Chapters 2, 5, 6 and 7; used in thousands of publications since the pioneering work of (Martin and Synge, 1941).
PD–$\overline{\text{Equil}}$ (sometimes called equilibrium dispersive model)	The column is approximated by a continuous medium in which plug flow and axial dispersion occur. Hydrodynamic dispersion and mass transfer limitations are modeled with a single parameter: the apparent Péclet number Pe^{app}. Lumped solid concentrations are used.	This is in general very similar to the MC model except under strongly non-linear conditions and low Pe^{app} values.	Chapters 2, 5, 6 and 7; has for instance been used by (Guiochon *et al.*, 1988), (Golshan-Shirazi and Guiochon, 1992b).
MC–$\overline{\text{LDF}}$	The effect of hydrodynamic dispersion is modeled by approximating the column as J mixing cells connected in series, and mass transfer limitations (external or internal) are modeled using constant characteristic times for mass transfer.	In "classical" situations (same mass transfer parameter(s) irrespective of the solute) this leads to chromatograms similar to those given by the MC–$\overline{\text{Equil}}$ model. Advantageous over the $\overline{\text{Equil}}$ models when strong kinetic limitations occur in a non-linear environment, or when different species are associated with significantly different mass transfer characteristics.	Chapters 2, 5 and 7; has for instance been used by (Nicoud and Schweich, 1989).

PD–$\overline{\text{LDF}}$ (sometimes called lumped kinetic model)	The column is approximated by a continuous medium in which plug flow and axial dispersion occur and are characterized by a Péclet number. Mass transfer limitations (external or internal) are modeled using constant characteristic times for mass transfer.	Very similar to MC–$\overline{\text{LDF}}$ except at very low efficiencies under strongly non-linear conditions.	Chapters 2, 5 and 7; has for instance been used by (Hiester *et al.*, 1956) and (Golshan-Shirazi and Guiochon, 1992b).
MC–$\overline{\text{Fick}}$ PD–$\overline{\text{Fick}}$	Hydrodynamic dispersion is modeled with a number of mixing cells or with a Péclet number. Mass transfer limitations are modeled by solving Fick's law in spherical coordinates in the lumped solid.	Usually considered more "rigorous" than the LDF models. However, the gain in precision compared to the LDF approach is questionable, so this model is generally of limited value in practical chromatographic processes design.	Chapters 2, 5 and 7; has for instance been used by (Weber and Charkravorti, 1974) and (Radeke *et al.*, 1981).
MC–$\overline{\overline{\text{Fick}}}$ PD–$\overline{\overline{\text{Fick}}}$ (sometimes called general rate model, GRM)	A more or less complex description of the particles is used, approximated by a porous medium. The diffusion inside the particle is typically modeled with Fick's law and equilibrium between solutes contained in the intragranular fluid and solutes adsorbed on the skeleton is normally assumed. Adsorbed diffusivity may be taken into account or not.	This model is not as general as the frequently used GRM terminology may suggest. It adds a significant degree of complexity and computational effort compared to the LDF models. It can only be used when the importance of the coupling between adsorption isotherms' non-linearity and intraparticular diffusion is such that the simpler models are incompatible with the required precision.	Chapters 5 and 7; has for instance been used by (Ma *et al.*, 1996) and (Hasnat and Jvekar, 1996).
MC–$\overline{\overline{\text{LDF}}}$ PD–$\overline{\overline{\text{LDF}}}$ (sometimes called pore model)	A simplification of the MC–$\overline{\overline{\text{Fick}}}$ and PD–$\overline{\overline{\text{Fick}}}$ models using differences instead of differentials to represent internal diffusion. Adsorbed diffusivity may be taken into account or not.	Mathematical expressions associated with this model can take different forms, as presented in Section 5.4.4.	Chapters 5 and 7; has for instance been used by (Morbidelli *et al.*, 1982), (Carta, 1998), (Carta and Cincotti, 1998).
More sophisticated	Influence of heat transfer, radial profiles, chemical equilibria can be taken into account, replacing Fick's law by the Maxwell–Stefan approach.	No limit!	Examples given in Section 3.7), in Chapter 6 and in following sections of Chapter 7.

operating conditions. By determining the surface under each chromatogram, one can define normalized curves $Y^1(t)$ and $Y^2(t)$, similarly to the approach in Section 2.1. By definition, the areas under $Y^1(t)$ and $Y^2(t)$ are each unity.

We define the *surface resolution* by:

$$R^{sf}(Y^1,Y^2) = \frac{\int_0^\infty \left|Y^1 - Y^2\right| dt}{\frac{1}{2}\int_0^\infty (Y^1 + Y^2) dt} = \int_0^\infty \left|Y^1 - Y^2\right| dt \tag{7.1}$$

A simple drawing should convince the reader that this is 0 if the two peaks are identical and 2 if they are totally disjoint (we have two separate peaks).

For illustration purposes, surface resolutions calculated for different peaks are given in Figure 7.1. If two curves $Y^1(t)$ and $Y^2(t)$ lead to a surface resolution lower than 0.05, these curves are almost indistinguishable.

We can now use surface resolution to compare the chromatograms obtained for a given solute A using two different models, M1 and M2. To do so, we define a dissimilarity criterion:

$$\Xi_{M1}^{M2} = R^{sf}\left(Y_A^{M1}, Y_A^{M2}\right) \tag{7.2}$$

where Y_A^{M1} and Y_A^{M2} represent the column outlet concentrations $C_A^{out}(\mathrm{M1})$ and $C_A^{out}(\mathrm{M2})$ as calculated by the two models, normalized to $S_A(\mathrm{M1})$ and $S_A(\mathrm{M2})$ the areas under the peaks.

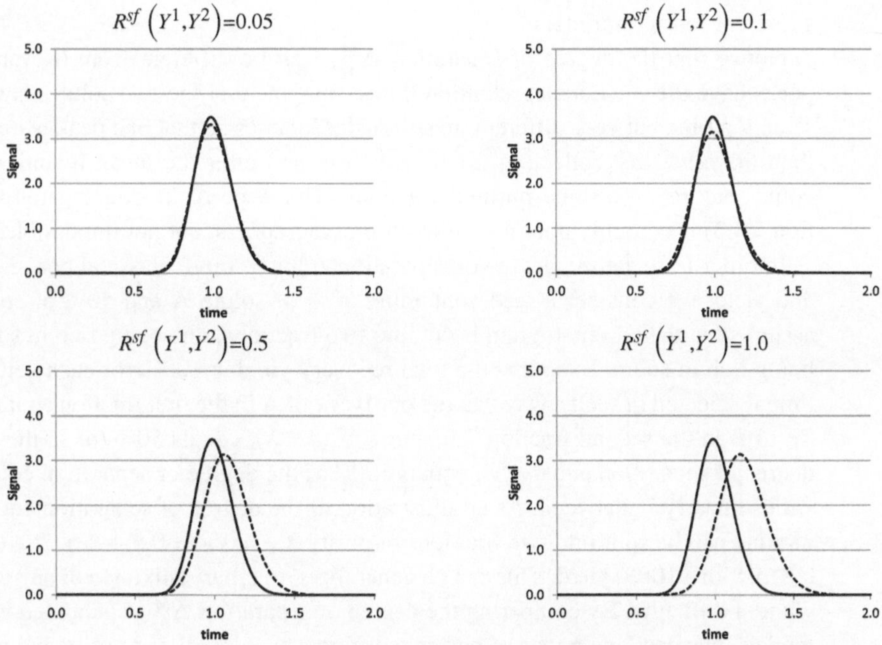

Figure 7.1 Influence of peak differences on the surface resolution. The solid line gives the chromatogram $Y^1(t)$ and the dashed line gives the chromatogram $Y^2(t)$.

Criterion Ξ_{M1}^{M2} compares the overall positioning of chromatograms and does not focus on front positioning or shape. Note that this criterion tends to decrease when the injected volume increases: at the limit, for very large injected volumes, both models would predict a very large plateau at feed concentration so that the dissimilarity criterion would tend towards zero. The criterion is thus adapted to peaks, not to breakthrough curves.

The value of the dissimilarity criterion allowing us to consider two models to be equivalent will be dictated by precision required by the industrial objectives. We know, however, that distinguishing chromatograms associated with a dissimilarity criterion lower than 0.05 is very difficult.

The surface resolution also allows an assessment of the extent to which two solutes A and B are separated. In order to do so, we define the *degree of separation* $\Delta_{A/B}^{M}$ between the two products A and B for a given model M by:[4]

$$\Delta_{A/B}^{M} = R^{sf}\left(Y_A^M, Y_B^M\right) \tag{7.3}$$

where Y_A^M and Y_B^M represent the normalized outlet concentrations $C_A^{out}(M)$ and $C_B^{out}(M)$ of solutes A and B as calculated by the model M.

By definition, the degree of separation $\Delta_{A/B}^{M}$ is 0 if the two peaks are identical and thus not separated, and 2 if they are perfectly separated (again we have two peaks ...). This definition is very simple and based on chromatograms. A more rigorous definition, the so-called *extent of separation*, has been proposed by (Rony, 1968) and generalized by (Valentin, 1981). The merit of the concept of extent of separation is that it can be used for multi-component separations and for any type of separation module. It is, however, mathematically rather involved, and the simple concept of degree of separation will be sufficient for our purposes.

Notice that the degree of separation $\Delta_{A/B}^{M}$ can be different from 0 even if the two peaks have the same mean retention times. Imagine that the two solutes have the same retention time but very different mass transfer kinetics so that one peak is much broader than the other. By collecting fractions before and after the mean retention time, one could thus recover some purified fractions. This type of kinetic separation (see Section 5.4.3) is certainly not familiar to chromatographers, but not impossible *a priori*.

In order to relate the degree of separation $\Delta_{A/B}^{M}$ to more classical concepts of purity and yield, we consider a feed containing 50% of solute A and 50% of solute B, and assume that the chromatogram is cut into two fractions, one being rich in solute A, one being rich in solute B, so that the total recovery yield is 100% for each solute. The cut time is selected in such a way that the purity X_A of A in the first fraction equals the purity X_B of B in the second fraction. The purity $X_A = X_B$ equals 50% (as in the feed) if the degree of separation equals 0; it equals 100% if the degree of separation equals 2 and it varies linearly in between. As an illustration, if the degree of separation equals 1.5, the mixture can be split into two fractions of purity $X_A = X_B = 0.5 + 1.5/2 \times (1 - 0.5) = 0.875$ with a 100% yield. This can be generalized to binary mixtures of any composition without difficulty. By comparing the degree of separation $\Delta_{A/B}^{M}$ calculated with the two models, one can get an idea of possible differences in predicted purity and yield.

4 The same concept can be used between two experimental chromatograms.

To design a chromatographic process, one needs to take into account simulated purity, yield and productivity, as discussed in Chapter 9. The required degree of precision of purity and yield is again dictated by the exact industrial demands. The criteria Ξ_{M1}^{M2} and $\Delta_{A/B}^{M}$ are certainly not sufficient to design a process and to select a model, but they give very useful information: if 99% purity and yield are required, one had better avoid using a model M1 known to be less precise than M2, if the dissimilarity criterion Ξ_{M1}^{M2} is a few percent or more and/or if the degree of separation $\Delta_{A/B}^{M}$ criteria for both the models differ by a few percent or more.

It was emphasized in Chapters 2, 4 and 5 that two models which lead to very similar or identical breakthrough curves or chromatograms under linear conditions can lead to very different ones when non-linearity is excited. We thus need to introduce a simple parameter for assessing (or at least giving an idea of) the non-linearity. The higher the feed concentration, the higher the non-linearity, but the impact of a high feed concentration can be mitigated by a small injection volume. That is why we define a *fictitious concentration* by:

$$C_i^{fic} = min\left(\frac{V_{inj}}{\varepsilon_e V_{col}} C_i^F, C_i^F\right)$$

This fictitious concentration C_i^{fic} is the concentration of an injection amount diluted in pure solvent so that the injection volume is identical to the extragranular porosity of the column. This varies between zero and C_i^F according to the injection volume. Other definitions following the same philosophy could be selected without impacting the methodology and further conclusions.

We then define a *non-linearity factor* Λ_i for each species by:

$$\begin{aligned} &\Lambda_i = \frac{\bar{f}_i'(C_1^{fic}, \ldots, C_n^{fic})}{\bar{K}_i} \\ &\text{for favorable adsorption isotherms, } \Lambda = min(\Lambda_i) \\ &\text{for unfavorable adsorption isotherms, } \Lambda = max(\Lambda_i) \end{aligned} \tag{7.4}$$

The non-linearity factor Λ indicates the maximum deviation between the slopes of the adsorption isotherms at concentrations C_i^{fic} and the lumped Henry's coefficients.[5] Λ is 1 for linear systems, less than 1 for favorable adsorption isotherms and greater than 1 for unfavorable adsorption isotherms. Note that Λ is not defined for S-shaped adsorption isotherms.

Λ is certainly not a perfect indicator,[6] as the same Λ can be associated with different situations (such as a large injection volume with a small concentration or a small injection volume with a large concentration). However, it is simple, able to capture increasing non-linearity when feed concentration or injection volume increases, and sufficient for the purposes of this chapter.

Having all our comparison criteria in hand, we are now ready to detail and compare different models, from the simple ones to the most complex ones. For most of the

[5] The definition can be extended without difficulty to columns not initially equilibrated with a solute-free solution.

[6] An alternative would be a so-called *loading factor* (Guiochon *et al.*, 2006). This is certainly an interesting concept, but limited to Langmuir adsorption isotherms.

simulations proposed below, we will use one of the following linearly modified Langmuir lumped adsorption isotherms:

- Iumped adsorption isotherm α:

$$\begin{array}{lll} \lambda_A = 0.5 \quad \bar{N} = 10\,\text{g/l} \quad \tilde{K}_A = 0.05\,\text{l/g} & & \\ \tilde{K}_B = 0.06\,\text{l/g} & \text{to get a selectivity } \alpha_{B/A} = \tilde{K}_B/\tilde{K}_A = 1.1 & \\ \tilde{K}_B = 0.15\,\text{l/g} & \text{to get a selectivity } \alpha_{B/A} = \tilde{K}_B/\tilde{K}_A = 2.0 & \end{array} \tag{7.5}$$

- Iumped adsorption isotherm β:

$$\begin{array}{ll} \lambda_A = 0.5 \quad \bar{N} = 50\,\text{g/l} \quad \tilde{K}_A = 0.2\,\text{l/g} & \\ \tilde{K}_B = 0.22\,\text{l/g} & \text{to get a selectivity } \tilde{K}_B/\tilde{K}_A = 1.1 \\ \tilde{K}_B = 0.41\,\text{l/g} & \text{to get a selectivity } \tilde{K}_B/\tilde{K}_A = 2.0 \end{array} \tag{7.6}$$

These two adsorption isotherms were used in Chapter 5 when we investigated the influence of non-linear adsorption on internal mass transfer.[7] Lumped adsorption isotherm α (resp. β) is associated with a Henry's coefficient that can be considered relatively small (resp. high) for many preparative chromatography applications. Moreover, $\tilde{K}_i$ being larger for lumped adsorption isotherm β, non-linearity is normally higher with lumped adsorption isotherm β.

We will, in general, arbitrarily fix the injection volume at $\Delta t_{inj} = 0.25\, t_0$, and we will vary the feed concentration from 0.01 g/l, ensuring linear behavior, to 100 g/l, ensuring non-linear behavior. In this range, the non-linearity factor Λ will vary from 0.999 to 0.59 with lumped adsorption isotherm α, and from 0.999 to 0.077 with adsorption isotherm β.

7.2 Modeling hydrodynamics and mass transfer with a single parameter

A simple way to obtain realistic chromatograms consists in using the so-called MC–$\overline{\text{Equil}}$ model, and thus in modeling the influence of hydrodynamic dispersion and mass transfer with a single parameter known as the *plate number*, N. Mass balance equations associated with this MC–$\overline{\text{Equil}}$ model have been presented in Chapter 2 and used in Chapters 3, 5 and 6. They are repeated as Eq. (7.7) for convenience:

$$C_i^{k-1} = C_i^k + \frac{t_0}{N}\frac{dC_i^k}{dt} + \frac{(1-\varepsilon_e)}{\varepsilon_e}\frac{t_0}{N}\frac{d\bar{C}_i^k}{dt} \qquad \left\{\begin{array}{l} i \in [1,n] \\ k \in [1,N] \end{array}\right. \qquad (\text{MC–}\overline{\text{Equil}}\text{ model}) \tag{7.7}$$

Adsorption equilibrium is assumed to have been reached in each cell k, so Eq. (7.7) can be rewritten as:

$$C_i^{k-1} = C_i^k + \frac{t_0}{N}\frac{dC_i^k}{dt} + \frac{1-\varepsilon_e}{\varepsilon_e}\frac{t_0}{N}\sum_{j=1}^{n}\frac{\partial \bar{C}_i^k}{\partial C_j^k}\frac{dC_j^k}{dt} \qquad i \in [1,n] \qquad k \in [1,N] \tag{7.8}$$

7 In Section 5 adsorption isotherms rather than lumped adsorption isotherms were used. Recal that adsorption capacities are related by $\bar{N} = (1-\varepsilon_i)\bar{\bar{N}}$.

The system of equations (7.8) is a set of $n \times N$ ordinary differential equations to be solved numerically.

The initial and boundary conditions are normally given by:

$$\begin{array}{ll} t = 0 & : C_i^k \text{ are uniform in the column and identical to } C_i^I \\ t \in [0, \Delta t_{inj}] & : C_i(k = 0) = C_i^F \end{array} \quad (7.9)$$

We start by comparing the results obtained with the MC–$\overline{\text{Equil}}$ model and with the equilibrium model (PF–$\overline{\text{Equil}}$). Some qualitative information has already been given in Chapter 3, and our goal here is to give some quantitative information by using the dissimilarity criterion Ξ_{MC}^{PF} and degree of separation $\Delta_{A/B}^{M}$ concepts.

Because of the variety of possible situations, proposing simple rules of thumb for quickly estimating the dissimilarity criterion Ξ_{MC}^{PF} is not feasible in general. In the case of linear adsorption isotherms, however, if one represents a chromatogram by a triangle whose width is imposed by the variance calculation in Chapter 2, one concludes that the dissimilarity criterion Ξ_{MC}^{PF} should depend primarily on $\Delta t_{inj}/t_R\sqrt{N}$. This is confirmed in Figure 7.2, where the results of simulations associated with very different experimental conditions are organized on a single curve. It is apparent that, under linear conditions, the dissimilarity criterion Ξ_{MC}^{PF} is below 0.1 if the parameter $\Delta t_{inj}/t_R\sqrt{N}$ is above 15. In the case of a very small injection such as $\Delta t_{inj} = 0.05\, t_0$, obtaining a dissimilarity criterion Ξ_{MC}^{PF} of 0.1 requires operating with a column equivalent to 9000 plates. Alternatively, 900 plates are sufficient for an injection of $\Delta t_{inj} = 0.5\, t_0$.

An idea for extending the comparison to non-linear system consists in plotting the dissimilarity criterion Ξ_{MC}^{PF} versus $\Delta t_{inj}/t_R\sqrt{N}$ for different values of the non-linearity factor Λ. Unfortunately, this plot is too scattered to be really useful: a single value of Λ can be associated with different experimental conditions, leading to different chromatograms.

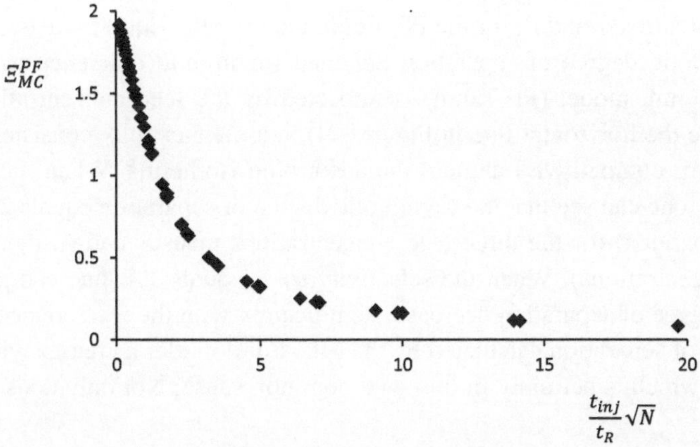

Figure 7.2 Influence of the parameter $t_{inj}/t_R\sqrt{N}$ on the dissimilarity criterion Ξ_{MC}^{PF} for linear systems. $\bar{K}_A$ varies between 1.5 and 6.5, Δt_{inj} between 0.1 t_0 and 2 t_0, N between 50 and 800.

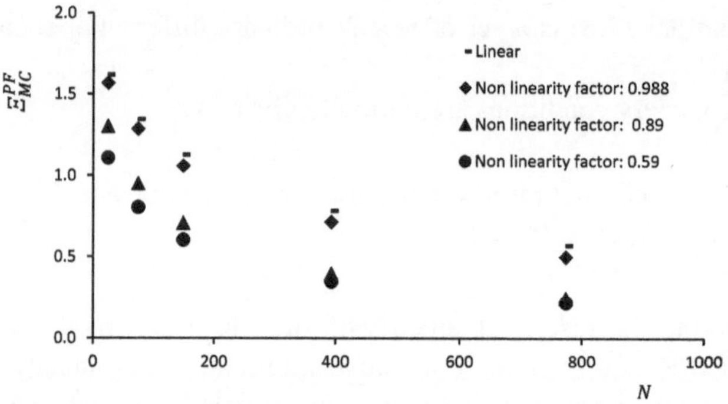

Figure 7.3 Influence of the number of plates and the non-linearity factor Λ on the dissimilarity criterion Ξ_{MC}^{PF}. $\varepsilon_e = 0.4$, $t_0 = 1$, adsorption isotherm A (Eq. (7.5)). Injection volume is constant and fixed at $\Delta t_{inj} = 0.25\, t_0$ for all simulations. Feed concentrations for A of 0.001, 1, 10 and 100 g/l lead to $\Lambda = 0.999$, 0.988, 0.89 and 0.59.

We are not aware of any extension of the plot in Figure 7.2 that would allow a quick comparison between the PF–$\overline{\text{Equil}}$ and the MC–$\overline{\text{Equil}}$ models.[8] However, as shown below, as non-linearity reduces the differences between the models, the curve in Figure 7.2 provides an overestimate of the dissimilarity criterion Ξ_{MC}^{PF}.

The effect of non-linearity on the dissimilarity criterion Ξ_{MC}^{PF} is illustrated for the adsorption isotherm A by analyzing the influence of the number of plates N and of the non-linearity factor Λ for a fixed injection time Δt_{inj} (Figure 7.3). It is clear that, for a given number of plates N, increasing non-linearity decreases the dissimilarity criterion Ξ_{MC}^{PF} compared to the value obtained under linear conditions. This result is easily understandable: when non-linearity increases, the broadening of the peak due to the adsorption isotherm increases, so that the relative influence of hydrodynamic dispersion and mass transfer limitations becomes relatively smaller.

We now compare the models in terms of separation of two solutes A and B. The results expressed in terms of degree of separation $\Delta_{A/B}^{M}$ are given in Figure 7.4 for adsorption isotherms α and β, for the PF–$\overline{\text{Equil}}$ and the MC–$\overline{\text{Equil}}$ models.

The asymptotic degree of separation obtained for infinite efficiency as calculated by the equilibrium model (PF–$\overline{\text{Equil}}$) is impacted by the feed concentration and the selectivity (see the horizontal line in Figure 7.4) in a manner fully consistent with our expectations for competitive Langmuirian adsorption isotherms. When the selectivity $\alpha_{B/A}$ equals 2, one can see that the asymptotic degree of separation equals 2 (the peaks are totally separated) for the three feed concentrations investigated (it then decreases at higher concentrations). When the selectivity $\alpha_{B/A}$ equals 1.1, one can see that the asymptotic degree of separation decreases significantly with the feed concentration.

The degree of separation calculated by the MC–$\overline{\text{Equil}}$ model increases with the number of plates, which is certainly in line with common sense. Not only does the asymp-

[8] Note, however, the so-called *Shirazi number*, presented, for instance, in (Guiochon *et al.*, 2006). While interesting, this number is limited to Langmuir adsorption isotherms, and does not allow quantitative comparisons.

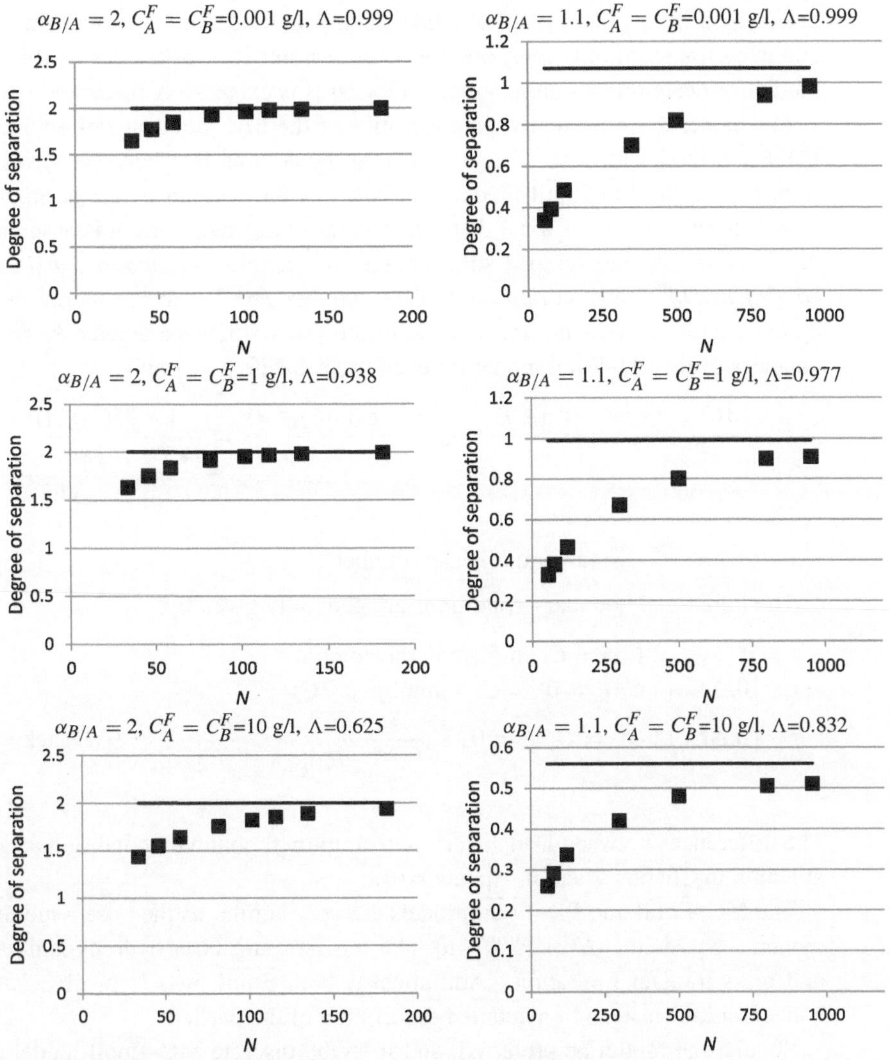

Figure 7.4 Influence of the number of plates and the non-linearity factor on the degree of separation calculated from the PF–$\overline{\text{Equil}}$ and MC–$\overline{\text{Equil}}$ models. $\Delta_{A/B}^{PF}$ given by the straight lines and $\Delta_{A/B}^{MC}$ by the points. $\varepsilon_e = 0.4$, $t_0 = 1$ min, adsorption isotherm α (Eq. (7.5)), $\Delta t_{inj} = 0.25\, t_0$.

totic degree of separation decrease when the selectivity decreases, but the influence of the number of plates becomes more drastic as well. In other words, for a given number of plates, the higher the selectivity, the closer the PF–$\overline{\text{Equil}}$ and the MC–$\overline{\text{Equil}}$ models. It is apparent that the degrees of separation predicted by the two models are very similar (let us say within 5–10%) provided that the number of plates is about 100 for a selectivity of 2, and about 1000 for a selectivity of 1.1.

This simple comparison confirms that the equilibrium model quickly gives an asymptotic behavior but the way to practically implement the separation must be determined by a more complex model taking into account hydrodynamic dispersion and mass transfer limitations, possibly via the number of plates. Using the equilibrium model for designing chromatographic processes is extensively presented in Chapter 9.

In Chapter 2 we presented an alternative to the MC–$\overline{\text{Equil}}$ model referred to as the PD–$\overline{\text{Equil}}$ model. Instead of approximating the column by a discrete series of equilibrium stages, the PD–$\overline{\text{Equil}}$ model approximates the column by a continuous medium in which plug flow plus axial dispersion takes place; equilibrium is assumed between the fluid mobile and lumped solid phases. By defining an *apparent axial dispersion coefficient* D_{ax}^{app}, thus an *apparent Péclet number* Pe^{app}, one can model the effects of hydrodynamic dispersion and mass transfer via a single parameter Pe^{app}. The mass balances of the PD–$\overline{\text{Equil}}$ model for each species i are given by:

$$\frac{\partial C_i(\xi,t)}{\partial \xi} + t_0 \frac{\partial C_i(\xi,t)}{\partial t} + t_0 \frac{(1-\varepsilon_e)}{\varepsilon_e} \frac{\partial \bar{C}_i(\xi,t)}{\partial t} = \frac{1}{Pe^{app}} \frac{\partial^2 C_i(\xi,t)}{\partial \xi^2} \quad \text{(PD–}\overline{\text{Equil}}\text{ model)} \tag{7.10}$$

with $Pe^{app} = \dfrac{u_e L_{col}}{D_{ax}^{app}}$ (apparent Péclet number).

The initial and boundary conditions are generally given by:

$$\begin{aligned}
&t = 0 &&: C_i \text{ and } \bar{C}_i \text{ uniform in the column} \\
&t \in [0,\Delta t_{inj}] &&: C_i(\xi = 0) = C_i^F \text{ (simplified BC)} \\
&t \in [0,\Delta t_{inj}] &&: C_A^F = C_A(\xi = 0) - \frac{1}{Pe^{app}} \left.\frac{dC}{d\xi}\right|_{\xi=0} ; \left.\frac{dC}{d\xi}\right|_{\xi=1} = 0 \text{ (Danckwerts BC)}
\end{aligned} \tag{7.11}$$

The differences between Danckwerts and simplified boundary conditions are discussed at length in Chapter 2 and in Appendix B.

The MC–$\overline{\text{Equil}}$ and PD–$\overline{\text{Equil}}$ models are very similar as they use a modified hydrodynamic parameter (N or Pe^{app}) to take into account both hydrodynamic dispersion and mass transfer limitations. Additionally, both $\overline{\text{Equil}}$ models involve *lumped solid* concentrations instead of a detailed description of the particles.

Should one model be preferred, and if so the discrete MC–$\overline{\text{Equil}}$ model or the continuous PD–$\overline{\text{Equil}}$ model? Provided that the adsorption isotherms are linear, it was shown in Chapter 2 that the models lead to almost identical results, provided that one uses $Pe^{app} = 2N$ and that N is greater than 10–20, which covers the vast majority of situations.[9]

A comparison between the discrete MC–$\overline{\text{Equil}}$ model and the continuous PD–$\overline{\text{Equil}}$ model, using the dissimilarity criterion Ξ_{MC}^{PD} with the parameters of Figure 7.3, is given in Figure 7.5. Confirming the linear chromatography results, only for a very small number of plates do significant differences occur. Let us point out that, in such a case, it is very hard to say that one model is more physically correct than the other. Should this situation be of practical relevance to a given industrial application, a more

[9] We saw in Chapter 2 that, according to the selected boundary conditions, more complex expressions than $Pe^{app} = 2N$ should in principle be used at low efficiencies. These more complex expressions do not materially change the conclusions of this study, so we will stick to the simplest ones.

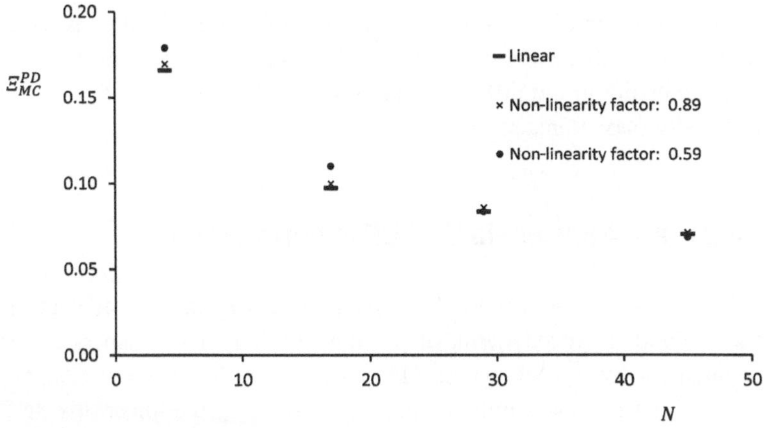

Figure 7.5 Influence of the number of plates and of the non-linearity factor on the dissimilarity criterion Ξ_{MC}^{PD}. Parameters: adsorption isotherm α (Eq. (7.5)). $\Delta t_{inj} = 0.25\, t_0$ for all simulations. Feed concentrations of 0.001, 10 and 100 g/l lead to $\Lambda = 0.999$, 0.89 and 0.59. The equivalence $Pe^{app} = 2N$ is used for the simulations.

sophisticated hydrodynamic description of the flow at the extremities of the column should be considered.

Additionally, in most practical situations, when the MC (resp. PD) model is associated with low N (resp. Pe^{app}) values, mass transfer limitations rather than hydrodynamic dispersion are dominant. Under these circumstances, discussing differences between MC and PD models is a bit meaningless: what really matters is modeling mass transfer adequately.

Thus the MC–$\overline{\text{Equil}}$ and PD–$\overline{\text{Equil}}$ models are in general equivalent for linear and non-linear systems. The decision to use one or the other is largely a matter of the habits of the development team.

The MC–$\overline{\text{Equil}}$ model, offering a "simple, ready-to-use, physical" discretization, appears, *prima facie*, a little bit less computer-demanding than the PD–$\overline{\text{Equil}}$ model. This is probably true for low-efficiency columns (let us say below 100 plates) but the progress made in numerical methods for solving PDEs (see further information in Section 7.9) is such that the PD–$\overline{\text{Equil}}$ model can now be solved with remarkably fast algorithms. Roughly speaking, the CPU time increases at least proportionally to N for the MC–$\overline{\text{Equil}}$ model but is relatively independent of Pe^{app} for the PD–$\overline{\text{Equil}}$ model. I tend to prefer the MC models at low efficiencies (let us say below 100 plates) and the PD models at high efficiencies. Philosophically this makes sense: the PD model being a deviation from plug flow, it looks odd using it to simulate a few mixing cells.

Finally, let us point out that, in some special cases of very strongly non-linear systems, the MC–$\overline{\text{Equil}}$ model may lead to strange and unacceptable behaviors such as a succession of plateaus or oscillating breakthrough curves. These situations are extremely rare and to some extent anecdotal, but they can occur and are addressed in Appendix I.

The simplicity of the $\overline{\text{Equil}}$ models does not prevent them from being used to simulate complex situations. Quite the opposite, they probably offer a unique platform, for instance, when one is willing to study the influence of complex mechanisms on chromatograms or to compare process modes: one can focus on the structure of the equilibria occurring in the different phases without bothering with undue refinements associated with mass transfer.

7.3 Modeling mass transfer via the LDF approximation

The $\overline{\text{Equil}}$ models can be refined by separately taking into account the influence of mass transfer and of hydrodynamic dispersion. Hydrodynamic dispersion can in general be accounted for by the MC or the PD models, and, as shown in Chapter 5, an easy way to model mass transfer limitations consists in using the linear driving force (LDF) approximation.

The mass-balance equations (7.7) are written for species j in mixing cell k for the MC–$\overline{\text{LDF}}$ model:

$$C_j^{k-1} = C_j^k + \frac{t_0}{J}\frac{dC_j^k}{dt} + \frac{(1-\varepsilon_e)}{\varepsilon_e}\frac{t_0}{J}\frac{d\bar{C}_j^k}{dt} \quad k \in [1,J]$$

- If internal diffusion is the rate-limiting step (MC–$\overline{\text{LDF}}_{\text{I}}$) :
$$\frac{d\bar{C}_j^k}{dt} = \frac{1}{t_j^i}\left(\bar{C}_j^{k*} - \bar{C}_j^k\right) \qquad \bar{C}_j^{k*} = \bar{f}_j\left(C_1^k,\ldots,C_n^k\right)$$
- If external diffusion is the rate-limiting step (MC–$\overline{\text{LDF}}_{\text{E}}$) : (7.12)
$$\frac{d\bar{C}_j^k}{dt} = \frac{1}{t_j^e}\left(C_j^k - C_j^{k*}\right) \qquad \bar{C}_j^k = \bar{f}_j\left(C_1^{k*},\ldots,C_n^{k*}\right)$$
- If external and internal diffusion are taken into account (MC–$\overline{\text{LDF}}_{\text{E+I}}$) :
$$\frac{d\bar{C}_j^k}{dt} = \frac{1}{t_j^e}\left(C_j^k - C_j^{k*}\right) = \frac{1}{t_j^i}\left(\bar{C}_j^{k*} - \bar{C}_j^k\right) \qquad \bar{C}_j^{k*} = \bar{f}_j\left(C_1^{k*},\ldots,C_n^{k*}\right)$$

The mass-balance equations (7.10) are written for the PD–$\overline{\text{LDF}}$ model:

$$\frac{\partial C_j(\xi,t)}{\partial \xi} + t_0\frac{\partial C_j(\xi,t)}{\partial t} + \frac{1-\varepsilon_e}{\varepsilon_e}t_0\frac{\partial \bar{C}_j(\xi,t)}{\partial t} = \frac{1}{Pe}\frac{\partial^2 C_j(\xi,t)}{\partial \xi^2}$$

- If internal diffusion is the rate-limiting step (PD–$\overline{\text{LDF}}_{\text{I}}$):
$$\frac{\partial \bar{C}_j(\xi,t)}{\partial t} = \frac{1}{t_j^i}\left(\bar{C}_j^*(\xi,t) - \bar{C}_j(\xi,t)\right) \qquad \bar{C}_j^* = \bar{f}_j(C_1,\ldots,C_n)$$
- If external diffusion is the rate-limiting step (PD–$\overline{\text{LDF}}_{\text{E}}$): (7.13)
$$\frac{\partial \bar{C}_i(\xi,t)}{\partial t} = \frac{1}{t_j^e}\left(C_j(\xi,t) - C_j^*(\xi,t)\right) \qquad \bar{C}_j = \bar{f}_j\left(C_1^*,\ldots,C_n^*\right)$$
- If external and internal diffusion are taken into account (PD–$\overline{\text{LDF}}_{\text{E+I}}$:)
$$\frac{\partial \bar{C}_i(\xi,t)}{\partial t} = \frac{1}{t_j^e}\left(C_j^k(\xi,t) - C_j^*(\xi,t)\right) = \frac{1}{t_j^i}\left(\bar{C}_j^*(\xi,t) - \bar{C}_j(\xi,t)\right)$$
$$\bar{C}_j^* = \bar{f}_j(C_1^*,\ldots,C_n^*)$$

The boundary conditions are similar to the ones given in Eq. (7.9) and Eq. (7.11).

Both MC–$\overline{\text{LDF}}$ and PD–$\overline{\text{LDF}}$ models are used extensively in the literature, and they lead to very similar results for comparable computational efforts. As with the $\overline{\text{Equil}}$ model, I tend to prefer the MC models at low efficiencies (low J or Pe) and the PD model at high efficiencies (high J or Pe).

We have shown that the MC and PD approaches are very similar in general, so we will focus on a comparison between the MC–$\overline{\text{Equil}}$ (resp. PD–$\overline{\text{Equil}}$) model and MC–$\overline{\text{LDF}}$ (resp. PD–$\overline{\text{LDF}}$) models.

Let us recall (Section 5.8) that, for linear systems, the MC–$\overline{\text{LDF}}$ model (using two parameters J and t_A^m to characterize hydrodynamic dispersion and mass transfer) can be approximated by the MC–$\overline{\text{Equil}}$ model (using a single parameter N to characterize hydrodynamic dispersion and mass transfer), provided that N is estimated by:

$$\begin{aligned} \frac{1}{N} &= \frac{1}{J} + \frac{2\bar{K}'_A}{1+\bar{K}'_A}\frac{t_A^m}{t_R} \\ t_R &= t_0\left(1+K'_A\right) \\ t_A^m &= t_A^i + \bar{K}_A t_A^e \end{aligned} \tag{7.14}$$

The Péclet number Pe to be used in the PD–$\overline{\text{LDF}}$ model and the Péclet number Pe^{app} to be used in the PD–$\overline{\text{Equil}}$ model are given by:

$$Pe = 2J \qquad Pe^{app} = 2N \tag{7.15}$$

Consequently, if the characteristic time t_A^m for mass transfer is known, the apparent number of plates (or Pe^{app}) can be estimated from Eq. (7.14). Recall that this equivalence has no theoretical basis for non-linear systems.

This works well for single-solute systems, but Eq. (7.14) shows that the equivalent number of plates N (or equivalent apparent Péclet number) is dependent on the solute. As a consequence, it is not possible to derive an equivalence condition between the MC–$\overline{\text{Equil}}$ and MC–$\overline{\text{LDF}}$ models when multi-solute systems are considered. In practice, a coarse approximation, consisting of a kind of average value of N, is often used.

We have shown in Chapter 5 that, under the linearity assumption, the $\overline{\text{Equil}}$ and $\overline{\text{LDF}}$ models lead to very similar chromatograms unless kinetic limitations are extremely strong. In the same chapter (Section 5.11), we have shown that significant differences can occur under non-linear conditions. In order to further quantify the differences between the models, we will again use the dissimilarity criterion Ξ_{Equil}^{LDF}.

Different simulations have been performed with the two models for a column equivalent to $J = 200$ mixing cells and various characteristic times for mass transfer, allowing N to vary according to Eq. (7.14). External mass transfer limitations are neglected so that internal diffusion is limiting (the LDF_I model is thus used), which is a frequent situation in chromatography, as explained in Section 5.6. The dissimilarity criterion Ξ_{Equil}^{LDF} is given in Figure 7.6 versus the ratio N/J for adsorption isotherms α (Eq. (7.5)) and β (Eq. (7.6)).

According to Eq. (7.14), when N is close to J, so that the ratio N/J is close to 1, the influence of mass transfer limitations is small compared to the influence of

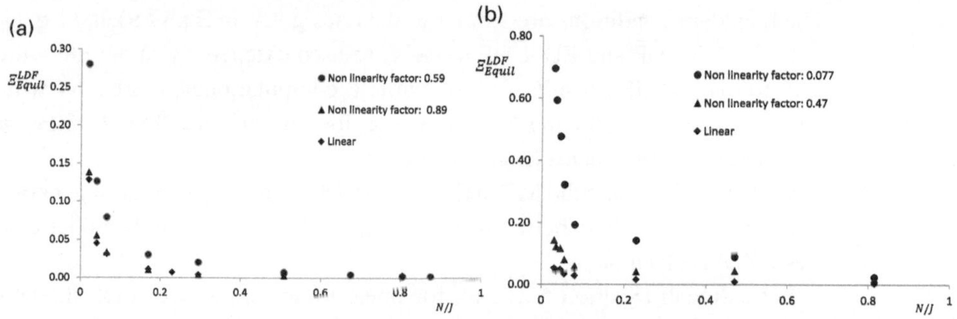

Figure 7.6 Influence of the number of plates and the non-linearity factor on the dissimilarity criterion Ξ^{LDF}_{Equil}. $\varepsilon_e = 0.4$, $t_0 = 1$ min, $J = 200$. N calculated from different values of t_i^A using Eq. (7.14). Concentrations $C_A^F = 0.001$, 10 and 100 g/l associated with the non-linearity factors given in the figure. (a) Adsorption isotherm α (Eq. (7.5)). (b) Adsorption isotherm β (Eq. (7.6)).

hydrodynamic dispersion. The $\overline{\text{Equil}}$ and $\overline{\text{LDF}}$ models must consequently be very similar when the ratio N/J approaches 1. This is confirmed in Figure 7.6 for both adsorption isotherms. The question is now to quantify the concept of "approaching 1" for the N/J ratio.

Under linear conditions, Figure 7.6 shows that chromatograms simulated by the MC–$\overline{\text{LDF}}_{\text{I}}$ and MC–$\overline{\text{Equil}}$ models are almost identical (dissimilarity criterion Ξ^{LDF}_{Equil} below 0.05) provided that the N/J ratio is greater than about 0.1. This means that the MC–$\overline{\text{Equil}}$ model can adequately represent mass transfer limitations, even if these limitations are strong enough to represent about 90% of the total dispersion represented by N. Consequently, under linear conditions, the $\overline{\text{Equil}}$ model is a good approximation of the $\overline{\text{LDF}}$ model unless mass transfer limitations represent 90% or more of the total dispersion.

In a non-linear situation, however, the difference between the two models can become visible even when the N/J ratio is about 0.5. This behavior is not a surprise considering what we learned from Section 5.11. For very strong mass transfer limitations, we have seen in Figure 5.23 that the shapes of the breakthrough curves predicted by the two models can be totally different. This is confirmed in Figure 7.6, which shows dissimilarity criteria not far from 1 (and thus very significant) when N/J is 0.1 or lower.

It is difficult to give precise quantitative rules, but we believe that the following semi-quantitative analysis is pertinent:

- When the ratio N/J is above 0.5, mass transfer plays a relatively secondary role compared to hydrodynamic dispersion, so the $\overline{\text{Equil}}$ and $\overline{\text{LDF}}$ models are very close, almost irrespective of the non-linearity of the system.
- When the ratio N/J is between 0.1 and 0.5, mass transfer limitations are important. The $\overline{\text{Equil}}$ and $\overline{\text{LDF}}$ models are very close if the system is linear but can show some significant differences under strong non-linearity (let us say $\Lambda > 1.5$ for unfavorable adsorption isotherms or $\Lambda < 0.5$ for favorable adsorption isotherms.
- When the ratio N/J is lower than 0.1, mass transfer limitations are extremely strong, and the two models lead to different chromatograms, even for linear systems, the differences being magnified by non-linearity.

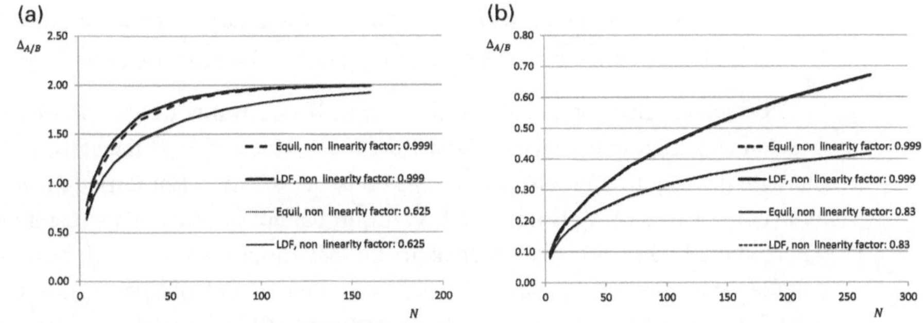

Figure 7.7 Influence of the number of plates and the non-linearity factor on the degrees of separation $\Delta_{A/B}^{Equil}$ and $\Delta_{A/B}^{LDF}$. $\varepsilon_e = 0.4$, $t_0 = 1$ min; adsorption isotherm A (Eq. (7.5)), $\Delta t_{inj} = 0.25\, t_0$. N calculated from different values of t_i^A using Eq. (7.14). $C_A^F = C_B^F = C^F = 0.01$ and 10 g/l leading to non-linearity factors Λ given in Figure 7.7 . (a) $J = 200$ and $\alpha_{B/A} = 2.0$. (b) $J = 400$ and $\alpha_{B/A} = 1.1$.

The same reasoning can be applied to the PD models: due to the equivalence between the MC and PD models, the ratio N/J can be replaced by the ratio Pe^{app}/Pe for assessing possible differences between the PD–$\overline{\text{LDF}}$ and PD–$\overline{\text{Equil}}$ models.

In Figure 7.7, we compare the ability of MC–$\overline{\text{Equil}}$ and MC–$\overline{\text{LDF}}$ models to predict the degrees of separation $\Delta_{A/B}^{Equil}$ and $\Delta_{A/B}^{LDF}$ between solutes A and B. As with the comparison between the PF and MC models, we limit ourselves to binary systems, and consider one situation of a rather small selectivity ($\alpha_{B/A} = 1.1$) and one situation of a more comfortable selectivity ($\alpha_{B/A} = 2.0$). Whereas the different models lead to dissimilarity criteria Ξ_{Equil}^{LDF} that can be very significant under very strong mass transfer limitations, it seems that the effect on the degrees of separation is very limited: curves associated with the two models are almost indistinguishable. The peaks for solutes A and B are deformed in "the same way" when moving from one model to the other, so that the surface resolution between the peaks for A and B is "relatively" insensitive to the model.

At this point one might conclude that, for many situations of interest in preparative chromatography, the use of the MC–$\overline{\text{Equil}}$ model (the famous plate model) is sufficient. We need to be precise, and identify situations where the above statement may be wrong. There are at least three situations for which the MC–$\overline{\text{Equil}}$ model shows significant deficiencies:[10]

- Extremely strong mass transfer limitations lead to columns operating at small N/J ratio. We will show in Chapter 10 that this situation is met when the SMB technology is used for some industrial applications.
- In multiple-solute systems, Eq. (7.14) may lead to numbers of plates which are significantly different for different species. This situation is incompatible with the mass balance structure of the MC approach, which requires the number of plates N to be independent of the species.

[10] These are in addition to the special situations presented in Appendix I.

- The characteristic times for internal diffusion vary with concentration, so a model similar to the $\overline{\overline{\mathrm{LDF}}}^{var}$ approach in Section 5.3.4 must be used.

The second situation is easy to deal with. If one wants to use the MC–$\overline{\mathrm{Equil}}$ model, one can check its sensitivity to the different estimates of N. If the sensitivity is compatible with the required precision, there is no issue: the MC–$\overline{\mathrm{Equil}}$ model can be used with an average value of N. If not, the MC–$\overline{\mathrm{Equil}}$ cannot be used. This results in a situation where the PD approach is superior to the MC approach as it offers more flexibility: whereas the number of plates N must be independent of species for the MC–$\overline{\mathrm{Equil}}$ model, nothing prevents the use of an apparent Péclet number Pe^{app} for each species in the PD–$\overline{\mathrm{Equil}}$ model. Consequently, a possibility could be to estimate a plate number per species N_i according to Eq. (7.14), then to estimate the corresponding apparent Péclet numbers by $Pe^{app}(i) = 2N_i$ and solve the PD–$\overline{\mathrm{Equil}}$ model with one apparent Péclet number per species. This approximation may allow the computational efforts to be reduced a bit compared to the $\overline{\mathrm{LDF}}$ models.

Situations of lumped diffusion coefficients, and consequently characteristic times for internal diffusion, varying with concentration have been presented in Chapter 5. In order to demonstrate the possible consequences of this variation, chromatograms based on the $\overline{\overline{\mathrm{LDF}}}_{\mathrm{I}}^{\mathrm{var}}$ approximation are given in Figure 7.8. We consider a single-solute situation together with the adsorption isotherm β (Eq. (7.6)). We further assume that the adsorbed solute is motionless, a situation which has been shown to maximize the impact of non-linearity on the characteristic time for internal mass transfer (Section 5.3.4). Assuming an intragranular porosity of $\varepsilon_i = 0.5$, Eq. (5.56) and Eq. (5.60) show that the characteristic time for diffusion of solute A is 10 times higher when the particles are in equilibrium with 100 g/l solution than when the particles are in equilibrium with a dilute solution. We can see in Figure 7.8 that differences between the MC–$\overline{\mathrm{LDF}}_{\mathrm{I}}$ and MC–$\overline{\overline{\mathrm{LDF}}}_{\mathrm{I}}^{\mathrm{var}}$ models are significant under non-linear conditions, while the two models are indistinguishable under linear conditions. Because of the

Figure 7.8 Chromatograms simulated with the MC–$\overline{\mathrm{LDF}}_{\mathrm{I}}$, MC–$\overline{\overline{\mathrm{LDF}}}_{\mathrm{I}}^{\mathrm{var}}$ and MC–$\overline{\mathrm{Equil}}$ models. $\varepsilon_e = 0.4$, $\varepsilon_i = 0.5$, $t_0 = 60$ s; adsorption isotherm β (Eq. (7.6)). $\Delta t_{inj} = 0.25\, t_0$. $J = 100$ and $t_A^i(0) = 30$ s, so $N = 16$ (Eq. (7.14)). t_A^{ivar} determined according to Eqs. (5.56) and (5.60). $C_A^F = 100$ g/l.

visible difference between chromatograms, there is no question that the variation of characteristic times for mass transfer can significantly impact the performance of the process.

In this specific example, a remarkable result should be highlighted: the chromatogram simulated with the MC–$\overline{\overline{\text{LDF}}}_{\text{I}}^{\text{var}}$ model is very different from the one simulated with the MC–$\overline{\overline{\text{LDF}}}_{\text{I}}$ model, but almost identical to the one obtained with the MC–$\overline{\text{Equil}}$ model! We mentioned in Section 5.3.4 that, for a given number of plates N, the predicted sharpening of the fronts is more important with the MC–$\overline{\text{Equil}}$ model than with the MC–$\overline{\overline{\text{LDF}}}_{\text{I}}$ model. The MC–$\overline{\overline{\text{LDF}}}_{\text{I}}^{\text{var}}$ model, including a decrease in the characteristic time for mass transfer when concentration increases, thus becomes closer to the MC–$\overline{\text{Equil}}$ model. This illustrates the fact that increasing complexity (in this case moving from the MC–$\overline{\text{Equil}}$ to the MC–$\overline{\text{LDF}}$ approach) does not necessarily bring simulation closer to reality. Conversely, similarity between predicted curves (in this situation, between the MC–$\overline{\text{Equil}}$ and MC–$\overline{\overline{\text{LDF}}}_{\text{I}}^{\text{var}}$ models) may be fortuitous, and result from unexpected compensation effects that are not reliable for extrapolation purposes.

In general we can certainly not bet on the fact that the MC–$\overline{\text{Equil}}$ model is "clever enough" to automatically account for the variation of characteristic times for mass transfer! Consequently, should characteristic times for mass transfer vary significantly with composition, and if significant precision is required, one should consider using the $\overline{\overline{\text{LDF}}}^{\text{var}}$ or similar approach proposed in Section 5.3.4. Another option, adding a considerable degree of complexity, consists in solving Fick's law in spherical geometry with the porous model, as shown in the next section.

7.4 Modeling mass transfer by solving Fick's law in spherical geometry

In Chapter 5 we discussed at length the differences between using the LDF approximation and solving Fick's law in spherical coordinates. Both models can be included in a column model, leading to the so-called MC–$\overline{\text{LDF}}$ and MC–$\overline{\text{Fick}}$ models if the MC model is used for describing hydrodynamic dispersion, and to the PD–$\overline{\text{LDF}}$ and PD–$\overline{\text{Fick}}$ models if the PD model is used for describing hydrodynamic dispersion. The terminologies $\overline{\text{LDF}}$ and $\overline{\text{Fick}}$ are replaced by $\overline{\overline{\text{Fick}}}$ and $\overline{\overline{\text{LDF}}}$ if a porous model instead of a lumped model is considered.

We know that, in general, the MC and PD approaches are very similar, so we now focus on the differences between the LDF and Fick approaches.

If one uses the PD model to describe column hydrodynamics, mass balances at the column level are identical to Eq. (7.13) and given by:

$$\frac{\partial C_j(\xi,t)}{\partial \xi} + t_0\frac{\partial C_j(\xi,t)}{\partial t} + \frac{1-\varepsilon_e}{\varepsilon_e}t_0\frac{\partial \bar{C}_j(\xi,t)}{\partial t} = \frac{1}{Pe}\frac{\partial^2 C_j(\xi,t)}{\partial \xi^2} \qquad j \in [1,n] \tag{7.16}$$

The lumped solid concentrations are obtained by volume averaging concentrations at the particle level:

$$\bar{C}_j(\xi,t) = \frac{\int_0^{R_0} r^2 \bar{c}_j(r,\xi,t)\,dr}{\int_0^{R_0} r^2 dr} = \frac{3}{R_0^3}\int_0^{R_0} r^2 \bar{c}_j(r,\xi,t)\,dr \tag{7.17}$$

- For the PD–$\overline{\overline{\overline{\text{Fick}}}}_{\text{E}}$ model:

  The mass balances inside the particle are given by Eq. (5.36) of Chapter 5:

$$\varepsilon_i \frac{\partial \grave{c}_j(r,\xi,t)}{\partial t} + (1-\varepsilon_i)\frac{\partial \bar{\bar{c}}_j(r,\xi,t)}{\partial t} = \varepsilon_i \grave{D}_k \nabla^2 \grave{c}_j(r,\xi,t) + (1-\varepsilon_i)\bar{\bar{D}}_j \nabla^2 \bar{\bar{c}}_j(r,\xi,t) \tag{7.18}$$

  The internal concentrations are related by:

$$\bar{c}_j(r,\xi,t) = \varepsilon_i \grave{c}_j(r,\xi,t) + (1-\varepsilon_i)\bar{\bar{c}}_j(r,\xi,t) \tag{7.19}$$
$$\bar{\bar{c}}_j(r,\xi,t) = \bar{\bar{f}}_j\left(\grave{c}_1(r,\xi,t),\ldots,\grave{c}_n(r,\xi,t)\right)$$

  The boundary conditions at the particle level are written:

$$\grave{c}_j(r = r_0,\xi,t) = C_j^*(\xi,t) \tag{7.20}$$

  The external mass transfer is modeled by:

$$\frac{\partial \bar{C}_j(\xi,t)}{\partial t} = \frac{C_j(\xi,t) - C_j^*(\xi,t)}{t_j^e} \tag{7.21}$$

- For the PD–$\overline{\text{Fick}}_{\text{E}}$ model:

  The mass balances inside the particle are given by:

$$\frac{\partial \bar{c}_j(r,\xi,t)}{\partial t} = \bar{D}_j \nabla^2 \bar{c}_j(r,\xi,t) \tag{7.22}$$

  The boundary conditions at the particle level are written:

$$\bar{C}_j(r = r_0,\xi,t) = \bar{f}_j\left(C_1^*(\xi,t),\ldots,C_n^*(\xi,t)\right) \tag{7.23}$$

  The external mass transfer is modeled by:

$$\frac{\partial \bar{c}_j(\xi,t)}{\partial t} = \frac{C_j(\xi,t) - C_j^*(\xi,t)}{t_j^e} \tag{7.24}$$

For further information on the relatively complex PD–$\overline{\overline{\overline{\text{Fick}}}}_{\text{E}}$ model (the subscript E refers to the fact that possible external limitations are taken into account), the reader is referred to (Weber and Charkravorti, 1974), (Lee *et al.*, 1989), (Liapis and Rippin, 1978) and (Ma *et al.*, 1996), representing some of the pioneering publications on the matter. (Lee *et al.*, 1989) compared chromatograms obtained by the PF–$\overline{\text{Equil}}$ and PD–$\overline{\overline{\overline{\text{Fick}}}}_{\text{E}}$ models for amino acids and proteins and discussed the influence of internal diffusion, external diffusion and axial dispersion. For more information on that family of models, the reader is referred to (Guiochon *et al.*, 2006, Section 16.3).

The MC–$\overline{\text{Fick}}_{\text{E}}$ and MC–$\overline{\overline{\text{Fick}}}_{\text{E}}$ models are derived in a very similar manner by replacing the mass-balance equation (7.16) by Eq. (7.7), and the ξ longitudinal variable by a cell index k.

What can we expect by moving from the LDF to the Fick type of approach?

Let us start with the simple case of lumped models, thus the $\overline{\text{Fick}}$ and $\overline{\text{LDF}}$ models. Under this condition, we have seen that diffusion inside the particles is a linear process, possible non-linearity occurring only at the fluid–lumped-solid interface. We have shown in Chapter 5 that, even for situations maximizing the differences between the two models (absence of external limitation, comparison of saturation/desaturation curves for which possible hydrodynamic dispersion smoothing is absent) the $\overline{\text{Fick}}$ and the $\overline{\text{LDF}}$ models give very similar behaviors. The value of the complex Fick approach instead of the simpler LDF approach is consequently to be found outside the lumped models.

With the porous model offering a more precise description of intragranular mass transfer, including internal equilibria and possible adsorbed-phase diffusivity, the situation may be different. We have seen in Chapter 5 that, with the porous model, the internal diffusion process may become non-linear, so apparent lumped diffusion coefficients may vary with concentration. Equations (7.16)–(7.21) describing the PD–$\overline{\overline{\text{Fick}}}$ model are often called the general rate model (GRM). In fact, as mentioned by (Guiochon *et al.*, 2006, Section 16.3), "GRM" is used for a family of models, taking into account the porous nature of the particles, external mass transfer and hydrodynamic dispersion, the differences being based on the mechanism of intragranular diffusion. We will avoid using the (presumptuous!) terminology GRM because, despite its complexity, a model based on Equations (7.16)–(7.21) is in fact:

- not general: factors including particle swelling/shrinking, the particle's intragranular geometry, distribution of properties, possible intragranular convection, steric hindrance, possible fluid-phase equilibria such as folding/unfolding and ionic dissociation, non-uniform velocity and temperature profiles are not taken into account;
- possibly not totally rigorous: after reading Section 5.12, the reader should be convinced that using Fick's law with constant diffusion coefficients to describe multi-solute systems is at least questionable.

There is no question that using a porous model, taking into account internal transfer mechanisms that we consider to be dominant, goes in the right direction. Using the relatively detailed description associated with a porous model does not necessarily require Fick's law to be solved in spherical geometry: averaging techniques at the particle level may allow retention of the essential features of the model and, at the same time, replacement of a complex PDE by a much simpler ODE. These are the $\overline{\overline{\text{LDF}}}$ approximations ($\overline{\overline{\text{LDF}}}^{\text{var}}$, $\overline{\overline{\text{LDF}}}^{1/5}$, $\overline{\overline{\text{LDF}}}^{\text{POR}}$, $\overline{\overline{\text{LDF}}}^{\text{NP}}$) presented in Sections 5.3.4 and 5.5.1.

The variety of possible situations is so wide that an extensive numerical comparison between the PD–$\overline{\overline{\text{Fick}}}_{\text{E}}$ model (or the MC–$\overline{\overline{\text{Fick}}}_{\text{E}}$) and the more or less classical $\overline{\overline{\text{LDF}}}$ models is out of scope. We know, however, that different $\overline{\overline{\text{LDF}}}$ models predict saturation/desaturation curves almost identical to those from the $\overline{\overline{\text{Fick}}}$ model (see, for instance,

Figure 5.11 and discussions in Sections 5.3.4 and 5.5.1). Agreement can only be better in chromatographic columns. This has been confirmed by (Morbidelli *et al.*, 1984) and (Morbidelli *et al.*, 1982), who have shown that the $\overline{\overline{\mathrm{LDF}}}^{\mathrm{POR}}$ model provides a very good approximation of the PD–$\overline{\overline{\mathrm{Fick}}}_{\mathrm{E}}$ model. This has been confirmed by (Kaczmarski and Antos, 1996), who showed that predictions of the PD–$\overline{\overline{\mathrm{Fick}}}_{\mathrm{E}}$ and $\overline{\overline{\mathrm{LDF}}}^{\mathrm{POR}}$ models were almost indistinguishable within their investigation range.

We thus have a body of evidence confirming that the complex PD–$\overline{\overline{\mathrm{Fick}}}_{\mathrm{E}}$ models can in general be approximated by simpler ones with limited, if any, loss of precision.

I am thus of the opinion that the significant computational effort needed to solve Fick's law in spherical geometry in a column model can only be justified provided that:

- the porous model predicts lumped diffusion coefficients varying significantly with concentration (and this being confirmed experimentally)
- there is a high enough degree of confidence that the porous model describes reality adequately
- the $\overline{\overline{\mathrm{LDF}}}$ approach does not work for some reason (not foreseen yet)
- industrial targets of purity, yield or productivity require this level of precision
- there is a decent degree of certainty that other peak-broadening effects (such as hydrodynamic dispersion, possibly due to non-uniform velocity profiles) are not going to dominate.

The problem is less replacing Fick's law in spherical geometry by some LDF approximation than properly modeling the intragranular mass transfer process. As far as multisolute systems are concerned, replacing Fick's law by the Maxwell–Stefan approach (Section 5.12) may lead to major improvements. An illustration of the major differences in experimental and breakthrough curves predicted by the PD–$\overline{\mathrm{LDF}}$, PD–$\overline{\overline{\mathrm{Fick}}}$ and PD–$\overline{\overline{\mathrm{MS}}}$ models is found in (Van Den Broeke and Krishna, 1995). The superiority of the $\overline{\overline{\mathrm{MS}}}$ model is clear. The improvement is not from taking the spherical geometry of the particle into account but from writing proper transfer equations via the Maxwell–Stefan approach. This being done, as mentioned in Section 5.12, nothing prevents us from averaging (or at least trying to!) the profile at the particle level, and deriving $\overline{\overline{\mathrm{LDF}}}^{\mathrm{MS}}$-or $\overline{\mathrm{LDF}}^{\mathrm{MS}}$-type models.

7.5 Influence of a non-uniform radial velocity profile

Though potentially significant, the effects of non-ideal hydrodynamic behavior, and especially radially non-uniform velocity profiles, are neglected in the vast majority of articles devoted to modeling chromatographic columns.[11] In Chapter 6, we discussed ways to model and investigate these effects. We will use this methodology to highlight

[11] For historical perspective, one of the first contributions addressing radially non-uniform profiles was written by (Le Van and Vermeulen, 1984), who proposed an extension of the Taylor work on diffusion to adsorbers.

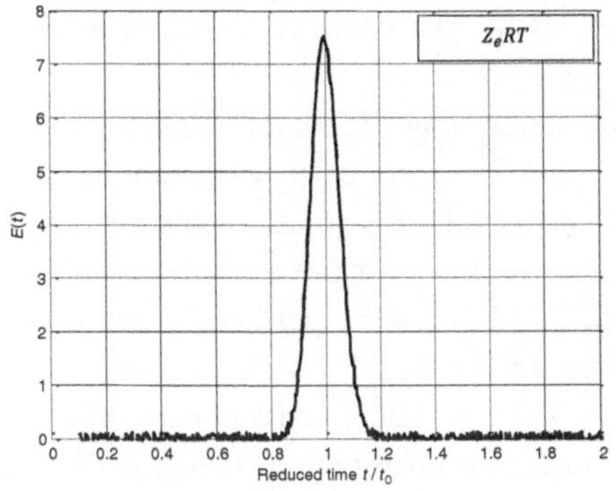

Figure 7.9 Experimental Z_eRTD obtained on a particular column.

the risks of using inappropriate (typically oversimplified) hydrodynamic models, and will propose a path to models closer to reality with reasonable computing effort.

We have stressed the known (but often forgotten) fact that knowledge of the ZRTD is not sufficient for unambiguously characterizing the hydrodynamics of a system in which non-linear processes will take place (see, for instance, Section 3.3.1, Exercise 3.3). A direct consequence is that approximating hydrodynamic dispersion by a number of mixing cells or by a Péclet number is not always sufficient for proper modeling.

As a simple example, consider the Z_eRTD given in Figure 7.9.[12] The Z_eRTD, obtained with an excluded non-interacting solute, characterizes hydrodynamic dispersion only and is unaffected by adsorption or mass transfer. A difficulty in modeling chromatographic columns arises from the fact that a given Z_eRTD can be represented equally well by different models. Let us consider two limiting cases:

- The dispersion observed on the Z_eRTD given in Figure 7.9 is entirely attributable to axial dispersion. The column is approximated by a set of J mixing cells connected in series. A simple graphical determination (width at the base or mid-height; see Chapter 2) shows that the Z_eRTD in Figure 7.9 can be represented by $J = 350$.
- The dispersion observed on the Z_eRTD given in Figure 7.9 is entirely attributable to the existence of a radial velocity profile u_r, using the definition introduced in Section 6.3.2. Recall that, as shown in Exercise 6.9, it is *always* possible to derive a radial velocity profile that reproduces *exactly* any experimental Z_eRTD.

The reality can be intermediate between these limiting situations, meaning that dispersion can result from both axial dispersion and a radial velocity profile. We have shown in Chapter 6 how to model that type of situation for linear chromatography, assuming either continuous or discrete radial velocity profiles. A simple way consists in subdividing the columns into multiple rings located between r and $r + dr$ and assuming that each ring can be modeled by J mixing cells connected in series. We further assumed the number of mixing cells to be independent of the ring, and also neglected possible

[12] The "experimental" Z_eRTD has in fact been simulated with the MC model and a bit of noise added . . .

radial diffusion. This model is called MC_{rad}. Alternatively PD_{rad}-type models can be built on the same principle.

For illustration purposes, we will consider that the radial velocity profile in the column associated with the Z_eRTD in Figure 7.9 is either flat or parabolic, as represented in Figure 7.10. This parabolic profile, as given by Eq. (6.89), is obtained with $\gamma = 0.95$, which means that the velocities u_r are within 5% of the average velocity u. We show in Figure 7.11 that the experimental Z_eRTD can be perfectly represented either by assuming a flat radial velocity profile and $J = 350$ or by assuming the parabolic velocity profile in Figure 7.10 but $J = 500$. The calculated and experimental curves are indistinguishable. Consequently, the available information does not allow us to select one model or the other to represent the experimental Z_eRTD. Valuable complementary information could be obtained, for instance, by determining internal dyes profiles or by studying the variation of the HETP with column length. These techniques give simple ways to assess the nature of radial velocity profiles, as explained in Chapter 6.

Let us now consider the response of the column to a large injection of a solute interacting non-linearly with the chromatographic medium according to the adsorption isotherm β of Eq. (7.6). Simulations performed with the MC–$\overline{\text{Equil}}$ and the

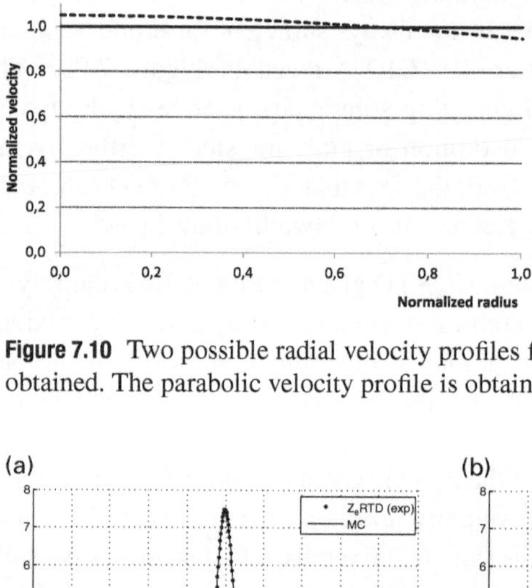

Figure 7.10 Two possible radial velocity profiles from which the Z_eRTD in Figure 7.9 can be obtained. The parabolic velocity profile is obtained with $\gamma = 0.95$ (Eq. (6.89)).

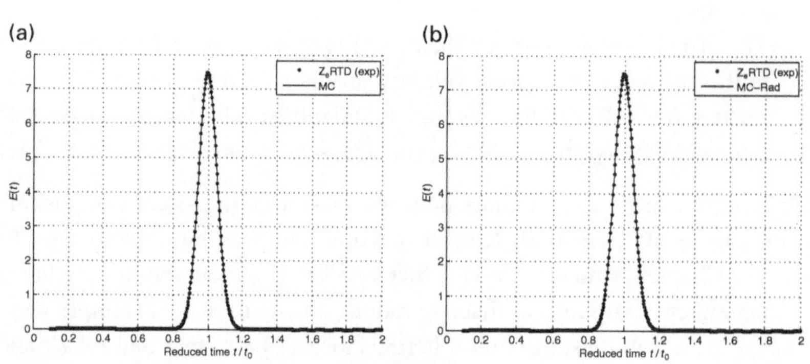

Figure 7.11 Comparison between the Z_eRTD given in Figure 7.9 and simulated curves. For both MC and MC_{Rad} models, simulated and experimental curves are indistinguishable. (a) MC model with a flat radial velocity profile; $N = 350$. (b) MC_{rad} model with parabolic velocity profile from Figure 7.10; $N = 500$ independently of radial position.

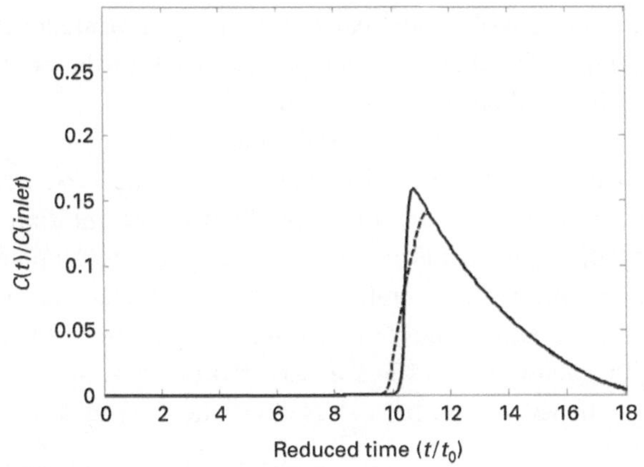

Figure 7.12 Influence of the hydrodynamic model on a non-linear system. $C_A^F = 10\,g/l\;\Delta t_{inj} = 0.5\,t_0$. Solid line: MC model (350 mixing cells). Dashed line: MC_{rad} model with velocity profile from Figure 7.10; $N = 500$ independently of radial position. Adsorption isotherm β (given by Eq. (7.6)) is assumed.

MC_{Rad}–$\overline{\text{Equil}}$ models in Figure 7.12 show that they lead to very different chromatograms. Under non-linear conditions, for a uniform radial velocity profile the MC model predicts a chromatogram saturation front that is remarkably sharp due to the non-linearity of the adsorption isotherm. By contrast, when a parabolic velocity profile is assumed, the MC_{rad} model predicts a front with a fair amount of broadening. In the presence of a non-uniform velocity profile, irrespective of the non-linearity of the adsorption isotherm, the outlet front cannot be too sharp because the different rings are associated with different velocities, thus different retention times. The rear parts of the chromatograms look more similar because of the broadening of the dispersive fronts.

For pedagogical reasons aimed at emphasizing the impact of non-linearity, we selected a Z_eRTD that is fully compatible with a plate model and showed that, even in that basic and simple case, more detailed information on hydrodynamics may be required to simulate non-linear chromatograms.

By experience, as mentioned in Chapter 6, we know that radial velocity profiles are often not parabolic: excluding situations of frit plugging, they are often relatively flat near the center, and show stronger variation near the periphery of the column. This leads to peak tailing, described in Section 6.3. Under these conditions, one must immediately recognize that the Z_eRTD is poorly represented by a plate model.

Should such a situation occur, one should either repack (not to say redesign) the column if a better one can be obtained, or use experience, or better, a *radial* model to estimate the compatibility of this tailing with the industrial constraints.

The impact of non-uniform radial velocity profiles is expected to play a significant role under strongly non-linear conditions, when low selectivities and high purities or yield are required.

Solving mass balances, and possibly heat balances, for a *radial* model is relatively demanding in terms of computational efforts. That is why a back-of-the-envelope calculation can be of interest. This method is presented below.

We assume that the concentration of solute i at the outlet of the column is $C_i^{out}(t,u)$, for a uniform radial velocity profile u. In order to account for the existence of a non-uniform radial velocity profile u_r, the column is subdivided into multiple rings. We neglect the possible impact of radial dispersion, following the development of Section 6.3.2. If one assumes uniform temperature profiles and variations of the radial velocity which are too small to significantly impact mass transfer, then the outlet chromatogram associated with a velocity u_r, expressed as $c_i^{out}(t,u_r)$, is simply related to the chromatogram obtained with velocity u, by a shift along the time axis. Consequently, one can write:[13]

$$c_i^{out}(t,u_r) = C_i^{out}(t\frac{u}{u_r},u) \tag{7.25}$$

Using interpolation formulas, Eq. (7.25) allows the chromatogram $c_i^{out}(t,u_r)$ to be obtained if a discrete chromatogram $C_i^{out}(t,u)$ for velocity u is known.

This property can then be used to determine the chromatogram $C_i^{out}(t)$ in a column with a non-uniform radial velocity, by averaging of all the local radial concentration profiles:

$$C_i^{out}(t) = \frac{2}{R_{col}^2}\int_0^{R_{col}} c_i^{out}(t,u_r)ru_r dr \tag{7.26}$$

We can thus obtain an approximation of the chromatogram obtained on a column with a non-uniform velocity profile by performing only one simulation assuming a flat velocity profile and then using Eq. (7.25) and Eq. (7.26). This avoids solving cumbersome two-dimensional mass-balance equations.

Notice that if Eq. (7.26) is used with shock concentration profiles $c_i^{out}(t,u_r)$ obtained with the equilibrium model for a velocity u_r, then Eq. (7.26) gives the asymptotic sharpest front that can be produced by the column as a response to a step change of inlet concentration due to the radial velocity profile.

This back-of-the-envelope calculation is based on two main assumptions:

- Negligible radial dispersion: we have seen in Chapter 6 that the radial dispersion diffusion coefficient is significantly smaller than the axial dispersion coefficient, making this assumption legitimate as a first approximation. Moreover, as radial dispersion would have the effect of homogenizing the radial concentration profiles, the back-of-the-envelope approximation gives an estimate by excess of the influence of the radial velocity profile.
- Uniform temperature profile: for a non-uniform temperature profile, the adsorption isotherms, thus retention times, are expected to vary with radial position, even at constant fluid velocity. In general, a higher temperature leads to a higher velocity (due to a lower viscosity) and to a lower retention (due to

[13] Similarly to the conventions used in Chapters 5 and 6, note the use of capital letters for average concentrations and lower-case letters for local concentrations.

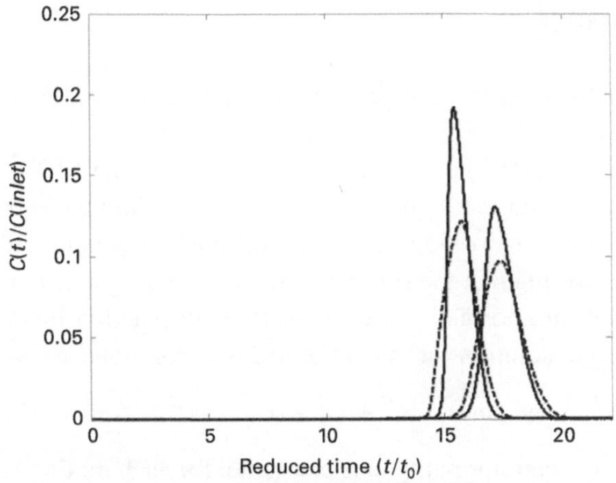

Figure 7.13 Influence of a non-uniform radial velocity profile on a typical preparative HPLC chromatogram. Solid line: reference chromatogram. Dashed line: chromatogram that would occur for a parabolic radial velocity profile ($\gamma = 0.05$). See other details in the text.

lower adsorption), so both effects lead to lower retention times. The consequence is that Eq. (7.25) is no longer valid, and that neglecting the temperature effect on the adsorption isotherms would lead to an underestimate of the impact of a non-uniform radial profile on the chromatogram. An idea (not further presented) consists in lumping the effects of temperature on fluid velocity and on retention time by writing $c_i^{out}(t,T_r) = C_i^{out}(t\alpha_r,T)$ instead of $c_i^{out}(t,u_r) = C_i^{out}(t\frac{u}{u_r},u)$, the parameter α_r expressing the fact that the concentration profiles of *all* the species are shifted by the same factor for a given temperature increment. In the case of multi-solute systems, this approximation makes sense if the temperature dependences of the retention times of the different solutes are similar.

Other attempts to derive simple solutions to this complex two-dimensional problem have been proposed, for instance by (Moate and Le Van, 2009) using perturbation methods. The results are limited to single-solute systems.

Finally, to emphasize the influence of non-uniform radial velocity profiles, we consider a separation performed under conditions close to the so-called "touching band" condition (the peaks are just separated). The associated hypothetical chromatogram is given in Figure 7.13.[14] For this chromatogram, the degree of separation $\Delta_{A/B}$ equals 1.79. Now we *superpose* the influence of a parabolic radial velocity profile ($\gamma = 0.95$) in the column. The resulting chromatograms in Figure 7.13 are obtained by transforming those obtained in the absence of non-uniform radial velocity profiles, using Eq. (7.25) and Eq. (7.26). The relatively moderate 5% velocity variation results in a very significant change in the chromatogram's shape. The degree of separation $\Delta_{A/B}$ drops from 1.79 for a flat velocity profile to 1.52 with γ value of 0.05.

[14] There is no need to know the associated parameters but, for readers willing to check simulations, this chromatogram was simulated with the MC–$\overline{\text{Equil}}$ model with $N = 1500$, the adsorption isotherm β, a selectivity of 1.1 and an injection time of 0.015 t_0.

7.6 The case of ion exchange

The models can be extended to ion exchange with limited adaptations. We will focus on cation exchange, the case of anion exchange being symmetrical.

As an illustration, we consider the constitutive equations for the MC–$\overline{\text{Equil}}$ model applied to ternary cation exchange (the presence of proton H^+ and two cations A and B is assumed). The generalization to higher number of ions presents no difficulty. For simplicity, we start by assuming that the solution contains one unique anion X.

We further assume constant normality and the absence of reaction in solution. The electroneutrality relations in solution for the initial and feed solutions are written:

$$H^I + C_A^I + C_B^I = X^I = N_0 \quad \text{and} \quad H^F + C_A^F + C_B^F = X^F = N_0 \quad \text{(eq/l)} \tag{7.27}$$

With the assumption of constant normality, one can write, for each mixing cell k:

$$X^k = N_0 \tag{7.28}$$

so that the concentration of anions is constant and uniform in the column. Consequently, the electroneutrality relation in solution can be written, for each mixing cell k:

$$H^k + C_A^k + C_B^k = N_0 = X^k \quad \text{(eq/l)} \tag{7.29}$$

Neglecting electrolyte sorption, the electroneutrality relation in the resin is written:

$$\bar{H}^k + \bar{C}_A^k + \bar{C}_B^k = \bar{N}^{iex} \quad \text{(eq/l)} \tag{7.30}$$

If a set of concentration C_A, C_B (resp. $\bar{C}_A, \bar{C}_B$) is known, a corresponding concentration H (resp. $\bar{H}$) is immediately calculated from Eq. (7.29) (resp. Eq. (7.30)).

The mass balances in cell k for the cations A and B are written:

$$\begin{aligned} C_A^{k-1} &= C_A^k + t_0\frac{dC_A^k}{dt} + t_0\frac{(1-\varepsilon_e)}{\varepsilon_e}\frac{d\bar{C}_A^k}{dt} \\ C_B^{k-1} &= C_B^k + t_0\frac{dC_B^k}{dt} + t_0\frac{(1-\varepsilon_e)}{\varepsilon_e}\frac{d\bar{C}_B^k}{dt} \end{aligned} \tag{7.31}$$

In each cell one has seven unknowns: X^k, H^k, C_A^k, C_B^k, $\bar{H}^k$, $\bar{C}_A^k$ and $\bar{C}_B^k$. There are seven equations: the electroneutrality relations in the solution and in the resin, two equilibrium relations (expressed via the selectivity coefficients), the conservation equation for anions X given by Eq. (7.28) and two mass balances given by Eq. (7.31). The system of equations (7.28)–(7.31) allows a complete simulation of the ion-exchange column.

Some simulation results are given in Figure 7.14 for the H, Na, Ca system. All simulations are performed assuming a column equivalent to $N = 100$ plates, selectivity coefficients $K_{Na/H} = 1.5$, $K_{Ca/H} = 1$ and $\bar{N}^{iex} = 2.1$ eq/l. The simulation in Figure 7.14a, performed at a normality of 0.01 eq/l, predicts that the Ca ions are significantly more retained on the resin than the Na ions are. This leads to the classical overshoot observed on the Na front. If the normality of the solution is increased from 0.01 eq/l to 1 eq/l, the ion-exchange equilibrium is much less in favor of the divalent Ca ions due to the structure of the mass action law for heterovalent ions (see Section 4.5.2). As a consequence, totally different breakthrough curves are obtained (Figure 7.14b).

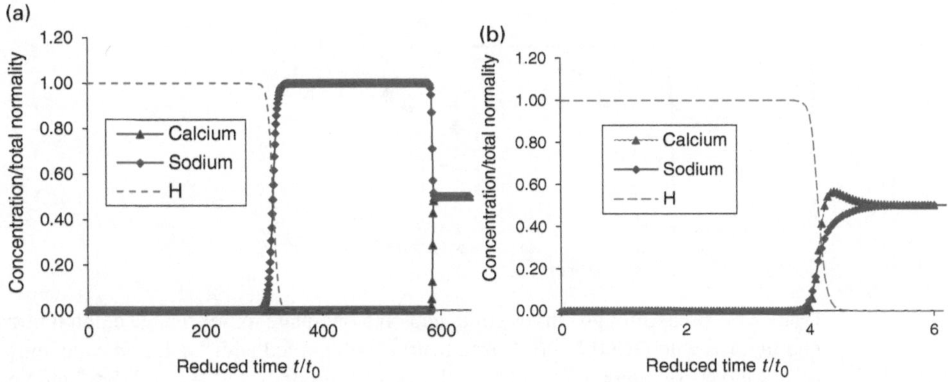

Figure 7.14 Simulated ion-exchange breakthrough curves for a H, Na, Ca system. For both figures: $N = 100$, $K_{Na/H} = 1.5$, $K_{Ca/H} = 1$ and $\bar{N}^{iex} = 2.1$ eq/l. (a) Column equilibrated with HCl 0.01 eq/l and then fed with a solution containing NaCl 0.005 eq/l and $CaCl_2$ 0.005 eq/l. (b) Column equilibrated with HCl 1 eq/l and then fed with a solution containing NaCl 0.5 eq/l and $CaCl_2$ 0.5 eq/l.

The modeling of ion exchange with constant normality is thus very similar to the modeling of adsorption, and the simple $\overline{\text{Equil}}$ model allows representation of the main features of typical ion-exchange processes.

When normality is not constant, Eq. (7.28) must be replaced by a mass balance on the anions having the form:

$$X^{k-1} = X^k + t_0 \frac{dX^k}{dt} + t_0 \frac{(1 - \varepsilon_e)}{\varepsilon_e} \frac{d\bar{X}^k}{dt} \tag{7.32}$$

In general, at low normality electrolyte sorption can be neglected, so the term $\frac{d\bar{X}^k}{dt}$ can be set to zero. If electrolyte sorption cannot be neglected, one has to relate $\bar{X}^k$ to X^k via an electrolyte sorption equilibrium law, as described in Sections 4.5.2 and 4.5.4. This does not add significant difficulty to the column modeling.

The above model can also be extended to take into account equilibria possibly occurring in solution. The simple but illuminating situation of the dissociation of water in H^+ and OH^- is presented below. More complex situations are discussed in Section 7.7.

At high pH, the hydroxide ions are present in such a concentration that they can no longer be neglected. In this situation, electroneutrality Eq. (7.29) must be replaced by:

$$H^k + C_A^k + C_B^k = X^k + OH^k \quad \text{(eq/l)} \tag{7.33}$$

The two concentrations H^k and OH^k are related by $K_e = H^k OH^k$, where K_e is the ionic product of water. One must notice that, if ionic dissociation of water is of significance in the considered process, the normality of the solution cannot be constant: when they leave the resin, protons H are consumed to form water.

In each cell one has eight unknowns: X^k, H^k, OH^k, C_A^k, C_B^k, $\bar{H}^k$, $\bar{C}_A^k$ and $\bar{C}_B^k$. We have eight equations: the electroneutrality relations in the solution and in the resin, the

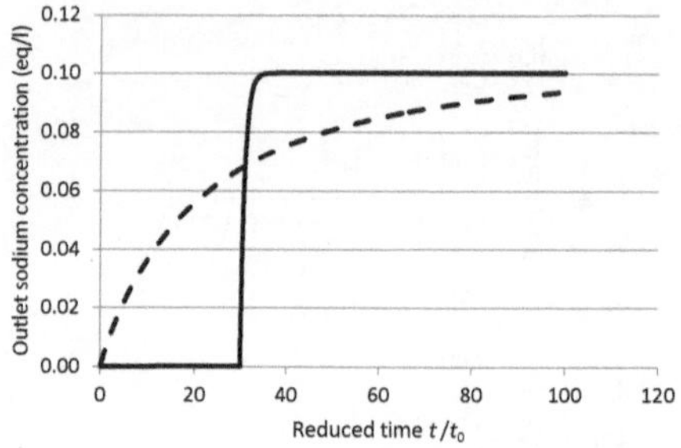

Figure 7.15 Breakthrough curves simulated on a one-plate ion-exchange column initially equilibrated with HCl 0.1 eq/l. Dashed curve: column fed with a solution containing NaCl 0.1 eq/l. Solid curve: column fed with a solution containing NaOH 0.1 eq/l. For both figures: $N = 1$, $K_{Na/H} = 1.5$ and $\bar{N}^{iex} = 2.1$ eq/l.

dissociation constant of water, two equilibrium relations (expressed via the selectivity coefficients), the conservation equation for anions X given by Eq. (7.28) (or possibly Eq. (7.32)) and two mass balances given by Eq. (7.31). The set of equations allows simulation of the column behavior.

An illustration of the influence of the dissociation of water is given in Figure 7.15 for the exchange between Na and H on a column equivalent to only $N = 1$ plate! When a column initially equilibrated with HCl 0.1 eq/l is fed with NaCl 0.1 eq/l, a very broad breakthrough curve is obtained. Certainly not a big surprise for a column equivalent to $N = 1$ plate! More surprisingly, starting with the same initial condition, when the column is fed with NaOH 0.1 eq/l, a very sharp front is obtained. This behavior was recognized and analyzed 40 years ago by (Rodrigues, 1974). The position of the very sharp front is identical to the average position of the front obtained with NaCl. The explanation for this extremely self-sharpening behavior is rather simple: the solution contains an excess of OH^- ions, so every H^+ released by the resin is consumed to produce water. The consequence is that the ionic fraction of Na in the solution is almost unity, which, according to the equilibrium law, requires the ionic fraction of Na in the resin to be almost unity. The ionic dissociation of water has the effect of transforming the H/Na ion-exchange system into an apparently irreversible process in favor of Na. Let us further notice that, should the column be equivalent to $N >> 1$ plates, but with a non-uniform radial velocity profile, the front obtained on the N-plate column may be less sharp than the front obtained on the one-plate column. Investigating strongly non-linear systems can also be an excellent way to investigate hydrodynamics.

This simple example illustrates the drastic influence that equilibria in solution may have on ion-exchange processes. The presence of reactions in solution is very frequent in ion exchange (think of the dissociation constants of amino acids or proteins, for instance), while they are seldom so in adsorption.

As with adsorption processes, the MC–$\overline{\text{Equil}}$ and PD–$\overline{\text{Equil}}$ models can be refined using the $\overline{\text{LDF}}$ approximation. An important thing to be aware of is that, in contrast to adsorption (where the eluent is a "phantom species"), the electroneutrality constraint does not allow us to use one characteristic time for mass transfer per species (see

Section 5.5). Additionally, because Fick's law must in general be replaced by the Nernst–Planck equations to model diffusion of ions, the mass transfer is not symmetric for saturation and desaturation: in other words, the characteristic times for diffusion typically exhibit strong composition dependence (see Section 5.5).

The next refinement consists in using the Fick or Nernst–Planck equations in spherical geometry to model the resin saturation in connection with the lumped model or even the more complex porous models for describing particles. Their integration in a column model does not lead to major difficulties, but it is certainly very demanding in terms of computational efforts. These approaches are presented in (Ma *et al.*, 1996) and (Hasnat and Jvekar, 1996).

Recall that techniques allowing us to replace complex resolutions of mass balances in spherical geometries by $\overline{\text{LDF}}$-or $\overline{\overline{\text{LDF}}}$-type approaches are available and have been introduced in Section 5.5.1.

7.7 Influence of reactions at equilibrium

It is common to see chromatographic processes influenced (desirably or undesirably) by chemical equilibria or reactions taking place in solution or in the solid phase. The topic has been discussed in Sections 3.7 and 4.7, and the simple illustration of water dissociation on ion exchange was presented in the previous section. Additional examples are given below; some have led to large industrial applications, while some are rather anecdotal.

For the sake of clarity, we will not discuss the field of *chromatographic reactors*, in which the objective is to achieve the chemical transformation of reactants into products, chromatography being used to displace equilibria by promoting a separation between products and reactants; we will merely discuss the situation of chromatographic separations that can be favored or hindered by the presence of reactions and equilibria. These situations can in general be modeled with the phenomenological models of Section 4.7.

- **Mixed-bed ion exchange** When cationic resins (in the H^+ form) and anionic resins (in the OH^- form) are mixed together in a column, cation and anion exchange occur simultaneously. These simultaneous ion-exchange processes are associated with H^+ and OH^- ion migration from the cationic and anionic resins to the solution, so they will react together to form water. As shown in Section 7.6, water formation allows a shifting of equilibria in favor of the ionic species which is different from H^+ and OH^- in water. This property is used to obtain highly demineralized water in mixed-bed ion exchangers (Grammont *et al.*, 1986).
- **Amino acid and peptide purification** If the pH is lower than their isoelectric point, amino acids or peptides are in a cationic form and can be fixed onto a cationic resin. If the pH increases, they become anionic and are easily removed from the cationic resin. Similar but opposite behavior is obtained with anionic resins. Ion exchange of amino acids is thus strongly influenced by equilibria in solution. Modeling has, for instance, been studied by (Carta *et al.*, 1988) and (Nagai and Carta, 2004).

- **Rare earth purification** Ion exchange offers the possibility of separating rare earths. Very often, ion exchange is associated with complexation in solution thanks to the introduction in the eluent of an adequate complexing agent like HEDTA (Strelow and Victor, 1990; Merciny *et al.*, 1986).
- **Hydrometallurgy** Even if their salts are cationic, metals like uranium or gold are fixed on anionic exchange resins because negatively charged complexes can be formed, for instance with sulfate ions (Streat, 1986).
- **Carbonates** When water is in contact with carbon dioxide, the carbonic equilibrium in solution produces carbonates and bicarbonates. This concept has been used for regenerating mixed beds of weak cation-exchange resins and strongly basic resins, and has led to the Carix process (Hoell and Feuerstein, 1986; Wenli *et al.*, 1994).
- **Optical isomers** The possibility of separating non-racemic mixtures on achiral chromatographic columns was mentioned in Chapter 4 and has been reported by (Jung and Schurig, 1992). This strange result can be interpreted in terms of complex formations between the enantiomers (Nicoud *et al.*, 1996; Baciocchi *et al.*, 2002, 2004).

The reader will find other examples (isomerization of dipeptides, protein dimerization, protein aggregation, ligand-assisted separation) presented experimentally and theoretically in (Ling and Wang, 2014).

In addition to equilibria occurring in the fluid phase, some equilibria can occur in the chromatographic medium itself, as shown in the two following situations:

- **Ligand-exchange chromatography** A well-known (but rather historical) application of ligand-exchange chromatography is the separation of optical isomers (Davankov, 1989). The adsorbent is prepared by grafting an optically pure amino acid (for instance L-proline) onto a polymeric material. If the eluent contains divalent metal ions (such as copper), complexes are formed between the grafted proline, copper and amino acids contained in the solution. Since this complexation is stereoselective, optical isomers of amino acids can be separated.
- **Weak ion-exchange resins** Weak cationic or anionic resins are extensively used. According to pH, their sites can be ionized or not, and this property allows ions to be fixed according to the eluent pH. This type of situation has been presented in Section 4.7.

Irrespective of the nature of chemical equilibria in solution, the modeling strategy is always the same: identify the dominant species and the possible equilibria and reactions, investigate whether and how the species involved in these equilibria/reactions are involved in the adsorption/ion-exchange process, and write and solve associated mass balances. As far as the column model is concerned, for obvious reasons of simplicity we recommend starting with the MC–$\overline{\text{Equil}}$ or the PD–$\overline{\text{Equil}}$ model.

Because of the variety of possible situations, we believe that the only general procedure for modeling chromatography in the presence of chemical equilibria is given by the phenomenological models presented in Section 4.7. We have already discussed the influence of the dissociation of water in Section 7.6. As a further illustration, we present below the case of ion exchange of amino acids. This is an extension of the MC–$\overline{\text{Equil}}$ model to situations treated in Section 4.7.

We assume that a solution containing anion X, sodium ions Na and a single amino acid AA is fed to a column packed with a cationic exchange resin. We further assume that the amino acid can be present in three different forms: cationic (AA^+), anionic (AA^-) and neutral ($AA^\pm$). The same approach can be extended to dissociation involving multivalent species such as lysine (Nagai and Carta, 2004).

The equations associated with the MC–$\overline{\text{Equil}}$ model are:

- Mass balance for Na, AA and X in cell k with $k \in [1,N]$:

$$\begin{aligned} AA^{k-1} &= AA^k + \frac{t_0}{J}\frac{dAA^k}{dt} + \frac{1-\varepsilon_e}{\varepsilon_e}\frac{t_0}{J}\frac{d\overline{AA}^k}{dt} \\ Na^{k-1} &= Na^k + \frac{t_0}{J}\frac{dNa^k}{dt} + \frac{1-\varepsilon_e}{\varepsilon_e}\frac{t_0}{J}\frac{d\overline{Na}^k}{dt} \\ X^{k-1} &= X^k + \frac{t_0}{J}\frac{dX^k}{dt} + 0 \end{aligned} \tag{7.34}$$

- Possible ionic forms of the amino acid and dissociation equilibria:

$$\begin{aligned} &AA^k = AA^{+k} + AA^{\pm k} + AA^{-k} \\ &K_+ = \frac{AA^{\pm k}H^k}{AA^{+k}} \quad K_- = \frac{AA^{-k}H^k}{AA^{\pm k}} \quad K_e = H^k OH^k \end{aligned} \tag{7.35}$$

- A single ionic form of the amino acid in the resin:

$$\overline{AA}^k = \overline{AA}^{+k} \tag{7.36}$$

- Electroneutrality constraints:

$$\begin{aligned} &AA^{+k} + H^k + Na^k = AA^{-k} + OH^k + X^k \\ &\overline{Na}^k + \overline{AA}^k + \overline{H}^k = \overline{N}^{iex} \end{aligned} \tag{7.37}$$

- Ion-exchange equilibria (with constant or variable selectivity coefficients):

$$\begin{aligned} K_{AA/H} &= \left[\frac{\overline{AA}^{+k}}{AA^{+k}}\right]\left[\frac{H^k}{\overline{H}^k}\right] \\ K_{Na/H} &= \left[\frac{\overline{Na}^k}{Na^k}\right]\left[\frac{H^k}{\overline{H}^k}\right] \end{aligned} \tag{7.38}$$

For each plate k, we have 11 unknowns: AA^k, $AA^{\pm k}$, AA^{+k}, AA^{-k}, H^k, Na^k, OH^k, X^k, $\overline{Na}^k$, $\overline{AA}^k = \overline{AA}^{+k}$, $\overline{H}^k$. For each plate k, we have three mass-balance equations (7.34), four dissociation equations (7.35), two electroneutrality equations (7.37) and two equilibrium equations (7.38), thus a total of 11 equations. It means that the set of equations (7.34) to (7.38) allows one to model the behavior of the amino acid on a strong cationic resin.

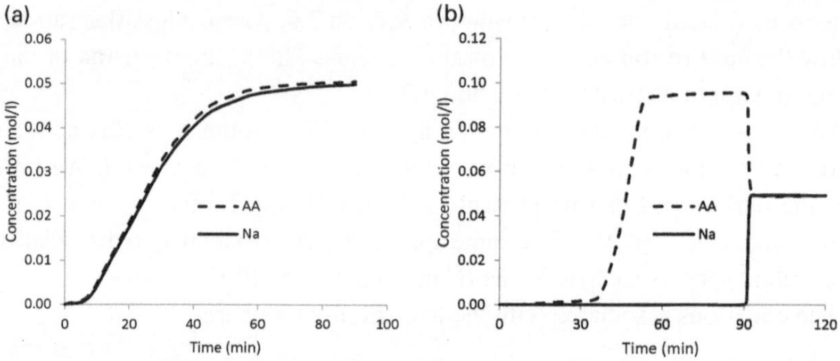

Figure 7.16 Breakthrough curves for cation exchange of amino acids. Resin initially in equilibrium with water and in the H^+ form. For both situations: $\bar{N}^{iex} = 3$ eq/l, $K_+ = 10^{-2}$ l/mol, $K_- = 10^{-9}$ l/mol , $K_e = 10^{-14}$ eq^{2}l^{-2}, $K_{AA/H} = 1.5$, $K_{Na/H} = 1.5$, $N = 5$, $t_0 = 1$ min, $\varepsilon_e = 0.4$. (a) Feed composition: X = 0.20, Na = 0.05, AA = 0.05 mol/l, thus pH = 1. (b) Feed composition: X = 0.00, Na = 0.05, AA = 0.05 mol/l, thus pH = 11.

As an illustration, let us investigate the influence of the anion concentration X^F in the feed solution using typical values for the equilibrium parameters. Results are given in Figure 7.16. Depending on this concentration X^F, the feed solution can be more or less acidic, and this feature has a strong influence on the breakthrough curves.

When the pH of the feed solution is acidic (Figure 7.16a), the amino acid is almost completely dissociated (in the cationic form), and the breakthrough curves of the amino acid and of sodium ion are very similar. This is understandable as identical selectivity coefficients $K_{AA/H}$ and $K_{Na/H}$ are assumed. When the pH of the solution is basic (Figure 7.16b), the amino acid, being in an anionic form, cannot in principle be captured by the cationic resin. However, due to the leakage of protons associated with sodium uptake, the solution in the column becomes acidic during a certain period. During this period, the amino acid can be fixed. After a while, the pH increases to that of the feed, and the amino acid is desorbed from the resin. This behavior is clearly shown by the significant overshoot of amino acid on its breakthrough curve.

This simple illustration is offered just to show that proper integration of key equilibria in a model as simple as the MC–$\overline{\text{Equil}}$ model allows the prediction and understanding of non-trivial phenomena. More sophisticated work including mass transfer modeling can be found in (Nagai and Carta, 2004).

However it is worth keeping in mind that, in rare situations of very strong non-linear coupling, the MC model can lead to physically incorrect results (Appendix I) so that a continuous PD model would be preferred.

7.8 Miscellaneous effects

In the previous sections, we discussed the methodology for modeling chromatographic columns in relatively ideal (even if possibly complex) situations. At least four possible perturbation factors were not considered:

- Fluid compressibility
- Bed compressibility
- Heat effects
- Viscous fingering and multi-phase situations.

These four items were discussed in Chapter 6 and multi-phase situations briefly mentioned in Section 3.7. While different in nature, they share the common feature of being neglected, for good reason (they are negligible) or not (they are difficult), in most contributions to chromatography modeling.

The difficulty in modeling chromatographic columns when one of the above perturbing factors plays a role is strongly connected to the validity of the decoupling postulate mentioned in the introduction to this chapter. Let us elaborate further, with a focus on the situations of practical interest that this book intends to cover.

7.8.1 Fluid compressibility

As presented in Section 6.1.3, when the eluent is a gas or a supercritical fluid, the fluid flow may be subject to significant variations in volumetric flow rate, density and viscosity along the axis of the column. Modeling becomes much more complex than in the liquid phase: fluid velocity influences pressure drop, thus pressure and density, and consequently adsorption.

In addition, a key question is to know if the adsorption of solutes is going to impact the fluid flow or not (decoupling postulate). There is certainly an impact in gas chromatography, where adsorption/desorption of solute can induce significant variations in the volumetric flow rate. The influence of gas compressibility and velocity variations on chromatography has been investigated by several authors including (Le Van *et al.*, 1988), (Scott, 1991), (Scott, 1993), (Sundaram and Wankat, 1988) and (Lu *et al.*, 1992).

As far as SFC is concerned, we have shown in Section 6.1.3 how to derive pressure, fluid density and volumetric flow rate profiles in the column. This requires an equation of state and an expression for viscosity. The characteristics of the equation of state and of the viscosity expression being, *a priori*, composition-dependent, the determination of pressure, fluid density and volumetric flow rate profiles is, *a priori*, coupled with the chromatographic adsorption process. The relatively low solute concentrations in the eluent are likely to have only a secondary influence on the equation of state and viscosity expressions. We will thus accept the decoupling postulate, keeping in mind that this may be questionable for very high solute concentrations.

Assuming that the pressure and volumetric flow rate profiles are known and calculated from the methodology presented in Section 6.1.3, the mass balances associated with the MC–$\overline{\text{Equil}}$ model are written for each species j in each cell k:

$$Q_{k-1} C_j^{k-1} = Q_k C_j^k + \frac{\varepsilon_e V_{col}}{N} \frac{dC_j^k}{dt} + \frac{(1-\varepsilon_e) V_{col}}{N} \frac{d\bar{C}_j^k}{dt} \tag{7.39}$$

where Q_k is the volumetric flow rate in cell k, calculated *independently of* the adsorption process.

In each cell, the lumped solid-phase concentrations are related to the fluid mobile phase concentrations by:

$$\bar{C}_j^k = \bar{f}_j^k(C_1^k, \ldots, C_n^k) \tag{7.40}$$

where $\bar{f}_j^k$ is the adsorption isotherm prevailing in cell k at current (P,T) for solute j.

Under the decoupling postulate, the MC–$\overline{\text{Equil}}$ model can thus be used to model situations involving compressible fluids, with minor adaptations compared to the situation of incompressible fluids.

7.8.2 Bed compressibility

We detailed the modeling of bed compressibility in Section 6.2. We showed that the effects of piston compression and pressure drop on compressible media can significantly impact the bed structure. The question now is to understand the extent to which bed compressibility can or must be coupled with the chromatographic process itself.

In general, as far as adsorption chromatography is concerned, the coupling is relatively weak and operates only via the dependence of fluid viscosity on concentration. Consequently, a decent approximation consists in using an average viscosity to model the bed compressibility; the bed structure being known, the mass balances presented in the previous sections then hold.

A much more complex and still not well described situation is associated with ion exchange when resin swelling/shrinking occurs in connection with the ionic form of the resins. The change in particle size, associated with the ion-exchange process, impacts pressure drop but also possibly induces a variation in the net fluid flow rate. Bed compressibility is unlikely to have an influence on adsorption or ion-exchange isotherms, but may influence mass transfer due to the change in particle and pore sizes.

Our modeling methodology can in principle be adapted, but this requires knowledge of the constitutive equations relating mass transfer to the physical state of the particles (e.g. characteristic time for mass versus particle and intragranular pore size). There are very few contributions on this effect in a dynamic environment, an exception being (Heinonen *et al.*, 2012).

7.8.3 Heat effects

So far we have assumed adsorption to be athermic. This assumption is usually valid in liquid chromatography, but it is well known in gas chromatography that adsorption is exothermic, and this exothermicity can strongly affect the chromatogram. For instance, the enthalpy of desorption of water on zeolites can be greater than the enthalpy of vaporization. In such cases, the propagation of the solutes in the column leads to a temperature front, and mass and energy balances are coupled. To put things in perspective, temperature variation can exceed 100°C in columns used for air drying processes.

Adiabatic equilibrium theory (as opposed to the isothermal equilibrium theory presented in Chapter 3) has been analyzed by (Amundson *et al.*, 1965), (Rhee *et al.*, 1970) and (Rhee *et al.*, 1972). This theory, which generalizes the isothermal equilibrium theory given in Chapter 3, provides the main features of concentration and propagation fronts. It is beyond the scope of this book to describe these developments: we focus on liquid and supercritical fluid chromatography, for which the influence of the exothermicity of adsorption is negligible.

In liquid chromatography, thermal effects are normally due to heat transfer with the column periphery and heat generation due to viscous forces. These have been discussed in Chapter 6. The main consequence of temperature profiles is velocity profiles, which have been discussed in Section 7.5. The absence of a significant impact of exothermicity of adsorption makes the decoupling postulate work: the velocity and temperature profiles can be determined independently of the chromatographic process. With knowledge of these profiles, the chromatographic column can then be modeled by writing a proper mass balance, as illustrated in Section 7.5.

In SFC, in addition to temperature elevation due to frictional forces, heat effects associated with decompression of the fluid while traveling inside the column can be observed (Rajendran, 2012).

7.8.4 Viscous fingering and multi-phase effects

This situation is certainly the toughest to model because the decoupling postulate clearly does not work. The existence of multiple phases affects the hydrodynamics, and prediction of the motion of the relative phases is difficult. In addition, the existence of the different phases is constrained by the dilution associated with hydrodynamic dispersion (axial and radial) and adsorption or ion exchange. As illustrations, ions leaving a resin can induce precipitation in solution, and adsorption of solutes on the chromatographic medium can induce phase disappearance by lowering the concentrations in solution.

Some information on these matters has been given in Sections 3.7 and 6.5. These situations, uncommon but certainly present in some production applications, are one of the main challenges in chromatographic column modeling.

7.9 Numerical methods

At this point in our presentation, we assume we can write a set of more or less complex equations allowing us to describe the behavior of our system. The "only" remaining task consists in solving these equations, and that can be quite challenging.

The equations must usually be solved numerically, even for simple models (single-solute Langmuirian adsorption in the MC–$\overline{\text{Equil}}$ model, for instance). Various commercial software is now available, some references being given in (Seidel-Morgenstern *et al.*, 2012, page 354). This software can simulate some, but certainly not all, of the models discussed in this book. The need to solve specific problems, or a determination to better understand, pushes many process developers to build their own software.

Various textbooks present numerical methods especially for chemical engineers, though it is beyond our scope to present them in detail. For readers not familiar with the subject, I recommend the book of (Rice and Do, 1995). We will limit our presentation to clarifying the key difficulties and options in navigating the forest of methods and algorithms.

Because of the complexity of the equations, relatively sophisticated numerical methods must be used. The sophistication of these methods should not lead the reader to believe that "auto-pilot" is set and nothing bad can happen. Prior to digging into the details, let us stress that systematic cross-checking is absolutely required. I have seen chromatograms simulated using sophisticated tools which do not respect mass balances.

Let us start with linear chromatography. While numerical tools for solving non-linear problems can obviously be used for linear systems as well, it would be a pity not to use the unique property of the Laplace transform and associated inversion algorithms. This has been presented at length in Chapter 2 and Appendix A. Chromatograms associated with the more detailed linear models can be calculated in a fraction of a second on a personal computer.

Much more challenging is the situation of non-linear systems. Irrespective of the type of adsorption isotherms or equilibria in solution, as far as numerical considerations are concerned the different models presented in this book can be organized into two categories:

- Models based on the MC approach (MC–$\overline{\text{Equil}}$, MC–$\overline{\text{LDF}}$, MC–$\overline{\overline{\text{LDF}}}$ etc.) involving only ordinary differential equations (ODEs) and algebraic equations
- Models based on the PD approach (PD–$\overline{\text{Equil}}$, PD–$\overline{\text{LDF}}$, PD–$\overline{\overline{\text{LDF}}}$ etc.) as well as the $\overline{\text{Fick}}$ and $\overline{\overline{\text{Fick}}}$ models, involving partial differential equations (PDEs) and algebraic equations (and also possibly ODEs).

Models taking into account radial velocity and temperature velocity can belong to either category according to the selected structure of mass balances.

The different properties of the equations will lead us to consider different types of numerical tools. Prior to presenting these tools, and for those who are a bit allergic to ODEs and PDEs but still interested in a first look at chromatographic processes performance, the historical Craig model (Craig, 1944) can be considered a first approximation. This is presented in the next section.

7.9.1 The Craig model

This is called the Craig model because it mimics the Craig machine proposed decades ago by (Craig, 1944). It gives a discontinuous version of the $\overline{\text{Equil}}$ model. The idea consists in representing the column as a series of N equilibrium stages. The fluid is transferred from stage to stage at different time steps; after each step equilibrium is assumed to be reached.

At time t, stage k contains concentrations C_i^k and $\bar{C}_i^k$ of solute i. A volume ΔV of fluid contained in stage $k-1$ is transferred to stage k. The operation takes a time:

$$\Delta t = \Delta V / Q \tag{7.41}$$

The volume ΔV must be less than the volume of extragranular liquid contained in one stage:

$$\Delta V \leq \varepsilon_e \frac{V_{col}}{N} \tag{7.42}$$

Thus

$$\Delta t \leq \frac{t_0}{N} \tag{7.43}$$

At time $t + \Delta t$, cell k contains q_i moles of species i, given by:

$$q_i = \varepsilon_e \frac{V_{col}}{N} C_i^k(t) + (1 - \varepsilon_e) \frac{V_{col}}{N} \bar{C}_i^k(t) + \Delta V \left(C_i^{k-1}(t) - C_i^k(t) \right) \tag{7.44}$$

The quantities q_i are known if all the concentrations are known at time t.

After equilibration, the concentrations are solutions of the system:[15]

$$\begin{aligned} q_i &= \varepsilon_e \frac{V_{col}}{N} C_i^k(t + \Delta t) + (1 - \varepsilon_e) \frac{V_{col}}{N} \bar{C}_i^k(t + \Delta t) \\ \bar{C}_i^k &= \bar{f}_i(C_1^k, \cdots, C_n^k) \end{aligned} \tag{7.45}$$

Equations (7.45) have to be solved cell by cell in $C_i^k(t + \Delta t)$, using standard methods for non-linear algebraic equations (Press *et al.*, 1989). This does not pose any difficulty in general, as the concentrations at time t normally provide a good initial guess. The set of concentrations C_i^k and $\bar{C}_i^k$ is thus known at time $t + \Delta t$. If this procedure is performed at successive time intervals, one obtains the time evolution of the concentrations.

The choice of the incremental volume and incremental time, which are linked by Eq. (7.41), is important. The maximum value of the incremental volume is given by Eq. (7.42). Moreover, the incremental time must be lower than the injection time, at least at the beginning of the calculation. If Δt_{inj} is less than t_0/N, a possibility consists of fixing $\Delta t = \Delta t_{inj}$ for the first iteration and then setting $\Delta t = t_0/N$. If $\Delta t_{inj} > t_0/N$, one can set $\Delta t = t_0/N$ for the complete calculation.

The Craig model, presented mainly for historical reasons, has the significant advantages of being robust and stable, and easily adaptable to complex situations involving equilibria in solution. However, it is not very fast, and is in any case limited to the stage model. More efficient tools are presented below. This is, however, the only method we can propose to avoid (or give the illusion of avoiding) differential equations!

7.9.2 Solving ordinary differential equations

The modeling of chromatographic columns with the MC models requires solving systems of ordinary differential equations (ODEs). Regardless of the exact model, the set of equations to be solved can be written:

[15] The mass-balance equation (7.44) holds even in the presence of equilibria in solution, but the mass-balance equation (7.45) should be modified to account for these equilibria, should they occur. Not all species are conserved, as explained in Section 4.7.

$$\frac{dY}{dt} = F(Y) \tag{7.46}$$

where Y is the vector containing all the unknown concentrations C_j^k, $\bar{C}_j^k$, possibly $\bar{\bar{C}}_j^k$, $\grave{C}_j^k$ of the different species in the different cells, and F is a function (linear or not) relating the time derivatives to the unknowns Y.

As an illustration, if one uses the MC–$\overline{\text{Equil}}$ model to describe the Na/H ion exchange at constant normality, presented in Section 7.6, the mass-balance equations are written:[16]

$$\frac{dC_{Na}^k}{dt} = \frac{1}{t_0} \frac{C_{Na}^k - C_{Na}^{k-1}}{1 + \dfrac{1-\varepsilon_e}{\varepsilon_e} \dfrac{d\bar{C}_{Na}^k}{dC_{Na}^k}} \tag{7.47}$$

The model is completed by writing the electroneutrality equations in the fluid moving phase and in the lumped solid phase. The derivative $\dfrac{d\bar{C}_{Na}^k}{dC_{Na}^k}$ can be calculated using the ion-exchange isotherm, so that Eq. (7.47) is of the form of Eq. (7.46). The MC–$\overline{\text{Equil}}$ model describing the Na/H ion exchange at constant normality is thus represented by a set of N ODEs and $2N$ algebraic (electroneutrality) equations. The reader can check that the LDF model would require $2N$ instead of N ODEs.

It is clear that, the higher the number of unknowns, the bigger the computational efforts. More subtle is the effect of column efficiency (number of plates with the MC models): increasing the number of plates increases the number of ODEs but it also makes the fronts steeper, which requires a decrease in the step time. As a consequence, computation times typically vary almost as the square of the number of plates. Other factors can make resolution difficult, and in order to present the matter one has to say a few words about *stiffness*.

A system of ODEs is said to be *stiff* if it involves processes having time constants of very different orders of magnitude. The process must be simulated for a time period of the order of the largest time constant, but the integration step is dictated by the smallest time constant. This may dramatically increase the computational time required for integration, make integration itself difficult and lead to numerical instabilities. Stiffness is a frequent difficulty when solving equations simulating chromatographic columns (Costa *et al.*, 1986). In order to become quickly familiar with stiffness, and how it can badly impact the resolution of an apparently simple problem, the reader is encouraged to solve the situation presented in Section 7.6 describing a monovalent ion exchange performed in a single-plate column at different values of feed pH. At acidic pH, resolution is a non-event; at basic pH, because of the extreme sharpness of the front it is an entirely different story. Another example is the use of LDF models when the characteristic times for mass transfer are small enough that lumped solid and fluid phase are very close to equilibrium. The smaller this time constant, the more difficult the integration, which can become

[16] Using $\dfrac{d\bar{C}_{Na}^k}{dt} = \dfrac{d\bar{C}_{Na}^k}{dC_{Na}^k}\dfrac{dC_{Na}^k}{dt}$.

almost impossible. Using a general model to simulate a system at equilibrium is in general a dead-end; selecting the right model with ad hoc considerations is mandatory.

Understanding how to address this difficulty requires a few words on discretization methods. The underlying idea in solving Eq. (7.46) is to use a discretization of the time derivative of the form:

$$\frac{dY}{dt} = \frac{Y(t + \Delta t) - Y(t)}{\Delta t} \tag{7.48}$$

so that with Eq. (7.46) one obtains:

$$\frac{Y(t + \Delta t) - Y(t)}{\Delta t} = F(Y) \tag{7.49}$$

This differentiation method is called Euler's method; it is a first-order method as it neglects higher orders of the Taylor expansion of the derivatives. Higher-order methods, like Runge–Kutta for instance, are available in various software packages, but this does not impact our discussion. The question is to decide at what time $F(Y)$ is to be estimated: at t or at $t + \Delta t$? If $F(Y)$ is estimated at time t, the method is called *explicit* and the estimates $Y(t + \Delta t)$ are calculated directly from Eq. (7.49) if one knows $Y(t)$. If $F(Y)$ is estimated at time $t + \Delta t$, the method is called *implicit* and estimating $Y(t + \Delta t)$ requires solving an implicit set of Equation (7.49).

Explicit methods are simpler to implement than implicit ones, but it can be shown that the latter are unconditionally stable,[17] whereas the former have strict stability limits (Costa *et al.*, 1986; Finlayson, 1980). This is why implicit methods are generally to be preferred unless simple and "smooth" problems are considered, for which explicit Euler or better higher-order methods are quicker.

Several numerical methods can be used for solving systems of ODEs (Costa *et al.*, 1986; Finlayson, 1980; Press *et al.*, 1989). The first efficient implicit integrators were LSODE, for solving differential systems of the form $y' = f(t, y)$ (Hindmarsh *et al.*, 2005), and DDASSL, for solving algebraic differential systems $g(t, y, y') = 0$ (Petzold, 1982; Breman *et al.*, 1989).

7.9.3 Solving partial differential equations

Solving partial differential equations (PDEs) is in principle more difficult than solving ODEs as they include at least two variables (time and one or two spatial variables such as column axis, particle or column radius) instead of one (time) variable. In addition, PDEs are typically associated with boundary conditions (see, for instance, the discussion in Appendix B), while ODEs in chromatography can be solved by specifying only initial conditions. As with the models based on ODEs, models based on PDEs can also include algebraic equations. Various resolution techniques have been presented, for instance by (Rice and Do, 1995) (for an introduction) and by (Finlayson, 1980; Costa *et al.*, 1986; Villadsen and Michelsen, 1978) (for more detailed presentations). Specific information

[17] Craig's machine corresponds to the implicit scheme, which is unconditionally stable. Craig's machine is physically realizable. This is a general property: all solving methods that correspond to a physically realizable process are intrinsically stable (unless the physical process is not!).

related to the simulation of efficient (high number of plates) chromatography columns can be found in (Golshan-Shirazi and Guiochon, 1992a) and (Guiochon *et al.*, 2006).

Just as solving ODEs requires a discretization of time, solving PDEs requires a discretization of space. There are basically two ways to achieve this, and thus two classes of methods can be distinguished:

- Finite-difference methods: continuous equations are replaced by discretized equations calculated at specific points (nodal points) of a grid. The grid is selected in such a way that the outermost points coincide with domain boundaries such as column inlet and outlet, column or particle periphery. This technique is similar in essence to the one used for ODEs and is detailed in Section 7.9.3.1.
- Finite-volume methods: this method originated for solving the equations of fluid dynamics. The domain is paved in adjacent subdomains (finite volumes). The solution is evaluated for the different subdomains (not at different points) and is thus an averaged value on the subdomain. Technically, volume integrals in a partial differential equation that contain divergence terms (the ∇ terms) are converted into surface integrals, using the divergence theorem (Gauss). Intuitively, this states that the sum of all sources minus the sum of all sinks gives the net flow out of a region.

The different discretization methods are schematically represented in Figure 7.17. The finite volumes paving the space can take any appropriate form and are not limited to the rectangles shown in Figure 7.17. The *finite-volume* method holds a lot of mystery for chromatographers and can certainly be associated with very significant mathematical complexity. For demystification purposes, simply note that, for a one-dimensional column, the MC model, paving the space with a series of cells or plates in which concentrations are homogeneous (thus averaged), can be seen as a simple finite-volume method (with a first-order upwind discretization for the specialists). In this case the above-mentioned Gauss divergence theorem simply ensures the mass balances in each cell. (Javeed *et al.*, 2011b) used finite-volume methods to model chromatography and (Cruz *et al.*, 2005) for simulating different separation processes, for example.

When the space is discretized with a grid having specific nodal points, the natural and classical next step consists in replacing the continuous derivatives with finite

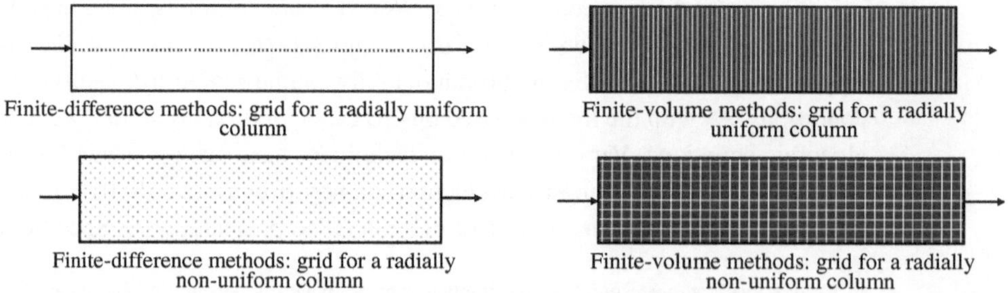

Figure 7.17 Discretization of PDEs (schematic). Finite-volume methods: the finite volumes are shown in dark gray, the surfaces separating the finite volumes, by white lines.

differences, as with the ODEs. There is, however, another route for solving the problem, which consists in assuming the solution to be represented analytically by an expression involving unknown functions and so-called "trial functions". The technique consists in determining the unknown functions so that the PDE is satisfied in the domain with a given degree of precision. This second option, which can be used either on a grid having specific nodal points or on a space paved with finite volumes, is called the *weighted residual method.*

We will briefly describe the two possible methods and then briefly compare them.

7.9.3.1 Finite-difference methods

We assume that a grid, including specific nodal points in time and geometric space, has been selected. The derivatives of the functions Y with respect to a variable x (time or space) are approximated at the different points of the grid by one of the following expressions:

$$\left.\frac{dY}{dx}\right|_k = \frac{Y(k+1) - Y(k)}{\Delta x} \quad \text{forward differences}$$

$$\left.\frac{dY}{dx}\right|_k = \frac{Y(k) - Y(k-1)}{\Delta x} \quad \text{backward differences} \tag{7.50}$$

$$\left.\frac{dY}{dx}\right|_k = \frac{Y(k+1) - Y(k-1)}{2\Delta x} \quad \text{central differences}$$

where $Y(k)$ represents the value of Y at the kth point of the grid. Two consecutive points of the grid are separated by Δx in the direction x. The second derivatives are approximated by:

$$\left.\frac{d^2Y}{dx^2}\right|_k = \frac{Y(k+1) - 2Y(k) + Y(k-1)}{\Delta x^2} \tag{7.51}$$

This scheme involves first-order approximations; higher schemes can be derived from a Taylor expansion ((Grosfils, 2009) used fifth-order approximations, for instance).

First-order schemes suffer from numerical diffusion (so a relatively low Δx must be used to ensure adequate precision) and higher-order schemes suffer from oscillations in the vicinity of steep gradients. Some methods have been proposed to mitigate these drawbacks (Ferziger and Peric, 2002, Chapter 10).

If the above discretization is performed only on the spatial terms of the PDEs, one obtains a set of ODEs that can be solved with the integration methods presented in Section 7.9.2. This method is attractive because it allows the use of available and established ODE solvers, and is called the *method of lines* (Schiesser, 1991). It has been used by (Haag *et al.*, 2001), for instance.

Now let us address techniques simultaneously using time and space grid discretization. As discussed by (Guiochon *et al.*, 2006) for the case of efficient chromatographic columns, the choice of the derivative approximation (forward differences, backward differences, central differences) is critical. An approximate numerical method in which physical dispersion is simulated by numerical dispersion has been proposed by (Rouchon *et al.*, 1987) and improved by (Kaczmarski and Antos, 1996). This algorithm summarized below is extremely fast, and usually leads to adequate precision for high-efficiency columns (Golshan-Shirazi and Guiochon, 1992a).

In the mass-balance equation associated with the PD–$\overline{\text{Equil}}$ model, Eq. (7.10), the dispersion term is neglected and its effect replaced by numerical dispersion:

$$\frac{\partial C_i(\xi,t)}{\partial \xi} + t_0 \frac{\partial C_i(\xi,t)}{\partial t} + \frac{1-\varepsilon_e}{\varepsilon_e} t_0 \frac{\partial \bar{C}_i(\xi,t)}{\partial t} = 0 \tag{7.52}$$

The space is discretized by a grid of points that are Δt apart along the time axis and $\Delta \xi$ apart along the space axis. The time derivatives are replaced by a backward two-point difference, while space derivatives are replaced by a forward two-point difference, leading to the algebraic equation:

$$\begin{aligned} \frac{C_i(\xi,t) - C_i(\xi - \Delta\xi,t)}{\Delta\xi} &+ t_0 \frac{C_i(\xi - \Delta\xi,t) - C_i(\xi - \Delta\xi,t - \Delta t)}{\Delta t} \\ &+ \frac{1-\varepsilon_e}{\varepsilon_e} t_0 \frac{\bar{C}_i(\xi - \Delta\xi,t) - \bar{C}_i(\xi - \Delta\xi,t - \Delta t)}{\Delta t} = 0 \end{aligned} \tag{7.53}$$

Dispersion is approximated by a proper choice of Δt and $\Delta \xi$.

From Eq. (7.53), $C_i(\xi,t)$ is calculated from knowledge of all concentrations prevailing at earlier values of time and abscissa:

$$\begin{aligned} C_i(\xi,t) &= C_i(\xi - \Delta\xi,t) - t_0 \frac{\Delta\xi}{\Delta t} \left[\Delta C_{\xi,t} + \frac{1-\varepsilon_e}{\varepsilon_e} \overline{\Delta C}_{\xi,t} \right] \\ \Delta C_{\xi,t} &= C_i(\xi - \Delta\xi,t) - C_i(\xi - \Delta\xi,t - \Delta t) \\ \overline{\Delta C}_{\xi,t} &= \bar{C}_i(\xi - \Delta\xi,t) - \bar{C}_i(\xi - \Delta\xi,t - \Delta t) \end{aligned} \tag{7.54}$$

In order to ensure the numerical stability of the algorithm, the ratio $\Delta t/\Delta\xi$ must be carefully selected. In the case of a linear single-solute problem, one defines:

$$\psi = \frac{1}{t_0 \left(1 + \frac{1-\varepsilon_e}{\varepsilon_e} \bar{K}_i\right)} \frac{\Delta t}{\Delta \xi} \tag{7.55}$$

and the algorithm is stable if $\psi > 1$. A good choice is to fix $\psi = 2$ and $\Delta\xi = \frac{\text{HETP}}{L}$. (Kaczmarski and Antos, 1996) have extended the methodology to non-linear $\overline{\overline{\text{LDF}}}$ models.

7.9.3.2 Weighted-residual methods

This very general approach is presented only briefly. Information on techniques of special relevance to chromatography can be found in (Finlayson, 1980) and (Villadsen and Michelsen, 1978).

The solution of the PDE in (*z*,*t*) is approximated by an expression of the form:

$$Y = Y_0 + \sum_l a_l(t) y_l(z) \tag{7.56}$$

where $a_l(t)$ are unknown and $y_l(z)$ are defined analytical functions, often called trial functions. The number of functions $a_l(t)$ is called the *order* of the method (not to be confused with the order associated with derivative approximation). The term Y_0 is chosen to satisfy the initial and boundary conditions (hence the term "boundary method" which is sometimes used). The trial functions are normally polynomials or trigonometric functions.

For illustration purposes, let us consider the PD–$\overline{\text{Equil}}$ model for a single-solute system. The associated mass balance given by Eq. (7.10) can be written:

$$L(C_A(z,t)) = \frac{\partial C_A(\xi,t)}{\partial \xi} + t_0 \frac{\partial C_A(\xi,t)}{\partial t} + t_0 \frac{(1-\varepsilon_e)}{\varepsilon_e} \frac{\partial \bar{C}_A(\xi,t)}{\partial t} - \frac{1}{Pe^{app}} \frac{\partial^2 C_A(\xi,t)}{\partial \xi^2} \tag{7.57}$$

The term $L(C_A)$ is zero if $C_A(z,t)$ is the solution of the PDE. If the solution is approximated by Eq. (7.56), the function *L* is not identical to zero; a residue *R*, a continuous function of (*z*,*t*), will remain:

$$L(Y(z,t)) = R \tag{7.58}$$

If one has selected a sufficiently large number of coefficients $a_l(t)$, these can be chosen such that *R* approaches zero over the computational domain. Technically this is achieved by requiring that an integral of the residue *R* times specific *weighting functions* equals zero over the domain. This is the origin of the term "weighted residual".

The different weighted-residual methods differ in the selection of the weighted integral, and the options generally used for chromatography are listed below:

- Subdomain methods: the residue is forced to zero in a subdomain. The finite-volume method can be viewed as a subdomain method.
- Collocation methods: the residue is forced to zero at *M* certain discrete points called collocation points. If the trial functions are a series of *M*th-degree orthogonal polynomials (normally Jacobi) and the *M* collocation points the roots of these orthogonal polynomials, the method is called the orthogonal collocation method (Villadsen and Michelsen, 1978). Notice that when the model involves two dimensions (such as particle radius and column axis) a double collocation must be performed. As an illustration, (Raghavan and Ruthven, 1983) used eight collocation points on the column axis and four in the particle radius, while (Liapis and Rippin, 1978) used eight collocation points on the column axis and eight in the particle radius. Incidentally, the latter reference shows that this method was already in use 35 years ago …
- Galerkin methods: the weighting functions belong to the same family of functions as the trial functions. This is the basis for a series of methods known as *finite-elements* methods. The term orthogonal collocation on finite elements (OCFE),

sometimes found in articles devoted to chromatography modeling, refers to this category of methods. One variation of the finite-elements method is the *moving finite-elements* methods that can be used for following sharp fronts. This has been used by (Lee *et al.*, 1989), (Sereno *et al.*, 1991), (Sereno *et al.*, 1992), (Ma *et al.*, 1996) and (Javeed *et al.*, 2011a).

Should one method be preferred for solving PDEs? In all honesty, I am unable to answer this question in general. However, there is a consensus that orthogonal collocation and/or orthogonal collocation on finite elements methods are more efficient than finite-difference methods, at least when efficient columns are to be simulated (Schmidt-Traub *et al.*, 2012; Ma *et al.*, 1996). This being said, I am not convinced that a general and valid conclusion can be drawn for the variety of situations met in chromatography.

Again, our objective is limited to helping the reader appreciate the differences between terms like methods of lines, orthogonal collocation, moving finite elements, finite difference and finite volume that are frequently found in the literature. Going further is either a matter of reading the specialized references mentioned above or comparing software packages.

7.10 Conclusion

Modeling chromatographic processes is associated with situations of extreme diversity: there is little in common between simulating, on the one hand, an HPLC column packed with reversed-phase material equivalent to 10 000 plates and used for separating very similar molecules, and, on the other hand, the desalting of sugar streams on ion-exchange resin columns equivalent to 10 plates. Neither situation is more difficult or more interesting than the other; they are simply different.

Irrespective of the chromatographic process, the required precision is dictated by industrial constraints: a high degree of precision may be required for the desalting example because of the need to maximize yield and minimize water consumption, while a simple initial go/no-go decision is sometimes expected for the HPLC example.

The problems to be solved are very different, so expecting a universal model, together with numerical resolution algorithms to cover all situations, is certainly chimeric and quite possibly useless. Importantly, however, the methodology – not to say the philosophy – remains the same.

Good practice is certainly to start with the simplest models (MC–$\overline{\text{Equil}}$ or PD–$\overline{\text{Equil}}$) and to add complexity only where and when required. A key to success is to start by addressing and understanding the predominant mechanisms, certainly not by using a "general" model that claims to solve every case like Harry Potter's magic wand.

Consider the separation of a relatively complex mixture containing amino acids and the identification of the pH zone that would favor the separation. In this case the key is to understand the equilibria in solution as well as the ion-exchange isotherms. At least in a first step, the hydrodynamic and mass transfer contributions are likely to be secondary, so they could be modeled with the simplest possible approach (MC). Risking the possibility of being out of date, in the absence of available software I would even start with the Craig model for simple resolution.

Now consider the case where two or three products need to be purified on a rather efficient column (a few thousand plates) with relatively modest non-linearity. Let us assume that initial information on the economic feasibility of the separation is required. Under these conditions, unless more sophisticated software is available, the Rouchon approach (Rouchon *et al.*, 1987) would be remarkably efficient.

A different situation is the quest for precise optimization of a separation. It is certainly wise to start with the PD–Equil or MC–Equil models, but the precision of these models may be incompatible with the required precision. Under these conditions, one must certainly investigate the influence on performance of kinetics (first with LDF approximations, then with more sophisticated methods) and/or of velocity and temperature profiles.

References

Amundson, N. R., Aris, R. and Swanson, R. (1965), *Proc. R. Soc. London, Ser. A* **286**(1404), 129–139.

Baciocchi, R., Mazzotti, M. and Morbidelli, M. (2004), *J. Chromatogr. A* **1024**(1–2), 15–20.

Baciocchi, R., Zenoni, G., Mazzotti, M. and Morbidelli, M. (2002), *J. Chromatogr. A* **944**(1–2), 225–240.

Breman, K., Campbell, S. and Petzold, L. (1989), *Numerical Solution of Initial-Value Problems in Differential Algebraic Equations*, Elsevier Science.

Carta, G. (1998), *Chem. Eng. Sci.* **43**, 2877.

Carta, G. and Cincotti, A. (1998), *Chem. Eng. Sci.* **53**, 3483.

Carta, G., Saunders, M. S., De Carli, J. P. and Vierow, J. B. (1988), *AIChE Symp. Ser.* **84**, 54–61.

Costa, C. A. V., Rodrigues, A. E. and Loureiro, J. M. (1986). In A. E. Rodrigues, ed., *Ion Exchange: Science and Technology*, vol. 107 of *NATO ASI Series, Series E: Applied Sciences*, Martinus Nijhoff, pp. 227–254.

Craig, L. C. (1944), *J. Biol. Chem.* **155**, 519–534.

Cruz, P., Santos, J., Magalhaes, F. and Mendes, A. (2005), *Comput. Chem. Eng.* **30**, 83–98.

Davankov, V. A. (1989), *Chromatographia* **27**(9–10), 475–482.

Ferziger, J. and Peric, M. (2002), *Computational Methods for Fluid Dynamics*, 3rd edn., Spring-Verlag.

Finlayson, B. (1980), *Nonlinear Analysis in Chemical Engineering*, McGraw-Hill.

Golshan-Shirazi, S. and Guiochon, G. (1992a). In F. Dondi and G. Guiochon, eds., Theoretical Advancement in Chromatography and Related Separation Techniques, vol. 383 of *NATO ASI Series, Series C: Mathematical and Physical Sciences*, Kluwer Academic Publishers, pp. 1–33.

Golshan-Shirazi, S. and Guiochon, G. (1992b), *J. Chromatogr. A* **603**(1–2), 1–11.

Grammont, P., Rothschild, W., Sauer, C. and Katsahian, J. (1986). In A. E. Rodrigues, ed., *Ion Exchange: Science and Technology*, vol. 107 of *NATO ASI Series, Series E: Applied Sciences*, Martinus Nijhoff, pp. 403–447.

Grosfils, V. (2009), Modelling and parametric estimation of simulated moving bed chromatographic processes. PhD thesis, Université Libre de Bruxelles.

Guiochon, G., Felinger, A., Golshan-Shirazi, S. and Katti, A. M. (2006), *Fundamentals of Preparative and Nonlinear Chromatography*, Elsevier Academic Press.

Guiochon, G., Golshan-Shirazi, S. and Jaulmes, A. (1988), *Anal. Chem.* **60**(18), 1856–1866.

Haag, J., Van de Wouwer, A., Lehoucq, S. and Saucez, P. (2001), *Control Eng. Pract.* **9**, 921–928.

Hasnat, A. and Jvekar, V. (1996), *AIChE J.* **42**(1), 161–175.

Heinonen, J., Rubiera Landa, H., Sainio, T. and Seidel-Morgenstern, A. (2012), *Sep. Purif. Technol.* **95**, 235–247.

Hiester, N. K., Radding, S. B., Nelson, R. L. and Vermeulen, T. (1956), *AIChE J.* **2**, 404–411.

Hindmarsh, A. C., Brown, P. N., Grant, K. E., *et al.* (2005), *ACM Trans. Math. Software* **31**(3), 363–396.

Hoell, W. and Feuerstein, W. (1986), *React. Polym., Ion Exch., Sorbents* **4**(2), 147–153.

Javeed, S., Qamar, S., Seidel-Morgenstern, A. and Warnecke, G. (2011a), *J. Chromatogr. A* **1218**, 7137–7146.

Javeed, S., Qamar, S., Seidel-Morgenstern, A. and Warnecke, G. (2011b), *Comput. Chem. Eng.* **35**, 2294–2305.

Jung, M. and Schurig, V. (1992), *J. Chromatogr. A* **605**(2), 161–166.

Kaczmarski, K. and Antos, D. (1996), *J. Chromatogr. A* **756**, 73–87.

Le Van, M. D., Costa, C. A. V., Rodrigues, A. E., Bossy, A. and Tondeur, D. (1988), *AIChE J.* **34**(6), 996–1005.

Le Van, M. D. and Vermeulen, T. (1984), *AIChE Symp. Ser.*, **80**, 34.

Lee, C. K., Yu, Q., Un Kim, S. and Wang, L. N.-H. (1989), *J. Chromatogr. A* **484**, 29–59.

Liapis, A. and Rippin, D. (1978), *Chem. Eng. Sci.* **33**, 593.

Ling, L. and Wang, N. (2014). In E. Grushka and N. Grinberg, eds., *Advances in Chromatography*, vol. 52, CRC Press.

Lu, Z., Loureiro, J. M., Le Van, M. D. and Rodrigues, A. E. (1992), *Ind. Eng. Chem. Res.* **31**(6), 1530–1540.

Ma, Z., Whitley, R. and Wang, N. (1996), *AIChE J.* **42**(5), 1244–1262.

Martin, A. and Synge, R. (1941), *Biochem. J.* **35**, 1358–1368.

Merciny, E., Desreux, J. F. and Fuger, J. (1986), *Anal. Chim. Acta* **189**(2), 301–311.

Moate, J. and Le Van, D. (2009), *Chem. Eng. Sci.* **64**, 1178–1184.

Morbidelli, M., Servida, A., Storti, G. and Carra, S. (1982), *Ind. Eng. Chem. Fundam.* **21**(2), 123–131.

Morbidelli, M., Storti, G., Carra, S., Niederjaufner, G. and Pontoglio, A. (1984), *Chem. Eng. Sci.* **39**, 384.

Nagai, H. and Carta, G. (2004), *Sep. Sci. Technol.* **39**(16), 3711–3738.

Nicoud, R. M., Jaubert, J.-N., Rupprecht, I. and Kinkel, J. N. (1996), *Chirality* **8**(3), 234–243.

Nicoud, R. M. and Schweich, D. (1989), *Water Resour. Res.* **25**(6), 1071–1082.

Petzold, L. (1982). In *Proceedings of the 10th IMACS World Congress*, August 8–13, Montreal IMACS.

Press, W. H., Flannery, B. P., Teukolsky, S. A. and Vetterling, W. T. (1989), *Numerical Recipes*, Cambridge University Press.

Radeke, K. H., Ortlieb, H. J. and Gelbin, D. (1981), *Chem. Eng. Sci.* **36**(1), 11–17.

Raghavan, N. and Ruthven, D. (1983), *AIChE J.* **29**(6), 922–925.

Rajendran, A. (2012), *J. Chromatogr. A* **1250**, 227–249.

Rhee, H. K., Heerdt, E. D. and Amundson, N. R. (1970), *Chem. Eng. J.* **1**(4), 279–290.

Rhee, H. K., Heerdt, E. D. and Amundson, N. R. (1972), *Chem. Eng. J.* **3**, 22–34.

Rice, R. and Do, D. (1995), *Applied Mathematics and Modeling for Chemical Engineers*, John Wiley & Sons.

Rodrigues, A. E. (1974), *J. Chromatography* **102**, 437–442.

Rony, P. (1968), *Sep. Sci.* **3**, 239.

Rouchon, P., Schonauer, M., Valentin, P. and Guiochon, G. (1987), *Sep. Sci. Technol.* **22**(8–10), 1793–1833.

Schiesser, W. (1991), *The Numerical Methods of Lines: Integration of Partial Differential Equations*, Academic Press.

Schmidt-Traub, H., Kaspereit, M., Engell, S., Susanto, A., Epping, A. and Jupke, A. (2012). In H. Schmidt-Traub, M. Schulte and A. Seidel-Morgenstern, eds., *Preparative Chromatography*, 2nd edn., Wiley-VCH, p. 425.

Scott, D. M. (1991), *Chem. Eng. Sci.* **46**(12), 2977–2982.

Scott, D. M. (1993), *Chem. Eng. Sci.* **48**(17), 3001–3006.

Seidel-Morgenstern, A., Schmidt-Traub, H., Michel, M., Epping, A. and Jupke, A. (2012). In H. Schmidt-Traub, M. Schulte and A. Seidel-Morgenstern, eds., *Preparative Chromatography*, 2nd edn., Wiley-VCH, p. 321.

Sereno, C., Rodrigues, A. E. and Villadsen, J. (1991), *Comput. Chem. Eng.* **15**, 25–23.

Sereno, C., Rodrigues, A. E. and Villadsen, J. (1992), *Comput. Chem. Eng.* **16**, 583–592.

Streat, M. (1986). In A. E. Rodrigues, ed., *Ion Exchange: Science and Technology*, vol. 107 of *NATO ASI Series, Series E: Applied Sciences*, Martinus Nijhoff, pp. 449–461.

Strelow, F. W. E. and Victor, A. H. (1990), *S. Afr. J. Chem.* **43**(3–4), 127–130.

Sundaram, N. and Wankat, P. C. (1988), *Chem. Eng. Sci.* **43**(1), 123–129.

Valentin, P. (1981). In A. E. Rodrigues and D. Tondeur, eds., *Percolation Processes Theory and Applications*, vol. 33 of *NATO ASI Series, Series E: Applied Sciences*, Sijthoff & Noordhof, pp. 141–195.

Van Den Broeke, L. and Krishna, R. (1995), *Chem. Eng. Sci.* **50**(16), 2507–2522.

Villadsen, J. and Michelsen, M. (1978), *Solution of Differential Equation Models by Polynomial Approximation*, Prentice-Hall.

Weber, T. W. and Charkravorti, R. K. (1974), *AIChE J.* **20**(2), 228–238.

Wenli, G., Wensheng, H. and Holl, W. (1994), *Aqua* **43**, 95–101.

8 Counter-current systems

8.1 Introduction

Partition chromatography resulted from the marrying of two techniques, that of chromatography and that of counter-current solvent extraction. All of the ideas are simple, and had peoples' minds been directed that way, the method would have flourished perhaps a century earlier. In fact the minds of laboratory workers seem to have been closed to counter-current procedures .... In industry the use of the counter-current principle led to great economies in heat and solvents, and its value was obvious.

A.J.P. Martin, Nobel Lecture, December 12, 1952

Even though chromatography is often considered, and used, as a discontinuous process, chemical engineers have conceived and implemented efficient continuous chromatographic systems.

The basic principle of continuous chromatographic processes is that they promote a counter-current contact between the fluid phase and the chromatographic solid-phase medium.[1] In such a system, the solid moves downwards, let us say thanks to gravity, whereas the fluid phase is pumped upwards. The feed mixture containing two solutes (or more generally two fractions to be separated) is injected continuously somewhere halfway along the system. Depending on their respective affinities for the chromatographic medium and on the choice of flow rates, the solutes may move upwards or downwards. This directional difference allows a continuous separation of one stream into two fractions. The counter-current contact is maintained thanks to the motion of the solid phase (the bed), so these systems are often called true moving bed (TMB) systems. The qualification *true* refers to the fact that the chromatographic medium is truly moving. We will use the expressions *true moving bed*, *moving bed* and *counter-current* systems equivalently for systems in which the solid and fluid phases are in contact counter-currently in a continuous manner.

[1] This probably explains the confusion between continuous and counter-current chromatographic processes. These terms are not synonymous: non-counter-current but continuous systems can be considered (Bridges and Barker, 1993; Nicoud, 2014).

Although these systems are attractive in principle, difficulties arising from solid handling have limited the applications of TMB, among them waste water treatment and uranium enrichment (Wesselingh and Van Der Meer, 1986; Streat, 1986a,b; Wankat, 1986; Eldridge, 1993). These technical difficulties led researchers from the company UOP (Broughton and Gerhold, 1961) to propose, more than 50 years ago, a system of fixed beds connected in such a way that it can mimic the behavior of a TMB. This system is called a *simulated moving bed* (SMB), because the solid motion is only simulated by astute column switching. We will use the term simulated counter-current and simulated moving bed interchangeably. These technologies have been used in the petrochemical industry for 50 years, in the sugar industry for 40 years and in the pharmaceutical industry for about 20 years, and their acceptance in bio-industries is growing (Broughton *et al.*, 1970; Johnson and Kabza, 1993; Ruthven, 1984; Wankat, 1986; Ruthven and Ching, 1989; Barker and Ganetsos, 1993; Nicoud, 2014). Hundreds of large-scale units are currently in operation.

The system of industrial interest is SMB, and it can be modeled by simulating a set of connected columns, so the methodology presented earlier in this book and summarized in Chapter 7 is, *a priori*, sufficient for simulating continuous chromatographic systems. Simulating a set of connected columns requires dealing with a complex system and is thus not convenient for deriving and understanding essential properties of these systems.

We will see that modeling and designing the TMB will considerably help in understanding complex SMB systems.

8.2 Single-zone counter-current systems: the equilibrium model

The first important contribution to the modeling of counter-current systems in the presence of non-linear adsorption is probably to be credited to (Rhee *et al.*, 1971). It is a bit involved mathematically, so it is advisable to read more recent contributions proposed by (Storti *et al.*, 1988), (Storti *et al.*, 1989), (Storti, Masi and Morbidelli, 1993) and (Migliorini *et al.*, 2000), which are perhaps a bit freer of detailed mathematics. Like most articles devoted to equilibrium theory, these are still considered very difficult by most chromatographers. Our ambition, as with fixed beds, is to propose a presentation that is at the same time complete and accessible, starting from the simplest situations and gradually increasing complexity.

It is convenient to start with a simple image. Consider the strange idea of people wanting to go upstairs using a downward-moving escalator: a typical counter-current situation. The escalator is the solid phase moving downwards, and the people are the solute molecules trying to move upwards. Now, let us study the possible steady-state situations. We assume that:

- everybody climbs stairs at the same speed
- the system is fed at the bottom by a constant flow of people
- there is nobody upstairs who wants to go downstairs ("pure escalator" is being fed in at the top).

There are then three possible steady-state situations:

- People climb faster than the escalator moves downwards: people move upwards, and the staircase is full of people at a uniform "concentration" imposed by the "inlet flow".
- People climb more slowly than the escalator moves downwards: nobody can be on the moving stairs, which are empty; the escalator has the composition of the top "feed stairs", i.e. there is nobody on it.
- People climb at exactly the speed that the escalator moves: the people on the stairs seem to be motionless; however, one may wonder how and when they reached their intermediate location on the escalator. It is obvious that they have reached this location because something happened at some earlier time, e.g. changes in the stairs or the person flow rate. In other words, this situation depends on a transient state reached prior to the steady state. Moreover, it is easily guessed that any tiny change in the escalator or person velocity will result, after some time, in one of the other situations: either an empty escalator or people moving upwards at constant concentration all along the escalator. In other words, this third case is unstable.

We will see that the single-zone TMB behaves in a very similar manner.

With this picture in mind, let us consider the simple single-zone system in Figure 8.1, in which the chromatographic medium and the fluid phase are in contact counter-currently.

We first assume that the system contains a single solute A. This solute is at concentration $\bar{C}_A^{inl}$ in the inlet lumped solid stream and at concentration C_A^{inl} in the inlet fluid stream. In this chapter we will use superscript *inl* to refer to the inlet conditions in the

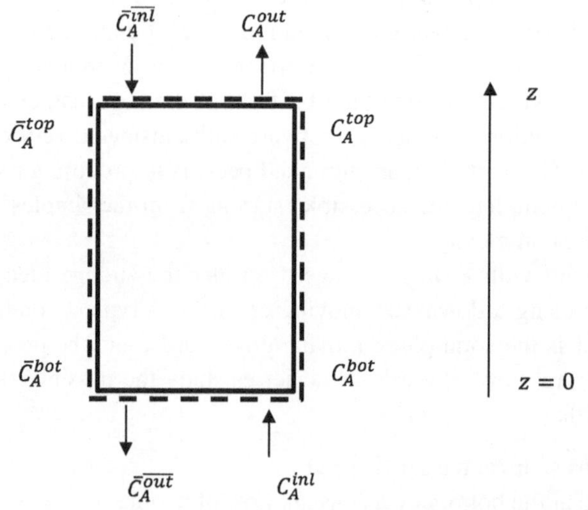

Figure 8.1 Counter-current system. The shaded rectangle around the counter-current system symbolizes the discontinuities that can occur at both inlets.

fluid stream, and superscript $\overline{inl}$ to refer to the inlet conditions in the lumped solid stream. The lumped solid and fluid concentrations of solute A exiting the system are respectively noted $\overline{C}_A^{\overline{out}}$ and C_A^{out}. Superscript *out* refers to the fluid stream exiting the system whereas superscript $\overline{out}$ refers to the lumped solid stream exiting the system. Modeling the system consists in determining C_A^{out} and $\overline{C}_A^{\overline{out}}$ and the internal concentration profiles knowing the fluid and lumped solid flow rates as well as inlet concentrations $\overline{C}_A^{\overline{inl}}$ and C_A^{inl}.

From the lumped solid and fluid flow rates, one can define fluid and solid superficial velocities:[2]

$$u = \frac{Q}{\Omega_{col}} \qquad \bar{u} = \frac{\overline{Q}}{\Omega_{col}} \tag{8.1}$$

and associated interstitial velocities:

$$u_e = u/\varepsilon_e \qquad \bar{u}_e = \bar{u}/(1-\varepsilon_e) \tag{8.2}$$

We will find that the ratio between fluid and lumped solid flow rates play a very important role; we thus define:

$$m = \frac{u}{\bar{u}} = \frac{Q}{\overline{Q}} \qquad m = \frac{\varepsilon_e}{1-\varepsilon_e}\frac{u_e}{\bar{u}_e} \tag{8.3}$$

Assuming steady state, the concentrations of solute in the fluxes entering and exiting the system are related by the following global mass balance:

$$mC_A^{inl} + \overline{C}_A^{\overline{inl}} = mC_A^{out} + \overline{C}_A^{\overline{out}} \tag{8.4}$$

Note that this mass balance is performed on the dashed rectangle surrounding the system in Figure 8.1.

We have two unknowns, $\overline{C}_A^{\overline{out}}$ and C_A^{out}, and only one equation, so any pair of concentrations (C_A^{out},$\overline{C}_A^{\overline{out}}$) satisfying the global mass-balance equation (8.4) is, *a priori*, an acceptable solution. Another constraint must be found.

Concentrations $\overline{C}_A^{\overline{out}}$ and C_A^{out}, being at opposite positions in the system, have little chance of being at equilibrium, so the equilibrium law cannot help us at this stage.

In order to determine the missing equation, let us determine the structure of the internal concentration profiles. Assuming steady state, in the absence of hydrodynamic dispersion and of kinetic limitations the local mass balance can be written:

$$m\frac{dC_A}{dz} - \frac{d\overline{C}_A}{dz} = 0 \tag{8.5}$$

with associated boundary conditions at $z = 0$ and $z = L_{col}$ that will be discussed later. Assuming local equilibrium in the system, Eq. (8.5) leads to:

$$m\frac{dC_A}{dz} = \frac{d\overline{C}_A}{dC_A}\frac{dC_A}{dz} \tag{8.6}$$

[2] To be consistent in our notation, $\overline{Q}$ is the flow rate of lumped particles (cf. Chapter 1). The volumetric flow rate of the geometrical volume containing the particles is $\overline{Q}/(1-\varepsilon_e)$.

Equation (8.6) accepts two types of solutions:

- trivial solutions for which $\frac{dC_A}{dz} = 0$ (and thus $\frac{d\overline{C}_A}{dz} = 0$)
- non-trivial solutions for which $\frac{dC_A}{dz}$ is *possibly* $\neq 0$ if:

$$m = \frac{d\overline{C}_A}{dC_A} \tag{8.7}$$

The trivial solutions associated with zero spatial derivatives describe uniform plateau compositions inside the system. Non-trivial solutions allow *possible* axial variations of concentration but, for a given value of m, Eq. (8.7) is satisfied for one concentration only, unless the adsorption isotherm has a constant slope identical to m. This shows that neither the trivial situation nor the non-trivial situation in general allows continuous spatial concentration variations.[3]

As a consequence, in order for C_A^{inl} to be different from C_A^{out} (otherwise the system would not be very useful ...), the internal profile must present at least one discontinuity. This discontinuity can occur either inside the system or, more frequently, at its boundaries, at either the inlet, the outlet, or both. The same holds for the lumped solid concentrations.

These boundary discontinuities were first recognized by (Rhee *et al.*, 1971). They require a more precise definition of the concentrations near the extremities of the system:

$$\begin{aligned} C_A^{top} &= C_A(z = L_{col}^-) \quad & C_A^{out} &= C_A(z = L_{col}^+) \\ \overline{C}_A^{bot} &= C_A(z = 0^+) \quad & \overline{C}_A^{\overline{out}} &= \overline{C}_A(z = 0^-) \end{aligned} \tag{8.8}$$

The superscripts *top* and *bot* thus refer to concentrations at the top and the bottom, but inside the system, whereas the superscript *out* refers to concentrations at the top and the bottom, but outside the system. Because of the boundary discontinuities, one has in general $C_A^{top} \neq C_A^{out}$, $C_A^{bot} \neq C_A^{inl}$, $\overline{C}_A^{bot} \neq \overline{C}_A^{\overline{out}}$ and $\overline{C}_A^{top} \neq \overline{C}_A^{\overline{inl}}$. The overall mass-balance equation (8.4) is thus written for the dashed line envelope *surrounding* the system.

The study of the simple steady-state single-zone counter-current system is thus a bit more complex than anticipated: we know that the internal concentration profiles are uniform and that discontinuities exist, but we are still unable to calculate the internal and outlet concentrations and to locate the discontinuities. Deriving a steady-state solution requires investigating the transient behavior of the system: solving transient mass balances will allow estimation of front propagation velocities; the directions of these propagation velocities will give us additional clues to the internal steady-state concentration profiles.

[3] We will show later that the non-trivial situation is in fact associated with a composition stabilizing a front velocity.

Neglecting hydrodynamic dispersion and kinetic limitations, the transient mass balance for a single solute can be written:

$$u\frac{\partial C_A}{\partial z} - \bar{u}\frac{\partial \overline{C}_A}{\partial z} + \left[\varepsilon_e\frac{\partial C_A}{\partial t} + (1-\varepsilon_e)\frac{\partial \overline{C}_A}{\partial t}\right] = 0 \tag{8.9}$$

and taking into account the equilibrium between the solid and fluid phases:

$$u\frac{\partial C_A}{\partial z} - \bar{u}\frac{d\overline{C}_A}{dC_A}\frac{\partial C_A}{\partial z} + \left[\varepsilon_e\frac{\partial C_A}{\partial t} + (1-\varepsilon_e)\frac{d\overline{C}_A}{dC_A}\frac{\partial C_A}{\partial t}\right] = 0 \tag{8.10}$$

Following the approach detailed in Chapter 3, it can be shown that Eq. (8.10) depends only on the parameter $\sigma = z/t$, and that it can be rewritten:

$$\left[u - \bar{u}\frac{d\overline{C}_A}{dC_A} - \sigma\left(\varepsilon_e + (1-\varepsilon_e)\frac{d\overline{C}_A}{dC_A}\right)\right]\frac{dC_A}{d\sigma} = 0 \tag{8.11}$$

Consequently, the problem has two possible types of solution:

- The concentration is uniform in the system:

$$\frac{dC_A}{d\sigma} = 0 \tag{8.12}$$

- A front forms, and moves at a propagation velocity given by:

$$\sigma = \frac{u - \bar{u}\dfrac{d\overline{C}_A}{dC_A}}{\varepsilon_e + (1-\varepsilon_e)\dfrac{d\overline{C}_A}{dC_A}} = \bar{u}\frac{m - \dfrac{d\overline{C}_A}{dC_A}}{\varepsilon_e + (1-\varepsilon_e)\dfrac{d\overline{C}_A}{dC_A}} \tag{8.13}$$

Equation (8.13) generalizes Eq. (3.18) in Chapter 3 for estimating the propagation velocity of fronts in fixed beds. But unlike with fixed beds, the propagation velocity here can be positive or negative:

- Case a: if $\sigma > 0$ the front moves upwards
- Case b: if $\sigma < 0$ the front moves downwards
- Case c: if $\sigma = 0$ the front is motionless.

This result is of great help in solving our steady-state problem: if $\sigma > 0$, the boundary discontinuities must be located at the top of the system, whereas they are located at the bottom if $\sigma < 0$; this is exactly the situation described for the escalator in the introduction to this section. Incidentally, note that the situation $\sigma = 0$ is reached for concentrations satisfying Eq. (8.7).

For simplicity, we will first consider the case of a linear adsorption isotherm. Under this assumption, Eq. (8.13) becomes:

$$\sigma = \frac{u - \bar{u}\overline{K}_A}{\varepsilon_e + (1-\varepsilon_e)\overline{K}_A} = \bar{u}\frac{m - \overline{K}_A}{\varepsilon_e + (1-\varepsilon_e)\overline{K}_A} \tag{8.14}$$

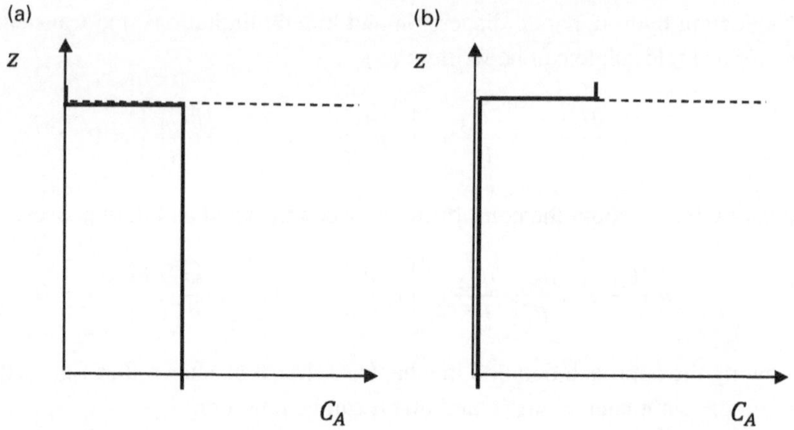

Figure 8.2 Possible steady-state profiles in a single-zone counter-current system. Linear behavior and $m > \overline{K}_A$ are assumed. (a) The fed lumped solid is solute-free. (b) The fed fluid is solute-free.

For a given lumped Henry's coefficient, the propagation velocity depends only on the fluid and lumped solid velocities. Consider further the possible situations.

- **Case a:** $\sigma > 0$
  According to Eq. (8.14), a positive front velocity imposes a ratio $m > \overline{K}_A$, thus a minimum fluid velocity given by $u > \bar{u}\overline{K}_A$. Possible steady-state profiles are illustrated in Figure 8.2, depending on the inlet conditions. Regardless of the inlet conditions, the front moves upwards until it reaches the top of the system; no discontinuity occurs at the bottom of the column. When the system is fed with a solute-free lumped solid (Figure 8.2a), the solute concentration is constant everywhere except at the upper boundary of the system, where it decreases sharply. When the system is fed with a solute-free fluid (Figure 8.2b), the solute concentration is constant everywhere except at the upper boundary of the system, where it increases sharply.
  The uniformity of the profiles and the absence of a discontinuity at the bottom allow one to write:

$$C_A^{bot} = C_A^{top} = C_A^{inl} \qquad \overline{C}_A^{bot} = \overline{C}_A^{top} = \overline{K}_A C_A^{inl} \tag{8.15}$$

Moreover, the lumped solid concentration $\overline{C}_A^{out}$ exiting the system equals $\overline{C}_A^{bot}$, again due to the absence of a discontinuity at the bottom of the system.

The last unknown is the fluid concentration C_A^{out} exiting the system, which is different from C_A^{top} because of the discontinuity at the top. This is obtained after substitution of $\overline{C}_A^{out} = \overline{K}_A C_A^{inl}$ in the overall mass-balance equation (8.4), which yields:

$$C_A^{out} = C_A^{inl} + \frac{1}{m}\left(\overline{C}_A^{inl} - \overline{K}_A C_A^{inl}\right) \tag{8.16}$$

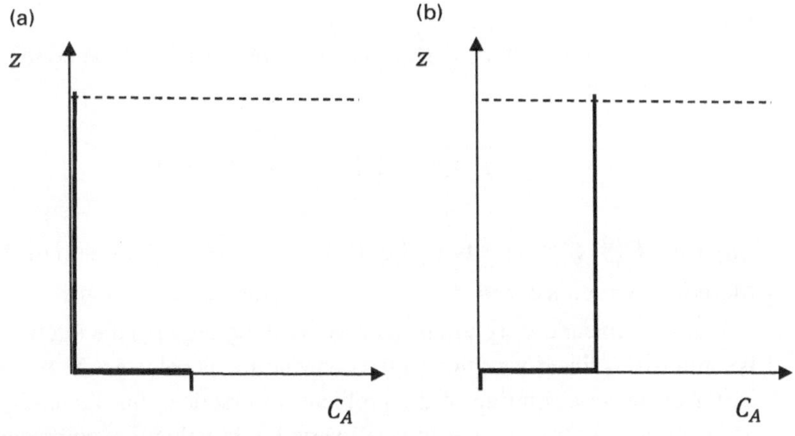

Figure 8.3 Possible steady-state profiles: linear behavior and $m < \overline{K}_A$ are assumed. (a) The fed lumped solid is solute-free. (b) The fed fluid is solute-free.

When the lumped solid velocity $\bar{u}$ equals zero, m reaches infinity and the standard behavior of a fixed bed is obtained from Eq. (8.16): the outlet fluid solute concentration C_A^{out} equals the inlet fluid solute concentration C_A^{inl}, when a steady state is reached.

- **Case b:** $\sigma < 0$

  According to Eq. (8.14), a negative front velocity imposes a ratio $m < \overline{K}_A$, thus a maximum fluid velocity given by $u < \bar{u}\overline{K}_A$. Possible steady-state profiles are given in Figure 8.3, depending on the boundary conditions. Regardless of the inlet conditions, the front moves downwards until it reaches the bottom of the system; no discontinuity occurs at the top of the column. When the system is fed with a solute-free lumped solid (Figure 8.3a), the solute concentration is constant everywhere except at the lower boundary of the system, where it increases sharply. When the system is fed with a solute-free fluid (Figure 8.3b), the solute concentration is constant everywhere except at the lower boundary of the system, where it decreases sharply. Profile uniformity and the absence of discontinuity at the top of the system allow one to write:

$$\overline{C}_A^{top} = \overline{C}_A^{bot} = \overline{C}_A^{inl} \qquad C_A^{top} = C_A^{bot} = \overline{C}_A^{inl}/\overline{K}_A \tag{8.17}$$

  Moreover, the absence of a discontinuity at the top of the system ensures that the fluid concentration C_A^{out} equals C_A^{top}.

  The last unknown is the lumped solid concentration $\overline{C}_A^{out}$ exiting the system. This is readily obtained with Eq. (8.4), and the fact that $C_A^{out} = C_A^{top} = \overline{C}_A^{inl}/\overline{K}_A$:

$$\overline{C}_A^{out} = \overline{C}_A^{inl} + m\left(C_A^{inl} - \overline{C}_A^{inl}/\overline{K}_A\right) \tag{8.18}$$

  When the fluid velocity u equals zero, m equals zero, and Eq. (8.18) confirms that the outlet lumped solid solute concentration $\overline{C}_A^{out}$ equals the inlet lumped solid solute concentration $\overline{C}_A^{inl}$, when a steady state is reached.

- **Case c:** $\sigma = 0$
  This situation refers to $m = \overline{K}_A$, thus $u = \bar{u}\overline{K}_A$. Under these conditions, Equation (8.4) gives:

$$\overline{K}_A C_A^{inl} + \overline{C_A^{inl}} = \overline{K}_A C_A^{out} + \overline{C_A^{out}} \tag{8.19}$$

  Any pair $(C_A^{out}, \overline{C_A^{out}})$ satisfying Eq. (8.19) is, *a priori*, a solution of the problem. Moreover, we notice with Eq. (8.7) that, under the condition $u = \bar{u}\frac{d\overline{C}_A}{dC_A}$ (thus $m = \overline{K}_A$ for linear adsorption isotherms), nothing requires the internal profiles to be uniform. A linear variation of the concentrations between the two boundaries is, for instance, a solution of the problem. In practice, this situation is unstable and cannot be observed: a very small variation in velocity would orient towards $\sigma < 0$ or $\sigma > 0$.

EXERCISE 8.1 *A small injection of solute A is performed into a 1 l column having an extragranular porosity of 0.4 and eluted at a 0.4 l/min flow rate. The measured retention time is 3 min. One wants to completely remove a solute A contained in a 10 l/min fluid stream using a single-zone counter-current system. Assuming linear adsorption, estimate the minimum solid flow rate required to achieve this task.*

We can now address the case of non-linear systems, which is more complex owing to the dependence of the front velocity on concentration, as implied by Eq. (8.13). Depending on the concentration, the front can move upwards or downwards, or it can be stabilized. While the velocity analysis is derived from transient mass balances, when a steady state is reached, the shape of the internal concentration profile is governed only by the inlet conditions and by the ratio m.

We have shown for fixed beds (see Section 3.3.1.1) that the fronts are compressive or dispersive as a consequence of the variation of the propagation velocity with concentration. In a fixed bed, for a given adsorption isotherm the fronts are compressive or dispersive according to the initial and final compositions: when the front velocity increases from initial to feed composition, the front is compressive; when the front velocity decreases from initial to feed composition, the front is dispersive. As a result, for favorable adsorption isotherms the saturation fronts are compressive while the regeneration fronts are dispersive. The same type of behavior is obtained for counter-current systems, the initial and final compositions being replaced by the boundary conditions. By analogy with fixed beds, we will speak of a *saturation* step when the solid captures some solutes due to contact with the fluid phase, and of a *regeneration* step when the solid releases some solutes due to contact with the fluid phase. Figure 8.4 illustrates situations leading to dispersive or compressive fronts for saturation and regeneration steps performed on a single-zone counter-current system when propagation velocities are positive.

The rules for identifying compressive and dispersive behaviors obtained for a fixed bed remain valid if one replaces the initial and final compositions by the boundary conditions. In the case of adsorption isotherms characterized by a one-signed curvature

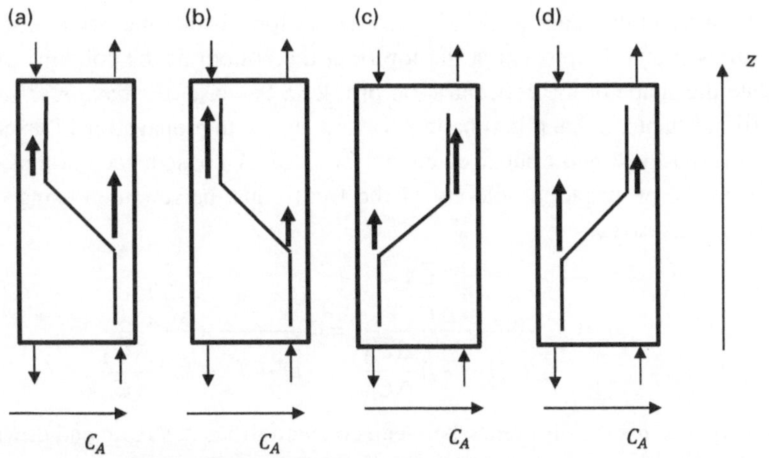

Figure 8.4 Dispersive and compressive character of fronts in a moving bed. The lengths of the heavy arrows inside the system are proportional to front velocity. (a) and (b) Saturation of the solid. (c) and (d) Regeneration of a presaturated solid.

(i.e. $\bar{f}''_A(C_A) \geq 0$ or $\bar{f}''_A(C_A) \leq 0$) one obtains:

$$\begin{array}{l}\text{If front velocity decreases from bottom to top, the front is compressive}\\ \text{If front velocity increases from bottom to top, the front is dispersive}\end{array} \tag{8.20}$$

Note that compressive behavior is illustrated by Figures 8.4a and d, and that dispersive behavior is illustrated by Figures 8.4b and c. The same conclusion would be drawn for negative propagation velocities. We can suggest a rule that is handier than Eq. (8.20) by defining $C_A^{\overline{inl}*}$, the solute concentration in the fluid phase which is in equilibrium with the concentration $\overline{C}_A^{\overline{inl}}$:

$$\overline{C}_A^{\overline{inl}} = \bar{f}_A\left(C_A^{\overline{inl}*}\right) \tag{8.21}$$

Concentration $C_A^{\overline{inl}*}$ is a reference concentration which is introduced for convenience. Equation (8.20) can be rewritten:

$$\begin{array}{ll}\text{If } \sigma\left(C_A^{\overline{inl}*}\right) > \sigma\left(C_A^{inl}\right) & \text{the front is dispersive}\\ \text{If } \sigma\left(C_A^{\overline{inl}*}\right) < \sigma\left(C_A^{inl}\right) & \text{the front is compressive}\end{array} \tag{8.22}$$

For favorable adsorption isotherms, the previous conditions become:

$$\begin{array}{ll}C_A^{inl} > C_A^{\overline{inl}*} & \text{saturation step; compressive behavior}\\ C_A^{inl} < C_A^{\overline{inl}*} & \text{regeneration step; dispersive behavior}\end{array} \tag{8.23}$$

while for unfavorable adsorption isotherms they become:

$$\begin{array}{ll}C_A^{inl} > C_A^{\overline{inl}*} & \text{saturation step; dispersive behavior}\\ C_A^{inl} < C_A^{\overline{inl}*} & \text{regeneration step; compressive behavior}\end{array} \tag{8.24}$$

Keeping in mind that concentrations must be uniform inside the column, and that discontinuities should be present at the top or at the bottom of the column, we will now elucidate the shape of the concentration profile in the case of a *compressive* front. We know Chapter 3 that this situation is associated with propagation of discontinuities, so continuous local mass balances cannot be used. A consequence is that Eq. (8.13), expressing the propagation velocity of the front, must be rewritten using differences instead of derivatives:

$$\sigma = \frac{u - \bar{u}\dfrac{\Delta \bar{C}_A}{\Delta C_A}}{\varepsilon_e + (1-\varepsilon_e)\dfrac{\Delta \bar{C}_A}{\Delta C_A}} = \bar{u}\frac{m - \dfrac{\Delta \bar{C}_A}{\Delta C_A}}{\varepsilon_e + (1-\varepsilon_e)\dfrac{\Delta \bar{C}_A}{\Delta C_A}} \tag{8.25}$$

where Δ represents the difference between concentrations upstream and downstream of the front, i.e. $C_A^{\overline{inl}^*}$ and C_A^{inl}.

Notice that the profile associated with a compressive front is in fact associated with a *double discontinuity*: one at the boundary resulting from the counter-current operation in an infinitely efficient system, and one due to the existence of the compressive front. These discontinuities are superimposed.

The possible steady-state internal profiles are the same as those given in Figure 8.2: one obtains a discontinuity either at the top or at the bottom of the system depending on the sign of σ in Eq. (8.25). If $\sigma > 0$, the solute concentration in the fluid equals C_A^{inl} everywhere except at the top boundary, where it equals C_A^{out}; if $\sigma < 0$, the lumped solid concentration equals $\bar{C}_A^{\overline{inl}}$ (so that the solute concentration in the fluid equals $C_A^{\overline{inl}^*}$) everywhere in the system except at the bottom boundary, where it equals $\bar{C}_A^{\overline{out}}$. The concentrations C_A^{out} and/or $\bar{C}_A^{\overline{out}}$ are calculated with the overall mass balance given by Eq. (8.4).

Determination of the concentration profiles associated with *dispersive* fronts is more subtle: due to the continuous variation of the propagation velocity with concentration according to Eq. (8.13), the propagation velocity can change sign with concentration, some concentrations moving up, others moving down.

Let us consider a concentration C_A^{stab} whose propagation velocity is zero. According to Eq. (8.13), this concentration must satisfy:

$$m = \left.\frac{d\bar{C}_A}{dC_A}\right|_{C_A^{stab}} \tag{8.26}$$

For a given value of m, this concentration may exist or not, may be located between C_A^{inl} and $C_A^{\overline{inl}^*}$ or not. In the case of adsorption isotherms characterized by a one-signed curvature, the derivative $\frac{d\bar{C}_A}{dC_A}$ varies continuously between $\left.\frac{d\bar{C}_A}{dC_A}\right|_{min}$ and $\left.\frac{d\bar{C}_A}{dC_A}\right|_{max}$. The possible solutions of the problem are:

- m is located outside the interval $\left[\left.\frac{d\bar{C}_A}{dC_A}\right|_{min}, \left.\frac{d\bar{C}_A}{dC_A}\right|_{max}\right]$: there is no concentration C_A^{stab} satisfying Eq. (8.26). All concentrations move in the same direction, given by the sign of the propagation velocity.

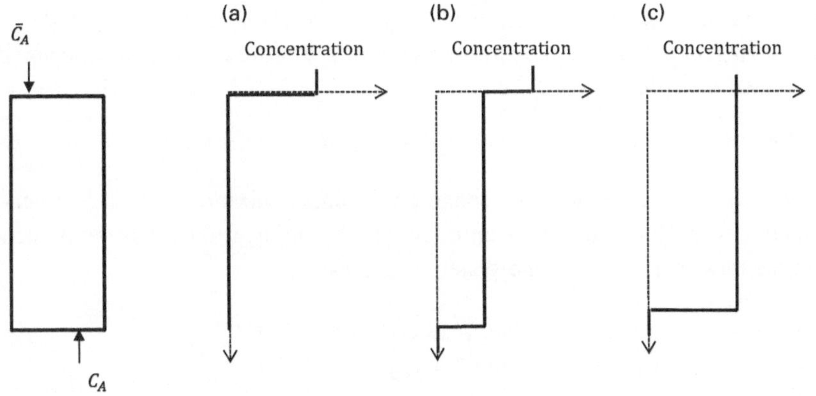

Figure 8.5 Influence of the m ratio on the steady-state internal concentration profile in a counter-current system operated in a regeneration mode. Favorable adsorption isotherm. The fed solid is solute-free. (a) $m > \left.\frac{d\bar{C}_A}{dC_A}\right|_{max}$; (b) $\left.\frac{d\bar{C}_A}{dC_A}\right|_{min} < m < \left.\frac{d\bar{C}_A}{dC_A}\right|_{max}$; (c) $m < \left.\frac{d\bar{C}_A}{dC_A}\right|_{min}$.

- m is located inside the interval $\left[\left.\frac{d\bar{C}_A}{dC_A}\right|_{min}, \left.\frac{d\bar{C}_A}{dC_A}\right|_{max}\right]$: there is one concentration C_A^{stab} satisfying Eq. (8.26). The system is filled with a fluid at concentration C_A^{stab} except at the boundaries. In the case of favorable adsorption isotherms, concentrations greater than C_A^{stab} move upwards and concentrations lower than C_A^{stab} move downwards. The opposite holds for unfavorable adsorption isotherms.

Possible internal profiles are given in Figure 8.5 for a system operated in a regeneration mode with a favorable adsorption isotherm. At small m ratio all concentrations move downwards, whereas all concentrations move upwards for large m. For intermediate m, a concentration C_A^{stab} exists which prevails in the system. In the case of m large or small enough to push all concentrations upwards or downwards, the unknown concentrations C_A^{out} and/or $\overline{C_A^{out}}$ are calculated using the global mass balance given by Eq. (8.4). The methodology is the same as the one used for linear and/or compressive fronts.

The case of an m ratio allowing some concentrations to move upwards and others downwards involves two discontinuities, one at the top and one at the bottom of the system. Noting that, under this circumstance, the fluid concentration in the system is uniform and identical to C_A^{stab}, one can determine the unknown concentrations C_A^{out} and/or $\overline{C_A^{out}}$ by writing a mass balance for the lower part of the system:

$$mC_A^{inl} + \overline{C}_A^{stab} = mC_A^{stab} + \overline{C_A^{out}} \tag{8.27}$$

and a mass balance for the upper part of the system:

$$mC_A^{stab} + \overline{C_A^{inl}} = mC_A^{out} + \overline{C}_A^{stab} \tag{8.28}$$

EXERCISE 8.2 *A solute A adsorbs according to a Langmuir lumped adsorption isotherm with parameters $\overline{N} = 10$ g/l, $\tilde{K}_A = 0.1$ l/g. Fluid and solid are in contact counter-currently in a single-zone system according to two different modes:*

- *Saturation: the fed solid is solute-free. The fluid contains 10 g/l of solute.*
- *Regeneration: the fed fluid is solute-free. The solute lumped concentration is 6 g/l on the solid.*

Determine the concentration profiles according to the value of the m ratio.

We will now extend our presentation to binary mixtures, which is a very important practical case for continuous chromatography applications. Assuming steady state, a simple mass balance over the system provides:

$$\begin{aligned} mC_A^{inl} + \overline{C}_A^{inl} &= mC_A^{out} + \overline{C}_A^{out} \\ mC_B^{inl} + \overline{C}_B^{inl} &= mC_B^{out} + \overline{C}_B^{out} \end{aligned} \tag{8.29}$$

The local steady-state mass balances are given by:

$$\begin{aligned} m\frac{dC_A}{dz} - \frac{d\overline{C}_A}{dz} &= 0 \\ m\frac{dC_B}{dz} - \frac{d\overline{C}_B}{dz} &= 0 \end{aligned} \tag{8.30}$$

Assuming local equilibrium between the fluid and the lumped solid phases, Eq. (8.30) becomes:

$$\begin{pmatrix} m - \dfrac{\partial \overline{C}_A}{\partial C_A} & \dfrac{\partial \overline{C}_A}{\partial C_B} \\ \dfrac{\partial \overline{C}_B}{\partial C_A} & m - \dfrac{\partial \overline{C}_B}{\partial C_B} \end{pmatrix} \begin{pmatrix} \dfrac{dC_A}{dz} \\ \dfrac{dC_B}{dz} \end{pmatrix} = \begin{pmatrix} 0 \\ 0 \end{pmatrix} \tag{8.31}$$

Equation (8.31) shows that the problem accepts two type of solutions:

- Trivial solutions given by $\dfrac{dC_A}{dz} = \dfrac{dC_B}{dz} = 0$, associated with plateau compositions
- Non-trivial solutions for which $\dfrac{dC_A}{dz} \neq 0$ and/or $\dfrac{dC_B}{dz} \neq 0$, associated with concentrations satisfying:

$$det \begin{bmatrix} m - \dfrac{\partial \overline{C}_A}{\partial C_A} & \dfrac{\partial \overline{C}_A}{\partial C_B} \\ \dfrac{\partial \overline{C}_B}{\partial C_A} & m - \dfrac{\partial \overline{C}_B}{\partial C_B} \end{bmatrix} = 0 \tag{8.32}$$

or, equivalently:

$$m^2 - \left(\frac{\partial \overline{C}_A}{\partial C_A} + \frac{\partial \overline{C}_B}{\partial C_B} \right) m + \frac{\partial \overline{C}_A}{\partial C_A}\frac{\partial \overline{C}_B}{\partial C_B} - \frac{\partial \overline{C}_A}{\partial C_B}\frac{\partial \overline{C}_B}{\partial C_A} = 0 \tag{8.33}$$

Equation (8.33) generalizes Eq. (8.7) obtained for the single-solute situation. As with single-component systems, trivial solutions, being associated with null spatial derivatives, describe uniform plateau compositions inside the system. Non-trivial solutions allow *possible* axial variations of concentrations provided that the concentrations belonging to the profile satisfy Eq. (8.33).[4]

In addition to the two internal concentration profiles $C_A(z)$ and $C_B(z)$, we have four unknowns C_A^{out}, C_B^{out}, $\overline{C}_A^{out}$ and $\overline{C}_B^{out}$. From our knowledge of single-solute and linear systems, we know that boundary discontinuities are likely to occur. This is addressed below.

We know that the steady-state regime of the system is associated with the presence of uniform concentration plateaus and boundary discontinuities (trivial solution), possibly with non-uniform internal concentration profiles (non-trivial solution), but we are not yet able to calculate these plateaus or profiles. As with the single-solute situation, solving the steady-state problem is greatly facilitated by investigating the transient behavior of the system.

A transient mass balance can be written for each species:

$$\begin{aligned} u\frac{\partial C_A}{\partial z} - \bar{u}\frac{\partial \overline{C}_A}{\partial z} + \left[\varepsilon_e\frac{\partial C_A}{\partial t} + (1-\varepsilon_e)\frac{\partial \overline{C}_A}{\partial t}\right] &= 0 \\ u\frac{\partial C_B}{\partial z} - \bar{u}\frac{\partial \overline{C}_B}{\partial z} + \left[\varepsilon_e\frac{\partial C_B}{\partial t} + (1-\varepsilon_e)\frac{\partial \overline{C}_B}{\partial t}\right] &= 0 \end{aligned} \tag{8.34}$$

Following the approach of Chapter 3 for binary systems, assuming equilibrium between the fluid and solid phases, it can be shown that these mass-balance equations depend on a single parameter $\sigma = z/t$, and that they can be rewritten:

$$\begin{pmatrix} u - \bar{u}\frac{\partial \overline{C}_A}{\partial C_A} - \sigma\left(\varepsilon_e + (1-\varepsilon_e)\frac{\partial \overline{C}_A}{\partial C_A}\right) & \bar{u}\frac{\partial \overline{C}_A}{\partial C_B} - \sigma(1-\varepsilon_e)\frac{\partial \overline{C}_A}{\partial C_B} \\ \bar{u}\frac{\partial \overline{C}_B}{\partial C_A} - \sigma(1-\varepsilon_e)\frac{\partial \overline{C}_B}{\partial C_A} & u - \bar{u}\frac{\partial \overline{C}_B}{\partial C_B} - \sigma\left(\varepsilon_e + (1-\varepsilon_e)\frac{\partial \overline{C}_B}{\partial C_B}\right) \end{pmatrix}\begin{pmatrix} \frac{dC_A}{d\sigma} \\ \frac{dC_B}{d\sigma} \end{pmatrix} = \begin{pmatrix} 0 \\ 0 \end{pmatrix} \tag{8.35}$$

In addition to plateaus associated with the trivial solutions $\frac{dC_A}{d\sigma} = \frac{dC_B}{d\sigma} = 0$, the above system shows that possible front propagation velocities are given by the roots of the characteristic equation:

$$det\begin{bmatrix} u - \bar{u}\frac{\partial \overline{C}_A}{\partial C_A} - \sigma\left(\varepsilon_e + (1-\varepsilon_e)\frac{\partial \overline{C}_A}{\partial C_A}\right) & \bar{u}\frac{\partial \overline{C}_A}{\partial C_B} - \sigma(1-\varepsilon_e)\frac{\partial \overline{C}_A}{\partial C_B} \\ \bar{u}\frac{\partial \overline{C}_B}{\partial C_A} - \sigma(1-\varepsilon_e)\frac{\partial \overline{C}_B}{\partial C_A} & u - \bar{u}\frac{\partial \overline{C}_B}{\partial C_B} - \sigma\left(\varepsilon_e + (1-\varepsilon_e)\frac{\partial \overline{C}_B}{\partial C_B}\right) \end{bmatrix} = 0 \tag{8.36}$$

Again, as in Chapter 3, one obtains:

$$\sigma_{\pm}(C_A,C_B) = \frac{u - \bar{u}\theta_{\pm}}{\varepsilon_e + (1-\varepsilon_e)\,\theta_{\pm}} = \bar{u}\frac{m - \theta_{\pm}}{\varepsilon_e + (1-\varepsilon_e)\,\theta_{\pm}} \tag{8.37}$$

[4] Generalizing the single-solute result, we will show later that the non-trivial situation is in fact associated with a composition stabilizing a front velocity, not with a continuous variation of the profiles.

where $\theta_{\pm}$ are the roots of the characteristic equation:[5]

$$\theta^2 - \left(\frac{\partial \bar{C}_A}{\partial C_A} + \frac{\partial \bar{C}_B}{\partial C_B}\right)\theta + \frac{\partial \bar{C}_A}{\partial C_A}\frac{\partial \bar{C}_B}{\partial C_B} - \frac{\partial \bar{C}_A}{\partial C_B}\frac{\partial \bar{C}_B}{\partial C_A} = 0 \tag{8.38}$$

Recall that, per our convention in Chapter 3, $\theta_+ < \theta_-$, so that $\sigma_- < \sigma_+$. Note that Eq. (8.38) is identical to Eq. (8.33) if θ is replaced by m.

We also know that, should the front be compressive, the expression for the velocities given by Eq. (8.37) must be replaced by a finite-difference expression:

$$\sigma_{\pm}(C_A,C_B) = \frac{u - \bar{u}\frac{\Delta \bar{C}}{\Delta C}}{\varepsilon_e + (1-\varepsilon_e)\frac{\Delta \bar{C}}{\Delta C}} = \bar{u}\frac{m - \frac{\Delta \bar{C}}{\Delta C}}{\varepsilon_e + (1-\varepsilon_e)\frac{\Delta \bar{C}}{\Delta C}} \tag{8.39}$$

where $\frac{\Delta \bar{C}}{\Delta C}$ represents the ratio of differences between concentrations upstream and downstream, the front being determined for either solute A or solute B (both options give the same result due to the coherence principle).

Provided that the adsorption isotherms are known, Eq. (8.37) and Eq. (8.38) allow an estimate the possible propagation velocities of a given composition (C_A,C_B). According to the values of θ and the ratio between fluid and solid velocities, fronts can move upwards or downwards or be stabilized, as with single-solute systems.

It is remarkable that Eq. (8.38) is the same as Eq. (3.44), derived in Chapter 3 for fixed beds. A direct consequence is that Eq. (3.48) defining the *hodograph plane* is still valid. We have thus shown that the concentrations (C_A,C_B) follow the same *characteristic paths* in a fixed bed and in a moving bed: thus the methodology derived in Chapter 3 for binary systems in fixed beds can be used for counter-current systems with little adaptation.

The first step consists in drawing the characteristic paths associated with the proper boundary conditions (see Figure 8.6).

The general methodology for determining the concentration profiles for adsorption isotherms characterized by a one-signed curvature is then as follows:

1. From composition (C_A^{inl},C_B^{inl}) take the *slow* path given by σ_-,[6] until one finds the intersection with the fast path given by σ_+ through the fluid composition $(C_A^{\overline{inl}^*},C_B^{\overline{inl}^*})$ in equilibrium with the inlet solid.
2. The intersection between the two above characteristics represents a composition plateau (*P*).
3. The response of the system to the inlet conditions results in two fronts, (*inl*,*P*) and ($P,\overline{inl}^*$).
4. When moving along a path in the direction of decreasing velocities, the corresponding front is a simple wave and otherwise a shock. For Langmuir

[5] The roots θ are often presented as so-called global derivatives (see the discussion of coherence in Section 3.4.1): $\theta = \frac{D\bar{C}_A}{DC_A} = \frac{\partial \bar{C}_A}{\partial C_A} + \frac{\partial \bar{C}_A}{\partial C_B}\frac{dC_B}{dC_A}\left(= \frac{D\bar{C}_B}{DC_B} = \frac{\partial \bar{C}_B}{\partial C_B} + \frac{\partial \bar{C}_B}{\partial C_A}\frac{dC_A}{dC_B}\right)$.

[6] The slow path must be taken because the bottom of the system is associated with low (even possibly negative) velocities.

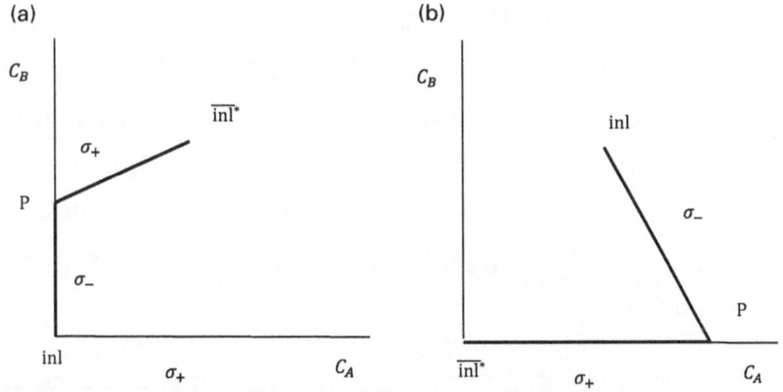

Figure 8.6 Schematic of the characteristic paths in a single-zone counter-current system. Langmuir adsorption isotherms. (a) Regeneration. (b) Saturation.

adsorption isotherms, the same straight paths are followed by self-sharpening and self-broadening fronts. For non-Langmuir adsorption isotherms, the path associated with self-broadening fronts must be determined according to the presentation in Section 3.4.1.

The problem is then solved by considering the sign of the propagation velocities according to the values of m (Migliorini *et al.*, 2000):

- Dispersive fronts:
  - Important composition propagation velocities are given by $\sigma_-(inl)$, $\sigma_-(P)$, $\sigma_+(P)$ and $\sigma_+(\overline{inl}^*)$, which are calculated using Eq. (8.37) with $\theta_+(inl)$, $\theta_+(P)$, $\theta_-(P)$ and $\theta_-(\overline{inl}^*)$, respectively.
  - Lower range: $m < \theta_+(inl)$. All compositions move downwards. The lumped solid concentration in the system is given by composition $\overline{inl}$; the fluid composition is uniform and given by $\overline{inl}^*$.
  - First transition range: $\theta_+(inl) < m < \theta_+(P)$. The system is filled with a composition located along characteristic path (inl,P). The exact composition is determined by solving Eq. (8.33).
  - Intermediate range: $\theta_+(P) < m < \theta_-(P)$. The system is filled with composition P.
  - Second transition range: $\theta_-(P) < m < \theta_-(\overline{inl}^*)$. The system is filled with a composition located along characteristic path $(P,\overline{inl}^*)$. The exact composition is determined by solving Eq. (8.33).
  - Upper range: $m > \theta_-(\overline{inl}^*)$. All compositions move upwards. The system is filled with composition *inl*.
- Compressive fronts:
  - The important propagation velocities are $\sigma_-(inl/P)$ and $\sigma_-(P/\overline{inl}^*)$, which represent the velocities of front (inl/P) and front $(P/\overline{inl}^*)$. They are

calculated using Eq. (8.39), where the finite-difference ratios calculated at the proper discontinuities are called $\left.\frac{\Delta\bar{C}}{\Delta C}\right|_{inl/P}$ and $\left.\frac{\Delta\bar{C}}{\Delta C}\right|_{P/\overline{inl}}$.

- If $m < \left.\frac{\Delta\bar{C}}{\Delta C}\right|_{inl/P}$, the lumped solid concentration in the system is given by composition $\overline{inl}$; the fluid composition is uniform and given by $\overline{inl}^*$.
- If $\left.\frac{\Delta\bar{C}}{\Delta C}\right|_{inl/P} < m < \left.\frac{\Delta\bar{C}}{\Delta C}\right|_{P/\overline{inl}^*}$, the system is filled with composition P.
- If $m > \left.\frac{\Delta\bar{C}}{\Delta C}\right|_{P/\overline{inl}^*}$, the fluid composition is uniform and given by composition *inl*.
- If the propagation velocity of a compressive front equals zero, the column is occupied partly by one of the states separated by the front and partly by the other. This front is unstable: any variation would push it upwards or downwards.

The sign of the velocity gives the location of the boundary discontinuities (at the bottom or at the top of the system). Knowing the location of the discontinuities and internal plateau compositions, the concentrations C_A^{out}, C_B^{out} and $\overline{C_A^{out}}, \overline{C_B^{out}}$ are determined by writing a mass balance either on the entire system or on its lower or upper part. This methodology is illustrated in Exercise 8.3.

EXERCISE 8.3 *Using the adsorption isotherms in Exercise 3.8, draw the internal profiles occurring in a counter-current system characterized by* $u = \bar{u} = 1$ *m/h for the following situations:*

- *Regeneration: the column is fed with a fluid of composition* α *and a solid in equilibrium with fluid composition* β.
- *Saturation: the column is fed with a fluid of composition* β *and a solid in equilibrium with fluid composition* α.

Composition α*:* $C_A = C_B = 0$*; composition* β*:* $C_A = C_B = 1$ *g/l.*

Readers willing to learn more on the use of the equilibrium model for single-zone counter-current systems and binary mixtures are referred to (Migliorini *et al.*, 2000), who used the above methodology for adsorption isotherms described by a bi-Langmuir equation or by the IAS theory, or to (Mazzotti, 2006a,b), who extended the methodology to generalized Langmuir adsorption isotherms, allowing one to represent a combination of favorable and unfavorable adsorption isotherms.

As with fixed-bed systems, the theory can be extended to multi-component mixtures only for Langmuirian adsorption isotherms. This work has been performed, thanks to the ω transform, by (Rhee *et al.*, 1971) and further refined and extended by (Storti *et al.*, 1989), (Storti, Mazzotti, Morbidelli and Carra, 1993), (Mazzotti *et al.*, 1994) and (Mazzotti *et al.*, 1996).

At this point in the presentation, one may be of the opinion that the steady counter-current problem is unambiguously solved for virtually all situations. This is only partly true. It appeared in the above presentation that using steady-state mass balances was not enough to fully characterize the behavior of the counter-current system, so we had to use transient mass balances. Using the transient equilibrium model for solving the steady-state problem allows us to find *one* solution, and not *the* solution, of the problem. To illustrate the matter, (Sundaresan *et al.*, 1987) used a PF–$\overline{\text{LDF}}$ model which tends towards the equilibrium model when all characteristic times for mass transfer tend towards zero. The authors showed that, in the case of binary-solute systems characterized by Langmuir adsorption isotherms, the asymptotic steady state reached when mass transfer limitations become negligible depends on the ratio between the characteristic times for mass transfer. The consequence for true moving beds is that the equilibrium model does not give an asymptotic steady state towards which all other models should tend. I am not totally sure whether the consequences are purely academic. In any case, from a practical point of view, we will only use the equilibrium model to determine m parameters allowing us to perform a given separation; precise internal concentration profiles will then be determined with more precise models like the MC and PD models.

8.3 Single-zone counter-current systems: the MC and PD models

In order to obtain a more realistic view of the behavior of single-zone systems, one can use the MC–$\overline{\text{Equil}}$ or the PD–$\overline{\text{Equil}}$ model, used for modeling fixed beds, with relatively little adaptation. Let us start with the MC–$\overline{\text{Equil}}$ model and assume that the system is equivalent to N theoretical stages, as shown in Figure 8.7.

For each cell j, the transitory mass balance is written, for each species i:

$$\bar{Q}\bar{C}_i^{j+1} + QC_i^{j-1} = \bar{Q}\bar{C}_i^j + QC_i^j + \varepsilon_e \frac{V_{col}}{N}\frac{dC_i^j}{dt} + (1-\varepsilon_e)\frac{V_{col}}{N}\frac{d\bar{C}_i^j}{dt} \tag{8.40}$$

$$j \in [1,N] \quad i \in [1,n]$$

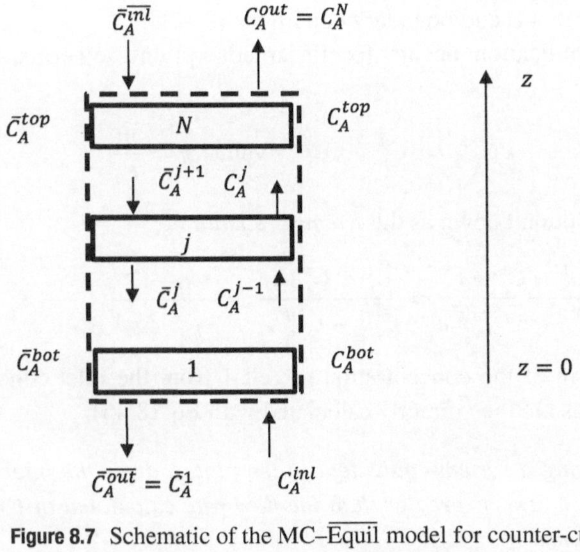

Figure 8.7 Schematic of the MC–$\overline{\text{Equil}}$ model for counter-current systems.

or, in a more normalized form:

$$\frac{1}{m}\overline{C}_i^{j+1} + C_i^{j-1} = \frac{1}{m}\overline{C}_i^j + C_i^j + \frac{t_0}{N}\frac{dC_i^j}{dt} + \frac{1-\varepsilon_e}{\varepsilon_e}\frac{t_0}{N}\frac{d\overline{C}_i^j}{dt} \tag{8.41}$$

By convention the boundary conditions are written:

$$\overline{C}_i^{N+1} = \overline{C}_i^{\overline{inl}} \quad \text{and} \quad C_i^0 = C_i^{inl} \tag{8.42}$$

The concentrations with superscript *top* or *bot* are the concentrations inside the system at its boundaries, defined by:

$$\overline{C}_i^N = \overline{C}_i^{top} \quad \text{and} \quad C_i^1 = C_i^{bot}$$

If the concentrations in the fluid and lumped solid phases are in equilibrium, following the methodology for fixed beds, the time derivatives of the lumped solid-phase concentrations can be related to the time derivatives of the fluid-phase concentration, so the transient behavior of the system can be simulated by solving $n \times N$ ordinary differential equations involving derivatives $\frac{dC_i^k}{dt}$.

The steady-state situation, which in general is the one of practical interest, is described by the following mass balances:

$$\overline{Q}\overline{C}_i^{j+1} + QC_i^{j-1} = \overline{Q}\overline{C}_i^j + QC_i^j \quad \text{with} \quad \overline{C}_i^j = \overline{f}_i\left(C_1^j, \ldots, C_n^j\right)$$
$$j \in [1,N] \quad i \in [1,n] \tag{8.43}$$

or, equivalently:

$$\overline{C}_i^{j+1} + mC_i^{j-1} = \overline{C}_i^j + mC_i^j \quad \text{with} \quad \overline{C}_i^j = \overline{f}_i\left(C_1^j, \ldots, C_n^j\right) \tag{8.44}$$

Steady-state internal profiles are consequently obtained by solving $n \times N$ algebraic equations given by Eq. (8.44) and boundary conditions (8.42).

An interesting simplification occurs for linear adsorption isotherms, for which Eq. (8.44) becomes:

$$C_i^{j+1} + \kappa C_i^{j-1} = C_i^j + \kappa C_i^j \quad \text{with} \quad \kappa_i = \frac{m}{\overline{K}_i} \tag{8.45}$$

System (8.45) has a solution known as the *Kremser equation*:

$$\frac{\overline{C}_i^1 - C_i^{inl}\overline{K}_i}{\overline{C}_i^{N+1} - C_i^{inl}\overline{K}_i} = \frac{\overline{C}_i^1 - C_i^{inl}\overline{K}_i}{\overline{C}_i^{\overline{inl}} - C_i^{inl}\overline{K}_i} = \frac{\kappa_i - 1}{\kappa_i^{N+1} - 1} \tag{8.46}$$

which allows calculation of the concentration in cell 1 from the inlet concentrations. The other concentrations are then directly calculated with Eq. (8.45).

EXERCISE 8.4 *Assuming a system equivalent to ten plates, draw the internal profile for a solid flow rate 10% less or greater than the flow rate calculated in Exercise 8.1. Linear behavior is assumed.*

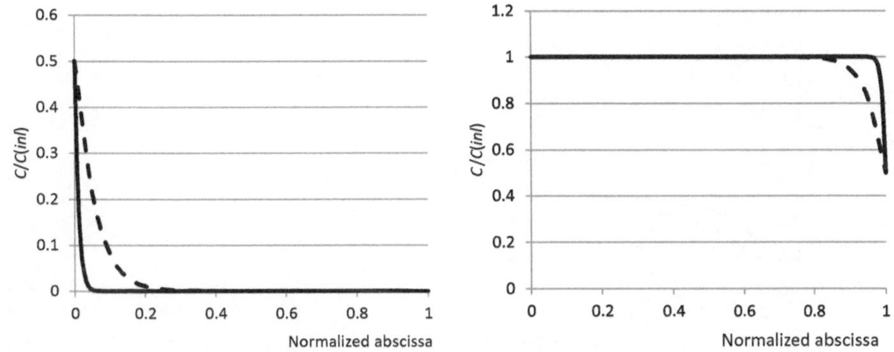

$m = 1/3$ $\qquad$ $m = 1.$

Figure 8.8 Internal concentration profiles (fluid phase) in a single-zone counter-current system simulated with the MC–$\overline{\text{Equil}}$ model: adsorption step, single-solute system. $C_A^{inl} = 10$ g/l, $\overline{C}_A^{inl} = 0$. Adsorption isotherm given in Exercise 8.2. Solid line: $N = 100$; dashed line: $N = 20$. Normalized abscissa: 0 at the bottom, 1 at the top of the zone.

When non-linear adsorption is considered, numerical simulation is required for solving Eq. (8.44).[7] Some typical behaviors are illustrated below, first for the single-solute system described in Exercise 8.2. Recall that the solute is adsorbed according to a Langmuir lumped adsorption isotherm described by parameters $\overline{\lambda}_A = 0$, $\overline{N} = 10$ g/l, $\tilde{K}_A = 0.1$.

The simulations associated with the single-solute saturation step are presented in Figure 8.8. In agreement with the results of Exercise 8.2, when the m ratio equals $1/3$ the front is pushed downwards and thus stabilized at the bottom of the system, whereas it is pushed upwards and stabilized at the top of the system when the m ratio equals 1. When the equivalent number of plates increases, the profiles simulated with the MC–$\overline{\text{Equil}}$ model become closer to those predicted by the equilibrium model, i.e. a uniform zero concentration for $m = 1/3$ and a uniform normalized concentration of unity for $m = 1$ associated with boundary discontinuities.

The simulations associated with the single-solute regeneration step are presented in Figure 8.9. In agreement with the results of Exercise 8.2, when $m = 2$ the front is pushed upwards and stabilized at the top of the system. When $m = 1/2$, a concentration of about 4 g/l is stabilized inside the system, this concentration being the one associated with a zero propagation velocity. When the equivalent number of plates increases, the profiles simulated with the MC–$\overline{\text{Equil}}$ model become closer to those predicted by the equilibrium model, i.e. a uniform internal concentration profile associated with boundary discontinuities.

[7] Newton–Raphson algorithms with numerous unknowns are often non-convergent. A robust alternative consists in simulating the transient behavior, by solving the transitory mass-balance equation (8.40) from a given initial state until the profiles are stabilized. An accurate integration method is not necessary (low-order scheme, large integration step allowed) because the steady state is the very same for the exact transient mass balance and for the discretized equations.

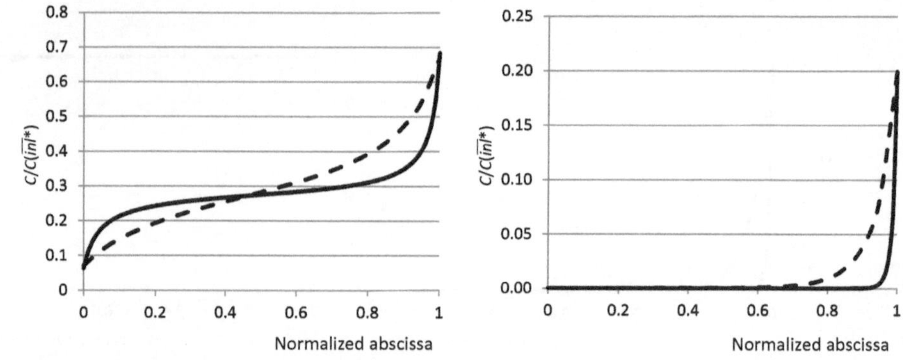

$m = 1/2$ $m = 2$

Figure 8.9 Internal concentration profiles (fluid phase) in a single-zone counter-current system simulated with the MC–$\overline{\text{Equil}}$ model: regeneration step, single-solute system. $C_A^{inl} = 0$, $\overline{C}_A^{\overline{inl}} = 6$ g/l (thus $C_A^{\overline{inl}*} = 15$ g/l). Adsorption isotherm given in Exercise 8.2. Solid line: $N = 100$; dashed line: $N = 20$. Normalized abscissa: 0 at the bottom, 1 at the top of the zone.

The profiles given in Figures 8.8 and 8.9 reveal a remarkable property of counter-current chromatographic systems: while the internal profiles are sensitive to the number of plates of the system, the outlet concentrations (given at normalized abscissa z of 0 or 1) are almost independent of the system efficiency. The reader can check that the outlet concentration obtained with a system equivalent to 20 plates is almost identical to those estimated in Exercise 8.2. This relative insensitivity of counter-current systems with respect to the number of plates largely explains, in my opinion, the possible superiority of SMB compared to elution chromatography.

I have to admit that the underlying reasons for the relative insensitivity of counter-current systems to the number of plates, when compared to elution chromatography, are not totally clear to me. I have simply learned to accept this behavior as a fact. For a tentative explanation, and for readers willing to learn more, the most convincing contribution that I am aware of is that of (Valentin, 1981), based on the concept of "contact time" between the solutes and the adsorbent. If one accepts that contact time is the key driver for efficiency, then the superiority of counter-current systems follows.

We now consider the situation of binary-solute saturation and regeneration analyzed in Exercise 8.3. Recall that the solutes are adsorbed according to Langmuir lumped adsorption isotherms described by the parameters $\overline{\lambda}_A = \overline{\lambda}_B = 0$, $\tilde{K}_A = 0.5$ l/g, $\tilde{K}_B = 1$ l/g and $\overline{N} = 2$ g/l.

As with the single-solute situation, the reader can easily check that the simulations in Figure 8.10 are in good agreement with the results of Exercise 8.3. For $m = 1$, we have shown that the regeneration step must lead to a plateau at composition P_2, whereas the saturation step must lead to a plateau concentration identical to the inlet stream composition. Note the presence of an overshoot on the less retained solute for the saturation step, the thickness of the overshoot decreasing when the number of plates increases, to reach zero for an infinitely efficient column. This is related to the double discontinuity mentioned in Section 8.2.

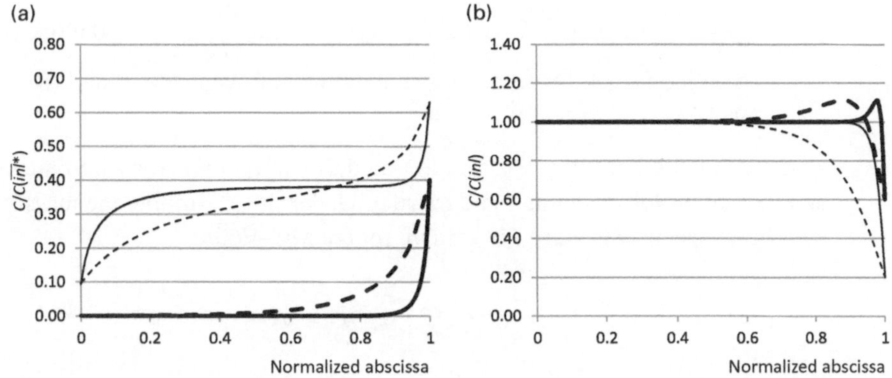

Figure 8.10 Internal concentration profiles (fluid phase) in a single-zone counter-current system simulated with the MC–$\overline{\text{Equil}}$ model: binary-solute system. See detailed conditions in Exercise 8.3; $m = 1$. Solid line: $N = 100$; dashed line: $N = 20$. Thick lines: solute A; thin lines: solute B. Normalized abscissa: 0 at the bottom, 1 at the top of the zone. (a) Regeneration. (b) Saturation.

Finally, confirming the result obtained for a single-solute system, Figure 8.10 and Exercise 8.3 show that the influence of the plate number on the outlet concentrations is relatively weak.

Let us now consider the possibility of using the PD–$\overline{\text{Equil}}$ model instead of the MC–$\overline{\text{Equil}}$ model to describe the counter-current system. If one assumes a plug flow of the solid, the transitory mass-balance equation (8.40) must be replaced by:

$$Q\frac{\partial C_i}{\partial z} - \overline{Q}\frac{\partial \overline{C}_i}{\partial z} + \Omega_{col}\left[\varepsilon_e\frac{\partial C_i}{\partial t} + (1-\varepsilon_e)\frac{\partial \overline{C}_i}{\partial t}\right] - \varepsilon_e\Omega_{col}D_{ax}^{app}\frac{\partial^2 C_i}{\partial z^2} = 0 \qquad (8.47)$$

where D_{ax}^{app} is an apparent axial dispersion coefficient, accounting for the effects of hydrodynamic dispersion and mass transfer limitations (see Section 7.2).

Equation (8.47) can be written in a more normalized form:

$$\frac{\partial C_i}{\partial \xi} - \frac{1}{m}\frac{\partial \overline{C}_i}{\partial \xi} + t_0\left[\frac{\partial C_i}{\partial t} + \frac{1-\varepsilon_e}{\varepsilon_e}\frac{\partial \overline{C}_i}{\partial t}\right] - \frac{1}{Pe^{app}}\frac{\partial^2 C_i}{\partial \xi^2} = 0 \qquad (8.48)$$

where Pe^{app} is an apparent Péclet number, accounting for the effect of hydrodynamic dispersion and mass transfer limitations (see Section 7.2). From Eq. (8.48), the steady-state mass balance for species i is readily obtained:

$$\frac{\partial C_i}{\partial \xi} - \frac{1}{m}\frac{\partial \overline{C}_i}{\partial \xi} - \frac{1}{Pe^{app}}\frac{\partial^2 C_i}{\partial \xi^2} = 0 \qquad (8.49)$$

with:

- $\overline{C}_i(\xi = 1) = \overline{C_i^{inl}}$ and $C_i(\xi = 0) = C_i^{inl}$ if one assumes simplified boundary conditions

- $\frac{1}{Pe^{app}}\frac{dC_A}{d\xi}\bigg|_{z=0}+\left(C_A^{inl}-C_A(\xi=1)\right)=0$ and $\frac{\partial C_A}{\partial\xi}\bigg|_{\xi=1}=0$ if one assumes the so-called Danckwerts boundary conditions (cf. Appendix B and (Ruthven, 1983, 1984)).

As with fixed beds, the MC and PD models can be refined by introducing an $\overline{\text{LDF}}$ approximation for modeling mass transfer. Under this condition, the mass steady-state balance of species i in stage j is written, for the MC–$\overline{\text{Equil}}$:

$$\begin{aligned}\bar{Q}\bar{C}_i^{j+1}&=\bar{Q}\bar{C}_i^j-\Delta N_i^j\\ QC_i^{j-1}&=QC_i^j+\Delta N_i^j\end{aligned}\tag{8.50}$$

where ΔN_i^j, representing the flux (mol/s or g/s) of species i transferred between the fluid phase and the solid phase in stage j of volume V_{cell}, is given by:

$$\begin{aligned}\Delta N_i^j&=(1-\varepsilon_e)\frac{V_{cell}}{t_i^e}\left(C_i^j-C_i^{j*}\right)\quad\text{for external mass transfer limitations}\\ \Delta N_i^j&=(1-\varepsilon_e)\frac{V_{cell}}{t_i^i}\left(\bar{C}_i^{j*}-\bar{C}_i^j\right)\quad\text{for internal mass transfer limitations}\end{aligned}\tag{8.51}$$

EXERCISE 8.5 *Using the theoretical framework presented in Chapter 5, justify the expressions in (8.51) for external limitations.*

In principle, as shown for fixed beds, the MC–$\overline{\text{LDF}}$ model becomes very similar to the MC–$\overline{\text{Equil}}$ model provided that the number of plates N is properly related to the number of mixing cells J and characteristic times for mass transfer (see Section 5.8), and that kinetic limitations are not too extreme (see Chapters 5 and 7). (Ruthven, 1983, 1984) has, however, shown that the equations relating the number of plates, N to the characteristic times for mass transfer must be adapted for true moving beds.

Nothing guarantees that the solid moves according to a perfect plug flow, so a key assumption of the model is highly questionable. Danckwerts boundary conditions in the fluid phase or detailed relationships between equivalent plate number and hydrodynamic dispersion and mass transfer coefficients probably is an undue refinement for a system (TMB) which will be used simply as a simple idealization of a real system (SMB).

I certainly recommend modeling TMB systems with the simple MC–$\overline{\text{Equil}}$ model and then using the number of plates N to obtain the required purity and yield to size the SMB systems. This is illustrated in Chapter 9. A more precise approach, detailing fluid- and solid-phase dispersion, complex boundary conditions and mass transfer, should probably be considered, should one be interested in implementing a real TMB system. I don't have enough experience of real TMB systems to address this subject.

8.4 Two-zone counter-current systems: the equilibrium model

The single-zone system allows the transfer of solutes from one moving phase to another, but cannot, in general, perform a complete separation of two solutes contained in a fluid phase. That is why these systems are normally used as elementary bricks, allowing us to build more complex processes. While the most classical implementation may be the four-zone system presented in Section 8.5, we will start with the two-zone system which is at the heart of the separation.

Let us consider the two-zone system schematically represented in Figure 8.11. A feed stream (Q_F) containing both solute A (less retained) and solute B (more retained) is injected between two single-zone counter-current systems so that it is mixed with the fluid coming out of the bottom zone. The system is also fed by a solute-free solid stream and a solute-free carrier-fluid stream. We will see in the next section that a simple way to deliver these solute-free streams is to add one zone at the bottom and one zone at the top of the system; that is why the two zones of the system described in Figure 8.11 are called zone 2 and zone 3.

This device, made of one saturation zone (zone 3, fluid flow rate Q_3) and one regeneration zone (zone 2 , fluid flow rate Q_2), connected by a unique solid flow rate $\bar{Q}$, is possibly the simplest device that allows us to perform a continuous binary separation. This system allows us to get pure solute A at the top of zone 3 and pure solute B at the bottom of zone 2 (adsorbed on the solid), provided that the following conditions are met:

- Zone 2: In order to recover pure B at the bottom, solute A must move upwards (and be directed towards zone 3) and solute B must move downwards.
- Zone 3: In order to recover pure solute A at the top, solute B must move downwards (and be directed towards zone 2) and solute A must move upwards.

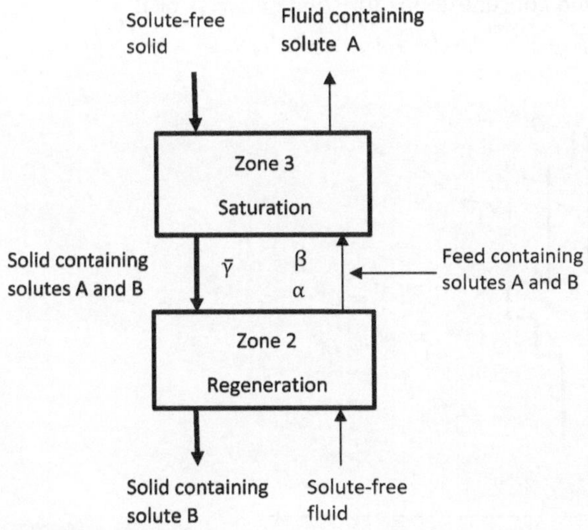

Figure 8.11 Principle of a two-zone TMB system: solute A less retained than solute B.

We know from the development proposed in Section 8.2 that the directions followed by the solutes depend on the ratio between the fluid and solid flow rates. This suggests defining:

$$m_k = \frac{Q_k}{\overline{Q}} = \frac{u_k}{\overline{u}} = \frac{1-\varepsilon_e}{\varepsilon_e}\frac{u_{ek}}{\overline{u}_e} \tag{8.52}$$

the ratio of the fluid to the solid flow rate in zone k.

The fluid flow rates in the two zones of the system are related by a simple mass balance:

$$Q_2 + Q_F = Q_3 \tag{8.53}$$

Prior to determining the set of flow rates allowing us to perform the separation, let us define more precisely the different concentrations in this two-zone system. We showed in Section 8.2 that, according to the equilibrium model, the internal concentration profiles in each single zone are uniform, regardless of the nature of the fronts, and that boundary discontinuities can occur at the inlet or at the outlet of the zones. Consequently, for each zone k, each species i is associated with one concentration entering the zone, one concentration inside the zone and one concentration exiting the zone. To simplify the notation, as shown in Figures 8.11 and 8.12, the fluid compositions exiting zone 2 and entering zone 3 are called, respectively, α and β. Additionally, the solid composition exiting zone 3 and entering zone 2 is denoted $\overline{\gamma}$. We also define γ, the fluid composition in equilibrium with the solid composition $\overline{\gamma}$, by $C_i^{\gamma} = C_i^{z3/\overline{out}*}$.

Noting that concentrations $C_i^{z2/inl}$ and $\overline{C}_i^{z3/\overline{inl}}$ are zero by definition of the two-zone system, and that the solid composition entering zone 2 is identical to the solid composition exiting zone 3, for a given set of m_k values the concentrations to be determined are:

- Five unknown fluid concentrations for solute i ($i =$ A or B): $C_i^{z2/inl} = 0$, C_i^{z2}, $C_i^{z2/out} (= C_i^{\alpha})$, $C_i^{z3/in} (= C_i^{\beta})$, C_i^{z3}, $C_i^{z3/out}$

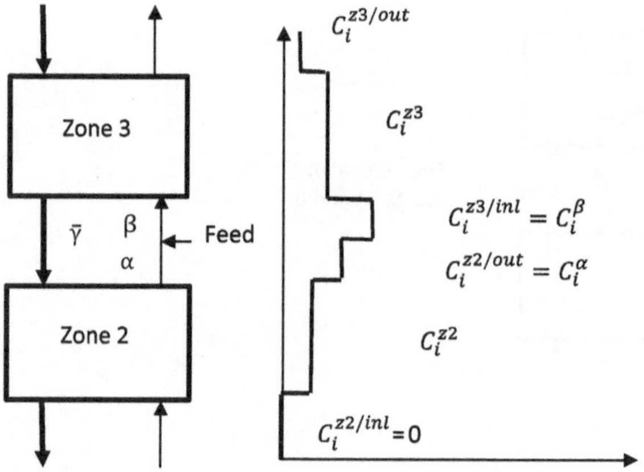

Figure 8.12 Schematic representation of the different concentrations in a two-zone system.

- Four solid concentrations for solute i (i = A or B): $\overline{C}_i^{z2/\overline{inl}}(=\overline{C}_i^{\gamma})$, $\overline{C}_i^{z2}$, $\overline{C}_i^{z2/\overline{out}}$, $\overline{C}_i^{z3/\overline{inl}}=0$, $\overline{C}_i^{z3}$, $\overline{C}_i^{z3/\overline{out}}(=\overline{C}_i^{\gamma})$

thus a total of 18 unknown concentrations.

A schematic representation of the different concentrations in the system is given in Figure 8.12. The fluid compositions α and β are related by mass balances written at the node between zone 2, zone 3 and the feed injection:

$$Q_2C_i^{z2/out}+Q_FC_i^F=Q_3C_i^{z3/inl}\quad\text{thus}\quad m_2C_i^{\alpha}+(m_3-m_2)C_i^F=m_3C_i^{\beta}\tag{8.54}$$

The different concentrations characterizing the system are related by mass balances written at the upper and lower parts of the two zones:

- Lower part of zone 2:

$$Q_2C_i^{z2}+\overline{Q}\overline{C}_i^{z2/\overline{out}}=Q_2\times 0+\overline{Q}\overline{C}_i^{z2}\quad\text{thus}\quad m_2C_i^{z2}+\overline{C}_i^{z2/\overline{out}}=\overline{C}_i^{z2}\tag{8.55}$$

- Upper part of zone 2:

$$Q_2C_i^{\alpha}+\overline{Q}\overline{C}_i^{z2}=Q_2C_i^{z2}+\overline{Q}\overline{C}_i^{\gamma}\quad\text{thus}\quad m_2C_i^{\alpha}+\overline{C}_i^{z2}=m_2C_i^{z2}+\overline{C}_i^{\gamma}\tag{8.56}$$

- Lower part of zone 3:

$$Q_3C_i^{z3}+\overline{Q}\overline{C}_i^{\gamma}=Q_3C_i^{\beta}+\overline{Q}\overline{C}_i^{z3}\quad\text{thus}\quad m_3C_i^{z3}+\overline{C}_i^{\gamma}=m_3C_i^{\beta}+\overline{C}_i^{z3}\tag{8.57}$$

- Upper part of zone 3:

$$Q_3C_i^{z3/out}+\overline{Q}\overline{C}_i^{z3}=Q_3C_i^{z3}+\overline{Q}\times 0\quad\text{thus}\quad m_3C_i^{z3/out}+\overline{C}_i^{z3}=m_3C_i^{z3}\tag{8.58}$$

Notice that the overall mass balance is a simple combination of mass balances (8.54)–(8.58):

$$Q_FC_i^F=Q_3C_i^{z3/out}+\overline{Q}\overline{C}_i^{z2/\overline{out}}\quad\text{thus}\quad(m_3-m_2)C_i^F=m_3C_i^{z3/out}+\overline{C}_i^{z2/\overline{out}}\tag{8.59}$$

Notice further that $\overline{C}_i^{z2}$ is in equilibrium with C_i^{z2} and $\overline{C}_i^{z3}$ is in equilibrium with C_i^{z3}.

We have a total of 18 unknown concentrations. We have four equilibrium relations (one per zone and per solute), two mass balances at the feed node (8.54), two mass balances at the lower part of zone 2 (8.55), two mass balances at the upper part of zone 2 (8.56), two mass balances at the lower part of zone 3 (8.57) and two mass balances at the upper part of zone 3 (8.58), thus a total of 14 equations. As with single-zone systems, the four missing relations will be provided by the equilibrium model.

We know from Section 8.2 that the different concentrations present in a given zone must follow specific paths in the hodograph space, as illustrated in Figure 8.13. Notice the presence of points α, β and γ introduced previously, as well as point SF, referring to a solute-free composition.

Let us start by discussing the behavior of zone 2. This zone, fed with a solute-free fluid and a solid in equilibrium with fluid at composition γ (to be determined later), works in a regeneration mode. The internal profiles in this zone are thus dispersive for favorable adsorption isotherms. We define composition P_2 as the composition at the

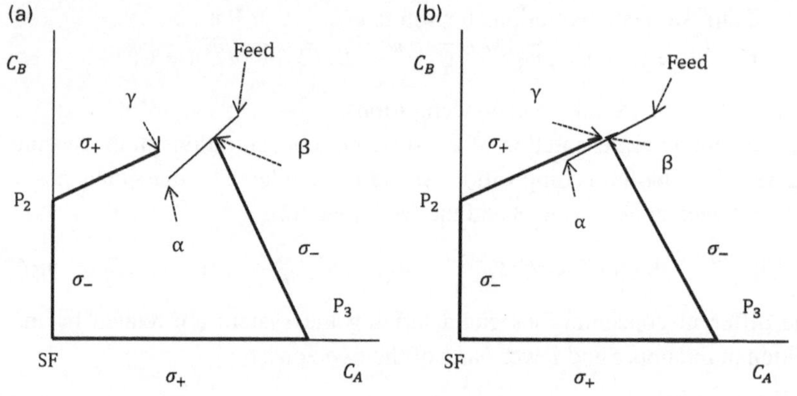

Figure 8.13 Representation of the two-zone true moving bed counter-current chromatography in the hodograph space; species A less retained than B. Favorable Langmuir adsorption isotherms. (a) Generic situation. (b) Compositions γ and β are identical.

intersection between the slow (−) characteristic path emanating from *SF* and the fast (+) characteristic path emanating from γ. We know from Section 8.2 that, according to the m_2 value, the composition inside the zone can be SF, γ or an intermediate value on one of the characteristic paths joining P_2 to SF or P_2 to γ.

Let us now consider the behavior of zone 3. This zone, fed with a solute-free solid (SF) and a fluid at composition β (to be determined later), works in a saturation mode. The internal profiles in this zone are thus compressive for favorable adsorption isotherms. We define composition P_3 as the composition at the intersection of the slow (−) characteristic path emanating from β and the fast (+) characteristic path emanating from SF. We know from Section 8.2 that, according to the m_3 value, the composition inside the zone can be SF, β or P_3.

With knowledge of the hodograph, and from the rules given in Section 8.2 for determining the single-zone compositions according to the m values, one can determine all the concentrations characterizing the system.

We now look at the (m_2,m_3) ratio allowing us to ensure a complete separation of the two solutes. (Migliorini *et al.*, 2000) have shown that a necessary and sufficient condition for the two solutes to be fully separated is for the two zones to operate at the above-mentioned P_2 and P_3 plateaus, so:

$$C_A^{z2} = 0 \quad C_B^{z2} = C_B^{P2} \quad \text{and} \quad C_A^{z3} = C_A^{P3} \quad C_B^{z3} = 0 \tag{8.60}$$

Important note: the plateau compositions P_2 and P_3 are defined from characteristic paths emanating from compositions β and γ, *not* from the feed composition.

The demonstration, a direct consequence of the front properties described in Section 8.2, is a bit too long to be repeated here. This result, as summarized in Eq. (8.60), is a fundamental property allowing a determination of the conditions for complete separation.

From Eq. (8.60), and from the rules given in Section 8.2 for determining the single-zone compositions according to the m values, we can set the constraints to be satisfied by the system to ensure complete separation of solutes A and B:

- For zone 2 (favorable adsorption isotherms):

$$m_2 \geq m_2^{min} = \left.\frac{\partial \overline{C}_A}{\partial C_A}\right|_{P_2}$$
$$m_2 \leq m_2^{max} = \left.\frac{\partial \overline{C}_B}{\partial C_B}\right|_{P_2} \tag{8.61}$$

- For zone 3 (favorable adsorption isotherms):

$$m_3 \geq m_3^{min} = \frac{\overline{C}_A^{P3}}{C_A^{P3}}$$
$$m_3 \leq m_3^{max} = \frac{\overline{C}_A^{\beta} - \overline{C}_A^{P3}}{C_A^{\beta} - C_A^{P3}} = \frac{\overline{C}_B^{\beta}}{C_B^{\beta}} \tag{8.62}$$

Under the assumption of complete separation, thus when conditions (8.60)–(8.62) are satisfied, the overall mass balance (8.59) reduces to:

$$(m_3 - m_2)C_A^F = m_3 C_A^{z3/out} + 0 \qquad (m_3 - m_2)C_B^F = 0 + \overline{C}_B^{z2/\overline{out}} \tag{8.63}$$

which gives immediate access to the compositions exiting the system, $C_A^{z3/out}$ and $\overline{C}_B^{z2/\overline{out}}$.

Determination of the ratios m_2 and m_3 is not straightforward in general, as equations Eq. (8.61) and Eq. (8.62) involve plateau compositions emanating from compositions β and γ that are unknown *a priori*. The α and β compositions are in general different, and coupled via the mass-balance equation (8.54). This mass balance means that points F, β and α are aligned in the hodograph space. This is schematically represented in Figure 8.13.

In the case of complete separation of solutes, the mass balances at the lower part of zone 3 (8.57) reduce to:

$$m_3 C_A^{\beta} + \overline{C}_A^{z3} = m_3 C_A^{z3} + \overline{C}_A^{\gamma}$$
$$m_3 C_B^{\beta} = \overline{C}_B^{\gamma} \tag{8.64}$$

which gives relations between γ and β compositions.

As a side comment, the mass-balance equation (8.64), together with Eq. (8.60) and Eq. (8.62), proves that β and γ are identical if $m_3 = m_3^{max}$. The situation $m_3 = m_3^{max}$ is shown in Figure 8.13b.

In general, compositions β and γ must be determined with an iterative procedure. For further illustration of the methodology, the reader is referred to (Migliorini *et al.*, 2000).

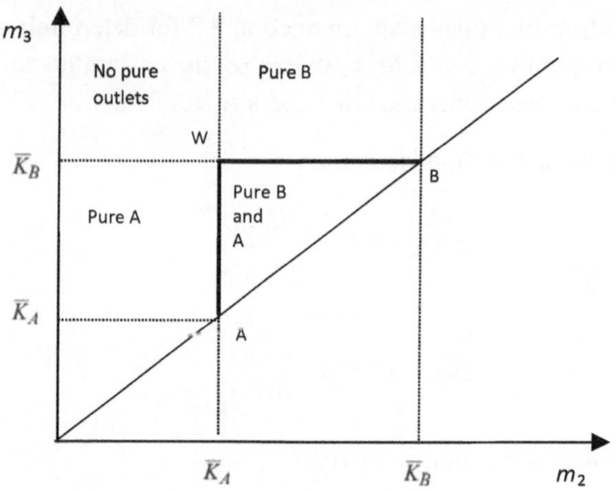

Figure 8.14 Regions of the (m_2,m_3) plane in connection with outlet purities. Linear adsorption isotherms (see Eqs. (8.65) and (8.66)).

As a simple illustration, consider the situation of linear adsorption isotherms associated with lumped Henry's coefficients $\overline{K}_A \leq \overline{K}_B$. Equations (8.61) and (8.62) reduce to:

$$\begin{aligned} m_2^{min} &= \overline{K}_A \\ m_2^{max} &= \overline{K}_B \end{aligned} \tag{8.65}$$

$$\begin{aligned} m_3^{min} &= \overline{K}_A \\ m_3^{max} &= \overline{K}_B \end{aligned} \tag{8.66}$$

The solution space can usefully be represented in a graphic form, as initially proposed by M. Morbidelli, M. Mazzotti and coworkers. From Figure 8.14, it is clear that the parameter region allowing us to get pure products is located within a triangle (A,B,W) in the case of linear adsorption isotherms.[8] Notice that the bisector $m_3 = m_2$ corresponds to a null feed flow rate and that the highest feed flow rate is obtained for point W. Operating at point W is certainly not robust because small variations in the flow rates would lead to impure products.

Recalling that the hodograph paths are orthogonal to the concentration axis for linear systems, one has $C_B^{\gamma} = C_B^{P2}$ and $C_B^{\beta} = C_B^{P3}$. From Eq. (8.60), we deduce that:

$$\begin{aligned} C_A^{z2} = 0 \quad &\text{and} \quad C_B^{z2} = C_B^{\gamma} \\ C_A^{z3} = C_A^{\beta} \quad &\text{and} \quad C_B^{z3} = 0 \end{aligned} \tag{8.67}$$

We can then determine all the unknown concentrations using the various mass balances presented earlier in this section. The problem is thus completely solved for linear

[8] Hence the terminology "triangle theory" often used in the literature. During a discussion in October 2013, Professor. M. Morbidelli expressed his surprise about the wide use of this terminology: "This is not a theory and unless the adsorption isotherms are linear, the result is not a triangle!"

adsorption isotherms: we know how to determine the (m_2,m_3) values that ensure complete separation, and, for a given set (m_2,m_3), we know how to determine the internal concentration profiles. This is demonstrated in Exercise 8.6.

EXERCISE 8.6 *Using the theoretical framework presented in this section, determine the main characteristics of the concentration profiles occurring in a two-zone system in which complete separation of two solutes A and B is realized. Linear adsorption isotherms are assumed.*

The two-zone problem can be solved analytically for the case of generalized Langmuir adsorption isotherms, defined in Section 3.2, by taking advantage of the properties of the ω transform, presented in Section 3.5.2. The case of Langmuir and generalized Langmuir adsorption isotherms has been addressed by Morbidelli and coworkers (Storti *et al.*, 1988, 1989; Storti, Mazzotti, Morbidelli and Carra, 1993; Mazzotti *et al.*, 1994, 1996; Mazzotti, 2006a,b), essentially for binary and ternary systems. A short summary for binary systems is given below.

The first step consists in solving the characteristic equation, derived from Eq. (3.69), for a binary mixture:

$$\left(1+p_A\tilde{K}_A C_A^F + p_B\tilde{K}_B C_B^F\right)\omega^2 - \left[\overline{K}_B\left(1+p_A\tilde{K}_A C_A^F\right)+\overline{K}_A\left(1+p_B\tilde{K}_B C_B^F\right)\right]\omega + \overline{K}_A\overline{K}_B = 0 \quad (8.68)$$

where $\overline{K}_i = \overline{N}_i\tilde{K}_i$ is the lumped Henry's coefficient of solute i and parameters p_i can take the values $+1$ or -1 (see Section 3.2).

It can be shown that Eq. (8.68) has two roots $0 < \omega_1^F < \omega_2^F$ satisfying the following inequalities:

- Case L: $0 < \omega_1^F \le \overline{K}_A \le \omega_2^F \le \overline{K}_B$
- Case A: $\overline{K}_A < \omega_1^F \le \overline{K}_B \le \omega_2^F < \infty$
- Case M_A: $0 < \omega_1^F \le \overline{K}_A < \overline{K}_B \le \omega_2^F < \infty$
- Case M_B: $\overline{K}_A \le \omega_1^F \le \omega_2^F \le \overline{K}_B$

All the information allowing us to derive the zone of complete separation is contained in ω_1^F and ω_2^F. As shown in Figure 8.15, the "triangle" defining the zone of complete separation in the (m_2,m_3) space is distorted due to non-linearity. Analytical expressions allowing us to position the different points in Figure 8.15 are given in Table 8.1. The results in Table 8.1 give the most general analytical solutions available for designing the two-zone TMB. While limited to generalized lumped Langmuir adsorption isotherms, they provide the interesting ability to propose solutions using a single frame for favorable and unfavorable adsorption isotherms. If the solutes follow linearly modified generalized lumped Langmuir adsorption isotherms with a linear term having the same coefficient $\overline{\lambda}$ for all species, one simply has to add $\overline{\lambda}$ to the m_k given in Table 8.1.

The equilibrium model has been extended and used to address various more or less complex situations. The associated references are presented in Section 8.5 on four-zone systems.

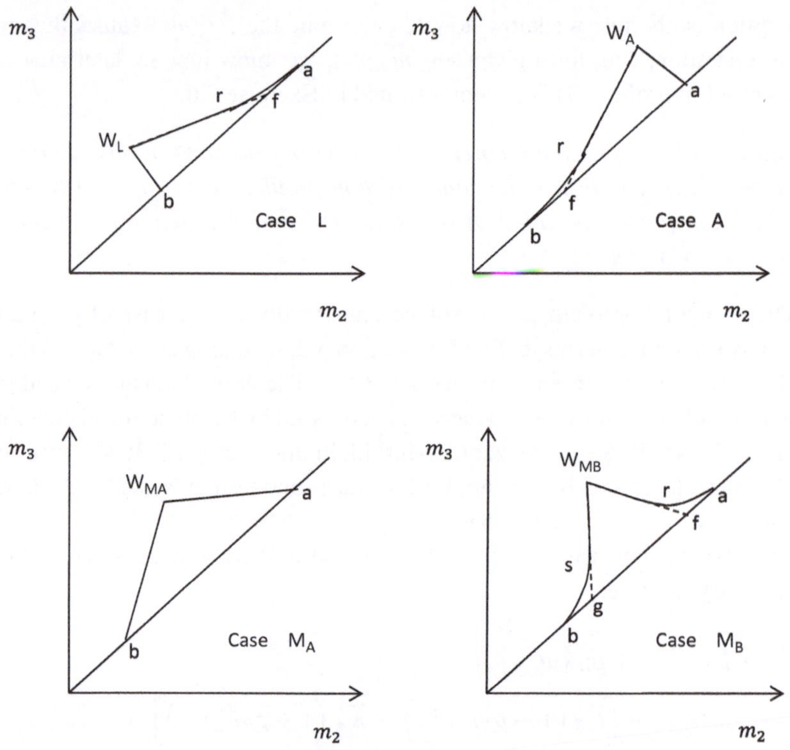

Figure 8.15 Regions of the (m_2,m_3) plane in connection with outlet purities. Generalized Langmuir lumped adsorption isotherms. Pure A and B are obtained within the "triangles".

8.5 Four-zone counter-current systems: the equilibrium model

The two-zone system is fed by a solute-free eluent stream and a solute-free solid stream. These can be delivered by adding one zone at the top and one zone at the bottom of the two-zone system, which leads to the classical four-zone TMB presented in Figure 8.16.

The four zones are denoted 1 to 4, from the bottom to the top. Between the zones, the feed and carrier fluids are injected at constant flow rates Q_F and Q_{El}, while extract and raffinate streams are collected at constant flow rates Q_{Ext} and Q_{Raf}. The solid flow rate $\overline{Q}$ is constant across the four zones, whereas the fluid flow rates take specific values Q_1, Q_2, Q_3 and Q_4 in the different zones. We will arbitrarily define the recycle fluid flow rate Q_R as the fluid flow rate taking place in zone 1.

The problem consists in determining the m values that allow us to get pure extract and raffinate streams, and in determining the internal concentration profiles for a given set of m values. The solution is a simple and straightforward extension of the solution of the two-zone problem.

The internal fluid flow rates are related to the inlet/outlet flow rates by mass balances:

$$\begin{aligned} Q_R &= Q_1 & Q_2 &= Q_1 - Q_{Ext} \\ Q_3 &= Q_2 + Q_F & Q_4 &= Q_3 - Q_{Raf} \end{aligned} \tag{8.69}$$

Table 8.1 Solution obtained using the equilibrium model to determine the (m_2, m_3) zone of complete separation, for generalized lumped Langmuir adsorption isotherms (Mazzotti, 2006a,b). $\omega_1^F < \omega_2^F$ are the roots of Eq. (8.68). The reference points are shown in Figure 8.15.

Point	m_2	m_3
a	$\bar{K}_B$	$\bar{K}_B$
b	$\bar{K}_A$	$\bar{K}_A$
f	ω_2^F	ω_2^F
g	ω_1^F	ω_1^F
r	$\dfrac{(\omega_2^F)^2}{\bar{K}_B}$	$\dfrac{\omega_2^F\left(\omega_2^F(\bar{K}_A-\omega_1^F)+\omega_1^F(\bar{K}_B-\bar{K}_A)\right)}{\bar{K}_A(\bar{K}_B-\omega_1^F)}$
s	$\dfrac{\omega_1^F\left(\omega_1^F(\omega_2^F-\bar{K}_B)+\omega_2^F(\bar{K}_B-\bar{K}_A)\right)}{\bar{K}_B(\omega_2^F-\bar{K}_A)}$	$\dfrac{(\omega_1^F)^2}{\bar{K}_A}$
W_0 (linear case)	$\bar{K}_A$	$\bar{K}_B$
W_L (case L)	$\dfrac{\omega_2^F\bar{K}_A}{\bar{K}_B}$	$\dfrac{\omega_2^F\left(\bar{K}_A(\bar{K}_A-\omega_1^F)+\omega_1^F(\bar{K}_B-\bar{K}_A)\right)}{\bar{K}_A(\bar{K}_B-\omega_1^F)}$
W_A (case A)	$\dfrac{\omega_1^F\left(\bar{K}_B(\omega_2^F-\bar{K}_B)+\omega_2^F(\bar{K}_B-\bar{K}_A)\right)}{\bar{K}_B(\omega_2^F-\bar{K}_B)}$	$\dfrac{\omega_1^F\bar{K}_B}{\bar{K}_A}$
W_{MA} (case M_A)	$\bar{K}_A\left(1+\dfrac{X}{Y}\right)$ $X=(\bar{K}_B-\omega_1^F)(\omega_2^F-\bar{K}_B)(\bar{K}_B-\bar{K}_A)$ $Y=\bar{K}_B(Y1+Y2)$ $Y_1=(\bar{K}_A-\omega_1^F)(\omega_2^F-\bar{K}_B)$ $Y_2=(\bar{K}_B-\omega_1^F)(\omega_2^F-\bar{K}_A)$	$\bar{K}_B\left(1+\dfrac{X}{Y}\right)$ $X=(\bar{K}_A-\omega_1^F)(\omega_2^F-\bar{K}_A)(\bar{K}_B-\bar{K}_A)$ $Y=\bar{K}_A(Y1+Y2)$ $Y_1=(\bar{K}_A-\omega_1^F)(\omega_2^F-\bar{K}_B)$ $Y_2=(\bar{K}_B-\omega_1^F)(\omega_2^F-\bar{K}_A)$
W_{MB} (case M_B)	$\dfrac{\omega_1^F\omega_2^F}{\bar{K}_B}=\dfrac{\bar{K}_A}{\delta_F}$ δ_F is the denominator of the lumped adsorption isotherm calculated at the feed composition	$\dfrac{\omega_1^F\omega_2^F}{\bar{K}_A}=\dfrac{\bar{K}_B}{\delta_F}$ δ_F is the denominator of the lumped adsorption isotherm calculated at the feed composition

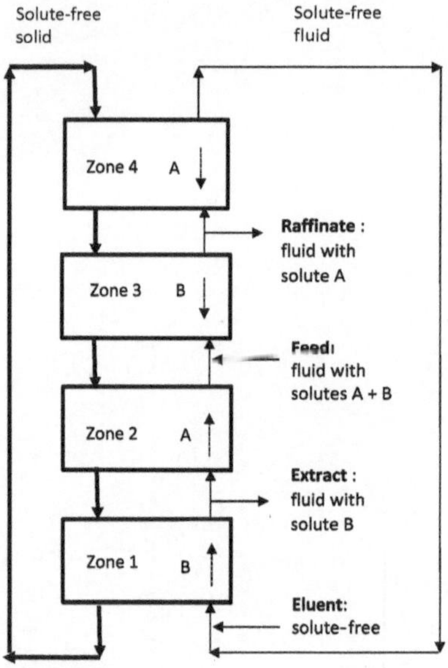

Figure 8.16 Principle of the four-zone TMB system. Solute A is less retained than solute B.

and the inlet/outlet flow rates are related by:

$$Q_F + Q_{El} = Q_{Raf} + Q_{Ext} \tag{8.70}$$

Let us assume that the feed contains two solutes A and B to be separated, B being more strongly retained than A. In order to recover A in the raffinate line and B in the extract line, various constraints must be fulfilled in the different zones. These constraints are qualitatively presented in Figure 8.16 and given by:

- Zone 1: Solutes A and B have to be pushed up in order to prevent a recycle of these solutes from zone 1 to zone 4 with the solid. As B is more strongly retained on the solid than A, this constraint is more difficult to fulfill for B. Consequently, a constraint is for the fluid flow rate in zone 1 to be sufficient to force the B front to go up.
- Zone 2: In order to recover pure B at the extract line, product A must be pushed up into zone 3. This constraint is the same as the one derived for the two-zone system.
- Zone 3: In order to recover pure A at the raffinate line, product B must be pushed down into zone 2. This constraint is the same as the one derived for the two-zone system.
- Zone 4: The liquid flow rate must be low enough to force products A and B to go down, so as to prevent a recycle of the solutes from zone 4 to zone 1 with the fluid stream. Note that if this constraint is satisfied for product A, it will also be satisfied for product B.

The system can be considered as being made of one regeneration zone of B (zone 1), one regeneration zone of A (zone 2), one saturation zone of B (zone 3) and one saturation zone of A (zone 4).

Assuming that zone 2 is fed by a solute-free fluid, and that zone 3 is fed by a solute-free solid, the constraints in zone 2 and 3 are exactly those presented in Section 8.4. Consequently, constraints on m_2 and m_3 to ensure complete separation of solutes A and B as expressed by Eq. (8.61) and Eq. (8.62) are still valid (favorable adsorption isotherms).

The constraints in zone 1 and 4 are simply aimed at delivering solute-free fluid and solid to the intermediate zones. These constraints are immediately written in terms of m constraints by noticing that zone 1, containing solute B only, is working in a regeneration mode (thus a dispersive front in the case of a favorable adsorption isotherm) and that zone 4, containing solute A only, is working in a saturation mode (thus a compressive front in the case of a favorable adsorption isotherm).

In the case of favorable adsorption isotherms, one obtains:

- Zone 1: solute B must move up at any composition:

$$m_1 \geq \left.\frac{\partial \bar{C}_B}{\partial C_B}\right|_{C_A=0, C_B=0}$$

- Zone 4: solute A is pushed down for the composition $(C_A^{z3},0)$ thus:

$$m_4 \leq \frac{\bar{C}_A^{z3} - 0}{C_A^{z3} - 0} = \frac{\bar{C}_A^{P3} - 0}{C_A^{P3} - 0} \qquad (8.71)$$

Equations (8.71), (8.61) and (8.62) allow the determination of the m_k values that allow complete solute separation. It is important to note that the (m_2,m_3) values are not influenced by the choice of (m_1,m_4). This very important result, which was not evident *a priori*, justifies focusing first on the set (m_2,m_3).

The concentration profiles are determined exactly as they were for the two-zone system. The profile associated with a complete separation of the two solutes is given in Figure 8.17.

For linear adsorption isotherms, Eqs. (8.71), (8.61) and (8.62) are simply written using the lumped Henry's coefficients. Consequently, one has:

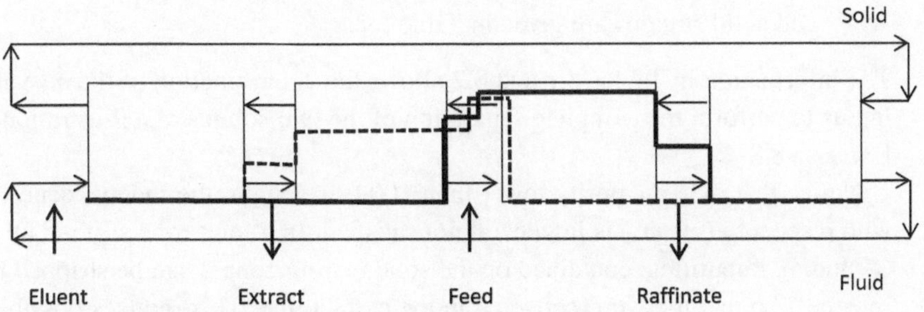

Figure 8.17 Internal profile in a four-zone system performing a binary separation: equilibrium model. Solid line: solute A; dashed line: solute B.

$$m_1 \geq \overline{K}_B \quad m_2 \geq \overline{K}_A \quad m_3 \leq \overline{K}_B \quad m_4 \leq \overline{K}_A \tag{8.72}$$

One way of satisfying these inequalities with some safety margin is to define a parameter $\vartheta \geq 1$ such that:

$$m_1 = \vartheta \overline{K}_B \quad m_2 = \vartheta \overline{K}_A \quad m_3 = \frac{\overline{K}_B}{\vartheta} \quad m_4 = \frac{\overline{K}_A}{\vartheta} \tag{8.73}$$

Note that there is a maximum value of ϑ ($\vartheta = \sqrt{\frac{\overline{K}_B}{\overline{K}_A}}$), for which the feed flow rate is zero ($m_2 = m_3$). If ϑ is set to 1, the inequalities of system (8.72) are just satisfied; the two solutes can be separated, but a significant efficiency (expressed by the number of plates, for instance) is likely to be required to get significant purities. If the value of ϑ is increased, the inequalities are satisfied with a higher safety margin, and a given purity is obtained with a more limited efficiency (number of plates, for instance), but at the expense of a lower feed flow rate relative to the other flow rates.

Taking into account the relationships between the flow rates given by Eq. (8.69), Eq. (8.73) can be solved for a given feed flow rate, and one obtains:

$$\begin{aligned}
\overline{Q} &= \frac{Q_F}{\overline{K}_B/\vartheta - \vartheta \overline{K}_A} \quad \text{with } 1 \leq \vartheta < \sqrt{\frac{\overline{K}_B}{\overline{K}_A}} \\
Q_1 &= \vartheta \overline{Q} \overline{K}_B \\
Q_{Ext} &= \vartheta \overline{Q} \left(\overline{K}_B - \overline{K}_A \right) \\
Q_{Raf} &= \overline{Q} \left(\overline{K}_B - \overline{K}_A \right) / \vartheta
\end{aligned} \tag{8.74}$$

EXERCISE 8.7 *Propose a design for a moving bed that performs the separation of two species adsorbing linearly with lumped Henry's coefficients $\overline{K}_A = 1$, $\overline{K}_B = 2$.*

The flow rates that ensure complete separation can be calculated analytically in the case of generalized Langmuir lumped adsorption isotherms:

- The (m_2,m_3) values can be calculated independently of zones 1 and 4 (assumption of solute-free feed streams), so the results in Table 8.1 can be used to design a four-zone system aimed at completely separating the solutes.
- The (m_1,m_4) values allowing us to feed the two-zone system with solute-free fluid and solid streams are given in Table 8.2.

The information in Tables 8.1 and 8.2 allows the determination of the m_k values, allowing us to perform the complete separation of the two solutes. An illustration is given in Exercise 8.8.

Notice that, when a purity lower than 100% is sought, the independence of (m_2,m_3) with respect to (m_1,m_4) is in general not satisfied. In a four-zone system, all the solutes (including impurities) contained on the solid exiting zone 2 can be stripped by the fluid in zone 1 so that they are recovered in the extract line. This induces possible pollution of the extract line, due to the regeneration of zone 1. This effect does not occur when

Table 8.2 Constraints to be satisfied by (m_1,m_4) for complete separation, as provided by the equilibrium model. Generalized lumped Langmuir adsorption isotherms (adapted from (Mazzotti, 2006a,b)). Expressions for (m_2,m_3) are given in Table 8.1.

Case	m_1	m_4
Linear case	$\overline{K}_B \le m_1$	$m_4 \le \overline{K}_A$
Case L	$\overline{K}_B \le m_1$	$m_4 \le \omega^F_{z4}$
Case A	$\omega^F_{z1} \le m_1$	$m_4 \le \overline{K}_A$
Case M_A	$\omega^F_{z1} \le m_1$	$m_4 \le \omega^F_{z4}$
Case M_B	$\overline{K}_B \le m_1$	$m_4 \le \overline{K}_A$

$$\omega^F_{z1} = \frac{B_1+\sqrt{B_1\times B_1-4C_1}}{2} \qquad \omega^F_{z4} = \frac{B_4-\sqrt{B_4\times B_4-4C_4}}{2}$$

$$B_1 = m_2 + \overline{K}_B - p_B\tilde{K}_B C^F_B(m_3 - m_2) \qquad B_4 = m_3 + \overline{K}_A + p_A\tilde{K}_A C^F_A(m_3 - m_2)$$

$$C_1 = m_2\overline{K}_B \qquad C_4 = m_3\overline{K}_A$$

100% purity is sought, as the solid entering zone 1 is free of solute A. The same applies between zone 3 and zone 4. The consequence is that using the two-zone system and thus the (m_2,m_3) representation for describing the four-zone system must be done with care when a purity lower than 100% is sought.

As mentioned for the two-zone system, if the compounds follow linearly modified lumped generalized Langmuir adsorption isotherms with a linear term having the same coefficient $\bar{\lambda}$ for all species, one has simply to add $\bar{\lambda}$ to the m_k given in Tables 8.1 and 8.2.

Finally, note that the introduction of a safety margin for non-linear systems is not trivial: as an illustration, taking a margin ϑ similar to the one for a linear system may lead to a point located *outside* the zone of complete separation. This will be further discussed in Chapter 9.

EXERCISE 8.8 *Using the adsorption isotherms* $\bar{\lambda}_A = 1$, $\bar{\lambda}_B = 1$, $\tilde{K}_A = 0.5$ *l/g,* $\tilde{K}_B = 1$ *l/g and* $\overline{N} = 2$ *g/l, estimate the flow rates of a TMB which allows the complete separation of solutes A and B with inlet concentrations* $C^F_A = C^F_B = 1$ *g/l. The feed flow rate is 10 l/h.*

The theoretical framework of the equilibrium model has been extended to adsorption isotherms following a bi-Langmuir representation (Migliorini *et al.*, 2000) and to those derived from IAS theory (Migliorini *et al.*, 2000); to obtaining partly purified fractions (Kaspereit *et al.*, 2007); and to multi-solute (n > 2) systems obeying Langmuir adsorption isotherms as shown by (Rhee *et al.*, 1971) and extended by (Storti *et al.*, 1989), (Storti, Mazzotti, Morbidelli and Carra, 1993), (Mazzotti *et al.*, 1994) and (Mazzotti *et al.*, 1996). The methodology does not, however, lead to explicit analytical solutions, some numerical work being required. Readers willing to investigate the *dynamics* of counter-current processes in depth are referred to (Hwang, 1987), (Hwang *et al.*, 1988), (Hwang and Helfferich, 1989) and (Hwang and Helfferich, 1990).

Finally, let us mention that TMB can be operated in a gradient mode, so that the adsorption isotherms vary in the different zones. This can lead to very significant improvements in performance, but often at the expense of significant technical difficulties. These situations have been investigated by (Abel *et al.*, 2002) and (Abel *et al.*, 2004) for liquid chromatography, and by (Di Giovanni *et al.*, 2001) for supercritical fluid chromatography.

We will see in Chapter 9 that the separation zone obtained in a system of infinite efficiency gives a very useful, not to say mandatory, starting point for optimizing industrial systems. When analytical solutions are not available, a simple method for determining the m_k zone leading to the desired separation is to simulate the system with the MC–$\overline{\text{Equil}}$ model.[9] Experience shows that using a few hundred plates per zone is sufficient to obtain results very close to those predicted by the equilibrium model.

8.6 Four-zone counter-current systems: the MC and PD models

In order to get more realistic profiles, let us model the entire four-zone TMB system with the MC–$\overline{\text{Equil}}$ model, assuming that the extract is recovered between stages j_{Ext} and $j_{Ext}+1$, that the feed is injected between stages j_F and j_F+1 and that the raffinate is recovered between stages j_{Raf} and $j_{Raf}+1$ (see Figure 8.18).

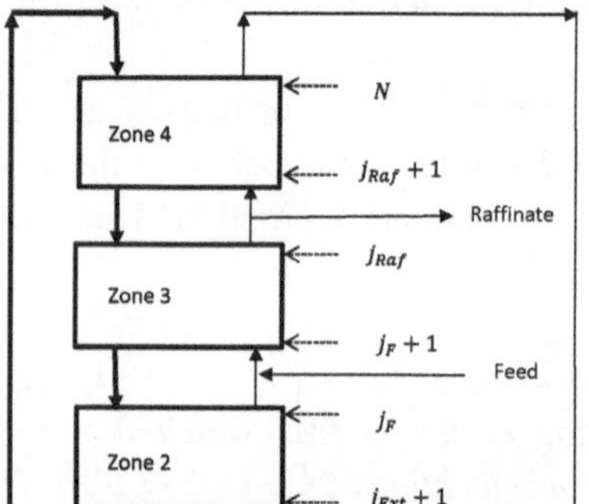

Figure 8.18 Representation of a four-zone TMB with the MC–$\overline{\text{Equil}}$ model.

[9] A good option is to work at fixed $m_3 - m_2$ values. Start with a very small $m_3 - m_2$ value, vary m_2 and check simulated outlet purities. The m_2 zone where products are pure gives the points a and b in Figure 8.15. Increase the $m_3 - m_2$ value until no m_2 permits the required purity/yield. This leads to point W.

The different zones are defined by:

$$\begin{aligned}
&\bullet \quad \text{Zone 1 } (1 \leq j \leq j_{Ext}) : \text{flow rate ratio } m_1 \\
&\bullet \quad \text{Zone 2 } (j_{Ext} + 1 \leq j \leq j_F) : \text{flow rate ratio } m_2 \\
&\bullet \quad \text{Zone 3 } (j_F + 1 \leq j \leq j_{Raf}) : \text{flow rate ratio } m_3 \\
&\bullet \quad \text{Zone 4 } (j_{Raf} + 1 \leq j \leq N) : \text{flow rate ratio } m_4.
\end{aligned} \tag{8.75}$$

For plate j located in zone k, the steady-state mass balances to be solved for species i are:

$$\begin{aligned}
&\bullet \quad \text{For } j \neq (1, j_F+1, N) : \overline{C}_i^{j+1} + m_k C_i^{j-1} = \overline{C}_i^j + m_k C_i^j \\
&\bullet \quad \text{For } j = 1 : \overline{C}_i^2 + m_1 C_i^{z1/inl} = \overline{C}_i^1 + m_1 C_i^1 \\
&\bullet \quad \text{For } j = j_F + 1 : \overline{C}_i^{j_F+2} + m_3 C_i^{z3/inl} = \overline{C}_i^{j_F+1} + m_3 C_i^{j_F+1} \\
&\bullet \quad \text{For } j = N : \overline{C}_i^{N+1} + m_4 C_i^{N-1} = \overline{C}_i^N + m_4 C_i^N \quad \text{with} \quad \overline{C}_i^{N+1} = \overline{C}_i^1
\end{aligned} \tag{8.76}$$

where $C_i^{z1/inl}$ and $C_i^{z3/inl}$ are the solute concentration in the fluid entering zones 1 and 3, respectively.

The flow rates in the different zones are related by Eq. (8.69), and the concentrations in the fluids feeding zones 1 and 3 are given by the following mass balances:

$$Q_1 C_i^{z1/inl} = Q_4 C_i^N + Q_{El} \times 0 \quad \text{thus} \quad m_1 C_i^{z1/inl} = m_1 C_i^0 = m_4 C_i^N \tag{8.77}$$

$$Q_3 C_i^{z3/inl} = Q_F C_i^F + Q_2 C_i^{j_F} \quad \text{thus} \quad m_3 C_i^{z3/inl} = (m_3 - m_2) C_i^F + m_2 C_i^{j_F} \tag{8.78}$$

Notice that the compositions of the fluid exiting stage j_F and entering stage $j_F + 1$ correspond to the states α and β presented in Section 8.4; composition γ corresponds to the fluid composition in equilibrium with the composition of the solid exit stage, $j_F + 1$.

Equations (8.76)–(8.78) can be solved via iterative procedures (such as Newton–Raphson) or, as mentioned for the single-zone system, by simulating a simple transient behavior.

An example of simulated concentration profiles is given in Figure 8.19. It is clear that the concentration profiles become sharper when the number of plates is increased and that, for very efficient systems, the internal concentration profiles tend towards the "step-like" concentration profile predicted by *Equilibrium Theory*.

Although the number of plates has a significant influence on the shape of the internal profiles, it must be noticed that the extract and raffinate concentrations are relatively insensitive to the number of plates unless this reaches an extremely low level. This is a direct consequence of the behavior of single-zone systems, as reported in Section 8.3. Once again, this relative insensitivity of TMB to the number of plates largely explains, in my opinion, the possible superiority of SMB over elution chromatography.

In practice, this TMB simulation is often used as the initialization step of a more sophisticated SMB model: the flow rates allowing us to get solute separation are calculated thanks to the equilibrium model, then the MC–$\overline{\text{Equil}}$ model is used to get an idea of the number of plates required to reach a given purity/yield. This will be discussed further in Chapter 9.

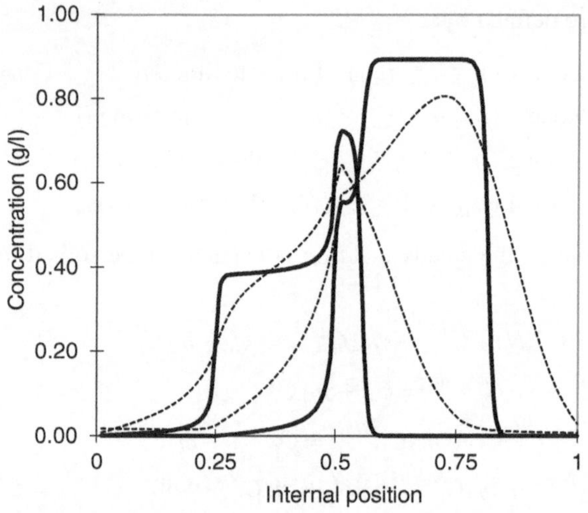

Figure 8.19 Influence of the number of plates on internal concentration profiles (MC–$\overline{\text{Equil}}$ model). The feed is injected at position 0.5, whereas the extract is recovered at position 0.25 and the raffinate at position 0.75. Dotted lines: the TMB is equivalent to 100 plates (25 per zone). Solid lines: the TMB is equivalent to 1000 plates (250 per zone). The adsorption isotherms, feed concentration and flow rates are those of Exercise 8.8.

More sophisticated models can be used to simulate the four-zone TMB systems. Confirming the comments at the end of Section 8.3, however, I recommend the MC model for that task.

8.7 Simulated moving beds

Presentations on how simulated moving bed or simulated counter-current allows practical implementation of counter-current contact between a chromatographic medium and a fluid phase have been proposed many times (Ruthven, 1984; Ruthven and Ching, 1989; Wankat, 1986; Nicoud, 1992; Seidel-Morgenstern *et al.*, 2012).

To explain the connection between TMB and SMB, we will suggest simple and consecutive modifications to TMB that will lead us to SMB. The first step is to consider a modified TMB system (Figure 8.20) in which the solid and fluid are counter-currently contacted in a circular column. The two parts of Figure 8.20 are clearly equivalent and simply correspond to two different representations of the same concept.

A third equivalent representation is given in Figure 8.21: the solid circulating in the circular system of Figure 8.20 (right) is now kept fixed, while the inlet/outlet lines are continuously displaced clockwise. The systems given in Figure 8.21a and in Figure 8.20b are equivalent, only the frame of reference has changed. For readers who are not convinced, imagine that, while one normally considers the station to be fixed (reference) and the train to move, nothing prevents us considering the train to be fixed (reference) and the station to move with respect to the train.

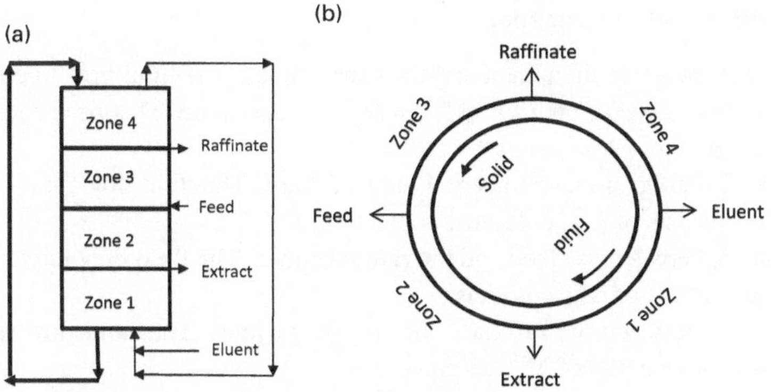

Figure 8.20 True moving bed: two equivalent representations.

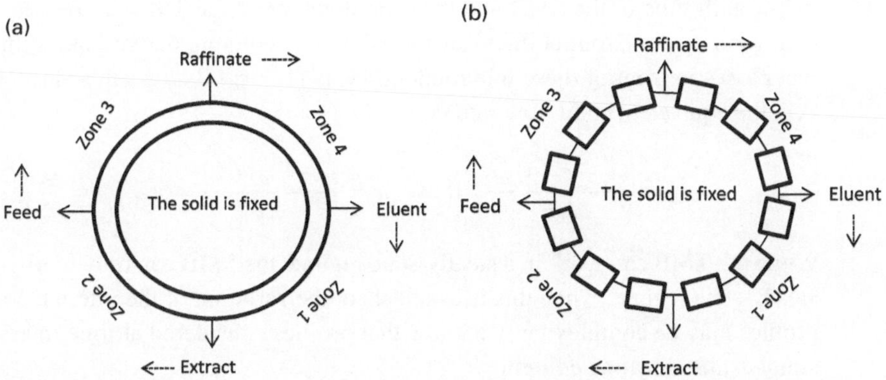

Figure 8.21 Principle of the simulated moving bed. (a) Concept: inlet/outlet lines are shifted continuously. (b) Implementation: inlet/outlet lines are shifted step by step.

Since it is extremely difficult to move the inlet/outlet lines in a continuous manner at the periphery of a continuous circular column (Figure 8.21a), a better way is to approximate the circular column by a series of N_{col} classical fixed-bed columns connected in a loop, and shifting the inlet/outlet lines stepwise between these N_{col} columns with a period ΔT (Figure 8.21b). This mode is referred to as SMB. The solid is no longer moving, its flow only simulated by the shift in inlet and outlet lines; the simulated solid flow rate downwards is directly related to the shift time ΔT.

An SMB is thus a collection of fixed-bed columns connected in series, between which an eluent and a feed are injected, and an extract and a raffinate are collected. The positions of the inlet/outlet lines are shifted periodically in order to simulate the solid motion with respect to the fluid. As with TMB, the key factor is the selection of flow rates that stabilize the different fronts in the different zones of the system, with the solid flow rate now replaced by the switching period.

The SMB is made of four zones:

- Zone 1, between the eluent and the extract lines. The fluid flow rate in zone 1 is called the *recycle flow rate*, $Q_R^{smb} = Q_1^{smb}$ by definition. This zone consists of N_{col}^1 columns.
- Zone 2, between the extract and the feed lines. The fluid flow rate is Q_2^{smb}. This zone consists of N_{col}^2 columns.
- Zone 3, between the feed and the raffinate lines. The fluid flow rate is Q_3^{smb}. This zone consists of N_{col}^3 columns.
- Zone 4, between the raffinate and the eluent lines. The fluid flow rate is Q_4^{smb}. This zone consists of N_{col}^4 columns.

For a given number of columns $N_{col} = \sum_k N_{col}^k$, the SMB is completely defined by the internal flow rates Q_k^{smb} in the different zones, and the shift time ΔT.

The shift time of the SMB and the solid flow rate of the TMB are related by noticing that, when the inlet/outlet lines are shifted by one column, the volume of lumped solid that passes in front of these inlet/outlet lines is $(1 - \varepsilon_e)\, V_{col}$. As this shift occurs every ΔT, the equivalent solid flow rate is:

$$\overline{Q} = \frac{(1 - \varepsilon_e)\, V_{col}}{\Delta T} \quad \text{thus} \quad \bar{u} = \frac{(1 - \varepsilon_e)\, L_{col}}{\Delta T} \quad \text{or} \quad \bar{u}_e = \frac{L_{col}}{\Delta T} \tag{8.79}$$

While the TMB can work in a steady-state regime, the SMB can only work in a periodic steady-state regime: when this is established, the variation of the internal concentration profiles may be complex but we know that profiles considered at time intervals ΔT are simply shifted by one column.

The relationships between the TMB and SMB fluid flow rates are derived by comparing Figure 8.20b, for which the inlet/outlet lines are fixed, and Figure 8.21a, for which the inlet/outlet lines are moving. Keeping a constant fluid velocity with respect to the inlet/outlet lines requires increasing the flow rates for the latter system compared to the former. The volume of fluid to be compensated because of the inlet/outlet lines' shift is given by $\varepsilon_e V_{col}$, leading to a relative fluid flow rate of $\varepsilon_e V_{col}/\Delta T$.

The internal SMB flow rates are thus related to the corresponding TMB flow rates by:

$$Q_k^{smb} = Q_k^{tmb} + \frac{\varepsilon_e V_{col}}{\Delta T} \quad k \in [1,4] \tag{8.80}$$

Taking into account the expression for the solid flow rate, one obtains:

$$Q_k^{smb} = Q_k^{tmb} + \frac{\varepsilon_e}{1 - \varepsilon_e}\overline{Q} \quad \text{thus} \quad u_k^{smb} = u_k^{tmb} + \frac{\varepsilon_e}{1 - \varepsilon_e}\bar{u} \quad \text{or} \quad u_{ke}^{smb} = u_{ke}^{tmb} + \bar{u}_e \tag{8.81}$$

Equations (8.79)–(8.81) are the fundamental equations that link the operating parameters of an SMB to those of the corresponding TMB.

From Eqs. (8.79) and (8.80) it is apparent that a TMB is strictly identical to an SMB made of an infinite number of columns of zero length, associated with a shift time of zero.

Finally, let us note that the classical SMB is characterized by constant inlet/outlet flow rates, constant feed composition and one unique shift time. Systems taking some liberties with the above constraints have been proposed (Kaspereit, 2009; Seidel-Morgenstern *et al.*, 2012; Nicoud, 2014). Their modeling offers no specific difficulties compared to SMB, and they are only briefly discussed in Chapter 9.

EXERCISE 8.9 *Calculate the SMB associated with the separation proposed in Exercise 8.8.*

The task of simulating the transient behavior of an SMB is quite straightforward provided that a model for simulating the columns is available. With this model in hand, all that remains is to connect the columns and properly include the inlet/outlet line shifting sequences. Any of the models presented in Chapter 7 can be used, selected according to the nature of the problem (number of solutes, adsorption or ion exchange, large particles with significant bed length or small particles with short bed length etc.) and the simulation precision required. The selected model can then be used to investigate the influence of various operating parameters on the system's behavior.

For illustration purposes, we have simulated the transitory behavior of the SMB calculated in Exercise 8.9. We consider:

- an SMB consisting of 16 columns of individual volume $V_{col} = 10$ l, so that the shift time is $\Delta T = 0.166$ h; each column is equivalent to 20 plates
- an SMB consisting of four columns of individual volume $V_{col} = 40$ l, so that the shift time is $\Delta T = 0.663$ h; each column is equivalent to 80 plates.

The number of plates of the different columns is set in such a way that both systems are equivalent to 320 plates. The systems are modeled by the MC–$\overline{\text{Equil}}$ model.

The transitory evolution of the concentrations obtained in the extract and raffinate lines are shown in Figure 8.22 for the 16-column and 4-column systems. Concentrations increase from zero (initial state) to a periodic steady state, where they exhibit some oscillations around an average value with a period identical to the shift time. The oscillations are obviously more significant for the four-column system, but the average concentrations remain similar to those predicted by the TMB model associated with the same number of plates.

For a given number of theoretical plates in the system, and for a given set of flow rates, an increase in the number of columns allows the extract and raffinate purities to be increased. Figure 8.23 demonstrates this for an SMB equivalent to 320 plates, which corresponds to the total number of plates of the systems in Figure 8.22.

As stated previously, the idealized TMB system is extremely useful as a reasonably good approximation of the SMB behavior: by using the flow rates as well as the number of plates determined for the TMB, one gets excellent starting conditions for designing an SMB. The SMB will deliver the same purities as the TMB, provided it is made of a sufficiently large number of columns. The important message is that the "sufficiently

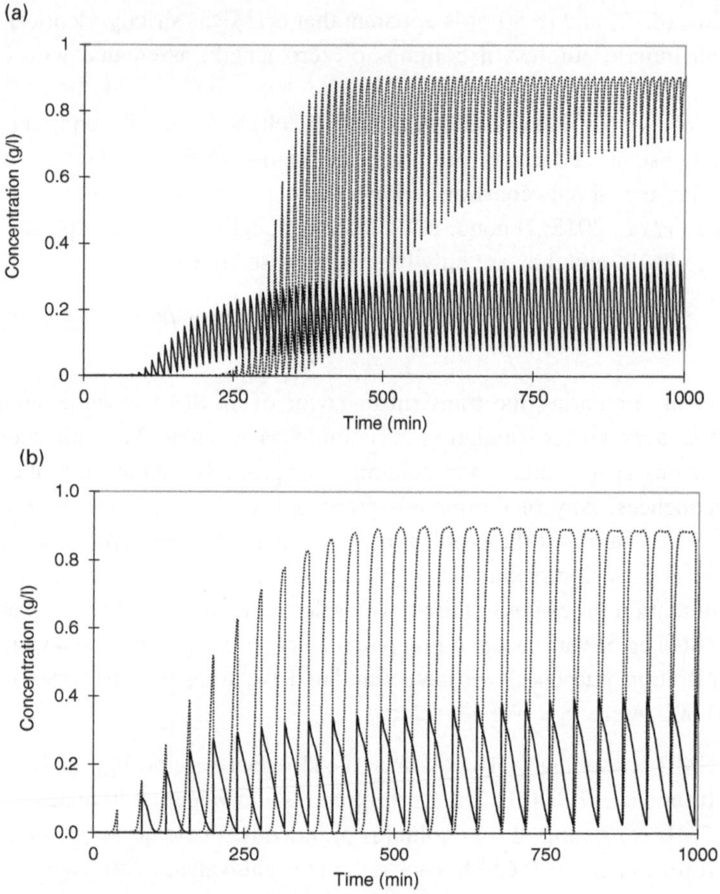

Figure 8.22 Time evolution of extract and raffinate concentrations exiting an SMB versus time. Detailed conditions are given in Exercise 8.9. Solute A in the raffinate: dotted line; solute B in the extract: solid line. (a) SMB consisting of 16 columns (4 columns per zone), each column being equivalent to 20 plates. (b) SMB consisting of 4 columns (1 column per zone), each column being equivalent to 80 plates.

high" qualification is in fact associated with a rather limited number of columns, let us say 8 to 12 according to the precision required. This is only a qualitative result, but in my experience, *for a given set of flow rates*, using more than 12 columns is unlikely to give meaningful improvement in purity/yield, and using less than 8 columns is normally associated with a visible decline in purity/yield. For practical and economic reasons, there is a trend to minimize the number of columns, down to four or six for some applications, as shown in (Nicoud, 2014) and Chapter 9.

Readers willing to learn more about the differences between SMB and TMB design and performance are referred to (Hashim and Chu, 1995), (Charton and Nicoud, 1995), (Migliorini *et al.*, 1998), (Biressi *et al.*, 2000) and (Seidel-Morgenstern *et al.*, 2012). Once again, when a high degree of precision or a detailed design is required, the TMB

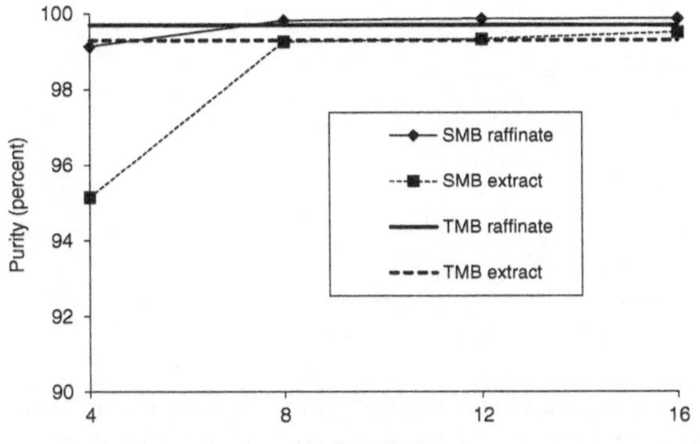

Figure 8.23 Influence of the number of columns on extract and raffinate purities. The total efficiency of the system is set to 320 plates. Same number of columns in the different zones. Other conditions are given in Exercise 8.9.

approximation is not sufficient and one has to use the more complex column models presented in Chapter 7, and thus to accept significantly increased computation times.

8.8 Analogy between TMB/SMB and elution chromatography

While different in principle, TMB/SMB and elution (batch) chromatography reveal intriguing similarities, as shown for the case of binary systems.

We consider the conditions just allowing complete separation, so the vertex of the separation zone is in the m space for TMB. We have shown how to determine the m ratio allowing us to reach these conditions of complete separation. The m ratios are fluid flow rates normalized to the solid flow rate, and we further define m_F, the ratio of the feed flow rate to the solid flow rate. One obtains immediately:

$$m_3 = m_2 + m_F \quad \text{with} \quad m_F = \frac{Q_F}{\overline{Q}} \tag{8.82}$$

For elution chromatography, the situation of complete separation is the so-called "touching-band" situation. In addition to the injection time Δt_{inj}, three important times characterize the chromatogram: t_{beg} at which the solute A appears, $t_A^{dis} = t_B^{app}$ at which the solute A disappears and the solute B appears, and t_{end} at which the solute B disappears. This is schematically represented in Figure 8.24.

For every time t of the chromatogram, the fluid volume passed onto the column is given by $V = t\,Q$. By analogy with the m ratios of the TMB, we normalize the volumes V to the volume of the lumped chromatographic medium in the column by defining:

$$b = \frac{V - \varepsilon_e V_{col}}{(1 - \varepsilon_e) V_{col}} \quad \text{thus} \quad b = \frac{t - t_0}{\frac{1-\varepsilon_e}{\varepsilon_e} t_0} \tag{8.83}$$

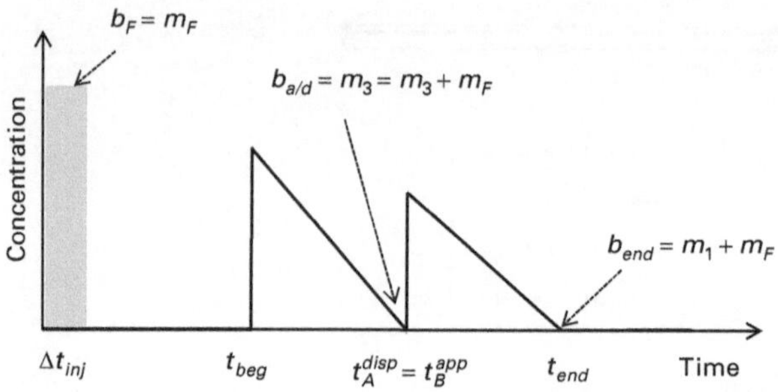

Figure 8.24 Representation of a touching-band chromatogram. The relation between chromatogram characteristic times and m ratio is calculated for TMB. Linearly modified Langmuir adsorption isotherm.

Note that, if one divides the numerator and the denominator of Eq. (8.83) by the time between two injections, the b ratios can be seen as ratios between fluid and solid flow rates, just like the m ratios for the TMB. Similarly to the definition of m_F, we define:

$$b_F = \frac{V_{inj}}{(1-\varepsilon_e)V_{col}} = \frac{\Delta t_{inj}}{\frac{1-\varepsilon_e}{\varepsilon_e} t_0} \tag{8.84}$$

which appears as a ratio of an "injection flow rate" to a "solid flow rate".

Thanks to simple definitions, we have been able to define ratios in a manner that is consistent between TMB and elution chromatography. How can we relate these ratios? For linearly modified Langmuir adsorption isotherms, the reader can check, using the results of Section 3.5.3 and Section 8.5, that:

- the situation of touching-band chromatography is obtained by selecting the injection time $\Delta t_{inj} = b_F \frac{1-\varepsilon_e}{\varepsilon_e} t_0$ with $b_F = m_F$. In other words, the normalized feed flow rate allowing us to deliver complete separation in the TMB is also the normalized feed flow rate allowing us to deliver the touching-band situation in elution chromatography. This is a very strong analogy!
- the time t_{end} is associated with the ratio $b_{end} = m_1 + m_F$
- the time $t_A^{dis} = t_B^{app}$ is associated with the ratio $b_{d/a} = m_3 = m_2 + m_F$
- the time t_{beg} is associated with a ratio b_{beg}. There is no simple way to relate b_{beg} and the m ratios. Note, however, that $b_{beg} = m_4$ for linear adsorption isotherms and $b_{beg} < m_4$ otherwise.

Some of the m ratios of the TMB are thus directly related to characteristics of the touching-band chromatogram (see Figure 8.24).

For illustration purposes, consider linear adsorption isotherms. We know that, for TMB, $m_1 = \overline{K}_B$, $m_2 = \overline{K}_A$, $m_3 = \overline{K}_B$ and $m_4 = \overline{K}_A$, so $m_F = \overline{K}_B - \overline{K}_A$. Now for elution chromatography, the maximum injection time allowing complete separation is $\Delta t_{inj} = t_0 \frac{1-\varepsilon_e}{\varepsilon_e}\left(\overline{K}_B - \overline{K}_A\right)$, so $b_F = \overline{K}_B - \overline{K}_A$ and consequently $m_F = b_F$. The time t_{beg} is given by $t_{beg} = t_0\left(1 + \frac{1-\varepsilon_e}{\varepsilon_e}\overline{K}_A\right)$, thus $m_{beg} = \overline{K}_A$ and consequently $b_{beg} = m_4$.

Using the results of Section 3.5.3 and Section 8.5, the reader can check that "some" equivalence between m ratios (TMB) and b ratios (elution) does exist for generalized Langmuir adsorption isotherms. In particular, the identity between m_F and b_F mentioned for Langmuir adsorption isotherms remains valid.

The above analogy has two consequences. First, the *thermodynamic* performance of TMB and elution chromatography being similar,[10] differences in performance must be found elsewhere; we have mentioned the influence of the plate numbers. The other consequence is that most of the m required to design a TMB can be read from a touching-band chromatogram.

8.9 Imperfect implementations of simulated moving beds

SMB being made of chromatographic columns, they are likely to suffer from all the possible imperfections of chromatographic columns, including heterogeneous beds, non-uniform velocity and/or temperature radial profiles, chromatographic media compressibility and so on, making the life of the chromatographer a little more difficult than he would like.

In addition to the above-mentioned technical difficulties, which can occur in any chromatographic process, two specific technical factors can affect the behavior of SMBs: asymmetry in the loop and differences between the columns. These two factors are briefly discussed below.

8.9.1 Asymmetry in the loop

The volumes of recycling pumps, sensors such as flow meters and pressure transducers, dampers, etc. can induce delays in the retention times of the solutes in various zones of the system. Such delays, varying from one part of the system to another, induce asymmetries that can significantly affect the efficiency of the SMB, especially when short columns and/or low retention factors are used. This decrease in efficiency can be overcome by various means, such as shorter columns or the asynchronous shift of inlets/outlets (Hotier and Nicoud, 1996; Hotier, 1996). These subjects present no modeling difficulties, so they will not be discussed further.

8.9.2 Differences between the columns

When modeling an SMB, one usually assumes all columns to be identical. What is the influence of heterogeneity in column characteristics? What is the maximum deviation between columns that allows a given separation to be performed? The only way to give precise answers is to simulate the behavior of SMBs made of different columns, but a few general remarks can be made:

- An SMB is certainly less sensitive to column heterogeneity than one might fear.

[10] For a numerical application, the reader is referred to Exercise 9.3.

- A proper arrangement of the columns can help to reduce the impact of a given heterogeneity.
- The lower the selectivity, the lower the acceptable heterogeneity.

To illustrate the matter, we consider an SMB made of eight columns of 40 plates each (this system will be called the *reference system*) and fed by pure eluent and feed composition $C_A^F = C_B^F = 1$ g/l. Two different adsorption isotherms, associated with low and medium selectivity ($\alpha = \overline{K}_B/\overline{K}_A$) are considered:

Selectivity	α	$\bar{\lambda}_A = \bar{\lambda}_B$	$\bar{K}_A$ (l/g)	$\bar{K}_B$ (l/g)	$\bar{N}$ (g/l)
Medium	1.5	1	0.5	1	2
Low	1.1	1	0.5	0.6	2

The flow rates and shift time used in the SMB are those given by the equilibrium model. When all the columns are identical, the system of eight columns of 40 plates each allows the following purities to be obtained:

Selectivity	Extract	Raffinate	Average
Medium	99.3%	99.8%	99.55%
Low	94.4%	95.4%	94.95%

For convenience we define an average purity as the arithmetic average of the extract and raffinate purities.

We start by considering that all the columns are identical, except for one which is 5% longer than the others. Under this circumstance, the average column length of the *heterogeneous* system is 0.625% longer than the average column length of the *reference* system.

With this column 5% longer than the others, keeping the same flow rates and shift time as those for the reference system, one obtains an average purity of about 94% in the heterogeneous system for the *medium selectivity case*, compared to 99.55% obtained in the reference system. The loss in purity is thus very significant.

Now let us increase the shift time by 0.625% in order to compensate for the increase in average column length. In this situation, the average purity is almost indistinguishable from the purity obtained with the reference system. We have thus shown that an SMB can work efficiently with one column different from the others, provided that the flow rates and shift time are calculated on the basis of the actual average column length. We have also shown that the key is the proper selection of the shift time, and that a very small 0.625% variation can make a significant difference in purity.

In order to further investigate the influence of column heterogeneity, we now assume that four columns out of eight are x% shorter than the average length and four are x% longer than the average length, and we consider two options for connecting the columns in the loop:

- one short column is followed by one long column ("alternate" configuration)
- four short columns followed by four long columns ("block" configuration).

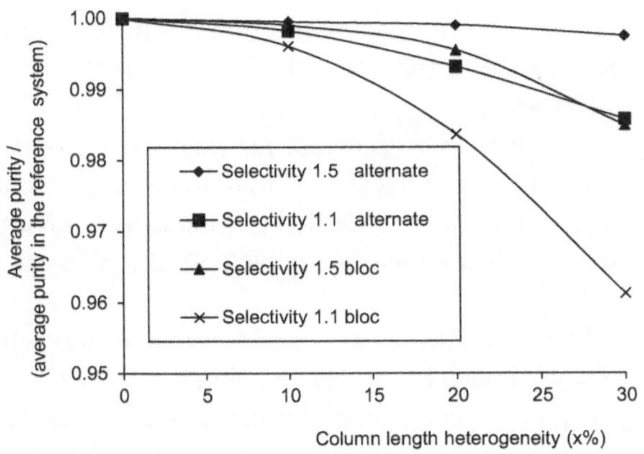

Figure 8.25 Influence of column length heterogeneity on the average extract plus raffinate purity in an eight-column SMB. See details in the text. The system is modeled with the MC–Equil model.

The results are shown in Figure 8.25. The SMB appears to be relatively accommodating with respect to column heterogeneity: even with a 30% difference in column length, the purity loss is "only" about 5% for the lowest selectivity.

The higher the selectivity, the lower the sensitivity of the system to column heterogeneity, and, finally, the "alternate" arrangement is less sensitive to column heterogeneity compared with the "block" configuration. It is thus possible, in principle, to minimize the influence of a column length heterogeneity by associating longer and shorter columns in order to build zones of relatively constant length.

8.10 Conclusion

While presenting significant practical differences, single-column elution chromatography and multi-column continuous chromatography systems, be they strictly SMB or not, can be modeled with the theoretical framework presented in the previous chapters.

It is worth recalling that most modeling of complex multi-column systems can be performed without modeling the columns! An approach based on the simple TMB concept allows a good approximation of the SMB parameters and performance to be quickly obtained. For a given elution chromatography system (and other derived techniques including displacement chromatography), performance and ability to deliver pure products are mainly influenced by injection volumes, regeneration volumes and cut times. For a given SMB chromatography system, performance is mainly imposed by internal fluid flow rates and shift time (lumped solid flow rate for TMB). These flow rates can be determined to a good accuracy using the equilibrium model and the TMB concept. It is remarkable that a very simple model used on an idealized process can deliver such a good approximation for real SMB parameters.

The determination of the flow rates of a four-zone TMB that ensure complete separation gives probably one of the most famous applications of the equilibrium model.

A remarkable result is that the set of flow rates ensuring complete separation can be calculated for the intermediate zones 2 and 3 first, the flow rates in the extremity zones 1 and 4 being calculated afterwards.

When detailed design or analysis is considered, for example determining column diameter and length, investigating the influence of dead volumes and column heterogeneity or searching for a precise optimum, detailed column models rather than TMB approximation must be used. An illustration of a complete SMB design will be given in Chapter 10.

An important property of SMB (related to its counter-current nature) is its moderate sensitivity to column efficiency compared to single-column elution chromatography. This relative insensitivity is quite general: full performance is delivered with a limited number of columns, the ability to accept column heterogeneity etc. This is probably at the very heart of the success of SMB.

8.11 Solutions of the exercises

Exercise 8.1

The retention time determined on the laboratory column allows a determination of the lumped Henry's coefficient:

$$\overline{K}_A = \left(\frac{t_R Q}{\varepsilon_e V_{\text{col}}} - 1\right)\frac{\varepsilon_e}{1-\varepsilon_e} = 1.33$$

In order to completely remove the solute from the fluid stream, the front of solute A should be pushed downwards, so $\sigma < 0$. This imposes $\bar{u} > u\overline{K}_A$ and finally:

$$\overline{Q} > \frac{Q}{\overline{K}_A} = 7.5 \text{ l/min.}$$

One can check that, according to Eq. (8.17), the concentration of solute A in the fluid stream exiting the system is given by $C_A^{top} = \overline{C}_A^{inl}\overline{K}_A$ and consequently equals zero.

Exercise 8.2

The adsorption isotherm is given by:

$$\overline{C}_A = \frac{\overline{N}\tilde{K}_A C_A}{1+\tilde{K}_A C_A} = \frac{C_A}{1+0.1\times C_A}$$

Its derivative is given by:

$$\frac{d\overline{C}_A}{dC_A} = \frac{\overline{N}\tilde{K}_A}{(1+\tilde{K}_A C_A)^2} = \frac{1}{(1+0.1\times C_A)^2}$$

Consequently, the sign of the front velocity given by Eq. (8.13) is determined by:

$$\text{Sign}(\sigma_{disp}) = \text{sign}\left(u - \bar{u}\frac{d\overline{C}_A}{dC_A}\right) = \text{sign}\left(u - \bar{u}\frac{1}{(1+0.1\times C_A)^2}\right)$$

Saturation

This situation is associated with $C_A^{inl} = 10$ g/l and $\overline{C}_A^{inl} = C_A^{\overline{inl}^*} = 0$ g/l. The front velocity Eq. (8.13) being higher at concentration C_A^{inl} than at concentration $C_A^{\overline{inl}^*}$, the front is compressive. The sign of the compressive front velocity is given by:

$$\text{Sign}(\sigma_{comp}) = \text{sign}\left(u - \bar{u}\frac{\Delta\overline{C}_A}{\Delta C_A}\right)$$

Noting that $\Delta\overline{C}_A = 0 - \frac{10}{1+0.1\times10}$ and $\Delta C_A = 0 - 10$, one obtains:

$$\text{Sign}(\sigma_{comp}) = \text{sign}\,(u - \bar{u}/2) = \text{sign}(m - 1/2)$$

Let us start by considering a ratio $m = 1$. Under this condition, $\text{sign}(\sigma_{comp}) = \text{sign}(1 - 1/2) > 0$. The front moves upwards and the boundary discontinuity occurs at the top of the system. Under this condition, the system is filled with uniform concentration $C_A^{inl} = 10$ g/l except at the top boundary. The concentration $\overline{C}_A^{\overline{out}}$ of the solid exiting the system is in equilibrium with the fluid concentration prevailing in the system and is thus given by:

$$\overline{C}_A^{\overline{out}} = \frac{10}{1 + 0.1 \times 10} = 5\text{g/l}$$

The concentration C_A^{out} of solute in the fluid exiting the system is obtained from the mass-balance equation (8.4), which is written:

$$C_A^{out} = C_A^{inl} + \frac{1}{m}\left(\overline{C}_A^{\overline{inl}} - \overline{C}_A^{\overline{out}}\right) = 10 + 1 \times (0 - 5) = 5\text{ g/l}$$

Let us now consider a ratio $m = 1/3$. Under this condition, $\text{sign}(\sigma_{comp}) = \text{sign}(1/3 - 1/2) < 0$. The front moves downwards; the boundary discontinuity occurs at the bottom of the system. Under this condition, the system is filled with uniform concentration $C_A^{\overline{inl}^*} = 0$ g/l except at the bottom boundary. The concentration C_A^{out} in the fluid exiting the system is in equilibrium with the lumped solid concentration prevailing in the system (0 g/l) and is thus given by:

$$C_A^{out} = C_A^{\overline{inl}^*} = 0\text{ g/l}$$

The concentration $\overline{C}_A^{\overline{out}}$ of solute in the lumped solid exiting the system is obtained from Eq. (8.4), which is written:

$$\overline{C}_A^{out} = \overline{C}_A^{inl} + m\left(C_A^{inl} - C_A^{out}\right) = 0 + 1/3 \times (10 - 0) = 3.33\text{ g/l}$$

If one now considers a ratio $m = 1/2$, the front velocity equals zero. A solution can, for instance, be associated with a system filled with a fluid containing the solute at concentration 10 g/l in the lower part and a solute-free fluid in the upper part.

Regeneration

This situation is associated with $C_A^{inl} = 0$ g/l and $\overline{C}_A^{\overline{inl}} = 6$ g/l, so $C_A^{\overline{inl}^*} = 15$ g/l. The front velocity (Eq. (8.13)) being lower at concentration C_A^{inl} than at concentration $C_A^{\overline{inl}^*}$, the front is dispersive. The sign of the dispersive front velocity is given by:

$$\mathrm{Sign}(\sigma_{disp}) = \mathrm{Sign}\left(u - \bar{u}\frac{d\bar{C}_A}{dC_A}\right) = \mathrm{Sign}\left(m - \frac{1}{(1+0.1\times C_A)^2}\right)$$

The derivative of the adsorption isotherm varies between 1 when $C_A = C_A^{inl} = 0$ and 0.16 when $\overline{C_A^{inl}}^* = 15$ g/l. By adjusting the ratio $\bar{u}/u$ one can stabilize a given concentration inside the system. Another way to look at this consists in saying that if the ratio $\bar{u}/u$ is between $1/1$ and $1/0.16$ a certain non-zero concentration will be stabilized in the system.

Let us start by considering a ratio $m = 1/2$. From the above equation, one sees that sign(σ_{disp}) becomes indeterminate if $\frac{1}{2} - \frac{1}{(1+0.1C_A)^2} = 0$, so $C_A = C_A^{stab} \approx 4$ g/l. One thus has a system filled with a fluid at concentration $C_A^{stab} = C_A^{top} = C_A^{bot}$. The lumped solid concentration is uniform as well, and equals $\bar{C}_A^{stab} = \frac{1}{1+0.1\times 4} \approx$ 2.85. Discontinuities are expected to occur at both boundaries. Outlet concentrations are obtained by writing the mass-balance equation (8.27) for the lower part of the system:

$$0 + 2\times 2.85 = 4 + 2\times \overline{C}_A^{out}$$

so $\overline{C_A^{out}} = 0.85$; and the mass-balance equation (8.28) for the upper part of the system:

$$4 + 2\times 6 = C_A^{out} + 2\times 2.85$$

so $C_A^{out} = 10.3$ g/l.

Now let us consider a ratio $m = 2$. One sees that the front velocity is positive, irrespective of the solute concentration. The front is thus pushed to the top of the system. The concentrations C_A^{out} and $\overline{C_A^{out}}$ are calculated using the global mass balance with a methodology identical to that for the compressive front.

Let us finally consider a ratio $m = 0.1$. One sees that the front velocity is negative, irrespective of the solute concentration in the range (C_A^{inl},$\overline{C_A^{inl}}^*$). The front is thus pushed to the bottom of the system. The concentrations C_A^{out} and $\overline{C_A^{out}}$ are calculated using the global mass balance with a methodology identical to that for the compressive front.

Exercise 8.3

The first thing to do is to evaluate the hodograph space (done in Exercise 3.8) and the propagation velocities associated with each composition. All concentrations are in g/l; all velocities are in m/h.

Composition α ($C_A = C_B = 0$):

$$\frac{\partial \bar{C}_A}{\partial C_A} = 1 \qquad \frac{\partial \bar{C}_A}{\partial C_B} = 0 \qquad \frac{\partial \bar{C}_B}{\partial C_A} = 0 \qquad \frac{\partial \bar{C}_B}{\partial C_B} = 2$$

Using Eq. (8.38), we find that $\theta_- = 2$ and $\theta_+ = 1$. The corresponding propagation velocities Eq. (8.37) with $u = \bar{u} = 1$ are $\sigma_-^\alpha = -0.625$ and $\sigma_+^\alpha = 0$.

Composition β ($C_A = C_B = 1$ g/l, which is associated with a composition on the lumped solid phase, $\overline{C}_A = 0.4$ and $\overline{C}_B = 0.8$ g/l):

$$\frac{\partial \overline{C}_A}{\partial C_A} = 0.32 \qquad \frac{\partial \overline{C}_A}{\partial C_B} = -0.16 \qquad \frac{\partial \overline{C}_B}{\partial C_A} = -0.16 \qquad \frac{\partial \overline{C}_B}{\partial C_B} = 0.48$$

At this composition, $\theta_- = 0.5789$ and $\theta_+ = 0.22$. Propagation velocities are $\sigma_-^\beta = 0.56$ and $\sigma_+^\beta = 1.46$.

In Exercise 3.8 we identified two points in the hodograph plane located at the intersection of characteristic paths starting from compositions α and β: P$_1$ ($C_A = 1.62$, $C_B = 0$) and P$_2$ ($C_A = 0$, $C_B = 0.38$). These compositions are associated with lumped solid-phase concentrations P$_1$ ($\overline{C}_A = 0.895$, $\overline{C}_B = 0$) and P$_2$ ($\overline{C}_A = 0$, $\overline{C}_B = 0.551$). Let us determine the front velocities at these two compositions.

Composition P$_1$ ($C_A = 1.62$, $C_B = 0$):

$$\frac{\partial \overline{C}_A}{\partial C_A} = 0.305 \qquad \frac{\partial \overline{C}_A}{\partial C_B} = -0.495 \qquad \frac{\partial \overline{C}_B}{\partial C_A} = 0 \qquad \frac{\partial \overline{C}_B}{\partial C_B} = 1.105$$

At this composition P$_1$, $\theta_- = 1.105$ and $\theta_+ = 0.305$; the propagation velocities are $\sigma_-^{(P1)} = -0.099$ and $\sigma_+^{(P1)} = 1.19$.

Composition P$_2$ ($C_A = 0$, $C_B = 0.38$):

$$\frac{\partial \overline{C}_A}{\partial C_A} = 0.7246 \qquad \frac{\partial \overline{C}_A}{\partial C_B} = 0 \qquad \frac{\partial \overline{C}_B}{\partial C_A} = -0.199 \qquad \frac{\partial \overline{C}_B}{\partial C_B} = 1.05$$

At this composition, $\theta_- = 1.05$ and $\theta_+ = 0.725$; the propagation velocities are $\sigma_-^{(P2)} = -0.049$ and $\sigma_+^{(P2)} = 0.33$.

We are now ready to determine the concentration profiles in the counter-current system.

Regeneration

The inlet fluid has composition α, $C_A = C_B = 0$, and the inlet solid is in equilibrium with composition β, $C_A = C_B = 1$ g/l. We start from composition α in the hodograph plane and select the slow moving, thus negative, characteristic path. The associated velocity at point α is σ_-^α and we move along the negative path and reach point P$_2$, where the velocity is $\sigma_-^{(P2)}$. One has $\sigma_-^\alpha < \sigma_-^{(P2)}$ and therefore the front F$_1$ is dispersive.

We then move from P$_2$ to composition β, following the only (positive) possible characteristic path. The associated velocity at point P$_2$ is $\sigma_+^{(P2)}$, while it is σ_+^β at point β. One has $\sigma_+^{(P2)} < \sigma_+^\beta$ and therefore the front F$_2$ is dispersive.

Given that front F$_1$ moves downwards (negative velocities) and front F$_2$ moves upwards (positive velocities), at steady state the bed will be uniformly loaded with a concentration corresponding to the plateau composition P$_2$.

The outlet compositions are then calculated using a mass balance for the lower part of the system:

$$uC_A^{inl} + \bar{u}\bar{C}_A^{P2} = uC_A^{P2} + \bar{u}\bar{C}_A^{out} \qquad uC_B^{inl} + \bar{u}\bar{C}_B^{P2} = uC_B^{P2} + \bar{u}\bar{C}_B^{out}$$

which leads to:

$$\bar{C}_A^{out} = 0 + 0 - 0 = 0 \qquad \bar{C}_B^{out} = 0 + 0.551 - 0.38 = 0.17$$

and for the upper part of the system:

$$uC_A^{P2} + \bar{u}\bar{C}_A^{\overline{inl}} = uC_A^{out} + \bar{u}\bar{C}_A^{P2} \qquad uC_B^{P2} + \bar{u}\bar{C}_B^{\overline{inl}} = uC_B^{out} + \bar{u}\bar{C}_B^{P2}$$

which leads to:

$$C_A^{out} = 0 + 0.4 - 0 = 0.4 \quad \text{g/l} \qquad C_B^{out} = 0.38 + 0.8 - 0.551 = 0.63$$

Saturation

The inlet fluid has composition β, $C_A = C_B = 1$, and the inlet solid is in equilibrium with composition α, $C_A = C_B = 0$. We start from composition β in the hodograph plane and select the slow moving, thus negative, characteristic path. The associated velocity at point β is σ_-^{β} and we move along the negative path and reach point P_1, where the velocity is $\sigma_-^{(P1)}$. One has $\sigma_-^{\beta} > \sigma_+^{(P1)}$ and therefore front F_1 is compressive.

We then move from P_1 to composition α following the only (positive) possible characteristic path. The associated velocity at point P_1 is $\sigma_+^{(P1)}$, while it is σ_+^{α} at point α. One has $\sigma_+^{(P1)} > \sigma_+^{\alpha}$ and therefore front F_2 is compressive.

The fronts being compressive, their velocities must be determined with finite differences. For front F_1:

$$\sigma(\beta \rightarrow P_1) = \frac{1 - 1 \times \dfrac{\Delta\bar{C}}{\Delta C}}{\varepsilon_e + (1 - \varepsilon_e)\dfrac{\Delta\bar{C}}{\Delta C}}$$

where $\frac{\Delta\bar{C}}{\Delta C}$ represents the ratio of the differences in lumped solid and fluid concentrations between compositions β and P_1. This ratio, calculated with either solute A or solute B, equals 0.8. Consequently:

$$\sigma(\beta \rightarrow P_1) = \frac{1 - 0.8}{0.4 + 0.6 \times 0.8} = 0.484$$

For front F_2:

$$\sigma(P_1 \rightarrow \alpha) = \frac{1 - 1 \times \dfrac{\Delta\bar{C}}{\Delta C}}{\varepsilon_e + (1 - \varepsilon_e)\dfrac{\Delta\bar{C}}{\Delta C}}$$

where $\frac{\Delta \overline{C}}{\Delta C}$ represents the ratio of the differences in lumped solid and fluid concentrations between compositions P_1 and α. This ratio, calculated with either solute A or solute B, equals 0.55. Consequently:

$$\sigma(P_1 \to \alpha) = \frac{1 - 0.55}{0.4 + 0.6 \times 0.55} = 0.616$$

Consequently, both fronts move upwards, so the system must be filled with a fluid at concentration (C_A^{inl}, C_B^{inl}). By calling $\overline{C}_A^{inl*}$ and $\overline{C}_B^{inl*}$ the lumped solid compositions in equilibrium with (C_A^{inl}, C_B^{inl}), a mass balance on the upper part of the system gives:

$$uC_A^{inl} + \bar{u}\overline{C}_A^{inl} = uC_A^{out} + \bar{u}\overline{C}_A^{inl*}$$

so that $C_A^{out} = 1 + 0 - 0.4 = 0.6$ and, for solute B:

$$uC_B^{inl} + \bar{u}\overline{C}_B^{inl} = uC_B^{out} + \bar{u}\overline{C}_B^{inl*}$$

so that $C_B^{out} = 1 + 0 - 0.8 = 0.2$.

Exercise 8.4

The system is characterized by $C_A^{inl} = 1$ g/l, $\overline{C}_A^{inl} = 0$, $Q = 10$ l/min, $\overline{Q}^{min} = 7.5$ l/min and $\overline{K}_A = 1.33$. We consider two fluid flow rates, $\overline{Q}^{high} = 1.1 \times \overline{Q}^{min} = 8.25$ l/min and $\overline{Q}^{low} = 0.9 \times \overline{Q}^{min} = 6.75$ l/min.

We can thus define $\kappa_A^{high} = \frac{Q}{\overline{Q}^{high}\overline{K}_A} = 0.9114$ and $\kappa_A^{low} = \frac{Q}{\overline{Q}^{low}\overline{K}_A} = 1.114$.

From the Kremser equation (8.46), we know that

$$\frac{\overline{C}_A^1 - C_A^{inl}\overline{K}_A}{\overline{C}_A^{N+1} - C_A^{inl}\overline{K}_A} = \frac{\kappa_A - 1}{\kappa_A^{N+1} - 1}$$

Using the numerical values of the inlet concentrations $C_A^{inl} = 1$ g/l, $\overline{C}_{N+1} = 0$, the Kremser equation allows an evaluation of $\overline{C}_A^1$. When $\kappa = \kappa^{high}$, one obtains $\overline{C}_A^{1high} = 1.146$ g/l; when $\kappa = \kappa^{low}$, one obtains $\overline{C}_A^{1low} = 1.2635$ g/l.

In addition, writing a balance on the column fraction including plates 1 to k, one obtains:

$$\overline{C}_A^{k+1} = \frac{Q}{\overline{Q}}C_A^k + \frac{\overline{Q}\overline{C}_A^1 - QC_A^{inl}}{\overline{Q}}$$

This defines an operating line having slope $\frac{Q}{\overline{Q}}$. When $\overline{Q} = \overline{Q}^{high}$, the operating line is given by $\overline{C}_A^{k+1} = 1.2121 \times C_A^k - 0.0661$; when $\overline{Q} = \overline{Q}^{low}$, the operating line is given by $\overline{C}_A^{k+1} = 1.4815 \times C_A^k - 0.218$. Using the equilibrium relation $\overline{C}_A^k = K_A C_A^k$, one can calculate the concentration profiles starting from plate 1. One obtains:

$\overline{Q} = \overline{Q}^{high}$			$\overline{Q} = \overline{Q}^{low}$		
Plate	C_A	$\overline{C}_A$	Plate	C_A	$\overline{C}_A$
0	1		0	1	
1	0.8617	1.146	1	0.95	1.2635
2	0.7356	0.9783	2	0.8944	1.1895
3	0.6207	0.8255	3	0.8323	1.107
4	0.5160	0.6862	4	0.7633	1.0151
5	0.4206	0.5593	5	0.6863	0.9128
6	0.3336	0.4437	6	0.6006	0.7988
7	0.2543	0.3382	7	0.5051	0.6718
8	0.1821	0.2422	8	0.3988	0.5303
9	0.1162	0.1546	9	0.2803	0.3728
10	0.0562	0.0748	10	0.1483	0.1972
11		0.0	11		0.0

The conclusion is that, for this system equivalent to only ten plates, about 95% of the solute is removed from the fluid stream when the lumped solid flow rate is high and about 85% when the lumped solid flow rate is low.

Exercise 8.5

The steady-state mass balances for species i in cell k are written:

$$\overline{Q}\overline{C}_i^{k+1} = \overline{Q}\overline{C}_i^k - \Delta N_i^k$$

$$QC_i^{k-1} = QC_i^k + \Delta N_i^k$$

where ΔN_i^k represents the flux (mol/s) for species i transferred from the fluid phase to the solid phase in stage k. In the case of external mass transfer limitations, ΔN_i^k is given by (cf. Chapter 5):

$$\Delta N_i^k = n_p D_i S_p \frac{C_i^k - C_i^{k*}}{\delta}$$

where D_i is a diffusion coefficient, δ is the thickness of a motionless layer surrounding the particles, S_p and n_p are respectively the peripheral surface of one lumped particle and the number of particles contained in the cell.

If V_p is the volume of one particle, the number of particles contained in cell k is related to the cell volume V_{cell} by $n_p = (1 - \varepsilon_e)\,\frac{V_{cell}}{V_p}$.

Consequently, $\Delta N_i^k = D_i\,(1 - \varepsilon_e)\,V_{cell}\frac{S_p}{V_p}\frac{C_i^k - C_i^{k*}}{\delta}$, which gives finally, with the definition of the characteristic time for external diffusion:

$$\Delta N_i^k = (1 - \varepsilon_e)\,\frac{V_{\text{cell}}}{t_e}\left(C_i^k - C_i^{k^*}\right)$$

Exercise 8.6

We assume that the m values have been selected for a complete separation of the two solutes (they thus satisfy the constraints given in Eq. (8.65) and Eq. (8.66)).

Consequently, solute A exits entirely at the top of the system and solute B entirely at the bottom. The overall mass balances (8.63) give:

$$\overline{C}_B^{z2/out} = (m_3 - m_2)C_B^F \qquad m_3 C_A^{z3/out} = (m_3 - m_2)C_A^F$$

The concentrations $\overline{C}_B^{z2/out}$ and $C_A^{z3/out}$ are thus known.

From the mass balance at the lower part of zone 2 (8.55) for solute B one obtains $\overline{C}_B^{z2} = m_2 C_B^{z2} + \overline{C}_B^{z2/out}$, which leads to $C_B^{z2} = \overline{C}_B^{z2/out}/(\overline{K}_B - m_2)$.

From the mass balance at the upper part of zone 3 (8.58) for solute A one obtains $\overline{C}_A^{z3} + m_3 C_A^{z3/out} = m_3 C_A^{z3}$, which leads to $C_A^{z3} = m_3 C_A^{z3/out}/(m_3 - \overline{K}_A)$.

At this point in the calculation, concentrations in zone 2 and 3 and concentrations in the streams exiting the system are known (note that $C_A^{z2} = \overline{C}_A^{z2/out} = C_B^{z3} = C_B^{z3/out} = 0$).

From Eq. (8.67) one determines $C_A^{\beta} = C_A^{z3}$ and $C_B^{\gamma} = C_B^{z2}$.

The next step consists in using the mass balance at the upper part of zone 2 (8.56) for solute B, $m_2 C_B^{z2} + \overline{C}_B^{\overline{\gamma}} = m_2 C_B^{\alpha} + \overline{C}_B^{z2}$, which gives, with Eq. (8.60) and the equilibrium inside zone 2, $C_B^{\alpha} = C_B^{z2}$. Concentration C_B^{α} is thus known.

The mass balance at the upper part of zone 2 (8.56) for solute A gives $m_2 C_A^{z2} + \overline{C}_A^{\overline{\gamma}} = m_2 C_A^{\alpha} + \overline{C}_A^{z2}$. Taking into account Eq. (8.60) and the equilibrium inside zone 2, one obtains $C_A^{\alpha} = \overline{K}_A C_A^{\overline{\gamma}}/m_2 = \overline{K}_A C_A^{z3}/m_2$. Concentration C_A^{α} is thus known.

Exercise 8.7

The system is characterized by $\overline{K}_A = 1$, $\overline{K}_B = 2$. ϑ is subject to $1 < \vartheta < \sqrt{\overline{K}_B/\overline{K}_A} = 1.414$. If, for example, we choose $\vartheta = 1.2$, we get:

$$\overline{Q} = \frac{Q_F}{\overline{K}_B/\vartheta - \vartheta\overline{K}_A} = \frac{Q_F}{2/1.2 - 1.2} = 2.143\, Q_F$$

$$Q_1 = \vartheta\overline{Q}\overline{K}_B = 1.2 \times 2 \times 2.14 \times Q_F = 5.136\, Q_F$$

$$Q_{Ext} = \vartheta\overline{Q}(\overline{K}_B - \overline{K}_A) = 1.2 \times 2.14 \times Q_F = 2.568\, Q_F$$

$$Q_2 = \vartheta\overline{Q}\overline{K}_A = 1.2 \times 1 \times 2.14 \times Q_F = 2.568\, Q_F$$

$$Q_3 = \frac{\overline{Q}\overline{K}_B}{\vartheta} = \frac{2.14 \times Q_F \times 2}{1.2} = 3.567\, Q_F$$

$$Q_{Raf} = \overline{Q}\frac{(\overline{K}_B - \overline{K}_A)}{\vartheta} = \frac{(2-1) \times 2.14 \times Q_F}{1.2} = 1.783\, Q_F$$

$$Q_4 = \frac{\overline{Q}\overline{K}_A}{\vartheta} = \frac{2.14 \times 1 \times Q_F}{1.2} = 1.783\, Q_F$$

$$Q_{El} = Q_1 - Q_4 = 3.353\, Q_F$$

Exercise 8.8

The lumped adsorption isotherms are given by $\bar{\lambda} = \bar{\lambda}_A = \bar{\lambda}_B = 1$, $\tilde{K}_A = 0.5$ l/g, $\tilde{K}_B = 1$ l/g and $\bar{N} = 2$ g/l and the feed concentrations are $C_A^F = C_B^F = 1$ g/l.

We first solve the problem with lumped adsorption isotherms *excluding* the $\bar{\lambda}$ terms. The lumped Henry's coefficients excluding the $\bar{\lambda}$ terms are $\bar{K}_A = 1$ and $\bar{K}_B = 2$. This situation corresponds to case L as defined in Tables 8.1 and 8.2.

The characteristic equation (8.68):

$$(1 + 0.51 + 1)\omega^2 - [2 \times (1 + 0.5) + 2 \times 0.5 \times (1 + 1)]\omega + 2 \times 0.5 \times 2 = 0$$

has two roots, $\omega_1^F = 0.5528$ and $\omega_2^F = 1.4472$. Using the equations given in Table 8.1, we get the following m values:

$$m_2 = \frac{\tilde{K}_A}{\tilde{K}_B}\omega_2^F = 0.7236$$

$$m_3 = \frac{(1 \times (1 - 0.5528) + 0.5528 \times (2 - 1)) \times 1.4472}{1 \times (2 - 0.5528)} = 1$$

Using Table 8.2, we get the following m values:

$$m_1 = \bar{K}_B = 2$$
$$B_4 = 1 + 1 + 0.5 \times 1 \times (1 - 0.7236) = 2.1382$$
$$m_4 = \omega_{z4}^F = \frac{1}{2}\left[B_4 - \sqrt{B_4 \times B_4 - 4 \times 1}\right] = 0.6909$$

We then need to add the $\bar{\lambda}$ terms to the m values, and one obtains:

$$m_1 = 3 \quad m_2 = 1.7236 \quad m_3 = 2 \quad m_4 = 1.6909$$

We have $Q_F = Q_3 - Q_2 = (m_3 - m_2)\bar{Q} = (2 - 1.7236)\,\bar{Q} = 10$ l/h, so $\bar{Q} = 36.18$ l/h.

Finally, one obtains:

$$Q_{\mathrm{R}} = Q_1 = 108.54 \text{ l/h}$$
$$Q_{\mathrm{Ext}} = Q_2 - Q_1 = (m_2 - m_1)\bar{Q} = 46.18 \text{ l/h}$$
$$Q_{\mathrm{Raf}} = Q_3 - Q_4 = (m_3 - m_4)\bar{Q} = 11.18 \text{ l/h}$$

Exercise 8.9

The flow rates obtained for the TMB in Exercise 8.8 are, for a feed flow rate of 10 l/h:

$$\bar{Q} = 36.18 \quad Q_R = 108.54 \quad Q_{Ext} = 46.18 \quad Q_{Raf} = 6.82 \text{ l/h}$$

Consequently, for an SMB consisting of 10 l columns:

$$\Delta T = \frac{(1 - \varepsilon_e)\,V_{col}}{\bar{Q}} = \frac{(1 - 0.4) \times 10}{36.181} = 0.166 \text{ h}$$

$$Q_R^{smb} = Q_R^{tmb} + \frac{\varepsilon_e}{1 - \varepsilon_e}\bar{Q} = 108.542 + \frac{0.4}{1 - 0.4} \times 36.18 = 132.6 \text{ l/h}$$

References

Abel, S., Mazzotti, M. and Morbidelli, M. (2002), *J. Chromatogr. A* **944**(1–2), 23–39.

Abel, S., Mazzotti, M. and Morbidelli, M. (2004), *J. Chromatogr. A* **1026**(1–2), 47–55.

Barker, P. E. and Ganetsos, G. (1993). In G. Ganetsos and P. E. Barker, eds., *Preparative and Production Scale Chromatography*, vol. 61 of *Chromatographic Science Series*, Marcel Dekker, pp. 173–186.

Biressi, G., Ludemann-Hombourger, O., Mazzotti, M., Nicoud, R. M. and Morbidelli, M. (2000), *J. Chromatogr. A* **876**(1–2), 3–15.

Bridges, S. and Barker, P. E. (1993). In G. Ganetsos and P. E. Barker, eds., *Preparative and Production Scale Chromatography*, vol. 61 of *Chromatographic Science Series*, Marcel Dekker, pp. 113–126.

Broughton, D. B. and Gerhold, C. G. (1961), US Patent 2 985 589.

Broughton, D. B., Neuzil, R. W., Pharis, J. M. and Brearley, C. S. (1970), *Chem. Eng. Prog.* **66**(9), 70–75.

Charton, F. and Nicoud, R. M. (1995), *J. Chromatogr. A* **702**(1–2), 97–112.

Di Giovanni, O., Mazzotti, M., Morbidelli, M., Denet, F., Hauck, W. and Nicoud, R. M. (2001), *J. Chromatogr. A* **919**(1), 1–12.

Eldridge, R. J. (1993). In G. Ganetsos and P. E. Barker, eds., *Preparative and Production Scale Chromatography*, vol. 61 of *Chromatographic Science Series*, Marcel Dekker, pp. 209–229.

Hashim, M. A. and Chu, K. H. (1995), *Chem. Eng. J. Bioch. Eng. J.* **56**(2), 59–65.

Hotier, G. (1996), *AIChE J.* **42**(1), 154–60.

Hotier, G. and Nicoud, R. M. (1996), US Patent 5 578 215.

Hwang, Y. L. (1987), *Chem. Eng. Sci.* **42**(1), 105–123.

Hwang, Y. L. and Helfferich, F. G. (1989), *Chem. Eng. Sci.* **44**(7), 1547–1568.

Hwang, Y. L. and Helfferich, F. G. (1990), *Chem. Eng. Sci.* **45**(9), 2907–2915.

Hwang, Y. L., Helfferich, F. G. and Leu, R. J. (1988), *AIChE J.* **34**(10), 1615–1626.

Johnson, J. A. and Kabza, R. G. (1993). In G. Ganetsos and P. E. Barker, eds., *Preparative and Production Scale Chromatography*, vol. 61 of *Chromatographic Science Series*, Marcel Dekker, pp. 257–271.

Kaspereit, M. (2009). In E. Grushka and N. Grinberg, eds., *Advances in Chromatography*, vol. 47, CRC Press.

Kaspereit, M., Seidel-Morgenstern, A. and Kiemle, A. (2007), *J. Chromatogr. A* **1162**, 2–13.

Mazzotti, M. (2006a), *Ind. Eng. Chem. Res.* **45**, 6311–6324.

Mazzotti, M. (2006b), *J. Chromatogr. A* **1126**, 311–322.

Mazzotti, M., Storti, G. and Morbidelli, M. (1994), *AIChE J.* **40**(11), 1825–1842.

Mazzotti, M., Storti, G. and Morbidelli, M. (1996), *AIChE J.* **42**(10), 2784–2796.

Migliorini, C., Mazzotti, M. and Morbidelli, M. (1998), *J. Chromatogr. A* **827**(2), 161–173.

Migliorini, C., Mazzotti, M. and Morbidelli, M. (2000), *AIChE J.* **46**(7), 1384–1398.

Nicoud, R. M. (1992), *LC-GC International* **5**(5), 43–47.

Nicoud, R. M. (2014), *Ind. Eng. Chem. Res.* **53**, 3755–3765.

Rhee, H.-K., Aris, R. and Amundson, N. R. (1971), *Philos. Trans. R. Soc., A* **269**, 187–215.

Ruthven, D. (1983), *Can. J. Chem. Eng.* **61**(6), 881–883.

Ruthven, D. M. (1984), *Principles of Adsorption and Adsorption Processes*, John Wiley & Sons.

Ruthven, D. M. and Ching, C. B. (1989), *Chem. Eng. Sci.* **44**(5), 1011–1038.

Seidel-Morgenstern, A., Schmidt-Traub, H., Michel, M., Epping, A. and Jupke, A. (2012). In H. Schmidt-Traub, M. Schulte and A. Seidel-Morgenstern, eds., *Preparative Chromatography*, 2nd edn., Wiley-VCH, p. 321.

Storti, G., Masi, M., Carra, S. and Morbidelli, M. (1988), *Prep. Chromatogr.* **1**(1), 1–27.

Storti, G., Masi, M., Carra, S. and Morbidelli, M. (1989), *Chem. Eng. Sci.* **44**(6), 1329–1345.

Storti, G., Masi, M. and Morbidelli, M. (1993). In G. Ganetsos and P. E. Barker, eds., *Preparative and Production Scale Chromatography*, vol. 61 of *Chromatographic Science Series*, Marcel Dekker, pp. 673–700.

Storti, G., Mazzotti, M., Morbidelli, M. and Carra, S. (1993), *AIChE J.* **39**(3), 471–492.

Streat, M. (1986a). In A. E. Rodrigues, ed., *Ion Exchange: Science and Technology*, vol. 107 of *NATO ASI Series, Series E: Applied Sciences*, Martinus Nijhoff, pp. 449–461.

Streat, M. (1986b). In A. E. Rodrigues, ed., *Ion Exchange: Science and Technology*, vol. 107 of *NATO ASI Series, Series E: Applied Sciences*, Martinus Nijhoff, pp. 289–335.

Sundaresan, S., Wong, J. and Jackson, R. (1987), *AIChE J.* **33**(9), 1466–1472.

Valentin, P. (1981). In A. E. Rodrigues and D. Tondeur, eds., *Percolation Processes: Theory and Applications*, vol. 33 of *NATO ASI Series, Series E: Applied Sciences*, Sijthoff & Noordhof, pp. 141–195.

Wankat, P. C. (1986), *Large-scale Adsorption and Chromatography, Vol II*, CRC Press.

Wesselingh, J. and Van Der Meer, A. (1986). In A. E. Rodrigues, ed., *Ion Exchange: Science and Technology*, vol. 107 of *Nato ASI Series E: Applied Science*, Martinus Nijhoff, pp. 289–318.

9 Chromatographic modes and their optimization

Optimizing for sure . . . but what and how?

9.1 Introduction

Options for implementing purification systems based on chromatographic processes are almost infinite in number, limited only by the creativity of the engineer. Even for a given chromatographic medium, one can vary the carrier fluid composition, use single-column or multi-column systems, connect columns in different ways, recycle partly purified fractions, change flow directions, take advantage of physical or chemical modulations and so on. These different ways of implementing chromatography for a given chromatographic medium are called *chromatographic modes*. A variety of articles presenting and comparing chromatographic modes have been published Bailly and Tondeur (1984); Wankat (1986a,b); Nicoud and Perrut (1991); Nicoud (1992); Tondeur and Bailly (1993a,b); Seidel-Morgenstern (1998); Kaspereit *et al.* (2012). Trying to find *the* best mode to systematically surpass the others is probably hopeless: if insulin is normally produced on batch chromatography columns, para-xylene on 24-column SMB systems and optical isomer separations on 6-column SMB or Varicol, there are certainly good reasons for that. The different chromatographic modes have to be considered as different tools belonging to the same toolbox: according to the problem to be solved, the right tool has to be used.

Prior to presenting our approach to optimization, let us describe a few process alternatives. The subsequent sections are not aimed at presenting a complete list of reported options, but rather at giving food for thought.

9.2 Presentation of chromatographic modes

9.2.1 Single-column systems

The simplest way of operating preparative chromatography is so-called elution chromatography, based on periodic injections of a mixture at the inlet of a single column, purified fractions being collected at the outlet. The basic principle is therefore equivalent to analytical chromatography, even if the objective is different.

The industrial efficiency of batch elution chromatography is certainly related to the ability to build high-efficiency columns (expressed in number of plates). The main evolution is from dynamic axial compression technology, permitting stable and efficient beds (Colin *et al.*, 1990). Detailed information on elution chromatography and associated techniques is given in (Toumi *et al.*, 2012) and (Kaspereit *et al.*, 2012). It is common knowledge that industrial quantities of insulin, paclitaxel, peptides, proteins, polyunsaturated fatty acid (PUFA) ethyl esters, steroids etc. are obtained using elution chromatography.

However, these references should not hide reality: while elution chromatography is widely used for purifying biomolecules (Carta and Jungbauer, 2011), there is still a high reluctance to use chromatography for small organic molecules in the fine chemical and pharmaceutical industries. The aphorism "avoid chromatography as far as possible" is very common, elution chromatography being normally considered expensive due to limited productivity and significant eluent consumption.

Dilution and limited productivity in elution chromatography are primarily linked to the non-linearity of the adsorption isotherms and to the broadening of the peaks due to low column efficiency (kinetic limitations or hydrodynamic dispersion). When one has exhausted the options for higher capacity, higher selectivity and improved kinetic characteristics, one still has a few cards to play.

One card is to change the nature of the eluent and move from liquids to supercritical fluids. General reviews of supercritical fluid chromatography (SFC) principles and applications are available in (Nicoud *et al.*, 1999), (Jusforgues *et al.*, 1998), (Guiochon and Tarafder, 2011) and (Rajendran, 2012). The invention of preparative SFC is normally credited to (Klesper *et al.*, 1962), with protection of eluent recycling credited to (Perrut, 1982), the latter invention being obviously mandatory for large-scale, economic implementation. Over 50 years after the basic invention, and 30 years after a key improvement, and despite numerous developments, there are very few examples of *large-scale* industrial separations performed by SFC (although laboratory-scale examples are numerous).

The main reasons for the lack of commercial success of large-scale SFC are the extra cost imposed by high-pressure technologies, difficulties in solubilizing polar compounds, the requirement for co-solvent, and … progress in liquid chromatography. The argument that "SFC is green" is weak because modern large-scale liquid chromatography systems include evaporation/recycling devices that allow the recycling of more than 99% of the eluent (Nicoud, 2014).

Accepting the need to work with liquid eluent, remaining possibilities for improving single-column chromatography include adapting piping and control, recycling of partially purified fractions, implementing chemical modulation and implementing physical modulation. These four options are briefly presented below.

9.2.1.1 Adapting piping and control

In this section we refer to relatively simple adaptations of the classical elution mode that can lead to substantial improvements. As an illustration, instead of using a gradient (step or continuous) for removing highly adsorbed solutes, one can reverse the eluent flow (back-flush) after collection of the product of interest at the outlet of the column:

the late eluting compounds are thus recovered at the inlet of the column. Assuming no complex non-linear behavior, the undesirable products are recovered at a time that is about twice the time at which the flow has been reversed (the products move forward and back at the same velocity). Using a back-flush instead of a gradient allows one to avoid the difficult recycling of complex solvent mixtures and lengthy equilibration times. This is discussed further in Section 10.4.

This type of implementation can be refined with the "flip-flop" concept proposed by (Colin *et al.*, 1991), in which the direction of the mobile phase flow is periodically reversed and products injected and recovered alternatively at the two extremities of the column. This shows some merit in decreasing cycle times when the mixture contains solutes associated with significantly different retention times. This procedure has been demonstrated for the separation of steroids (Colin *et al.*, 1991).

9.2.1.2 Recycling of partially purified fractions

In order to maximize the productivity of batch chromatography, some authors (Dingenen and Kinkel, 1994) have proposed injecting more than the maximum amount, allowing them to obtain a touching-band situation, collect the purified part of the chromatogram and recycle the non-separated fraction to the inlet of the column. This technique is often called *recycling and peak shaving*. Recycling the unresolved part of the chromatogram at the inlet of the column permits working with a virtual column that is longer than the real column. Hydrodynamic dispersion associated with the recycling loop must be carefully controlled, using an adapted system with small size tubing or so-called *external recycling* (Grill, 1999).

The recycling and peak shaving mode certainly helps to improve the recovery yield for low-selectivity separations. Comparison with classical elution in terms of productivity is not that clear, and may vary case by case. For a specific example, it has been shown (Seidel-Morgenstern, 1998) that productivities obtained for elution and recycling and peak shaving chromatography are very similar, but that recycling and peak shaving allows reduction of eluent consumption by about 30%. A more systematic and theoretical approach has shown that productivity is almost always lower for recycling and peak shaving than for elution chromatography (Seidel-Morgenstern and Guiochon, 1993).

Various modifications of this basic recycling mode have been proposed in order to get a steady-state recycling process. The idea consists in re-injecting fresh feed in association with the recycled fraction, the approaches differing in how the fresh feed is re-injected with respect to the recycled fraction:

- *Recycling and mixing with fresh feed*: the recycled fraction is mixed with fresh feed before re-injection
- *Segmented recycle*: the fraction to be recycled is collected in different fractions, these fractions being re-injected after reversal of their elution order[1]

[1] This may look awkward because it ruins the partial separation already achieved. The idea of reversing the fraction order is to take advantage of the non-linear behavior of the system and to promote what could be called "self displacement of the products", i.e. the recompression of the self-broadened peaks.

- *Mono-segmented recycling*: a pulse of feed is injected immediately after the injection of the recycled fraction
- *Steady-state recycling (Cyclojet)*: the feed is injected at the middle of the recycled fraction.

Detailed descriptions of the first three modes can be found in (Bailly and Tondeur, 1982), (Bailly and Tondeur, 1984) and (Tondeur and Bailly, 1993a), while detailed descriptions of *steady-state recycling* can be found in (Grill and Hampton, 1997), (Grill, 1998), (Grill, 2000) and (Quinones *et al.*, 2000). More recently, (Siitonen and Sainio, 2011) have suggested concentrating the recycled fraction prior to re-injection and compared different options for practical implementation.

9.2.1.3 Chemical modulation

As with analytical chromatography, a variation of the eluent strength (gradient or step-wise evolution of the eluent composition) can be implemented at production scale. Its value is in minimizing the total elution time while maintaining separation ability. While gradient elution is an extremely useful tool in bio-chromatography, it must be kept in mind that it makes solvent recycling very difficult. Solvent recycling being a key to designing cost-effective chromatographic processes, use of gradients must normally be limited to situations where no other choices are available.

Another way to make use of modulation of the eluent composition is so-called *displacement chromatography* (Carta and Jungbauer, 2011). To understand the concept, recall that, owing to the non-linearity of the adsorption isotherms, when a strongly retained solute replaces a weakly retained solute one obtains a sharp (compressive) front, a broadening (dispersive) front being obtained when a weakly retained solute replaces a strongly retained solute. Solutes are normally less adsorbed than eluent, so one typically obtains characteristic asymmetric overloaded peaks with one sharp front and a broadening rear. The basic principle of displacement chromatography consists in using a displacer which is more strongly adsorbed than the solutes for desorbing them. Instead of getting successive broadening peaks, one obtains a succession of concentrated bands of solutes.

Displacement chromatography has been known for a long time and was extensively used for the purification of rare earth elements in the Manhattan Project (Spedding *et al.*, 1947). Other examples are separations of resorcinol, catechol and phenol (Frenz and Horvath, 1985); of peptides and related products (Cramer and Horvath, 1988; Liao *et al.*, 1987) and of peptides, antibiotics and proteins (Subramanian *et al.*, 1988).

While very attractive *prima facie*, this chromatographic mode presents a significant drawback: after separation, the column is equilibrated with the displacer and must be regenerated by the eluent. This regeneration step involves the desorption of a strongly adsorbed component (the displacer) by a weakly adsorbed product (the eluent), so it is in general highly dispersive, ruining most if not all of the benefits of the separation step. Unfortunately the influence of this regeneration step is seldom addressed in articles on displacement chromatography.

A detailed comparison between elution and displacement chromatography is provided by (Felinger and Guiochon, 1993). Their main conclusion is that productivity is normally higher in elution chromatography than in displacement chromatography, and that, especially due to the constraint of regenerating the displacer, "for the large-scale production of chemicals whose adsorption behavior is described by a Langmuir or related isotherm, the preferred mode of implementation of preparative chromatography should be elution."

However, there are noticeable exceptions allowing a displacement effect to be easily implemented. This is, for instance, the case when a pH modulation allows modification of the affinity of the species for the chromatographic medium. As an illustration, amino acids can be fixed on cationic ion-exchange resins at a pH lower than their isoelectric points, where they are in cationic form. The column is then easily regenerated by increasing the pH because the amino acids, now in an anionic form, are immediately released from the resin. Such behavior has been modeled in Chapter 7. Examples of amino acids separated using this procedure can be found in (De Carli *et al.*, 1990). A well-known bio-industrial illustration is given by the use of protein A to purify monoclonal antibodies (Carta and Jungbauer, 2011): the procedure involves passage of clarified cell culture supernatant over the column at pH 6–8; under these conditions the antibodies bind and unwanted components such as host cell proteins, cell culture media components and putative viruses flow through the column. The procedure is continued with an optional wash step to remove non-specifically bound impurities from the column, and finally the product is recovered by elution at pH 3–7.

9.2.1.4 Physical modulation

Regeneration of the column is normally performed by using the *chemical energy* brought by the eluent. An idea aimed at minimizing the solvent requirement is to bring *physical energy* (for instance thermal or electrical) into the system. This idea, known as *parametric pumping*, has been presented by (Tondeur *et al.*, 1985) and reviewed by (Grevillot, 1986).

In gas-phase adsorption processes, a classical way of implementing a physical modulation consists in performing the adsorption at a high pressure (the solutes being strongly adsorbed) and the regeneration at a lower pressure (the solutes being weakly adsorbed). This mode of operation, known as *pressure swing adsorption* (PSA), has led to several applications including air drying, enrichment of oxygen from air and hydrogen purification (Grevillot, 1986).

Halfway between gases and liquids, the use of supercritical fluids exhibiting adsorption properties that are strongly affected by pressure looks like an interesting option for parametric pumping. Using pressure modulation in SFC is illustrated in Section 9.2.2.3.

Pressure having virtually no influence on adsorption in the liquid phase, it cannot be used as a way to modulate adsorption, but temperature can be used instead. This mode is known as *temperature-swing adsorption* (TSA). The pioneering work published by (Wilhelm *et al.*, 1966) and (Wilhelm and Sweed, 1968) on the separation of toluene and n-heptane has opened the way for applications such as benzene–n-hexane separation, removal of NaCl from water, separation of cations or of gases like ethane–propane, argon–propane and propylene–propane (Grevillot, 1986). While interesting in principle, practical application of TSA is limited because the time required for thermal equilibration can easily ruin the benefits expected from this technique. To overcome this limitation, attempts have been made to use inductive heating to limit the heating to the surface of the adsorbent where the adsorption takes place (Hauck, 1997).

For separation of ionic species, advantage can be taken of electrical modulation (Chen *et al.*, 1981; Oren and Soffer, 1983a,b): according to the intensity of an electric/magnetic field applied to the column, solutes can be more or less strongly adsorbed. Again, these ideas seem to have a limited range of industrial applicability.

9.2.2 Multi-column systems

Just as we considered elution chromatography to be the reference for single-column systems, the four-zone SMB system, being widely used, is certainly the reference for multi-column systems.

The principles of the true moving bed (TMB) and the simulated moving bed (SMB) as invented by (Broughton and Gerhold, 1961) have been explained in Chapter 8 and presented in different articles (Kaspereit, 2009; Kaspereit *et al.*, 2012; Gomes and Rodrigues, 2012). This chromatographic mode has been extensively used in petrochemical applications and sugar separations (Ruthven and Ching, 1989; Wankat, 1986b), and has given rise to a strong interest in the fine chemical and pharmaceutical industries since the 1990s (Nicoud, 2014).

If the production of para-xylene on zeolites represents the largest application of chromatography, the separation of fructose and glucose on cationic resins is the second-largest one, and has been extensively studied (Ganetsos and Barker, 1993). SMBs packed with cationic resins in calcium form have also been used for obtaining other monosaccharides such as xylose and arabinose (Balannec and Hotier, 1993) and for performing the separation of mono- and disaccharides and between disaccharides (Kishihara *et al.*, 1989; Nicoud, 2000).

Desalting is another simple and interesting application of SMB in biotechnology, different mechanisms being possible such as ion exclusion, hydrophobic interaction and size exclusion (Hashimoto *et al.*, 1988).

The first attempt to purify proteins by SMB can probably be credited to (Huang *et al.*, 1986), who performed the purification of trypsin from extract of porcine pancreas. Almost 30 years later, the first industrial SMB purification of therapeutic protein is still awaited, possibly not for long.

SMB technology has also been used to purify different natural products, both ionic ones like lysine (Van Walsem and Thompson, 1997) and betaine (Kampen, 1990) and apolar ones like polyunsaturated fatty acids (Perrut *et al.*, 1998).

A major breakthrough in recent years is certainly the separation of optical isomers by SMB (MacCoy, 2000; Perrin and Nicoud, 2001; Nicoud, 2014). For a review, the reader is referred to (Rajendran *et al.*, 2009).

Modifications of the reference four-zone SMB are aimed either at minimizing eluent consumption and maximizing productivity or at obtaining multiple purified fractions. This subject has been discussed by (Nicoud, 2000) and by (Nicoud, 2014).

9.2.2.1 Adapting piping and control

This section covers relatively small technical adaptations to the reference four-zone system (SMB) that can lead to meaningful improvements.

The Varicol process is a multi-column continuous chromatographic process proposed by (Ludemann-Hombourger, Nicoud and Bailly, 2000) and (Adam *et al.*, 2000): it uses SMB hardware but with a different mode of control. While the inlet and outlet lines are shifted synchronously in SMB, they are shifted asynchronously in Varicol so that the zone lengths vary with time. This allows one to work with a non-integer average number of columns per zone. This offers fine tuning opportunities, making Varicol more productive than SMB, as confirmed by (Toumi *et al.*, 2002), (Ludemann-Hombourger, Nicoud and Bailly, 2000) and (Ludemann-Hombourger *et al.*, 2002). Various large-scale optical isomer separations are now performed with Varicol technology.

Instead of varying the zone length over time, varying feed concentration or inlet/outlet flow rates have been proposed (Kaspereit, 2009; Seidel-Morgenstern *et al.*, 2012; Nicoud, 2014).

Another alternative to the four-zone system consists in withdrawing the liquid output of zone 1 and concentrating it prior to entering zone 2. The concentration of the highly retained solutes provokes a displacement of the less retained solutes, which leads to productivity or purity improvement (Abdelmoumen *et al.*, 2006; Kaspereit, 2009).

A further alternative consists in using a three-zone system, meaning a system in which zone 4 of the system described in Figure 8.16 has been removed. The entire flux exiting zone 3 is thus collected in the raffinate stream and sent to the concentration unit. The value of the three-zone scheme is in decreasing the overall bed length of the system and thus maximizing productivity in a simpler system. These productivity gains are, however, mitigated by a higher eluent flow rate to the raffinate concentration unit. The three-zone SMB process is an economical alternative to the four-zone one when eluent consumption is not an important concern relative to chromatographic media or investment costs, for instance. This may happen in specific cases, for example in removing salts from biological products or sucrose from molasses.

In order to take the best of two worlds offered by the three-zone and four-zone configurations, one can consider systems oscillating between the two configurations. This is the case for so-called I-SMB and S-SMB, proposed by Mitsubishi and Novasep (Nicoud, 2014) and widely used in the sugar industry. For information on the I-SMB design and performance, the reader is referred to (Katsuo and Mazzotti, 2010a), (Katsuo and Mazzotti, 2010b) and (Katsuo *et al.*, 2011).

For very simple separations associated with a very high selectivity between solutes, for instance the so-called capture step of proteins, one can even consider continuous two-zone systems. In general, two-zone systems are better adapted for enriching than for highly purifying mixtures. For a presentation of possible two-zone configurations the reader is referred to (Tondeur, 1981) and to (Rodrigues *et al.*, 2008).

There is a quest, more or less justified, to decrease the number of columns used in SMB-type systems (Nicoud, 2014). As illustrations, (Rodrigues *et al.*, 2008) have described a two-column SMB and (Jin and Wankat, 2005, 2006) have proposed a two-zone SMB which incorporates a storage tank to hold desorbent for a certain time before further use. Stretching the limits of the exercise, (Abunasser *et al.*, 2003) and (Mota and Araujo, 2005) have proposed a one-column system analogous to a four-zone SMB using additional tanks. These examples require us to stress the difference between the number of columns and the number of zones. The number of columns is intuitively higher than the number of zones, as one should in principle have at least one column per zone. However, by using asynchronous port switching and/or intermittent feeding and withdrawal, one can work with a number of columns lower than the number of zones.

Instead of decreasing the number of zones from four down to three or two, one can consider adding zones and designing, for instance, a five-zone system. There are basically two reasons for adding a fifth zone to an SMB:

- *Recovery of a third fraction:* The addition of a fifth zone simply allows three fractions to be obtained, two being pure, one containing an enriched mixture. For instance, by adding a fifth zone at the top of the TMB system in Figure 8.16, from a feed containing three solutes A, B and C (A less retained than B, B less retained than C), one can obtain one fraction containing pure A, the second one a mixture A+B and the third one pure C.
- *Removal of strongly retained products:* By adding a fifth zone at the bottom of the TMB system in Figure 8.16 that is eluted with a stronger desorbent than the one used in zones 1 to 4, the system simultaneously performs the separation of A and B and the removal of the strongly retained impurities.

The classical four or even five-zone system can only deliver a pure fraction if the target solute is the least or the most retained. If one is interested in purifying a fraction containing a solute having an intermediate retention while keeping the classical TMB/SMB concept, one needs to add further zones. Systems including up to nine zones have been proposed (Nicolaos *et al.*, 2001a,b). The value of these modes is rather limited, because of their complexity but also because the productivity obtained using two classical four-zone SMBs connected in series is normally higher.

In order to get more than two pure fractions with a limited number of columns, one can again use asynchronous port switching and/or intermittent feeding and withdrawal. As an illustration, (Ando *et al.*, 1990) proposed a system operating partly in counter-current mode and partly in fixed-bed mode to achieve ternary separation. This system has been further studied by (Borges da Silva and Rodrigues, 2008) and (Lee and Wankat, 2010). Different systems aimed at performing ternary separations with two-zone SMBs have been proposed by (Hur and Wankat, 2005) and (Hur and Wankat, 2006). Systems involving gradients are presented in the following sections.

Finally, let us mention other multi-column schemes, including *moving feed* and *moving port*, that typically combine elution and port movement (Wankat, 1986b) and are not aimed at simulating counter-current contact between the fluid and the chromatographic medium.

9.2.2.2 Recycling of partially purified fractions and implementing gradients

Deviations of the systems presented in this section from a reference four-zone SMB are more important as they involve either a solvent composition gradient, internal recycling of partially purified fractions or internal by-pass.

As with elution chromatography, the concept of recycling partly purified fractions can be extended to multi-column systems (continuous or not). Continuously recycling some internal fractions can, for instance, be beneficial for binary separations when very high purities are required as shown by (Hotier, 1993).

Additionally, nothing prevents one from operating a classical four-zone SMB in a gradient mode. A steady-state spatial gradient composition can sometimes be easily obtained: using solvents of different compositions in the feed and eluent lines can promote the existence of an eluent composition gradient in the system that can lead to performance enhancement (Jensen *et al.*, 2000; Abel *et al.*, 2002; Antos and Seidel-Morgenstern, 2002; Houwing *et al.*, 2002; Abel *et al.*, 2004).

The significance of the spatial gradient is easily understood by considering the case of a linear system. We have shown in Chapter 8 that normalized flow rates must satisfy $m_3 - m_2 < \overline{K}_B - \overline{K}_A$ for complete separation. More precisely, assuming that the different zones are associated with different eluent compositions, the above inequality can be written $m_3 - m_2 < \overline{K}_B(3) - \overline{K}_A(2)$, where $\overline{K}_S(k)$ represents the lumped Henry's coefficient of solute S in zone k. By changing the eluent strength in the different zones, one can increase the difference $m_3 - m_2$ and thus the feed flow rate.

Multi-column systems can also be operated with temporal composition gradients, possibly associated with fraction recycling and/or by-pass. These ideas have been used for multi-solute bioseparation in systems referred to as MCSGP (multi-column counter-current solvent gradient purification), developed by ETH (Strohlein *et al.*, 2006), and GSSR (gradient with steady-state recycling), developed by Novasep (Silva *et al.*, 2010). Using astute sequencing, but losing the continuous feature, these have been practically implemented in systems containing column numbers as low as three (Aumann and Morbidelli, 2007).

A specific need for biomolecules is associated with the so-called capture step, illustrated by the use of protein A to purify monoclonal antibodies, presented in Section 9.2.1.3. A typical batch purification scheme involves different steps to be performed in sequence: loading, washing, elution, regeneration, equilibration. Systems allowing continuous performance of these tasks in a multi-column environment have been proposed: the BioSc concept of Novasep (www.novasep.com), capture SMB from ETH (www.chromacon.ch) and periodic counter-current packed bed chromatography (3C-PCC or 4C-PCC) from GE (www.gelifesciences.com).

While the variety of proposed systems is a source of complexity, not to say confusion, it must be noticed that many can be seen as systems oscillating between two or three simple states (Nicoud, 2014). In addition to I-SMB, S-SMB, MCSGP, GSSR, 3C-PCC, BioSc and Capture-SMB discussed and presented in (Nicoud, 2014), systems like those described by (Ando *et al.*, 1990; Hur and Wankat, 2005) and (Hur and Wankat, 2006) obey the same logic. As a further illustration of systems oscillating between different states, let us mention the so-called Relay-SMB, which replaces the continuous fluid split required at the points of collection of the raffinate and extract with three-way valves, allowing the system to oscillate between two configurations (Silva *et al.*, 2012).

9.2.2.3 Physical modulation

In Section 9.2.1.4 we emphasized the potential value of physical gradients, as opposed to chemical gradients. Supercritical fluids are well suited to this purpose because the adsorption isotherms vary strongly with pressure (Chapter 4). Supercritical fluid SMB consists not only in replacing the liquid eluent by a supercritical fluid but in implementing a pressure gradient in the different zones of the bed (Nicoud *et al.*, 1992, 1999; Clavier *et al.*, 1996; Mazzotti *et al.*, 1997).

This technology has been investigated theoretically by (Mazzotti *et al.*, 1997) and (Di Giovanni *et al.*, 2001)[2] and tested on a separation of fatty acid ethyl esters (Clavier *et al.*, 1996), of isomers of phytol (Depta *et al.*, 1999) and of optical isomers (Denet *et al.*, 2001). Significant productivity gains have been confirmed, but no industrial applications have been reported at the time of writing.

9.2.3 Combining different modes or steps

For complex mixtures containing a large number of solutes with possibly different physico-chemical characteristics such as hydrophobicity, ionizability or size, it is common practice to associate different chromatographic steps in series. These can, for instance, be associated with different chromatographic media. As an illustration, protein purification often requires affinity chromatography, ion exchange, size exclusion and hydrophobic interaction. Another example is lysine purification, which associates a strong cationic step with a weak anionic step (Van Walsem and Thompson, 1997), both steps being performed continuously. Other examples involving SMB are the purification of cyclosporine (Nicoud, 2000), paclitaxel (Wu *et al.*, 1999) and an ascomycin derivative (Küsters *et al.*, 2000).

When the mixture contains many solutes having similar characteristics, it may be appropriate to connect two or three systems packed with the same chromatographic media, and to perform the purification in different subsequent steps. This is, for instance, the case with EPA purification (Perrut *et al.*, 1998), associating batch and SMB, or SMB and SMB. Notice that there are many ways to connect SMB systems in series, as illustrated in (Kessler and Seidel-Morgenstern, 2006), (Kim *et al.*, 2003) and (Kim and Wankat, 2004).

9.2.4 Conclusion

Options for implementing chromatography are only limited by the creativity of the engineer, and it is consequently difficult to find a way through this very rich environment. I am unable to give even simple rules to allow one to select the optimum process for a given separation process. In the past, I was convinced that, at small scale,[3] elution/batch should be favored for simplicity reasons, whereas continuous processes

[2] It is interesting to note that the behavior of a linear SF-SMB leads to unrealistic behavior as it allows solute concentrations to increase indefinitely. This example shows that non-linearity absolutely must be taken into account to obtain realistic behavior.

[3] The exact definition of "small" does not matter.

should be preferred at large scale for cost reasons. I am not even sure about these very simple qualitative rules of thumb:

- I have seen some laboratory managers very happy to operate micro-SMB (with columns of 4.6 mm diameter) because it saved liters of eluent per day in a very constrained laboratory environment.
- While the switch to continuous was the way to think big (scale), the biopharmaceutical industry is interested in continuous systems for thinking small (size).
- In pharmaceutical industries, uncertainties about the long-term needs for a given product may tend to orient towards batch systems. The *perception* that a large-scale batch system is more easily reusable than a highly optimized but specific continuous system may compensate for other drawbacks.

A key is understanding what an optimum process can be. We will see in Section 9.3 that defining an objective function requires rigor, as the chromatographic system must not be considered alone but as part of a broader environment. I am convinced that there is not a single system surpassing the others. Depending on whether one is interested in minimizing cost, eluent consumption, chromatographic media inventory, recovery yield, system compactness, footprint, quality controls, flexibility, ease of starting and stopping or cleaning, very different systems may be considered.

9.3 Presentation of the optimization problem

Assuming a reasonable knowledge of physico-chemical parameters, the tools given in the previous chapters allow simulation of the behavior of chromatographic systems. These tools can now be used to answer the following question: for a given separation problem, how can we determine the chromatographic system that will maximize the performance according to criteria selected by the user?

Because of the variety of possible chromatographic implementations and of possible industrial situations, answering this question is a task of tremendous difficulty. Many contributions aimed at optimizing chromatographic processes have been published and will be referred to later in this chapter. These contributions are normally based on simple assumptions about the objective function, the constraints, the chromatographic mode, the chromatographic model etc., so they often have more academic than immediate industrial applicability. Instead of proposing detailed results only valid for specific situations, we will propose a methodology that, for a problem presented in its full complexity, will give a path allowing the derivation of reasonable solutions.

Properly solving an optimization problem requires that we proceed via four steps:

- recognizing the set of "tuning parameters" on which it is possible to act; such parameters include column length, fluid velocity, injection time and others
- recognizing the constraints of the problem, which may be associated with a "constrained parameter" (for instance: column length must be less than 1 m) or a "constrained result" (for instance: yield must be greater than 75%)

- selecting an objective function to be minimized or maximized
- selecting an optimization algorithm.

These steps are discussed in the following sections.

9.3.1 Variables, tuning parameters, constrained parameters and results

We will reserve the term "variable" for time, axial abscissa and possibly radial abscissa, used to generate chromatograms via mass balances.

In order to optimize the chromatographic process, one can act on several "tuning parameters" which can be organized in different categories and subcategories. A list, possibly not exhaustive, is given below.

- Discrete tuning parameters:
  - Chemical: nature of the carrier fluids and of the chromatographic medium
  - Chromatographic mode: batch with or without recycling, multi-column like SMB, more generally one of the modes presented in Section 9.2, number of columns per zone for SMB
  - Solvent processing devices: decision to recycle the eluent or not; if yes, the choice of technology
  - Particle shape: spherical or irregular
- Continuous tuning parameters:
  - Physical: temperature, pressure
  - Geometrical: particle size, column diameter and length
  - Chemical: feed concentration and purity, composition of the carrier fluid
  - Chromatographic process parameters: injection volume, cycle time, cut times, flow rate for elution flow rates and period for SMB, etc.
  - Solvent processing parameters: recycling ratio, composition of the recycled streams etc.

The feed concentration C_{tot}^F and solute purity X_i^F deserve some specific comments:

- C_{tot}^F is defined as the total amount (mole or kg) of solutes of interest or of impurities per liter of solution.
- The process just upstream of the chromatographic unit delivers an effluent that is rarely adapted to optimal separation (Murphy's law):
  - The effluent may be too dilute to ensure decent productivity, and must be preconcentrated.
  - Alternatively, it may be too concentrated (thus subsequent miscibility issues with the carrier fluid, precipitations etc.), thus requiring dilution with a suitable solvent.
  - The solvent in the feed may even be incompatible with the chromatographic medium, requiring its elimination and dilution in an appropriate solvent.

- The preconcentration or predilution affects C^F_{tot} and thus the shape of the chromatogram, which in turn affects the cost of the separation. Consequently the concentration C^F_{tot} of the feed stream must be considered a tuning parameter.

The purity X_i of a given solute i is defined as the ratio of the concentration C_i of this solute to the total concentration C_{tot}:

$$X_i = \frac{C_i}{C_{tot}} \qquad X_i^F = \frac{C_i^F}{C_{tot}^F} \quad \text{in the feed} \tag{9.1}$$

It is important to note that the purities are defined *without* taking the solvent into consideration. We will see later that purities X_i^F are in general tuning parameters.

The distinction between discrete and continuous tuning parameters is intuitively important as these two groups of parameters cannot be treated in the same way in an optimization algorithm.

It is important to realize that process optimization requires *simultaneous* optimization of all the tuning parameters. As an illustration, everything else being fixed, the optimum particle size for elution chromatography and for SMB are different. This demonstrates that selecting chromatographic conditions from simple educated observation of laboratory chromatograms without using adequate simulation software is almost certain to result in options that are not optimized.

Let us now present the constraints. Assuming that the problem consists in purifying one solute (*) contained in a feed mixture, at least one constraint is expected to prevail: the purity of the product to be recovered should exceed a given target. This can be expressed as:

$$X_* > X_*^{min} \tag{9.2}$$

where X_* is the purity of the compound of interest in the purified fraction and X_*^{min} is the minimum purity to be reached. The purity X_* is our first "constrained result".

It is common to see constraints other than purity prevailing. For instance, one may want to reach a minimum recovery yield:

$$\theta_* > \theta_*^{min} \tag{9.3}$$

or to force certain impurities to be present at levels lower than a given value:

$$X_{imp} < X_{imp}^{max} \tag{9.4}$$

In addition to constraints related to solutes, other technical constraints may prevail:

- Geometry: column length and diameter limited by the available technology
- Pressure: limited because of chromatographic media stability
- Temperature: limited because of chromatographic media or product stability
- Solvent choice and composition: frequently limited due to chromatographic media or product stability as well as incompatibility with regulatory constraints.

These constraints may apply to the "constrained results" (purity, yield, pressure drop etc.) or to the "constrained parameters" (solvent composition, column geometry etc.).

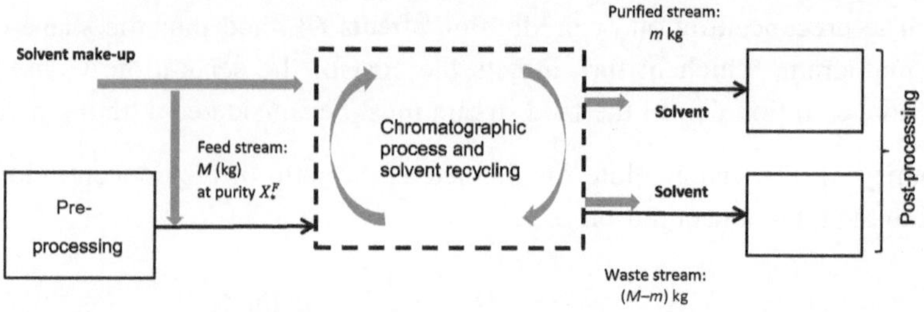

Figure 9.1 Principle of the general production/purification process. The flux of solutes is represented by black arrows. Solvent fluxes are represented by gray arrows. Quantities in kg represent the masses of solutes.

The most important constraint, which is not always clearly expressed, comes from the environment: chromatography being normally (not to say systematically) embedded in a more complex process as described in Figure 9.1, it cannot be optimized independently of this environment.

Let us define the amount M (kg) of feed material (excluding the solvent) to be processed by the chromatographic system. The feed stream, at concentration C_{tot}^F, contains the target solute at purity X_*^F, and it is produced by an *upstream* process possibly including chemical synthesis, fermentation, extraction from biomass or various purification processes. At the outlet of the chromatographic step, one obtains a stream containing a mass m of purified material containing the target solute at purity X_*. The mass of pure solute in the purified material is m_*.

A simple mass balance on the targeted solute gives:

$$m_* = \theta_* M X_*^F \qquad \text{(kg of pure solute)}$$
$$m = \frac{m_*}{X_*} \qquad \text{(kg of purified material)} \tag{9.5}$$

where θ_* is the recovery yield of the targeted solute.

The quantities M, m and m_* refer to products entering and exiting the dashed box in Figure 9.1. Possible recycling of partly purified fractions at the inlet of the chromatographic column is included in the dashed box. In parallel with the purified stream, one obtains waste streams, which are for convenience lumped in one single stream in Figure 9.1.

In order to avoid confusion with biopharmaceutical terminology,[4] instead of using the terms *upstream* and *downstream*, we will speak of processes located *pre* or *post* the chromatographic step.

The chromatographic process requires solvents that are partly delivered via the feed stream and partly via the eluent stream. A solvent can be partly recycled using devices directly connected with the chromatographic process and thus located inside the dashed box in Figure 9.1. The non-recycled fraction of solvent is recovered either in the purified

[4] In biopharmaceutical industries, *upstream* typically refers to fermentation and cell culture steps excluding purification steps, while *downstream* refers to the purification steps, be they chromatographic or not.

stream or in waste streams. The solvent streams are symbolized with gray arrows in Figure 9.1; they contain eluent mixtures of constant or variable composition.

A mass balance on the quantity of solvent entering and exiting the chromatographic dashed line box gives:[5]

$$\Upsilon^{inl} = \Upsilon_p^{out} + \Upsilon_w^{out} \qquad \text{volume of eluent to obtain } m \text{ kg of purified material} \tag{9.6}$$

where Υ^{inl} represents the volume (l) of eluent contained in the feed and fresh eluent make-up streams to obtain m kg of purified material; Υ_p^{out} and Υ_w^{out} represent respectively the volumes of eluent contained in the purified and waste streams.

Because of the possible recycling of solvent with devices (evaporation, for instance) directly coupled with the chromatographic equipment, the net quantity of solvent Υ^{inl} entering/exiting the chromatographic process (the dashed box of Figure 9.1) is in general much lower than the net quantity of solvent Υ^{chr} passing through the chromatographic unit. We define the *solvent recycling ratio* S^R by:

$$\begin{aligned} \Upsilon^{rec} &= S^R \Upsilon^{chr} \\ \Upsilon^{inl} &= (1 - S^R)\Upsilon^{chr} \end{aligned} \tag{9.7}$$

which allows expression for the different solvent volumes as a function of the volume Υ^{chr} passed onto the chromatographic system and of the solvent recycling ratio S^R.

If the recycling ratio S^R equals 0.9, 90% of the solvent used for chromatography is recycled in a closed loop associated with the chromatography unit, so that only 10% of the solvent requirement is brought by the feed and eluent make-up streams.

In addition to purity, for large-scale production a *production rate* $\dot{m}^*$ is normally imposed in order to fulfill production requirements:

$$\dot{m}^* = \frac{m}{H} \qquad \text{kg of purified material obtained per unit time} \tag{9.8}$$

where H represents the duration of operation. Note that this productivity is often related to the quantity m of purified material (thus having adequate purity) rather than to the quantity m_* of recovered solute in the purified stream.

A special word about feed and targeted purities is appropriate here. Because of the complex process in which the chromatographic step may be embedded, the purity constraint is not always unambiguously present at the chromatographic stage: for instance, products coming out of a chromatographic system may be further purified by crystallization (for instance, in the large-scale, Eluxyl process for obtaining pure para-xylene or in continuous industrial separation of optical isomers (Nicoud, 2014)) or by a subsequent chromatographic step, in order to reach a final desired purity. Similar considerations can also apply to the feed mixture, which may be more or less enriched prior to the chromatographic step. As an illustration, industrial chromatographic purification of polyunsaturated fatty acids is not performed on raw fish oil but on an enriched mixture obtained after urea extraction and vacuum distillation. The higher the pre-purification, the higher the cost of the feed, but the lower the chromatography cost. In general, the purity of the streams entering and exiting the chromatographic step are not constraints, but can be parameters of a more general optimization problem.

[5] Rigorously, the mass-balance equation (9.6) should be written in kilograms.

9.3.2 Objective function (simplified)

When chromatographers deal with optimization of chromatography, their natural inclination is to maximize the productivity per unit volume of the system, the volume of the system being defined as the total column volume:

$$\phi_v = \frac{\dot{m}^*}{\text{system volume}} = \frac{\dot{m}^*}{V_{syst}} \quad \text{kg/(h m}^3\text{)} \tag{9.9}$$

or the productivity per unit mass of chromatographic medium used to perform the separation:

$$\phi_m = \frac{\dot{m}^*}{\text{mass of chromatographic medium}} = \frac{\dot{m}^*}{\overline{M}} \quad \text{kg/(h kg)} \tag{9.10}$$

The productivities ϕ_v and ϕ_m are related via the density of the chromatographic medium. The fact that the chiral stationary phases are normally sold on a price/kg basis makes ϕ_m a parameter of practical relevance to chiral separations. For these applications, chromatographers often think in terms of KKD (kilograms of purified material per kilogram of chromatographic media per day).

In many situations, and especially for high-pressure systems using dynamic axial compression technology, the system cost is mainly imposed by the column diameter and only marginally by the column length.[6] As a consequence, a third productivity based on column diameter is sometimes used:

$$\phi_D = \frac{\dot{m}^*}{\text{column diameter}} \quad \text{kg/(h m)} \tag{9.11}$$

While productivity, irrespective of the exact definition, is a relevant parameter for process optimization, it is not the only one. For instance, eluent consumption often plays a critical role in the process economy. We will define the specific eluent consumption by:

$$\Upsilon_* = \frac{\Upsilon^{chr}}{\text{mass of purified material product}} = \frac{\Upsilon^{chr}}{m} \quad (\text{m}^3/\text{kg}) \tag{9.12}$$

Large-scale chromatography generally requires recycling the solvent in order to be economical,[7] the biopharmaceutical industry being an exception for technical and regulatory reasons. Recycling the solvent has the effect of replacing solvent costs by investment and energy costs when evaporation/distillation is the technique selected. We will show later that a trade-off between maximizing productivity and minimizing eluent consumption is unavoidable.

For large-scale production, functions to be optimized are frequently expressed in units such as \$ or €. Deriving an economic function to be optimized means recognizing that chromatography is only a piece of a more complete (and complex) production process, as illustrated in Figure 9.1.

[6] For very large-scale low-pressure systems, the column length can be imposed by the standardized width of stainless steel sheets that are then rolled and welded. Deviating from commercially available standard sheets can lead to unacceptable extra costs, so column length is imposed by considerations which are pretty remote from chromatography.

[7] (Nicoud, 2014) mentioned a 99.9% solvent recycling for a large-scale separation of optical isomers performed at UCB Pharma.

The economic objective function is derived in detail in the next section. For the time being, we focus on a simple expression, obtained largely from common-sense considerations. The cost per kilogram of purified material is seen as the addition of four contributions:

$$\psi^{tot} = A_F \frac{1}{\theta_*} + A_{CP} \frac{1}{\phi_v} + A_S \Upsilon_* + c_p(X_*) \quad (\text{€/kg of purified material}) \tag{9.13}$$

- $A_F \frac{1}{\theta_*}$, a "quality term". The materials processed in chromatography are often expensive and one has to work with a decent yield in order to conserve it. A_F is expressed in €/kg and is related to the feed cost.
- $A_{CP} \frac{1}{\phi_v}$, the "chromatography productivity term". Increasing the productivity has a positive influence on the purification cost. A_{CP} is expressed in €/(h m^3) and is related to system characteristics (one hour of operation of a complex multi-column high-pressure system is more expensive than one hour of operation of a low-pressure single column of the same volume).
- $A_S \Upsilon_*$, the cost associated with solvent processing. A_S is expressed in €/m^3 and is related to eluent purchasing, recycling and destruction costs. The impact of the recycling ratio S^R is embedded in A_S.
- $c_p(X_*)$, the cost of post-processing the purified material. This is not influenced by chromatographic performance but possibly by the outlet purity X_*^{out}.

Equation (9.24) is a general expression for the cost per unit mass of purified product. It is extremely useful because it bridges economic and chromatographic performance. It is an extension of an expression originally published by (Nicoud and Colin, 1989).

The optimum design will be influenced by the set of technico-economic parameters A_F, A_{CP}, A_S and $c_p(X_*)$. If the cost of feed is very high (for instance if you consider the last step of a complex protein purification), yield can be your major concern. In sugar separations performed on low-pressure SMB filled with ion-exchange resins, the costs associated with water evaporation are the main ones. For optical isomer separations with high-performance SMB using optically active chromatographic media, the main contributor is often productivity.

Section 9.3.3 presents in detail how to derive the technico-economic parameters. Readers who want to skip detailed calculations can go directly to Section 9.4.

9.3.3 Objective function (detailed)

Defining an economic objective function is certainly a task requiring rigor. In order to appear credible, let us use key terms favored by financial controllers: *Capex* (capital expenditure) and *Opex* (operational expenditure). (This will be our only tribute to accounting in this book.) The distinction between Capex and Opex will only be made for the chromatographic part, the costs associated with processing before or after chromatography being presented in a lumped manner.

We express the total cost Ψ^{tot} to obtain a quantity m of purified material as:

$$\Psi^{tot} = \Psi^{pre} + \Psi^{chr/ev}_{op} + \Psi^{chr/ev}_{cap} + \Psi^{post} \quad (\text{€}) \tag{9.14}$$

where Ψ^{pre} and Ψ^{post} represent the costs associated with pre-processing and post-processing. The terms $\Psi_{op}^{chr/ev}$ and $\Psi_{cap}^{chr/ev}$ represent respectively the Opex and Capex costs of chromatography, including possible eluent recycling. The different contributions are detailed below.

- Ψ^{pre} (cost of pre-processing) is the cost associated with the materials required to perform the chromatographic separation, i.e. the feed material to be purified and, per our convention, the solvent make-up.
  - As a decent approximation, the feed material contribution can be considered proportional to the amount of feed M (kg) engaged in the chromatographic unit. The feed cost per unit mass, $a_F(X_*^F)$, expressed in €/kg of feed material is, *a priori*, a function of the purity of the product of interest (the higher the purity, the higher the cost per unit mass).
  - Similarly, the solvent make-up costs are proportional to the solvent cost per unit volume, c_{sol} (€/m^3 of solvent), and to the quantity of solvent used, Υ^{inl}.

  Consequently, one can write:

$$\Psi^{pre} = a_F(X_*^F)M + c_{sol}\Upsilon^{inl} \qquad (€) \tag{9.15}$$

- Ψ^{post} (cost of post-processing) is the cost associated with processing the purified and waste streams.
  - The cost for further purification to final purity (for instance with a crystallization step) is likely to be proportional to the quantity of product m (kg) to be processed, the proportionality factor $c(X_*)$ (€/kg of purified stream) being, *a priori*, a function of the purity X_* (the higher this purity, the easier the work in possible subsequent steps).
  - The cost to process and possibly dispose of or destroy the solvent exiting the chromatography is likely to be proportional to the quantity of solvent, $\Upsilon_p^{out} + \Upsilon_w^{out} = \Upsilon^{inl}$ (cf. Eq. (9.6)) via the proportionality factor $c_{sol/dis}$ (€/m^3 of solvent).

  Consequently, one can write:

$$\Psi^{post} = c_p(X_*)m + c_{sol/dis}\Upsilon^{inl} \qquad (€) \tag{9.16}$$

- $\Psi_{cap}^{chr/ev}$: Capex is normally expressed in €/h after proper definition of a depreciation policy.[8] Main contributors are:
  - buildings and infrastructure, including solvent storage, utilities etc.
  - the chromatographic system itself
  - possible solvent recycling unit and associated utilities
  - maintenance costs, as a first approximation proportional to the investment, can be included in depreciation costs without loss of generality.

[8] We will simply consider a Y-year depreciation period, no actualization and H_y hours of operation per year, so the hourly cost contribution of a piece of equipment acquired for € X is $(X/(Y H_y))$ €/h.

The above items are directly impacted by the type of process selected (for instance, elution or SMB), its size and associated eluent needs.

We will divide depreciation costs into two parts, the first associated with the solvent recycling device, the second with all other investment items.[9]

The investment associated with the chromatographic system (and associated buildings) is taken to be proportional to a certain power x_c of the system volume, while the investment associated with the solvent recycling unit is expected to be proportional to a certain power x_r of the flow rate. The contribution of depreciation being proportional to the duration H of the production, one can write:

$$\Psi_{cap}^{chr/ev} = d_{chr}(V_{syst})^{x_c} H + d_{rec}\left(\frac{\Upsilon^{rec}}{H}\right)^{x_r} H \qquad (€) \qquad (9.17)$$

where d_{chr} is expressed in (€/(h m^{3x_c})) and d_{rec} in (€/(h$^{1-x_r}$ m^{3x_r})). For a given operating pressure, the cost of chromatographic units varies *roughly* proportionally to the diameter of the column and is moderately impacted by length. In general, $x_c \simeq 1/2$ offers a simple first approximation. The costs of concentration and evaporation units vary roughly as the square root of the eluent flow rate to be processed, so $x_r \simeq 1/2$.

- $\Psi_{op}^{chr/ev}$: these operational costs can be of very different natures; a simple list is given below.
  - Fixed costs associated with the start-up and cleaning of the unit: these can be negligible for large-scale continuous operations, but can represent the lion's share in some biopharmaceutical applications where batch changes can require complex and lengthy cleaning operations. This explains the interest in single-use devices, and possibly switching from batch to continuous operation. The influence of chromatography performance (productivity, yield, eluent consumption) on these costs being normally secondary, they will not be taken into account in this presentation, which does not mean that they are negligible in practice.
  - Labor costs: these are proportional to the duration H of the operation and slightly dependent on the unit size. The proportionality factor is c_{lb} (€/h). Notice that *quality control* costs are included in this contribution. These costs are only secondarily impacted by the size of the unit but are dependent on the chromatographic mode (for instance, fewer controls required for a continuous automated unit compared to a batch manual one).
  - Recycled solvent costs: these are proportional to the quantity of solvent recycled in the separation via the coefficient c_{rec} (€/l of eluent), and are often mainly energy costs.
  - Chromatographic media costs:[10] these are proportional to the system volume V_{syst}, to the cost of chromatographic media per unit bed volume

[9] These depend on many factors, from unit type and size to legal constraints such as anti-seismic norms in Japan or California ...

[10] These costs are arbitrarily located in this section. I have seen companies treating them either as an investment or as consumables (like solvent), depending on the lifetime.

c_{cmd} (€/m^3), to the reciprocal of the lifetime of the chromatographic medium, $\overline{H}$, and to the duration H of the operation. We define $\bar{c}_{cmd} = c_{cmd}/\overline{H}$, which corresponds to the cost per volume of chromatographic medium and per hour, taking into account the lifetime of the medium.

- Other costs depending on the situation, on the creativity of the controlling department etc.

Neglecting the costs associated with start-up and cleaning, we will thus write:

$$\Psi_{op}^{chr/ev} = c_{lb}H + c_{rec}\Upsilon^{rec} + \bar{c}_{cmd}V_{syst}H \qquad (€) \tag{9.18}$$

We have thus expressed the various contributions to Eq. (9.14) for the total cost to produce a quantity m of purified material.

By dividing the cost contributions by the quantity m of purified material, one obtains the total production cost per unit mass of purified product, and by introducing the yield θ_* (Eq. (9.5)), the production rate $\dot{m}^*$ (Eq. (9.8)), the productivity per unit volume ϕ_v (Eq. (9.9)), the specific eluent consumption Υ_* (Eq. (9.12)) and the recycle ratio S^R (Eq. (9.7)), one obtains:

$$\psi^{tot} = \frac{\Psi^{tot}}{m} = \psi^{pre} + \psi_{cap}^{chr/ev} + \psi_{op}^{chr/ev} + \psi^{post} \quad (\text{€/kg of purified material}) \tag{9.19}$$

$$\psi^{pre} = \frac{\Psi^{pre}}{m} = a_F(X_*^F)\frac{X_*}{X_*^F}\frac{1}{\theta_*} + c_{sol}(1 - S^R)\Upsilon_* \quad (\text{€/kg}) \tag{9.20}$$

$$\psi_{cap}^{chr/ev} = \frac{\Psi_{cap}^{chr/ev}}{m} = d_{chr}\frac{V_{syst}^{x_c-1}}{\phi_v} + d_{rec}(S^R)^{x_r}\Upsilon_*{}^{x_r}\dot{m}_*^{x_r-1} \quad (\text{€/kg}) \tag{9.21}$$

$$\psi_{op}^{chr/ev} = \frac{\psi_{op}^{chr/ev}}{m} = c_{lb}\frac{1}{\dot{m}_*} + c_{rec}S^R\Upsilon_* + \frac{\bar{c}_{cmd}}{\phi_v} \quad (\text{€/kg}) \tag{9.22}$$

$$\psi^{post} = c_p(X_*) + c_{sol/dis}(1 - S^R)\Upsilon_* \quad (\text{€/kg}) \tag{9.23}$$

These equations can be rewritten in a format showing the influence of feed material, solvent, chromatography and post-processing and thus using the parameters of Eq. (9.13). Noting that $\dot{m}_* = V_{syst}\phi_v$, one obtains:

$$\psi^{tot} = A_F\frac{1}{\theta_*} + A_{CP}\frac{1}{\phi_v} + A_S\Upsilon_* + c_p(X_*) \quad (\text{€/kg of purified material}) \tag{9.13}$$

- Feed contribution:

$$A_F = a_F(X_*^F)\frac{X_*}{X_*^F} \tag{9.24}$$

- Chromatography contribution:

$$A_{CP} = \frac{c_{lb}}{V_{syst}} + d_{chr}V_{syst}{}^{x_c-1} + \bar{c}_{cmd} \qquad V_{syst} = \frac{\dot{m}_*}{\phi_v} \tag{9.25}$$

- Solvent contribution:

$$A_S = (c_{sol} + c_{sol/dis})(1 - S^R) + c_{rec}\left(S^R\right)^{x_r} + d_{rec}\left(S^R\right)^{x_r}(\Upsilon_* \dot{m}_*)^{x_r - 1} \quad (9.26)$$

- Post-processing contribution:

$$c_P(X_*) \quad (9.27)$$

Equations (9.13) and (9.24)–(9.27) are certainly not fully general,[11] and are based on a few approximations, such as cost inventory and allocation, depreciation policy, "elasticity" and so on, but they can be adapted by the reader to the specific situations he has to deal with. They clearly show the necessary trade-off between purity, yield, productivity and investment strategy regarding solvent cost; more complex and possibly precise expressions would exhibit the same characteristics.

EXERCISE 9.1 *A mixture containing 50% of a solute of interest must be processed in order to get 10 t/year of this solute at 99.0% purity.*

- *The cost of the crude mixture is 25 €/kg.*
- *Chromatographic system selling prices (excluding evaporation and solvent recycling) are given by (operating pressure around 50 bar, L_{col}=100 cm):*
  - *20 cm internal diameter: 1.5 M€*
  - *40 cm internal diameter: 2.8 M€*
  - *60 cm internal diameter: 4.0 M€.*
- *The eluent and chromatographic media have been chosen, and the associated parameters are:*
  - *Solvent: € 2/l; energy cost for recycling the solvent € 0.1/l*
  - *Chromatographic media: € 10 000/l; lifetime: 2 years.*
- *The solvent is recycled in evaporators able to deliver dry solids and to recycle 99.5% of the solvent stream. The energy cost for recycling one liter of solvent is about € 0.1. Typical costs for such evaporation units (including utilities) are:*
  - *100 l/h: € 7×10^5*
  - *1000 l/h: € 2×10^6*
- *The cost of buildings and associated facilities is not taken into account.*
- *The cost to post-process the purified material at 99.0% purity is € 10/kg.*
- *The cost to destroy the residual solvent is € 0.5/l.*
- *In this factory, the policy is to depreciate equipment over 7 years.*
- *Labor costs associated with the operation of the units (including quality control) are estimated to be € 6.4×10^5/year.*

Derive the objective function allowing one to minimize the production cost of the purified material.

[11] In particular, the coefficients d_{chr} and $\bar{c}_{cmd}$ embedded in A_{CP} can only be considered constant in a certain range of system size. The term $\frac{c_{lb}}{V_{syst}}$ in the coefficient A_{CP} simply shows that the influence of labor costs decreases with production scale.

9.4 Algorithmic part of the optimization procedure

9.4.1 General considerations

With our objective function well defined and our tuning and constrained parameters identified, we can now address the algorithmic part of the optimization procedure. One need not be an expert in numerical optimization to realize that the complete optimization of a chromatographic process is a task of a serious complexity owing to the number of continuous and discrete parameters. Additionally, a detailed chromatographic (especially multi-column) process model is computationally expensive, which becomes an issue for rigorous model-based optimization.

The art of optimization has made substantial progress over recent years, and algorithms which can work with a large number of parameters and constraints, continuous or discrete, are now available. Stochastic approaches, as well as gradient-based approaches, have been used to optimize chromatographic processes. Stochastic optimization methods such as generic algorithms, adaptive random search and simulated annealing are robust and less sensitive to non-convexities in both the objective function and constraints than gradient-based methods. They have been applied with success to the optimization of different classes of chromatographic processes (Araujo *et al.*, 2006). Gradient-based methods require the values of the partial derivatives of the objective function and of the constraints with respect to the design parameters, and may converge towards local optima. The latter method may, however, converge more quickly than the former, should a preliminary parameter study have been performed in order to provide adequate initial guesses (Araujo *et al.*, 2006).

Irrespective of the progress in numerical techniques, this black-box type of approach suffers from major deficiencies: there is usually no certainty that the absolute optimum (vs. local optima) has been determined and it gives little if any information on the "drivers" allowing maximization of the efficiency of the process. To be a little bit provocative, this is sort of a "plug and pray" approach for the chromatographer. With all due respect to modern optimization algorithms, our approach is based on the derivation of a rational structure for the algorithm, by including as much knowledge of chromatography as possible. This approach leads to the definition of key loops and sub-loops.[12] This may be considered a drawback compared with global optimization algorithms which can sometimes reach an optimum more quickly, but it has the advantage of a path to guide the convergence which takes into account the physics of the problem.

In order to make the best use of our modeling skills, we will organize the continuous tuning parameters presented in Section 9.3.1 into three categories:

- Purities: often constrained parameters, but can be tuning parameters in general
- Continuous tuning parameters at the local scale: tuning parameters such as temperature, pressure, carrier-fluid composition (constant or variable), particle size and thickness of an adsorbing coating. All are independent of the nature of the

[12] By "loop" we mean a subpart of the algorithm in which one or more parameters are determined by an iterative procedure, all other parameters being fixed.

chromatographic system (elution, SMB etc.) and its size. In line with the introduction to Chapter 1, these parameters define the *elementary chromatographic processes* – in other words, the parameters impacting *physico-chemistry*.

- Continuous tuning parameters at the process scale: injection volume and composition, column diameter, collection time etc. Flow rate is in this category, although it influences dispersion and external mass transfer at the local scale (this is a reminder that the boundary between *local* and *process* may be somewhat fuzzy).

Using this classification, we will distinguish three types of optimization problems, as presented in Figure 9.2:

- Type II: determining the optimum of the objective function, assuming that all parameters, discrete or continuous, can be tuned. This is the most general problem.
- Type I: determining the optimum of the objective function with imposed inlet/outlet purities, assuming that only continuous parameters with no impact on physico-chemistry can be tuned. In other words, temperature, pressure, eluent composition and chromatographic media are fixed.
- Type 0: determining the optimum of the Type I problem using shortcuts, science, intuition and experience

Type I problems, aimed at optimizing the process for a given set of (X_*, X_*^F) purities and given physico-chemical parameters, can be considered the core of the optimization procedure. Solving a Type 0 problem can be seen as a step to estimate parameters for a preliminary design, or simply to provide a sound initialization for a Type I problem.

As far as discrete parameters are concerned, moving from Type I to Type II problems is normally a matter of comparing a few process modes and chromatographic media. Having solved a Type I problem, taking account of the continuous parameters impacting the physico-chemistry (particle size, carrier-fluid composition, temperature etc.) in order to optimize a Type II problem can require a fair amount of simulation. The reader will

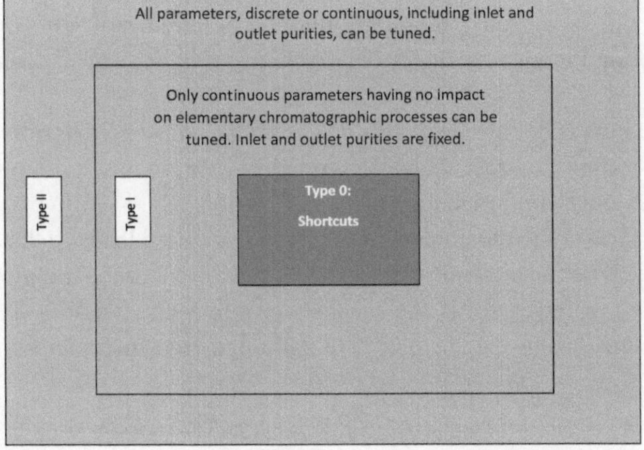

Figure 9.2 Organization of the optimization problem by level of complexity.

find more general information on algorithms and case studies in (Zhang *et al.*, 2003), (Araujo *et al.*, 2006), (Toumi *et al.*, 2007), (Rodrigues *et al.*, 2007), (Mota *et al.*, 2007), (Kawajiri and Biegler, 2008), (Gomes and Rodrigues, 2012) and (Agrawal and Kawijari, 2012).

Rather than further discussing Type II problems and complex multi-parameter algorithms, we will concentrate on what I believe is the core of optimization: solving Type I problems.

9.4.2 Type I problems

Our objective is to propose a strategy for solving Type I problems that is as independent as possible of the chromatographic system and of the model used to simulate it. We will also propose a strategy for estimating the relevant tuning parameters.

9.4.2.1 Structure

Let us consider the situation of ideal chromatographic systems defined in Chapter 1, so that the scale-up is linear versus column cross section. Under these conditions, we can normalize the flow rates by the cross section Ω_{col}; in other words, we can work with fluid velocities instead of flow rates. When the fluid velocities are known, the required cross section is then determined so as to fulfill the required production rate (see Section 9.4.2.2). In the following, we will thus primarily work with velocities.

However, recall that while assuming chromatography to be scalable linearly with respect to Ω_{col} is in general an excellent starting approximation, hydrodynamic considerations can make this approximation unsatisfactory for compressible media or in case of non-uniform radial velocity or temperature profiles. These situations are presented in Chapter 6 but they will not be considered further in the present chapter.

In the case of elution chromatography,[13] parameters to be optimized in the Type I problem are C^F_{tot}, u_e, Δt_{inj}, L_{col}, t_{cycle}, t_{c1} and t_{c2}, where t_{cycle} is the time between successive injections and t_{c1} and t_{c2} are respectively the cut times at which collection of the product of interest is started and stopped (Figure 9.3).

In production conditions, injections are normally superposed in order to minimize the cycle time, so the recovered fraction (*) may be polluted not only by impurities in the current injection but also by impurities from previous and subsequent injections (Figure 9.4).[14]

The seven parameters (C^F_{tot}, u_e, Δt_{inj}, L_{col}, t_{cycle}, t_{c1}, t_{c2}) must be determined in order to satisfy the constraints (at least on purity for Type I problems, possibly on yield, column length, etc.) and to optimize the objective function.

A more detailed look at the parameters of the Type I problem leads to the conclusion that they should not all be treated equivalently. From a given set of the tuning parameters (C^F_{tot}, u_e, Δt_{inj}, L_{col}), referred to as the *primary group*, what freedom do we really have for the *secondary group* (t_{cycle}, t_{c1}, t_{c2})? Although free in principle, the tuning

[13] Recall that carrier-fluid composition, constant or variable, is imposed in Type I problems.

[14] Notice the analogy with pollution from zones 1 and 4 in TMB/SMB systems.

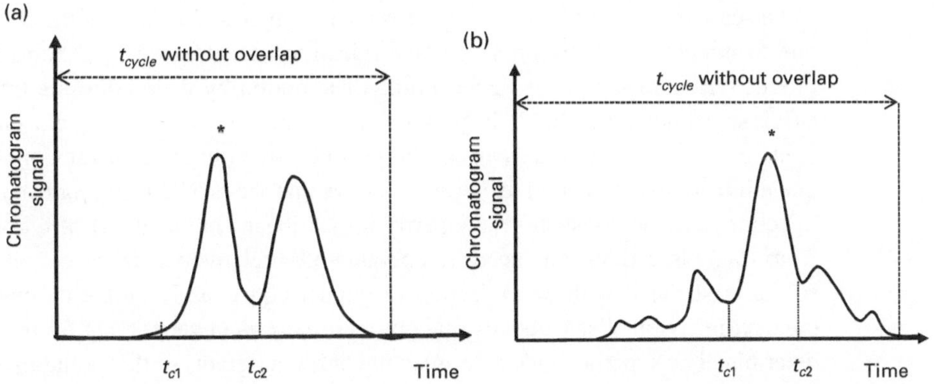

Figure 9.3 Typical chromatogram: (a) binary; (b) multi-component. The recovered fraction is indicated by *.

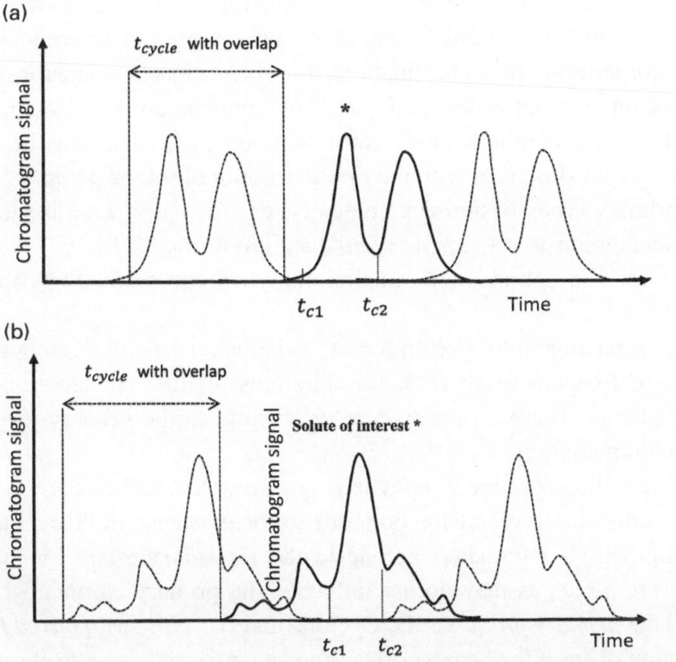

Figure 9.4 Typical chromatogram: superposition of successive injections: (a) binary; (b) multi-component. The recovered fraction is indicated by *.

parameters (t_{c1}, t_{c2}) do not influence the chromatogram and can be determined using common sense: cut times are determined so that the quantities of recovered purified product are maximized. This determination can be more or less complex in the case of multi-solute systems,[15] but a simple sub-loop is normally adequate for quickly determining (t_{c1}, t_{c2}) from a given chromatogram.

[15] A rigorous approach to determining cut times has been proposed by (Sreedhar *et al.*, 2013). Additionally, note that there is a difference between simulation and industrial life. With a simulated chromatogram, one

The case of t_{cycle} is a bit more subtle as it can influence the shape of the chromatogram due to interactions between successive injections. Decreasing t_{cycle} helps to increase productivity and minimize carrier fluid consumption by decreasing the time between successive injections, but if it goes too far, it may decrease the yield due to cross-contamination: the product $(*)$ can be polluted by solutes from previous and subsequent injections. In general, experience shows that the profile resulting from successive injections can be considered a superposition (linear combination) of chromatograms from single injections: one needs to operate under relatively extreme conditions for the solutes associated with one injection to impact non-linearly on the chromatogram of the next injection. Consequently, the optimum t_{cycle} is in general not far from the value determined by superposing single-injection chromatograms so that pollution of the product $(*)$ from previous and subsequent injections is just avoided. Determination of t_{cycle} from a single-injection chromatogram is trivial when a binary mixture is processed (see Figure 9.4a): for instance, t_{cycle} can be determined as the period between t_{beg}, at which the least retained solute has grown to $x\%$ of its maximum, and t_{end}, at which the most retained solute product has shrunk to $x\%$ of its maximum. The determination is a bit more complex for multi-component situations (to avoid headaches, superpose individual chromatograms on transparencies, and you will immediately get t_{cycle}!). Under the assumption of a linear combination of chromatograms, t_{cycle} can thus be determined from a chromatogram simulated with the *primary group* of tuning parameters.

When the primary group of tuning parameters (C_{tot}^F, u_e, Δt_{inj}, L_{col}) is set, the single-injection chromatogram is set, so the secondary group (t_{cycle}, t_{c1}, t_{c2}) is determined with a simple sub-loop optimization, maximizing yield and productivity for this chromatogram.

For elution chromatography performed at variable carrier-fluid composition, additional degrees of freedom result from the durations of different eluent composition plateaus or gradients. These tuning parameters belong to the primary group, as they influence the chromatogram.

The case of recycling chromatography is very similar to the classical batch situation, but one has additional degrees of freedom due to the existence of "recycling periods". These recycling-periods parameters belong to the *secondary group* for the *recycling and peak shaving* mode, as they do not influence the primary chromatograms, while they belong to the *primary group* for the Cyclojet and *recycling with mixed feed* modes, as the re-injection of fresh feed has a critical impact on the steady-state chromatograms.

The case of TMB is at the same time similar and different. It is similar because, as with elution chromatography, there are seven tuning parameters: C_{tot}^F, u_{e1}^{tmb}, u_{e2}^{tmb}, u_{e3}^{tmb}, u_{e4}^{tmb}, $\bar{u}_e$ and L_{col}^{tmb}, where L_{col}^{tmb} represents the total TMB bed length. Note that L_i, representing the length of zone i, is not part of the Type I problem, by analogy with SMB in the next section. It is different because, in contrast to elution chromatography, all the tuning parameters influence the internal concentration profiles. Differentiating between a primary group and a secondary group of tuning parameters is thus not that straightforward. From Chapter 8 we know, however, that flow rates in zones 1 and 4 can

can, a posteriori, identify some very small fractions that are then pooled in an optimal way. This ideal pooling is unrealistic at industrial scale, and (Westerberg *et al.*, 2010) showed how the pooling strategy can impact performance.

in general be determined after we have designed zones 2 and 3. We will thus consider that $(C^F_{tot}, u^{tmb}_{e2}, u^{tmb}_{e3}, \bar{u}_e, L^{tmb}_{col})$ belong to the primary group of tuning parameters, while $(u^{tmb}_{e1}, u^{tmb}_{e4})$ belong to the secondary group.

For SMB, one has to consider seven parameters as well: $C^F_{tot}, u^{smb}_{e1}, u^{smb}_{e2}, u^{smb}_{e3}, u^{smb}_{e4}$, ΔT and L_{col}. By convention, L_{col} is the length of one column, the total SMB length thus being $N_{col}L_{col}$. Note that the number of columns N_{ci} in zone i are not tuning parameters of the Type I problem, as they are discrete tuning parameters. Similarly to the choice made with TMB, we will consider that $(C^F_{tot}, u^{smb}_{e2}, u^{smb}_{e3}, L_{col}, \Delta T)$ belong to the primary group of tuning parameters, while $(u^{smb}_{e1}, u^{smb}_{e4})$ belong to the secondary group.

In general, we define the secondary group as the set of tuning parameters that:

- can mainly be determined as a consequence of the choices made for the primary group (such as cut times in elution chromatography)
- offer fine tuning rather than fundamental changes (for example, t_{cycle} in elution chromatography, fluid velocities u_{e1} and u_{e4} in TMB) ... at least as long as the user does not explore excessively "creative" areas.

This distinction between primary and secondary group is not absolutely mandatory (and possibly not fully scientific) but it allows one to rationalize the optimization algorithm. Extending the philosophy to other process modes is not difficult.

Finally, the principle structure of the optimization algorithm is given in Figure 9.5.

At this point of the presentation, the reader may feel that separation optimization is a cumbersome mathematical problem. Furthermore, anybody knows that any optimization algorithm needs an initialization – simply speaking, an initial guess of the tuning

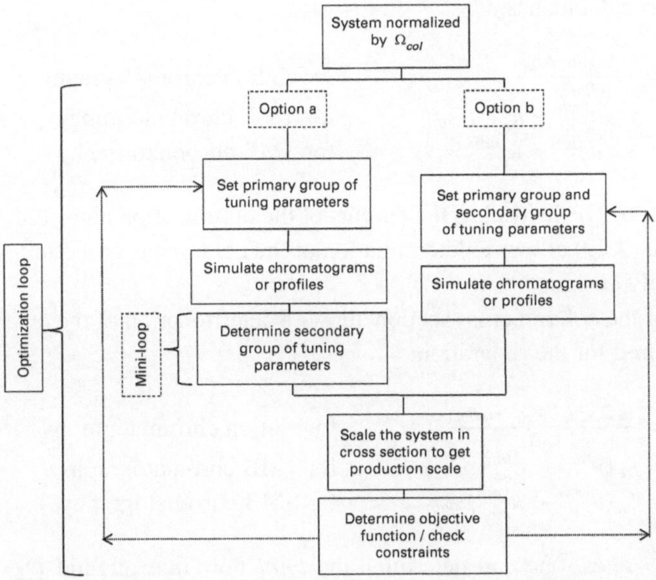

Figure 9.5 Principle of the optimization procedure: Type I problem. Option a: primary and secondary groups of tuning parameters are optimized separately. Option b: primary and secondary groups of tuning parameters are optimized simultaneously.

parameters. Further physical insight into the optimization problem is thus necessary to satisfy at least three objectives:

- a better understanding of the effects of various parameters
- a better structure for the optimization problem, ensuring simple and fast calculations; this is already partially done with the primary and secondary groups
- initial guesses which are close to the optimum; if they are good enough, they may in some cases be considered the (pseudo-) optimal solution.

These three objectives are discussed in the sections following.

9.4.2.2 Scaling the system

As a result of the optimization, the fluid velocities (and solid velocities in the case of TMB) and column length are known. The missing information is the column section.

We assume that the system is designed to deliver a given production rate $\dot{m}^*$ of purified material. Let us define Q_F, the feed flow rate, expressed in m^3/s. For TMB, Q_F is the continuous feed flow rate presented in Chapter 8. For elution chromatography, the feed flow rate Q_F is given by $Q\Delta t_{inj}/t_{cycle}$.

The feed flow rate is related to the production rate by:

$$Q_F C_{tot}^F \theta_* \frac{X_*^F}{X_*} = \dot{m}^* \qquad \text{(kg purified material per unit time)} \tag{9.28}$$

which allows Q_F to be calculated from the production rate requirement, the feed and outlet purities, and the yield, which is an output of the optimization algorithm.

On the other hand, one has:

$$\begin{aligned} Q_F &= \frac{\varepsilon_e \Omega_{col} u_e \Delta t_{inj}}{t_{cycle}} && \text{for elution chromatography} \\ Q_F &= \varepsilon_e (u_{e3}^{tmb} - u_{e2}^{tmb}) \Omega_{col} && \text{for TMB chromatography} \\ Q_F &= \varepsilon_e (u_{e3}^{smb} - u_{e2}^{smb}) \Omega_{col} && \text{for SMB chromatography} \end{aligned} \tag{9.29}$$

From knowledge of Q_F, as well as the outputs of the optimization algorithm (the tuning parameters), Eq. (9.29) allows a determination of the column cross section. The system is thus completely designed.

Knowledge of the column cross section allows a determination of the eluent flow rate Q_E^{eq} (m^3/s) required for the separation:

$$\begin{aligned} Q_E &= \frac{\varepsilon_e \Omega_{col} u_e (t_{cycle} - \Delta t_{inj})}{t_{cycle}} && \text{for elution chromatography} \\ Q_E &= \varepsilon_e (u_{e1}^{tmb} - u_{e4}^{tmb}) \Omega_{col} && \text{for TMB chromatography} \\ Q_E &= \varepsilon_e (u_{e1}^{smb} - u_{e4}^{smb}) \Omega_{col} && \text{for SMB chromatography} \end{aligned} \tag{9.30}$$

From these flow rates, one can determine the *total* flow rate of fluid $Q_{EF} = Q_F + Q_E$ (m^3/s) required for the separation. Notice that Q_{EF} corresponds to the fluid flow rate Q eluting the column in elution chromatography, white it corresponds to the total inlet/outlet flow rate for TMB/SMB (thus not the recycling flow rate).

9.4.2.3 Reference to the equilibrium model

In order to further structure the optimization algorithm, making reference to the equilibrium model results will prove to be very helpful. First, we will specify the possible inputs and outputs of the equilibrium model.

Let us start with elution chromatography, and consider a chromatographic column of length L_{col}, operated in the elution mode with fluid velocity u_e. From the tuning parameters L_{col} and u_e one can determine $t_0 = \varepsilon_e L_{col}/u_e$, and from t_0 and C^F_{tot}, the equilibrium model allows a determination of Δt^{eq}_{inj}, t_{c1}, t_{c2}, and t_{cycle} that give the required purity at 100% yield. Notice that the equilibrium model has two input tuning parameters (t_0 and C^F_{tot}) and delivers four output tuning parameters ($\Delta t^{eq}_{inj}, t^{eq}_{c1}, t^{eq}_{c2}, t^{eq}_{cycle}$). Column length and fluid velocity appear only as a ratio (t_0); the column is assumed to be infinitely efficient.

The chromatograms obtained for the equilibrium model and for the optimized chromatogram at the same t_0 and C^F_{tot} are given in Figure 9.6 for the simple situation of a linear system. It is pretty clear that, at given t_0 and C^F_{tot}, the optimized tuning parameters ($\Delta t_{inj}, t_{c1}, t_{c2}, t_{cycle}$) cannot be very far from ($\Delta t^{eq}_{inj}, t^{eq}_{c1}, t^{eq}_{c2}, t^{eq}_{cycle}$) derived from the equilibrium model. It is equally clear that this reasoning is valid irrespective of the structure of the adsorption isotherms as well as of the number of solutes. The only information that we had, *a priori*, about the tuning parameters was that they are located between 0 and ∞, so the above reference to the equilibrium model is thus very useful.

Let us now consider the TMB process. From the tuning parameters C^F_{tot} and $\bar{u}_e$, the equilibrium model allows a determination of the four fluid velocities u^{eq}_{ei} that allow us to get the target purity at 100% yield. Again, we have two input tuning parameters (in this example, $\bar{u}_e$ and C^F_{tot}) and four output tuning parameters (in this example, u^{eq}_{ei}). Column length is neither an input nor an output of the equilibrium model: the column is simply assumed to be infinitely efficient. The extension to SMB and other modes presents no difficulty.

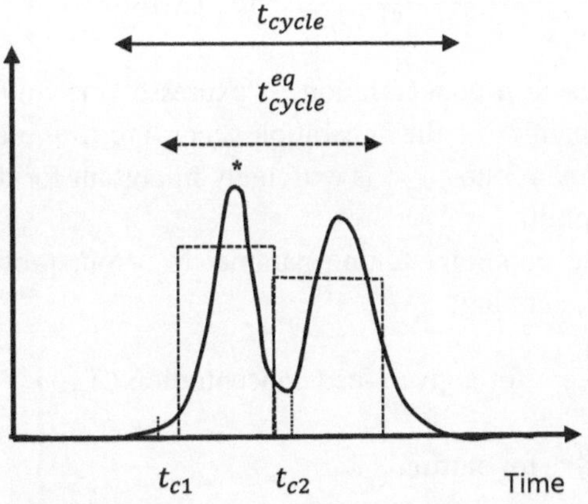

Figure 9.6 Chromatograms: obtained from the equilibrium model and after optimization at identical t_0 and C^F_{tot}. Linear system.

We will now quantify the relation between the actual set of parameters and those provided by the equilibrium model. While our algorithm depends only on velocities, for convenience let us consider the flow rates for a moment; we will see that only ratios of flow rates will then be used, and thus ratios of the corresponding velocities. Let us again assume that C_{tot}^F, t_0 (for elution) and $\bar{u}_e$(for TMB) are given. The equilibrium model then allows determination of the feed flow rate Q_F^{eq} (m^3/s), the eluent flow rate Q_E^{eq} (m^3/s) and consequently the *total* flow rate of fluid $Q_{EF}^{eq} = Q_F^{eq} + Q_E^{eq}$ (m^3/s) required for getting the target purity at 100% yield in the absence of hydrodynamic dispersion and kinetic limitations. These flow rates are related to fluid velocities (as obtained from the equilibrium model) and column section by equations similar to Eqs. (9.29) and (9.30).

In order to make full use of our reference to the equilibrium model, we introduce various ratios:

$$\left.\begin{aligned} \varrho^{eq} &= \frac{Q_F^{eq}}{Q_E^{eq}} \\ \text{thus} \quad \varrho^{eq} &= \frac{t_{inj}^{eq}}{t_{cycle}^{eq} - t_{inj}^{eq}} \quad \text{for elution} \\ \text{or} \quad \varrho^{eq} &= \frac{u_{e3}^{eq} - u_{e2}^{eq}}{u_{e1}^{eq} - u_{e4}^{eq}} \quad \text{for TMB/SMB} \end{aligned}\right\} \tag{9.31}$$

which is the ratio between feed and eluent flow rate from the equilibrium model.

Another definition is also very useful:

$$\left.\begin{aligned} \chi_F^{eq} &= C_{tot}^F \frac{Q_F^{eq}}{Q_{EF}^{eq}} \\ \chi_F^{eq} &= C_{tot}^F \frac{\varrho^{eq}}{1+\varrho^{eq}} \quad (\text{kg/m}^3) \\ \text{thus} \quad \chi_F^{eq} &= C_{tot}^F \frac{t_{inj}^{eq}}{t_{cycle}^{eq}} \quad \text{for elution} \\ \text{or} \quad \chi_F^{eq} &= C_{tot}^F \frac{u_{e3}^{eq} - u_{e2}^{eq}}{u_{e3}^{eq} - u_{e2}^{eq} + u_{e1}^{eq} - u_{e4}^{eq}} \quad \text{for TMB/SMB} \end{aligned}\right\} \tag{9.32}$$

The parameter χ_F^{eq} is analogous to a concentration: it expresses the dilution of the feed imposed by the thermodynamics of the separation according to the equilibrium model. We will show in Section 9.4.3 that χ_F^{eq} is extremely important for determining productivity and eluent consumption.

In addition, we normalize the important tuning parameters of our problem to the output of the equilibrium model, defining:

$$\left.\begin{aligned} \eta_F^{eq} &= \frac{Q_F}{Q_F^{eq}} \quad \text{for a given feed concentration } C_{tot}^F \\ \text{thus} \quad \eta_F^{eq} &= \frac{t_{inj}}{t_{inj}^{eq}} \quad \text{for elution} \\ \text{or} \quad \eta_F^{eq} &= \frac{u_{e3} - u_{e2}}{u_{e3}^{eq} - u_{e2}^{eq}} \quad \text{for TMB/SMB} \end{aligned}\right\} \tag{9.33}$$

η_F^{eq} is thus the ratio between the actual feed flow rate and the feed flow rate that generates the required purity at 100% yield according to the equilibrium model.[16]

We define a normalized eluent consumption by:

$$\left.\begin{aligned} \eta_E^{eq} &= \frac{Q_E}{Q_E^{eq}} \quad \text{for a given feed concentration } C_{tot}^F \\ \text{thus} \quad \eta_E^{eq} &= \frac{t_{cycle} - \Delta t_{inj}}{t_{cycle}^{eq} - \Delta t_{inj}^{eq}} \quad \text{for elution} \\ \text{or} \quad \eta_E^{eq} &= \frac{u_{e1} - u_{e4}}{u_{e1}^{eq} - u_{e4}^{eq}} \quad \text{for TMB} \end{aligned}\right\} \tag{9.34}$$

η_E^{eq} is thus the ratio between the eluent flow rate and the eluent flow rate that generates the required purity at 100% yield according to the equilibrium model.

Finally, we define a normalized carrier fluid flow rate η_{EF}^{eq} relative to the equilibrium model by:

$$\left.\begin{aligned} \eta_{EF}^{eq} &= \frac{Q_{EF}}{Q_{EF}^{eq}} = \frac{Q_E + Q_F}{Q_E^{eq} + Q_F^{eq}} \quad \text{for a given feed concentration } C_{tot}^F \\ \text{thus} \quad \eta_{EF}^{eq} &= \frac{t_{cycle}}{t_{cycle}^{eq}} \quad \text{for elution} \\ \text{or} \quad \eta_{EF}^{eq} &= \frac{u_{e3} - u_{e2} + u_{e1} - u_{e4}}{u_{e3}^{eq} - u_{e2}^{eq} + u_{e1}^{eq} - u_{e4}^{eq}} \quad \text{for TMB} \end{aligned}\right\} \tag{9.35}$$

The parameter η_{EF}^{eq} compares the total carrier fluid (brought by either the feed or eluent stream) consumption to the minimum carrier fluid consumption obtained by the equilibrium model. Extension of the definition of η_F^{eq} or η_{EF}^{eq} to other modes is straightforward.

Allowing a safety margin relative to the equilibrium model design means taking $\eta_F^{eq} < 1$ and/or $\eta_E^{eq} > 1$. These two margins are embedded in η_{EF}^{eq}, which is related to them by:

$$\eta_{EF}^{eq} = \frac{\eta_F^{eq} + \eta_E^{eq}/\varrho^{eq}}{1 + 1/\varrho^{eq}} \quad \text{for a given feed concentration } C_{tot}^F \tag{9.36}$$

With a given set of $\eta_F^{eq} < 1$ and/or $\eta_E^{eq} > 1$, η_{EF}^{eq} can be less or greater than 1. Nothing prevents us from working with $\eta_F^{eq} > 1$ but under these conditions the yield will be lower than 100%, even for a system equivalent to an infinity of plates.

Determination of the parameters η_F^{eq}, η_E^{eq}, η_{EF}^{eq}, ϱ^{eq} and χ_F^{eq} requires access to the solution provided by the equilibrium model. We presented at length the equilibrium

[16] Recall that the fluid velocities in a TMB are related to the m values by (Chapter 8):

$$m_i = \frac{Q_i}{\bar{Q}} = \frac{u_i}{\bar{u}} = \frac{\varepsilon_e}{1 - \varepsilon_e} \frac{u_{ei}}{\bar{u}_e}$$

so that, for continuous processes, η_F^{eq} can be written

$$\eta_F^{eq} = \frac{m_3 - m_2}{m_3^{eq} - m_2^{eq}}$$

and thus corresponds to the dimensionless distance between the vertex of the separation zone and the diagonal in (m_2, m_3) space.

model for single-column systems in Chapter 3, and in Chapter 8 for TMB and thus SMB. Extensions to various specific situations involving binary mixtures have been described in the literature:

- Limited purities: (Siitonen *et al.*, 2013) for elution chromatography and (Kaspereit *et al.*, 2007) for SMB
- Recycling modes: (Bailly and Tondeur, 1982) for 100% purities, extended by (Sainio, 2009) and (Kaspereit and Sainio, 2011) for arbitrary purities and by (Siitonen and Sainio, 2011) for systems including solvent removal
- Modified SMB (I-SMB): (Katsuo and Mazzotti, 2010a), (Katsuo and Mazzotti, 2010b), (Katsuo *et al.*, 2011)
- Gradient SMB: (Abel *et al.*, 2002), (Abel *et al.*, 2004)
- Supercritical fluid SMB: (Mazzotti *et al.*, 1997), (Di Giovanni *et al.*, 2001).

Despite real progress, analytical or quasi-analytical solutions are not available (and will probably not be available for the foreseeable future) for specific processes (steady-state recycling, for instance) and for complex adsorption isotherms, especially for multi-component systems.

For the general situation, a numerical approach must be used: solutions of the equilibrium model are approximated numerically by a system equivalent to a number of plates which is significantly higher than the maximum number of plates that can practically be considered for a real system (for instance a few hundreds to a few thousands for SMB and a few thousands to a few tens of thousands for elution). System simulation can be performed using, for instance, the Rouchon algorithm presented in Chapter 7, which allows very fast calculations. With a few simulations using MC or PD models, a good approximation of the solution of the equilibrium model is thus determined.

We will specify in the next section how solutions of the equilibrium model can be used to facilitate our optimization task.

9.4.3 Type 0 shortcuts: estimating tuning parameters from scratch

Shortcuts consist in determining a reasonable set of tuning parameters without performing the complete optimization of the Type I problem. Depending on the required precision, the obtained tuning parameters can then be directly used for the process design, or as an excellent initialization set for detailed optimization.

We focus on the primary group of parameters. Determination of the secondary group of parameters is presumed to be a more or less direct consequence of the choice for the primary group; alternatively, variations of the secondary group relative to the predictions of the equilibrium model plays a lesser role, so their fine tuning is not of major importance, at least for shortcut purposes. All other parameters are outside the scope of the Type I problem.

As already mentioned, an important starting point consists in guessing η_F^{eq} (to a lesser extent η_{EF}^{eq}), which cannot be too far from 1. All parameters will then be determined in sequence.

9.4.3.1 Estimating all tuning parameters except column length and one velocity

In the first part of this section, we will assume that C^F_{tot} is known. We will subsequently see how to determine C^F_{tot}.

We know that productivity per unit volume ϕ_v is an important performance factor that must be taken into account in a complete economic optimization; ϕ_v is the ratio of the recovered purified fraction to the system volume (see Eq. (9.9)). After introducing the associated feed flow rate (Eq. (9.28)), one obtains:

$$\phi_v = \frac{\dot{m}_*}{V_{syst}} = \theta_* \frac{C^F_{tot} Q_F}{V_{syst}} \frac{X^F_*}{X_*} \quad \text{kg/(s m}^3\text{)}$$

After rearranging, one obtains:

$$\left.\begin{aligned}
\phi_v &= \theta_* \frac{\eta^{eq}_F}{\eta^{eq}_{EF}} \frac{\chi^{eq}_F}{t^\chi_\phi} \frac{X^F_*}{X_*} \quad \text{with} \quad t^\chi_\phi = \frac{V_{syst}}{Q_{EF}} \quad \text{s} \\
\text{with} \quad t^\chi_\phi &= \frac{L_{col}}{\varepsilon_e} \frac{1}{u_e} \left(= \frac{t_0}{\varepsilon_e} \right) \quad \text{for elution} \\
\text{or} \quad t^\chi_\phi &= \frac{L^{tmb}_{col}}{\varepsilon_e} \frac{1}{u^{tmb}_{e3} - u^{tmb}_{e2} + u^{tmb}_{e1} - u^{tmb}_{e4}} \quad \text{for TMB} \\
\text{or} \quad t^\chi_\phi &= \frac{L_{col} N_{col}}{\varepsilon_e} \frac{1}{u^{smb}_{e3} - u^{smb}_{e2} + u^{smb}_{e1} - u^{smb}_{e4}} \quad \text{for SMB}
\end{aligned}\right\} \tag{9.37}$$

Extension to other chromatographic modes poses no difficulty.

The specific eluent consumption Υ_*, defined in Eq. (9.12), is immediately obtained:

$$\Upsilon_* = \frac{1}{\phi_v t^\chi_\phi} = \frac{1}{\theta_* \chi^{eq}_F} \frac{\eta^{eq}_{EF}}{\eta^{eq}_F} \frac{X_*}{X^F_*} \quad \text{m}^3\text{/kg} \tag{9.38}$$

Equations (9.37) and (9.38) are remarkably simple and compact expressions relating productivity and specific eluent consumption to our tuning parameters (constrained or not). The different contributors are:

- Purities, which are assumed to be fixed in the Type I problem
- The parameter χ^{eq}_F, which is analogous to a concentration, expresses the dilution imposed by the thermodynamic ability to separate the feed mixture: a large χ^{eq}_F is not an absolute guarantee of high productivity, but it is a good start.
- The parameters η^{eq}_F and η^{eq}_{EF} are, in a sense, safety margins compared to the predictions of the equilibrium model. They cannot be very far from 1; this will be explained later in this section.
- The time t^χ_ϕ, the ratio of the total fluid flow rate required to perform the separation to the system volume, is simply connected to column length and fluid interstitial velocities. The higher t^χ_ϕ, the larger the ratio of t^χ_ϕ to characteristic times for mass transfer, the lower the mass transfer limitations.
- Yield θ_*: in contrast to the above parameters, yield cannot be estimated, and must be determined by simulation (using one of the models in Chapters 7 and 8), for example to obtain the target purity with the tuning parameters embedded in η^{eq}_F, η^{eq}_{EF}, χ^{eq}_F and t^χ_ϕ.

As an exercise, let us estimate the productivity per unit volume ϕ_v associated with the equilibrium model. χ_F^{eq} is determined according to the adsorption isotherms to the feed concentration, and, by definition, one has $\eta_F^{eq} = \eta_{EF}^{eq} = \theta_* = 1$ for the equilibrium model. In order to deliver this "performance", the column must be equivalent to an infinite number of plates. The absence of kinetic limitations requires $L_{col} = \infty$ in the case of elution chromatography, so $t_\phi^X = \infty$; consequently $\phi_v = 0$, which means that the equilibrium model cannot be used for estimating productivity. The same conclusion is reached for other modes. Instead of productivity per unit volume, the specific eluent consumption can be unambiguously calculated with the equilibrium model: $\Upsilon_*^{eq} = \frac{1}{\theta_* \chi_F^{eq}} \frac{X_*}{X_*^F}$.

It is worth noting that the different parameters appearing in Eqs. (9.37) and (9.38) indicate a competition between different factors. As an illustration, if one uses lower values of η_F^{eq}, the peaks or steady-state profiles are better separated because of the safety margin taken. As a consequence, for a given purity, at constant t_ϕ^X, a better yield can be delivered. Alternatively, with the same safety margin, one can decide to work at a lower yield and a lower t_ϕ^X. The aim of the following sections is to propose reasonable compromises.

EXERCISE 9.2 *A separation of two solutes A and B, linearly adsorbed, must be performed in elution or in the TMB mode. One has $C_A^F = C_B^F = 5$ g/l, $\overline{K}_A = 1$ and $\overline{K}_B = 3$. Use Eq. (9.37) to express productivity associated with the production of 100% pure fractions.*

Determining the feed concentration

Feed concentration should *in general* (at least for Langmuirian systems) be taken as high as possible to maximize productivity and minimize eluent consumption (Knox and Pyper, 1986; Guiochon *et al.*, 2006). While this is appropriate in many situations, exceptions occur. As an illustration, in the case of selectivity reversal occurring at some concentration, it is probably not advisable to inject the feed at a concentration above the one at which this selectivity reversal occurs.

Fortunately, the qualitative evolution of productivity (and eluent consumption) versus feed concentration can be quickly apprehended with Eq. (9.37). The yield θ_* and the parameters η_F^{eq}, η_{EF}^{eq} (thus η_E^{eq}) appearing in Eq. (9.37) will be the results of the optimization. We know, however, that the chromatographic system will probably not be operated at yield θ_* and η_F^{eq}, η_{EF}^{eq} (thus η_E^{eq}) too far from 1. As a consequence, purities being fixed, we learn from Eq. (9.37) that, in the design space of practical interest, the order of magnitude of ϕ_v is given by $\frac{\chi_F^{eq}}{t_\phi^X}$. We have thus identified the two parameters, χ_F^{eq} and t_ϕ^X, which are the main drivers for productivity.

We know that the feed concentration directly affects χ_F^{eq} according to the equilibrium model. As the feed concentration C_{tot}^F tends towards zero, the concentration χ_F^{eq} tends towards zero. When C_{tot}^F increases, χ_F^{eq} normally increases to a plateau determined by the equilibrium model. That at least is the situation for typical competitive adsorption isotherms. In the case of linear adsorption isotherms, χ_F^{eq} increases linearly with feed concentration; an example associated with synergistic adsorption will be given in Section 10.1.

How can feed concentration impact t_ϕ^χ as a result of the optimization? The value of t_ϕ^χ influences the system efficiency: the higher t_ϕ^χ compared to characteristic times for mass transfer, the lower the kinetic limitations. Everything else being fixed, the choice of t_ϕ^χ results from a compromise: it must be low enough to impose a short cycle time, but high enough to allow the column to be efficient enough to deliver the required purity at sufficient yield.

The feed concentration can certainly play a role in the required/optimum number of plates, and thus influences the optimum value of t_ϕ^χ. While possibly real, the influence of the feed concentration on the required t_ϕ^χ must be secondary compared to its influence on concentration χ_F^{eq}, which is directly associated with the non-linearity of the adsorption isotherms. We will thus consider that the influence of feed concentration on productivity is essentially captured in the χ_F^{eq} term.

From the preceding considerations, concentrations χ_F^{eq} and productivity ϕ_v must exhibit similar variations with feed concentration. We also know that injecting a too highly concentrated feed material can lead to difficulties such as miscibility issues, very abrupt viscosity changes and thus pressure excursions, and to very small injection times (elution) or feed flow rates (TMB/SMB), which could be problematic for operating the system. As a consequence, the feed concentration will be selected so as to:

- maximize χ_F^{eq} as calculated by the equilibrium model (Eq. (9.32))
- ensure Δt_{inj} ($u_{e3} - u_{e2}$ for TMB/SMB) compatible with operational constraints; for instance $\Delta t_{inj} \geq 0.05\ t_0$ and $(u_{e3} - u_{e2}) \geq 0.05\ u_{e1}$ can be considered reasonable
- avoid miscibility and viscosity risks. This is to be assessed case-by-case: a 5 bar pressure excursion is a non-event on an HPLC column but can be detrimental with very compressible media.

An illustration of the determination of feed concentration is given in Exercise 9.3.

EXERCISE 9.3 *The separation described in Exercise 9.1 is to be performed on a chromatographic system having the following characteristics:*

- *Lumped adsorption isotherms: linearly modified Langmuir with* $\bar{\lambda}_A = \bar{\lambda}_B = 0.5$, $\tilde{K}_A = 0.01$ *l/g*, $\tilde{K}_B = 0.015$ *l/g*, $\overline{N} = 100$ *g/l*
- *Extragranular porosity: 0.4*
- *Particle size:* $d_p = 20\ \mu m$
- *Pressure drop:* $\frac{\Delta P}{L} = 2\ u_e$ *bar/cm* (u_e *in cm/s*)
- *Number of plates:* $HETP_A(cm) \approx HETP_B(cm) \approx 3\ d_p + 0.15\ u_e$ (u_e *in cm/s and* d_p *in cm*)
- *No solubility and no viscosity issues.*

A is the solute of interest. Propose a reasonable feed concentration C_{tot}^F *for the separation. Do we need all the above information?*

Determining parameters other than C^F_{tot}, L_{col} and u_e ($\bar{u}_e$ for TMB)

At this point, we have determined C^F_{tot}. We know that the solution of the equilibrium model can be derived for any given t_0 (for elution) or $\bar{u}_e$ (for TMB). Common sense tells us again that optimal η^{eq}_F parameters are normally not far from 1. When η^{eq}_F is substantially less than 1, injected quantities are too small for the chromatographic medium to be used in a decent manner. When η^{eq}_F is substantially greater than 1, required purities can still be reached but at the expense of very small yields.

The case of η^{eq}_E or η^{eq}_{EF}, which are related via Eq. (9.36), is similar in nature but their influence on the process performance is normally lower than that of η^{eq}_F. Notice that η^{eq}_F involves only primary tuning parameters, whereas η^{eq}_E (resp. η^{eq}_{EF}) involves only secondary parameters for elution chromatography (resp. TMB/SMB). This drives us to make a distinction between elution chromatography and TMB/SMB:

- Elution chromatography: the cycle time t_{cycle} may have to be increased with respect to t^{eq}_{cycle} because of the sensitivity of chromatograms to system efficiency (plates). Adequate t_{cycle} can be derived simply from the simulation of one chromatogram with primary tuning parameters; this leads to η^{eq}_{EF}. Alternatively, η^{eq}_{EF} can be guessed: experience shows that η^{eq}_{EF} from 1.1 up to about 1.5 for systems equivalent to a few hundred plates only are reasonable values. The parameter η^{eq}_E is then obtained from η^{eq}_{EF} via Eq. (9.36).
- TMB or SMB: The parameter η^{eq}_E is known when secondary tuning parameters u_{e1} and u_{e4} have been determined. Because of the moderate sensitivity of TMB/SMB to system efficiency (plates), experience shows that η^{eq}_E can be taken very close to 1 (1.2 is already a very significant safety margin). The parameter η^{eq}_{EF} is then obtained from η^{eq}_E via Eq. (9.36).

With η^{eq}_F, and to a lesser extent η^{eq}_E and η^{eq}_{EF}, we have thus identified a very important set of parameters whose *order of magnitude* can simply be guessed; this is a very interesting way to initialize and also to structure the optimization algorithm. As a simple illustration, Δt_{inj} is simply given by $\Delta t_{inj} = \eta^{eq}_F \Delta t^{eq}_{inj}$ according to Eq. (9.33). All our parameters are thus expressed as functions of L_{col} and u_e.

The case of TMB is more complex. Equations (9.33) and (9.34) relate $u^{tmb}_{e2} - u^{tmb}_{e3}$ as well as $u^{tmb}_{e1} - u^{tmb}_{e4}$ to the solid velocity $\bar{u}_e$. The feed concentration being fixed, at given η^{eq}_F and η^{eq}_E we have still four free parameters, for instance two fluid velocities in addition to $\bar{u}_e$ and L^{tmb}_{col}. Consequently, we need two additional constraints in order to express our problem as a function of $\bar{u}_e$ and L^{tmb}_{col} only. We know that the safety margin η^{eq}_E is likely to play a secondary role (don't forget that we are deriving shortcuts). Consequently, we don't take a risk if we split this margin evenly between zones 1 and 4, for instance using the parameter ϑ introduced in Eq. (8.73). This allows the expression for u^{tmb}_{e1} and u^{tmb}_{e4} as functions of $\bar{u}_e$. Splitting the η^{eq}_F margin between zones 2 and 3 requires more attention due to the shape of the separation zone in (m_2,m_3) space. Further explanation will be given in Section 9.6.2. With these approximations, all our parameters are expressed as functions of $\bar{u}_e$ and L^{tmb}_{col} only. Extending the above TMB approximation to the SMB is straightforward. The algorithm structure is presented in Figure 9.7.

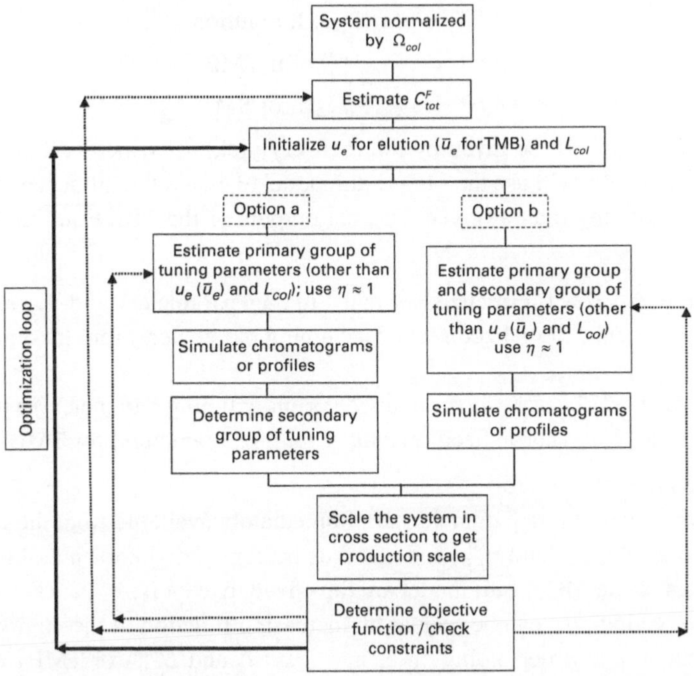

Figure 9.7 Principle of the optimization procedure (Type I problem): estimate of all tuning parameters except column length and one velocity. If estimates are precise enough, the loops associated with dotted lines are optional. η means η_F^{eq}, η_E^{eq} or η_{FE}^{eq}.

9.4.3.2 Determining length and velocity

Our objective is now to determine the system length and the remaining velocity (u_e for elution, $\bar{u}_e$ for TMB). The feed concentration has been selected, and reasonable values for η_F^{eq} and η_E^{eq} (thus η_{EF}^{eq}) have been postulated. The last tuning parameters u_e and L_{col} ($\bar{u}_e$ and L_{col}^{tmb} for TMB) are obtained by requiring a reasonable yield and pressure drop.

A physically meaningful constraint often consists in requiring the system to operate at a given pressure. In any chromatographic optimization, pressure drop plays a critical role. Not only must this parameter stay within limits compatible with the chromatographic medium's stability, but it seriously impacts the cost of the equipment; this is a very serious contributor to the parameter d_{chr} defined in Section 9.3.3. On the other hand, the higher the operating pressure, the smaller the system diameter (see further details in Section 9.5.2), so the optimum operating pressure results from an economic compromise. Instead of determining the operating pressure that minimizes the economic objective function, we will arbitrarily fix the pressure drop at ΔP^{tg}. This latter choice is in fact no more restrictive than the former, but it allows simpler thinking.

The required pressure is given by experience: a few bar maximum for very compressible gels, 5–10 bar for resins, about 50 bar for high-performance systems. Darcy's law requires a relation of the form:

$$\Delta P^{tg} = g(u_e, L_{col}) \text{ for elution}$$
$$\Delta P^{tg} = g(\bar{u}_e, L_{col}^{tmb}) \text{ for TMB} \quad (9.39)$$
$$\Delta P^{tg} = g(\Delta T, L_{col}) \text{ for SMB}$$

The other constraint will be given by the recovery yield θ_*, which is impacted by the column efficiency. We will use the MC–$\overline{\text{Equil}}$ model to assess the influence of efficiency. It is worth mentioning that, at given C_{tot}^F, the outputs of the MC–$\overline{\text{Equil}}$ model depend only on:

- the ratio $\Delta t_{inj}/t_0$ (normalized primary tuning parameter) and the ratios t_{c1}/t_0, t_{c2}/t_0, t_{cycle}/t_0 (normalized secondary tuning parameters) for elution chromatography
- the ratios $u_{2e}^{tmb}/\bar{u}_e$, $u_{3e}^{tmb}/\bar{u}_e$ (normalized primary tuning parameters) and the ratios $u_{1e}^{tmb}/\bar{u}_e$, $u_{4e}^{tmb}/\bar{u}_e$ (normalized secondary tuning parameters) for TMB

and on the number of plates.

These normalized tuning parameters are immediately available from the equilibrium model and from η_F^{eq}, η_E^{eq} and η_{EF}^{eq}, and can thus be considered known at this point. All the parameters of the MC–$\overline{\text{Equil}}$ model are thus fixed, except N.

The recovery yield, θ_*, can be seen as an increasing function of the number of plates N, the number of plates depending on u_e and L_{col} ($\bar{u}_e$ and L_{col}^{tmb} for TMB) via the Van Deemter curve.[17] One can thus write $\theta_* = f(u_e, L_{col})$ (or $\theta_* = f(\bar{u}_e, L_{col}^{tmb})$ for TMB), where θ_* represents the maximum yield at which the target product can be recovered at purity X_*. If one sets a yield target θ_*^{tg}, one obtains:

$$\theta_*^{tg} = f(u_e, L_{col}) \text{ for elution}$$
$$\theta_*^{tg} = f(\bar{u}_e, L_{col}^{tmb}) \text{ for TMB} \quad (9.40)$$
$$\theta_*^{tg} = f(\Delta T, L_{col}) \text{ for SMB}$$

Solving Eqs. (9.39) and (9.40), one obtains u_e and L_{col} ($\bar{u}_e$ and L_{col}^{tmb} for TMB) and most of the problem is then solved.

Based on the above considerations, the structure of the complete shortcut algorithm is given in Figure 9.8. More detailed information and illustrations are given in the following sections.

9.5 Application to elution chromatography

9.5.1 Type I problems

For elution chromatography, solving a Type I problem consists in determining seven parameters (C_{tot}^F, u_e, Δt_{inj}, L_{col}, t_{cycle}, t_{c1}, t_{c2}). In order to simplify the presentation, we will segregate these into four primary parameters (C_{tot}^F, u_e, Δt_{inj}, L_{col}) and three secondary parameters (t_{cycle}, t_{c1}, t_{c2}) that can be determined if the former are known.

[17] Note that the above reasoning is not limited by the fact that the Van Deemter equation is derived only for linear systems. The system is simulated under possibly non-linear conditions with the MC–$\overline{\text{Equil}}$ model. Based on the simulation, a certain number of plates is required to reach certain performance criteria. The Van Deemter equation is simply a convenient way to estimate the length and velocity for this number of plates.

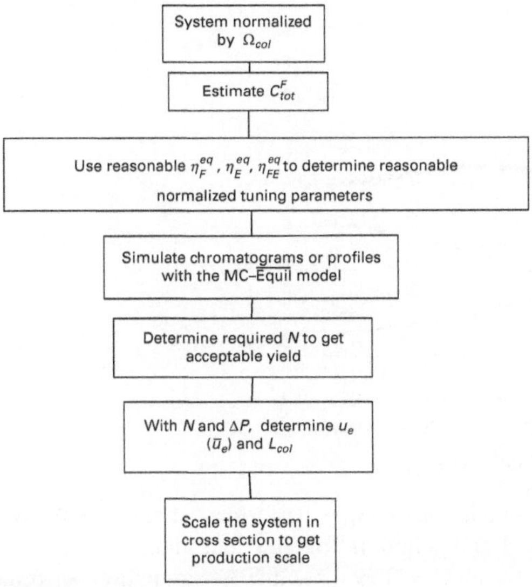

Figure 9.8 Principle of the complete shortcut: all tuning parameters are estimated.

In this section, we will neglect the influence on chromatogram i of pollution from chromatograms $i-1$ and $i+1$, i.e. successive chromatograms are exactly superpositions of single-injection chromatograms. Following the presentation given in Section 9.4.2.1, the secondary variable t_{cycle} is defined as the period between t_{beg}, at which the least retained solute has grown, for instance, to 0.1% of its maximum, and t_{end}, at which the most retained solute product has shrunk, for instance, to 0.1% of its maximum.

Optimizing the four primary parameters simultaneously may be a little bit computer-intensive but poses no specific difficulties. Instead, we will use the methodology presented in Section 9.4.2.

For illustration purposes, we will simulate the problem of Exercise 9.3 for C_{tot}^F = 100 g/l with $\Delta P^{tg} = 50$ bar. The pressure drop being fixed, the column length and fluid velocity are thus related by Darcy's law,[18] and we will select L_{col} as the independent parameter. Note that the fluid velocity varies as the reciprocal of L_{col} due to the pressure-drop constraint. The column will be simulated with the MC–$\overline{\text{Equil}}$ model; using more sophisticated non-equilibrium models poses no difficulty and would lead to the same qualitative conclusions.

Let us follow the algorithm presented in Figure 9.7. For a given L_{col}, the interstitial fluid velocity u_e is determined according to the pressure-drop constraint. This allows a determination of t_0 and, with the Van Deemter equation, of the number of plates of the column. For a given L_{col}, the injection time is given by $\Delta t_{inj} = \eta_F^{eq} \Delta t_{inj}^{eq}(L_{col}, C_{tot}^F)$, and we assume for the time being that $\eta_F^{eq} = 1$. Chromatograms can thus be unambiguously

[18] Should a pressure drop not be imposed, one needs to initialize the algorithm with decent initial values of u_e and L_{col}. We know from experience that optimum column lengths are typically a few tenths of a centimeter for HPLC systems packed with particles of size 10–50 μm, and a few meters for particle sizes of a few hundred μm. Using the Van Deemter equation, one can then estimate fluid velocity in order to get a few hundred plates for comfortable selectivity (let us say about 2) and of a few thousand for small selectivity (let us say 1.1).

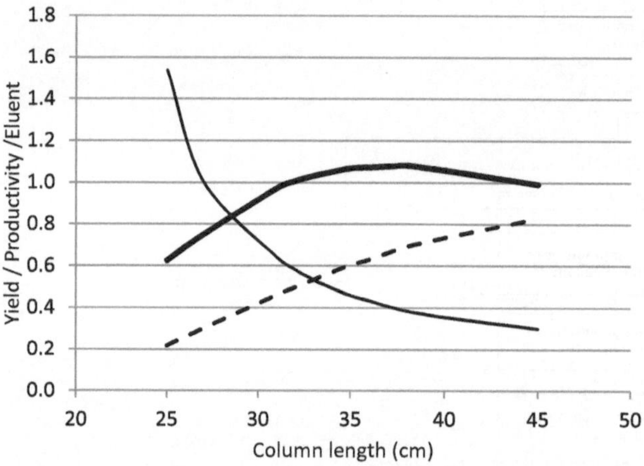

Figure 9.9 Influence of L_{col} on elution chromatography performance; parameters given in Exercise 9.3. $C_{tot}^{F} = 100$ g/l, $\eta_F^{eq} = 1$ and pollution from previous and subsequent chromatograms neglected, $\Delta P = 50$ bar, $X^* = 0.99$ for solute A. Simulation performed with the MC–Equil model. Productivity φ_V (g/(l min)) (heavy solid line), eluent consumption Υ_* (l/g) (light solid line), yield θ_* (dashed line).

simulated for any column length. The secondary parameters are then determined, and consequently the parameters η_{EF}^{eq} and η_E^{eq}, the productivity per unit volume φ_V, the eluent consumption Υ_* and the yield θ_*. The results are given in Figure 9.9.

It is apparent that productivity reaches a maximum with respect to column length, the result of a compromise between a good yield (thus a significant column length to get a high resolution) and minimum column size (thus small length and poor yield).

The system can be optimized in length with very elementary algorithms since L_{col} is the only tunable parameter, and the general optimization framework of Figure 9.7 is not necessary. However, this example, presented for illustrative purposes, is not realistic. Let us progress further towards realism with two tuning parameters, namely Δt_{inj} and L_{col}. We will consider successively an unconstrained and a constrained problem: either we optimize productivity or we optimize productivity in addition to satisfying $\theta_* \geq 0.95$. In Figure 9.10, the influence of η_F^{eq} on productivity φ_V, eluent consumption Υ_* and yield θ_* is given; this generalizes the results of Figure 9.9.

When optimum productivity is sought without yield constraint, the simulation results show that the productivity reaches a maximum for $\eta_F^{eq} \approx 1.5$ (Figure 9.10a). When productivity is optimized and yield kept free, yield decreases when the η_F^{eq} parameter increases, and at the optimum *productivity* the yield is quite small (around 0.5 or less).

When a significant yield is sought (95% in our example), the simulation shows that the productivity reaches a maximum for η_F^{eq} slightly below 1 (Figure 9.10b). Note that, in this example, the optimum productivity is obtained when yield equals the minimum target value (95%). Moreover, with or without yield constraint, the eluent consumption systematically decreases when η_F^{eq} increases.

These results confirm that the equilibrium model gives a very reasonable starting point for a precise optimization: taking η_F^{eq} about 1 is not far from optimum.

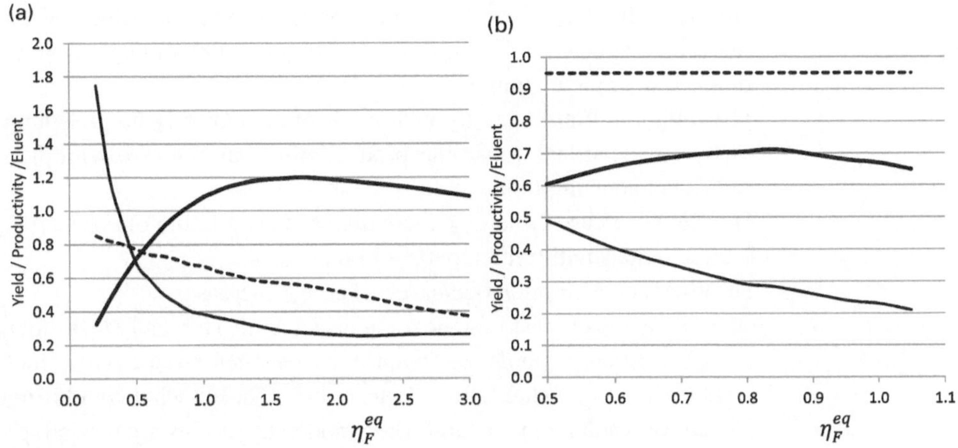

Figure 9.10 Influence of the η_F^{eq} parameter on elution chromatography performance; parameters given in Exercise 9.3. $C_{tot}^F = 100$ g/l; pollution from previous and subsequent chromatograms neglected, $\Delta P = 50$ bar, $X^* = 0.99$ for solute A. Simulation performed with the MC–$\overline{\text{Equil}}$ model. Productivity φ_V (g/(l min)) (heavy solid line), eluent consumption Υ_* (l/g) (light solid line), yield θ_* (dashed line). (a) For each η_F^{eq} column length has been chosen in order to maximize productivity. (b) For each η_F^{eq} column length has been chosen in order to maximize productivity and to ensure $\theta_* \geq 0.95$.

Additionally, it is worth noting that the optimized column length L_{col} (not shown in Figure 9.10) increases with η_F^{eq}, from about 30 cm to about 60 cm when η_F^{eq} varies from 0.2 to 2.0 when productivity is optimized, and from about 52 cm to 65 cm when η_F^{eq} varies from 0.75 to 1.1 at a constant yield of 95%. This confirms the role of η_F^{eq}: when η_F^{eq} increases, the safety margin decreases, so that a greater column length is needed for decent work.

These results also confirm that complete optimization results in a trade-off between yield, eluent consumption and productivity and that they cannot be optimized simultaneously. An illustration of how to optimize an objective function by compromising between productivity, eluent consumption and yield is given in Exercise 9.4.

EXERCISE 9.4 *Using the information given in Exercise 9.3 and the results in Figure 9.10a, propose a system design operating at 50 bar allowing one to minimize production cost under the conditions of Exercise 9.1. Is it a true optimum? Compare this with the results in Figure 9.10b.*

The detailed behavior of chromatographic systems with respect to η_F^{eq} varies according to the number of solutes, the targeted solute, the exact adsorption isotherms etc. However, the following qualitative conclusions are quite general (to the best of my knowledge):

- The productivity reaches a maximum for η_F^{eq} between 1 and 2 in the absence of yield constraints.

- The productivity reaches a maximum for η_F^{eq} between 0.75 and 1 when significant yield constraints exist. Significance is certainly problem-dependent, but let us say above 95% to pick a number.
- The variation of productivity with η_F^{eq} is often relatively flat, meaning that different systems can lead to similar productivities but obtained with different yields and eluent consumptions.
- The yield decreases when η_F^{eq} increases. At the optimum of *productivity*, the yield is often quite small (typically 0.5 or below).
- The eluent consumption decreases when η_F^{eq} increases.
- Optimizing the process (and not only productivity) typically leads to η_F^{eq} values in the 0.7–1.0 range in order to compromise between productivity, yield and eluent consumption. η_F^{eq} values above 1 can only be chosen when yield is not a concern.
- The above results remain valid when moderate purities are needed, provided that η_F^{eq} is, according to the definition, referred to the injection time t_{inj}^{eq}, allowing one to get the target purity at 100% yield on a column of infinite efficiency.

9.5.2 Type 0 shortcuts: influence of pressure drop and particle size

We will now follow the shortcut methodology in Section 9.4.3.2 and Figure 9.8 to get simple and reasonable information on how productivity is impacted by pressure drop, and consequently by particle size, for elution chromatography systems. After determining the feed concentration and selecting η_E^{eq} and η_F^{eq}, the *normalized tuning parameters* are known, and one determines L_{col} and u_e in order to obtain a given pressure drop and yield.

Darcy's law, given by Eq. (6.3), is written:

$$\frac{\Delta P^{tg}}{L_{col}} = \delta_e u_e \tag{9.41}$$

where ΔP^{tg} is the target pressure drop. We know that, for soft particles, more complex expressions should be used. This poses no difficulty.

A few simulations allow determination of the column length (fluid velocity being imposed by the pressure-drop constraint), allowing us to reach the target yield. Specifically, using the MC–$\overline{\text{Equil}}$ model,[19] one can determine the number of plates allowing one to reach the target purity at the targeted yield. Let us call this number of plates N^{tg}.

We know from Section 5.8 that the number of plates N can in general be expressed by an equation of the form:

$$N^{tg} = \frac{L_{col}}{HETP} = \frac{L_{col}}{\frac{A_e}{u_e} + B_e + C_e u_e} \tag{9.42}$$

Equations (9.41) and (9.42) allow a determination of L_{col} and u_e required to obtain ΔP^{tg} and N^{tg}. This allows the unit to be quickly sized. The column diameter is then obtained, as explained in Section 9.4.2.2.

[19] Considering number of plates and HETP is a convenient way of presenting things. The same reasoning could be used with the more sophisticated kinetic models.

We now investigate the influence of Darcy's and Van Deemter's coefficients on our two parameters u_e and L_{col}, thus on system characteristics. In liquid or supercritical fluid chromatography, the term A_e/u_e of the Van Deemter curves is always negligible with respect to the others. Under these conditions, Eqs. (9.41) and (9.42) lead to:

$$u_e = \frac{B_e}{2C_e}\left[-1+\sqrt{1+\frac{4\Delta P^{tg} C_e}{\delta_e N^{tg} {B_e}^2}}\right] \qquad L_{col} = \frac{\Delta P^{tg}}{\delta_e u_e} \tag{9.43}$$

For given hydrodynamic (δ_e) and mass transfer (B_e,C_e) characteristics, Eq. (9.43) allows the unit to be sized.

Furthermore, systems are frequently operated in the high-velocity range of the Van Deemter curve, so the term $C_e u_e$ dominates B_e. Under these common circumstances, Eq. (9.43) leads to:

$$L_{col} = \sqrt{\frac{N^{tg} C_e \Delta P^{tg}}{\delta_e}} \qquad u_e = \sqrt{\frac{\Delta P^{tg}}{\delta_e N^{tg} C_e}} \tag{9.44}$$

In the rest of this section, we will assume that the system is operated in the high-velocity range of the Van Deemter curve so that Eq. (9.44) holds.

Purities, yield, feed concentration (thus χ_F^{eq}) and the parameters η_F^{eq} and η_{EF}^{eq} being fixed, productivity expressed in Eq. (9.37) is thus proportional to the reciprocal of t_ϕ^χ, so one writes $\phi_v \sim \frac{1}{t_\phi^\chi}$. This means that:

$$\phi_m \sim \phi_v \sim \frac{1}{t_\phi^\chi} \sim \frac{u_e}{L_{col}} \sim \frac{1}{N^{tg} C_e} \tag{9.45}$$

The proportionality factors in Eq. (9.45) depend on thermodynamics, feed concentration, η_F^{eq} and η_{EF}^{eq} as well as yield, but not on hydrodynamics or mass transfer.

Equation (9.45) shows that the productivity per unit volume (or mass) is independent of the pressure-drop target but proportional to the reciprocal of the required number of plates.

The productivity per unit diameter is determined by noting that $\phi_D = \phi_v \frac{V_{col}}{D_{col}}$ and $\phi_v \sim \frac{u_e}{L_{col}}$, so $\phi_D \sim u_e D_{col}$. The fluid flow rate, determined from the equilibrium model and parameters η_F^{eq} and η_{EF}^{eq}, is, according to this shortcut, independent of ΔP^{tg} and N^{tg}. This leads to $D_{col} \sim {u_e}^{-1/2}$, so that one obtains:

$$\phi_D \sim {u_e}^{1/2} \sim \left[\frac{\Delta P^{tg}}{\delta_e N^{tg} C_e}\right]^{1/4} \tag{9.46}$$

Instead of productivity per unit volume, the productivity per unit diameter is thus impacted by the pressure-drop target.

With Eqs. (9.45) and (9.46), we have related productivities to the pressure-drop target and the number of plates (related to yield). We can go further and analyze the influence of particle size on the system characteristics.

The coefficient δ_e in Darcy's law (see Eq. (9.41)) is proportional to the reciprocal of the square of the particle diameter, so we write $\delta_e \sim d_p^{-2}$. The proportionality coefficient depends on the fluid viscosity and on the column porosity.

In addition, when the internal diffusion limitation prevails, the coefficient C_e is proportional to the square of particle size, so we write $C_e \sim d_p^2$. The proportionality coefficient depends on diffusion coefficients, particle porosity and adsorption isotherms. Introducing the dependence of parameters δ_e and C_e on particle diameter in Eqs. (9.45) and (9.46), one obtains:

$$\begin{aligned} \phi_v \sim \phi_m &\sim \frac{1}{N^{tg} C_e} \sim \frac{1}{{d_p}^2} \\ \phi_D &\sim \left[\frac{\Delta P^{tg}}{\delta_e N^{tg} C_e}\right]^{-1/4} \sim {d_p}^0 \end{aligned} \tag{9.47}$$

In addition, other proportionality relations are directly derived and summarized below:

$$\begin{aligned} u_e \sim \sqrt{\Delta P^{tg}} \quad & L_{col} \sim \frac{d_p^2 \Delta P^{tg}}{u_e} \sim {d_p}^2 \sqrt{\Delta P^{tg}} \\ t_{cycle} \sim \frac{L_{col}}{u_e} \sim d_p^2 \quad & D_{col} \sim (\Delta P^{tg})^{-1/4} \end{aligned} \tag{9.48}$$

This simple derivation shows that, for classical chromatography operation (significant internal mass transfer limitations):

- productivities per unit mass or volume are not influenced by the target pressure but vary as the reciprocal of the square of the particle diameter
- productivity per column diameter is not influenced by the particle size but varies as the 1/4 power of the pressure drop
- the cycle time varies as the square of the particle diameter
- the column length varies as the square of the particle diameter and as the square root of the pressure target.

These results were probably not totally intuitive ...

The methodology can be extended without difficulty to situations involving external mass transfer limitations or pressure drop, exhibiting complex variations with column length, diameter and fluid velocity.

EXERCISE 9.5 *A separation is performed using 100 μm silica particles in a column (L_{col} = 100 cm, D_{col} = 50 cm) operated at 10 bar. A pulse test performed on the unit shows that the number of plates drops from 500 to 200 when the flow rate is tripled. An operation with 25 μm particles in an HPLC system operated at 30 bar is considered. Assuming that the adsorption isotherms are not affected by particle size or operating pressure, what is the expected size of the new unit?*

9.6 Application to TMB

We will now use our optimization methodology for the design of TMB systems.

9.6.1 Type I problems

Solving a Type I problem for a TMB consists in determining seven parameters that can be organized into two groups: the primary group (C_{tot}^F, u_{e2}^{tmb}, u_{e3}^{tmb}, $\bar{u}_e$, L_{col}^{tmb}) and the secondary group (u_{e1}^{tmb}, u_{e4}^{tmb}), all other parameters being fixed.

Let us follow the algorithm presented in Figure 9.7. The concentration C_{tot}^F is estimated as explained in Section 9.4.3. The strategy consists in expressing the entire problem in terms of $\bar{u}_e$ and L_{col}, which represent the key parameters of the optimization loop. From knowledge of C_{tot}^F, all the ratios m_i^{eq} can be determined with the equilibrium model. Recall that the m_i (m_i^{eq} when those are derived from the equilibrium model) are simple ratios of fluid to solid velocities:

$$u_{ei}^{tmb} = m_i \frac{1-\varepsilon_e}{\varepsilon_e}\bar{u}_e \qquad u_{ei}^{eq} = m_i^{eq}\frac{1-\varepsilon_e}{\varepsilon_e}\bar{u}_e \tag{9.49}$$

For given $\bar{u}_e$, working in terms of u_{ei}^{tmb} or in terms of m_i are equivalent. We will use one or the other, as appropriate.

Let us start by estimating the u_{ei}^{tmb}. As mentioned in Section 9.4, the secondary group of parameters is determined in a sub-loop for the given primary group of parameters. This sub-loop does not pose specific difficulties provided that initialization is done adequately. The secondary group can be initialized using the m_1^{eq}, m_4^{eq} of the equilibrium model; in other words, it means that $\eta_E^{eq} = 1$. The optimum parameters are in general slightly, but not significantly, different from those given by the equilibrium model. For instance, it can sometimes be beneficial to work with an m_1 parameter slightly lower than m_1^{eq}: the highly retained component will be partly recycled from zone 1 to zone 4; the purity at the raffinate will certainly drop, but this choice may allow an increase in productivity and a decrease in eluent consumption. Similar considerations apply for m_4.

Let us now consider zones 2 and 3. By definition of η_F^{eq}, one has:

$$\begin{gathered} m_3 - m_2 = \eta_F^{eq}\left(m_3^{eq} - m_3^{eq}\right) \text{ or equivalently} \\ \frac{u_{e3}^{tmb} - u_{e2}^{tmb}}{\bar{u}_e} = \frac{1-\varepsilon_e}{\varepsilon_e}\eta_F^{eq}\left(m_3^{eq} - m_3^{eq}\right) \end{gathered} \tag{9.50}$$

This shows that, at given $\bar{u}_e$, selecting η_F^{eq} imposes a relationship between u_{e2}^{tmb} and u_{e3}^{tmb}. By requiring a given η_F^{eq} (that cannot be too far from 1), our problem is thus reduced to a three-parameter problem in u_{e2}^{tmb}, $\bar{u}_e$ and L_{col}^{tmb} (whereas it was limited to a two-parameter problem in u_e and L_{col} for elution chromatography).

As with elution chromatography, for illustration purposes, instead of considering the pressure drop to be part of a complex optimization function, it can simply be imposed.

The TMB is composed of different zones having different lengths, defined by:

$$\begin{aligned} &L_i = l_i L_{col}^{tmb} \text{ for } i \in [1,4] \\ &l_i \text{ representing fractions of the total length} \\ &\sum_{i=1}^{i=4} l_i = 1 \end{aligned} \tag{9.51}$$

The pressure drop in the system is related to the different zone lengths and associated fluid velocities by:

$$\frac{\Delta P^{tmb}}{L_{col}^{tmb}} = \delta_e \sum_{i=1}^{i=4} (u_{ei}^{tmb} + \bar{u}_e) l_i \tag{9.52}$$

Note that Darcy's law has been modified to take into account the fluid velocity relative to the solid, $u_{ei} + \bar{u}_e$. Incidentally, this relative fluid velocity is also the fluid velocity that must be considered in an SMB (see Section 9.7.2).

Using the relation between the interstitial fluid velocities and the m_i, Eq. (9.52) becomes:

$$\begin{aligned} \frac{\Delta P^{tmb}}{L_{col}^{tmb}} &= \bar{\delta}_e \bar{u}_e \\ \bar{\delta}_e &= \delta_e \left[1 + \frac{1-\varepsilon_e}{\varepsilon_e} \sum_{i=1}^{i=4} m_i l_i \right] \approx \delta_e \left[1 + \frac{1-\varepsilon_e}{\varepsilon_e} \sum_{i=1}^{i=4} m_i^{eq} l_i \right] \end{aligned} \tag{9.53}$$

Note that replacing the m_i by m_i^{eq} in the above expressions is in general an excellent approximation: not only are the m_i and m_i^{eq} close but, in addition, compensations occur in the weighted sum in Eq. (9.53). Equation (9.53) is formally identical to the one derived for elution chromatography, provided that u_e is replaced by $\bar{u}_e$ and δ_e by $\bar{\delta}_e$. This shows that, if pressure drop is imposed, $\bar{u}_e$ and L_{col}^{tmb} are related. We will then use L_{col}^{tmb} as the independent parameter.

For simplicity, we will use the MC–$\overline{\text{Equil}}$ model to determine internal concentration profiles and thus purity and yield.[20] The number of plates N_i^{tmb} associated with each zone i are determined from u_{ei}^{tmb} and L_i with the Van Deemter equation:

$$N_i^{tmb} = \frac{L_i}{HETP_i} = L_{col}^{tmb} \frac{l_i}{\frac{A_e}{(u_{ei}^{tmb} + \bar{u}_{ei})} + B_e + C_e(u_{ei}^{tmb} + \bar{u}_e)} \tag{9.54}$$

Note that fluid velocities relative to the solid are used in the Van Deemter equation. Following the discussion in Section 8.3, there is no certainty that Eq. (9.54) adequately describes the TMB efficiency, but, as with pressure drop, the expression will readily be usable for SMB (see Section 9.7.2).

As far as initialization is concerned, we know from experience that optimum zone lengths are typically a few tens of centimeters for systems packed with particles of size 10–50 μm, and a few meters for particles of a few hundred μm in size.

20 The same methodology can be used with more complex models.

We can now use the data of Exercise 9.3 and follow the algorithm presented in Figure 9.7. C_F^{tot} is fixed at 100 g/l. For any given system length, a solid velocity is associated with it due to the pressure-drop constraint. Prior to addressing optimization, let us illustrate key system characteristics by fixing $m_1 = m_1^{eq}$, $m_4 = m_4^{eq}$ (thus $\eta_E^{eq} = 1$) and $\eta_F^{eq} = 1$. Under these conditions, one velocity (let us pick u_{e2}) is free and can be optimized. Our remaining free parameters are thus u_{e2} and L_{col}^{tmb}; for normalization purposes, we will use m_2 and L_{col}^{tmb} instead, which is equivalent. The numbers of plates N_i^{tmb} associated with each zone i are determined with the Van Deemter equation (9.54).

For a given column length, the outlet purity (raffinate in this example) is strongly influenced by the precise value of m_2; this is illustrated in Figure 9.11. If one sets $m_2 = m_2^{eq}$, for the considered system length one obtains a purity for solute A which is a bit below the target of Exercise 9.3 (99%). A way to increase the raffinate purity is to work with a slightly lower m_2 ratio (and thus m_3 according to Eq. (9.50)). The fronts are moved slightly towards the lower zones in the TMB: this helps to increase the raffinate purity but at the expense of yield (see Figure 9.11). The important message is that, even when all the parameters but m_2 are fixed, m_2 can be so tuned as to get the targeted purity. Tuning the m_i is an important occupation of SMB developers and operators ...

Now, as we did for elution chromatography, let us investigate the influence of η_F^{eq} on system performance for two situations: either productivity is optimized, or productivity is optimized subject to a yield constraint ($\theta_* \geq 95\%$). All simulations are performed for an imposed pressure drop of 50 bar (Figure 9.12). For a given η_F^{eq}, the parameters to be optimized, m_2 and L_{col}^{tmb}, have been determined so as to obtain the target purity at maximum productivity, either with or without yield constraints. For simplicity, we impose $m_1 = m_1^{eq}$, $m_4 = m_4^{eq}$ for all simulations.

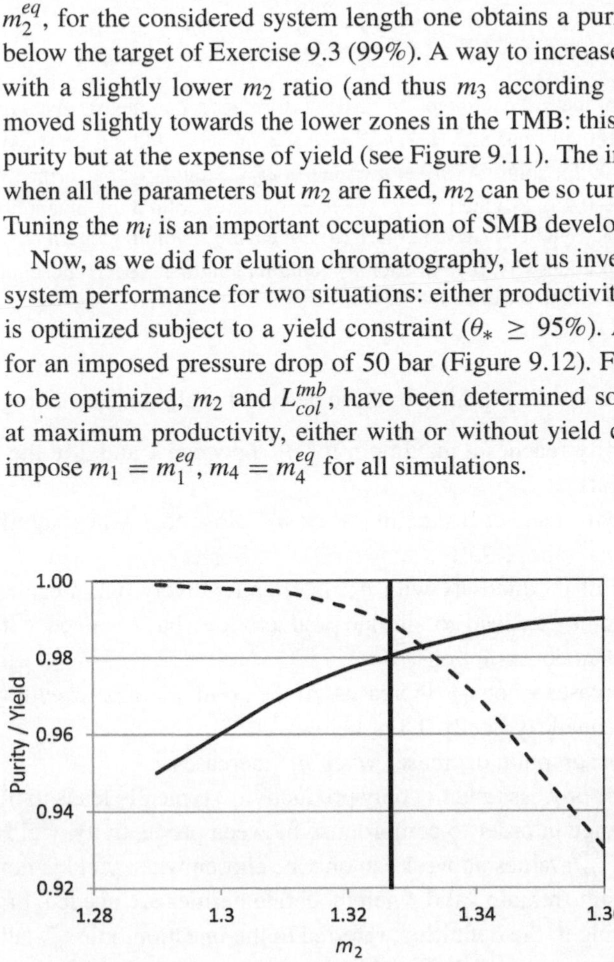

Figure 9.11 Influence of the m_2 ratio on the raffinate purity and yield; parameters given in Exercise 9.3. $C_{tot}^F = 100$ g/l (thus $m_1^{eq} = 2$, $m_2^{eq} = 1.3268$, $m_3^{eq} = 1.4423$, $m_4^{eq} = 1.2597$), $l_i = 0.25$, $\eta_F^{eq} = 1$, $\eta_E^{eq} = 1$, $L_{col}^{tmb} = 40$ cm, $\Delta P = 50$ bar; $m_1 = m_1^{eq}$, $m_4 = m_4^{eq}$ thus $\eta_E^{eq} = 1$; m_3 is calculated from m_2 according to Eq. (9.50). Simulation performed with the MC–Equil model. Dashed line: purity of solute A at the raffinate; solid line: recovery yield of solute solid A; vertical line: position of m_2^{eq}.

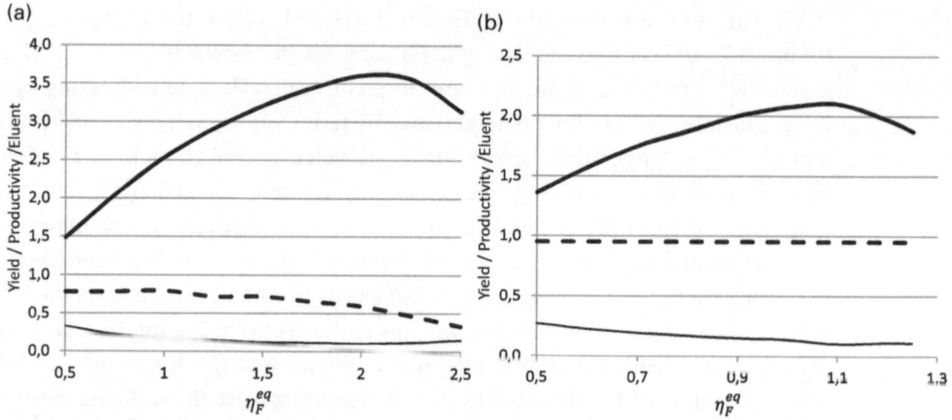

Figure 9.12 Influence of the η_F^{eq} parameter on TMB performance; parameters given in Exercise 9.3. $C_{tot}^F = 100$ g/l (thus $m_1^{eq} = 2$, $m_2^{eq} = 1.327$, $m_3^{eq} = 1.442$, $m_4^{eq} = 1.260$); $\eta_F^{eq} = 1$, $\Delta P = 50$ bar, $X^* = 0.99$ for solute A; $m_1 = m_1^{eq}$, $m_4 = m_4^{eq}$. Simulation performed with the MC–Equil model ($l_i = 0.25$). Productivity φ_V (g/(l min)) (heavy solid line), eluent consumption Υ_* (l/g) (light solid line), yield θ_* (dashed line). (a) For each η_F^{eq} column length has been chosen in order to maximize productivity. (b) For each η_F^{eq} column length has been chosen in order to maximize productivity and to ensure $\theta_* \geq 0.95$.

The general trends are very similar to those derived for elution chromatography:

- The productivity reaches a maximum for η_F^{eq} between 1 and 2 in the absence of yield constraints.
- The productivity reaches a maximum for η_F^{eq} close to 1 when significant yield constraints exist (above 95%).
- The variation of productivity with η_F^{eq} is often relatively flat, meaning that very different systems can lead to similar productivities but obtained with different yields and eluent consumptions.
- The yield decreases when η_F^{eq} increases. At the optimum of *productivity*, the yield is often quite small (typically 0.5 or below).
- The eluent consumption decreases when η_F^{eq} increases.
- Optimizing the process (and not only productivity) typically leads to η_F^{eq} values in the 0.7–1.0 range in order to compromise between productivity, yield and eluent consumption. η_F^{eq} values above 1 can only be chosen when yield is not a concern.
- The above results remain valid when moderate purities are needed, provided that η_F^{eq} is, according to the definition, referred to the injection time t_{inj}^{eq}, allowing one to get the *target purity* at 100% yield on a column of infinite efficiency.

Note that typical optimum η_F^{eq} values allowing one to reach high yield are a bit higher for TMB than they are for elution chromatography. This means that a TMB (thus SMB) can normally be operated more closely to the maximum thermodynamic possibilities. This is consistent with our finding in Chapter 8 that the TMB is less sensitive to efficiency (number of plates) than elution is.

9.6.2 Type 0 shortcuts: influence of pressure drop and particle size

As with the elution process, we will follow the shortcut methodology presented in Section 9.4.3, and thus estimate all the parameters. We have shown how to estimate C_{tot}^F in Section 9.4.3.1.

We know that (m_1,m_4) values can be set close to those determined by the equilibrium model; in other words $\eta_E^{eq} \approx 1$ in general. Should a η_E^{eq} margin be taken, we mentioned that this margin could be evenly split between zones 1 and 4. "Evenly" could mean taking $u_{e1}^{tmb} = u_{e1}^{eq} + \frac{\eta_E^{eq} - 1}{2}(u_{e1}^{eq} - u_{e4}^{eq})$ and $u_{e4}^{tmb} = u_{e4}^{eq} - \frac{\eta_E^{eq} - 1}{2}(u_{e1}^{eq} - u_{e4}^{eq})$. Alternatively, one can make use of the parameter ϑ introduced in Eq. (8.73) and one obtains: $u_{e1}^{tmb} = \vartheta u_{e1}^{eq}$ and $u_{e4}^{tmb} = u_{e4}^{eq}/\vartheta$, so as to ensure that $\eta_E^{eq} = \frac{u_{e1}^{eq}\vartheta - u_{e4}^{eq}/\vartheta}{u_{e1}^{eq} - u_{e4}^{eq}}$. There is no specific difficulty; this is fine tuning, so we will simply assume $\eta_E^{eq} = 1$.

Much more complex is the situation associated with zones 2 and 3. For illustration let us consider a system for which $m_2^{eq} = 2$ and $m_3^{eq} = 3$. If one selects $\eta_F^{eq} = 0.8$, the difference between m_3 and m_2 should now equal 0.8 instead of 1. The problem is to decide where to allocate this safety margin, as shown in Figure 9.13. For instance, do we want $m_2 = 2$ and $m_3 = 2.8$ or $m_2 = 2.2$ and $m_3 = 3.0$? The answer is case-specific and depends on the shape of the separation region in the (m_2,m_3) space determined by the equilibrium model. In general, one must ensure that the operating point (m_2,m_3) is inside the complete separation zone; otherwise, in selecting $\eta_F^{eq} < 1$, we would inject less than theoretically possible according to the equilibrium model and fail to get pure products. With the situation presented in Figure 9.13, virtually all the margin must be allocated to m_2 (shifted to the right), while m_3 must be kept constant. If there is any doubt as to the proper selection of (m_2,m_3) associated with $\eta_F^{eq} < 1$, the best option is η_F^{eq}=1.

At this point in the presentation, all the (m_1, m_2, m_3, m_4) parameters are set, as well as the various length ratios l_i, which are imposed. The system is thus fully defined if one determines the solid velocity $\bar{u}_e$ and the system length L_{col}^{tmb}. We saw with Eq. (9.53) how

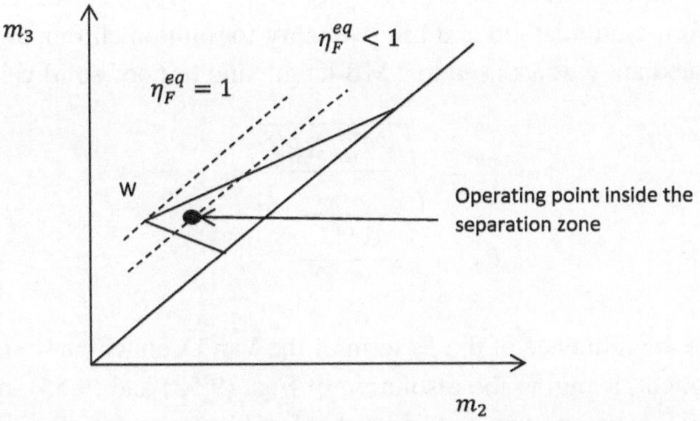

Figure 9.13 Representation of the influence of η_F^{eq} parameter in the (m_2, m_3) plane.

the pressure-drop constraint allows the TMB length L_{col}^{tmb} and the average solid velocity $\bar{u}_e$ to be related. We can thus use L_{col}^{tmb} as our single variable. The column length is then determined in order to make sure that the target purity is obtained at the targeted yield. For simplicity, let us use the MC–Equil model (once again, using more complex kinetic models presents no difficulty).

Each zone i is equivalent to a number of plates N_i^{tmb}, which is related to the zone length and velocities via the Van Deemter equation (9.54). Using the definition of m_i, the total number of plates N^{tmb} equivalent to the TMB can thus be written:

$$\begin{aligned}\frac{N^{tmb}}{L_{col}^{tmb}} &= \sum_{i=1}^{i=4} \frac{l_i}{\dfrac{A_e}{(1+\frac{1-\varepsilon_e}{\varepsilon_e}m_i)\bar{u}_e} + B_e + C_e(1+\frac{1-\varepsilon_e}{\varepsilon_e}m_i)\bar{u}_e} \\ \frac{N^{tmb}}{L_{col}^{tmb}} &\approx \sum_{i=1}^{i=4} \frac{l_i}{\dfrac{A_e}{(1+\frac{1-\varepsilon_e}{\varepsilon_e}m_i^{eq})\bar{u}_e} + B_e + C_e(1+\frac{1-\varepsilon_e}{\varepsilon_e}m_i^{eq})\bar{u}_e}\end{aligned} \tag{9.55}$$

A few simulations allow a determination of the N^{tmb} required to deliver adequate purity and yield. By requiring a certain pressure drop, one can determine L_{col}^{tmb} and $\bar{u}_e$ from Eq. (9.53) and Eq. (9.55). All velocities are then determined with Eq. (9.49).

For illustration purposes, as with the development for elution chromatography, let us assume that the system is operated under significant mass transfer limitations so that the term in $C_e\bar{u}_e$ dominates the Van Deemter equation. Under these conditions, Eq. (9.55) gives:

$$\begin{aligned}\frac{N^{tmb}}{L_{col}^{tmb}} &\approx \frac{1}{\overline{C}_e\bar{u}_e} \\ \overline{C}_e &= C_e\left[\sum_{i=1}^{i=4}\frac{l_i}{1+\frac{1-\varepsilon_e}{\varepsilon_e}m_i^{eq}}\right]^{-1}\end{aligned} \tag{9.56}$$

Under these common circumstances, we have shown that the expressions for pressure drop and number of plates which are valid for elution chromatography can be used for TMB provided that one substitutes u_e, C_e and δ_e for $\bar{u}_e$, $\overline{C}_e$ and $\bar{\delta}_e$. A direct consequence is that the results for column length and fluid velocity for elution chromatography (see Eq. (9.44)) are immediately transposed to TMB length and lumped solid velocity:

$$\begin{aligned}L_{col}^{tmb} &= \sqrt{\frac{N^{tg}\overline{C}_e\Delta P^{tg}}{\bar{\delta}_e}} \\ \bar{u}_e &= \sqrt{\frac{\Delta P^{tg}}{\bar{\delta}_e N^{tg}\overline{C}_e}}\end{aligned} \tag{9.57}$$

If one needs to take the influence of the B_e term of the Van Deemter curve into account, another simple shortcut, avoiding the resolution of Eqs. (9.53) and (9.55), replaces the zone-dependent HETPs by an averaged value. Under this approximation, one can use Eq. (9.43) by substituting u_e, B_e, C_e and δ_e for $\bar{u}_e$, $\overline{B}_e$, $\overline{C}_e$ and $\bar{\delta}_e$. One obtains:

$$\bar{u}_e = \frac{\bar{B}_e}{2\bar{C}_e}\left[-1 + \sqrt{1 + \frac{4\Delta P^{tg}\bar{C}_e}{\delta_e N^{tg}\bar{B}_e^{\,2}}}\right]$$
$$\bar{C}_e = C_e\left[\sum_{i=1}^{i=4} \frac{l_i}{1 + \frac{1-\varepsilon_e}{\varepsilon_e} m_i^{eq}}\right]^{-1} \qquad \bar{B}_e = B_e \tag{9.58}$$
$$L_{col}^{tmb} = \frac{\Delta P^{tg}}{\delta_e \bar{u}_e}$$

The qualitative influence of pressure drop and particle size on productivity and system geometry is the same as for elution chromatography.

EXERCISE 9.6 *Propose a reasonable TMB design to perform the separation described in Exercises 9.1 and 9.3. The target operating pressure is 50 bar. The TMB system is made of four zones of identical length. Feed concentration C_{tot}^F is set to 160 g/l. Additional information: simulations have shown that about 160 plates are required to get 95% recovery when the m_i are set to m_i^{eq}.*

9.7 Application to SMB and multi-column systems

The task of structuring multi-column systems to maximize performance is an ongoing area of research. Schemes aimed at comparing different structures have been proposed by (Toumi *et al.*, 2002), (Araujo *et al.*, 2006), (Kawajiri and Biegler, 2006), (Kawajiri and Biegler, 2008), and (Agrawal and Kawijari, 2012). A comparison of the performance of various multi-column systems can be found in (Schmidt-Traub *et al.*, 2012). While these approaches are certainly of key interest for the global optimization of complex processes, they hardly allow us to draw general conclusions. Once again, we will stick to our objective of explaining how Type I problems (i.e. for given system schemes) can be solved with a simple approach and physical arguments; more general situations can be dealt with by comparing different Type I solutions.

The methodology for designing SMB systems derives directly from that for TMB systems in Section 9.6. The approach detailed here is similar to those published by (Charton and Nicoud, 1995), (Ludemann-Hombourger, Bailly and Nicoud, 2000) and (Biressi *et al.*, 2000).

At the outset, let us recall Eqs. (8.79), (8.80) and (8.81) of Chapter 8 relating TMB to SMB parameters:

$$\bar{u} = \frac{(1-\varepsilon_e)\,L_{col}}{\Delta T} \quad \text{thus} \quad \bar{u}_e = \frac{L_{col}}{\Delta T} \tag{9.59}$$

$$u_{ie}^{smb} = u_{ie}^{tmb} + \bar{u}_e \tag{9.60}$$

9.7.1 Type I problems

Solving a Type I problem for an SMB consists in determining seven parameters that can be organized in two groups: the primary group (C_{tot}^F, u_2^{smb}, u_3^{smb}, L_{col}, ΔT) and the secondary group (u_1^{smb}, u_4^{smb}). All other parameters (including the number of columns per zone, N_{ci}) are fixed. The core of the optimization procedure thus consists in determining the parameters (C_{tot}^F, u_2^{smb}, u_3^{smb}, L_{col}, ΔT).

While the detailed behavior varies with the number of columns per zone and exact system conception (e.g. SMB, Varicol, I-SMB and so on), the main features for elution and TMB remain: the η_F^{eq} parameter plays a critical role in process optimization, optimum systems being associated with η_F^{eq} not far from 1. The optimum design depends on the objective function: maximum productivity is obtained at the expense of poor yield or eluent consumption, high yields are obtained at lower η_F^{eq} parameters, eluent consumption is minimized at higher η_F^{eq} and so on.

Experience shows that the optimum parameters for SMB and TMB are very similar when the number of columns is high enough (let us say at least two columns per zone), differences appearing at lower numbers of columns, as shown by (Pais and Rodrigues, 2003). In the latter situation, and for fine-tuned optimization, nothing replaces complete optimization based on a simulation of the actual system.

It is certainly wise to initialize the optimization algorithm of an SMB with information from the TMB optimization, which is much less computer-intensive. In order to select a TMB for initialization which is as close as possible to the SMB to be optimized, one needs to relate the zone lengths L_i of the TMB to the numbers of columns per zone, N_{ci}, of the SMB.

The lengths of the various zones of the TMB are expressed as:

$$\begin{gathered} L_i = N_{ci} L_{col} \text{ for } i \in [1,4] \\ \sum_{i=1}^{i=4} N_{ci} = N_{col} \end{gathered} \tag{9.61}$$

which can be written:

$$L_i = l_i L_{col}^{tmb} \quad L_{col}^{tmb} = N_{col} L_{col} \text{ for } i \in [1,4] \tag{9.62}$$

Consequently, the number of columns per zone (integer for SMB but possibly real for Varicol) suggests that we define a *relative zone length*, $l_i = N_{ci}/N_{col}$, for the TMB.

With known relative zone lengths, the TMB used for initialization can be optimized in terms of total length, velocities etc., according to the methodology in Section 9.6. Equations (9.59) and (9.60) are then used to determine the initial SMB parameters from the TMB ones.

9.7.2 Type 0 shortcuts: influence of pressure drop and particle size

Let us specify how the shortcut design of the TMB can be directly transferred to the SMB. The pressure drop in the SMB is related to the different zone lengths and associated fluid velocities by:

$$\frac{\Delta P^{smb}}{L_{col}} = \delta_e \sum_{i=1}^{i=4} u_{ie}^{smb} N_{ci} \tag{9.63}$$

Using Eqs. (9.60) and (9.62) to express the l_i, and introducing the m_i, one obtains:

$$\frac{\Delta P^{smb}}{N_{col}L_{col}} = \frac{\Delta P^{smb}}{L_{col}^{smb}} = \bar{u}_e\delta_e \sum_{i=1}^{i=4}(1 + \frac{1-\varepsilon_e}{\varepsilon_e} m_i) l_i = \bar{\delta}_e \bar{u}_e \tag{9.64}$$

Equation (9.64) is similar to Eq. (9.53), which means that, due to the selected expression for Darcy's law for TMB, one can simply use the pressure-drop target of the SMB (real system) to design the TMB (idealized system).

The reader can check that the same applies to the expression for the number of plates: the estimated number of plates of the SMB as a function of length and fluid velocities is given by Eq. (9.55), which was derived for the TMB.

This means that, if one uses identical N^{tg} and ΔP^{tg} for TMB and SMB, the solid flow rate and length determined for the TMB can be used without adaptation to design the SMB. Therefore the shortcut results obtained for TMB are readily applicable to size the SMB, using the correspondence equations (9.59), (9.60) and (9.62).

As with elution and TMB chromatography, this simplified method quickly gives a decent approximation of potential systems. For the sake of completeness, let us mention the article by (Charton and Nicoud, 1995), which estimates the number of plates required to operate an SMB as a function of the selectivity and feed concentration in the case of Langmuir adsorption isotherms. For a preliminary design, this allows one to avoid the use of simulation for estimating N^{tg}.

EXERCISE 9.7 *Propose a reasonable SMB design for the separation studied in Exercise 9.6.*

The influence of parameters like pressure and particle size on productivity and system geometry is the same as the one derived for elution chromatography.

9.8 Conclusion

The possibilities for chromatographic processes are almost infinite, and that is what makes this technology so fascinating. First, although this is not the subject of this book, the variety of existing chromatographic media offers a considerable range of opportunities. Once a chromatographic medium is selected, playing with "columns, valves and pumps", the so-called chromatographic modes, offers a further layer of possibilities.

Can we propose a mode that is likely to be superior to the others in most situations? Certainly not, if one mode maximizes productivity, another minimizes eluent consumption, a third offers the best long-term economy but at the expense of high initial investment, a fourth offers ideal robustness and the selected one may minimize the "volumetric footprint" because HVAC[21] costs a fortune ...

This is a reminder that defining an objective function, even assuming that all desires can be translated into technico-economic terms, requires a fair amount of rigor. It is

[21] Heating, ventilation and air conditioning.

worth keeping in mind that a chromatographic process is never isolated, and that taking into account the environment is of primary importance. One of many consequences is that the inlet and outlet purities might be considered as optimization parameters, rather than as unambiguously defined constraints.

The number of parameters which the chromatographer or process engineer can adjust is a great confirmation of the possibilities offered by chromatography. Numerical optimization tools are certainly making great progresses, and allow more and more complex problems to be solved. However, optimizing chromatography is still, and will probably remain for a long period, a task of significant difficulty owing to the number of parameters to be considered. That is why, presenting a structure of an optimization algorithm including as much as possible of our knowledge of chromatography is of key interest.

In addition to the initial and general distinction between discrete and continuous parameters, organizing the parameters by nature (do they influence physico-chemistry or not?) and role (*primary* or *secondary*?) have allowed us to identify the core of the optimization process: for a given choice of mode, chromatographic medium and solvent, how can we determine optimum column geometry, fluid velocity and feed concentration?

Along this chapter we have presented various optimization strategies from purely numerical ones based on "blind" computer algorithms that ignore chromatographic knowledge, to shortcuts based either on engineer practice (admissible pressure drop for instance) or more sophisticated reasoning (admissible feed concentration for instance). We have shown that using the equilibrium model results (analytical or simulated) gives remarkable hint for designing very reasonable systems, with limited calculation efforts. Further detailed optimizations can be performed using the models presented in the previous chapters of this book. It is user's responsibility to make sure that the selected model is adapted to the problem to be solved, and that relevant physico-chemical parameters (especially adsorption isotherms) have been determined with enough precision.

Once again, ability to invent and develop new and efficient chromatographic process is primarily limited by the creativity of the process engineer . . . including mine. That is why I don't want to be more specific in terms of recommending chromatographic modes; this is to avoid taking the risk of orientating towards suboptimal directions.

9.9 Solutions of the exercises

Exercise 9.1

One needs to process a feed characterized by $X_*^F = 0.5$ in order to get $\dot{m}_* = 10$ t/year (1.26 kg/h assuming 8000 hours per year operation) of purified material at purity $X_* = 0.99$. The function to be minimized is given by Eq. (9.13), written here again for convenience:

$$\psi^{tot} = A_F \frac{1}{\theta_*} + A_{CP} \frac{1}{\phi_v} + A_S \Upsilon_* + c_p(X_*) \quad (\text{€ /kg of purified material})$$

- Term A_F: in this problem, the composition of the feed is fixed (50% of desired product) and the feed cost is set to € 25/kg. As a consequence:

$$A_F = a_F(X_*^F)\frac{X_*}{X_*^F} = \frac{25 \times 0.99}{0.5} = 49.5\ €/\text{kg}$$

- Term $A_{CP} = \dfrac{c_{lb}\phi_v}{\dot{m}_*} + d_{chr}\left(\dfrac{\dot{m}_*}{\phi_v}\right)^{x_c-1} + \bar{c}_{cmd}$:

  - Labor costs amount to € 6.4×10^5/year, thus $c_{lb} =$ € 80/h.
  - The reported cost of chromatographic systems (in €) can be approximated by $8 \times 10^6\ V_{syst}{}^{x_c}$ with $x_c = 1/2$ and V_{syst} expressed in m^3. With a depreciation policy of 7 years and 8000 h operation per year, one obtains: $d_{chr} =$ € 142/(h m$^{3/2}$)
  - The chromatographic medium contribution is $\bar{c}_{cmd} = 10\,000 \times 1000/(2 \times 8000) = 625$€/(h m^3).

  We obtain $A_{CP} = \dfrac{80}{1.26}\phi_v + 142\left(\dfrac{1.26}{\phi_v}\right)^{-1/2} + 625$€/(h m^3).

- Term $A_S = (c_{sol} + c_{sol/dis})(1 - S^R) + c_{rec}\left(S^R\right)^{x_r} + d_{rec}\left(S^R\right)^{x_r}\left(\Upsilon_*\dot{m}_*\right)^{x_r-1}$:

  - Solvent costs are $c_{sol} =$ € 2000/m^3
  - Solvent disposal costs are $c_{sol/dis} =$ € 500/m^3
  - Solvent recycling costs are $c_{rec} =$ € 100/m^3
  - Solvent recycling depreciation costs: the cost of the evaporation unit (€) can be approximated by $2.1 \times 10^6\ Q^{x_r}$ (€) with $x_r = 1/2$ if the flow rate is measured in m^3/h. Taking into account a depreciation policy of 7 years, and 8000 h operation per year, one obtains $d_{rec}=$ € 37.5/(h$^{1/2}$m$^{3/2}$).
  - Recycling ratio $S^R = 0.995$.

  We obtain $A_S = 2500 \times 0.05 + 100 \times 0.995^{1/2} + 37.5 \times (0.995)^{1/2}\left(\Upsilon_* \times 1.26\right)^{-1/2}$, so:
  $A_S = 224.75 + 33.3\Upsilon_*{}^{-1/2}$ €/m^3

- Term $c_p(X_*)$: the purity of the outlet stream is fixed at 99% and processing this stream costs € 10/kg, $c_p(X_*) = 10$ € /kg.

Consequently, the objective function, expressed in € /kg of purified material, is:

$$\psi^{tot} = \frac{49.5}{\theta_*} + \frac{63.5\phi_v + 142\left(\dfrac{1.26}{\phi_v}\right)^{-1/2} + 625}{\phi_v} + \left(224.75 + 33.3 \times \Upsilon_*{}^{-1/2}\right)\Upsilon_* + 10$$

where the productivity per unit volume, ϕ_v, is expressed in kg/(h m^3) and Υ_* in m^3/kg.

Exercise 9.2

Let us consider elution chromatography first. The sets of parameters are (C_{tot}^F, u_e, Δt_{inj}, L_{col}) (primary group) and (t_{cycle}, t_{c1}, t_{c2}) (secondary group). For a set (u_e, L_{col}), one has $t_0 = \varepsilon_e L_{col}/u_e$. According to the equilibrium model, the chromatogram is given by:

- for solute A, a peak starting at $t_0\left(1+\frac{1-\varepsilon_e}{\varepsilon_e}\overline{K}_A\right)$ and ending at $t_0\left(1+\frac{1-\varepsilon_e}{\varepsilon_e}\overline{K}_A\right)+\Delta t_{inj}$
- for solute B, a peak starting at $t_0\left(1+\frac{1-\varepsilon_e}{\varepsilon_e}\overline{K}_B\right)$ and ending at $t_0\left(1+\frac{1-\varepsilon_e}{\varepsilon_e}\overline{K}_B\right)+\Delta t_{inj}$
- $\Delta t_{inj}^{eq}=t_0\frac{1-\varepsilon_e}{\varepsilon_e}\left(\overline{K}_B-\overline{K}_A\right)$ to ensure complete separation of the two peaks
- $t_{cycle}^{eq}=2t_0\dfrac{1-\varepsilon_e}{\varepsilon_e}\left(\overline{K}_B-\overline{K}_A\right)$, the difference between the beginning of peak A and the end of peak B.

Consequently, one obtains:

$$t_\phi^\chi=\frac{t_0}{\varepsilon_e}\ \mathrm{s}^{-1}\qquad \eta_F^{eq}=\frac{\Delta t_{inj}}{t_0\frac{1-\varepsilon_e}{\varepsilon_e}\left(\overline{K}_B-\overline{K}_B\right)}$$

$$\eta_{EF}^{eq}=\frac{t_{cycle}}{2t_0\frac{1-\varepsilon_e}{\varepsilon_e}\left(\overline{K}_B-\overline{K}_B\right)}\qquad \chi_F^{eq}=\frac{(C_A^F+C_B^F)\Delta t_{inj}^{eq}}{t_{cycle}^{eq}}=5\ \mathrm{g/l}$$

Let us now consider the case of TMB. The seven parameters to be determined for a TMB are C_{tot}^F, u_{e2}^{tmb}, u_{e3}^{tmb}, $\bar{u}_e$, L_{col}^{tmb} (primary group) and u_{e1}^{tmb}, u_{e4}^{tmb} (secondary group). According to the equilibrium model, the complete separation is obtained for $m_1^{eq}=\overline{K}_B$, $m_2^{eq}=\overline{K}_A$, $m_3^{eq}=\overline{K}_B$, $m_4^{eq}=\overline{K}_A$, with $u_{ei}^{eq}=m_i^{eq}\frac{1-\varepsilon_e}{\varepsilon_e}\bar{u}_e$. Consequently, one obtains:

$$t_\phi^\chi=\frac{\varepsilon_e L_{col}^{tmb}}{u_{e3}^{tmb}-u_{e2}^{tmb}+u_{e1}^{tmb}-u_{e4}^{tmb}}\ \mathrm{s}\qquad \eta_F^{eq}=\frac{u_{e3}^{tmb}-u_{e2}^{tmb}}{\frac{1-\varepsilon_e}{\varepsilon_e}\bar{u}_e(\overline{K}_B-\overline{K}_A)}$$

$$\eta_{EF}^{eq}=\frac{u_{e3}^{tmb}-u_{e2}^{tmb}+u_{e1}^{tmb}-u_{e4}^{tmb}}{2\bar{u}_e\frac{1-\varepsilon_e}{\varepsilon_e}(\overline{K}_B-\overline{K}_B)}\qquad \chi_F^{eq}=\frac{(C_A^F+C_B^F)(\overline{K}_B-\overline{K}_A)}{2(\overline{K}_B-\overline{K}_A)}=5\ \mathrm{g/l}$$

From Eqs. (8.79) and (8.81) one has $u_{ei}^{smb}=u_{ei}^{tmb}+\bar{u}_e$ and $\bar{u}_e=\dfrac{L_{col}}{\Delta T}$, so extension to SMB is straightforward:

$$t_\phi^\chi=\frac{\varepsilon_e N_{col} L_{col}^{smb}}{u_{e3}^{smb}-u_{e2}^{smb}+u_{e1}^{smb}-u_{e4}^{smb}}\ \mathrm{s}\qquad \eta_F^{eq}=\frac{u_{e3}^{smb}-u_{e2}^{smb}}{\frac{1-\varepsilon_e}{\varepsilon_e}\frac{L_{col}}{\Delta T}(\overline{K}_B-\overline{K}_A)}$$

$$\eta_{EF}^{eq}=\frac{u_{e3}^{smb}-u_{e2}^{smb}+u_{e1}^{smb}-u_{e4}^{smb}}{2\frac{1-\varepsilon_e}{\varepsilon_e}\frac{L_{col}}{\Delta T}(\overline{K}_B-\overline{K}_B)}\qquad \chi_F^{eq}=\frac{(C_A^F+C_B^F)(\overline{K}_B-\overline{K}_A)}{2(\overline{K}_B-\overline{K}_A)}=5\ \mathrm{g/l}$$

Notice the remarkable symmetry between the expressions for elution and for TMB/SMB. In particular, the parameters χ_F^{eq} have the same value.

Exercise 9.3

The feed concentration will be determined so as to maximize χ_F^{eq}, which is estimated from simple knowledge of the adsorption isotherms. The information associated with pressure drop and HETP are not required. We will use the analytical expressions given in Chapters 3 and 8, valid for generalized Langmuir adsorption isotherms, thus for $\bar{\lambda}_A=\bar{\lambda}_B=0$ knowing that we have $\bar{\lambda}_A=\bar{\lambda}_B=0.5$ in this exercise. The two λ being identical, they simply require a shift in the profiles so that adapting the analytical results obtained for generalized Langmuir adsorption isotherms is straightforward. We can do this even more simply: the two λ being identical, they have no impact on the influence of concentration on χ_F^{eq}. We will thus simply neglect them and consider $\overline{K}_i = \overline{N}\tilde{K}_i$.

We start with elution chromatography. In order to estimate an adequate feed concentration, we need to estimate the parameter $\chi_F^{eq} = \frac{\Delta t_{inj}^{eq}}{t_{cycle}^{eq}} C_{tot}^F$.

The expressions in Table 3.2 allow determination of the abscissa Z_{resol} at which complete resolution occurs in elution chromatography. For Langmuir systems (no added linear term), after noting that $L_{col} = Z_{resol}$ one obtains:

$$\frac{t_0}{\Delta t_{inj}^{eq}} = \frac{(\bar{K}_B - \omega_1^F)}{\rho \omega_1^F \omega_2^F (\bar{K}_B - \bar{K}_A)^2}$$

The cycle time is the difference between t_{beg}, at which the front (compressive) peak of solute A appears, and t_{end}, the end of the peak (dispersive) of solute B. The time t_{end} is given by $t_{end} = \Delta t_{inj} + t_0(1 + \frac{1-\varepsilon_e}{\varepsilon_e}\bar{K}_B)$. The calculation of t_{beg}, the position of the front of the peak of the first solute, is obtained from $t_R^{max}(A) = t_0\left(1 + Fe\frac{\bar{C}_A^{P1}}{C_A^{P1}}\right)$ when the plateau P_1 exists, and by (3.77) when it is eroded.

Determination of the ω_i^F by solving the characteristic equation for different feed concentrations C_{tot}^F immediately gives the influence of C_{tot}^F on the ratio χ_F^{eq}. The results are given in Table 9.1.

The concentration χ_F^{eq} increases linearly with the feed concentration at low concentrations, but there is not much to gain after 100 g/l. Working at 100 g/l allows capture of most of the possible productivity while being compatible with an "agreeable" $\Delta t_{inj}^{eq}/t_0$ ratio of about 17% (working at 1000 g/l, for instance, would require a ratio of 1.7%, which can be technically more challenging). In addition, in practice, the probability of a situation allowing us to work at 1000 g/l without solubility or viscosity issues is pretty small!

The same methodology is used with TMB chromatography. The concentration χ_F^{eq} is given by:

$$\chi_F^{eq} = C_{tot}^F \frac{u_{e3}^{eq} - u_{e2}^{eq}}{u_{e1}^{eq} - u_{e4}^{eq}}$$

Table 9.1 For Exercise 9.3, elution.

C_{tot}^F (g/l)	0.1	1	10	100	1000
C_A^{P1} (g/l)	0.05	0.5	5.7	85	975
ω_1^F	0.999	0.995	0.946	0.537	0.092
ω_2^F	1.499	1.489	1.409	1.24	1.204
t_{beg}/t_0	2.5	2.49	2.42	1.91	1.71
t_{end}/t_0	4.0	3.98	3.85	3.42	3.27
$\Delta t_{inj}^{eq}/t_0$	0.74	0.73	0.6	0.17	0.02
χ_F^{eq}	0.049	0.49	4.2	11.45	12.65

The expressions given in Tables 8.1 and 8.2 give m_i, thus u_{ei}^{eq}. For a system obeying a Langmuir adsorption isotherm, one has $m_1^{eq} = \overline{K}_B$, $m_2^{eq} = \omega_2^F \overline{K}_A/\overline{K}_B$, $m_3^{eq} = \frac{\omega_2^F(\overline{K}_A(\overline{K}_A-\omega_1^F)+\omega_1^F(\overline{K}_B-\overline{K}_A))}{\overline{K}_A(\overline{K}_B-\omega_1^F)}$, $m_4^{eq} = \omega_{z4}^F$. Determination of the ω_i by solving the characteristic equation for different feed concentrations C_{tot}^F immediately gives the influence of C_{tot}^F on the ratio χ_F^{eq}. The results are given in Table 9.2. The results are similar to those obtained for elution chromatography (a consequence of the presentation given in Section 8.8). The preferred feed concentration is 100 g/l, which allows a significant χ_F^{eq} and an "acceptable" $\frac{u_{e3}^{eq} - u_{e2}^{eq}}{u_{e1}^{eq}}$ ratio of about 8%.

Exercise 9.4

The objective function as determined in Exercise 9.1 is given (in € /kg) by:

$$\psi^{tot} = \frac{49.5}{\theta_*} + \frac{63.5\phi_v + 142\left(\frac{1.26}{\phi_v}\right)^{-1/2} + 625}{\phi_v} + \left(224.75 + 33.3 \times \Upsilon_*^{-1/2}\right)\Upsilon_* + 10$$

where the productivity per unit volume ϕ_v is expressed in kg/(h m^3) and Υ_* in m^3/kg.

Using the curves in Figure 9.10a, the results for different values of η_F^{eq} are given in Table 9.3.

With our set of parameters, and with the precision associated with the reading of the figures, one sees that the cost decreases continuously with η_F^{eq} and reaches a plateau at $\eta_F^{eq} \simeq 1$. This is not a real optimum because the curves in Figure 9.10a maximize productivity and not the objective function for each η_F^{eq}.

Table 9.2 For Exercise 9.3, TMB.

C_{tot}^F (g/l)	0.1	1	10	100	1000
ω_1^F	0.999	0.995	0.946	0.537	0.092
ω_2^F	1.499	1.489	1.409	1.24	1.204
χ_F^{eq}	0.049	0.49	4.2	13.5	15.7
$\frac{u_{e3}^{eq} - u_{e2}^{eq}}{u_{e1}^{eq}}$	0.33	0.32	0.27	0.08	0.009

Table 9.3 For Exercise 9.4, free yield.

η_F^{eq}	0.5	0.75	1.0	1.25	1.5	2.0
θ_*	0.77	0.74	0.69	0.60	0.57	0.52
ϕ_v (g/(min l))	0.71	0.92	1.08	1.17	1.19	1.18
ϕ_v (kg/(h m^3))	42.6	55.2	64.8	70.2	71.4	70.8
Υ_* (m^3/kg)	0.69	0.5	0.38	0.35	0.33	0.26
ψ^{tot} (€/kg)	353	304	277	278	272	267

Table 9.4 For Exercise 9.4, imposed yield.

η_F^{eq}	0.5	0.75	1.0	1.05
θ_*	0.95	0.95	0.95	0.95
ϕ_v (g/(min 1))	0.6	0.7	0.67	0.65
ϕ_v (kg/(hm^3))	36	42	40.2	39
Υ_* (m^3/kg)	0.49	0.32	0.23	0.21
ψ^{tot} (€ /kg)	298	251	231	228

Using the curves in Figure 9.10b (obtained when yield was required to be greater than 95%), the results for different values of η_F^{eq} are given in Table 9.4.
The production cost is lower than the one obtained with the "free yield" because the negative impact of lower productivity obtained at the imposed 95% yield is more than compensated by the positive impact of lower eluent consumption and better yield.

Conclusion: productivity is not everything!

Exercise 9.5

At low flow rate, the number of plates is 500, so the HETP is 2000 μm, leading to a reduced plate height of $2000/100 = 20$. This indicates a significant kinetic limitation. This is confirmed by the strong decrease in plate number with the flow rate. The fact that this separation is performed on silica makes internal diffusion likely to be the kinetic limiting step.

These considerations make us comfortable using the results given in Eq. (9.48). The length of the new system operated at 30 bar with the small particles is thus:

$$L_{col} = 100\frac{25^2}{100^2}\frac{\sqrt{30}}{\sqrt{10}} = 10.8 \text{ cm}$$

and its diameter is:

$$D_{col} = 50\frac{10^{1/4}}{30^{1/4}} = 38 \text{ cm}$$

Exercise 9.6

Using the expressions in Exercise 9.3, one obtains, for a feed concentration of 100 g/l, $\omega_1^F = 0.537$ and $\omega_2^F = 1.24$. Assuming $\bar{\lambda}_A = \bar{\lambda}_B = 0$, this leads to $m_1^{eq} = 1.5$, $m_2^{eq} = 0.827$, $m_3^{eq} = 0.942$, $m_4^{eq} = 0.759$ and $\chi_F^{eq} = 13.5$ g/l.

In the present situation, we have $\bar{\lambda}_A = \bar{\lambda}_B = 0.5$. The two $\bar{\lambda}$ being identical, they simply require a shift of the chromatogram, thus of the m_i values, so that we finally have $m_1^{eq} = 2$, $m_2^{eq} = 1.327$, $m_3^{eq} = 1.442$, $m_4^{eq} = 1.259$ and $\chi_F^{eq} = 13.5$ g/l. We will take $m_i = m_i^{eq}$ (this imposes $\eta_F^{eq} = \eta_F^{eq} = 1$).

The system is assumed to consist of four zones of identical length, so one has $l_i = 0.25$. From Eq. (9.56) one obtains:

$$\overline{C}_e = C_e \left[\sum_{i=1}^{i=4} \frac{l_i}{1 + \frac{1-\varepsilon_e}{\varepsilon_e} m_i^{eq}}\right]^{-1} = 3.23 \times C_e = 0.48 \text{ s} \tag{9.65}$$

From Eq. (9.53) one obtains:

$$\overline{\delta_e} = \delta_e \left[1 + \frac{1-\varepsilon_e}{\varepsilon_e} \sum_{i=1}^{i=4} m_i^{eq} l_i \right] = 3.28 \times \delta_e = 6.52 \text{ bar s/cm}^2 \tag{9.66}$$

With Eq. (9.57), one obtains:

$$\begin{aligned} L_{col}^{tmb} &= \sqrt{\frac{N^{tg}\overline{C}_e \Delta P^{tg}}{\bar{\delta}_e}} = \sqrt{\frac{160 \times 0.48 \times 50}{6.52}} = 24.3 \text{ cm} \\ \bar{u}_e &= \sqrt{\frac{\Delta P^{tg}}{\bar{\delta}_e N^{tg} \overline{C}_e}} = \sqrt{\frac{50}{6.52 \times 160 \times 0.48}} = 0.32 \text{ cm/s} \end{aligned} \tag{9.67}$$

Finally, the various $u_{ei}^{tmb} = m_i \frac{1-\varepsilon_e}{\varepsilon_e} \bar{u}_e$ are determined: $u_{e1}^{tmb} = 0.947$, $u_{e2}^{tmb} = 0.628$, $u_{e3}^{tmb} = 0.683$, $u_{e4}^{tmb} = 0.596$ (cm/s). The system is thus designed.

One can determine the productivity per unit volume with Eq. (9.37), and the eluent consumption with (9.38):

$$\begin{aligned} t_\phi^X &= \frac{L_{col}^{tmb}}{\varepsilon_e} \frac{1}{u_{e3}^{tmb} - u_{e2}^{tmb} + u_{e1}^{tmb} - u_{e4}^{tmb}} = 150 \text{ s} \\ \phi_v &= \theta_* \frac{\eta_F^{eq}}{\eta_{EF}^{eq}} \frac{\chi_F^{eq}}{t_\phi^X} \frac{X_*^F}{X_*} = 0.043 \text{ kg/(s m}^3) \\ \Upsilon_* &= \frac{1}{\phi_v t_\phi^X} = 0.15 \text{ m}^3\text{/kg} \end{aligned} \tag{9.68}$$

Exercise 9.7

Let us assume an SMB of eight columns, with two columns per zone. From Eq. (9.62), one has $l_i = N_{ci}/N_{col} = 0.25$, which corresponds to the geometry of the TMB of Exercise 9.6.

From the results of Exercise 9.6, one obtains $L_{col} = L_{col}^{tmb}/N_{col} = 3$ cm and, with Eq. (9.59), $\Delta T = L_{col}/\bar{u}_e = 9.6$ s. Obtaining very short column lengths is not uncommon with this type of separation (fairly good selectivity, small particle size). More information on this important matter is given in Section 10.2. Finally, from Eq. (9.60) one obtains $u_{1e}^{smb} = 1.26$, $u_{2e}^{smb} = 0.944$, $u_{3e}^{smb} = 0.998$ and $u_{4e}^{smb} = 0.911$ (cm/s).

References

Abdelmoumen, S., Muhr, L., Bailly, M. and Ludemann-Homburger, O. (2006), *Sep. Sci. Technol.* **41**, 2639.

Abel, S., Mazzotti, M. and Morbidelli, M. (2002), *J. Chromatogr. A* **944**(1–2), 23–39.

Abel, S., Mazzotti, M. and Morbidelli, M. (2004), *J. Chromatogr. A* **1026**(1–2), 47–55.

Abunasser, N., Wankat, P., Kim, Y. and Koo, Y. (2003), *Ind. Eng. Chem. Res.* **42**(21), 5268–5279.

Adam, P., Nicoud, R. M., Bailly, M. and Ludemann-Hombourger, O. (2000), US Patent 6 136 198.

Agrawal, G. and Kawijari, Y. (2012), *J. Chromatogr. A* **1238**, 105–113.

Ando, M., Tanimura, M. and Talura, M. (1990), US Patent 4 970 002.

Antos, D. and Seidel-Morgenstern, A. (2002), *J. Chromatogr. A* **944**(1–2), 77–91.

Araujo, J., Rodrigues, R. and Mota, J. (2006), *J. Chromatogr. A* **1132**, 76–89.

Aumann, L. and Morbidelli, M. (2007), *Biotechnol. Bioeng.* **99**(3), 728–733.

Bailly, M. and Tondeur, D. (1982), *Chem. Eng. Sci.* **37**(8), 1199–1212.

Bailly, M. and Tondeur, D. (1984), *Chem. Eng. Process.* **18**(6), 293–302.

Balannec, B. and Hotier, G. (1993). In G. Ganetsos and P. E. Barker, eds., *Preparative and Production Scale Chromatography*, vol. 61 of *Chromatographic Science Series*, Marcel Dekker, pp. 301–357.

Biressi, G., Ludemann-Hombourger, O., Mazzotti, M., Nicoud, R. M. and Morbidelli, M. (2000), *J. Chromatogr. A* **876**(1–2), 3–15.

Borges da Silva, E. and Rodrigues, A. E. (2008), *Sep. Sci. Technol.* **43**, 533.

Broughton, D. B. and Gerhold, C. G. (1961), US Patent 2 985 589.

Carta, G. and Jungbauer, A. (2011), *Protein Chromatography*, Wiley-VCH.

Charton, F. and Nicoud, R. M. (1995), *J. Chromatogr. A* **702**(1–2), 97–112.

Chen, H. T., Ahmed, Z. M. and Rollen, V. (1981), *Ind. Eng. Chem. Fundam.* **20**(2), 171–174.

Clavier, J. Y., Nicoud, R. M. and Perrut, M. (1996). In R. von Rohr and C. Trepp, eds., *High Pressure Chemical Engineering*, Elsevier Science, pp. 429–434.

Colin, H., Hilaireau, P. and De Tournemire, J. (1990), *LC-GC* **8**(4), 302–312.

Colin, H., Hilaireau, P. and Martin, M. (1991), *J. Chromatogr. A* **557**, 137–153.

Cramer, S. M. and Horvath, C. (1988), *Prep. Chromatogr.* **1**, 29–49.

De Carli, J. P., Carta, G. and Byers, C. H. (1990), *AIChE J.* **36**(8), 1220–1228.

Denet, F., Hauck, W., Nicoud, R. M., *et al.* (2001), *Ind. Eng. Chem. Res.* **40**, 4603–4609.

Depta, A., Giese, T., Johannsen, M. and Brunner, G. (1999), *J. Chromatogr. A* **865**(1–2), 175–186.

Di Giovanni, O., Mazzotti, M., Morbidelli, M., Denet, F., Hauck, W. and Nicoud, R. M. (2001), *J. Chromatogr. A* **919**(1), 1–12.

Dingenen, J. and Kinkel, J. N. (1994), *J. Chromatogr. A* **666**(1–2), 627–650.

Felinger, A. and Guiochon, G. (1993), *Biotechnol. Bioeng.* **41**, 134–147.

Frenz, J. and Horvath, C. (1985), *AIChE J.* **31**(3), 400–409.

Ganetsos, G. and Barker, P. E. (1993). In G. Ganetsos and P. E. Barker, eds., *Preparative and Production Scale Chromatography*, vol. 61 of *Chromatographic Science Series*, Marcel Dekker, pp. 233–255.

Gomes, P. and Rodrigues, A. E. (2012), *Chem. Eng. Technol.* **35**(1), 17–34.

Grevillot, G. (1986). In N. P. Cheremisinoff, ed., *Handbook of Heat and Mass Transfer. Volume 2: Mass Transfer and Reactor Design*, Gulf Publishing, pp. 1429–1474.

Grill, C. M. (1998), *J. Chromatogr. A* **796**(1), 101–113.

Grill, C. M. (1999), *Isol. Purif.* **3**, 29–50.

Grill, C. M. (2000), US Patent 6 063 284.

Grill, C. M. and Hampton, W. T. (1997), US Patent 5 630 943.

Guiochon, G., Felinger, A., Golshan-Shirazi, S. and Katti, A. M. (2006), *Fundamentals of Preparative and Nonlinear Chromatography*, Elsevier Academic Press.

Guiochon, G. and Tarafder, A. (2011), *J. Chromatogr. A* **1218**, 1037–1114.

Hashimoto, K., Adachi, S. and Shirai, Y. (1988), *Agric. Biol. Chem.* **52**(9), 2161–2167.

Hauck, W. (1997), Procede d'adsorption sur charbon actif avec regeneration thermique de la colonne par induction electromagnetique. Application au traitement d'air charge en composes organiques volatils, unpublished PhD thesis, Institut National Polytechnique de Lorraine, Nancy.

Hotier, G. (1993). In R. M. Nicoud, ed., *Simulated Moving Bed: Basics and Applications*, Separex (ISBN 2-905-267-21-6), pp. 4–34.

Houwing, J., Billiet, H. A. H. and van der Wielen, L. A. M. (2002), *J. Chromatogr. A* **944**(1–2), 189–201.

Huang, S. Y., Lin, C. K., Chang, W. H. and Lee, W. S. (1986), *Chem. Eng. Commun.* **45**(1–6), 291–309.

Hur, H. and Wankat, P. (2005), *Ind. Eng. Chem. Res.* **44**, 1906.

Hur, H. and Wankat, P. (2006), *Ind. Eng. Chem. Res.* **45**, 1426.

Jensen, T. B., Reijns, T. G. P., Billiet, H. A. H. and van der Wielen, L. A. M. (2000), *J. Chromatogr. A* **873**(2), 149–162.

Jin, W. and Wankat, P. (2005), *Ind. Eng. Chem. Res.* **44**, 1565.

Jin, W. and Wankat, P. (2006), *Ind. Eng. Chem. Res.* **45**, 2793–2807.

Jusforgues, P., Shaimi, M. and Barth, D. (1998). In K. Anton and C. Berger, eds., *Supercritical Fluid Chromatography with Packed Columns: Techniques and Applications*, vol. 75 of *Chromatographic Science Series*, Marcel Dekker, pp. 403–427.

Kampen, W. H. (1990), US Patent 5 177 008.

Kaspereit, M. (2009). In E. Grushka and N. Grinberg, eds., *Advances in Chromatography*, vol. 47, CRC Press.

Kaspereit, M. and Sainio, T. (2011), *Chem. Eng. Sci.* **66**, 5428–5438.

Kaspereit, M., Schulte, M., Wekenborg, K. and Wavers, W. (2012). In H. Schmidt-Traub, M. Schulte and A. Seidel-Morgenstern, eds., *Preparative Chromatography*, Wiley-VCH.

Kaspereit, M., Seidel-Morgenstern, A. and Kiemle, A. (2007), *J. Chromatogr. A* **1162**, 2–13.

Katsuo, S., Langel, C., Sandre, A. and Mazzotti, M. (2011), *J. Chromatogr. A* **1218**, 9345–9352.

Katsuo, S. and Mazzotti, M. (2010a), *J. Chromatogr. A* **1217**, 1354–1361.

Katsuo, S. and Mazzotti, M. (2010b), *J. Chromatogr. A* **1217**, 3067–3075.

Kawajiri, Y. and Biegler, L. T. (2006), *AIChE J.* **52**(4), 1343–1350.

Kawajiri, Y. and Biegler, L. T. (2008), *Comput. Chem. Eng.* **32**(1–2), 135–144.

Kessler, L. and Seidel-Morgenstern, A. (2006), *J. Chromatogr. A* **1126**, 323–337.

Kim, J. K. and Wankat, P. C. (2004), *Ind. Eng. Chem. Res.* **43**, 1071–1080.

Kim, J. K., Zang, Y. and Wankat, P. C. (2003), *Ind. Eng. Chem. Res.* **42**(20), 4849–4860.

Kishihara, S., Horikawa, H., Tamaki, H., Fujii, S., Nakajima, Y. and Nishio, K. (1989), *J. Chem. Eng. Jpn.* **22**(4), 434–436.

Klesper, E., Corwin, A. H. and Turner, D. A. (1962), *J. Org. Chem.* **27**, 700–701.

Knox, J. H. and Pyper, H. M. (1986), *J. Chromatogr.* **363**(1), 1–30.

Küsters, E., Heuer, C. and Wieckhusen, D. (2000), *J. Chromatogr. A* **874**(2), 155–165.

Lee, J. and Wankat, P. (2010), *J. Chromatogr. A* **1217**, 3418.

Liao, A., El Rassi, Z., Lemaster, D. and Horvath, C. (1987), *Chromatographia* **24**, 881–885.

Ludemann-Hombourger, O., Bailly, M. and Nicoud, R. M. (2000), *Sep. Sci. Technol.* **35**(9), 1285–1305.

Ludemann-Hombourger, O., Nicoud, R. M. and Bailly, M. (2000), *Sep. Sci. Technol.* **35**(12), 1829–1862.

Ludemann-Hombourger, O., Pigorini, G., Nicoud, R. M., Ross, D. S. and Terfloth, G. (2002), *J. Chromatogr. A* **947**(1), 59–68.

MacCoy, M. (2000), *Chem. Eng. News* **78**(25), 17–19.

Mazzotti, M., Storti, G. and Morbidelli, M. (1997), *J. Chromatogr. A* **786**(2), 309–320.

Mota, J. and Araujo, J. (2005), *AIChE J.* **51**(6), 1641.

Mota, J., Esteves, I. and Eusebio, M. (2007), *AIChE J.* **53(5)**, 1192–1203.

Nicolaos, A., Muhr, L., Gotteland, P., Nicoud, R. M. and Bailly, M. (2001a), *J. Chromatogr. A* **908**(1–2), 71–86.

Nicolaos, A., Muhr, L., Gotteland, P., Nicoud, R. M. and Bailly, M. (2001b), *J. Chromatogr. A* **908**(1–2), 87–109.

Nicoud, R. M. (1992), *LC-GC International* **5**(5), 43–47.

Nicoud, R. M. (2014), *Ind. Eng. Chem. Res.* **53**, 3755–3765.

Nicoud, R. M. (2000). In S. Ahuja, ed., *Handbook of Bioseparations*, vol. 2 of *Separation Science and Technology*, Academic Press, pp. 475–509.

Nicoud, R. M., Clavier, J. Y. and Perrut, M. (1999). In M. Caude and D. Thiebaut, eds., *Practical Supercritical Fluid Chromatography and Extraction*, Harwood Academic Publishers, pp. 397–433.

Nicoud, R. M. and Colin, H. (1989), *LC-GC International* **3**(2), 28–36.

Nicoud, R. M. and Perrut, M. (1991). In C. A. V. Costa and J. S. Cabral, eds., *Chromatographic and Membrane Processes in Biotechnology*, vol. 204 of *NATO ASI Series, Series E: Applied Sciences*, Kluwer Academic Publishers, pp. 381–413.

Nicoud, R. M., Perrut, M. and Hotier, G. (1992), French Patent 9 205 304.

Oren, Y. and Soffer, A. (1983a), *J. Appl. Electrochem.* **13**(4), 473–487.

Oren, Y. and Soffer, A. (1983b), *J. Appl. Electrochem.* **13**(4), 489–505.

Pais, L. S. and Rodrigues, A. E. (2003), *J. Chromatogr. A* **1006**(1–2), 33–44.

Perrin, S. R. and Nicoud, R. M. (2001), In G. Subramanian, ed., *Chiral Separation Techniques: A Practical Approach*, Wiley-VCH, pp. 253–285.

Perrut, M. (1982), French Patent 2 527 934.

Perrut, M., Nicoud, R. M. and Breivik, H. (1998), US Patent 5 719 302.

Quinones, I., Grill, C. M., Miller, L. and Guiochon, G. (2000), *J. Chromatogr. A* **867**(1–2), 1–21.

Rajendran, A. (2012), *J. Chromatogr. A* **1250**, 227–249.

Rajendran, A., Paredes, G. and Mazzotti, M. (2009), *J. Chromatogr.* **1216**, 709–738.

Rodrigues, C., Araujo, J. and Mota, J. (2007), *J. Chromatogr. A* **1162**(1), 14–23.

Rodrigues, R., Canhoto, T. J., Araujo, J. and Mota, J. (2008), *J. Chromatogr.* **1180**, 42–52.

Ruthven, D. M. and Ching, C. B. (1989), *Chem. Eng. Sci.* **44**(5), 1011–1038.

Sainio, T. and Kaspereit, M. (2009), *Sep. Purif. Technol.* **66**, 9–18.

Schmidt-Traub, H., Schulte, M. and Seidel-Morgenstern, A. (2012), *Preparative Chromatography*, 2nd edn., Wiley-VCH.

Seidel-Morgenstern, A. (1998), *Analusis* **26**(7), 46–55.

Seidel-Morgenstern, A. and Guiochon, G. (1993), *AIChE J.* **39**(5), 809–819.

Seidel-Morgenstern, A., Schmidt-Traub, H., Michel, M., Epping, A. and Jupke, A. (2012). In H. Schmidt-Traub, M. Schulte and A. Seidel-Morgenstern, eds., *Preparative Chromatography*, 2nd edn., Wiley-VCH, p. 321.

Siitonen, J. and Sainio, T. (2011), *J. Chromatogr. A* **1218**, 6379–6387.

Siitonen, J., Sainio, T. and Rajendran, A. (2013), *J. Chromatogr. A* **1286**, 55–68.

Silva, R., Rodrigues, R. and Mota, J. (2012), *J. Chromatogr. A* **1260**, 132–142.

Silva, R., Rodrigues, R., Osuna-Sanchez, H., Bailly, M., Valery, E. and Mota, J. (2010), *J. Chromatogr. A* **1217 (52)**, 8257–8269.

Spedding, F. H., Fulmer, E. I., Butler, T. A., *et al.* (1947), *J. Am. Chem. Soc.* **69**(11), 2812–2818.

Sreedhar, B., Wagler, A, Kaspereit, M. and Seidel-Morgenstern, A. (2013), *Comput. Chem. Eng.* **49**, 158–169.

Strohlein, G., Aumann, L., Mazzotti, M. and Morbidelli, M. (2006), *J. Chromatogr. A* **1126**, 338–346.

Subramanian, G., Phillips, M. W. and Cramer, S. M. (1988), *J. Chromatogr. A* **439**(2), 341–351.

Tondeur, D. (1981). In A. E. Rodrigues and D. Tondeur, eds., *Percolation Processes: Theory and Applications*, vol. 33 of *NATO ASI Series, Series E: Applied Sciences*, Sijthoff & Noordhoff, pp. 517–538.

Tondeur, D. and Bailly, M. (1993a). In G. Ganetsos and P. E. Barker, eds., *Preparative and Production Scale Chromatography*, vol. 61 of *Chromatographic Science Series*, Marcel Dekker, pp. 79–109.

Tondeur, D. and Bailly, M. (1993b). In R. M. Nicoud, ed., *Simulated Moving Bed: Basics and Applications*, Separex, pp. 97–117.

Tondeur, D., Migault, G. and Wankat, P. C. (1985), *Entropie* **21**(123), 39–55.

Toumi, A., Dingenen, J., Genolet, J., *et al.* (2012). In H. Schmidt-Traub, M. Schulte and A. Seidel-Morgenstern, eds., *Preparative Chromatography*, Wiley-VCH.

Toumi, A., Engell, S., Diehl, M., Bock, H. and Schlöder, J. (2007), *Chem. Eng. Process.* **46**, 1067–1084.

Toumi, A., Hanisch, F. and Engell, S. (2002), *Ind. Eng. Chem. Res.* **41**(17), 4328–4337.

Van Walsem, H. J. and Thompson, M. C. (1997), *J. Biotechnol.* **59**, 127–132.

Wankat, P. C. (1986a), *Large-Scale Adsorption and Chromatography, Vol I*, CRC Press.

Wankat, P. C. (1986b), *Large-Scale Adsorption and Chromatography, Vol II*, CRC Press.

Westerberg, K., Degerman, B. and Nilsson, B. (2010), *Bioprocess Biosyst. Eng.* **33**, 375.

Wilhelm, R. H., Rice, A. W. and Bendelius, A. R. (1966), *Ind. Eng. Chem. Fundam.* **5**(1), 141–144.

Wilhelm, R. H. and Sweed, N. H. (1968), *Science* **159**(3814), 522–524.

Wu, D. J., Ma, Z. and Wang, L. N.-H. (1999), *J. Chromatogr. A* **855**(1), 71–89.

Zhang, Z., Mazzotti, M. and Morbidelli, M. (2003), *J. Chromatogr. A* **989**(1), 95–108.

10 Addressing a few industrial problems

Time to harvest . . .

In Chapter 9, the tools allowing us to optimize and to rationally design chromatographic processes have been presented. Rules of thumb like "continuous processes must be preferred at large scale" are dangerous, and only a rigorous approach based on a detailed definition of tuning parameters, constrained parameters, objective function and algorithms guarantees the best option among those considered. The rigorous approach makes extensive use of numerical simulations and of the more or less detailed models presented in the first nine chapters of this book. We are, however, aware that a complete numerical design approach does not say a lot about the key drivers for performance. This chapter is therefore a series of situations designed as pedagogical examples, to be solved using the framework presented in the previous chapters. Our objective is not to make the reader an expert in specific processes, some of which have benefited from 40 years of continuous improvement, nor to give recipes for ready-to-cook meals, but to illustrate a methodology, to give decent orders of magnitude and, it is hoped, ideas to help the reader solve his own problems. Instead of proposing detailed process design, we will therefore favor shortcuts to better highlight features that we consider essential. We will keep the mathematics and detailed modeling to a minimum, in order to focus on a few messages that we consider key. With the essence of the problem well in mind, the previous chapters give the reader the methodologies and techniques for addressing specific topics requiring further detail and precision.

For confidentiality reasons, while the problems are real, the laboratory results and design parameters are not exact industrial data; they are, however, reasonable and precise enough for our conclusions to be realistic. Having orders of magnitude and references in mind is very important for performing an engineer's work; that is why typical characteristics of chromatographic processes are given in Appendix J.

10.1 Sugar separation

We will start by investigating the separation of glucose and fructose on an SMB, which will also be addressed in Section 10.3.2. Scientific information on this separation can be found in (Nicoud, 2000), (Azevedo and Rodrigues, 2000), (Azevedo and Rodrigues, 2001), (Subramani *et al.*, 2003) and (Seidel-Morgenstern *et al.*, 2012). More practical information can be found in resin supplier brochures.[1] This separation is often studied because it is relatively simple, usually linear and of practical industrial interest. We will show that, despite its alleged simplicity, this subject offers many important research directions at the frontiers of our current scientific knowledge.

10.1.1 Problem description

It is well known that the separation of fructose and glucose can be performed on cationic ion-exchange resins (typical particle size 320 μm) in calcium (Ca) form using water (65°C) as eluent. Our problem is to design an SMB which can process a flow rate of 4 m^3/h of a solution containing about 58% glucose and 42% fructose at a total concentration of 800 g/l.[2] We will eventually set $C_{gl}^F = 465$ g/l and $C_{fr}^F = 335$ g/l. The objective is to obtain fructose at 90% purity and at 90% yield on an SMB system.

10.1.2 Laboratory data

Small injections performed at 65°C on a laboratory column filled with the cation-exchange resin selected for production have allowed the determination of Henry's coefficients: $\overline{K}_{gl} = 0.27$ and $\overline{K}_{fr} = 0.46$.

The lumped adsorption isotherms are then measured by frontal chromatography. One obtains a relatively moderate non-linear and synergistic influence of the concentrations. For fitting purposes, we will simply use the expression proposed by (Seidel-Morgenstern *et al.*, 2012):

$$\begin{aligned} \overline{C}_{gl} &= 0.27 \times C_{gl} + 0.00012 \times C_{gl}^2 + 0.0001 \times C_{gl}C_{fr} \\ \overline{C}_{fr} &= 0.46 \times C_{fr} + 0.00012 \times C_{fr}^2 + 0.00024 \times C_{gl}C_{fr} \end{aligned} \tag{10.1}$$

all concentrations being expressed in g/l.

Note that one could have used the IAS approach to model the competitive adsorption isotherms from the single-solute adsorption isotherms, but the quadratic representation in Eq. (10.1) has the merit of being very simple and accurate enough for our purposes. A nasty question: are the adsorption isotherms defined in Eq. (10.1) thermodynamically consistent?[3]

Additionally, the Van Deemter curves have been obtained by determining the variance of chromatograms (see method description in Section 5.10.4) associated with small injections performed under linear conditions. They can be represented by (results compatible with those of (Nicoud, 2000)):

$$\begin{aligned} &HETP = B_e + C_e \times u_e \\ &\text{glucose: } B_e = 0.001, C_e = 0.0035 \qquad \text{fructose: } B_e = 0.001, C_e = 0.005 \end{aligned} \tag{10.2}$$

u_e being expressed in m/h, C_e in h, and B_e and $HETP$ in m.

[1] For instance, Purolite, Dow, Finex and Mitsubishi.

[2] In industrial reality, the feed contains about 5% of other solutes that are neglected in this study.

[3] See Chapters 3 and 4 for definitions and constraints.

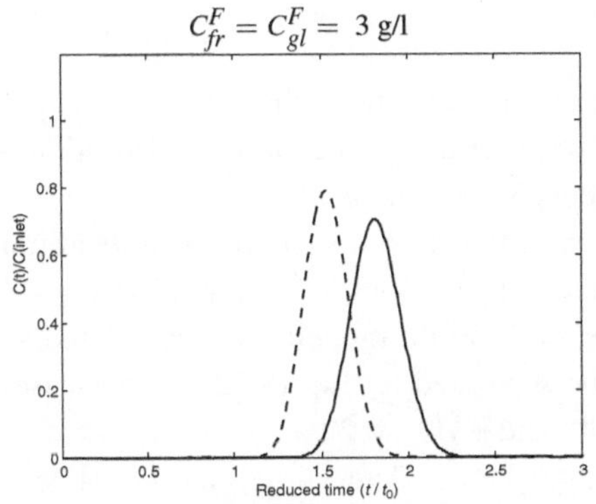

Figure 10.1 Simulated chromatograms obtained with the MC–Equil model. $t_{inj} = 0.25t_0$; $N = 200$. Adsorption isotherms given by Eq. (10.1). Glucose: dashed lines; fructose: solid lines.

The influence of concentration on mass transfer has not been investigated.

From the expressions given in Section 5.8, the Van Deemter curves allow determination of the hydrodynamic contribution to dispersion as well as the characteristic times for mass transfer (limited by intragranular diffusion in this situation).

With the above thermodynamic, hydrodynamic and kinetic information, one can, for instance, use the MC–$\overline{\text{Equil}}$, MC–$\overline{\text{LDF}}$, PD–$\overline{\text{Equil}}$ or PF–$\overline{\text{LDF}}$ model to simulate the system. For simplicity, we will use the MC–$\overline{\text{Equil}}$ model with an average Van Deemter curve characterized by $B_e = 0.001$ m and C_e =0.0043 h.

Typical simulated chromatograms are given in Figure 10.1 for one dilute and one concentrated injection. Notice that the non-linear behavior is not immediately visible on the chromatograms, as peak asymmetry remains very limited. However, note the slight *increase* in retention times with concentration as well as the change in relative peak heights between fructose and glucose.

Finally, the pressure drop has been determined at different fluid velocities on the laboratory column. In an industrial system, fluid viscosity is expected to exhibit significant fluctuations, from about 0.5 cP for pure water at 65°C to about 5–10 cP at the level of the injection of the concentrated feed mixture. For simplicity, some pressure-drop measurements have been carried out with a water/sugar mixture containing a total sugar concentration of about 200 g/l, which *approximately* represents the average concentration likely to be obtained in an industrial unit.[4] The results can be represented by:

$$\frac{\Delta P}{L} \approx \delta_e u_e \text{ bar/m} \qquad \delta_e = 0.04 \text{ bar/(m}^2\text{ h)}$$

where u_e is expressed in m/h. It is worth mentioning that the operations are systematically performed down-flow; otherwise they would be accompanied by a significant bed expansion. The effect of possible bed compression on pressure drop due to the down-flow operation is not taken into account in this study (see Chapter 4 for information).

[4] This is sufficient for our purposes, but, for a precise design, having access to a precise pressure-drop estimate is critical.

10.1.3 Process design: linear

In this section, we will neglect the quadratic terms in the lumped adsorption isotherms in Eq. (10.1) and consider only the linear terms, so the lumped adsorption isotherms can be described with knowledge of Henry's coefficients.

For a linear system, determination of the m_i ratios (fluid flow rate normalized by the lumped solid flow rate) defining the zone of complete separation for 100% purity at 100% yield is immediate using the equilibrium model: they are related to the Henry's coefficients of the two solutes to be separated by Eq. (8.72). In our case, one obtains $m_1^{eq} = 0.46$, $m_2^{eq} = 0.27$, $m_3^{eq} = 0.46$ and $m_4^{eq} = 0.27$.

The target fructose purity being only 90%, determining the m_i values given by the equilibrium model for *90% fructose purity at 100% yield* is important, as we know from Section 9.4 that this reference gives an excellent starting point for process design.

Let us analyze the additional freedom given in the m_i space by the fact that the target purity is 90% instead of 100%, the fructose yield being fixed at 100% in both cases. Obtaining a 100% yield can only be achieved if fructose is totally stripped out of zone 1, so that $m_1 \geq 0.46$ and thus $m_1^{eq} = 0.46$. Additionally, fructose must be totally pushed down in zone 3 so that $m_3 \leq 0.46$ and thus $m_3^{eq} = 0.46$. The additional freedom can thus only come from zones 2 and 4. We will not tune m_4, which is considered to be a secondary tuning parameter, as explained in Chapter 9. Under the linearity assumption, using the theoretical framework presented in Section 8.4, one can determine an analytical expression for the fructose purity according to the m_2 and m_3 values. Assuming that m_4 is low enough to prevent recycling from zone 4 to zone 1, we will look at m_2 values lower than $\overline{K}_A$ in order to maximize the feed injection flow rate. Under these conditions, the purity of fructose at the extract is given by:

$$\text{fructose purity} = \frac{C_{fr}^{Ext}}{C_{gl}^{Ext} + C_{fr}^{Ext}} = \frac{m_3 - m_2}{m_3 - m_2 + (\overline{K}_A - m_2)C_{gl}^F / C_{fr}^F} \quad (10.3)$$

$$\text{assuming no pollution from zone 4 and with} \quad m_2 \leq \overline{K}_A \text{ and } m_3 \leq \overline{K}_B$$

It may be a good idea to derive this expression yourself.[5] Note that Eq. (10.3) forecasts a fructose purity of 100% if $m_2 = \overline{K}_A$, which corresponds to the border of the zone of complete separation. Equation (10.3) can be used to determine pairs (m_2,m_3) yielding a given purity. As shown in Figure 10.2, the (m_2,m_3) pairs allowing one to get 90% purity are very close to those delivering 100% purity. This means that, in terms of the equilibrium model prediction, the zone of partial separation is very similar to the zone of complete separation.

An expression similar to Eq. (10.3) is in general not available for non-linear systems, so numerical simulations must be used to determine the zone of partial separation. By setting m_1 and m_4 to 100 and 0.01, respectively, so that recycling between zones 1 and 4 is completely avoided, the zone of partial separation predicted by the MC–$\overline{\text{Equil}}$ model is indistinguishable from the prediction of Eq. (10.3), provided that the TMB is equivalent to at least 200 plates per zone. We assume that purity and yield are reached

[5] Hint: under these conditions, the concentration profile of glucose is uniform in zones 2 and 3.

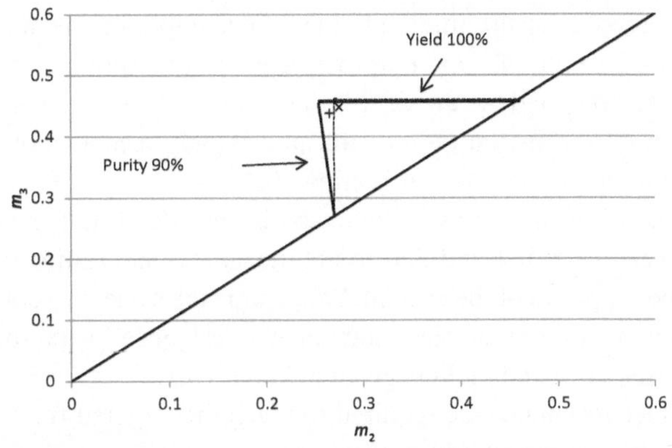

Figure 10.2 Separation zone in (m_2,m_3) space for the glucose–fructose separation to obtain at least 90% fructose purity at a yield of 100%. Linear behavior is assumed. $C_{fr}^F = 335$ g/l, $C_{gl}^F = 465$ g/l. Solid line: zone of partial separation determined from Eq. (10.3) or using the MC–Equil model (four-zone TMB with 200 plates per zone, $m_1 = 100$ and $m_4 = 0.01$). Dashed line: solution of the equilibrium model providing 100% purity and 100% yield. +: set of selected (m_2,m_3) for process design (Table 10.1); ×: set of selected (m_2,m_3) for process design (Table 10.2).

if they are within 0.5% of the target. At the vertex of the separation zone in Figure 10.2, one has $m_2^{eq} = 0.253$ and $m_3^{eq} = 0.46$ for 90% fructose purity at a yield of 100%. In addition, one has $m_1^{eq} = 0.46$ and $m_4^{eq} = 0.27$. With this information, one can determine the parameters χ_F^{eq} and ϱ^{eq} defined in Eq. (9.32) and Eq. (9.31), known to be important for estimating system performance:

$$\begin{aligned} \chi_F^{eq} &= C_{tot}^F \frac{u_{e3}^{eq} - u_{e2}^{eq}}{u_{e3}^{eq} - u_{e2}^{eq} + u_{e1}^{eq} - u_{e4}^{eq}} \\ &= C_{tot}^F \frac{m_3^{eq} - m_2^{eq}}{m_3^{eq} - m_2^{eq} + m_1^{eq} - m_4^{eq}} = 417 \quad \text{(g/l)} \end{aligned} \tag{10.4}$$

$$\varrho^{eq} = \frac{u_{e3}^{eq} - u_{e2}^{eq}}{u_{e1}^{eq} - u_{e4}^{eq}} = \frac{m_3^{eq} - m_2^{eq}}{m_1^{eq} - m_4^{eq}} = 1.09 \tag{10.5}$$

In the absence of a clear economic objective function, one cannot fully optimize the system, but must simply make a decent guess. We know from Chapter 9 that the optimum generally results from a compromise between productivity, eluent consumption and yield, and that, according to the results in Section 9.4.2, taking $\eta^{eq} \approx 1$ is often a good compromise (at least in the absence of precise optimization criteria). Following the route proposed in Section 8.5 for linear systems (Eq. (8.73)), we will allocate the safety margins with respect to the equilibrium model via a parameter $\vartheta = 1.03$ such that $m_i = m_i^{eq} \times 1.03$ for $i \in [1,2]$ and $m_i = m_i^{eq}/1.03$ for $i \in [3,4]$. These m_i lead to $\eta_F^{eq} = 0.9$, $\eta_E^{eq} = 1.11$ and $\eta_{EF}^{eq} = 1.002$. Note that taking a 3% margin on the m_i, thus on fluid velocities, leads to a 10% margin on η_F^{eq} and an 11% margin on η_E^{eq}.

In order to ensure minimal eluent consumption, one would need to work at higher values of η^{eq} and η_F^{eq} and thus to accept a lower margin on the fluid velocities compared to those predicted by the equilibrium model. Investigating the influence of η^{eq} on the system design is straightforward, and is only a matter of a few simulations. Further illustration will be given in Section 10.1.4.

This set of m_i values is then used in the MC–$\overline{\text{Equil}}$ model (Eq. (8.43)) to simulate the internal profiles and thus to get the extract and raffinate purities with respect to the number of plates of the system. We assume the system to consist of four zones of similar efficiency in terms of plate number; the influence of the total number of plates on the fructose purity and yield is given in Table 10.1.

About 160 plates are required to reach the targeted fructose purity (> 90%) but this would be associated with a yield of 95%, which exceeds our target. It is apparent that, with the above set of parameters, delivering the target purity requires a yield higher than the target (we mentioned in Chapter 9 that allocating a safety margin between zones 2 and 3 is not trivial). This means that our set of (m_2, m_3) could be increased slightly. The influence of the number of plates on a system for which m_2 and m_3 have been increased by 0.01 is given in Table 10.2. The set of m_i in Table 10.2 is associated with the same set of parameters $\eta_F^{eq} = 0.83$, $\eta_E^{eq} = 1.19$, $\eta_{EF}^{eq} = 1.003$ as those in Table 10.1.

Table 10.2 shows that, with the same margin as in the design based on the equilibrium model, the new parameters provide a better balance between purity and yield, so that only about 120 plates are required. This shows the extreme sensitivity of SMB/TMB to the precise set of m_i (the respective operating points associated with Tables 10.1 and 10.2 are given in Figure 10.2). We mentioned in Chapter 9 that a good fraction of an SMB user's time consists in tuning m_i. Certainly, again tuning the m_i values would make our estimate of N and therefore of the system design more precise. This fine tuning would not, however, change our conclusion. Also, don't forget that we are deriving an

Table 10.1 Influence of the number of plates N for four-zone TMB on extract and raffinate purities. Simulations performed with the MC–$\overline{\text{Equil}}$ model. System of four zones equivalent to $N/4$ plates. Linear behavior assumed. Feed: $C_{fr}^F = 335\ g/l$, $C_{gl}^F = 465\ g/l$; $m_1 = 0.474$, $m_2 = 0.261$, $m_3 = 0.447$, $m_4 = 0.262$.

Number of plates	80	100	120	140	160	200
Fructose purity (%)	83	86	88	89	90	91
Fructose yield (%)	87	90	92	94	95	97

Table 10.2 Influence of the equivalent number of plates N for four-zone TMB on extract and raffinate purities. Same conditions as for Table 10.2 except that m_2 and m_3 have been increased by 0.01, so: $m_1 = 0.474$, $m_2 = 0.271$, $m_3 = 0.457$, $m_4 = 0.262$.

Number of plates	80	100	120	140	160	200
Fructose purity (%)	87	89	91	93	94	95
Fructose yield (%)	85	88	90	91	93	94

Table 10.3 Key design parameters of the TMB and associated SMB for separating fructose and glucose. Linear adsorption isotherms. Feed mixture: $C_{fr}^F = 335$ g/l, $C_{gl}^F = 465$ g/l, target: purity fructose 90%, yield fructose 90%. Velocities are expressed in m/h, time in h, lengths in m, pressure in bar, concentrations in kg/m^3, ϕ_V in kg/(h m^3) and Υ_* in m^3/kg.

Zone	1	2	3	4
m_i^{eq}	0.46	0.253	0.46	0.27
m_i	0.478	0.271	0.457	0.262
u_{ei}^{smb}	24.6	20.2	24.2	20.0

$\bar{u}_e$	L_{col}^{tmb}	N_{col}	L_{col}	ΔT
14.35	11.3	8	1.4	0.098

η_F^{eq}	η_E^{eq}	η_{EF}^{eq}	χ_F^{eq}	t_ϕ^χ
0.9	1.11	1.002	417	3.3

ϕ_V	Υ_*	θ_*	ΔP	N
47.6	6.4 10^{-3}	0.9	10	120

D_{col}	Q_F	Q_E	Q_{ext}	Q_{raf}	Q_R
1.78	4	4.55	4.37	4.18	24.5

approximation: thinking in terms of plate numbers, independent of species, in a TMB is a coarse approximation of reality.

In the following, we will use the m_i in Table 10.2, and assume the required number of plates to be 120 and the pressure drop to be 10 bar. Based on these constraints, one can shortcut the SMB design according to the methodology presented in Section 9.6.2 and demonstrated in Exercises 9.6 and 9.7. The key features of such a TMB and thus SMB are given in Table 10.3. The productivity per unit volume is obtained with Eq. (9.37) and the eluent consumption with Eq. (9.38). Knowing the velocities in the different zones, the flow rates are finally determined for accommodating a feed flow rate of 4 m^3/h. This shortcut predicts that, despite the fact that an elution chromatogram does not show significant resolution (see Figure 10.2 obtained with $N = 200$), the glucose–fructose separation can be performed efficiently on an SMB system equivalent to only 120 plates. The productivity and eluent consumption for this glucose–fructose separation certainly represents some of the highest performance available for this particle size. Notice that, as a consequence of operations under linear conditions, the extract and raffinate flow rates are similar but a bit higher than the feed flow rate. Should η_F^{eq} and η_F^{eq} be set to 1, these flow rates would be identical. As a side comment, note that the HETP is about 10 cm when the particle size is 320 μm; kinetic limitations are thus very strong.

We have thus determined a first design of an SMB with very simple equations and considerations based on the TMB. In order to refine the design, a few considerations can be proposed:

- An SMB is not strictly equivalent to a TMB unless the number of columns is significant. We know from experience that about eight columns is adequate for this separation. Simulations confirm that the agreement between TMB and SMB performance is good when $N_{col} = 8$. Additionally, the way the columns are allocated between the different zones certainly impacts performance (there is a trend to put more columns in zones 2 and 3, fewer in zone 1 and 4). Information on this topic can be found in (Subramani *et al.*, 2003).

- Using the MC–$\overline{\text{Equil}}$ model is certainly a rough description of reality. At least, the PD–$\overline{\text{Equil}}$ and $\overline{\text{LDF}}$ models can represent the fact that mass transfer limitations are stronger for fructose than for glucose. The approach can even be refined using bi-LDF models (Subramani *et al.*, 2003; Azevedo and Rodrigues, 2001).

The reader who is brave enough has all the means to simulate different SMB units taking into account different numbers of columns per zone and different kinetic models. There are some differences from one option to another, but orders of magnitude are conserved, and there is no real "game changer" in the above considerations.

While the glucose–fructose separation is almost universally considered to be linear in the literature, we will see that investigating the influence of slight deviations from non-linearity is rewarding.

10.1.4 Process design: non-linear

We will now briefly present the design of the system when the adsorption isotherms are characterized by the non-linear expressions given in Section 10.1.2. First, we determine the hodograph space (C_{gl},C_{fr}) as defined in Chapter 4, the result being given in Figure 10.3.

One immediately recognizes the impact of non-linearity: for a linear system, the hodograph would consist of straight lines, parallel to the axes and passing through the feed and initial compositions. One also notices a difference from the hodographs

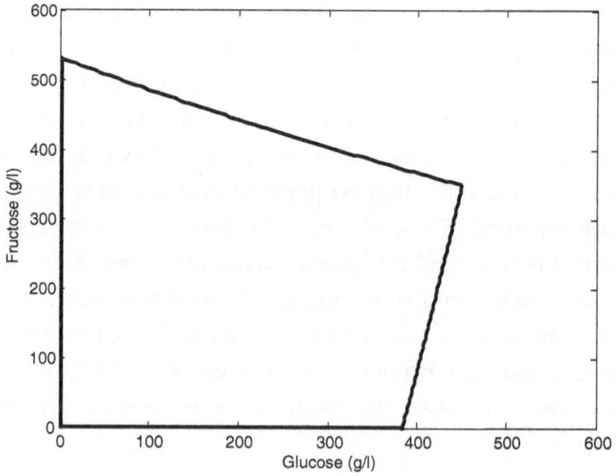

Figure 10.3 Hodograph for the glucose–fructose system. Characteristics which pass through solute-free solution and feed mixture $C_{fr}^F = 335$ g/l, $C_{gl}^F = 465$ g/l are shown. Non-linear adsorption isotherms are given by Eq. (10.1).

in Chapter 3 for Langmuir adsorption isotherms: the plateau composition obtained for pure component B having the highest affinity for the resin is *above* its concentration in the feed. This means that, in contrast to the behavior with Langmuir adsorption isotherms, the most retained solute (rather than the least retained) can be concentrated on a breakthrough curve obtained by injecting the feed mixture into a column initially equilibrated with a solute-free fluid. Can we use this property to improve our separation process?

Let us determine the m_i^{eq}, and again recall that, when a purity lower than 100% is sought, using a two-zone system, and thus the (m_2,m_3) representation, to investigate the four-zone system must be done with care, because (m_2,m_3) cannot be determined independently of (m_1,m_4) (see Section 8.5).

The zone of separation has been determined in the (m_2,m_3) space for a four-zone system having a set of (m_1,m_4) which prevents any recycling between zones 1 and 4. The zone of separation has been obtained by simulating the outlet purities of a system equivalent to 800 plates for different m_2 and m_3 values, and the result is given in Figure 10.4. We have checked that a higher system efficiency does not significantly influence the determined zone of separation. While an analytical expression for the zone of separation might be obtained with these simple adsorption isotherms, we have used a numerical approach for the sake of generality. The region of separation is significantly impacted by the non-linearity of the adsorption isotherms. From Figure 10.4, one obtains, from the vertex of the region of partial separation, $m_2^{eq} = 0.335$ and $m_3^{eq} = 0.55$. Notice that the difference $m_3^{eq} - m_2^{eq} = 0.215$ is "symbolically" greater than the difference $m_3^{eq} - m_2^{eq} = 0.207$ obtained with the linear adsorption isotherms. This is due

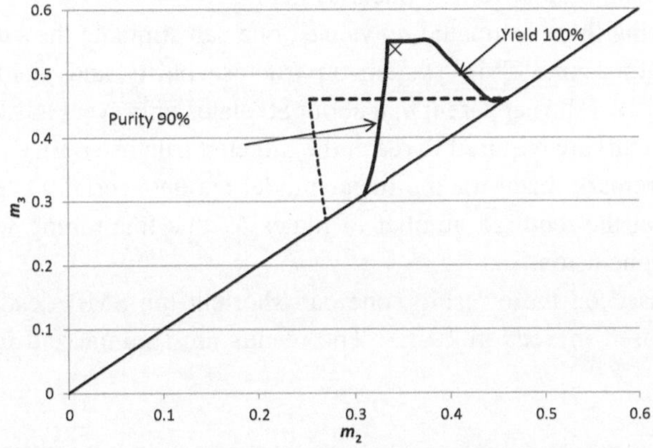

Figure 10.4 Separation zone in (m_2,m_3) space for the glucose–fructose separation to obtain 90% fructose purity at a yield of 100%. $C_{fr}^F = 335$ g/l, $C_{gl}^F = 465$ g/l. Solid line: zone of partial separation determined using the MC–Equil model (four-zone TMB with 200 plates per zone, $m_1 = 100$ and $m_4 = 0.01$) with non-linear adsorption isotherms given by Eq. (10.1). Dashed line: zone of partial separation determined with the linear adsorption isotherms having a slope given by the lumped Henry's coefficient. ×: set of selected (m_2,m_3) for process design (see Table 10.5).

to the synergistic characteristics of the adsorption isotherms (for a Langmuir adsorption isotherm, $m_3^{eq} - m_2^{eq}$ would decrease with concentration).

The ratio m_4^{eq} is given by the lumped Henry's coefficient of glucose ($\overline{K}_{gl}$ =0.27) as a consequence of the presentation given in Chapter 8 on the design of TMB with unfavorable adsorption isotherms. The m_1^{eq} value has been determined in order to just avoid leakage of fructose from zone 1 to zone 4 on a system of high efficiency (800 plates) operated with ratios m_2^{eq}, m_3^{eq} and m_4^{eq}; one obtains m_1^{eq}=0.51.

From these m_i^{eq}, one can determine $\chi_F^{eq} = 378$ g/l and $\varrho^{eq} = 0.90$. The values of these two parameters are similar to those obtained under the linearity assumption. In contrast to the example in Exercise 9.3, χ_F^{eq} increases almost linearly with feed concentration up to very high concentrations; there is no plateau of χ_F^{eq} versus feed concentration. The parameters χ_F^{eq} and ϱ^{eq} are, however, slightly lower than they are under the linearity assumption. This means that, despite the synergistic effect mentioned above, taking into account the non-linearity of the adsorption isotherm is going to slightly increase the eluent consumption. I believe this is a general rule: adding non-linearity requires an increase in eluent consumption.

Additionally, the difference $m_3^{eq} - m_2^{eq} = 0.215$ (feed) is greater than the difference $m_1^{eq} - m_2^{eq} = 0.175$ (extract), which confirms the possibility of purifying the fructose and at the same time concentrating it. Unfavorable adsorption isotherms do not mean unfavorable performance!

As in the linear case, we will take a 2.5% margin on the m_i values compared to those determined with the equilibrium model. We will thus take $m_4 = 0.263$ and $m_1 = 0.523$. The case of (m_2,m_3) is a bit more subtle (see Figure 9.13) and we will take $m_2 = 0.343$ and $m_3 = 0.523$. The set of parameters (m_2,m_3) is indicated by a cross in Figure 10.4. With this set of parameters, one obtains $\eta_F^{eq} = 0.9$ and $\eta_E^{eq} = 1.08$, which are almost identical to those for the linear system.

Using these estimated m_i values, one can simulate the influence of the plate number of a four-zone TMB system on fructose purity and yield. The results are given in Table 10.4. It is apparent that about 80 plates only (versus 120 with the linear adsorption isotherm) are required to reach the targeted fructose purity (> 90%) and yield (> 90%). The remark made for the linear model applies: certainly tuning the m_i, values would impact the required number of plates N; this fine tuning would not, however, change our conclusion.

Based on these results, one can shortcut the SMB design following the methodology used in Section 10.1.3. The results are summarized in Table 10.5. Compared to

Table 10.4 Influence of the equivalent number of plates N of the TMB on the extract and raffinate purities. The system is made of four zones equivalent to $N/4$ plates with $m_1 = 0.523$, $m_2 = 0.343$, $m_3 = 0.537$, $m_4 = 0.263$.

Number of plates	60	80	100	120
Fructose purity	87	90	92	94
Fructose yield	85	90	94	96

Table 10.5 Key design parameters of TMB and associated SMB for separating fructose and glucose. Non-linear adsorption isotherms. Feed mixture: $C_{fr}^F = 335\ g/l$, $C_{gl}^F = 465\ g/l$, target: purity fructose 90%, yield fructose 90%. Velocities are expressed in m/h, time in h, lengths in m, pressure in bar, ϕ_V in kg/(h m^{-3}) and Υ_* in m^3/kg.

Zone	1	2	3	4
m_i^{eq}	0.51	0.335	0.55	0.27
m_i	0.523	0.343	0.537	0.263
u_{ei}^{smb}	29.8	25.3	30.1	23.3

$\bar{u}_e$	L_{col}^{tmb}	N_{col}	L_{col}	ΔT
16.7	9.2	8	1.15	0.069

η_F^{eq}	η_E^{eq}	η_E^{eq}	χ_F^{eq}	t_ϕ^χ
0.9	1.08	0.995	378	2.03

ϕ_V	Υ_*	θ_*	ΔP	N
70.5	$7.0\ 10^{-3}$	0.9	10	80

D_{col}	Q_F	Q_E	Q_{ext}	Q_{raf}	Q_R
1.62	4	5.37	3.71	5.66	24.6

Table 10.6 Influence of the equivalent number of plates N of the TMB on the extract and raffinate purities. The system is made of four zones equivalent to $N/4$ plates with $m_1 = 0.515$, $m_2 = 0.338$, $m_3 = 0.545$, $m_4 = 0.267$.

Number of plates	80	100	120
Fructose purity	88	90	91
Fructose yield	88	92	94

Table 10.3, it is apparent that the design summarized in Table 10.5 is more productive by about 50%. The improvement is mainly due to the differences in the m_i values, and to the fact that the non-linear behavior is compatible with operation under a reduced number of plates.

The extract flow rate is slightly lower than the feed flow rate, which means that fructose is purified and at the same time concentrated. We have thus confirmed the ability to concentrate the purified fructose fraction by a bit less than 10%. Our immediate and initial guess, after a quick look at the hodograph, was thus perfectly correct! Possibly even more interesting is that this ability is used in large-scale industrial production units.

Can we further improve our design? Again, a complete optimization would require a detailed economic objective function, but we know that minimizing the eluent consumption is pretty important for this application. Let us try to minimize the eluent consumption by lowering the safety margin on the m_i from 2.5% to 1%.

The influence of the number of plates on fructose purity and yield is given in Table 10.6. The target purities and yield can be obtained with a system equivalent to 100 plates, which leads to the design in Table 10.7: compared to the previous design,

Table 10.7 Key design parameters of TMB and associated SMB for separating fructose and glucose; minimized eluent consumption. Non-linear adsorption isotherms. Feed mixture: $C_{fr}^F = 335\ g/l$, $C_{gl}^F = 465\ g/l$, target: purity fructose 90%, yield fructose 90%. Velocities are expressed in m/h, time in h, lengths in m, pressure in bar, ϕ_v in kg/h m^3 and Υ_* in m^3/kg.

Zone	1	2	3	4
m_i^{eq}	0.51	0.335	0.55	0.27
m_i	0.515	0.338	0.545	0.267
u_{ei}^{smb}	26.5	22.5	27.1	20.9

$\bar{u}_e$	L_{col}^{tmb}	N_{col}	L_{col}	ΔT
14.9	10.3	8	1.3	0.086

η_F^{eq}	η_E^{eq}	η_E^{eq}	χ_F^{eq}	t_ϕ^χ
0.96	1.03	0.998	378	2.53

ϕ_v	Υ_*	θ_*	ΔP	N
60.2	$6.6\ 10^{-3}$	0.9	10	100

D_{col}	Q_F	Q_E	Q_{ext}	Q_{raf}	Q_R
1.66	4	4.8	3.4	5.4	22.9

the productivity is about 17% lower, but the eluent consumption is 6% lower and the fructose concentration about 8% higher. Is it worth it? Possibly yes, due to the high costs associated with water evaporation.

10.1.5 Conclusion

For simplicity, we used the TMB approach and shortcuts to approximate an SMB able to separate fructose and glucose. Although following a very simplified approach, we have been able to extract important and unique features in good agreement with industrial experience:

- Although chromatograms obtained by elution chromatography exhibit poor resolution between the two sugars, the separation can be performed efficiently using SMB technology.
- Performance, expressed in terms of productivity and eluent consumption, is at the high end of what can be expected with chromatography.
- The adsorption behavior (slight deviation from linearity) of the sugars on the resin is such that one can select a feed concentration as high as possible, compatible with technical constraints, and there is also a possibility of purifying and at the same time concentrating the target product (fructose).

A complete optimization of the process requires intensive use of numerical tools, in order to investigate the precise influence of the number of columns, number of columns per zone, column length, fluid velocity etc. on the objective function. The way to achieve this task is to follow the algorithms given in Chapter 9 and to use the models presented in the previous chapters.

Readers will find more detail on the optimization of the glucose–fructose separation on SMB and Varicol in (Subramani *et al.*, 2003). Simulation (and industrial experience) shows that some modes described in Chapter 9, such as Varicol or, even better, I-SMB or S-SMB, are more efficient than SMB for this separation (Nicoud, 2014). This result is not a surprise: Varicol, I-SMB and S-SMB offer additional degrees of freedom compared to SMB, so they offer means for deeper optimization.

If one wants to go further, and better understand the performance of the systems, one needs to take into account different technical factors that make the system non-ideal according to the definition in Chapter 1. An important effect is the swelling and shrinking of resins in contact alternatively with concentrated sugar solutions and water. Experience shows that the bed length can vary by about 10% over a cycle for large columns, the variation being more limited at laboratory scale (remember the so-called wall effect). It is also known that resins located at the bottom of the bed can be deformed and become ellipsoid. The reader will find information on how to describe shrinking beds in Section 6.2.

Can one still calculate how a dead volume representing 10% of the column bed volume would impact an apparent number of plates and whether this is likely to impact the glucose–fructose separation? I am sure the answer is yes (see Section 2.5)! If you are not sure that this dead volume is equivalent to a mixing cell (I am not sure either), and believe that at least some investigation is required, the linear chromatography theory presented in Chapter 2 will give you great tools for that. More generally, this will give you tools to investigate imperfectly filled (or designed) columns.

This separation is performed at 65°C. The possible existence and influence of temperature profiles on column efficiency can be investigated with the approach given in Section 6.3.

Other important effects are associated with the high viscosity of the concentrated feed injected. Viscous fingering can and does impact the efficiency of the column, as described in Section 6.5. Additionally, one may wonder how the influence of concentration on swelling/shrinking can impact mass transfer. This is investigated in (Heinonen *et al.*, 2012).

Finally, with one example which is too often considered to be a simple separation (true under dilute conditions), one touches the limits of what simulation is currently able to deliver in terms of hydrodynamics and kinetics. We know that our shortcut allowed us to derive correct orders of magnitude, but I also know that the scale-up of these systems is not trivial, large-scale efficiency being in general not as good as at small scale.

10.2 Optical isomer separation

This illustration addresses the very important industrial application of optical isomer separation. More information on this application can be found in (Rajendran *et al.*, 2009), and in (Nicoud, 2014) for a historical perspective. A "model" separation will be studied in order to highlight the possibilities and limits of SMB technology for this application, as well as possible gains versus the elution chromatography mode.

Most of the data associated with this problem have been given in (Ludemann-Hombourger *et al.*, 2000). They are used below in a broader perspective.

10.2.1 Problem description

The problem consists in separating the two optical isomers of methyl mandelate. The objective is to split a feed flow rate of 125 l/h containing 8 g/l of racemate into two streams, one containing principally isomer A and one containing principally isomer B, the target purity in the enriched product being the same for both streams. Let us call the purity obtained in both streams the *purity split*. Although expressed in a different way, this problem is not different in principle from targeting a certain purity and yield for a given isomer; the yield associated with each purified fraction is simply identical to the purity split. Different purity splits will be considered, as this could be considered a tuning parameter in a more global optimization scheme, as discussed in Chapter 9.

A screening has shown that a possible chromatographic medium is Chiralcel OD (Chiral Technologies, Illkirch, France) associated with a heptane/isopropanol mixture (90/10 v/v) as eluent. The temperature is fixed at 30°C. The chiral stationary phase is expected to be available in four particle sizes: 10 μm, 25 μm, 50 μm and 100 μm. The problem is to compare elution and SMB chromatography and to select the particle size optimizing productivity.

10.2.2 Laboratory data

All tests have been performed on a column having a diameter of 4.6 mm and a length of 250 mm, eluted with an eluent flow rate of 1 ml/min at 30°C. The fluid density is 0.7 and its viscosity is 0.55 cP. The influence of feed viscosity is very small because of the dilute conditions used in this study.

The adsorption isotherms are almost independent of particle size.[6] They can be approximated by the linearly modified Langmuir adsorption isotherms in Eq. (3.10), but we will assume that we operate at concentrations low enough to be in the linear range so that the adsorption isotherms can be represented by their lumped Henry's coefficients, $\overline{K}_A = 2.20$ and $\overline{K}_B = 3.65$.

Using particle sizes of 25 and 51 μm, (Ludemann-Hombourger *et al.*, 2000) have shown that:

- The pressure drop is represented by Darcy's law:

$$\frac{\Delta P}{L} = \delta_e u_e \quad \text{with} \quad \delta_e = \frac{5.57 \times 10^{-10}}{d_p^2}$$

  Pressure is expressed in bar, lengths in m and velocity in m/h.

[6] This is expected, but not totally evident: for instance, the process of binding/grafting ligands on silica can be influenced by the particle geometry. This could then influence the properties of the chromatographic medium.

- The HETP (the average of the results obtained with the two species) varies with fluid velocity and particle diameter as:

$$HETP = B_e + C_e u_e \quad \text{with} \quad B_e = 6d_p \quad C_e = 1.98 \times 10^4 d_p^2$$

Lengths are expressed in m and velocity in m/h.

We assume that the above expressions can be used for the particle diameter range 10–100 μm.

At this point, one has gathered the required information to use models such as MC–$\overline{\text{Equil}}$, PD–$\overline{\text{Equil}}$, MC–$\overline{\text{LDF}}$ or PD–$\overline{\text{LDF}}$.

10.2.3 Equilibrium model analysis

We start by using the information from the adsorption isotherms to derive critical information on our system. The adsorption isotherms are independent of particle size (at least for this example), so the results from the equilibrium model for both elution and TMB/SMB are independent of particle size. We further assume the adsorption isotherms to be linear in our concentration range. Should one be willing to investigate the non-linear region, the adsorption isotherms could be approximated by a linearly modified Langmuir expression, so that one could use the analytical solutions in Tables 3.2, 8.1 and 8.2. Taking into account the impact of non-linearity would not substantially impact our conclusions.

TMB-SMB chromatography

Our first step consists in determining the reference provided by the equilibrium model, and we will use the set of parameters allowing us to obtain 100% purity and 100% yield, irrespective of the targeted purity split. The purity splits to be considered will be 93% or more, obtained by adapting the number of plates.

The m_i values defining the zone of complete separation are given for this linear problem by (see Eq. (8.72)): $m_1^{eq} = 3.65$, $m_2^{eq} = 2.2$, $m_3^{eq} = 3.65$, $m_4^{eq} = 2.2$. From this set of m_i^{eq}, one determines $\chi_F^{eq} = 4$ g/l and $\varrho^{eq} = 1$ (see Eqs. (9.32) and (9.31)).

Our aim is to quickly investigate the influence of operating modes and particle sizes on system productivity. We are not looking for a fine-tuned design, so we will simply take $m_i = m_i^{eq}$ and consequently $\eta_F^{eq} = \eta_E^{eq} = \eta_{EF}^{eq} = 1$. Tuning the m_i values according to the target purity would certainly allow a small improvement in performance but, again, this would not change our conclusions.

Elution chromatography

The injection time that allows the two solutes to be just separated on a column of infinite efficiency is related to the Z_eRT t_0 by $\Delta t_{inj}^{eq} = \frac{1-\varepsilon_e}{\varepsilon_e} t_0 \left[\overline{K}_B - \overline{K}_A\right]$, and the cycle time is $t_{cycle}^{eq} = 2\frac{1-\varepsilon_e}{\varepsilon_e} t_0 \left[\overline{K}_B - \overline{K}_A\right]$. Deriving these expressions is straightforward for linear systems. They can also be seen as special cases of the results in Chapter 3, specifically those in Section 3.5.3 for Langmuir adsorption isotherms.

With the present adsorption isotherms, this leads to $\Delta t_{inj}^{eq} = 2.174\, t_0$ and $t_{cycle}^{eq} = 4.35\, t_0$. As with the methodology in Section 10.1, one then determines $\chi_F^{eq} = 4$ g/l and $\varrho^{eq} = 1$ (see Eqs. (9.32) and (9.31)). These parameters are identical to those determined for the SMB. It is a simple confirmation of what we know from Section 8.8.

We will set $\Delta t_{inj} = \Delta t_{inj}^{eq}$ so that $\eta_F^{eq} = 1$. As explained in Section 9.5.1, the secondary tuning parameter t_{cycle} is determined by numerical simulation as the period between t_{beg}, at which the least retained solute has grown to 1% of its maximum, and t_{end}, at which the most retained solute product has shrunk to 0.1% of its maximum. The parameters η_E^{eq} and η_{EF}^{eq} are then determined from t_{cycle}. The cut time is selected so as to obtain the target purity.

10.2.4 MC–$\overline{\text{Equil}}$ model analysis

In order to assess the sensitivity of elution and TMB/SMB systems to efficiency, we will simulate the separation of the two optical isomers using the MC–$\overline{\text{Equil}}$ model. For both systems, the feed stream is divided into two streams containing one or the other isomer at the purity split. The purity split is an increasing function of the system efficiency expressed in number of plates. Results associated with $\eta_F^{eq} = 1$ are given in Table 10.8. As we take $\eta_F^{eq} = 1$, there is no margin with respect to the equilibrium model and consequently we exaggerate the influence of system efficiency. Recall that the yield associated with the two purified fractions is identical to the purity split. It is clear that, to obtain a given purity split, elution chromatography requires a higher number of theoretical plates than TMB/SMB does. This cannot be a surprise after reading Chapter 8; it is a major advantage of counter-current systems. Let us further notice that TMB/SMB is truly working according to the equilibrium model set of m_i, whereas a safety margin is taken on elution chromatography via t_{cycle}. Requiring $t_{cycle} = t_{cycle}^{eq}$ would dramatically increase the number of plates required to obtain a given purity split, and would not make practical sense.

Let us consider two different scenarios:

- Situation a): a high purity of 96.5% is sought. From Table 10.8, we see that 1400 plates would be required for the elution system, while only 450 plates are required for the TMB/SMB system.

Table 10.8 Influence of system efficiency on the purity split. Elution chromatography: $t_{inj} = 2.174\, t_0$; t_{cycle} (thus η_E^{eq} and η_{EF}^{eq}) is determined so as to collect solutes if their concentration exceeds 1% of peak maximum. TMB chromatography: $m_1 = 3.65$, $m_2 = 2.2$, $m_3 = 3.65$, $m_4 = 2.2$; $\chi_i = 0.25$, $i \in [1,4]$. For TMB, the purity split is defined as the average of the extract and raffinate purities.

N	100	200	300	400	500	600	800	1000	1400
TMB	87	93.0	95.0	96.2	96.9	97.5	98.0	> 98.5	> 98.5
Elution	88.6	91.6	93.0	93.9	94.4	94.9	95.6	96.0	96.5
η_{EF}^{eq} (elution)	1.58	1.41	1.33	1.28	1.25	1.23	1.2	1.18	1.15

Table 10.9 Influence of particle size on the elution and SMB systems, both delivering 96.5% split purities. Operating pressure: 50 bar. Processing of 125 l/h of feed at 8 g/l. Productivity per unit volume ϕ_V is expressed in kg/(h m^3). No technical constraints other than maximum pressure drop are considered.

Particle diameter (μm)		10	25	50	100
Elution (1400 plates)	u_e (m/h)	56.9	56.9	56.9	56.9
	L_{col} (m)	0.16	0.99	3.9	15.8
	D_{col} (m)	0.13	0.13	0.13	0.13
	ϕ_V	251	40	10	2.5
TMB/SMB (450 plates)	$\bar{u}_e$ (m/h)	19	19	19	19
	$N_{col}L_{col}$ (m)	0.09	0.55	2.2	8.7
	D_{col} (m)	0.10	0.10	0.10	0.10
	ϕ_V	756	121	30	7.5

- Situation b): a moderate purity of 93% is sought. From Table 10.8, we see that 300 plates would be required for the elution system, while only 200 plates are required for the TMB/SMB.

Using the shortcut approach proposed in Section 9.4.3, and requiring a pressure drop of 50 bar, one can solve Darcy's and the Van Deemter equations with respect to length and velocity for the different particle sizes and systems. The results are given in Table 10.9 for a purity split of 96.5%. Recall that, for the MC–Equil model, hydrodynamic dispersion and mass transfer limitations are characterized by a single parameter, the number of plates, N. It does not matter how the plates are obtained: significant length with large particles, shorter length with smaller particles, more or less important fluid velocity impacting mass transfer limitations etc. From Table 10.9, one sees that the fluid phase velocity and system diameter are independent of particle size but that the column length varies as the square of particle size, and productivity as the reciprocal of the square of particle size. This is totally consistent with the results in Section 9.4.3.

The productivity obtained with TMB/SMB is always higher (about three times) than with elution chromatography. From these results, it is clear that SMB is much more productive than elution, which is consistent with many published articles.

Let us now consider the results given in Table 10.10 for a 93% rather than 96.5% purity split. The productivity is significantly increased when the target purity decreases, and again, this is no surprise. Note that this increase is entirely attributable to the requirement of number of plates, as all other parameters are kept constant.

Whereas the TMB/SMB was about three times more productive than elution for a 96.5% purity split, it is only 70% more productive than elution for a 93% purity split. This is again associated with the influence of the number of plates: fewer plates being required at lower purity, the advantage of SMB/TMB over elution decreases.

Excluding technical constraints other than a maximum pressure drop, we have confirmed with this example that productivity varies as the reciprocal of the square of the particle diameter, and that the advantage of SMB over elution is mainly related to the number of plates required for the separation. The consequence is that, in principle, SMB should be favored for difficult separations (high purities, limited selectivities etc.) for which a significant number of plates is required.

Should we consider the results in Tables 10.9 and 10.10 as a ready-to-implement design? No: some of the technical characteristics are frightening!

10.2.5 Comparison of elution and SMB: practical considerations

Let us first consider the fluid velocities presented in Tables 10.9 and 10.10. They are very high compared with orders of magnitude given in Appendix J. Readers used to thinking in terms of t_0 or shift times will find these in Table 10.11. For the 96.5% purity split, when 10 μm particles are used, one obtains t_{cycle} as low as 57 s for elution chromatography, and ΔT as low as 2 s for an eight-column SMB. One obtains even smaller values for the 93% purity split: t_{cycle} is about 12 s and ΔT about 1 s. One thus obtains very small t_{cycle} and ΔT when small particles are used.

It is difficult to suggest a minimum acceptable t_{cycle} value, and there may be no sound technical arguments supporting a precise threshold definition. Because of vibrations, pressure excursion when injecting feed, valves and more generally system response

Table 10.10 Influence of particle size on the elution and SMB systems, both delivering 93% split purities. Operating pressure: 50 bar. Processing of 125 l/h of feed at 8 g/l. Productivity per unit volume ϕ_V is expressed in kg/(h m^3). No technical constraints other than maximum pressure drop.

Particle diameter (μm)		10	25	50	100
Elution (300 plates)	u_e (m/h)	122.9	122.9	122.9	122.9
	L_{col} (m)	0.07	0.46	1.83	7.3
	D_{col} (m)	0.09	0.09	0.09	0.09
	ϕ_V	1012	162	40	10
TMB/SMB (200 plates)	$\bar{u}_e$ (m/h)	28.5	28.5	28.5	28.5
	$N_{col}L_{col}$ (m)	0.06	0.36	1.46	5.8
	D_{col} (m)	0.08	0.08	0.08	0.08
	ϕ_V	1700	272	68	17

Table 10.11 Influence of particle size on cycle and shift times. Operating pressure: 50 bar. See details in the text and in Tables 10.9 and 10.10. No technical constraints other than maximum pressure drop.

Particle diameter (μm)		10	25	50	100
Elution	t_{cycle} (h)	1.6×10^{-2}	1.0×10^{-1}	4.0×10^{-1}	1.6
SMB (8 columns)	ΔT (h)	5.8×10^{-4}	3.6×10^{-3}	1.4×10^{-2}	5.8×10^{-2}
Purity split: 96.5%					

Particle diameter (μm)		10	25	50	100
Elution	t_{cycle} (h)	3.4×10^{-3}	2.1×10^{-2}	8.6×10^{-2}	3.4×10^{-1}
SMB (8 columns)	ΔT (h)	2.6×10^{-4}	1.6×10^{-3}	6.4×10^{-3}	2.6×10^{-2}
Purity split: 93%					

time, or even habits, we will consider that the minimum acceptable value for the cycle time is 2 min for elution (I am not aware of any production system operated with such a short cycle time). In our case, this translates to a t_0 that should not be less than about 30 s. As far as SMB is concerned, the smallest shift time ΔT that I am aware of is about 30 s, which corresponds in our case to a t_0 calculated on one of the SMB columns of only about 5 s.

Another, related concern is the calculated column length. When small particles are used, and especially with limited purity requirements, very short column lengths (of a few centimeters or less) are calculated for TMB/SMB implementation. At industrial scale, columns are at least about 50 cm long for low-pressure design, and can be as short as 10 cm or a bit less when DAC technology is used. It is again difficult to propose a minimum value that is backed by experimental or theoretical evidence, but let us consider that 10 cm per column is a minimum at large scale; again, this requires DAC technology. Assuming that the SMB consists of eight columns (typical for this type of application), the minimum acceptable system length $N_{col}L_{col}$ is 0.8 m.

In addition, we could require a minimum value for the L_{col}/D_{col} ratio. Although such a limitation is sometimes mentioned, it depends a lot on the column technology: I have seen L_{col}/D_{col} ratios well below 0.1 in large-scale SMB made of DAC columns delivering an adequate and stable efficiency.

Another question is whether one should consider a maximum bed length. It is certainly not uncommon to see columns a few meters long operated at low–medium pressure and packed with particles a few hundred in size. In our case, the use of particle sizes below 100 μm points to DAC technology. The maximum length for a stable and compact bed is probably impacted by the L_{col}/D_{col} ratio (see Chapter 6). I have seen DAC columns with diameters in excess of 1 m packed with particles a few tens of μm in size having bed lengths of a few meters (the classical length being closer to half a meter). In Table 10.9, the maximum calculated length is 15.8 m for elution chromatography, which is certainly too much, especially for DAC technology. An idea for implementing this configuration would be to separate the column into a few columns connected in series, but then I would probably try to inject and collect between the columns, so the mode would no longer be elution!

Why is our model predicting unreasonably high interstitial fluid velocities, or unreasonably low t_0, ΔT or L_{col}? The number of plates has been determined so as to obtain the required purity split; it is thus not really negotiable! In addition, we decided to take the parameters derived from the equilibrium model: any safety margin would decrease the required number of plates and thus make things even worse. The other constraint is to operate at the maximum operating pressure; this looks very reasonable if one wants to get the best out of the system. The selectivity of the investigated separation is relatively comfortable (thanks to the great job performed by colleagues in charge of screening!), so a relatively limited number of plates (a few hundred) is required to deliver the purity. In addition, one uses small particles, so that even if the HETP is 100 particle diameters because of strong mass transfer limitations, obtaining the required number of plates is achievable with very short column lengths, too short to be practical. In addition, as the carrier fluid has a low viscosity, nothing prevents the fluid velocity from increasing dramatically while the operating pressure drop is kept below the threshold. Using small particle size and low-viscosity carrier fluid to perform a separation requiring a limited number of plates is like a car without brakes: nothing prevents the car from speeding up. We have to put on the brakes by taking technical constraints other than maximum pressure drop into account.

The results in Tables 10.9 and 10.10 have been reprocessed in order to require $t_0 >$ 30 s (or $\Delta T >$ 30 s) and $L_{col} >$ 10 cm as additional technical constraints. The exact values of these thresholds can be fine-tuned, as discussed later, but the order of magnitude is correct for current technology. Taking these additional technical constraints into account poses no difficulty, but will lead to drastic consequences, as detailed below.

One first determines the system without technical constraints (except for pressure drop), and then:

- If $L_{col} <$ 10 cm is calculated, one sets $L_{col} =$ 10 cm and determines u_e such that the maximum pressure drop is not exceeded. Note that any increase in L_{col} has a significant impact on system volume, thus on productivity: not only is the length increased, but the diameter must be increased as well, in order to decrease the fluid velocity so as to keep pressure drop constant.
- With the revised L_{col} and u_e values (resp. $\bar{u}_e$), one determines t_0 (resp. ΔT). If t_0 (resp. ΔT) is below the threshold, a new set of L_{col} and u_e (resp. $\bar{u}_e$) is determined, satisfying the minimum t_0 (resp. ΔT) as well as the pressure-drop constraint.
- Taking into account the above constraints, one oversizes the system compared to "strict needs", and productivity can decrease dramatically. A consequence is that the number of plates N associated with the revised system is higher than the number required to reach the purity split with the set of injection parameters determined in Section 10.2.3.
- Having more plates than strictly necessary offers an opportunity to inject more with respect to the equilibrium model design, or to work at higher yield. In general, this cannot compensate for the productivity loss: let us say that taking advantage of the extra plates can lead to productivity increases on the order of a few times 10%, while the productivity decrease due to the L_{col}, t_0 and ΔT constraints can be 100-fold when small particles are used.

The systems required to deliver either 96.5% or 93.0% purity split have been redesigned, taking into account the three additional technical constraints, and the results are given in Figure 10.5. Figures 10.5a and 10.5c reproduce the results already given in Tables 10.9 and 10.10: only a constraint on pressure drop is taken into account. In Figures 10.5b and 10.5d the constraints on t_0, ΔT and L_{col} are taken into account.

An important qualitative observation is that taking into account the additional constraints dramatically impacts the shape of the productivity versus particle size curves. Whereas productivity decreases continuously with particle size with only a constraint on pressure drop, it exhibits a maximum and a plateau when additional constraints are considered. Let us explain why. At "large" particle sizes, the t_0, ΔT and L_{col} parameters are above their thresholds; consequently the new constraints have no impact,

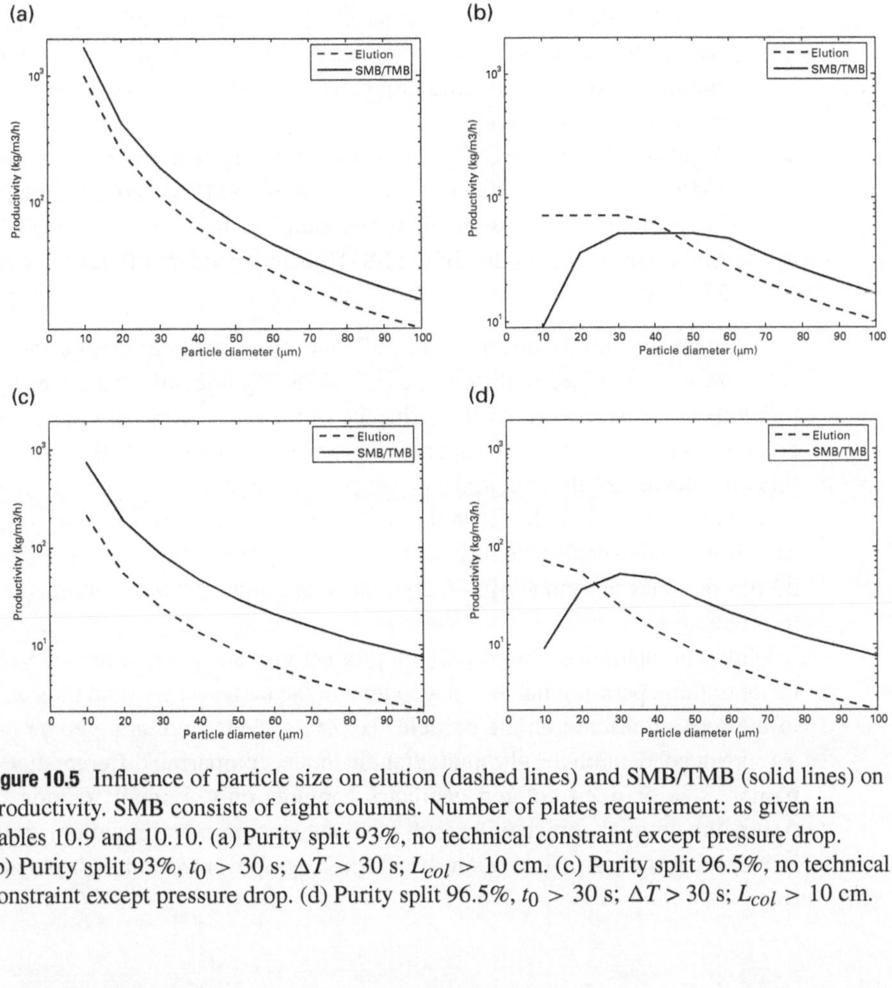

Figure 10.5 Influence of particle size on elution (dashed lines) and SMB/TMB (solid lines) on productivity. SMB consists of eight columns. Number of plates requirement: as given in Tables 10.9 and 10.10. (a) Purity split 93%, no technical constraint except pressure drop. (b) Purity split 93%, $t_0 > 30$ s; $\Delta T > 30$ s; $L_{col} > 10$ cm. (c) Purity split 96.5%, no technical constraint except pressure drop. (d) Purity split 96.5%, $t_0 > 30$ s; $\Delta T > 30$ s; $L_{col} > 10$ cm.

and productivity decreases with particle size. At very "small" particle sizes, L_{col} is fixed and set at 10 cm. Association of very small particles and a minimum imposed bed length leads to a number of plates N which is significantly greater than required. In order to keep the pressure drop constant, one needs to decrease the velocity, thus to increase the column diameter when particle size is decreased. In this region, decreasing the particle size decreases productivity! The plateau is obtained for intermediate particle sizes, and is due to the threshold on t_0 or ΔT. The concept of being "small" or "large" depends on the situation (purity sought, mode).

An important quantitative observation is that the impact of the constraints on t_0, ΔT and L_{col} can be dramatic: the productivity of an SMB operated with 10 μm particles to obtain a 93% purity split is decreased by about a hundred times.

A final observation is that optimal particle sizes and modes are totally problem-dependent:

- At a "moderate" 93% purity split, elution appears to be more productive than SMB if particles of 45 μm or less are used; SMB is more productive otherwise.

The rationale for this result is that SMB is more affected by the additional technical constraints, especially the one associated with minimum column length, than elution is. Additionally, the productivity of SMB is independent of particle size in the range 25–60 μm.

- At a "significant" 96.5% purity split, elution appears to be more productive than SMB if particles of 20 μm or less are used; SMB is more productive otherwise. The rationale for this result is the same as the one for the 93% purity split. Additionally, the productivity of SMB is independent of particle size in the range 25–45 μm.

Different thresholds taken for t_0, ΔT and L_{col}, different purity splits or different selectivities would lead to different conclusions, significantly impacting system design.

Let us say a final word on the influence of the number of columns in SMB. Adding columns is certainly a way to operate more closely to ideal TMB system performance; this also decreases the column unit length. The likelihood of being constrained by a minimum column length threshold is thus more important at higher column number. That is why industrial systems used for separating optical isomers and filled with about 20 μm particles are made of 5–6 columns maximum. Adding columns is totally counterproductive.

With this separation, based only on productivity considerations, we have shown that the optimum operating mode is dependent on the target purity (and thus on the difficulty of the separation) and on the particle size. As optimum column geometry and particle size are dependent on the chromatographic mode, comparing different modes for a given particle size or fixed column geometry captures only a small fraction of the reality. Again, the optimal choice will result from an economic optimization, and certainly the influence of the particle size on the chromatography medium cost plays a significant role.

10.2.6 Conclusion

Results obtained in the preceding sections show that comparing SMB and elution modes is hardly compatible with general considerations. We know that, in the theoretical absence of hydrodynamic dispersion and mass transfer limitations, thermodynamics and the equilibrium model predict similar performance for the two systems, as shown by the analogy in Section 8.8. The difference in behavior, and thus in performance, is intimately related to the influence of hydrodynamic dispersion and mass transfer limitations. These considerations are strongly dependent on selectivity, adsorption isotherms, mass transfer, particle size, viscosity, etc., so a fair amount of information is required for a sensible comparison between these two modes. An important result is that the comparison depends on the particle size and that optimal particle size is different for each system. Qualitatively, SMB is favored for low selectivities, moderate non-linearities, significant purity or yield constraints and significant particle sizes.

In terms of economy, besides the impact of eluent consumption, which can be major, the dependence of the chromatographic medium cost on particle size is important: when chromatographic media are available in large particle sizes at low cost, SMB may be ideally suited; this is exactly the situation for glucose–fructose separation. The success of SMB/Varicol for optical isomers is intimately related to the technical ability to build columns with very short L/D ratios packed with small particles (typically 20 μm). It is worth noting that these systems are associated with a limited number of columns (typically five), as more columns would lead to even lower and more problematic column lengths.

In this section, we have more or less arbitrarily set some thresholds. Let us discuss the rationale and consequences from a modeling and simulation perspective.

Extra-column factors requiring a minimum column length

Irrespective of the chromatographic mode, chromatographic beds are surrounded by distributors, pipes, sensors, valves etc., each device associated with its own dynamics. We have shown in Chapters 2 and 6 how a linear chromatography analysis and system dynamics provide great tools for analysis and investigation. The "surrounding pieces" being largely independent of column length, their relative influence increases when column length decreases. At some point, their influence is such that the system is no longer able to perform. The detailed analysis and practical solutions are very much dependent on the precise system design: as an illustration, the presence of flow meters and pumps in the recycling loop of the SMB induces an asymmetry that pushed (Hotier *et al.*, 1996) to propose using a first column shorter than the others and (Hotier and Nicoud, 1996) to suggest period desynchronization.

Consequences of working with very short columns

Working with very short bed lengths has various implications from a modeling standpoint:

- Representing mass transfer limitations: Besides the fact that the MC–$\overline{\text{Equil}}$ model is not able to account for species-dependent mass transfer limitations, it may further differ substantially from the models that explicitly take into account mass transfer limitations. We have seen in Chapters 1 and 7 that when N/J is below 0.1 the chromatograms obtained using MC–$\overline{\text{Equil}}$ models can significantly differ from those using the MC–$\overline{\text{LDF}}$ model. One can check, using results from Tables 10.10 and 10.11, and the Van Deemter equation, that the ratio N/J can be as low as 0.03 for 100 μm particles. This means that using a plate model to represent the column can be a rough approximation of reality so that the MC–$\overline{\text{LDF}}$, PD–$\overline{\text{LDF}}$ or other detailed kinetic model must be considered.
- Boundary conditions/inlet signal: Modeling hydrodynamic dispersion requires taking a serious look at boundary conditions (see Chapter 1 and Appendix B). It is not only a matter of knowing whether simplified boundary conditions or Danckwerts boundary conditions must be used: one needs to precisely investigate the hydrodynamics of the distributor that will actually be used, the key lying in the ability to deliver a plug flow over a very short distance. Computational fluid dynamics can help for the conception, and linear chromatography tools presented

in Chapters 2 and 6 must be used for investigation purposes. In particular, the existence of non-uniform injection or radial velocity profiles must be assessed, as we know that they can significantly influence chromatograms, as explained in Chapters 6 and 7.

From this, it is apparent that the tools derived from the linear chromatography methodology, including transfer functions, must be at the center of investigation methods.

10.3 Chromatography as a Lego piece

10.3.1 Problem description

We explained in Chapter 9 that chromatography is in general a simple part of a more complex process. As an illustration, interest in coupling SMB and crystallization, racemization or membranes has been discussed in (Blehaut and Nicoud, 1998), (Adam *et al.*, 2005), (Strohlein *et al.*, 2003), (Lorenz *et al.*, 2001) and (Nimmig and Kaspereit, 2013). It should be noticed that most large-scale applications of SMB for the pharmaceutical industry associate chromatography and crystallization, and that the very large-scale separations of xylene isomers also take advantage of this coupling (Nicoud, 2014).

Coupling chromatography and racemization is important for maximizing the overall yield. Coupling chromatography and crystallization is often a good way to improve productivity, gain being related to the shape of the crystallization curve and especially the position of the eutectic, if any.

Coupling chromatography, isomerization and crystallization has been extensively discussed in the literature, so we will focus on two situations that are probably less known.

10.3.2 Process design: HFCS and by-pass strategy

This example can be seen as a generalization of the presentation given in Section 10.1. High-fructose corn syrup (HFCS) is produced by milling corn to produce corn starch, then processing that starch to yield corn syrup, which is almost entirely glucose. Glucose is then converted by enzymes that change some of the glucose into fructose. The resulting syrup (after enzyme conversion) contains approximately 42% fructose and 53% glucose, and is referred to as HFCS 42. Some of the HFCS 42 must then be purified to yield HFCS 55 (a sweetener used mostly in soft drinks) containing approximately 55% fructose and 42% glucose. Both HFCS 42 and HFCS 55 contain about 5% impurities that are not separated from glucose or fructose on the SMB.

Complete processes use carbon adsorption for impurity removal and numerous filtration, ion-exchange and evaporation steps, but we will focus on the chromatographic and isomerization steps and neglect solutes different from glucose and fructose.

In order to simulate the entire process, we will assume that the isomerization step yields HFCS 42, irrespective of the inlet fructose/glucose ratio, and that the data given in Section 10.1 associated with the non-linear adsorption isotherms are accurate enough for designing the SMB. Furthermore, the composition of HFCS 42 injected into the

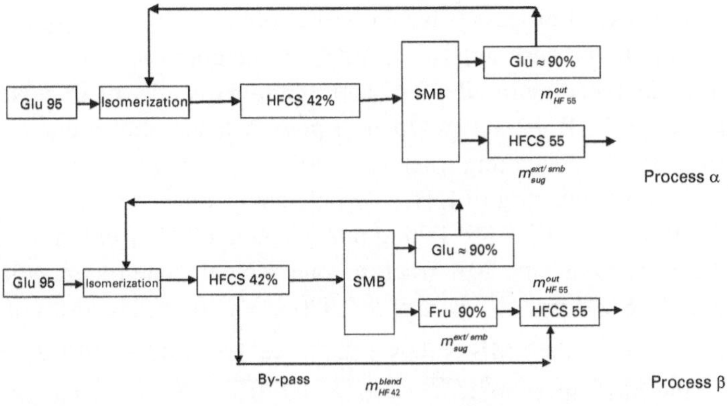

Figure 10.6 Schematic of two possible production processes to obtain high-fructose syrup.

SMB is still given by $C_{gl}^F = 465$ g/l and $C_{fr}^F = 335$ g/l. We will consider two possible processes for obtaining HFCS 55, as illustrated in Figure 10.6. According to process α, the chromatographic system (assumed to be SMB) is designed to directly produce HFCS 55 from HFCS 42. From the retention order of the sugars, we know that HFCS 55 is recovered in the extract, and we want the associated yield to be 90%. This is the most obvious process to produce HFCS 55 (55% fructose) from HFCS 42 (42% fructose) by SMB.

According to process β, the chromatographic system (again assumed to be SMB) is designed to produce HFCS 90 (90% fructose) from HFCS 42, the two streams then being blended to produce HFCS 55. From the retention order of the sugars, we know that HFCS 90 is recovered in the extract, and we want the associated yield to be 90%. This process is less intuitive, but this is the path followed industrially. Can we understand why?

We will assume that the isomerization step is cheap compared to chromatography, so the primary goal is to minimize the chromatographic system size, as well as water consumption, since evaporation typically represents about 90% of Opex. The isomerization step is followed by an evaporation step, so the concentration of HFCS 42 entering the SMB is constant and independent of the volume of the raffinate fraction recycled.

The production rate of HFCS 55 is assumed to be $\dot{m}^* = 1000$ t/y for normalization purposes. This corresponds to a production of $m_{fr}^{out} = 550$ t/y of pure fructose in the final HFCS 55 stream.

Let us start by detailing process α. Taking into account a 90% yield, a fructose mass balance allows the quantity of dry fructose (kg/h) that must be processed on the SMB to be obtained: $m_{fr}^{inl/smb} = 550/0.9 = 611$ t/y. Taking into account the fructose content in the HFCS 42, one obtains the mass of HFCS 42 that must be processed on the SMB: $m_{HF42}^{inl/smb} = 611/0.42 = 1454$ t/y. The 1454 t/y of sugar fed onto the SMB is split into $m_{sug}^{ext/smb} = 1000$ t/y exiting at the extract (HFCS 55) and $m_{sug}^{raf/smb} = 554$ t/y of the glucose-rich fraction sent to isomerization. Finally, the feed concentration being 800 g/l, assuming 8000 h of operation per year, the feed flow rate of HFCS 42 to be processed in the TMB/SMB is 227 l/h.

The case of process β is a bit more subtle due to the by-pass. The quantity of sugar exiting the process, $m_{HF55}^{out} = 1000$ t/y including $m_{fr}^{out} = 550$ t/y of fructose, results from the blending of HFCS 42 and a certain quantity of fructose at 90% purity coming from the SMB extract. A fructose mass balance allows one to obtain the quantity of sugar (kg/h) containing fructose at 90% purity coming from the SMB ($m_{sug}^{ext/smb} = 270$ t/y) and the quantity of HFCS 42 delivered from the by-pass ($m_{HF42}^{blend} = 730$ t/y) to get the 1000 t of HFCS 55. Taking into account a 90% yield in the SMB, and the fact that the extract contains 90% fructose, one can calculate the quantity of fructose to be fed into the SMB: $m_{fr}^{inl/smb} = 270 \times 0.9/0.9 = 270$ t/y. The mass of HFCS 42 that must be processed on the SMB is thus $m_{HF42}^{inl/smb} = 270/0.42 = 643$ t/y. A simple mass balance on the SMB gives $m_{sug}^{raf/smb} = 643 - 270 = 373$ t/y of the glucose-rich fraction sent to isomerization. Finally, the feed concentration being 800 g/l, the feed flow rate to be processed on the SMB is 100 l/h.

The key flow rates characterizing both processes are summarized in Table 10.12. This analysis shows that the choice to be made is between:

- Process α: processing 611 kg/h of fructose contained in HFCS 42, thus 227 l/h of HCFS 42 to produce 55% pure fructose at 90% yield
- Process β: processing 270 kg/h of fructose contained in HFCS 42, thus 100 l/h of HCFS 42 to produce 90% pure fructose at 90% yield.

A priori, not an easy choice!

We start by considering the zones of partial separation in the (m_2,m_3) space. The zone associated with the (90% purity, 100% yield) region has been given in Figure 10.4. The zone associated with the (55% purity, 100% yield) region is given in Figure 10.7. For both situations, the (m_2,m_3) zone has been determined with m_1 and m_4 ratios avoiding any recycling of fructose or glucose between zones 1 and 4.

One immediately sees that the zone of separation associated with the 55% pure fructose target is significantly wider than the zone associated with the 90% fructose purity given in Figure 10.4: that cannot be a surprise, and this is true irrespective of the choice of a linear or non-linear representation. The impact of non-linearity on the separation zones is also clear. Two non-obvious results are, however, obtained:

- It is possible to simultaneously increase purity and productivity. For instance, consider coordinates $m_2 = 0.2$ and $m_3 = 0.3$. From Figure 10.7 it is clear that the target purity is not met. One way to reach the purity target consists in increasing

Table 10.12 Different quantities (t/y) of dried sugars associated with the production of 1000 t/y of HFCS 55. The yield of fructose is assumed to be 90% on the SMB step (thus 90% of the fructose goes into the extract stream). Process α: $Q_F = 227$ l/h; Process β: $Q_F = 100$ l/h.

Process	$m_{fr}^{inl/smb}$	$m_{HF42}^{inl/smb}$	$m_{sug}^{ext/smb}$	$m_{sug}^{raf/smb}$	m_{HF42}^{blend}	m_{HF55}^{out}	m_{fr}^{out}
α	611	1454	1000	454	0	1000	550
β	270	643	270	373	730	1000	550

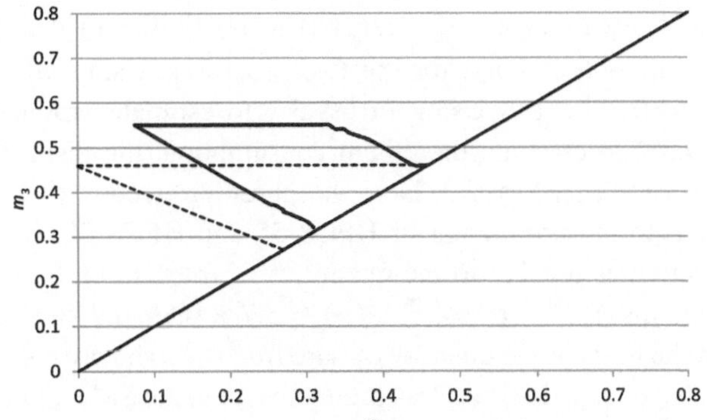

Figure 10.7 Separation zone in (m_2,m_3) space for the glucose–fructose separation to obtain at least 55% fructose purity at a yield of 100%. $C_{fr}^F = 335$ g/l, $C_{gl}^F = 465$ g/l. Solid line: zone of partial separation determined using the MC–Equil model (four-zone TMB with 200 plates per zone, $m_1 = 100$ and $m_4 = 0.01$) with non-linear adsorption isotherms given by Eq. (10.1). Dashed line: zone of partial separation determined with the linear adsorption isotherms having a slope given by the lumped Henry's coefficient.

m_3 above 0.4, but below 0.55 for the non-linear adsorption isotherms. In other words, at constant m_2 one can, at the same time, increase purity and feed flow rate. This ability, under certain circumstances, to increase purity by increasing productivity is rare enough to be worth mentioning.

- The ratio m_2 can be almost zero (and even zero for a linear system) while still allowing the delivery of fractions at the target purity. This implies that the extract flow rate equals the eluent make-up flow rate and that there is no fluid flow rate in zone 2. This feature is rare enough to be worth mentioning.

Recall that the vertex of the separation zone associated with 90% purity is given by $m_2^{eq} = 0.335$ and $m_3^{eq} = 0.55$ (non-linear adsorption isotherms). Additionally, we also determined $m_1^{eq} = 0.51$ and $m_4^{eq} = 0.27$. The vertex of the separation zone associated with 55% purity given in Figure 10.7 is: $m_2^{eq} = 0.075$ and $m_3^{eq} = 0.55$. Additionally, we have $m_4^{eq} = 0.27$ and one can determine $m_1^{eq} = 0.51$.

The m_i^{eq} allow the determination of the concentration χ_F^{eq} and the ratio ϱ^{eq} for both processes. One obtains $\chi_F^{eq}(\alpha) = 378$ g/l, $\chi_F^{eq}(\beta) = 531$ g/l, $\varrho^{eq}(\alpha) = 0.9$ and $\varrho^{eq}(\beta) = 1.98$. The fact that a lower-purity target is associated with a higher normalized feed flow rate is reflected in the higher values of χ_F^{eq} and ϱ^{eq} for SMB process β compared to SMB process α. In order to derive orders of magnitude, let us consider situations for which $m_i = m_i^{eq}$ so that $\eta_F^{eq} = \eta_E^{eq} = \eta_{EF}^{eq} = 1$ for both SMB processes. Under these conditions, productivity is proportional to the reciprocal of t_ϕ^χ (see Eq. (9.37)). The time t_ϕ^χ must be determined by process simulation, but we can "guesstimate" an order of magnitude for the ratio $t_\phi^\chi(\beta)/t_\phi^\chi(\alpha)$. Using SMB technology, and the same set of η^{eq} for both processes, the number of plates required by each process to deliver the target purity and yield should not be too different. For a given pressure drop, this requires the fluid velocity and column length to be similar for both processes. In other words, one can expect $t_\phi^X(\beta)/t_\phi^X(\alpha)$ to be similar to the ratio of the algebraic sums $[m_4 - m_1 + m_3 - m_2]$ for the two processes. Using m_i^{eq} (because $\eta = 1$) leads to $t_\phi^X(\beta)/t_\phi^X(\alpha) \approx 1.6$. Using Eq. (9.37) to estimate productivity per unit volume and Eq. (9.38) to estimate the eluent consumption, the result is that $\phi_v(\alpha)/\phi_v(\beta) \approx 3.7$ and $\Upsilon_*(\alpha)/\Upsilon_*(\beta) \approx 2.3$. Note that these productivities and eluent consumptions are referred to the production of HFCS 55 and HFCS 90, respectively. If one expresses productivities relative to the quantities of fructose in the feed for both SMB systems, one obtains $\phi_v^{purefru}(\alpha)/\phi_v^{purefru}(\beta) \approx 3.7 \times 0.55/0.9 \approx 2.26$.

On the basis of the quantity of pure fructose to be processed, the SMB of process α is expected to be 2.26 times less productive than the SMB of process β. On the other hand, we have shown that the quantity of pure fructose to be processed is $611/270 \approx 2.27$ times greater for process α than for process β. With this simple shortcut, we have thus shown that the two SMB systems must be of similar size. With the level of precision of our approach, the only conclusion we can draw is that over-purifying and mixing with fresh feed is a route worth considering.

Process α is associated with one technical difficulty: operating with a quasi-zero flow rate in zone 2 requires some very significant velocity variations during one shift. With the above m_i values, it turns out that the flow rate must vary from 1 to $0.51/0.075 = 6.8$ when a column moves from zone 2 to zone 1. Practical implementation is thus associated with technical difficulties, and the smoother operation of process β is preferable.

Another industrial consideration is associated with process stability. The fructose composition of the HFCS 42 in fact varies from 40 to 44% over time. Operating process α at its maximum capacity would thus require a continuous tuning of the internal flow rate that is difficult in practice. On the other hand, should a variation in feed composition impact the outlet fructose purity of process β, this could immediately be compensated by adjusting the blending ratio in order to keep the purity of HFCS 55 adequate.

The two preceding comments, together with the fact that productivities of the two systems are in principle equivalent, explain why process β is preferred industrially.

Should we want more precision and a better comparison between the two processes, we could simply use the methodology presented several times (select η_F^{eq} not far from 1), and perform simulations with the MC–$\overline{\text{Equil}}$ or, better, the MC–$\overline{\text{LDF}}$ model.

10.3.3 Process design: elimination of radioactive pollutants

Another illustration of how the environment can qualitatively impact the design of a chromatographic system is given by the Nymphea process. The information in this section is taken largely from (Pierlas *et al.*, 1986), and the problem is briefly presented below. In the 1980s, the French authorities decided to build a facility which could store and process up 1600 t/year of radioactive waste (from nuclear power auxiliary units). This radioactive waste, stored in pools, produces a certain flux of contaminants, including radioactive cesium 137 and cobalt 60. We will focus on cobalt 60, which is the predominant species in terms of radioactivity. Notice that the resin cannot distinguish between isotopes, so the total cobalt flux matters, and we define ϕ_s^{Co}, the flux (eq/s) of cobalt ions released by the waste (sources). The cobalt must be continuously captured by ion exchangers (the Nympheas) in order to maintain the radioactivity level of the water in the pools below a certain threshold. In addition, let us define ϕ_s^{Na} and ϕ_s^{Ca}, the fluxes (eq/s) of sodium and calcium ions released by the waste and by the water added to compensate for evaporation. Na and Ca ions are not radioactive, but we will see that their fluxes play a key role. Note that, for neutrality reasons, there must be a flux of anions, identical to the sum of the cations fluxes (in eq/s). The concentration of cobalt 60 must be maintained below $C_{Co}^{max} = 1.5 \times 10^{-11}$ eq/l. The water in the pools is demineralized, its normality being typically about 10^{-6} eq/l. The radioactive species are thus at trace levels compared to non-radioactive cations like Ca and Na. The reader will find some underlying information on the modeling and simulation of the H/Na/Co/Ca ion exchange in (Nicoud and Schweich, 1989).

The radioactive contaminants (and other ions) are captured by ion exchangers immersed in the pools. A schematic of the system is given in Figure 10.8. Immersion of the system (Figure 10.8a) avoids the use of external piping and thus possible leakages; for modeling purposes, however, it is equivalent to consider that the Nympheas and the pool are inserted in a recycling loop (Figure 10.8b). The ion exchangers are filled with mixed beds composed of cationic resins initially in H^+ and OH^- forms respectively. Adding anion exchange to cation exchange is important: if anions were not captured simultaneously with cations, the pH of the water would decrease continuously and corrosion would take place. The fact that the different resins are in mixed or separated beds is relatively secondary.

The problem is to design an ion exchanger that can continuously fix the radioactive cobalt ions. Important questions have been raised: what is the lifetime of these ion exchangers and can it be maximized? Can an optimal flow rate for a given ion exchanger volume be defined, and thus an optimal design? Can a possible abrupt increase in water radioactivity be anticipated?

The affinity sequence of the species for the cationic resins is H<Na<Co<Ca. In the absence of a specific selectivity of the ion-exchange resins towards Co, and because of the very low concentration of Co with respect to Na and Ca, the cationic resins mainly capture the predominant ions Na and Ca.

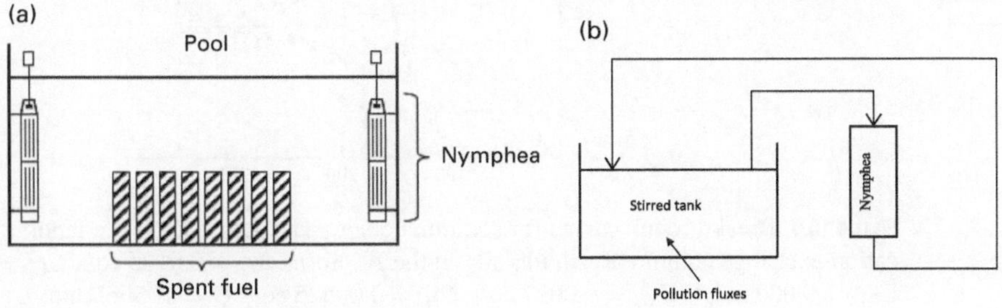

Figure 10.8 Schematic of the pool (a) and model (b).

Three different periods can be recognized during the operation of one Nymphea:

1 First, a short, transient stabilization period (a few hours) due to the hydrodynamic time constant of the system, on the order of $\frac{V_{pool}}{QN_{ny}}$, where V_{pool} is the volume of the pool (about 10 000 m^3), N_{ny} is the number of Nympheas in the pool (normally six) and Q is the flow rate of water processed per Nymphea (about 400 m^3/h).

2 Second, a long "quasi-steady state" (a few months) during which the flux of ions released by the waste equals the flux of ions captured by the ion exchanger. Ion concentrations in the water are almost constant, but as far as cations are concerned, Na, Ca, Co (etc.) concentrations in the resin continuously increase, while the concentration of H in the resins continuously decreases.

3 Finally, an increase in the cobalt concentration in the pool up to the maximum level, C_{Co}^{max} (eq/l), due to progressive resin saturation. This last period is never experienced.

Ion exchange capacity and selectivity coefficients have been determined by the adsorption/desorption method (Section 4.8.2.1): $\overline{N}^{iex} = 2.1$ eq/l, $K_{Na/H} = 1.8$, $K_{Co/H} = 0.8$, $K_{Ca/H} = 1.4$. Be aware that the last two selectivity coefficients are associated with heterovalent ion exchange; we have seen in Section 4.5 that this implies the affinity of the resin for the divalent ions with respect to H to increase when normality decreases. Despite the fact that $K_{Co/H} \leq 1$, under the normality conditions prevailing in the Nymphea environment, Co is preferred by the resins to H and Na (but not to Ca). At low normality, the affinity sequence being H<Na<Co<Ca, the breakthrough curves are expected to exhibit first a front of sodium, then a front of cobalt, then a front of calcium. This is confirmed in Figure 10.9.

Under the operating conditions of the Nymphea, external mass transfer is limiting. The characteristic times t_i^e for external mass transfer have been determined with the shallow-bed technique presented in Section 5.10.3 and have been found to be about 1.2 s.

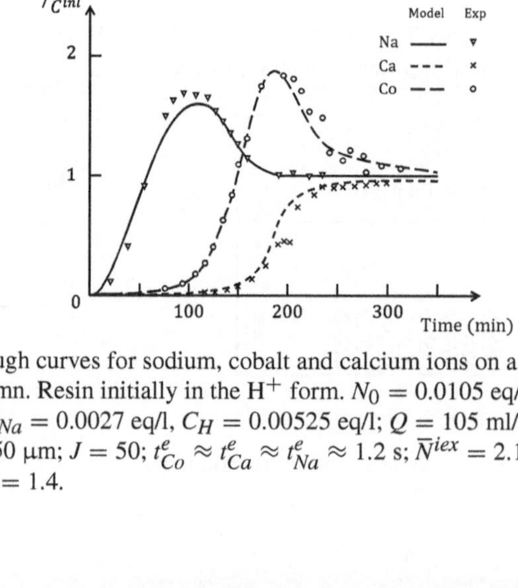

Figure 10.9 Breakthrough curves for sodium, cobalt and calcium ions on a laboratory cation-exchange column. Resin initially in the H^+ form. $N_0 = 0.0105$ eq/l, $C_{Ca} = 0.002$ eq/l, $C_{Co} = 0.0005$ eq/l, $C_{Na} = 0.0027$ eq/l, $C_H = 0.00525$ eq/l; $Q = 105$ ml/min; $L_{col} = 30$ cm, $D_{col} = 1$ cm; $d_p = 550$ µm; $J = 50$; $t_{Co}^e \approx t_{Ca}^e \approx t_{Na}^e \approx 1.2$ s; $\overline{N}^{iex} = 2.1\ eq/l$, $K_{Na/H} = 1.8$, $K_{Co/H} = 0.8$, $K_{Ca/H} = 1.4$.

The breakthrough curves have been simulated using the MC–$\overline{\text{LDF}}_{\text{E}}$ model. Following the presentation given in Section 5.5.2, the mass balance equations associated with the MC–$\overline{\text{LDF}}_{\text{E}}$ model on the cationic resin are:

$$C_i^{k-1} = C_i^k + \frac{t_0}{J}\frac{C_i^k}{dt} + \frac{t_0}{J}\frac{1-\varepsilon_e}{\varepsilon_e}\frac{\overline{C}_i^k}{dt}$$

$$\frac{d\overline{C}_i^k}{dt} = \frac{C_i^k - C_i^{k*}}{t_e^i} \qquad i = \text{Co, Na, Ca}; i \neq \text{H} \tag{10.6}$$

C_i^{k*} in equilibrium with the lumped solid-phase concentrations

The agreement with experimental results is satisfactory, as shown in Figure 10.9.

This model can then be used to simulate a Nymphea immersed in a pool which is assumed to be perfectly mixed and fed with a prescribed and constant flux of cations. For modeling purposes, the pool is assumed to be in a recycling loop, as shown in Figure 10.8b.

The average concentration C_i^{pool} of ion i in the pool equals the concentration C_i^{inl} entering the Nymphea. The outlet concentrations C_i^{out} exiting the Nymphea being the concentrations entering the pool, the mass balances for the different species i are written:

$$QC_i^{out} = QC_i^{inl} + \frac{V_{pool}}{N_{ny}}\frac{dC_i^{inl}}{dt} - \frac{\phi_s^i}{N_{ny}} \qquad C_i^{inl} = C_i^{pool} \tag{10.7}$$

This mass balance, coupled with a model of the Nymphea (see Eq. (10.6)), allows a calculation of the transient evolution of the different concentrations. As a first approximation, one can get rid of the simulation of the anion exchangers for reasons explained below.

The pool being in contact with ambient air, its pH is largely imposed by the dissolution equilibrium of carbon dioxide in water. Electroneutrality requires:

$$N_0 = \sum_i C_i^{inl} + H^{inl} = OH^{inl} + \sum_i A_i^{inl} + HCO_3^- + CO_3^{2-} \quad i \neq \text{OH}, \text{HCO}_3^-, \text{CO}_3^{2-} \tag{10.8}$$

We know that the dissolution equilibrium of carbon dioxide in demineralized water requires pH $\approx$ 5.5 and a predominance of HCO_3^- and CO_3^{2-} anions. This approximation is obviously valid as long as the ion exchangers capture enough ions for the water to be "demineralized". Assuming pH $\approx$ 5.5 allows a determination of all the cation concentrations at the inlet of the Nymphea (identical to the pool concentrations), and thus simulation of its behavior independently of anion exchange.[7]

In the Nymphea situation, requiring equilibrium with the atmosphere is mandatory. This is rarely taken into account when chromatography processes are modeled, but what happens to water-rich eluents stored in tanks not flushed with nitrogen?. Certainly the phenomenological approach presented in Section 4.7 finds another interesting application here.

Let us start by studying the "quasi-steady state" mentioned above. Under this assumption, Eq. (10.7) becomes:

[7] In fact, in a mixed bed, water formation impacts the front shape. This is a second-order phenomenon for our purposes.

$$\phi_s^i = N_{ny}Q\left(C_i^{inl} - C_i^{out}\right) \qquad C_i^{inl} = C_i^{pool}$$

At steady state, assuming uniform concentrations in the pool, the flux of cations captured by the Nymphea, $N_{ny}Q\left(C_i^{inl} - C_i^{out}\right)$, simply equals the pollution flux. If one defines the leakage F_i as the ratio between the outlet and inlet concentrations in the Nymphea, the steady-state concentration of ion i in the pool is given by:

$$F_i = \frac{C_i^{out}}{C_i^{inl}} \qquad i = \text{Co, Na, Ca; } i \neq \text{H}$$
$$C_i^{pool} = \frac{\phi_s^i}{N_{ny}Q\left(1 - F_i\right)} \tag{10.9}$$

Let us focus on the cobalt ion (i = Co). On the one hand, increasing the flow rate increases the flux of cobalt seen by the ion exchanger; on the other hand, increasing the flow rate increases the leakage F_{Co} due to increasing kinetic limitations. Intuitively, if the flow rate tends towards infinity, the mass transfer limitations become very strong, so the leakage F_{Co} tends towards 1. The ingredients for the presence of an optimum are thus met.

We have seen in Section 5.10.3 that, for a strong affinity of the species for the chromatographic medium, and in the presence of external mass transfer limitations, after a few t_0, the breakthrough curve reaches a concentration plateau such that:

$$F_{Co} = \frac{C_{Co}^{plat}}{C_{Co}^{inl}} = \frac{1}{\left[1 + \frac{1}{J}\frac{1 - \varepsilon_e}{\varepsilon_e}\frac{t_0}{t_{Co}^e}\right]^J} \tag{10.10}$$

The underlying assumptions in Section 5.10.3 for establishing Equation (10.10) are satisfied: the concentrations are very low so the affinity of the H^+ form resin for the divalent cobalt cation is extremely high ($\overline{K}_{Co} \approx K_{Co}^H \overline{N}^{iex}/N_0 \approx 2 \times 10^6$); low concentrations and large affinity clearly drives towards external mass transfer limitations for ion exchange.

Equation (10.10) is valid for a bed composed of cationic resins. In the case of mixed beds containing a volumetric fraction f_{cat} of cationic resins, Eq. (10.10) simply becomes:

$$F_{Co} = \frac{C_{Co}^{plat}}{C_{Co}^{inl}} = \frac{1}{\left[1 + \frac{1}{J}\frac{1 - \varepsilon_e}{\varepsilon_e}\frac{f_{cat}t_0}{t_{Co}^e}\right]^J} \tag{10.11}$$

From the presentation in Section 5.1, we know that the characteristic time for external diffusion must vary with fluid velocity via the Sherwood number. Being interested in significant fluid velocities, we will assume that $Sh \gg 2$, so that the characteristic time for external mass transfer can be written:

$$t_{Co}^e \sim Re^{-1/2} \sim u_e^{-1/2}$$

Let us write $t_{Co}^e = a_e u_e^{-1/2}$. From results published by (Nicoud and Schweich, 1989), one obtains $a_e \sim 0.03\ \text{m}^{1/2}\text{s}^{1/2}$. Noting that $t_0 = L_{col}/u_e$ and $Q = \varepsilon_e u_e \Omega_{col}$, Eq. (10.9) and (10.11) can be rewritten:

$$C_{Co}^{pool} = \frac{\phi_s^{Co}}{N_{ny}\varepsilon_e\Omega_{col}u_e} \frac{1}{1 - \frac{1}{\left[1 + \frac{1}{J}\frac{1-\varepsilon_e}{\varepsilon_e}\frac{f_{cat}L_{col}}{a_e}u_e^{-1/2}\right]^J}} \tag{10.12}$$

From Eq. (10.12) it is apparent that, for a given geometry of the column defined by Ω_{col} and L_{col}, the stationary concentration of cobalt in the pool decreases continuously with fluid velocity, thus with the flow rate increase. There is thus no optimum: the benefit of processing a larger flow rate is greater than the drawback associated with the increase in leakage F_{Co} due to increased mass transfer limitations. The influence of the number of plates is insignificant provided that J is at least about 5 or 10, which is very easily obtained in practice. The consequence is that the ion exchanger design must allow maximization of the flow rate under minimal pressure drop. This is why an annular geometry has been selected for the Nymphea. The bed height is about 4.5 m, the external diameter 0.75 m and internal diameter 0.15 cm, so total bed volume is about 1.8 m^3. The flow rate being 400 m^3/h, the Z_eRT is only about 5 s. A very strange animal, operated successfully for about 30 years!

Let us now address the transient regime of the system composed of the pool and the Nympheas. This requires solving the set of ODEs presented in Eqs. (10.6) and (10.7). Note that a complete simulation requires simulating anion exchange in the mixed bed. There are no specific difficulties, except that the system of ODEs becomes stiff, so again reading the discussion of numerical methods in Chapter 7 may be worthwhile.

The system must be stopped prior to any leakage of cobalt from the ion exchanger. Fortunately for process efficiency, but unfortunately for process monitoring, the cobalt concentration in the pool is not sensitive to the degree of saturation of the resins: it remains almost constant until complete saturation, and then increases very quickly. The simulated curves in Figure 10.10 show that the evolution of the sodium concentration, which increases continuously, is an excellent indicator of what is going to happen with the expected leakage of cobalt. Additionally, note the initial leakage F_{Co}, which is about 5–10% under the conditions of simulation of Figure 10.10, due to significant mass transfer limitations.

Finally, let us address the influence of the flow rate on the lifetime of the Nympheas, thus on the evolution of the cobalt concentration in the pool. We see in Figure 10.11 that the lifetime of the ion exchangers is independent of the water flow rate: it is primarily imposed by the flux ϕ_s^{Ca} of calcium ions in water, which is the predominant species for resin saturation. Operating an ion exchanger whose lifetime is independent of the water flow rate is rare enough to be worth mentioning.

10.3.4 Conclusion

The production of HFCS 55 and the elimination of radioactive pollutants are processes having, a priori, very little in common, although the two processes use ion-exchange resins. They are presented in the same section because they perfectly illustrate how

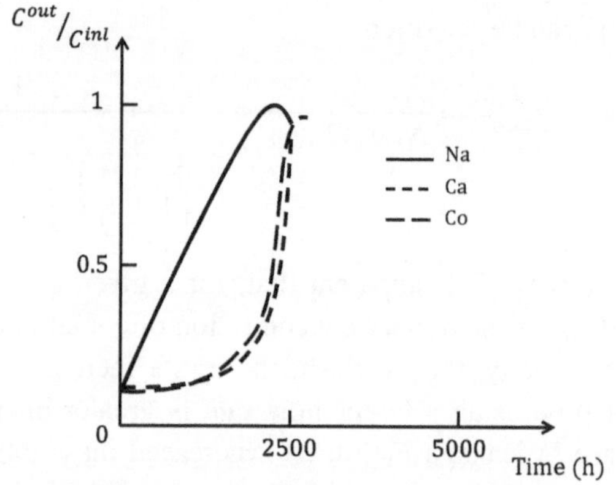

Figure 10.10 Simulated evolution over time of the ratio of ion concentrations exiting and entering a Nymphea. $\phi_s^{Co} = 9.33\times10^{-7}$ eq/s, $\phi_s^{Ca} = 1.1\times 10^{-4}$ eq/s, $\phi_s^{Na} = 6.0\times10^{-5}$ eq/s; $V_{col} = 1.8$ m^3 containing 40% cation exchange resins and 60% anion exchange resins; $V_{pool} = 10\,000$ m^3, $Q = 400$ m^3/h, $J = 50$, $N_{ny} = 4$; $t^e_{Co} \approx t^e_{Ca} \approx t^e_{Na} \approx 1.5$ s. Selectivity coefficients are given in Figure 10.9.

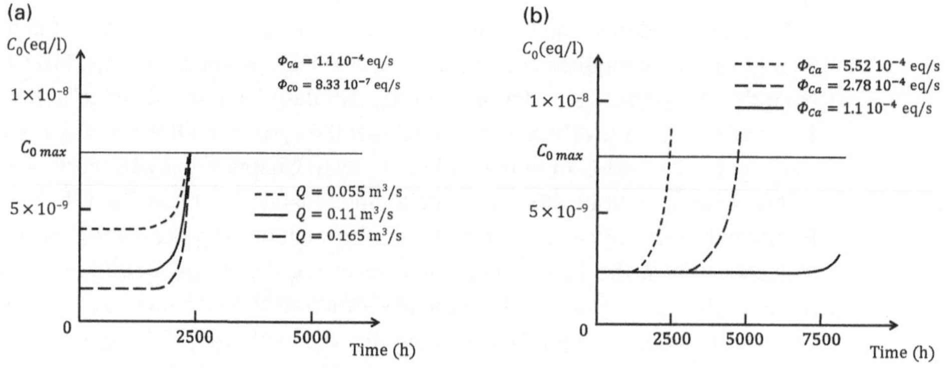

Figure 10.11 Influence of the flow rate passing into one Nymphea and of the calcium flux on the evolution of the cobalt concentration in the pool. (a) Influence of the water flow rate; parameters identical to those given in Figure 10.10 except for the water flow rate. (b) Influence of the calcium flux; parameters identical to those given in Figure 10.10 except for the calcium flux.

the constraints of the environment in which the chromatographic system is placed can dramatically impact the process design:

- The production of HFCS 55 illustrates that the inlet and outlet purities of a chromatographic process can sometimes be tuning instead of constrained parameters. In this specific situation, this has led industry to design processes in which the target product is over-purified and then blended with fresh feed. For a more scientific approach to the concept of over-purifying and blending, the reader is referred to the work of (Siitonen *et al.*, 2012) for elution chromatography.

- Because of its integration in a loop, the design of the Nymphea leads to a system with extremely strange characteristics: very short bed length, significant cross section, very low Z_eRT, significant leakage due to kinetic limitations and a lifetime independent of the flow rate.

10.4 Complex mixtures

Many industrial separations involve complex multi-component mixtures, and the spectrum of situations is extremely wide. To illustrate the matter, one could take one or two examples, such as the purification of PUFAs, cyclosporine, paclitaxel, insulin or Mabs, estimate as precisely as possible the fluid-phase equilibria and mass transfer characteristics and propose more or less detailed process simulation results. We might obtain precise information on a very specific problem, but the spectrum of situations is so wide that the probability for it to be close to the reader's needs is very small. We are now at the end of this book, so the reader should know at least as much as the author in terms of modeling and simulation. Consequently, I will simply offer "food for thought", and let readers adapt it to their own problems and perform the detailed work.

In the development of a multi-component separation process, two questions systematically cross my mind:

- When solutes associated with very different retention times are present, should we consider a gradient operation or not?
- For a given chromatographic medium and carrier fluid, should we try to perform the chromatographic work in one step or proceed in two or more steps?

These two questions are addressed in the following sections.

10.4.1 Gradient operations

Purification of biomolecules is often very complex because the mixtures typically contain a large number of different solutes associated with a wide spectrum of properties. Some solutes may be more or less excluded from the chromatographic medium, others may have retention times similar to the solute of interest, while a final category could have retention times which are dramatically greater. Such a separation is often performed in a gradient mode, so the eluent strength is modulated in order to get decent selectivity between the target solute and the closely eluted solutes, and rapid elution of the strongly retained solutes. The gradient profile is normally either linear or in steps. The reader will find a presentation of gradient elution chromatography in (Carta and Jungbauer, 2011). It is worth mentioning that, in elution chromatography, two types of gradients exist: a temporal gradient dictated by the time variation of the eluent composition at the inlet of the column, and an axial gradient that results from the former when the carrier fluid is traveling through the column. The latter gradient induces peak compression effects. In TMB chromatography, only the spatial gradient prevails.

How can our modeling methodology be adapted to simulating gradient operations? It is at the same time straightforward in principle and possibly difficult: straightforward because there is no difference in principle from the isocratic modeling of multi-solute systems, except that the boundary conditions at the inlet of the column must be changed. Whereas isocratic systems are often modeled by considering the solvent to be in large excess and keeping this "phantom" species away from the mass balances, modeling gradient operations requires at least identifying and modeling the behavior of one of the constituents of the carrier fluid. Let us call this component the *modifier*. The propagation of the modifier in the column is described by writing appropriate mass balances together with a description of its interaction with other solutes and with the chromatographic medium.

It is possibly difficult because the interactions and competition between the solutes to be separated and the modifier must be properly described. Different situations can be distinguished:

- The gradient is a salt gradient or a pH gradient associated with operations on ion-exchange resins. In this situation, if the system is described by the phenomenological models in Section 4.7, modeling is not difficult. The system behavior and the influence of the composition gradient are directly determined from knowledge of the mass action laws.
- The gradient results from the variation of the modifier concentration in adsorption chromatography. The different solutes, as well as the modifier, are adsorbed on the chromatographic medium skeleton, and compete for available adsorption sites. This situation can be modeled with multi-solute adsorption laws, possibly more or less similar to Langmuir adsorption isotherms. The influence of the modifier's concentration on retention is a consequence of competition between modifier and solutes for available adsorption sites.
- The gradient is associated with a change in modifier composition, for example in reversed-phase chromatography. The mechanism of retention is not based only on competition for adsorption sites (4.3.3), so Langmuir-type expressions are likely to be inadequate to describe the influence of the modifier concentration on retention. Very often, one postulates a variation of the lumped Henry's coefficient $\overline{K}_i$ versus modifier concentration according to a power or exponential function of the eluent composition, as given for instance in (Abel *et al.*, 2002), (Silva *et al.*, 2010) and (Carta and Jungbauer, 2011). Adsorption isotherms are then represented by more or less empirical formulas based on this variation of Henry's coefficient with composition (see, for instance, (Abel *et al.*, 2004)).

Assuming the solutes and modifier adsorption isotherms to be known, the system is simulated by solving one mass balance per solute and one for the modifier.

For process design, as for isocratic conditions, using the equilibrium model prediction as a reference is extremely helpful. Because analytical expressions are unlikely to be available for gradient operations, don't hesitate to use simulations performed on highly efficient columns with a Rouchon type of algorithm (see Section 7.9.3.1) to get this reference.

In general, gradient operations are unlikely to have a significant impact on mass transfer and hydrodynamics. An exception may be associated with possible swelling/shrinking of the particles with composition.

Finally let us mention possible thermal effects due to the mixing of solvents. This is especially the case with alcohol/water and acetonitrile/water mixtures: thermal effects must in general be strictly compensated by adequate system design, ensuring isothermal operation, unless one decides to exploit this thermal effect in some creative processes.

10.4.2 Flow reversal

In general, gradient operations have two drawbacks:

- They make solvent recycling more costly, more difficult, possibly impossible. To be a bit provocative, I used to say that, irrespective of the chromatographic separation, the optimum acetonitrile/water composition was 84/16, as it corresponds to the position of the azeotrope, making the solvent recycling very easy.
- They require a regeneration step to go from the end of the gradient composition to its initial state. The modifier being normally strongly adsorbed, the associated regeneration front is strongly dispersive in general. An exception can be found with ion exchange when taking advantage of dissociation equilibria.

The cycle time, defined as the duration between two successive injections, is the sum of the duration of the chromatogram itself and of the regeneration step, and the second could be higher than the first.

As mentioned in Section 9.2, there is another simple way to handle the elution of very strongly retained products in the column: instead of using a gradient, simply reverse the flow. Let us consider a solute mixture containing solutes relatively close to the target solute (*) and a very strongly retained impurity, referred to as *str*, as depicted in Figure 10.12. The product of interest is collected between times t_{c1} and t_{c2}. For simplicity, we assume the presence of only one strongly retained solute. In reality, many solutes can be located between * and *str*. In isocratic elution chromatography, the cycle time is pretty close to t_R^{str}, the retention time of solute *str*; we write $t_{cycle}^{Iso} \approx t_R^{str}$.

In Figure 10.12, we also present the case of gradient chromatography. Because of the increase in eluent strength, solute *str* elutes earlier, and the total cycle time must include the time required for regenerating the column t_{reg} in addition to the duration t_{ch} of the chromatogram itself, so $t_{cycle}^{Gr} = t_{ch} + t_{reg}$.

A last option is also depicted in Figure 10.12. Immediately after collection time t_{c2}, the flow is reversed. The solutes are collected at the extremity, which was initially the inlet of the column. Let us define t_{back}, the time required for the complete solute elution after flow reversal. The cycle time is given by $t_{cycle}^{FR} = t_{c2} + t_{back}$.

It is intuitive that, in general, a gradient mode can only be interesting if t_{cycle}^{Gr} is significantly lower than t_{cycle}^{Iso} and/or t_{cycle}^{FR}; otherwise one would have the drawbacks of gradient without the benefits of short cycle times.

The cycle time t_{cycle}^{Gr} is a complex function of the gradient profile and the regeneration time. It can be determined and possibly optimized without specific difficulties, but there

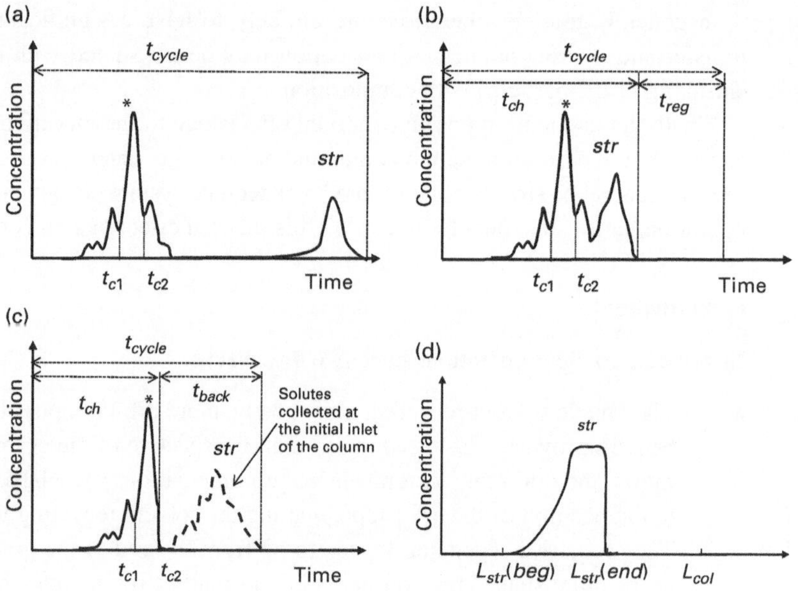

Figure 10.12 Chromatograms associated with isocratic and gradient-elution modes. Chromatograms and internal concentration profile obtained at t_{c2} for the flow-reversal mode. (a) Isocratic chromatogram. (b) Gradient chromatogram. (c) Flow-reversal chromatogram; flow is reversed at time t_{c2}. (d) Flow reversal: concentration profile of solute *str* at time t_{c2}.

is little to say other than that these are typically of the order of an hour in industrial applications. By assumption, the feed mixture contains very strongly retained solutes, so the condition $t^{Gr}_{cycle} \ll t^{Iso}_{cycle}$ is normally met without difficulty (otherwise a gradient would not be considered).

How can we estimate t^{FR}_{cycle}? For the sake of simplicity, and for illustration purposes, we assume that solute *str* is the only one present inside the column after solute *, that it follows a non-competitive Langmuir adsorption isotherm described with parameters $\overline{N}$ and $\tilde{K}_{str}$ so that the lumped Henry's coefficient is given by $\overline{K}_{str} = \tilde{K}_{str}\overline{N}$, and that the injection time is very small compared to the retention time of solute *str*.

Let us first consider the situation of isocratic chromatography. In terms of the equilibrium model, the time $t_R(str)$ at which solute *str* would completely elute from the column is given by the end of the dispersive part of the peak: $t_R(str) = \Delta t_{inj} + t_0\left[1 + F_e\overline{K}_{str}\right]$, so $t_R(str) \approx t_0\left[1 + F_e\overline{K}_{str}\right]$ because the injection time is likely to be very small compared to $t_R(str)$. In practical situations, $\overline{K}_{str}$ can easily be 100 or more, so $t_R(str)$ can be very significant. The cut time t_{c2} is determined with the methodology followed for the classical (without reversed flow) isocratic mode.

In order to determine t_{back}, the time required to complete the solute elution after flow reversal, let us consider the profiles occurring in the column at time t_{c2} and given in Figure 10.12. At time t_{c2}, solute *str* is located between $L_{str}(beg)$ and $L_{str}(end)$ from the inlet of the column. $L_{str}(end)$ is given by the position of the compressive part of the peak, so $L_{str}(end) = \frac{t_{c2}u_e}{1+F_e\frac{C^{max}_{str}}{C^{max}_{str}}}$, and C^{max}_{str}, the concentration at the maximum of the peak, can be calculated with the methodology in Chapter 3. $L_{str}(beg)$ is given by the position of the rear of the dispersive part of the peak, which, neglecting the feed injection time, is $L_{str}(beg) = \frac{t_{c2}u_e}{1+F_e\overline{K}_{str}}$. If the adsorption constant is very high, the distance traveled by str is much shorter than the column length L_{col}. This means that it may be efficient to force the solute to travel back (a short distance) and to exit at the initial inlet of the column.

Let us now see what happens after flow reversal. The downstream and sharp part of the str peak (near $L_{str}(end)$) will generate a dispersive front, while the upstream and dispersed part of the peak (near $L_{str}(beg)$) will be "compressed". An important effect of flow reversal is thus to reverse the respective positioning of dispersive and compressive parts of the peaks. Using the development in Chapter 3, the reader has the means, at least for the single-solute situation, to calculate the shape of the peak collected at the initial inlet of the column after flow reversal. For our purposes, we will simply admit that the rear part of this peak is given by the rear part of the dispersive front generated by the elution of the front initially located at $L_{str}(end)$. This is associated with time $t_{back} = \frac{L_{str}(end)}{u_e}\left[1+F_e\overline{N}\overline{K}_{str}\right]$.

Using the expression for $L_{str}(end)$, one obtains:

$$t_{cycle}^{FR} = t_{c2} + t_{back} = t_{c2} + t_{c2}\frac{1+F_e\overline{K}_{str}}{1+F_e\frac{\overline{C}_{str}^{max}}{C_{str}^{max}}}$$

In the case of a linear adsorption isotherm, this expression shows that $t_{cycle}^{FR} = t_{c2}+t_{c2} = 2t_{c2}$: this is pretty obvious as the fronts travel at the same velocity in one direction as in the other. Under the assumption of linear behavior, the time t_{back} is independent of Henry's coefficient $\overline{K}_{str}$: flow reversal is thus a great way to decrease cycle time in the presence of strongly adsorbed solutes. In the case of a non-linear favorable adsorption isotherm, things are less favorable; t_{cycle}^{FR} is greater than $2t_{c2}$, depending on the ratio between the chord of the adsorption isotherm at concentration C_{str}^{max} and its initial slope. The importance of flow reversal thus depends on the structure of the adsorption isotherm and feed concentration. Fine tuning can easily be added to our simplistic approach, but the bottom line is that flow reversal is potentially a great way to dramatically decrease cycle time without using a gradient.

I am convinced that flow-reversal schemes can advantageously replace gradient operations in many cases. As mentioned by (Carta and Jungbauer, 2011), however, notice that gradients can have an advantage in situations where protein retention is highly sensitive to composition, so highly that operating at constant solvent composition would require a very high degree of eluent composition precision.

10.4.3 Multi-step strategies

It is common practice to link chromatographic steps of different natures in series for the purification of complex mixtures. An illustration is the use of affinity chromatography followed by ion exchange and reversed-phase chromatography for purifying proteins.

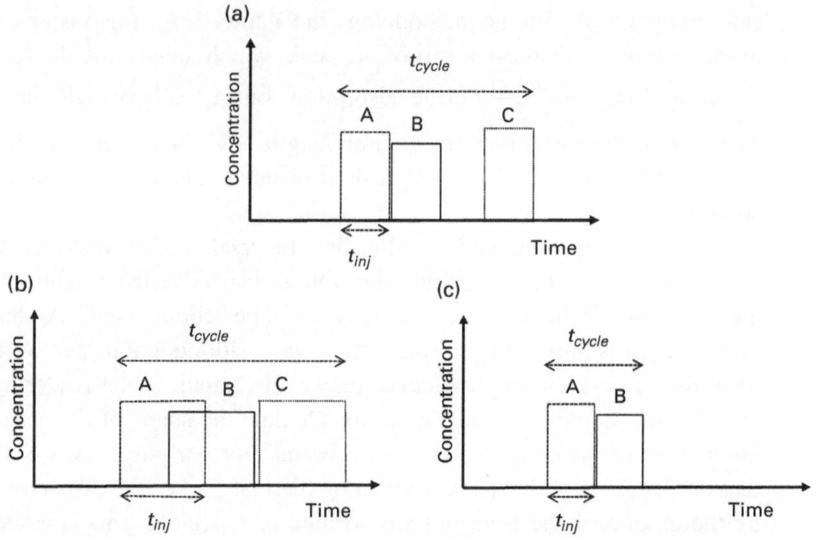

Figure 10.13 Handling a ternary separation in one step or in two steps (linear system). (a) One-step process (process α): equilibrium model solution for the touching-band chromatogram. (b) Two-step process (process β): first step; equilibrium model solution for the touching-band chromatogram. (c) Two-step process (process β): second step; equilibrium model solution for the touching-band chromatogram.

Alternatively, some processes link chromatographic steps of the same nature in series. For instance, we have seen in Section 9.2 that a good way to isolate one solute from a complex mixture is to use two SMBs in series. A systematic study has been carried out by (Kim and Wankat, 2004) for quaternary mixtures. Associating different systems in series is probably mandatory for multi-solute separation via SMB, but what prevents us from doing the same for elution chromatography? Consider the schematic chromatogram in Figure 10.13: two solutes, A and B, are closely eluted, and a third one, C, is eluted later and fairly well separated from the others. The mass fractions of the solutes in the feed are X_A^F, X_B^F and X_C^F. The target product is B. The difference in retention times between B and C is significant, but probably not enough for the gradient method to be worth considering.

Let us compare two simple strategies. According to the first strategy (process α), the injection and cycle times are set in such a way that the target product, B, is obtained at the required purity in one step (Figure 10.13a). This is the classical elution process.

Another option, process β, is a two-step strategy: in a first step (Figure 10.13b), the injection is set so as to obtain, on the one hand, a fraction (A+B) and, on the other hand, a pure fraction C. The injection time associated with this first step is imposed by the selectivity between B and C, which is assumed to be relatively large, so this step is expected to be very productive. The (A+B) fraction is then processed in a second step (after possible concentration) to get pure A and pure B (Figure 10.13c). The two processes can easily be compared by process simulation, but straightforward calculations allow a simple comparison for linear systems.

We have seen in Chapter 9 that the concentration $\chi_F^{eq} = \frac{\Delta t_{inj}}{t_{cycle}} C_{tot}^F$ plays a very important role in determining productivity and eluent consumption. For process α, one has:

$$\Delta\, t_{inj} = t_0 \frac{1-\varepsilon_e}{\varepsilon_e}\left(\overline{K}_B - \overline{K}_A\right) \qquad t_{cycle} = t_0 \frac{1-\varepsilon_e}{\varepsilon_e}\left(\overline{K}_B + \overline{K}_C - 2\overline{K}_A\right)$$

so $\chi_F^{eq}(\alpha) = \dfrac{\overline{K}_B - \overline{K}_A}{\overline{K}_B + \overline{K}_C - 2\overline{K}_A} C_{tot}^F$.

For the first step of process β, one has:

$$\Delta\, t_{inj} = t_0 \frac{1-\varepsilon_e}{\varepsilon_e}\left(\overline{K}_C - \overline{K}_B\right) \qquad t_{cycle} = t_0 \frac{1-\varepsilon_e}{\varepsilon_e}\left(2\overline{K}_C - \overline{K}_A - \overline{K}_B\right)$$

so $\chi_F^{eq}(\beta_1) = \dfrac{\overline{K}_C - \overline{K}_B}{2\overline{K}_C - \overline{K}_A - \overline{K}_B} C_{tot}^F$.

For the second step of process β, one has:

$$\Delta\, t_{inj} = t_0 \frac{1-\varepsilon_e}{\varepsilon_e}\left(\overline{K}_B - \overline{K}_A\right) \qquad t_{cycle} = 2t_0 \frac{1-\varepsilon_e}{\varepsilon_e}\left(\overline{K}_B - \overline{K}_A\right)$$

so $\chi_F^{eq}(\beta_2) = \frac{1}{2} C_{tot}^{S2}$, where C_{tot}^{S2} is the concentration of fraction (A+B) collected after step 1 and sent to step 2. One immediately obtains $C_{tot}^{S2} = (X_A^F + X_B^F) C_{tot}^F$.

In order to compare these two strategies, we will assume 100% yield and purities, and $\eta_F^{eq} = 1$, and we will focus on the eluent consumption, which is given, according to Eq. (9.38), by:

$$\Upsilon_* = \frac{1}{\chi_F^{eq}} \frac{1}{X_*^F} \quad (\text{m}^3/\text{kg})$$

where X_*^F is the purity of the fraction in the feed material and to be purified on the considered step (solute B for process α, mixture (A+B) for process β step 1 and solute B for process β step 2). The specific eluent consumption is expressed in m^3 per mass of purified fraction (i.e. solute B for process α, mixture (A+B) for process β step 1 and solute B for process β step 2). These specific eluent consumptions will be called $\Upsilon_*(\alpha)$ for process α, and $\Upsilon_*(\beta_1)$ and $\Upsilon_*(\beta_2)$ respectively for the two steps of process β.

The eluent consumption can be expressed in terms of the purified quantity of solute B, which is the target solute. These eluent consumptions will be called Υ_*^B and are given for process α:

$$\Upsilon_*^B(\alpha) = \Upsilon_*(\alpha) = \frac{1}{\chi_F^{eq}(\alpha)} \frac{1}{X_B^F} \quad (\text{m}^3/\text{kg})$$

and, after a simple mass balance, for process β:

$$\Upsilon_*^B(\beta) = \Upsilon_*(\beta_1) \frac{X_A^F + X_B^F}{X_B^F} + \Upsilon_*(\beta_2) = \frac{1}{X_B^F}\left[\frac{1}{\chi_F^{eq}(\beta_1)} + \frac{1}{\chi_F^{eq}(\beta_2)}\right] \quad (\text{m}^3/\text{kg})$$

These two equations allow a comparison of eluent consumptions associated with the two processes on the basis of the equilibrium model.

Depending on the respective positioning of the $\overline{K}_i$ coefficients, one process can be better than the other. For instance, process β is favored when $\overline{K}_C$ is sufficiently greater than $\overline{K}_A$ and $\overline{K}_B$. This very simple approach is certainly not sufficient to make a final choice, but it allows the identification of situations of possible interest. The MC–$\overline{\text{Equil}}$ model is in general sufficient for an adequate comparison of the two options.

10.5 Conclusion

With selected illustrations, we have showed how simplified assumptions and more or less idealistic models allow the identification and explanation of important features of industrial processes. For all these illustrations, the first nine chapters of this book provide many tools for refining the preliminary design, taking into account, for instance, mass transfer limitations and specific hydrodynamic situations.

Many other topics can be addressed in the same way. To mention a few:

- Investigating new system concepts and comparing the many options described in Chapter 9, beyond simple elution and SMB modes
- Investigating the possibility of impurity formation during the separation process, due, for instance, to product degradation
- Deriving ways to identify the best chromatographic medium from laboratory experiments. A simple look at resolution is certainly not sufficient: the same resolution can be obtained with a moderate selectivity and a significant number of plates, or vice versa
- Handling the so-called capture step for protein chromatography: do we need continuous systems? Counter-current systems?

It seems that the same strategy always wins: start with simple descriptions based on knowledge of fluid-phase equilibria, use the equilibrium model as a reference, start the design with an MC–$\overline{\text{Equil}}$ type of simple model and then refine when the limiting factors (mass transfer, hydrodynamics, heterogeneities etc.) have been identified.

References

Abel, S., Mazzotti, M. and Morbidelli, M. (2002), *J. Chromatogr. A* **944**(1–2), 23–39.

Abel, S., Mazzotti, M. and Morbidelli, M. (2004), *J. Chromatogr. A* **1026**(1–2), 47–55.

Adam, P., Ludemann-Hombourger, O., Ndzie, E., Ross, D., Schaeffer, M. and Suteu, C. (2005), WO Patent 2 005 040 140.

Azevedo, D. and Rodrigues, A. E. (2001), *AIChE J.* **47**(9), 2042–2051.

Azevedo, D. S. and Rodrigues, A. E. (2000), *Sep. Sci. Technol.* **35**(16), 2561–2581.

Blehaut, J. and Nicoud, R. M. (1998), *Analusis* **26**(7), M60–M70.

Carta, G. and Jungbauer, A. (2011), *Protein Chromatography*, Wiley-VCH.

Heinonen, J., Rubiera Landa, H., Sainio, T. and Seidel-Morgenstern, A. (2012), *Sep. Purif. Technol.* **95**, 235–247.

Hotier, G., Cohen, C., Couenne, N. and Nicoud, R. M. (1996), US Patent 5 578 216.

Hotier, G. and Nicoud, R. M. (1996), US Patent 5 578 215.

Kim, J. K. and Wankat, P. C. (2004), *Ind. Eng. Chem. Res.* **43**, 1071–1080.

Lorenz, H., Sheehan, P. and Seidel-Morgenstern, A. (2001), *J. Chromatogr. A* **908**(1–2), 201–214.

Ludemann-Hombourger, O., Bailly, M. and Nicoud, R. M. (2000), *Sep. Sci. Technol.* **35**(9), 1285–1305.

Nicoud, R. M. (2014), *Ind. Eng. Chem. Res.* **53**, 3755–3765.

Nicoud, R. M. (2000). In S. Ahuja, ed., *Handbook of Bioseparations*, vol. 2 of *Separation Science and Technology*, Academic Press, pp. 475–509.

Nicoud, R. M. and Schweich, D. (1989), *Water Resour. Res.* **25**(6), 1071–1082.

Nimmig, S. and Kaspereit, M. (2013), *Chem. Eng. Process.* **67**, 89– 98.

Pierlas, R., Dozol, J., Kalimbadjian, M., Nicoud, R. M. and Schweich, D. (1986). In *World Congress III Of Chemical Engineering*, September 21–25, Tokyo 1986.

Rajendran, A., Paredes, G. and Mazzotti, M. (2009), *J. Chromatogr.* **1216**, 709–738.

Seidel-Morgenstern, A., Schmidt-Traub, H., Michel, M., Epping, A. and Jupke, A. (2012). In H. Schmidt-Traub, M. Schulte and A. Seidel-Morgenstern, eds., *Preparative Chromatography*, 2nd edn., Wiley-VCH, p. 321.

Siitonen, J., Sainio, T. and Rajendran, A. (2012), *J. Chromatogr. A* **1320**, 77–92.

Silva, R., Rodrigues, R., Osuna-Sanchez, H., Bailly, M., Valery, E. and Mota, J. (2010), *J. Chromatogr. A* **1217**(52), 8257–8269.

Strohlein, G., Schulte, M. and Strube, J. (2003), *Sep. Sci. Technol.* **28**(14), 3353–3383.

Subramani, H. J., Hidajat, K. and Ray, A. K. (2003), *Comput. Chem. Eng.* **27**(12), 1883–1901.

11 Conclusion

Let us end the journey...

Starting with relatively basic definitions and a description of ideal chromatographic systems, we have explored thermodynamics, hydrodynamics and mass transfer to illustrate different models of chromatography. Similarly, we have explored different techniques, from transfer functions to methods of characteristics, from numerical methods for solving ODEs to optimization algorithms for simulating chromatographic processes. All these models and techniques have then been linked in order to explore systems design. Modeling, simulating and designing chromatographic processes require many ingredients that must be combined in due proportions, according to the problem to be solved.

Modeling is neither a simple fitting of experimental results nor the use of statistics for representing data without physical knowledge. It should be a tool, allowing us to identify dominant mechanisms and predict behaviors in unexplored regions, a way to investigate, to think, to improve, to gain knowledge and understanding, and finally to stimulate creativity.

It is amazing that chromatography has for so long avoided one of the main factors of progress of the 20th century. Today, when a new airplane is developed, it is numerically designed, taking into account constraints such as the desired number of passengers, range, speed, fuel consumption, noise and so on, and based on this numerical design, it is built and it flies. This example is certainly simplified, but the test pilots of today are much safer than their ancestors were, when planes were designed by pure trial and error.

At the beginning of the 21st century, many chromatographic processes are still developed by trial-and-error methodologies like those used by the pioneers of aviation. I see no reason in principle (but many practical difficulties) why science cannot produce greater and greater numerical tools for simulating (and stimulating ...) chromatographic processes.

The modeling of chromatographic processes is neither as difficult as it may seem to many practitioners nor as well understood and ready to use as modeling specialists may dream. Let us conclude with three key messages.

Modeling chromatographic processes is possible

The key lies in the decoupling of thermodynamic, hydrodynamic and mass transfer processes, and then their integration into adequate mass and possibly heat balances. The most difficult task is probably the identification of the dominant mechanisms: complexity is not synonymous with efficiency. The experience, and maybe the art, of the scientist must lead him always to adapt the complexity of the model to the required precision, and the available information.

Modeling chromatographic processes is useful

Designing and optimizing chromatographic processes is a difficult task because of the very large number of parameters that can be tuned. Modeling cannot in general replace experimentation, but can help to minimize the need for it; that is already a lot. How many months, not to say years, would a chromatographer need to design a complex multi-column process with a pure experimental trial-and-error methodology, without a minimum of theoretical knowledge? How many ideas are still unexplored because the time required to investigate them is not affordable? What is the value associated with the identification of rational critical parameters for operating better controlled and thus safer processes?

Furthermore, modeling that allows a better understanding of processes is fully aligned with Quality By Design requirements.

Modeling chromatographic processes offers great research opportunities

At various points we have touched what can probably be considered the limits of our current modeling skills. Let us mention the modeling of multi-component adsorption equilibria, multi-dimensional velocity profiles, compressible and generally non-uniform chromatographic beds, multi-phase flows, multi-component mass transfer and the limits of Fick's law, the need to use all the available information and thus to solve the "inverse problem", the possibility of deriving general qualitative rules for multi-component systems ("phenomenological models"), complex multi-variable algorithms . . .

Fascinating areas for further research!

Appendix A Some important properties of the Laplace transform

The aim of this appendix is to give essential properties of the Laplace transform for the theory of linear chromatography. More detailed presentations are available in several books, including (Varma and Morbidelli, 1997).

A.1 Definition

The Laplace transform of a function $x(t)$ is defined as:

$$\hat{x}(s) = \mathscr{L}(x(t)) = \int_0^\infty e^{-st}x(t)dt \tag{A.1}$$

The definition involves an improper integral, which is given by the limit of the corresponding proper (bounded) integral:

$$\mathscr{L}(x(t)) = \lim_{T\to\infty} \int_0^T e^{-st}x(t)dt$$

Existence of the Laplace transform requires this proper integral to be defined for all T and to converge when $T \to \infty$. In practice, the existence of the Laplace transform will not be an obstacle for chromatography applications.

It can also be shown that, with minor exceptions, two different functions cannot have the same Laplace transform, so one can define the inverse of the Laplace transform:

$$\mathscr{L}^{-1}(\hat{x}(s)) = x(t) \tag{A.2}$$

which is unique.

Many Laplace transforms can be calculated; see, for instance, (Rice and Do, 1995) and Table A.1.

A.2 Linearity

From the definition of the Laplace transform, one obtains readily:

$$\mathscr{L}\left(\sum a_i x_i(t)\right) = \sum a_i \mathscr{L}(x_i(t)) = \sum a_i \hat{x}_i(t) \tag{A.3}$$

Table A.1 Some Laplace transforms.

$x(t)$	$\hat{x}(s)$
Dirac delta function of surface a at time τ	$ae^{-\tau s}$
Step of height a at time τ	$a\frac{e^{-\tau s}}{s}$
Constant a	$\frac{a}{s}$
Rectangular pulse of duration t_{inj}	$\frac{1}{st_{inj}}(1-\exp(-st_{inj}))$
$e^{-t/\tau}$	$\frac{\tau}{1+\tau s}$
at	$\frac{a}{s^2}$
$\sin(at)$	$\frac{a}{s^2+a^2}$
$t^n, \quad n=0,1,2,..$	$\frac{n!}{s^{n+1}}$
$\cos(at)$	$\frac{s}{s^2+a^2}$

A.3 Transforms and derivatives of integrals

Let us consider the Laplace transform of the time derivative of a function $x(t)$. Using Eq. (A.1), one has:

$$\mathcal{L}\left(\frac{dx(t)}{dt}\right)=\int_0^\infty e^{-st}\left(\frac{dx(t)}{dt}\right)dt$$

so, integrating by parts:

$$\mathcal{L}\left(\frac{dx(t)}{dt}\right)=\int_0^\infty \frac{d\left(e^{-st}x(t)\right)}{dt}dt-\int_0^\infty \frac{x(t)d\left(e^{-st}\right)}{dt}dt$$

which leads to:

$$\mathcal{L}\left(\frac{dx(t)}{dt}\right)=\left[e^{-st}x(t)\right]_0^\infty+s\int_0^\infty x(t)e^{-st}dt$$

Assuming that $\lim_{t\to\infty}\left[e^{-st}x(t)\right]=0$, it follows immediately that the Laplace transform of the time derivative of $x(t)$ is given by:

$$\mathcal{L}\left(\frac{dx(t)}{dt}\right)=s\hat{x}(s)-x(0) \tag{A.4}$$

Differentiation with respect to time in the time domain thus translates into multiplication by s in the Laplace domain. This is a very interesting property of the Laplace transform, as it allows the transformation of ODEs into algebraic equations.

We can obtain a similar relation for integrals. Let us define $y(t) = \int_0^t x(\tau)d\tau$. Applying Eq. (A.4), one obtains:

$$\mathscr{L}\left(y'(t)\right) = s\mathscr{L}\left(y(t)\right) - y(0)$$

By the definition of $y(t)$, one has $y(0) = 0$. The above equation yields:

$$\mathscr{L}\left(\int_0^t x(\tau)d\tau\right) = \frac{1}{s}\hat{x}(s) \tag{A.5}$$

Thanks to (A.4) and (A.5), operating in the Laplace domain allows the transformation of ODEs into algebraic equations and PDEs into ODEs.

Example:
Let us determine the Laplace transform of $\cos(at)$ knowing the Laplace transform of $\sin(at)$. From Table A.1 we know that the Laplace transform of $\sin(at)$ is $\frac{a}{s^2+a^2}$. According to Eq. (A.4), $\mathscr{L}\left(\frac{d\sin(at)}{dt}\right) = s\frac{a}{s^2+a^2} - \sin(a \times 0) = \frac{as}{s^2+a^2}$. The derivative of $\sin(at)$ is $a\cos(at)$, so $\mathscr{L}\left(\cos(at)\right) = \frac{1}{a}\mathscr{L}\left(\frac{d\sin(at)}{dt}\right)$. Consequently, $\mathscr{L}\left(\cos(at)\right) = \frac{1}{a}\frac{as}{s^2+a^2} = \frac{s}{s^2+a^2}$, which is the expression given in Table A.1.

A.4 Differentiation and integration in the Laplace domain

Let us denote by $\hat{x}'(s)$ the derivative of the Laplace transform of $x(t)$. By definition:

$$\hat{x}'(s) = \frac{d}{ds}\int_0^\infty e^{-st}x(t)dt = \int_0^\infty -te^{-st}x(t)dt$$

so that:

$$\hat{x}'(s) = \frac{d}{ds}\int_0^\infty e^{-st}x(t)dt = \mathscr{L}\left(-tx(t)\right) \tag{A.6}$$

The derivative of the Laplace transform of $x(t)$ is simply the Laplace transform of $-tx(t)$.

Reciprocally, it can be shown that:

$$\int_s^\infty \hat{x}(s)ds = \mathscr{L}\left(\frac{x(t)}{t}\right) \tag{A.7}$$

so integration of the Laplace transform corresponds to division of the initial function $x(t)$ by t in the real domain.

Example:
Using Table A.1, let us determine the Laplace transform of $te^{-t/\tau}$. From Eq. (A.6), one has $\mathcal{L}\left(te^{-t/\tau}\right) = -\left[\mathcal{L}(e^{-t/\tau})\right]'$, so $\mathcal{L}\left(te^{-t/\tau}\right) = \left[\frac{\tau}{1+\tau s}\right]' = \frac{-\tau^2}{(1+\tau s)^2}$.

A.5 Initial-value and final-value theorems

The limits of both sides of Eq. (A.4) can be taken:

$$\lim_{s\to\infty} \int_0^\infty e^{-st} \frac{dx(t)}{dt} dt = \lim_{s\to\infty} \left(s\hat{x}(s) - x(0)\right)$$

The left-hand side of the above equation equals 0 since:

$$\lim_{s\to\infty} \int_0^\infty e^{-st} \frac{dx(t)}{dt} dt = \int_0^\infty \lim_{s\to\infty} e^{-st} \frac{dx(t)}{dt} dt = 0$$

Consequently, the right-hand side equals zero, so:

$$\lim_{s\to\infty} \left(s\hat{x}(s) - x(0)\right) = 0$$

or, equivalently:

$$\lim_{t\to 0} x(t) = \lim_{s\to\infty} \left[s\hat{x}(s)\right] \tag{A.8}$$

Similarly, it can be shown that:

$$\lim_{t\to\infty} x(t) = \lim_{s\to 0} \left[s\hat{x}(s)\right] \tag{A.9}$$

Note that that these equalities only hold provided that *both* limits exist.

These theorems are useful for evaluating some limiting values in the real time domain without inverting the Laplace transform.

Example:
Let us address the response of a chromatography column to a step input of inlet temperature and/or external temperature. For simplicity, we assume a uniform temperature in the column. Under this assumption, an energy balance can be written in the form:

$$T^{inl} = T + \tau \frac{dT}{dt} + \alpha(T - T_{ext})$$

At time $t = 0$, the system is assumed to be at equilibrium, so:

$$T^{inl}(t=0) = T(t=0) + \alpha\left[T(t=0) - T_{ext}(t=0)\right]$$

Subtracting the two above equations, one obtains:

$$\Delta T^{inl} = \Delta T + \tau \frac{d\Delta T}{dt} + \alpha(\Delta T - \Delta T_{ext})$$
$$\text{with } \Delta X = X - X(t=0)$$

After transformation into the Laplace domain, one obtains:

$$\widehat{\Delta T} = \frac{\widehat{\Delta T}^{inl} + \alpha \widehat{\Delta T}_{ext}}{1 + \alpha + \tau s}$$

Assuming that the inlet and outside temperature have step inputs ΔT^{inl} and ΔT_{ext}, one has:

$$\widehat{\Delta T} = \frac{1}{s} \frac{\Delta T^{inl} + \alpha \Delta T_{ext}}{1 + \alpha + \tau s}$$

Using Eq. (A.8) one obtains:

$$\lim_{t \to 0} \Delta T = \lim_{s \to \infty} \frac{s}{s} \frac{\Delta T^{inl} + \alpha \Delta T_{ext}}{1 + \alpha + \tau s} = 0$$

which is obvious according to the definition of ΔT.

Using Eq. (A.9) one obtains:

$$\lim_{t \to \infty} \Delta T = \lim_{s \to 0} \frac{s}{s} \frac{\Delta T^{inl} + \alpha \Delta T_{ext}}{1 + \alpha + \tau s} = \frac{\Delta T^{inl} + \alpha \Delta T_{ext}}{1 + \alpha}$$

A.6 Convolution theorem

Let us define the function $z(t)$ as the convolution of functions $x(t)$ and $y(t)$ over the interval $[0,t]$:

$$z(t) = \int_0^t x(\tau) y(t - \tau) d\tau \tag{A.10}$$

The Laplace transform of $z(t)$ is given by the product of the Laplace Transforms:

$$\hat{z}(s) = \hat{x}(s)\hat{y}(s) \tag{A.11}$$

Equation (A.11) allows us to substitute a simple product in the Laplace domain for a convolution product.

This is an essential property as it proves that, in the Laplace domain, the outlet signal of a chromatography column is the product of the inlet signal and the Transfer Function of the system.

A.7 Shifting of the independent variable

Consider the function $\vec{x}(t)$, obtained by shifting $x(t)$ along the t axis by an amount a, and which equals zero for $t < a$:

$$\vec{x}(t) = u(t-a)x(t-a)$$
$u(t-a)$ is the unit step function, equaling 0 for $t < a$ and 1 for $t \geq a$ (A.12)

The Laplace transform of the shifted function is given by:

$$\mathscr{L}\left(\vec{x}(t)\right) = \int_0^\infty e^{-st}u(t-a)x(t-a)dt = \int_a^\infty e^{-st}x(t-a)dt$$

Substituting $\tau = t - a$, one obtains:

$$\mathscr{L}\left(\vec{x}(t)\right) = e^{-sa}\int_0^\infty e^{-s\tau}x(\tau)d\tau$$

so that:

$$\mathscr{L}(\vec{x}(t)) = e^{-sa}\hat{x}(s) \tag{A.13}$$

The Laplace transform of the shifted function is given by the product of the Laplace transform of the initial function and e^{-sa}.

Conversely, a shift of the transform along the s axis leads to:

$$\hat{x}(s-a) = \mathscr{L}\left(e^{at}x(t)\right) \tag{A.14}$$

Example:
One can directly calculate the Laplace transform of the unit step function $u(t-a)$:

$$\mathscr{L}(u(t-a)) = \int_0^\infty e^{-st}u(t-a)dt = \int_a^\infty e^{-st}dt = \frac{e^{-as}}{s}$$

The same result is obtained with Eq. (A.14) using $x(t) = 1$. Equation (A.13) gives:

$$\mathscr{L}(\vec{x}(t)) = \mathscr{L}(u(t-a)\times 1) = e^{-sa}\mathscr{L}(1) = e^{-sa}\frac{1}{s}$$

A.8 Moment determination

Using the Maclaurin expansion of $e^{-st} = 1 - st + \frac{1}{2!}s^2t^2 + \cdots$ for small s, one can write:

$$\mathscr{L}(x(t)) = \int_0^\infty x(t)dt - s\int_0^\infty x(t)tdt + \frac{1}{2!}s^2\int_0^\infty x(t)t^2dt + \cdots$$

so that:

$$\mathscr{L}(x(t)) = \hat{x}(s) = \mu_0 - s\mu_1 + s^2\mu_2 + \cdots$$

After equating this with the expansion of $\hat{x}(s)$ for small s, one obtains Van der Laan's theorem:

$$\mu_n = (-1)^n \left. \frac{d\hat{x}(s)}{dt} \right|_{s=0} \quad \text{(A.15)}$$

A.9 Inverse transform

Inverting the Laplace expressions to derive expressions in the real domain is certainly the most difficult part of the work. For simple situations, one can use analytical expressions such as those in Table A.1. For more complex situations (for example, when transient diffusion processes are involved), more general methods are required.

A general methodology known as the Heaviside expansion theorem (Varma and Morbidelli, 1997), allowing one to obtain analytical expressions, is available when the Laplace transform can be expressed as a ratio of two polynomials with real coefficients. Obtaining transfer functions as a ratio of polynomials with real coefficients is relatively common.

For more complex situations, a numerical inversion must be performed. Numerical methods aimed at inverting the Laplace transform are discussed in (Press *et al.*, 2007; Davies and Martin, 1979; Duffy, 1993). The preferred methods are based on what is sometimes called a Fast Fourier transform (FFT), in which certain numerical concepts are used to accelerate the calculation speed.

(Honig and Hirdes, 1984) has detailed a sophisticated inversion method based on Fourier expansion, and included Fortran code in their paper. (Chen and Hsu, 1987; Hsu and Dranoff, 1987) have detailed the use of FFT for chromatography. (Mees *et al.*, 1989) have detailed the use of the Laplace transform for modeling non-isothermal chromatography, taking into account axial dispersion, external and internal mass and heat transfer.

The above-mentioned algorithms are extremely quick (two to three orders of magnitude quicker than conventional methods, as suggested by (Chen and Hsu, 1987)), and numerical inversion does not take more than a few milliseconds in a modern computer. This speed is extremely useful when included in an optimization algorithm.

However, it must be noted that these tools may lead to numerical problems for specific ranges of parameters, so adaptations may be required, as discussed by (Seidel-Morgenstern, 1991).

References

Chen, L. and Hsu, J. (1987), *AIChE J.* **33**(8), 1387.

Davies, B. and Martin, B. (1979), *J. Comput. Phys.* **33**(1), 1–35.

Duffy, D. (1993), *ACM Trans. Math. Software* **19**(3), 333–359.

Honig, G. and Hirdes, U. (1984), *J. Comput. Appl. Math.* **10**, 113–132.

Hsu, J. and Dranoff, J. (1987), *Comput. Chem. Eng.* **11**, 101.

Mees, P., Gerritsen, A. and Verheijen, P. (1989), *AIChE J.* **35**(8), 1380–1384.

Press, W., Teukolsky, S., Vetterling, W. and Flannery, B. (2007), *Numerical Recipes: The Art of Scientific Computing*, Cambridge University Press.

Rice, R. and Do, D. (1995), *Applied Mathematics and Modeling for Chemical Engineers*, John Wiley & Sons.

Seidel-Morgenstern, A. (1991), *Chem. Eng. Sci.* **46**(10), 2567–2571.

Varma, A. and Morbidelli, M. (1997), *Mathematical Methods in Chemical Engineering*, Oxford University Press.

Appendix B Inlet and outlet boundary conditions

In Chapter 2 we mentioned possible concerns with the way boundary conditions should be written for the PD model, and this appendix is aimed at clarifying the matter.

The mass balance associated with the PD model is given in the Laplace domain by Eq. (2.23):

$$\frac{d\hat{C}_A(\xi,s)}{d\xi} + t_O s \hat{C}_A(\xi s) = \frac{1}{Pe}\frac{d^2 C_A(\xi,s)}{d^2\xi} \tag{B.1}$$

If one defines:

$$\lambda_- = \frac{Pe}{2}\left(1 - \sqrt{1 + \frac{4t_O s}{Pe}}\right) \qquad \lambda_+ = \frac{Pe}{2}\left(1 + \sqrt{1 + \frac{4t_O s}{Pe}}\right) \tag{B.2}$$

the general solution of Eq. (2.23) is given by:

$$\hat{C}_A(\xi) = \alpha_- \exp(\lambda_- \xi) + \alpha_+ \exp(\lambda_+ \xi) \tag{B.3}$$

where α_- and α_+ are independent of ξ and determined by the boundary conditions.

We defined simplified boundary conditions (simplified BC) in Chapter 2 by Eqs. (2.24):

$$\hat{C}_A(s)\Big|_{\xi=0} = \hat{C}_A^{inl} \text{ and } \hat{C}_A(s)\Big|_{\xi=1} = \hat{C}_A^{out} \text{ must be finite} \tag{B.4}$$

Using these conditions, from Eq. (B.3) one obtains:

$$\alpha_- = \hat{C}_A^{inl} \qquad \alpha_+ = 0 \tag{B.5}$$

so Eq. (2.23) has the solution:

$$\hat{C}_A(s)\Big|_{\xi=1} = \hat{C}_A(s)\Big|_{\xi=0} \exp\left[\frac{Pe}{2}\left(1 - \sqrt{1 + \frac{4t_O s}{Pe}}\right)\right] \tag{B.6}$$

This relation, in association with the simplified BC in Eq. (B.4), shows that the transfer function for the PD model is given by:

$$\hat{E}_e(s) = \exp\left[\frac{Pe}{2}\left(1 - \sqrt{1 + \frac{4t_O s}{Pe}}\right)\right] \tag{B.7}$$

These conditions, assuming continuity of the concentrations at the inlet and outlet of the column, would be totally rigorous for a double measurement experiment, in which the injection is performed on a hypothetical negative axis, $\xi < 0$, and signals measured at both $\xi = 0$ and $\xi = 1$.

The subject of suitable boundary conditions has been extensively addressed in the literature after the pioneering work of (Danckwerts, 1953) and (Wehner and Wilhelm, 1956), as presented in (Wen and Fan, 1975). A reactor, or more specifically a chromatography column, is normally closed to diffusion at the inlet: the inlet pipe diameter being drastically smaller than the column diameter, solutes cannot diffuse from the very beginning of the column ($\xi = 0$) back into the inlet pipe. The same applies to the outlet.

We consequently need to assume a discontinuity in flow pattern at the inlet and outlet of the column, as the axial dispersion effect is assumed to play a role only inside the column, and not immediately before or after. Writing a flux conservation at the inlet and outlet, one obtains (Danckwerts, 1953; Wehner and Wilhelm, 1956):

$$C_A^{inl} = C_A(\xi = 0) - \frac{1}{Pe} \left. \frac{dC}{d\xi} \right|_{\xi=0} \qquad \left. \frac{dC}{d\xi} \right|_{\xi=1} = 0 \tag{B.8}$$

We will call these conditions B.8 the *closed–closed BC*, as they describe a system closed to diffusion at the inlet and at the outlet, or as the Danckwerts conditions.

Using (B.8) instead of (B.4), from (B.2) one obtains:

$$\begin{aligned} \alpha_- &= \hat{C}_A^{inl} \frac{\lambda_+ \exp(\lambda_+)}{\lambda_+ \exp(\lambda_+) \left(1 - \frac{\lambda_-}{Pe}\right) - \lambda_- \exp(\lambda_-) \left(1 - \frac{\lambda_+}{Pe}\right)} \\ \alpha_+ &= \hat{C}_A^{inl} \frac{-\lambda_- \exp(\lambda_-)}{\lambda_+ \exp(\lambda_+) \left(1 - \frac{\lambda_-}{Pe}\right) - \lambda_- \exp(\lambda_-) \left(1 - \frac{\lambda_+}{Pe}\right)} \end{aligned} \tag{B.9}$$

so that the concentration profile in the Laplace domain and thus the transfer function are known. The expression for the transfer function is less handy than it is with the simplified BC and does not allow us to obtain an analytical solution of $E(t)$.

Obviously, if the dispersion is "small enough" or, in other words, if the *Pe* number is "large enough", the boundary conditions (B.4) and (B.8) become equivalent, and one can approximate the Danckwerts conditions by (B.4). This approximation is typically considered acceptable for *Pe* greater than about 20.

The Danckwerts conditions can be better understood if one uses the MC model with back-flow, as presented, for instance by (Wen and Fan, 1975). The column is approximated by a series of J mixing cells between which a back-flow can occur, as presented in Figure B.1. This approach will be called the MC_{BF} model.

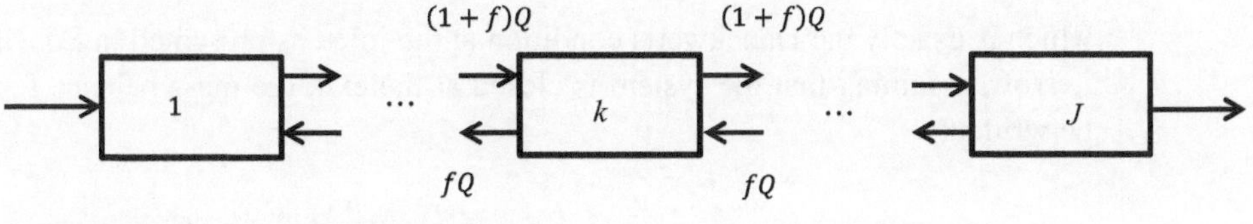

Figure B.1 Schematic of the MC_{BF} model.

A mass balance can be written for cell k:

$$(1+f)QC_A^{k-1} + fQC_A^{k+1} = (1+2f)QC_A^k + \frac{\varepsilon_e V_{col}}{J}\frac{dC_A^k}{dt} \tag{B.10}$$

Defining $\delta z = \frac{L_{col}}{J}$ and noting that $\frac{L_{col}Q}{V_{col}} = u$, one obtains:

$$\frac{dC_A^k}{dt} = \frac{uf\delta z}{\varepsilon_e}\frac{\left(C_A^{k+1} - 2C_A^k + C_A^{k-1}\right)}{(\delta z)^2} + \frac{u}{\varepsilon_a}\frac{C_A^{k-1} - C_A^k}{\delta z} \tag{B.11}$$

so that, if one takes J large enough:

$$\frac{\partial C_A}{\partial t} \simeq \frac{uf\delta z}{\varepsilon_e}\frac{\partial^2 C_A}{\partial z^2} + \frac{u}{\varepsilon_e}\frac{\partial C_A}{\partial z} \tag{B.12}$$

Comparing this with the mass-balance equation (2.21) of the PD model one obtains:

$$D_{ax} = \frac{uf\delta z}{\varepsilon_e} \quad \text{so} \quad f = \frac{J}{Pe}$$

Consequently, if one takes J large enough, the $\mathrm{MC_{BF}}$ model is similar to the PD model, provided that $f = \frac{J}{Pe}$.

Now, assuming that the system is closed to diffusion at the entrance, the mass-balance equation for cell 1 must be written:

$$QC_A^{inl} + fQC_A^2 = (1+f)C_A^1 + \frac{\varepsilon_e V_{col}}{J}\frac{dC_A^1}{dt} \tag{B.13}$$

so that:

$$\delta z\frac{dC_A^1}{dt} = \frac{uf}{\varepsilon_e}(C_A^2 - C_A^1) + \frac{u}{\varepsilon_e}(C_A^{inl} - C_A^1) \tag{B.14}$$

and, recalling that $D_{ax} = \frac{uf\delta z}{\varepsilon_e}$, one obtains:

$$\delta z\frac{\partial C_A^1}{\delta t} \simeq D_{ax}\left.\frac{\partial C_A}{\partial z}\right|_{z=0} + \frac{u}{\varepsilon_e}(C_A^{inl} - C_A^1) \tag{B.15}$$

Taking the limit $\delta z = 0$ in Eq. (B.15), one obtains:

$$D_{ax}\left.\frac{\delta C_A}{\delta z}\right|_{z=0} + \frac{u}{\varepsilon_e}(C_A^{inl} - C_A(z=0)) = 0$$

which is exactly the Danckwerts condition at the inlet, as presented in Eq. (B.8).

Now, assuming that the system is closed at the exit, the mass balance for cell J can be written:

$$\delta z\frac{dC_A^J}{dt} = \frac{uf}{\varepsilon_e}(C_A^{J-1} - C_A^J) + \frac{u}{\varepsilon_e}(C_A^{J-1} - C_A^J)$$

so that:

$$\delta z \frac{\partial C_A^J}{\delta t} = D_{ax} \left. \frac{\partial C_A}{\partial z} \right|_{z=L} + \frac{u}{\varepsilon_e}(C_A^{J-1} - C_A^J)$$

When $\delta z = 0$, the left-hand side of the above equation equals zero, just like the second term on the right-hand side. Consequently one obtains $\left. \frac{\partial C_A}{\partial z} \right|_{z=L} = 0$, which is the Danckwerts condition at the exit, as presented in Eq. (B.8). The rather theoretical Danckwerts conditions are thus given a simple physical meaning with the MC_{BF} model.

References

Danckwerts, P. (1953), *Chem. Eng. Sci.* **2**, 1–9.

Wehner, J. and Wilhelm, R. (1956), *Chem. Eng. Sci.* **6**, 89–93.

Wen, C. and Fan, L. (1975), *Chemical Processes and Engineering. Vol. 3. Models of Flow Systems and Chemical Reactors*, Marcel Dekker.

Appendix C Equilibrium theory: single-solute chromatograms

In order to understand how dispersive and compressive fronts interfere, let us consider a point of concentration C_A^i at position (z_i,t_i) belonging to both fronts (Figure C.1).

The concentration C_A^i of the interference varies as it travels in the column, and finally reaches C_A^{max} at the outlet (Figure C.2).

The abscissa z_{0i} and time t_{0i} at which the interference starts (Figure C.1) are obtained by writing that, at this moment, C_A^i equals C_A^F and this concentration belongs to both the compressive and the dispersive fronts. Since entering the column, the compressive front has traveled at a velocity given by:

$$\frac{u_e}{1+F_e\dfrac{\bar{f}_A(C_A^F)}{C_A^F}}$$

so that:

$$t_{0i}=z_{0i}\frac{1+\dfrac{\varepsilon_e}{1-\varepsilon_e}\dfrac{\bar{f}_A(C_A^F)}{C_A^F}}{u_e} \tag{C.1}$$

The concentration C_A^F belonging to the dispersive front travels into the column at a velocity given by:

$$\frac{u_e}{1+F_e\bar{f}'_A\left(C_A^F\right)}$$

so that:

$$t_{0i}=z_{0i}\frac{1+F_e\bar{f}'_A\left(C_A^F\right)}{u_e}+\Delta t_{inj} \tag{C.2}$$

Eliminating t_{0i} between the two above equations, one obtains:

$$z_{0i}\left[\frac{\bar{f}_A(C_A^F)}{C_A^F}-\bar{f}'_A\left(C_A^F\right)\right]=\frac{\varepsilon_e}{1-\varepsilon_e}\Delta t_{inj}u_e \tag{C.3}$$

Equation (C.3), together with (C.1) or (C.2), allows us to determine the position (z_{0i}, t_{0i}) at which interference starts for any given adsorption isotherm.

The question is now to determine the evolution of concentration C_A^i in the (z,t) plane beyond (z_{0i}, t_{0i}). The difficulty with the interference calculation lies in the fact that

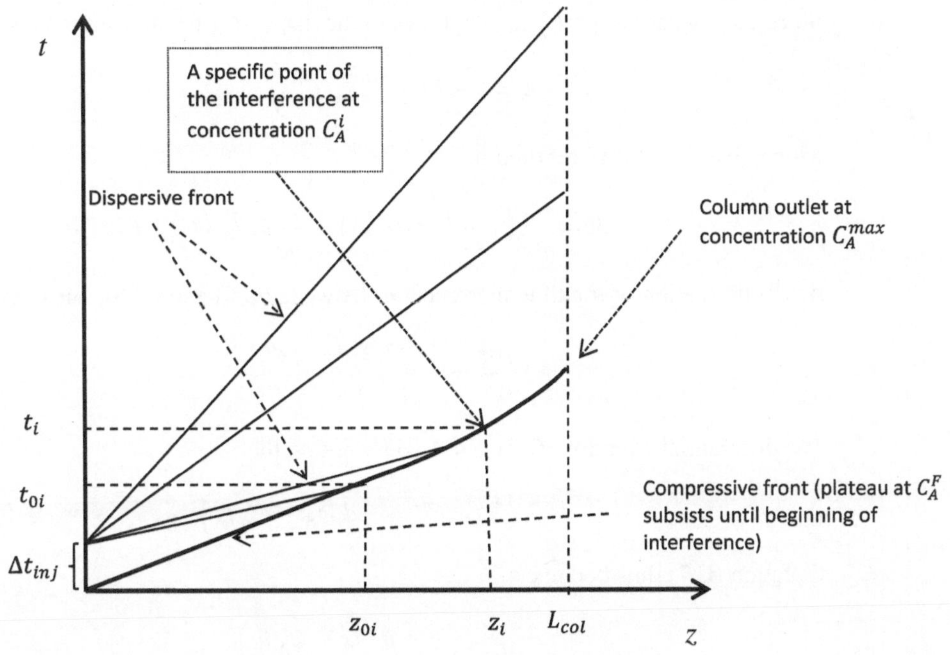

Figure C.1 Representation of interference (favorable adsorption isotherm).

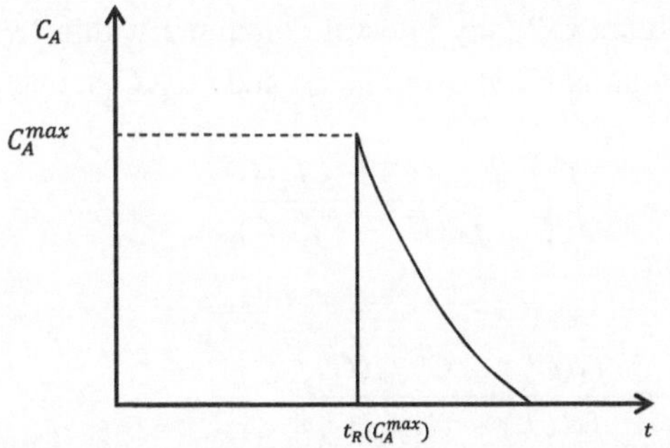

Figure C.2 Chromatogram obtained at the column outlet as the result of the interference (favorable adsorption isotherm).

after (z_{0i}, t_{0i}) the compressive front travels through the column at a variable velocity (Figure C.1) due to the erosion in concentration C_A^i.

Let us describe how the "shrinking" compressive front velocity evolves. During time increment dt, the compressive front moves from abscissa z_i by a distance dz such that:

$$dz = dt \frac{u_e}{1 + F_e \dfrac{\bar{f}_A(C_A^i)}{C_A^i}} \quad \text{(C.4)}$$

Since the concentration C_A^i also belongs to the dispersive front, it must satisfy:

$$t_i = \frac{z_i}{u_e}\left(1 + F_e\bar{f}'_A(C_A^i)\right) + \Delta t_{inj} \tag{C.5}$$

which gives, after differentiation:

$$dt = \frac{dz}{u_e}\left(1 + F_e\bar{f}'_A(C_A^i)\right) + \frac{z_i}{u_e}F_e\bar{f}''_A\left(C_A^i\right)dC_A^i \tag{C.6}$$

After elimination of the time increment dt between (C.4) and (C.6), one obtains:

$$\frac{dz}{z} = \frac{C_A^i\bar{f}''_A\left(C_A^i\right)dC_A^i}{\bar{f}_A\left(C_A^i\right) - C_A^i\bar{f}'_A(C_A^i)} \tag{C.7}$$

The differential equation (C.7) is solved by noting that:

$$C_A^i\bar{f}''_A dC_A^i = -d\left(\bar{f}_A - C_A^i\bar{f}'_A\right)$$

Equation (C.7) thus becomes:

$$\frac{dz}{z} = -\frac{d\left(\bar{f}_A - C_A^i\bar{f}'_A\right)}{\bar{f}_A\left(C_A^i\right) - C_A^i\bar{f}'_A(C_A^i)} \tag{C.8}$$

The differential equation (C.8) can be integrated over the regions where interference occurs, meaning, as shown in Figure C.2, any interval contained within (z_{0i},L_{col}) and (C_A^F,C_A^{max}). In particular, by integrating (C.8) over (z_{0i},z_i) and (C_A^F,C_A^i), one obtains:

$$\int_{z_{0i}}^{z_i}\frac{dz}{z} = -\int_{C_A^F}^{C_A^i}\frac{d(\bar{f}_A(C) - C\bar{f}'_A(C))}{\bar{f}_A(C) - C\bar{f}'_A(C)} \tag{C.9}$$

so that:

$$\frac{z_i}{z_{0i}} = \frac{\bar{f}_A(C_A^F) - C_A^F\bar{f}'_A(C_A^F)}{\bar{f}_A(C_A^i) - C_A^i\bar{f}'_A(C_A^i)} \tag{C.10}$$

Equation (C.10) describes the path followed by the interference in the (z,C) plane. For every concentration C_A^i, one can calculate an associated abscissa z_i; with this abscissa one can then associate a time t_i, thanks to Eq. (C.5) which describes the motion of the dispersive front. Consequently, the path followed by the interference in the (z,t) plane is known.

In order to determine what happens at the column outlet, one has to set $z_i = L_{col}$ in Eq. (C.10), so:

$$\frac{L_{col}}{z_{0i}} = \frac{\bar{f}_A(C_A^F) - C_A^F\bar{f}'_A(C_A^F)}{\bar{f}_A(C_A^{max}) - C_A^{max}\bar{f}'_A(C_A^{max})} \tag{C.11}$$

Notice that if $z_{0i} = L_{col}$, Eq. (C.11) requires that $C_A^{max} = C_A^F$. This is fully consistent with the definition of z_{0i}: when the interference begins, $C_A^i = C_A^F$. If this interference begins just at the outlet of the column, one must have $C_A^i = C_A^F = C_A^{max}$. Replacing z_{0i} by the expression given by Eq. (C.3), one finally obtains:

$$\Delta t_{inj} C_A^F = t_0 F_e \left[\bar{f}_A(C_A^{max}) - C_A^{max} \bar{f}'_A(C_A^{max}) \right]$$

which is Eq. (3.32) from Chapter 3.

An alternative and simpler way of deriving Eq. (3.32), without calculating the path followed by the interference in the (z,t) plane, consists in writing the conservation of the injected quantity of solute A:

$$\Delta t_{inj} C_A^F = \int_{t_R(C_A^{max})}^{\infty} C(t)dt \tag{C.12}$$

The integral on the right-hand side of Eq. (C.12) is the area under the chromatogram of Figure C.2. Instead of determining this area by integrating over time, one can integrate over concentration. A simple drawing should convince the reader that:

$$\int_{t_R(C_A^{max})}^{\infty} C(t)dt = \int_0^{C_A^{max}} t(C)dC - C_A^{max} t(C_A^{max}) \tag{C.13}$$

Introducing the expression for the retention time:

$$t(C) = t_0 \left(1 + F_e \bar{f}'_A(C) \right) + \Delta t_{inj}$$

into Eq. (C.12) and Eq. (C.13), one obtains:

$$\Delta t_{inj} C_A^F = \int_0^{C_A^{Max}} \left[t_0 \left(1 + F_e \bar{f}'_A(C) \right) + \Delta t_{inj} \right] dC$$
$$- C_A^{max} \left[t_0 \left(1 + F_e \bar{f}'_A(C_A^{max}) \right) + \Delta t_{inj} \right]$$

so that:

$$\Delta t_{inj} C_A^F = t_0 F_e \left[\bar{f}_A(C_A^{max}) - C_A^{max} \bar{f}'_A(C_A^{max}) \right]$$

which is again Eq. (3.32).

Appendix D Equilibrium theory: binary chromatograms

The aim of this appendix is to present a methodology allowing us to derive Z_{plat} and Z_{resol} for *favorable* lumped adsorption isotherms.

D.1 Calculation of Z_{plat}

The propagation velocities of the two compressive fronts are given by:

$$\sigma_{F_c^2} = \frac{u_e}{1 + F_e \dfrac{\bar{C}_B^{P1} - 0}{C_B^{P1} - 0}} = \frac{u_e}{1 + F_e \dfrac{\bar{C}_A^{P1} - 0}{C_A^{P1} - 0}}$$

$$\sigma_{F_c^1} = \frac{u_e}{1 + F_e \dfrac{\bar{C}_B^F - \bar{C}_B^{P1}}{C_B^F - C_B^{P1}}} = \frac{u_e}{1 + F_e \dfrac{\bar{C}_A^F - \bar{C}_A^{P1}}{C_A^F - C_A^{P1}}} \tag{D.1}$$

The propagation velocities of the two dispersive fronts are estimated using Eq. (3.45), where $\theta_\pm$ are the roots of Eq. (3.44) calculated at the appropriate concentrations:

$$\sigma_\pm = \frac{u}{1 + F_e \theta_\pm} \tag{D.2}$$

The feed plateau exactly disappears when the characteristic representing front F_c^2 crosses the characteristic representing front F_d^1 for the feed composition. At this point $(z = Z_{plat}, t = t_{plat})$ one has simultaneously:

$$t_{plat} = \Delta\, t_{inj} + \frac{Z_{plat}}{\sigma_{F_d^1}(F)} \quad (F_d^1) \quad \text{and} \quad t_{plat} = \frac{Z_{plat}}{\sigma_{F_c^2}(F)} \quad (F_c^2) \tag{D.3}$$

and consequently:

$$Z_{plat} = \frac{\Delta\, t_{inj}}{1/\sigma_{F_c^2(F)} - 1/\sigma_{F_d^1(F)}} \tag{D.4}$$

where t_{plat} is given by one of the equations (D.3).

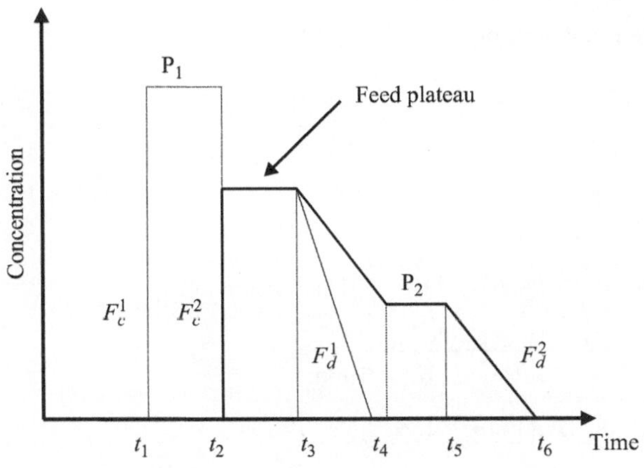

Figure D.1 Expected chromatogram for $Z_{inj} < Z_{plat}$.

D.2 Calculation of the chromatogram for $Z < Z_{plat}$

If $Z < Z_{plat}$, the feed plateau still exists, and the shape of the chromatogram is given in Figure D.1.

The compositions of plateaus P_1 and P_2 are determined as explained in Chapter 7. One still has to evaluate the times $t_1, \ldots, t_6$. These times are calculated from the associated propagation velocities of the different fronts:

$$\begin{aligned} t_1 &= L_{col}/\sigma_{F_c^1} \\ t_2 &= L_{col}/\sigma_{F_c^2} \\ t_3 &= \Delta\, t_{inj} + L_{col}/\sigma_{F_d^1}(\mathrm{F}) \\ t_4 &= \Delta\, t_{inj} + L_{col}/\sigma_{F_d^1}(\mathrm{P}_2) \\ t_5 &= \Delta\, t_{inj} + L_{col}/\sigma_{F_d^2}(\mathrm{P}_2) \\ t_6 &= \Delta\, t_{inj} + L_{col}/\sigma_{F_d^2}(0) \end{aligned} \tag{D.5}$$

D.3 Calculation of Z_{resol}

Since the zone between t_4 and t_6 contains pure B regardless of the injected volume, the formulas allowing us to calculate t_4, t_5 and t_6 in Eq. (D.5) are still valid, and one has:

$$\begin{aligned} t_4 &= \Delta\, t_{inj} + Z_{resol}/\sigma_{F_d^1}(\mathrm{P}_2) \\ t_5 &= \Delta\, t_{inj} + Z_{resol}/\sigma_{F_d^2}(\mathrm{P}_2) \\ t_6 &= \Delta\, t_{inj} + Z_{resol}/\sigma_{F_d^2}(0) \end{aligned} \tag{D.6}$$

The maximum volume of feed that can be injected for a complete resolution can be obtained by performing a mass balance on compound B. A mass balance written for the

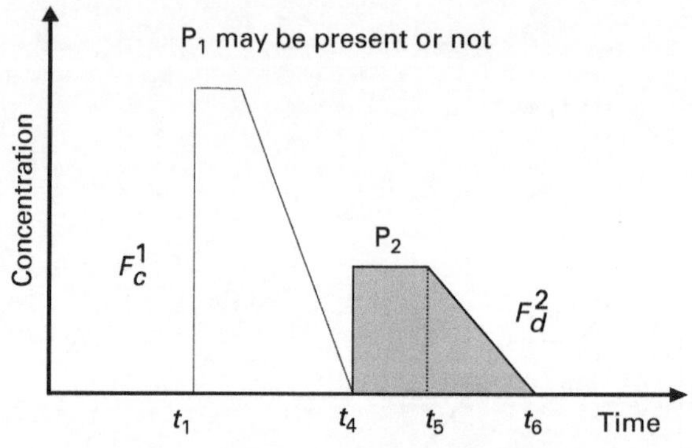

Figure D.2 Chromatogram obtained for $Z_{inj} = Z_{resol}$.

more retained compound B gives:

$$t_{inj} C_B^F = (t_5 - t_4)\, C_B^{P2} + \int_{t_5}^{t_6} C_B(t)dt \tag{D.7}$$

One can check in Figure D.2 that:

$$\int_{t_5}^{t_6} C_B(t)dt = \int_0^{P_2} t(C)dC - t_5 C_{Feed}^{P2}$$

Consequently, after substitution of the times t_4, t_5 and t_6 by their expressions in Eq. (D.6), one obtains:

$$\begin{aligned} t_{inj} C_B^F &= Z_{resol} \left(1/\sigma_{F_d^2(P2)} - 1/\sigma_{F_d^1(P2)}\right) C_B^{P2} \\ &+ \int_0^{P_2} \left(\Delta\, t_{inj} + Z_{resol}/\sigma_{F_d^2(C)}\right) dC - \left(\Delta\, t_{inj} + Z_{resol}/\sigma_{F_d^2(P2)}\right) C_B^{P2} \end{aligned} \tag{D.8}$$

which leads to an analytical expression for Z_{resol}:

$$Z_{resol} = \frac{\Delta\, t_{inj} C_B^F}{C_B^{P2} \left(\dfrac{1}{\sigma_{F_d^2(P2)}} - \dfrac{1}{\sigma_{F_d^1(P2)}} \right) + \int_0^{P_2} \dfrac{dC}{\sigma_{F_d^2}(C)} - \dfrac{C_B^{P2}}{\sigma_{F_d^2(P2)}}} \tag{D.9}$$

Appendix E The influence of the porosity determination on chromatographic modeling

Respective ways to measure extragranular, intragranular and total porosity, and difficulties associated with these measurements, are discussed in Chapter 4. Experience shows that some imprecision in the porosities does not significantly affect the main chromatogram characteristics, provided that a consistent modeling scheme is used for laboratory and production-scale columns.

For illustration purposes we assume that an adsorption isotherm is determined via frontal chromatography on a laboratory column for which porosity has been characterized by injecting a tracer. The tracer is frequently a product which can enter the intragranular porosity of the particles, ideally not adsorbed but frequently weakly adsorbed.

The retention time of the tracer is given by:

$$t_R^{lab}(tracer) = \left[\varepsilon_e^{lab} + \left(1 - \varepsilon_e^{lab}\right)\overline{K}_{tracer}\right]\left[\frac{V_{col}}{Q}\right]_{lab} \tag{E.1}$$

Notice that the quantity $\mathscr{R}(lab) = t_R^{lab}(tracer)\left[\frac{Q}{V_{col}}\right]$ is a dimensionless number based on experimental values only, and is not affected by the porosity choice. One has:

$$\mathscr{R}(lab) = \varepsilon_e^{lab} + (1 - \varepsilon_e^{lab})\overline{K}_{tracer} \tag{E.2}$$

If ε_e^{lab} represents exactly the extragranular porosity of the column and if the tracer is simply able to enter the intragranular porosity without interacting with the chromatographic medium, then $\overline{K}_{tracer}$ should equal the intragranular porosity ε_i of the particle.

If ε_e^{lab} is set to a more or less arbitrary value, from Eq. (E.2) it is clear that $\overline{K}_{tracer}$ is affected by this external porosity choice according to:

$$\overline{K}_{tracer} = \frac{\mathscr{R}(lab) - \varepsilon_e^{lab}}{1 - \varepsilon_e^{lab}} \tag{E.3}$$

If a breakthrough curve (assumed to be compressive) of a solute A is determined on this column, the associated average retention time $t_R^{lab}(A)$ is given by:

$$t_R^{lab}(A) = \left[\varepsilon_e^{lab} + \left(1 - \varepsilon_e^{lab}\right)\frac{\overline{C}_A}{C_A}\right]\left[\frac{V_{col}}{Q}\right]_{lab} \tag{E.4}$$

One point of the adsorption isotherm is thus known, and calculated from:

$$\begin{aligned}\overline{\mathscr{R}}(lab) &= t_R^{lab}(A)\left[\frac{Q}{V_{col}}\right]_{lab}\\ \frac{\overline{C}_A(C_A)}{C_A} &= \frac{\overline{\mathscr{R}}(lab)-\varepsilon_e^{lab}}{1-\varepsilon_e^{lab}}\end{aligned} \tag{E.5}$$

$\overline{\mathscr{R}}(lab)$ is a dimensionless number based on experimental values only and is not impacted by the initial porosity choice. Equation (E.5) expresses that the calculated lumped solid-phase concentration is impacted by the initial extragranular porosity choice.

Let us use these results to calculate the retention time of a front of the same solute A on a production column. This production column has been characterized by a tracer injection (the same tracer as for the lab column) which leads to a retention time $t_R^{prod}(tracer)$ given by:

$$t_R^{prod}(tracer) = \left[\varepsilon_e^{prod} + \left[1-\varepsilon_e^{prod}\right]\overline{K}_{tracer}\right]\left[\frac{V_{col}}{Q}\right]_{prod} \tag{E.6}$$

Similarly to the laboratory column, the quantity $\mathscr{R}(prod) = t_R^{prod}(tracer)\left[\frac{Q}{V_{col}}\right]$ is a dimensionless number based on experimental values only and is not affected by the porosity choice. This allows a calculation of the extragranular porosity of the preparative column thanks to knowledge of $\overline{K}_{tracer}$:

$$\varepsilon_e^{prod} = \frac{\mathscr{R}(prod)-\overline{K}_{tracer}}{1-\overline{K}_{tracer}} \tag{E.7}$$

Note: $\overline{K}(tracer)$ determined on the laboratory column must be as small as possible and in any case different from 1.

With Eq. (E.7) and (E.3) one obtains:

$$\varepsilon_e^{prod} = \frac{\mathscr{R}(prod)-\mathscr{R}(lab)+\varepsilon_e^{lab}\left[1-\mathscr{R}(prod)\right]}{1-\mathscr{R}(lab)} \tag{E.8}$$

Equation (E.8) allows a determination of the porosity of the preparative column, taking into account the result of the tracer injection performed on this column, and in a manner which is consistent with the porosity choice made for the analytical column.

The retention time of a front of solute A on the production column is calculated by:

$$t_R^{prod}(A) = \left[\varepsilon_e^{prod} + \left(1-\varepsilon_e^{prod}\right)\frac{\overline{C}_A}{C_A}\right]\left[\frac{V_{col}}{Q}\right]_{prod} \tag{E.9}$$

After substitution of ε_e^{prod} and $\frac{\overline{C}_A}{C_A}$ by their expressions in (E.9), one obtains:

$$t_R^{prod}(A) = \frac{1}{1-\mathscr{R}(lab)}\left[\mathscr{R}(prod)-\mathscr{R}(lab)+(1-\mathscr{R}(prod))\overline{\mathscr{R}}(lab)\right]\left[\frac{V_{col}}{Q}\right]_{prod} \tag{E.10}$$

Equation (E.10) expresses that the retention time of the front is exclusively linked to experimental retention times (via the parameters $\mathscr{R}(lab)$, $\mathscr{R}(prod)$ and $\overline{\mathscr{R}}(lab)$) obtained with the tracer on both columns or with the product A on the laboratory column, and is *independent of the assumptions made on the porosity of the laboratory column*. This result is achieved because the porosities of the different columns are defined in a consistent manner.

Appendix F Useful physico-chemical data and orders of magnitude

F.1 Density and transport coefficients: orders of magnitude

	Density (kg/m^3)	Viscosity, μ (Pa s)	Diffusivity (m^2/s)
Gases	1	10	10^{-6}–10^{-5}
Supercritical fluids	100–1000	50–100	10^{-8}–10^{-7}
Liquid solvent	1000	500–1000	10^{-9}

Note that the viscosities of highly concentrated feeds injected in chromatography can be an order of magnitude higher than the viscosities of liquid solvents.

For properties of solvents of interest in chromatography, see (Schmidt-Traub *et al.*, 2012) and (Guiochon *et al.*, 2006). For general information on fluids and fluid mixtures, see (Poling *et al.*, 2001).

F.2 Supercritical fluids

Various equations of state can be found in (Poling *et al.*, 2001). The simple Peng–Robinson equation of state (Peng and Robinson, 1976), which is good enough for first estimates, is given by:

$$P = \frac{RT}{v-b} - \frac{a}{v\,(v+b)+b(v-b)}$$

$$a = 0.457\,24\,\frac{R^2T_c^2}{P_c}\left[1+m\left(1-\left(\frac{T}{T_c}\right)^{1/2}\right)\right]^2 \qquad b = 0.077\,80\,\frac{RT_c}{P_c}$$

$$m = 0.374\,64 + 1.542\,26\,\omega - 0.269\,92\,\omega^2$$

where $T_c = 304.21$ K, $P_c = 73.795$ bar and $\omega = 0.239$ for pure carbon dioxide.

The Peng–Robinson equation leads to density via $\rho = \frac{M}{V} = \frac{MP}{ZRT}$, where the compressibility factor Z is a solution of the implicit equation:

$$Z^3 - (1-B)\,Z^2 + \left(A - 3B^2 - 2B\right)Z - \left(AB - B^2 - B^3\right) = 0$$

with $A = \frac{aP}{R^2T^2}$ and $B = \frac{bP}{RT}$.

Note that $\rho_c \approx 468$ kg/m^3 for carbon dioxide.

Viscosities at high pressure (in micropoise, 10^{-7} Pa s) are estimated by the correlations of:

- (Stiel and Thodos, 1964) for polar solvents:

$$\log_{10}\left[4 - \log_{10}\left(\left(\mu\left(T,P\right) - \mu_0\left(T\right)\right)\xi\right)\right] = 0.6439 - 0.1005\rho_r \quad \text{for} \quad 0.9 < \rho_r < 2.2$$

- (Jossi *et al.*, 1962) for non-polar solvents:

$$\left[(\mu(T,P) - \mu_0(T))\,\xi + 1\right]^{0.25} = 1.0230 + 0.23364\rho_r + 0.58533\rho_r^2 - 0.40758\rho_r^3 + 0.093324\rho_r^4 \quad \text{for} \quad 0.1 < \rho_r < 3.0$$

with $\xi = T_c^{1/6} M^{-1/2} P_c^{-2/3}$. M is the molecular weight (g/mol) and the critical parameters are given in K and atm. $\mu_0(T)$ is the viscosity at low pressure (atmospheric), given by (Guiochon and Tarafder, 2011):

$$\mu_0(T) = \frac{1.0069\, T^{1/2}}{f(T)} \quad \text{(micropoise, } 10^{-7} \text{ Pa s)}$$

$$f(T) = 0.235\,156 - 0.491\,266\left(\ln\frac{kT}{\epsilon}\right) + 5.211\,155\times10^{-2}\left(\ln\frac{kT}{\epsilon}\right)^2 + 5.347\,906\times10^{-2}\left(\ln\frac{kT}{\epsilon}\right)^3 - 1.537\,102\times10^{-2}\left(\ln\frac{kT}{\epsilon}\right)^4$$

F.3 Heat transfer coefficients

Heat capacity of liquids

Many organic compounds	2–2.5
Water/milk	4
Silica	0.7
Steel	0.4

Heat capacity in kJ/(kg K)

Thermal conductivity

Many organic compounds	0.15
Water/milk	6
Carbon steel	54
Glass	1.05
Iron	80
Plastics	0.25–0.5
Sand, dry	0.15–0.25
Stainless steel	16

Thermal conductivity in W/(m K)

References

Guiochon, G., Felinger, A., Golshan-Shirazi, S. and Katti, A. M. (2006), *Fundamentals of Preparative and Nonlinear Chromatography*, Elsevier Academic Press.

Guiochon, G. and Tarafder, A. (2011), *J. Chromatogr. A* **1218**, 1037–1114.

Jossi, J., Stiel, L. and Thodos, G. (1962), *AIChE J.* **8**, 59.

Peng, D.-Y. and Robinson, D. B. (1976), *Ind. Eng. Chem. Fundam.* **15**(1), 59–64.

Poling, B., Prausnitz, J. and O'Connell, J. (2001), *The Properties of Gases and Liquids*, 5th edn., McGraw-Hill.

Schmidt-Traub, H., Schulte, M. and Seidel-Morgenstern, A. (2012), *Preparative Chromatography*, 2nd edn., Wiley-VCH.

Stiel, L. and Thodos, G. (1964), *AIChE J.* **10**, 275.

Appendix G Fick and Maxwell–Stefan approaches to diffusion

G.1 Introduction

Prior to discussing diffusion and diffusion coefficients, we need to introduce some fundamental definitions regarding flux and velocities.

We consider a system made of n species, each one being associated with a molar fraction x_i and a velocity u_i with respect to a given fixed reference (the laboratory for instance). The molar flux N_i of species i with respect to this reference is given by:

$$N_i = x_i C_{tot} u_i \qquad (\mathrm{mol/m^2\,s^1}) \tag{G.1}$$

and the total flux is defined by:

$$N_{tot} = \sum_i N_i \qquad (\mathrm{mol/m^2\,s^1}) \tag{G.2}$$

We define the molar average velocity $\langle u \rangle$ by:

$$\langle u \rangle = \sum_i x_i u_i \tag{G.3}$$

In the absence of external forces and at uniform P and T, the molar average velocity $\langle u \rangle$ must be constant (according to Galileo Galilei ...).

In general, there is no reason for the velocities u_i to be identical to the molar average velocity $\langle u \rangle$ (Brownian motion), so one can define a flux relative to the average velocity by:

$$J_i = x_i C_{tot} (u_i - \langle u \rangle) \tag{G.4}$$

This flux is called the *diffusional flux* of species i.

From these definitions, one sees immediately that fluxes J_i and N_i are related by:

$$J_i = N_i - x_i C_{tot} \langle u \rangle = N_i - x_i N_{tot} \tag{G.5}$$

and that:

$$\sum_i J_i = 0 \tag{G.6}$$

Fick's law postulates that the diffusional flux is proportional to the molar fraction gradient, so, for each species:

$$J_i = -D_i C_{tot} \nabla x_i \approx -D_i \nabla C_i \tag{G.7}$$

the diffusion coefficient D_i appearing to be a simple proportionality factor. Notice that only $n-1$ fluxes and molar fractions are independent in Eq. (G.7): the nth component flux is obtained by setting the sum of the fluxes to zero, and the nth component molar fraction is obtained by setting the sum of the molar fractions to one.

In chromatography, molecular diffusion is the driving force of mass transfer through the boundary layer or inside the particle. Over the thickness of the boundary layer or the diameter of a particle, it is unlikely that the total concentration varies significantly. Consequently, replacing the molar fraction gradient by the concentration gradient is unlikely to cause any meaningful loss of precision. That is why Fick's law is normally expressed in terms of a concentration gradient.

In Chapter 5, when we presented single-component systems (single solute A), we wrote Fick's law as:

$$J_A = -D_A \nabla C_A \tag{G.8}$$

This equation is generally written so automatically and systematically that one easily forgets the underlying assumptions and definitions: a diffusion process refers to the motion of one species relative to the other(s) ($J_i = 0$ if there is only one species), so it involves at least a binary mixture.

Consequently, the term *single-component system* used in many articles and books is misleading, and a better wording is certainly *single-solute system*, the solvent being the second component. A direct consequence is that the diffusion coefficient D_A is in fact the diffusion coefficient $D_{A,S}$ of A in the solvent, and is not an intrinsic property of solute A, not a kind of auto-diffusion coefficient.

Fick's law should be written $J_A = -D_{A,S} C_{tot} \nabla x_A \approx -D_{A,S} \nabla C_A$ and, by symmetry, $J_S = -D_{S,A} C_{tot} \nabla x_S \approx -D_{S,A} \nabla C_S$. By definition, these two diffusional fluxes must be opposite, and, taking into account that $x_A + x_S = 1$, one must have $D_{A,S} = D_{S,A}$.

When using Fick's law, Eq. (G.7), to describe multi-component mixtures, one needs to keep in mind that the diffusion coefficient D_i in fact represents the diffusion coefficient of i in "some mixture".

Fick's law, Eq. (G.7), is only valid in the absence of electrostatic and centrifugal force field and, as we will show using the Maxwell–Stefan equations, under the following conditions (Krishna and Wesselingh, 1997):

- binary mixtures
- diffusion of dilute species in a multi-component mixture

If one takes the view that Eq. (G.7) provides a definition of Fickian diffusivity in multi-component mixtures, then this parameter shows a complex behavior with respect to composition. It is known for very simple multi-component systems that results almost incompatible with Fick's law are obtained (Krishna and Wesselingh, 1997): nitrogen can move *against* its gradient in N_2, H_2, CO_2 mixtures (inducing negative Fickian diffusion coefficients), H^+ and Cl^- ions have very different diffusion coefficients but diffuse together, and so on. In short, when multi-component systems are involved, Fickian diffusivities typically show a strong dependence on composition, and can even be negative.

Analyzing Fick's law a little bit further will allow us to understand its limitations. The main issue relates to the lack of a physical meaning for coefficient D_i. In order to illustrate the matter let us consider a ternary system ($n = 3$) and the use of Eq. (G.7) for component 1. From the binary mixture presentation given above, it is clear that, in the absence of component 3, the diffusion coefficient D_1 represents D_{12}, the diffusivity of species 1 in species 2. Similarly, in the absence of component 2, the diffusion coefficient D_1 represents D_{13}, the diffusivity of species 1 in species 3. This means that, when dealing with a three-component mixture, the diffusion coefficient D_1 should tend towards D_{12} or D_{13} according to the missing species and take different values when all species are present.

This simple example shows that, unless the diffusivities of species 1 in species 2 and in species 3 are identical, the Fickian diffusivity D_1 as defined by Eq. (G.7) *cannot* be constant with respect to composition. Expecting constant diffusion coefficients in using Fick's law is thus a chimera for multi-component systems.

One can argue that a more rigorous writing of Fick's law would use a matrix form, often called the generalized Fick's law:

$$\begin{bmatrix} J_1 \\ \\ J_{n-1} \end{bmatrix} = -C_{tot}\left[D_{i,j}\right]\begin{bmatrix} \nabla x_1 \\ \\ \nabla x_{n-1} \end{bmatrix} \quad \text{for an } n\text{-component system} \tag{G.9}$$

in which the flux of species i is related to the gradients of the various species. The diffusion coefficient $D_{i,j}$ relates the flux of species i to the molar fraction gradient of species j.

By adding free parameters, the generalized Fick's law certainly represents the experimental data better than Fick's law. The gain in rigor/efficiency is questionable, and the fundamental difficulty, that the coefficient D_i ($D_{i,i}$ in the generalized Fick's law) cannot in general be constant, remains. Using Eq. (G.9) may even lead to some fallacious statements such as "$D_{i,j}$ is the *binary* diffusion coefficient dealt with above and $\left[D_{i,j}\right]$ is a symmetric matrix". This is wrong: $D_{i,j}$ in Eq. (G.9) is a "cross-diffusivity" relating the diffusion flux of i to the gradient of j. It will be defined in Section G.3.

There is thus a need for a more general and physically sound theory involving better-defined parameters. This is probably to be obtained using the Maxwell–Stefan equations. For an introduction, the reader is strongly encouraged to read (Wesselingh and Krishna, 2006). More detailed developments are found in (Krishna and Wesselingh, 1997), (Taylor and Krishna, 1993) and (Gandhi, 2012).

G.2 The Maxwell–Stefan approach

The Maxwell–Stefan equations can be derived by writing a force balance expressing the fact that frictional forces on solute i equilibrate with driving forces (deriving from a potential) exerted on the same species. Friction forces exerted by solute j on solute i are assumed to be proportional to the molar fraction of solute j and to the difference in diffusional velocities between the two species:

$$F_i^{drag} = \sum_{j \neq i} \zeta_{i,j} x_j \left(u_i - u_j\right) \tag{G.10}$$

One notices that, should two species have different velocities, a frictional force would immediately appear tending to minimize this velocity difference.

One can replace velocities by fluxes using Eq. (G.1), and one obtains:

$$F_i^{drag} = \frac{1}{x_i C_{tot}} \sum_{j \neq i} \zeta_{i,j} \left(x_j N_i - x_i N_j\right) = \frac{1}{x_i C_{tot}} \sum_{j \neq i} \zeta_{i,j} \left(x_j J_i - x_i J_j\right) \tag{G.11}$$

The so-called *Maxwell–Stefan diffusivities* $Đ_{i,j}$ are related to the coefficients of friction by $Đ_{i,j} = \frac{RT}{\zeta_{i,j}}$, so that:

$$F_i^{drag} = \frac{RT}{x_i C_{tot}} \sum_{j \neq i} \frac{x_j J_i - x_i J_j}{Đ_{i,j}} \tag{G.12}$$

Note that the coefficient $Đ_{i,i}$ is not defined and never appears in the Maxwell–Stefan approach (it would be related to the relative motion of i with respect to i, which does not make a lot of sense). Consequently, there is no ambiguity as with Fick's law. Additionally, as shown below, considering a binary mixture leads immediately to $Đ_{i,j} = Đ_{j,i}$.

If the driving force derives from a chemical potential gradient, one has (at constant T and P and in the absence of external forces):

$$F_i^{drive} = -\nabla \mu_i = -RT \nabla \ln \left(a_i\right) = -\frac{RT}{a_i} \nabla a_i \tag{G.13}$$

which, for an ideal solution at constant T,P and in the absence of external forces, becomes:

$$F_i^{drive} \approx -RT \nabla \ln \left(x_i\right) = -\frac{RT}{x_i} \nabla x_i \tag{G.14}$$

The Maxwell–Stefan equation expresses that frictional forces equilibrate with driving forces, so:

$$F_i^{drive} = F_i^{drag} \tag{G.15}$$

Assuming that the driving force derives from a chemical potential and that the solution is ideal, one obtains:

$$C_{tot}\nabla x_i = \sum_{j \neq i} \frac{x_j J_i - x_i J_j}{Đ_{i,j}} \tag{G.16}$$

at constant T and P and in the absence of external forces

This is a frequently used version of the Maxwell–Stefan equation. It does not look very handy at first glance: instead of expressing flux versus gradient, it expresses gradient versus flux. This little inconvenience will largely be compensated by important advantages.

G.3 Relationship between Maxwell–Stefan and Fick's law

The Maxwell–Stefan and Fick's laws can be simply related for binary systems (including solvent as the case may be). When only two species (including solvent) are involved, the Maxwell–Stefan equation is written:

$$C_{tot}\nabla x_1 = \frac{x_2 J_1 - x_1 J_2}{Đ_{1,2}} \tag{G.17}$$

Because $J_1 + J_2 = 0$ and $x_1 + x_2 = 1$, this leads to:

$$J_1 = -Đ_{1,2} C_{tot} \nabla x_1 \tag{G.18}$$

which is Fick's law for a binary mixture. We have thus shown that the Maxwell–Stefan and Fick's laws are equivalent for a binary mixture and that one has $D_{1,2} = Đ_{1,2}$ (assuming an ideal solution).

There are two limiting situations for which the Maxwell–Stefan equations reduce to Fick's law for multi-component systems:

- If all $Đ_{i,j}$ are identical to $Đ$, Eq. (G.16) reduces to Fick's law (G.7), all Fickian diffusion coefficients D_i being identical to $Đ$.
- If x_i is very small compared to the other molar fractions, Eq. (G.16) again gives Fick's law (G.7) but with:[1]

$$\frac{1}{D_i} = \sum_{j \neq i} \frac{x_j}{Đ_{i,j}} \tag{G.19}$$

Consequently, we have shown that the Maxwell–Stefan equations reduce to Fick's law in the case of binary systems, systems for which all the $Đ_{i,j}$ are identical or for a compound having a molar fraction very small compared to the others.

[1] Equation (G.19) is rather obvious. The $Đ_{i,j}$ are "abilities" for displacement; thus the $1/Đ_{i,j}$ are resistances (i.e. drag) to displacement, as shown by Eq. (G.12). Equation (G.19) then stipulates that the resistance to displacement of species i is due to the weighted sum of the resistance of each species j.

Outside these situations, the two approaches are different, as explained below. The Maxwell–Stefan equations (G.16) can be written in matrix form:

$$C_{tot}\begin{bmatrix}\nabla x_1\\ \\ \nabla x_{n-1}\end{bmatrix}=-\left[B_{i,j}\right]\begin{bmatrix}J_1\\ \\ J_{n-1}\end{bmatrix}$$

$$B_{i,i}=\sum_{k\neq i}\frac{x_k}{Đ_{i,k}} \tag{G.20}$$

$$B_{i,j}=-\frac{x_i}{Đ_{i,j}}\quad i\neq j$$

After inversion, one obtains expressions having the form of the generalized Fick's law (for ideal systems):

$$\begin{bmatrix}J_1\\ J_{n-1}\end{bmatrix}=-C_{tot}\left[B_{i,j}\right]^{-1}\begin{bmatrix}\nabla x_1\\ \\ \nabla x_{n-1}\end{bmatrix}=-C_{tot}\left[D_{i,j}\right]\begin{bmatrix}\nabla x_1\\ \\ \nabla x_{n-1}\end{bmatrix} \tag{G.21}$$

The Fickian diffusion coefficients are related to the Maxwell–Stefan diffusivities by $\left[D_{i,j}\right]=\left[B_{i,j}\right]^{-1}$. As matrix $\left[B_{i,j}\right]$ is dependent on the molar fractions via Eq. (G.20), if the $Đ_{i,j}$ are constant with composition then the $D_{i,j}$ must vary with composition, and vice versa. Notice that this composition dependence is not related to any type of non-ideality: it has been derived for ideal systems!

This experience shows that the Maxwell–Stefan diffusivities $Đ_{i,j}$ are much more constant with respect to composition than the Fickian diffusivities $D_{i,j}$ are. This is not a surprise as Maxwell–Stefan addresses the inherent limitations of Fick's law. As an illustration, let us consider the ternary system discussed above and assume that species 1 is present in very small quantities, so that x_1 is smaller than the two other molar fractions. We saw that the Fickian diffusivity D_1 must vary from $D_{1,2}$ (in the absence of 3) to $D_{1,3}$ (in the absence of 2), so it cannot be constant in the ternary mixture. This variation is at least qualitatively given by Eq. (G.19) using the constant Maxwell–Stefan diffusivities $Đ_{1,2}$ and $Đ_{1,3}$.

Finally, let us say a few words on non-ideal systems in which the driving force is expressed by:

$$F_i^{drive}=-\nabla\mu_i=-\frac{RT}{a_i}\nabla a_i=-\frac{RT}{\gamma_i x_i}\nabla\gamma_i x_i \tag{G.22}$$

where γ_i is the activity coefficient of i in the solution. For a binary system, the Maxwell–Stefan equation (G.16) becomes:

$$-C_{tot}\nabla\ln\left(\gamma_1 x_1\right)=\frac{1}{x_1}\frac{x_2J_1-x_1J_2}{Đ_{1,2}}=\frac{1}{x_1}\frac{J_1}{Đ_{1,2}} \tag{G.23}$$

Comparing this with the flux predicted by Fick's law, one obtains:

$$\frac{D_{1,2}}{Đ_{1,2}}=1+x_1\frac{d\ln(\gamma_1)}{dx_1} \tag{G.24}$$

One can say that non-idealities are embedded in the expression for the driving force when using Maxwell–Stefan, while they are embedded in the diffusion coefficients with Fick's law.

G.4 Perspectives for chromatography modeling

Let us consider the lumped model for describing the particles. The driving force exerted on species i in the lumped particle is given by $F_i^{drive} = -\nabla \bar{\mu}_i = -RT\nabla \ln(\bar{a}_i)$, where $\bar{a}_i$ is the activity of species i in the lumped particle. At equilibrium, the chemical potential in the lumped solid phase ($\bar{\mu}_i = \bar{\mu}_i^0 + RT \ln(\bar{a}_i)$) equals the chemical potential in the fluid mobile phase ($\mu_i = \mu_i^0 + RT \ln(C_i)$ for an ideal solution and constant total concentration).

Consequently one has $\bar{F}_i^{drive} = -\nabla\bar{\mu}_i = -RT\nabla \ln(C_i)$, and, using the lumped equilibrium law and the chain rule, one obtains, for a binary mixture:

$$\bar{F}_1^{drive} = -\frac{RT}{C_1}\frac{dC_1}{d\bar{C}_1}\nabla\bar{C}_1 \tag{G.25}$$

With our notation the frictional force, Eq. (G.12), is expressed as:

$$\bar{F}_1^{drag} = \frac{RT}{\bar{C}_1}\frac{\bar{x}_2\bar{J}_1 - \bar{x}_1\bar{J}_2}{\bar{Đ}_{1,2}} = \frac{RT}{\bar{C}_1}\frac{\bar{J}_1}{\bar{Đ}_{1,2}} \tag{G.26}$$

Equating frictional and driving forces, and equating the flux to the one derived from Fick's law $\bar{J}_1 = \bar{D}_{1,2}\nabla\bar{C}_{1,2}$, one obtains $\bar{D}_{1,2} = \bar{Đ}_{1,2}\frac{\bar{C}_1}{C_1}\frac{dC_1}{d\bar{C}_1}$, which relates Fickian diffusivities to Maxwell–Stefan diffusivities. This very simple model predicts lumped Fickian diffusivities which are dependent on the adsorption isotherm.

References

Gandhi, K. (2012), *AIChE J.* **58**(11), 3601–3605.

Krishna, R. and Wesselingh, J. (1997), *Chem. Eng. Sci.* **52**(6), 861–911.

Taylor, R. and Krishna, R. (1993), *Multicomponent Mass Transfer*, John Wiley & Sons.

Wesselingh, J. and Krishna, R. (2006), *Mass Transfer in Multicomponent Mixtures*, VSSD.

Appendix H Non-linear LDF for multi-solute systems

This appendix is an attempt to generalize the $\overline{\overline{\text{LDF}}}^{\text{var}}$ model to multi-solute systems.

By integrating the mass-balance equation (5.36) over the entire particle, one obtains, for each species k:

$$\frac{d\bar{C}_k}{dt} = \frac{3}{r_0}\left(\varepsilon_i \grave{D}_k \left.\frac{\partial \grave{c}_k}{\partial r}\right|_{r_p} + (1-\varepsilon_i)\bar{\bar{D}}_k \left.\frac{\partial \bar{\bar{c}}_k}{\partial r}\right|_{r_p}\right) \tag{H.1}$$

so, using the properties of the derivative and of the adsorption isotherms:

$$\frac{d\bar{C}_k}{dt} = \sum_j \bar{D}_{k,j}^{loc*} \left.\frac{\partial \bar{c}_j}{\partial r}\right|_{r_p} \tag{H.2}$$

with:

$$\bar{D}_{k,j}^{loc*} = \varepsilon_i \grave{D}_k \left.\frac{d\grave{C}_k}{d\bar{C}_j}\right|_{\bar{c}(c^*)} + (1-\varepsilon_i)\bar{\bar{D}}_k \left.\frac{d\bar{\bar{C}}_k}{d\bar{C}_j}\right|_{\bar{c}(c^*)} \tag{H.3}$$

Using the derivative of the expression $\bar{c}_k = \varepsilon_i \grave{c}_k + (1-\varepsilon_i)\,\bar{\bar{c}}_k$ one obtains:

$$\bar{D}_{k,j}^{loc*} = \varepsilon_i\left(\grave{D}_k - \bar{\bar{D}}_k\right) \left.\frac{d\grave{C}_k}{d\bar{C}_j}\right|_{c^*} \quad \text{if } j \neq k$$

$$\bar{D}_{k,k}^{loc*} = \varepsilon_i \grave{D}_k \left.\frac{d\grave{C}_k}{d\bar{C}_k}\right|_{\bar{c}(c^*)} + \left(1 - \varepsilon_i \left.\frac{d\grave{C}_k}{d\bar{C}_k}\right|_{\bar{c}(c^*)}\right)\bar{\bar{D}}_k$$

Equation (H.2) expresses that the time variation of the lumped concentration $\bar{C}_k$ is proportional to the different peripheral gradients, the lumped diffusion coefficients $\bar{D}_{k,j}^{loc*}$ being calculated using the derivatives of the adsorption isotherm at the periphery of the particle.

On the other hand, we know that the local lumped diffusion coefficients are likely to play an important role in the description of the saturation process. By analogy with the above expressions derived at the periphery, and with the expressions derived for the quasi-linear situation, we define:

$$\bar{D}_{k,j}^{loc} = \varepsilon_i\left(\grave{D}_k - \bar{\bar{D}}_k\right) \left.\frac{d\grave{C}_k}{d\bar{C}_j}\right|_{\bar{C}(C)} \quad \text{if } j \neq k$$

$$\bar{D}_{k,k}^{loc} = \varepsilon_i \grave{D}_k \left.\frac{d\grave{C}_k}{d\bar{C}_k}\right|_{\bar{C}(C)} + \left(1 - \varepsilon_i \left.\frac{d\grave{C}_k}{d\bar{C}_k}\right|_{\bar{C}(C)}\right) \bar{\bar{D}}_k$$

where the derivatives are calculated at the average lumped concentration ($\bar{C}_1, \ldots, \bar{C}_n$).

By analogy with the approach successfully used for single-component systems, Eq. (H.2) suggests writing:

$$\frac{d\bar{C}_k}{dt} = \sum_j \frac{1}{t_{j,k}^{i\,var}} \left(\bar{c}_j^* - \bar{C}_j\right) \tag{H.4}$$

the variable characteristic time for diffusion $t_{j,k}^{i\,var}$ being calculated from:

$$t_{k,j}^{i\,var} = \frac{d_p^2}{60\bar{D}_{k,j}^{var}} \tag{H.5}$$

Again, by analogy with the approximation proposed for the single-component $\overline{\overline{\overline{\text{LDF}}}}^{\text{var}}$ model, one can consider using:

$$\bar{D}_{k,j}^{var} = \frac{\bar{D}_{k,j}^{loc*} + \bar{D}_{k,j}^{loc}}{2} \tag{H.6}$$

Equations (H.4), (H.5) and (H.6) give a generalization of the $\overline{\overline{\overline{\text{LDF}}}}^{\text{var}}$ approximation to multi-component systems.

Appendix I Situations that make the use of the MC model problematic

The MC model can lead to physically unacceptable solutions when used for simulating chromatographic columns in extreme situations involving strongly non-linear interactions.

A first illustration is associated with cationic exchange performed in basic solutions with strong external mass transfer limitations (Nicoud, 1994). A cationic resin column in the H^+ form is fed with a solution containing Na^+ and anions X^-. The equations associated with the MC–$\overline{\text{LDF}}_{\text{E}}$ model are:

- Mass balance for sodium ions and anion X^- in cell k:

$$Na^{k-1} = Na^k + \frac{t_0}{J}\frac{dNa^k}{dt} + \frac{1-\varepsilon_e}{\varepsilon_e}\frac{t_0}{J}\frac{d\overline{Na}^k}{dt} \qquad X^{k-1} = X^k + \frac{t_0}{J}\frac{dX^k}{dt} + 0 \quad \text{(I.1)}$$

- Ion-exchange and dissociation equilibria:

$$K_{Na/H} = \left[\frac{\overline{Na}^k}{Na^k}\right]^{z1}\left[\frac{H^k}{\overline{H}^k}\right] \qquad K_e = H^k OH^k \quad \text{(I.2)}$$

- Electroneutrality constraints:

$$H^k + Na^k = OH^k + X^k \qquad \overline{H}^k + \overline{Na}^k = \overline{N}^{iex} \quad \text{(I.3)}$$

- External mass transfer limitations modeled with the LDF approximation:

$$\frac{d\overline{Na}^k}{dt} = \frac{Na^k - Na^{k*}}{t^e_{Na}} \quad \text{(I.4)}$$

We first consider the situation of a column equivalent to one mixing cell fed with a basic solution ($Na^F > X^F$). As explained in Chapter 4, the fact that H^+ ions exiting the resin immediately react with OH^- ions contained in the solution to form water considerably displaces the ion-exchange equilibrium in favor of sodium ions. We assume in addition that the system is operated under strong external mass transfer limitations. We saw in Section 5.10.3 that, under these circumstances, the sodium concentration at the outlet reaches a plateau whose height is imposed by the ratio of t^e_{Na}, the characteristic time for external mass transfer, and t_0. Qualitatively, the response of the one mixing cell column to the inlet front of sodium is: a first part having a duration of t_0 during which Na ions are absent, a plateau imposed by mass transfer, and the plateau ending when the resin is totally saturated.

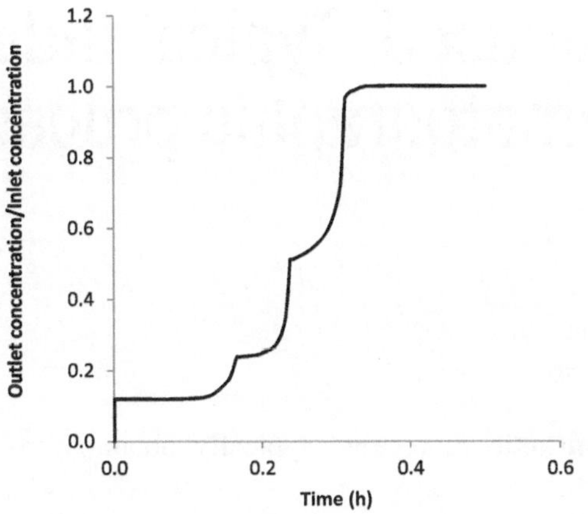

Figure I.1 Response of an ion-exchange column in the H form to a step of a basic solution of Na ions. $Na^F = 0.02$ eq/l, $X^F = 0.01$ eq/l; MC–$\overline{\text{LDF}}_{\text{E}}$ model with $t_0 = 3$ s, $t^e_{Na} = 2$ s, $J = 3$.

If the parameters are properly chosen, when the column is equivalent to a few mixing cells, one can expect a chromatogram that looks like a stairway. This is confirmed by the simulation in Figure I.1: when the column is approximated by $J = 3$ mixing cells, the breakthrough curve is represented by a stairway of four steps. This result is certainly incompatible with the smooth breakthrough curve expected from a chromatographic column.

Another situation associated with a failure of the MC model to deliver physically acceptable breakthrough curves has been reported by (Nicoud, 1994) for ion exchange of amino acids. A combination of a pH excursion and ion-exchange selectivity coefficients varying with composition and represented by the expression from (Dye *et al.*, 1990) given in Eq. (4.75) can, with a specific set of parameters, lead to breakthrough curves showing an oscillatory behavior, one oscillation being obtained per plate.

References

Nicoud, R. M. (1994), Contribution à la modélisation et à la conception de procédés chromatographiques, Habilitation á diriger des recherches (thesis), Institut Polytechnique de Lorraine, Nancy.

Appendix J Typical industrial chromatographic processes

Small synthetic molecules: typically obtained from chemical synthesis or hemi-synthesis:

- Production scale: a few kg to about 10^5 kg/y
- Chromatographic media: typically silica-based materials
- Batch systems, typically operated in the HPLC mode:

d_p (μm)	ΔP (bar)	L_{col} (m)	D_{col} (m)	u (m/s)	t_{cycle} (s)
10–50	10–100	0.1–1	0.1–1.5	$\sim 10^{-3}$	300–3600

- Continuous systems, mainly used for optical isomer separation

d_p (μm)	ΔP (bar)	L_{col} (m)	D_{col} (m)	u (m/s)	ΔT (s)	N_{col}
10–100	10–50	0.1–0.3	0.1–1.5	$\sim 1\text{–}3 \times 10^{-3}$	30–300	4–8

Biopharmaceutical molecules: typically obtained from fermentation or cell culture processes:

- Production scale: a few kg to a few 10^3 kg/y
- Chromatographic media: typically polymeric (natural or synthetic), exceptionally silica-based
- Batch systems:
  - HPLC: same as for small molecules. Mainly used for insulin polishing
  - LPLC: widely used for capture steps

d_p (μm)	ΔP (bar)	L_{col} (m)	D_{col} (m)	u (m/s)	t_{cycle} (s)
50–500	1–10	0.1–1	0.1–2.0	$\sim 0.5\text{–}2 \times 10^{-3}$	1500–15000

  When very compressible gels are used, velocities as low as a few $\times 10^{-4}$ m/s are used with a pressure drop of about 1 bar.

- Continuous systems: first applications under development

Industrial biomolecules: feed, food additives, amino acids etc., typically obtained from fermentation or from biomass:

- Production scale: from a few 10^5 kg/y to 10^8 kg/y
- Chromatographic media: typically polymeric media

- Batch systems: very uncommon
- Continuous systems:

d_p (μm)	ΔP (bar)	L_{col} (m)	D_{col} (m)	u (m/s)	ΔT (s)	N_{col}
200–500	1–10	0.5–2.0	0.5–5.0	$\sim 1\text{–}3 \times 10^{-3}$	300–3000	4–8

Notation

Symbol	Description	MKSA units
A_e, B_e, C_e	Coefficients of the Van Deemter curve	
$\overline{A}_e, \overline{B}_e, \overline{C}_e$	Coefficients of the Van Deemter curve based on lumped solid velocity	
$\overline{\overline{A}}$	Surface area of the adsorbent	m^2
$\overline{\overline{A}}_{sp}$	Specific surface area of the adsorbent ($\overline{\overline{A}}/\overline{\overline{V}}$)	m^2/m^3
$\breve{A}, \breve{B}$	Coefficients of the Ergun equation	m^2
$\grave{c}_A(r,t)$	Local concentration of solute A in the intragranular fluid	mol/m^3, eq/m^3 *or* kg/m^3
$\bar{c}_A(r,t)$	Lumped concentration of solute A in the solid phase (local)	mol/m^3, eq/m^3 *or* kg/m^3
$\bar{\bar{c}}_A(r,t)$	Concentration of solute A adsorbed on the solid skeleton (local)	mol/m^3, eq/m^3 *or* kg/m^3
$c_A\ (r,t)$	Concentration of solute A in the fluid mobile phase at radial position r from the symmetry axis of the column	mol/m^3, eq/m^3 *or* kg/m^3
$c_A^{inl}\ (r,t)$	Concentration of solute A in the fluid mobile phase at the inlet of a ring between radial positions r and $r + dr$ from the symmetry axis of the column	mol/m^3, eq/m^3 *or* kg/m^3
C_A	Concentration of solute A in the fluid mobile phase	mol/m^3, eq/m^3 *or* kg/m^3
$\grave{C}_A$	Concentration of solute A in the fluid of the intragranular pores (volume-averaged)	mol/m^3, eq/m^3 *or* kg/m^3
$\overline{C}_A$	Lumped concentration of solute A in the solid phase (volume-averaged)	mol/m^3, eq/m^3 *or* kg/m^3

$\overline{\overline{C}}_A$	Concentration of solute A adsorbed on the solid skeleton (volume-averaged)	mol/m^3, eq/m^3 *or* kg/m^3
C_A^k	Concentration of solute A in mixing cell k	mol/m^3, eq/m^3 *or* kg/m^3
C_A^F	Feed (inlet) concentration of solute A	mol/m^3, eq/m^3 *or* kg/m^3
C_A^I	Initial concentration of solute A	mol/m^3, eq/m^3 *or* kg/m^3
C_A^{Inj}	Concentration of solute A in the injection pulse	mol/m^3, eq/m^3 *or* kg/m^3
C_A^{Final}	Final concentration of solute A	mol/m^3, eq/m^3 *or* kg/m^3
C_A^{inl}	Concentration of solute A in the fluid mobile phase at the inlet of a column or a moving bed	mol/m^3, eq/m^3 *or* kg/m^3
C_A^{out}	Concentration of solute A in the fluid mobile phase at the outlet of a column or a moving bed	mol/m^3, eq/m^3 *or* kg/m^3
$C_A^{\overline{inl}}$	Concentration of solute A in the lumped solid mobile phase at the inlet of a moving bed	mol/m^3, eq/m^3 *or* kg/m^3
$C_A^{\overline{out}}$	Concentration of solute A in the lumped solid mobile phase at the outlet of a moving bed	mol/m^3, eq/m^3 *or* kg/m^3
C_A^{max}	Maximal concentration of solute A	mol/m^3, eq/m^3 *or* kg/m^3
C_A^{stab}	Concentration of solute A in the fluid mobile phase that is stabilized inside a moving bed	mol/m^3, eq/m^3 *or* kg/m^3
C_A^{bot}	Concentration of solute A in the fluid mobile phase at the bottom and inside a moving bed	mol/m^3, eq/m^3 *or* kg/m^3
C_A^{top}	Concentration of solute A in the fluid mobile phase at the top and inside a moving bed	mol/m^3, eq/m^3 *or* kg/m^3

C_A^{zi}	Concentration of solute A in the fluid mobile phase in zone i of a moving bed	mol/m^3, eq/m^3 *or* kg/m^3
$C_A^{zi/in}$	Concentration of solute A in the fluid mobile phase entering zone i of the moving bed	mol/m^3, eq/m^3 *or* kg/m^3
$C_A^{zi/out}$	Concentration of solute A in the fluid mobile phase exiting zone i of the moving bed	mol/m^3, eq/m^3 *or* kg/m^3
C_A^0	Fictitious concentration of solute A used in IAS theory	mol/m^3 *or* kg/m^3
C_{tot}^F	Total feed concentration	mol/m^3, eq/m^3 *or* kg/m^3
C_p	Heat capacity	J/kg K
d_p	Particle diameter	m
D_A	Diffusion coefficient in solution of solute A	m^2/s
$\grave{D}_A$	Diffusion coefficient of solute A in the intra-granular pore solution	m^2/s
$\overline{D}_A$	Lumped internal diffusion coefficient of solute A	m^2/s
$\overline{D}_A(0)$	Lumped internal diffusion coefficient of solute A when solute concentration equals zero	m^2/s
$\overline{\overline{D}}_A$	Adsorbed phase diffusivity of solute A	m^2/s
$\overline{D}_A^{Augm}$	Augmented internal diffusion due to internal convection	m^2/s
$\overline{D}_A^{var}$	Lumped diffusion coefficient of solute A, concentration-dependent	m^2/s
$\overline{D}_A^{loc}$	Local lumped diffusion coefficient of solute A	m^2/s
$\overline{D}_A^{loc*}$	Local lumped diffusion coefficient of solute A at the fluid-solid interface	m^2/s
D_{ax}	Axial dispersion coefficient (hydrodynamic) $(Pe = 2J = \frac{uL_{col}}{D_{ax}})$	m^2/s
D_{ax}^{app}	Apparent axial dispersion coefficient in which hydrodynamic and kinetic contributions are lumped $(Pe^{app} = 2N = \frac{uL_{col}}{D_{ax}^{app}})$	m^2/s
D_{col}	Column diameter	m
D_{rad}	Radial dispersion coefficient	m^2/s
E	Young modulus	N/m^2
E_ε	Modified Young modulus	N/m^2
E_{fo}	Modified Young modulus	N/m^2
$E(t)$	Retention times distribution RTD	s^{-1}

$E_e(t)$	Zero retention times distribution (access to extragranular porosity) Z_eRTD	s^{-1}
$E_t(t)$	Zero retention times distribution (access to total porosity) Z_tRTD	s^{-1}
$\hat{E}(s)$	Laplace transform of the RTD	
$\hat{E}_e(s)$	Laplace transform of the Z_eRTD	
$\hat{E}_t(s)$	Laplace transform of the Z_tRTD	
$\bar{f}_A$	Lumped adsorption isotherm for solute A	mol/m^3, eq/m^3 *or* kg/m^3
$\bar{\bar{f}}_A$	Adsorption isotherm for solute A	mol/m^3, eq/m^3 *or* kg/m^3
$\grave{f}_A$	Partition law relating extragranular fluid concentration to intragranular fluid concentration for solute A	mol/m^3, eq/m^3 *or* kg/m^3
F_e	Phase ratio $\frac{1-\varepsilon_e}{\varepsilon_e}$	
F	Faraday constant (96 485)	C/mol
$F(t)$	Normalized response of the column to a step input of concentration	
F_{stress}, F_{fluid}, F_{fric}	Forces acting on a chromatographic bed	N
$\bar{g}_A$	Ratio of the lumped diffusion coefficient to the molecular diffusion coefficient, $\overline{D}_A/D_A$	
$\bar{g}_A(0)$	Ratio $\bar{g}$ calculated with $\overline{D}_A(0)$	
h_u	Heat generated per unit volume of bed due to viscous flow	W/m^3
h_B	Burke–Plummer coefficient	
h_k	Kozeny–Kármán coefficient	
H	Duration of production	s
$\overline{H}^{Fick}(s)$	Transfer function relating the average lumped solid concentration to the lumped solid concentration at the periphery of the particle (Fick model)	
$\overline{H}^{LDF}(s)$	Transfer function relating the average lumped solid concentration to the lumped solid concentration at the periphery of the particle (LDF model)	
j	James–Martin correction factor (compressible fluids)	
J	Number of mixing cells	
k_A	Mass transfer coefficient	m/s

$\bar{\bar{K}}_A$	Initial slope of the adsorption isotherm, $\bar{\bar{C}}_A = \bar{\bar{f}}_A(C_A)$ (Henry's coefficient)	
$\bar{K}_A$	Initial slope of the lumped adsorption isotherm, $\bar{C}_A = \bar{f}_A(C_A)$ (lumped Henry's coefficient)	
$\bar{K}'_A$	Lumped retention factor $\frac{1-\varepsilon_e}{\varepsilon_e}\bar{K}_A$	
k'_A	Retention factor $\frac{1-\varepsilon_T}{\varepsilon_T}\bar{\bar{K}}_A$	
$\tilde{K}_A$	Adsorption coefficient for the Langmuir adsorption isotherm	
$\grave{K}_A$	Initial slope of the partition law, $\grave{f}_A(C_A)$	
$K_{C/H}$	Selectivity coefficient of cation C with respect to proton H, expressed in concentration	
$K^x_{C/H}$	Selectivity coefficient of cation C with respect to proton H, expressed in ionic fraction	
l_i	Relative length of zone i in a TMB	
L_{col}	Column length	m
L_{col0}	Column length under no stress (gravity-settled bed)	m
L^{tmb}_{col}	Total length of a TMB	m
$\breve{L}$	Characteristic length for heat exchange	m
$\hat{L}(s)$	Transfer function relating the quantity of solute contained in the mobile (δn_A) and lumped solid ($\delta \bar{n}_A$) phases	
m	Ratio of fluid to lumped solid flow rate in a counter-current system ($m = Q/\bar{Q} = u/\bar{u}$)	
m_i	Value of ratio m in zone i ($m = Q_i/\bar{Q} = u_i/\bar{u}$)	
m	Mass of purified material	kg
m_*	Mass of purified solute	kg
$\dot{m}^*$	Production rate	kg/s
M	Molecular weight	kg/mol
M	Mass of feed material	kg
n	Number of solutes or number of components	
n_A	Quantity of solute A in the extragranular fluid	mol *or* kg
$\grave{n}_A$	Quantity of solute A in the intragranular fluid	mol *or* kg
$\bar{n}_A$	Quantity of solute A in the lumped solid	mol *or* kg
$\bar{\bar{n}}_A$	Quantity of solute A adsorbed on the solid skeleton	mol *or* kg
n^{tot}_A	Total quantity of solute A in the system	mol *or* kg
n_f	Number of species in the fluid phase (phenomenological models)	
$n_{\bar{s}}$	Number of species in the lumped solid phase (phenomenological models)	

$n_{\bar{\bar{s}}}$	Number of species in the solid phase (phenomenological models)	
n_R	Number of concentric rings	
n_T	Total number of species (phenomenological models)	
N	Number of plates	
N_{col}	Number of columns	
N_0	Normality	eq/m^3
$\overline{N}$	Saturation capacity (lumped adsorption isotherm)	eq/m^3
$\overline{\overline{N}}$	Saturation capacity (adsorption isotherm)	mol/m^3 *or* eq/m^3
$\overline{N}^{iex}$	Ion-exchange capacity (lumped resin)	eq/m^3
$\overline{\overline{N}}^{iex}$	Ion-exchange capacity (resin skeleton)	eq/m^3
P	Pressure	Pa
p	Parameter for modeling compressible beds	m^2/s
p_A	Parameter associated with solute A in generalized Langmuir adsorption isotherm, taking values $+1$ or -1	
P^{inl}	Pressure at the inlet of the column	Pa
P^{out}	Pressure at the outlet of the column	Pa
P_{mech}	Mechanical pressure acting at the inlet of the chromatographic bed due to a compression piston	N
q	Parameter used for modeling compressible beds	m^2/s
Q	Fluid flow rate	m^3/s
Q^{out}	Fluid flow rate at the outlet of the column	m^3/s
Q^{mass}	Mass flow rate	m^3/s
$\overline{Q}$	Lumped solid flow rate	m^3/s
$\overline{\overline{Q}}$	Solid skeleton flow rate	m^3/s
$\grave{Q}$	Intragranular fluid flow rate	m^3/s
Q_i	Fluid flow rate in zone i of solute A in a counter-current system	m^3/s
Q_R, Q_{Ext}, Q_F, Q_{Raf}	Recycling, extract, feed and raffinate flow rates in a four-zone TMB/SMB	
R	Ideal gas constant ($\approx$8.314)	J/(k mol)
r	Radius inside particle or inside column	m
r_p	Particle radius	m
R_{col}	Column radius	m
R^{sf}	Surface resolution	
S	Stress acting on a chromatographic bed	Pa

$S^{inl}(t)$	Inlet signal	arbitrary
$S^{out}(t)$	Outlet signal	arbitrary
S_{col}	Column surface ($\pi D_{col} L_{col}$)	m^2
S_p	Peripheral surface of a particle	
S^R	Recycling ratio	
t	Time	s
t_{cycle}	Cycle time	s
t_{ci}	Cut time	s
t_0	Excluded zero retention time Z_eRT ($\varepsilon_e V/Q$)	s
t_0^{out}	t_0 calculated with the conditions prevailing at the outlet of the column (compressible fluids)	s
t_{0R}	Total zero retention time Z_tRT ($\varepsilon_T V/Q$)	s
t_{0R}^{out}	t_{0R} calculated with the conditions prevailing at the outlet of the column (compressible fluids)	s
t_ϕ^χ	Ratio of total fluid flow rate to system volume	s
t_A^F	Fickian diffusion time for solute A ($r_p^2/\overline{D}_A$)	s
t_A^m	Global mass transfer time for solute A	s
t_A^e	Characteristic time for external diffusion for solute A	s
t_A^i	Characteristic time for internal diffusion for solute A	s
$t_A^{i\,var}$	Characteristic time for internal diffusion for solute A, varying with concentration	s
t_A^{ads}	Adsorption time for solute A	s
t_R	Retention time	s
t_R^{max}	Retention time of the peak maximum	s
T	Temperature	K
T^{inl}	Inlet temperature	K
T^{ext}	External temperature	K
$\langle T(z=\infty)\rangle$	Radially averaged temperature in the column at long distances from the inlet	K
$\langle T(r)\rangle$	Axially averaged temperature in the column	K
u	Superficial fluid velocity (Q/Ω_{col})	m/s
$\bar{u}$	Superficial lumped solid velocity ($\overline{Q}/\Omega_{col}$)	m/s
u_e	Interstitial fluid velocity (u/ε_e)	m/s
$\bar{u}_e$	Interstitial lumped solid velocity ($\bar{u}/(1-\varepsilon_e)$)	m/s
u_i	Fluid velocity in ring i (column approximated by concentrated rings)	m/s
u_i^{tmb}, u_i^{smb}	Superficial fluid velocity in zone i of a TMB or SMB	m/s
u_{ei}^{tmb}, u_{ei}^{smb}	Interstitial fluid velocity in zone i of a TMB or SMB	m/s

u_r	Short notation for $u(r)$, the fluid velocity at radius r	m/s
u^{out}	Fluid velocity prevailing at the outlet of the column	m/s
$\tilde{u}$	Reduced fluid velocity $(\mu_e d_p / D_A)$	
$U^{Fick}(t)$	Normalized fractional uptake of solute-free particles suddenly immersed in a solution of constant concentration (Fick model)	
$U^{LDF}(t)$	Normalized fractional uptake of solute-free particles suddenly immersed in a solution of constant concentration (LDF model)	
V_{col}	Column volume	m^3
$\grave{V}$	Volume of the intragranular pores	m^3
$\overline{V}$	Volume of the lumped solid	m^3
$\overline{\overline{V}}$	Volume of the skeleton	m^3
V_{inj}	Injection volume	m^3
V_p	Volume of the particle	m^3
w	Maxwell–Boltzmann factor	
X_A	Massic or molar fraction of solute A in the extragranular solution	
$\overline{\overline{X}}_A$	Massic or molar fraction of solute A on the solid skeleton	
$\overline{X}_A$	Massic or molar fraction of solute A in the lumped solid	
X_*^{out}	Massic purity of the target solute in the outlet stream	
X_*^{min}	Target massic purity of the target solute in the outlet stream	
$X(t)$	Surface normalized inlet signal	s^{-1}
$Y(t)$	Surface normalized outlet signal	s^{-1}
$\hat{X}(s)$	Laplace transform of $X(t)$	
$\hat{Y}(s)$	Laplace transform of $Y(t)$	
z	Axial position	m
z	Valence	
Z	Compressibility factor	
Z^{out}	Compressibility factor at the outlet of the column	

GREEK SYMBOLS

$\alpha_e, \beta_e, \gamma_e$	Coefficients of the Van Deemter curve (reduced units)

α_i	Fraction of the total flow rate associated with ring i (discrete modeling of solute axial velocity profiles)	
β	Friction factor	
$\breve{\beta}$	Geometric friction factor ($\beta \frac{L_{col}}{D_{col}}$)	
γ	Normalized parameter used in parabolic radial velocity profiles	
$\Delta^M_{A/B}$	Degree of separation of solutes A and B according to model M	
ΔT	Shift time in an SMB	s
Δt_{inj}	Injection time	s
$\Delta \breve{T}$	Asymptotic temperature increase (heat mass balance)	K
ΔT_{adiab}	Temperature increase in an adiabatic column due to viscous flow	K
ΔP	Pressure drop	Pa
ΔP^{tg}	Pressure-drop target	Pa
ΔP_0	Pressure drop in the absence of bed stress	Pa
δ	Parameter for relating pressure drop to gravitational acceleration ($+1$, 0 or -1)	
δn_A	Quantity of solute contained in the mobile phase in a small volume of column where the composition is assumed to be uniform	mole
$\delta \bar{n}_A$	Quantity of solute contained in the stationary phase in a small volume of column where the composition is assumed to be uniform	mole
δ_s	Proportionality coefficient in Darcy's law between pressure drop per unit length and superficial velocity	Pa/(m^2 s)
δ_e	Proportionality coefficient in Darcy's law between pressure drop per unit length and interstitial velocity	Pa/(m^2 s)
$\bar{\delta}_e$	Proportionality coefficient in Darcy's law between pressure drop per unit length and lumped solid interstitial velocity	Pa/(m^2 s)
ε_e	External (extragranular) porosity	
ε_{e0}	External (extragranular) porosity in the absence of bed stress	
ε_i	Internal (intragranular) porosity	
ε_T	Total porosity	
ε_{T0}	Total porosity in the absence of bed stress	

η_F^{eq}	Ratio of feed flow rate to feed flow rate allowing delivery of target purity at 100% yield according to the equilibrium model	
η_E^{eq}	Ratio of eluent flow rate to eluent flow rate allowing delivery of target purity at 100% yield according to the equilibrium model	
η_{EF}^{eq}	Ratio of carrier fluid flow rate to carrier fluid flow rate allowing delivery of target purity at 100% yield according to the equilibrium model	
θ	Yield	
ϑ	Safety margin for the calculation of TMB flow rates	
$\bar{\bar{\theta}}_A$	Ratio of surface diffusivity to the diffusion coefficient in the intragranular pores for solute A	.
$\bar{\bar{\Theta}}_A$	Ratio of the concentration $\bar{\bar{C}}_A$ to the saturation capacity $\bar{\bar{N}}_A$ for solute A	
κ	Permeability of the bed	m^2
κ_0	Permeability of the bed in the absence of stress	m^2
κ_A	Parameter of the Kremser equation ($m/\bar{K}_A$)	m^2
λ	Thermal conductivity	W/(m K)
λ_{bed}	Bed thermal conductivity	W/(m K)
λ_{comp}	Bed compression factor	
λ_{comp}^{crit}	Critical bed compression factor	
$\bar{\bar{\lambda}}_A$	Coefficient of the linear term added to the adsorption isotherm of solute A	
$\bar{\lambda}_A$	Coefficient of the linear term added to the lumped adsorption isotherm of solute A	
Λ	Non-linearity factor	
μ	Viscosity	Pa s
μ_i	ith-order moment of solute A distribution	depending
μ_{ci}	ith-order centered moment of solute A distribution	depending
$\Xi_A^{M1/M2}$	Dissimilarity criterion for solute A between models M1 and M2	
χ_F^{eq}	Solute concentration expressing the dilution due to thermodynamics	mol/m^3 *or* kg/m^3
ν	Kinematic viscosity (η/ρ)	m^2/s
ν	Intraparticular Péclet number ($\frac{\xi u d_p}{6 D_A}$)	
ν	Enhancement factor for diffusivity	

$\nu_{i,j}$	Stoichiometric coefficient of species j in reaction i	
ξ	Dimensionless distance	
ξ	Dimensionless parameter characterizing intraparticular convection	
ξ_p	Dimensionless particle radius (r/r_p)	
ξ_c	Dimensionless column radius (r/R_{col})	
ρ	Density	kg/m^3
π	Spreading pressure	N/m
Π	Modified spreading pressure ($\frac{\bar{\bar{A}}_{sp}}{RT}\pi$)	
σ	Variance	s^2
σ	Front/composition velocity	m/s
σ	Normalized stress	
σ_P	Shielding factor for protein ion exchange	
τ	Tortuosity	
φ	Electrical potential	V
φ_m	Productivity per unit mass	kg/(kg s)
φ_v	Productivity per unit volume	kg/(m^3 s)
φ_D	Productivity per unit diameter	kg/(m s)
φ_k^{num}	Particle size distribution (number)	
Υ	Volume of solvent to obtain a given mass of purified material	m^3/kg
φ_k^{vol}	Particle size distribution (volume)	
ψ	Dimensionless factor describing the influence of concentration on the lumped diffusion coefficient	
ψ	Production cost per unit mass of purified fraction	€/kg
Ψ	Production cost	€
ω_i	Fraction of the column section Ω associated with ring i	
ω_i^y	Solution number i of the characteristic equation in the ω space associated with concentration conditions y	
Ω_{col}	Column cross section	m^2

DIMENSIONLESS NUMBERS

Pe	Column Péclet number for axial dispersion ($\frac{u_e L_{col}}{D_{ax}} = \frac{u L_{col}}{\varepsilon_e D_{ax}}$)
Pe^{app}	Column Péclet number for apparent axial dispersion ($\frac{u_e L_{col}}{D_{ax}^{app}}$)

Pe_{ax}	Particle Péclet number for axial dispersion $(\frac{u_e d_p}{D_{ax}})$	
Pe_{rad}	Particle Péclet number for radial dispersion $(\frac{u_e d_p}{D_{rad}})$	
Re	Reynolds number $(\frac{d_p u}{\nu} = \frac{\rho d_p u}{\mu})$	
Sc	Schmidt number $(Sc = \frac{\nu}{D_m} = \frac{\eta}{\rho D_m})$	
Sh	Sherwood number $(Sh = \frac{d_p}{\delta})$	

ACRONYMS

DAC	Dynamic axial compression	
CFD	Computational fluid dynamics	
LDF	Linear driving force	
HPLC	High-performance (-pressure) liquid chromatography	
LPLC	Low-pressure liquid chromatography	
RTD	Retention times distribution	
Z_eRT	Zero (excluded) retention time	s
Z_tRT	Zero (total) retention time	s
Z_eRTD	Zero (excluded) retention time distribution	
Z_tRTD	Zero (total) retention time distribution	

Index

Adsorbed phase diffusivity 238
Adsorption isotherm 7, 63, 144
Adsorption kinetics 257
Amino acid 116, 166, 178, 185
Axial dispersion 14, 28, 40, 322

Band broadening 83
Batch method 193, 231, 292
Bed compression factor 348
Breakthrough curve 11, 71
Buffer 190

Capture 503, 505
Cation exchanger 161, 266
Characteristic time 37, 219, 222, 225
Characteristic path 89
Characteristics 89
Coherence 94, 95
Column efficiency 39, 282
Competitive adsorption 155, 199
Complexation 159, 175, 420
Compressibility factor 325
Compressive front 70
Critical bed compression factor 348
Critical velocity 348

Danckwerts 400
Darcy 318
Dead volumes 21, 42
Detector 43
Dispersive wave 69
Displacement 62, 116, 373, 500

Eddy diffusion 321
Effective diffusion 228
Electrical double layer 169, 268
Equilibrium model 67, 417
Equilibrium theory 67
External diffusion 216
External mass transfer 217
External porosity 3
Extra column 43, 583
Extract 468
Extragranular porosity 3

Fast Fourier transform 612
Favorable lumped adsorption isotherms 64
Fick 14, 217, 221
Fluid mobile phase 3
Fronting 62, 120

Generalized lumped adsorption isotherm 64, 66
Gradient 190, 505, 595

Henry's coefficient 9
HETP 39
Hodograph 88
Hydrodynamic dispersion 14, 27, 321

Ideal model 17
Ideal system 18
Injection signal 43
Interference 77
Internal diffusion 216
Internal porosity 3
Intragranular fluid 3
Intragranular porosity 3
Invariant 182
Inversed size exclusion 142

James–Martin 327

Kozeny–Kármán 318

Langmuir 64, 147
Laplace transform 25
Linear driving force (LDF) 225
Linearly modified Langmuir adsorption isotherm 65
Loading factor 110
Lumped concentration 5
Lumped diffusion coefficient 222
Lumped Henry's coefficient 35
Lumped model 140
Lumped solid 5

Macropores 253
Maxwell–Boltzmann 169
Maxwell–Stefan 301
Mercury intrusion porosimetry 143

Micropores 253
Mixing cells 14, 27
Moments 23

Nitrogen sorptometry 143
Non-linearity factor 395
Number of mixing cells 14
Number of plates 14

Particle size distribution 370
Partition law 8
Péclet number 28, 40, 256
Perfusion 256
Permeability 318
Phenomenological model 178, 419
Pore size 143, 177, 258
Porosity 140
Porous model 140
Pressure drop 318
Primary group 520
Productivity 512
Purity 511

Radial dispersion 324, 377
Radial temperature profiles 357, 364
Radial velocity profiles 360, 362, 364
Rate-controlling step 271
Reduced units (HETP) 283
Reduced velocity 283
Refolding 178
Residence time distribution 23
Retention factor 12
Retention time distribution 23
Reynolds number 219

Saturation capacity 147
Schmidt number 219
Secondary group 520
Selectivity 107, 113
Self-broadening front 69
Sensor 42
Sherwood number 219
Shielding factor 167
Shift time 477
Shock wave 70, 71
Simple wave 69
Simulated moving bed 439
Size exclusion 8, 142, 177
Skewness 25
Solid-phase skeleton 3
Steric hindrance 167, 237, 409
Stiffness 428
Stoichiometric 117, 118, 178
Superficial fluid velocity 28
Surface resolution 393

Tortuosity 237
Total porosity 7, 140
Transfer function 20, 25
True moving bed 438

Unfavorable lumped adsorption isotherm 64

Van Deemter 47, 280
Viscous fingering 373